Undergraduate Texts in Mathematics

T0178259

For other titles published in this series, go to
http://www.springer.com/series/666

Kenneth R. Davidson
Allan P. Donsig

Real Analysis and Applications

Theory in Practice

 Springer

Kenneth R. Davidson
Department of Pure Mathematics
University of Waterloo
Ontario N2L 3G1
Canada
krdavids@uwaterloo.ca

Allan P. Donsig
Department of Mathematics
University of Nebraska-Lincoln
Lincoln, NE 68588
USA
adonsig1@unl.edu

ISSN 0172-6056
ISBN 978-1-4614-9900-8   ISBN 978-0-387-98098-0 (eBook)
DOI 10.1007/978-0-387-98098-0
Springer New York Dordrecht Heidelberg London

Mathematics Subject Classification (2000): 26Exx, 26E40

Springer is part of Springer Science+Business Media (www.springer.com)

To Virginia and Stephanie

Preface

This book provides an introduction both to real analysis and to a range of important applications that depend on this material. Three-fifths of the book is a series of essentially independent chapters covering topics from Fourier series and polynomial approximation to discrete dynamical systems and convex optimization. Studying these applications can, we believe, both improve understanding of real analysis and prepare for more intensive work in each topic. There is enough material to allow a choice of applications and to support courses at a variety of levels.

This book is a substantial revision of *Real Analysis with Real Applications*, which was published in 2001 by Prentice Hall. The major change in this version is a greater emphasis on the latter part of the book, focussed on applications. A few of these chapters would make a good second course in real analysis through the optic of one or more applied areas. Any single chapter can be used for a senior seminar.

The first part of the book contains the core results of a first course in real analysis. This background is essential to understanding the applications. In particular, the notions of limit and approximation are two sides of the same coin, and this interplay is central to the whole book. Several topics not needed for the applications are not included in the book but are available online, at both this book's official website www.springer.com/978-0-387-98097-3 and our own personal websites, www.math.uwaterloo.ca/~krdavids/ and www.math.unl.edu/~adonsig1/.

The applications have been chosen from both classical and modern topics of interest in applied mathematics and related fields. Our goal is to discuss the theoretical underpinnings of these applied areas, showing the role of the fundamental principles of analysis. This is not a methods course, although some familiarity with the computational or methods-oriented aspects of these topics may help the student appreciate how the topics are developed. In each application, we have attempted to get to a number of substantial results, and to show how these results depend on the theory.

This book began in 1984 when the first author wrote a short set of course notes (120 pages) for a real analysis class at the University of Waterloo designed for students who came primarily from applied math and computer science. The idea was to get to the basic results of analysis quickly, and then illustrate their role in a variety of applications. At that time, the applications were limited to polynomial approximation, Newton's method, differential equations, and Fourier series.

A plan evolved to expand these notes into a textbook suitable for a one- or two-semester course. We expanded both the theoretical section and the choice of applications in order to make the text more flexible. As a consequence, the text is not uniformly difficult. The material is arranged by topic, and generally each chapter gets more difficult as one progresses through it. The instructor can omit some more difficult topics in the early chapters if they will not be needed later.

We emphasize the role of normed vector spaces in analysis, since they provide a natural framework for most of the applications. So some knowledge of linear algebra is needed. Of course, the reader also should have a reasonable working knowledge of differential and integral calculus. While multivariable calculus is an asset because of the increased level of sophistication and the incorporation of linear algebra, it is not essential. Some of this background material is outlined in the review chapter.

By and large, the various applications are independent of each other. However, there are references to material in other chapters. For example, in the wavelets chapter (Chapter 15), it seems essential to make comparisons with the classical approximation results for Fourier series and for polynomials.

It is possible to use an application chapter on its own for a student seminar or topics course. We have included several modern topics of interest in addition to the classical subjects of applied mathematics. The chapter on discrete dynamical systems (Chapter 11) introduces the notions of chaos and fractals and develops a number of examples. The chapter on wavelets (Chapter 15) illustrates the ideas with the Haar wavelet. It continues with a construction of wavelets of compact support, and gives a complete treatment of a somewhat easier continuous wavelet. In the final chapter (Chapter 16), we study convex optimization and convex programming. Both of these latter chapters require more linear algebra than the others.

We would like to thank various people who worked with early versions of this book for their helpful comments, in particular, Robert André, John Baker, Jon Borwein, Ola Bratteli, Brian Forrest, John Holbrook, Stephen Krantz, Michael Lamoureux, Leo Livshits, Mike McAsey, Robert Manning, John Orr, Justin Peters, Gabriel Prajitura, David Seigel, Ed Vrscay, and Frank Zorzitto. We also thank our students Geoffrey Crutwell, Colin Davidson, Sean Desaulniers, Masoud Kamgarpour, Michael Lipnowski, and Alex Wright for working through parts of the book and solving many of the exercises. We also thank the students in various classes at the University of Waterloo and at the University of Nebraska, where early versions of the text were used and tested.

We welcome comments on this book.

Waterloo, ON & Lincoln, NE   *Kenneth R. Davidson*
March, 2009   *Allan P. Donsig*

Contents

Part B   Applications

Part A
Analysis

Chapter 1
Review

Since we use results from calculus and linear algebra regularly, we review the key definitions and theorems here. If something seems unfamiliar, reviewing this material would be wise. We list a few good books in each subject in the bibliography. For the theoretical part of calculus, there is a detailed development in Chapter 6 and in the supplementary materials for this book available online. Finally, we give a brief treatment of equivalence relations.

1.1 Calculus

To read and understand this book, you are expected to have taken and understood a full course on calculus, although it need not be a proof-oriented course. In general, you should have an understanding of functions, and the mechanics of differentiation and integration. We will make use of these tools to analyze examples before we get to Chapter 6, where the theory of differentiation and integration is developed carefully, with complete proofs.

The first part of this book is a careful treatment of the basic ideas of real analysis. These ideas are illustrated by a wide variety of examples, most of which are based on knowledge of calculus. In particular, we expect the reader to be familiar with the standard functions such as logarithm, exponential, trigonometric and inverse trigonometric functions. We rely on your ability to sketch graphs of functions and compute extrema, asymptotes, and inflection points as needed.

The basic theory that underlies calculus is generally not taught in a first course, because the ideas are difficult and subtle. The basic structure of the real numbers and various formulations that express the important property of completeness are only tacitly assumed. A couple of centuries ago, serious mathematicians did the same thing, but developments in the nineteenth century forced them to examine the basics and put them on a better footing. This is what we do in the next chapter.

K.R. Davidson and A.P. Donsig, *Real Analysis and Applications: Theory in Practice*,
Undergraduate Texts in Mathematics, DOI 10.1007/978-0-387-98098-0_1,
© Springer Science + Business Media, LLC 2010

This treatment of the real numbers goes hand in hand with a careful discussion of limits. Although the material is developed from scratch, it is useful for the reader to have a working knowledge of how to compute basic limits.

The theory of the derivative is developed in Chapter 6. We assume a working knowledge of differentiation, and do not spend time on methods or applications here. This includes various methods for calculating derivatives. You should also understand the relationship between the derivative and tangent lines.

We also develop the theory of integration. We don't spend time on the techniques of integration using the various tricks of the trade such as substitution and integration by parts. We assume that the reader is comfortable with these methods. They are used when the need arises throughout the book. If you have seen a proof-oriented development of calculus, then most of Chapter 6 may safely be omitted.

There are two ideas from calculus that you need to be aware of now, to understand some exercises and material in the first few chapters.

One central fact from differential calculus that we make use of frequently is the Mean Value Theorem (6.2.2). Intuitively, this says that if f is a differentiable function on (a, b), then the line through the endpoints is parallel to a tangent line to the curve at some interior point.

1.1.1. MEAN VALUE THEOREM.
Suppose that f is a function that is continuous on $[a, b]$ and differentiable on (a, b). Then there is a point $c \in (a, b)$ such that

$$f'(c) = \frac{f(b) - f(a)}{b - a}.$$

In a calculus course, most integrals are actually computed by finding antiderivatives. But if you think that integration *is* antidifferentiation, then Chapter 6 will show you that integration really is the computation of area. The connection to antidifferentiation is a *theorem*. This is the Fundamental Theorem of Calculus, Theorems 6.4.2 and 6.4.3, which connects the notions of tangent line and area in a surprising way. You should be aware that integrals can be computed even when no simple antiderivative exists.

1.1.2. FUNDAMENTAL THEOREM OF CALCULUS, PART 1.
Let f be an integrable function on $[a, b]$. If $F(x) = \int_a^x f(t)\, dt$ for $a \le x \le b$, then F is a continuous function. If f is continuous at x_0, then F is differentiable at x_0 and $F'(x_0) = f(x_0)$.

1.1.3. FUNDAMENTAL THEOREM OF CALCULUS, PART 2.
Let f be integrable on $[a, b]$. If there is a continuous function g on $[a, b]$ that is differentiable with $g'(x) = f(x)$ on (a, b), then

$$\int_a^b f(x)\, dx = g(b) - g(a).$$

1.2 Linear Algebra

Many of our applications are naturally set up in the context of normed vector spaces. So it is worth reviewing carefully the definition of a vector space and the basic results about them. We use $\mathbf{v}$ for vectors and r for real numbers.

1.2.1. DEFINITION. A (real) **vector space** consists of a set V with elements called **vectors** and two operations with the following properties:

vector addition: for each pair $\mathbf{u}, \mathbf{v} \in V$, there is a vector $\mathbf{u} + \mathbf{v} \in V$. This satisfies

(1) **commutativity:** $\mathbf{u} + \mathbf{v} = \mathbf{v} + \mathbf{u}$ for all $\mathbf{u}, \mathbf{v} \in V$
(2) **associativity:** $\mathbf{u} + (\mathbf{v} + \mathbf{w}) = (\mathbf{u} + \mathbf{v}) + \mathbf{w}$ for all $\mathbf{u}, \mathbf{v}, \mathbf{w} \in V$
(3) **zero:** there is a vector $\mathbf{0} \in V$ such that $\mathbf{0} + \mathbf{u} = \mathbf{u} = \mathbf{u} + \mathbf{0}$ for all $\mathbf{u} \in V$
(4) **inverses:** for each $\mathbf{u} \in V$, there is a vector $-\mathbf{u}$ such that $\mathbf{u} + (-\mathbf{u}) = \mathbf{0}$

scalar multiplication: for each vector $\mathbf{v} \in V$ and real number $r \in \mathbb{R}$, there is a vector $r\mathbf{v} \in V$. This satisfies, for all $\mathbf{u}, \mathbf{v} \in V$ and all $r, s \in \mathbb{R}$,

(1)	$(r+s)\mathbf{v} = r\mathbf{v} + s\mathbf{v}$	(4)	$1\mathbf{v} = \mathbf{v}$
(2)	$r(s\mathbf{v}) = (rs)\mathbf{v}$	(5)	$0\mathbf{v} = \mathbf{0}$
(3)	$r(\mathbf{u} + \mathbf{v}) = r\mathbf{u} + r\mathbf{v}$	(6)	$(-1)\mathbf{v} = -\mathbf{v}$

1.2.2. DEFINITION. A **subspace** of a vector space V is a nonempty subset W of V that is a vector space using the operations of V.

A nonempty subset of a vector space is a subspace if and only if it is closed under addition and scalar multiplication, that is, for all $\mathbf{w}_1, \mathbf{w}_2 \in W$ and $r \in \mathbb{R}$, we have $\mathbf{w}_1 + \mathbf{w}_2, r\mathbf{w}_1 \in W$.

1.2.3. DEFINITION. If S is a subset of a vector space V, the **span** of S is the smallest subspace containing S, denoted by span S. A vector $\mathbf{w}$ is a **linear combination** of S if there are $\mathbf{v}_1, \ldots, \mathbf{v}_k \in S$ and $r_1, \ldots, r_k$ such that $\mathbf{w} = r_1\mathbf{v}_1 + \cdots + r_k\mathbf{v}_k$.

Phrases like *the smallest* are dangerous, because they assume that there is a unique smallest subspace. After making such a definition, we should prove that there *is* such a subspace; this process of showing that the definition of an object makes sense is known as showing that the object is **well defined**. It comes up often.

For a nonempty set $S \subset V$, it is a theorem that span S is exactly the set of all linear combinations of elements of S.

1.2.4. DEFINITION. A subset S of a vector space V is **linearly independent** if whenever vectors $\mathbf{v}_1, \ldots, \mathbf{v}_k \in S$ and scalars $r_1, \ldots, r_k \in \mathbb{R}$ satisfy $r_1\mathbf{v}_1 + \cdots + r_k\mathbf{v}_k = \mathbf{0}$, this implies that $r_1 = \cdots = r_k = 0$. We say S is **linearly dependent** if it is not linearly independent. A **basis** for a vector space V is a linearly independent set that spans V.

Saying that $B = \{\mathbf{v}_1, \ldots, \mathbf{v}_n\}$ is a basis for V means that each element of V can be written *uniquely* as a *finite* linear combination of elements of B. For example, let $\mathbb{P}$ be the vector space of polynomials over the real numbers. Then the infinite set $B = \{1\} \cup \{x^j : j \geq 1\}$ is a basis for $\mathbb{P}$. However, if we enlarge our vector space by adding in even very nice power series, like $1 + x/2 + x^2/4 + \cdots + x^n/2^{n+1} + \cdots$, then B is no longer a basis. This power series is not a *finite* linear combination of elements of B.

1.2.5. THEOREM. *Let V be a vector space with a basis having finitely many elements. Then every basis for V has the same (finite) number of elements, called the **dimension** of V and denoted by $\dim V$. We say V is **finite-dimensional**.*

A **linear transformation** A from a vector space V to a vector space W is a function $A : V \to W$ satisfying

$$A(r_1 \mathbf{v}_1 + r_2 \mathbf{v}_2) = r_1 A \mathbf{v}_1 + r_2 A \mathbf{v}_2 \quad \text{for all } \mathbf{v}_1, \mathbf{v}_2 \in V \text{ and } r_1, r_2 \in \mathbb{R}.$$

We use $\mathscr{L}(V, W)$ to denote the set of all linear transformations from V to W and $\mathscr{L}(V)$ for $\mathscr{L}(V, V)$.

A linear transformation is determined by what it does to a basis. If $\mathbf{e}_1, \ldots, \mathbf{e}_m$ is a basis for V, then each element of V has the form $r_1 \mathbf{e}_1 + \cdots + r_m \mathbf{e}_m$, and A applied to such an element yields $r_1 A \mathbf{e}_1 + \cdots + r_m A \mathbf{e}_m$. If $\mathbf{f}_1, \ldots, \mathbf{f}_n$ is a basis for W and $A \mathbf{e}_j = a_{1j} \mathbf{f}_1 + \cdots + a_{nj} \mathbf{f}_n$, then

$$A \left(\sum_{j=1}^{m} r_j \mathbf{e}_j \right) = \sum_{j=1}^{m} r_j \left(\sum_{i=1}^{n} a_{ij} \mathbf{f}_i \right) = \sum_{i=1}^{n} \left(\sum_{j=1}^{m} a_{ij} r_j \right) \mathbf{f}_i.$$

The $n \times m$ matrix $\left[a_{ij} \right]$ is the **matrix representation** of A with respect to the bases $\mathbf{e}_1, \ldots, \mathbf{e}_m$ and $\mathbf{f}_1, \ldots, \mathbf{f}_n$.

The space $\mathscr{L}(V, W)$ is a vector space with the two operations $A + B$ and rA for A and B in $\mathscr{L}(V, W)$ and scalars r, defined by $(A + B)\mathbf{v} = A\mathbf{v} + B\mathbf{v}$ and $(rA)\mathbf{v} = r(A\mathbf{v})$ for $\mathbf{v} \in V$. In $\mathscr{L}(V)$ we also have a multiplication: for $A, B \in \mathscr{L}(V)$, define $BA \in \mathscr{L}(V)$ by $(BA)(\mathbf{v}) = B(A\mathbf{v})$. The matrix representation of BA is the product of the matrix representations of B and A.

1.2.6. DEFINITION. The **kernel** of a linear transformation $A \in \mathscr{L}(V, W)$ is $\ker A = \{\mathbf{v} \in V : A\mathbf{v} = \mathbf{0}\}$, which is a subspace of V. The **range** of A is $\operatorname{ran} A = \{A\mathbf{v} : \mathbf{v} \in V\}$, which is a subspace of W. The **rank** of A is $\operatorname{rank} A = \dim \operatorname{ran} A$.

1.2.7. THEOREM. *Let V, W be vector spaces with V finite-dimensional. For $A \in \mathscr{L}(V, W)$, $\dim \ker A + \operatorname{rank} A = \dim V$.*

1.2.8. COROLLARY. *For $A \in \mathscr{L}(V)$, for V as before, A is invertible if and only if A is one-to-one (i.e., $\ker A = \{\mathbf{0}\}$) if and only if A is onto (i.e., $\operatorname{ran} A = V$).*

1.3 Appendix: Equivalence Relations

Equivalence relations occur frequently in mathematics and will appear occasionally later in this book.

1.3.1. DEFINITION. Let X be a set, and let R be a subset of $X \times X$. Then R is a **relation** on X. Let us write $x \sim y$ if $(x, y) \in R$. We say that R or $\sim$ is an **equivalence relation** if it is

(1) **(reflexive)** $x \sim x$ for all $x \in X$.
(2) **(symmetric)** if $x \sim y$ for any $x, y \in X$, then $y \sim x$.
(3) **(transitive)** if $x \sim y$ and $y \sim z$ for any $x, y, z \in X$, then $x \sim z$.

If $\sim$ is an equivalence relation on X and $x \in X$, then the **equivalence class** $[x]$ is the set $\{y \in X : y \sim x\}$. By $X/\sim$ we mean the collection of all equivalence classes.

1.3.2. EXAMPLES.

(1) Equality is an equivalence relation on any set. Verify this.

(2) Consider the integers $\mathbb{Z}$. Say that $m \equiv n \pmod{12}$ if 12 divides $m - n$. Note that 12 divides $n - n = 0$ for any n, and thus $n \equiv n \pmod{12}$. So it is reflexive. Also if 12 divides $m - n$, then it divides $n - m = -(m - n)$. So $m \equiv n \pmod{12}$ implies that $n \equiv m \pmod{12}$ (i.e., symmetry). Finally, if $l \equiv m \pmod{12}$ and $m \equiv n \pmod{12}$, then we may write $l - m = 12a$ and $m - n = 12b$ for certain integers a, b. Thus $l - n = (l - m) + (m - n) = 12(a + b)$ is also a multiple of 12. Therefore, $l \equiv n \pmod{12}$, which is transitivity.

There are twelve equivalence classes $[r]$ for $0 \leq r < 12$ determined by the remainder r obtained when n is divided by 12. So $[r] = \{12a + r : a \in \mathbb{Z}\}$.

(3) Consider the set $\mathbb{R}$ with the relation $x \leq y$. This relation is reflexive ($x \leq x$) and transitive ($x \leq y$ and $y \leq z$ implies $x \leq z$). However, it is **antisymmetric**: $x \leq y$ and $y \leq x$ both occur if and only if $x = y$. This is not an equivalence relation.

When dealing with functions defined on equivalence classes, we often define the function on an equivalence class in terms of a representative. In order for the function to be well defined, that is, for the definition of the function to make sense, we must check that we get same value regardless of which representative is used.

1.3.3. EXAMPLES.

(1) Consider the set of real numbers $\mathbb{R}$. Say that $x \equiv y \pmod{2\pi}$ if $x - y$ is an integer multiple of 2π. Verify that this is an equivalence relation. Define a function $f([x]) = (\cos x, \sin x)$. We are really defining a function $F(x) = (\cos x, \sin x)$ on $\mathbb{R}$ and asserting that $F(x) = F(y)$ when $x \equiv y \pmod{2\pi}$. Indeed, we then have $y = x + 2\pi n$ for some $n \in \mathbb{Z}$. Since sin and cos are 2π-periodic, we have

$$F(y) = (\cos y, \sin y) = (\cos(x + 2\pi n), \sin(x + 2\pi n)) = (\cos x, \sin x) = F(x).$$

It follows that the function $f([x]) = F(x)$ yields the same answer for every $y \in [x]$. So f is well defined. One can imagine the function f as wrapping the real line around the circle infinitely often, matching up equivalent points.

(2) Consider $\mathbb{R}$ modulo 2π again, and look at $f([x]) = e^x$. Then $0 \equiv 2\pi \pmod{2\pi}$ but $e^0 = 1 \neq e^{2\pi}$. So f is not well defined on equivalence classes.

(3) Now consider Example 1.3.2 (2). We wish to define multiplication modulo 12 by $[n][m] = [nm]$. To check that this is well defined, consider two representatives $n_1, n_2 \in [n]$ and two representatives $m_1, m_2 \in [m]$. Then there are integers a and b such that $n_2 = n_1 + 12a$ and $m_2 = m_1 + 12b$. Then

$$n_2 m_2 = (n_1 + 12a)(m_1 + 12b) = n_1 m_1 + 12(am_1 + n_1 b + 12ab).$$

Therefore, $n_2 m_2 \equiv n_1 m_1 \pmod{12}$, and multiplication modulo 12 is well defined.

Exercises for Section 1.3

A. Put a relation on $C[0,1]$ by $f \sim g$ if $f(k/10) = g(k/10)$ for k with $0 \leq k \leq 10$.

    (a) Verify that this is an equivalence relation.
    (b) Describe the equivalence classes.
    (c) Show that $[f] + [g] = [f + g]$ is a well-defined operation.
    (d) Show that $t[f] = [tf]$ is well defined for all $t \in \mathbb{R}$ and $f \in C[0,1]$.
    (e) Show that these operations make $C[0,1]/\sim$ into a vector space of dimension 11.

B. Consider the set of all infinite decimal expansions $x = a_0.a_1 a_2 a_3 \ldots$, where a_0 is any integer and a_i are digits between 0 and 9 for $i \geq 1$. Say that $x \sim y$ if x and y represent the same real number. That is, if $y = b_0.b_1 b_2 b_3 \ldots$, then $x \sim y$ if (1) $x = y$, or (2) there is an integer $m \geq 1$ such that $a_i = b_i$ for $i < m - 1$, $a_{m-1} = b_{m-1} + 1$, $b_i = 9$ for $i \geq m$ and $a_i = 0$ for $i \geq m$, or (3) there is an integer $m \geq 1$ such that $a_i = b_i$ for $i < m - 1$, $a_{m-1} + 1 = b_{m-1}$, $a_i = 9$ for $i \geq m$ and $b_i = 0$ for $i \geq m$. Prove that this is an equivalence relation.

C. Define a relation on the set $PC[0,1]$ of all piecewise continuous functions on $[0,1]$ (see Definition 5.2.3) by $f \approx g$ if $\{x \in [0,1] : f(x) \neq g(x)\}$ is finite.

    (a) Prove that this is an equivalence relation.
    (b) Decide which of the following functions are well defined.

    (i) $\varphi([f]) = f(0)$ (ii) $\psi([f]) = \int_0^1 f(t)\, dt$ (iii) $\gamma([f]) = \lim_{x \to 1^-} f(x)$

D. Let $d \geq 2$ be an integer. Define a relation on $\mathbb{Z}$ by $m \equiv n \pmod{d}$ if d divides $m - n$.

    (a) Verify that this is an equivalence relation, and describe the equivalence classes.
    (b) Show that $[m] + [n] = [m + n]$ is a well-defined addition.
    (c) Show that $[m][n] = [mn]$ is a well-defined multiplication.
    (d) Let $\mathbb{Z}_d$ denote the equivalence classes modulo d. Prove the distributive law:
        $[k]([m] + [n]) = [k][m] + [k][n].$

E. Say that two real vector spaces V and W are **isomorphic** if there is an invertible linear map T of V onto W.

    (a) Prove that this is an equivalence relation on the collection of all vector spaces.
    (b) When are two finite-dimensional vector spaces isomorphic?

Chapter 2
The Real Numbers

2.1 An Overview of the Real Numbers

Doing analysis in a rigorous way starts with understanding the properties of the
real numbers. Readers will be familiar, in some sense, with the real numbers from
studying calculus. A completely rigorous development of the real numbers requires
checking many details. We attempt to justify one definition of the real numbers
without carrying out the proofs.

Intuitively, we think of the real numbers as the points on a line stretching off to
infinity in both directions. However, to make any sense of this, we must label all the
points on this line and determine the relationship between them from different points
of view. First, the real numbers form an algebraic object known as a field, meaning
that one may add, subtract, and multiply real numbers and divide by nonzero real
numbers. There is also an order on the real numbers compatible with these algebraic
properties, and this leads to the notion of distance between two points.

All of these nice properties are shared by the set of rational numbers:

$$\mathbb{Q} = \left\{ \frac{a}{b} : a, b \in \mathbb{Z}, b \neq 0 \right\}.$$

The ancient Greeks understood how to construct all fractions geometrically and
knew that they satisfied all of the properties mentioned above. However, they were
also aware that there were other points on the line that could be constructed but
were not rational, such as $\sqrt{3}$. While the Greeks were focussed on those numbers
that could be obtained by geometric construction, we have since found other reasonable numbers that do not fit this restrictive definition. The most familiar example is
perhaps π, the area of a circle of radius one. Like the Greeks, we accept the fact that
$\sqrt{3}$ and π are bona fide numbers that must be included on our real line.

We will define the real numbers to be objects with an infinite decimal expansion.
A subtle point is that an infinite decimal expansion is used only as a name for a point
and does mean the sum of an infinite series. It is crucial that we do not use limits to
define the real numbers because we deduce properties of limits from the definition.

K.R. Davidson and A.P. Donsig, *Real Analysis and Applications: Theory in Practice*,
Undergraduate Texts in Mathematics, DOI 10.1007/978-0-387-98098-0_2,
© Springer Science + Business Media, LLC 2010

This construction of the real numbers appears to be strongly dependent on the choice of 10 as the base. We are left with the nagging possibility that the number line we construct depends on the number of digits on our hands. For this reason, some purists prefer a base-independent method of defining the real numbers, albeit a more abstract one. (See Exercise 2.8.L.) Our construction does yield the same object, independent of choice of base; but the proof requires considerable work.

2.2 The Real Numbers and Their Arithmetic

We define a real number using an infinite decimal expansion such as

$$\frac{1}{3} = 0.3333333333333333333333333333333333333333333333333333\ldots$$
$$\sqrt{3} = 1.7320508075688772935274463415058723669428052538 1038\ldots$$
$$\pi = 3.1415926535897932384626433832795028841971693993 7510\ldots$$

In general, an infinite decimal expansion has the form

$$x = a_0.a_1a_2a_3a_4a_5a_6a_7a_8a_9a_{10}a_{11}a_{12}a_{10}a_{11}a_{12}a_{13}a_{14}a_{15}a_{16}a_{17}a_{18}\ldots.$$

Formally, an **infinite decimal expansion** is a function $x(n) = a_n$ from $\{0\} \cup \mathbb{N}$ into $\mathbb{Z}$ such that for all $n \geq 1$, $a_n \in \{0, 1, \ldots, 9\}$.

Be warned that, by this construction, the point usually thought of as $-5/4$ will be denoted by 2.75, for example, because we think of this as $-2 + .75$. The notation is simpler if we do this. After we have finished the construction, we will revert to the standard notation for negative decimals.

To relate infinite decimal expansions to our geometric idea of the real line, start with a line and mark two points on it; and call the left one 0 and the right one 1. Then we can construct points for every integer $\mathbb{Z}$, equally spaced along the line. Now divide each interval from an integer n to $n+1$ into 10 equal pieces, marking the cuts as $n.1$, $n.2$, $\ldots$, $n.9$. Proceed in this way, cutting each interval of length 10^{-k} into 10 equal intervals of length 10^{-k-1} and mark the endpoints by the corresponding number with $k+1$ decimals. In this way, all finite decimals are placed on the line.

To obtain a geometric version of the line, we postulate that for every infinite decimal $x = a_0.a_1a_2a_3\ldots$, there will be a point (also called x) on this line with the property that for each positive integer k, x lies in the interval between the two rational numbers $y = a_0.a_1\ldots a_k$ and $y + 10^{-k}$. For example,

$$3.141592653589 \leq \pi \leq 3.141592653590.$$

One difficulty with using infinite decimal expansions to define the real numbers is that some points have two names. For example consider the expansions $1.000000000\ldots$ and $0.999999999\ldots$. Call them 1 and z, respectively. Clearly these are *different* infinite decimal expansions. However, for each positive integer k,

$$1 - 10^{-k} = 0.\underbrace{99999999999999}_{k} \le z \le 1.$$

Thus the difference between z and 1 is arbitrarily small. It would create quite an un-intuitive line if we decided to make z and 1 different real numbers. To fit in with our intuition, we must agree that $z = 1$. That means that some real numbers (precisely all those numbers with a finite decimal expansion) have two different expansions, one ending in an infinite string of zeros, and the other ending with an infinite string of nines. For example, $0.12500\ldots$ and $0.12499999\ldots$ are the same number.

Formally, this defines an equivalence relation on the set of infinite decimals by pairing off each decimal expansion ending in a string of zeros with the corresponding decimal expansion ending in a string of nines:

$$a_0.a_1 a_2 \ldots a_{k-1} a_k 000 \ldots = a_0.a_1 a_2 \ldots a_{k-1}(a_k - 1)999 \ldots,$$

where $a_k \ne 0$. Each real number is an equivalence class of infinite decimal expansions given by this identification. The set of all real numbers is denoted by $\mathbb{R}$.

To recognize the rationals as a subset of the reals, we need a function F that sends a fraction a/b to an infinite decimal expansion. This is accomplished by long division, as you learned in grade school. For example, to compute $27/14$, divide 14 into 27 to obtain

$$F\left(\tfrac{27}{14}\right) = 1.92857142857142857142857142857142857714\ldots.$$

Notice that this decimal expansion is *eventually periodic* because after the initial 1.9, the six-digit sequence 285714 is repeated ad infinitum. In the exercises, hints are provided to show that an infinite decimal represents a rational number if and only if it is eventually periodic.

We have a built-in order on the real line given by the placement of the points which extends the natural order on the finite decimals. When two infinite decimals $x = a_0.a_1 a_2 \ldots$ and $y = b_0.b_1 b_2 \ldots$ represent *distinct* real numbers, we say that $x < y$ if there is some integer $k \ge 0$ such that $a_i = b_i$ for $i < k$ and $a_k < b_k$. For example, if

$$x = 2.7342118284590452354000064338325028841971693993\ldots,$$
$$y = 2.7342118284590452353999928747135224977572470936\ldots,$$

then $y < x$ because

$$y < 2.73421182845904523539999 < 2.73421182845904523540000 < x.$$

For two real numbers x and y, either $x < y$, $x = y$, or $x > y$.

Next we extend the addition and multiplication operations on $\mathbb{Q}$ to all of $\mathbb{R}$. The basic idea is to extend addition and multiplication on finite decimals to $\mathbb{R}$ respecting the order properties. That is, if $w \le x$ and $y \le z$, then $w + y \le x + z$, and if $x \ge 0$, then $xy \le xz$. Some of the subtleties are explored in the exercises.

A basic fact about the order and these operations is known as the **Archimedean property** of $\mathbb{R}$: *for $x, y > 0$, there is always some $n \in \mathbb{N}$ with $nx > y$.* It is not hard to show this is equivalent to the following almost-obvious fact: *if $z > 0$, then there is some integer $k \geq 0$ so that $10^{-k} < z$.* To see this fact, observe that a decimal expansion of $z = z_0.z_1 z_2 \ldots$ has a first nonzero digit, z_{k-1} and, since $z_{k-1} \geq 1$, we have $z \geq 10^{-(k-1)} > 10^{-k}$.

Finally, consider the distance between two points. The **absolute value function** is $|x| = \max\{x, -x\}$. Define the distance between x and y to be $|x - y|$. This is always nonnegative, and $|x - y| = 0$ only if $x - y = 0$, namely $x = y$.

Exercises for Section 2.2

A.　Why, in defining the order on $\mathbb{R}$, did we insist that x and y be distinct real numbers?
HINT: consider a real number with two decimal expansions.

B.　Prove that $|xy| = |x| \, |y|$ and $|x^{-1}| = |x|^{-1}$.

C.　(a) Prove the **triangle inequality**: $|x + y| \leq |x| + |y|$.
　　　　HINT: Consider x and y of the same sign and different signs as separate cases.
　　　(b) Prove by induction that $|x_1 + x_2 + \cdots + x_n| \leq |x_1| + |x_2| + \cdots + |x_n|$.
　　　(c) Prove the **reverse triangle inequality**: $\bigl| |x| - |y| \bigr| \leq |x - y|$.

D.　(a) Prove that if $x < y$, then there is a rational number r with a finite decimal expansion and an integer k so that $x < r < r + 10^{-k} < y$.
　　　(b) Prove that if $x < y$, then there is an irrational number z such that $x < z < y$.
　　　HINT: Use (a) and add a small multiple of $\sqrt{2}$ to r.

E.　(a) Explain how $x + y$ is worked out for
$$x = 2.1357\underbrace{999999\ldots999999}_{10^7 \text{ nines}}0123456789\underbrace{\ldots}_{10^{19} \text{ repetitions}}0123456789\,34524\ldots,$$
$$y = 3.8642\underbrace{999999\ldots999999}_{10^7 \text{ nines}}9876543210\underbrace{\ldots}_{10^{19} \text{ repetitions}}9876543210\,39736\ldots.$$
　　　(b) How many digits of x and y must we know to determine the first 6 digits of $x + y$?
　　　(c) How many digits of x and y must we know to determine the first 10^8 digits of $x + y$?

F.　Describe an algorithm for adding two infinite decimals. You should work from 'left to right', determining the decimal expansion in order, as much as possible. When are you assured that you know the integer part of the sum? In what circumstance does it remain ambiguous?
HINT: Given infinite decimals a and b, define a carry function $\gamma : \{0\} \cup \mathbb{N} \to \{0, 1\}$ and then define the decimal expansion of $a + b$ in terms of $a(n) + b(n) + \gamma(n)$.

G.　Show that if x and y are known up to k decimal places, then the $x + y$ is known to within $2 \cdot 10^{-k}$, i.e., there is a finite decimal r with $r \leq x + y \leq r + 2 \cdot 10^{-k}$.

H.　An infinite decimal $x = a_0.a_1 a_2 \ldots$ is *eventually periodic* if there are positive integers n and k such that $a_{i+k} = a_i$ for all $i > n$. Show that any decimal expansion which is eventually periodic represents a rational number.　HINT: Compute $10^{n+k} x - 10^n x$.

I.　Prove that the decimal expansion of a rational number p/q is eventually periodic. We will use the Pigeonhole Principle, which states that if $n + 1$ items are divided into n categories, then at least two of the items are in the same category.
　　　(a) Assume $q > 0$. Let r_k be the remainder when 10^k is divided by q. Use the Pigeonhole Principle to find two different exponents $k < k + d$ with the same remainder.
　　　(b) Express $p/q = 10^{-k}\bigl(a + b/(10^d - 1)\bigr)$ with $0 \leq b < 10^d - 1$.
　　　(c) Write b as a d-digit number $b = b_1 b_2 \ldots b_d$ even if it starts with some zeros. Show that the decimal expansion of p/q ends with the infinitely repeated string $b_1 b_2 \ldots b_d$.

J. Explain how the associative property of addition for real numbers: $x + (y+z) = (x+y) + z$ follows from knowing it for for finite decimals.

K. Show that if r is rational and x is irrational, then $r + x$ and, if $r \neq 0$, rx are irrational.

L. Show that the two formulations of the Archimedean property of $\mathbb{R}$ are equivalent.

2.3 The Least Upper Bound Principle

After defining the least upper bound of a set of real numbers, we prove the Least Upper Bound Principle (2.3.3). This result depends crucially on our construction of the real numbers. It will be the basis for the deeper properties of the real line.

2.3.1. DEFINITION. A set $S \subset \mathbb{R}$ is **bounded above** if there is a real number M such that $s \leq M$ for all $s \in S$. We call M an **upper bound** for S. Similarly, S is **bounded below** if there is a real number m such that $s \geq m$ for all $s \in S$, and we call m a **lower bound** for S. A set that is bounded above and below is called **bounded**.

Suppose a nonempty subset S of $\mathbb{R}$ is bounded above. Then L is the **supremum** or **least upper bound** for S if L is an upper bound for S that is smaller than all other upper bounds, i.e., for all $s \in S$, $s \leq L$, and if M is another upper bound for S, then $L \leq M$. It is denoted by $\sup S$.

Similarly, if S is a nonempty subset of $\mathbb{R}$ which is bounded below, the **infimum** or **greatest lower bound**, denoted by $\inf S$, is the number L such that L is an lower bound *and* whenever M is another lower bound for S, then $L \geq M$.

The supremum of a set, if it exists, is unique. We have not defined suprema or infima for sets that are not bounded above or bounded below, respectively. For example, $\mathbb{R}$ itself has neither a supremum nor an infimum. For a nonempty set $S \subseteq \mathbb{R}$, sometimes we write $\sup S = +\infty$ if S is not bounded above and $\inf S = -\infty$ if S is not bounded below. Finally, by convention, $\sup \varnothing = -\infty$ and $\inf \varnothing = +\infty$.

Note that $\sup S = L \in \mathbb{R}$ if and only if L is a upper bound for S and for all $K < L$, there is $x \in S$ with $K < x < L$. There is an equivalent characterization for $\inf S$.

Recall that the **maximum** of a set $S \subset \mathbb{R}$, *if it exists*, is an element $m \in S$ such that $s \leq m$ for all $s \in S$. Thus, when the maximum of a set exists, it is the least upper bound. The situation for the **minimum** of a set and its infimum is the same. We use $\max S$ and $\min S$ to denote the maximum and minimum of S.

2.3.2. EXAMPLES.

(1) If $A = \{4, -2, 5, 7\}$, then any $L \leq -2$ is a lower bound for A and any $M \geq 7$ is an upper bound. So, $\inf A = \min A = -2$ and $\sup A = \max A = 7$.

(2) If $B = \{2, 4, 6, \ldots\}$, then $\inf B = \min B = 2$ and $\sup B = +\infty$.

(3) If $C = \{\pi/n : n \in \mathbb{N}\}$, then $\sup C = \max C = \pi$. However, for any element of C, say π/n, we have a smaller element of C, such as $\pi/(2n)$. So C does not have a

minimum. Clearly, 0 is a lower bound and for all $x > 0$, there is some $\pi/n \in C$ with $\pi/n < x$, showing that 0 is the greatest lower bound.

(4) If $D = \{(-1)^n n/(n+1) : n \in \mathbb{N}\}$, then D has neither a maximum nor a minimum. However, D has upper and lower bounds, and $\inf D = -1$ and $\sup D = 1$. Neither 1 nor -1 belongs to D.

In proving the Least Upper Bound Principle, the definition of the real numbers as *all* infinite decimals is essential. The principle is not true for some subsets of the rational numbers. For example, $\{s \in \mathbb{Q} : s^2 < 2\}$ is bounded above but has no least upper bound in $\mathbb{Q}$.

2.3.3. LEAST UPPER BOUND PRINCIPLE.
Every nonempty subset S of $\mathbb{R}$ that is bounded above has a supremum. Similarly, every nonempty subset S of $\mathbb{R}$ that is bounded below has an infimum.

PROOF. We prove the second statement first, since it is more convenient. Let M be some lower bound for S with decimal expansion $M = m_0.m_1 m_2 \ldots$. Let s be some element of S with decimal expansion $s = s_0.s_1 s_2 \ldots$. Notice that since $m_0 \le M$, we have that m_0 is a lower bound for S. On the other hand, $s < s_0 + 2$. So $s_0 + 2$ is not a lower bound. There are only finitely many integers between m_0 and $s_0 + 1$. Pick the largest of these that is still a lower bound for S, and call it a_0. Since $a_0 + 1$ is not a lower bound, we may also choose an element x_0 in S such that $x_0 < a_0 + 1$.

Next pick the greatest integer a_1 such that $y_1 = a_0 + 10^{-1} a_1$ is a lower bound for S. Since $a_1 = 0$ works and $a_1 = 10$ does not, a_1 belongs to $\{0, 1, \ldots, 9\}$. To verify our choice, pick an element x_1 in S such that $a_0.a_1 \le x_1 < a_0.a_1 + 0.1$. Continue in this way recursively. Figure 2.1 shows how a_2 and x_2 would be chosen.

FIG. 2.1 The second stage ($k = 2$) in the proof.

At the kth stage, we have a lower bound $y_{k-1} = a_0.a_1 \ldots a_{k-1}$ and an element $x_{k-1} \in S$ such that $y_{k-1} \le x_{k-1} < y_{k-1} + 10^{1-k}$. Select the largest integer a_k in $\{0, 1, \ldots, 9\}$ such that $y_k = a_0.a_1 a_2 \ldots a_k$ is a lower bound for S. Since $y_k + 10^{-k}$ is not a lower bound, we also pick an element x_k in S such that $x_k < y_k + 10^{-k}$ to verify our choice.

We claim that $L = a_0.a_1 a_2 \ldots$ is $\inf S$. If $L = y_k$ for some k, then L is a lower bound for S. Otherwise, $L > y_k$ for all k and, in particular, for each k there is $l > k$ with $y_l > y_k$. If $s = s_0.s_1 s_2 \ldots$ is in S, then it follows that $s > y_k$ for each k. By the definition of the order, either $s_i = a_i$ for $1 \le i \le k$ or there is some j, $0 \le j \le k$, with $s_i = a_i$ for $1 \le i < j$ and $s_j > a_j$. If the latter occurs for some k, then $s > L$; if the former occurs for every k, then $s = L$. Either way, L is a lower bound for S.

To see that L is the greatest lower bound, suppose $M = b_0.b_1b_2\ldots > L$. By the definition of the ordering, there is some first integer k such that $b_k > a_k$ and $b_i = a_i$ for all i with $0 \le i < k$. But then

$$M \ge a_0.a_1\ldots a_{k-1}b_k \ge y_k + 10^{-k} > x_k.$$

So M is not a lower bound for S. Hence L is the greatest lower bound.

A simple trick handles upper bounds. Notice that $S \subset \mathbb{R}$ is bounded above if and only if $-S = \{-s : s \in S\}$ is bounded below and that L is an upper bound for S precisely when $-L$ is a lower bound for $-S$. Further, $M < L$ if and only if $-M > -L$, so M is an upper bound of S less than L exactly when $-M$ is a lower bound of $-S$ greater than $-L$. Thus $\sup S = -\inf(-S)$, so $\sup S$ exists.   ∎

Exercises for Section 2.3

A. Suppose that $S \subset \mathbb{R}$ is bounded above. When does S have a maximum? Your answer should be expressed in terms of $\sup S$.

B. A more elegant way to develop the arithmetic properties of the real numbers is to prove the results of this section first and then define addition and multiplication using suprema. Let $\mathscr{D}$ denote the set of all finite decimals.
(a) Let $x, y \in \mathbb{R}$. Prove that $x + y = \sup\{a + b : a, b \in \mathscr{D}, a \le x, b \le y\}$.
(b) Suppose that $x, y \in \mathbb{R}$ are positive. Show that $xy = \sup\{ab : a, b \in \mathscr{D}, 0 \le a \le x, 0 \le b \le y\}$.
(c) How do we define multiplication in general?

C. With $\mathscr{D}$ as in the previous exercise, show that $\sup\{a \in \mathscr{D} : a^2 \le 3\} = \sqrt{3}$.

D. For the following sets, find the supremum and infimum. Which have a max or min?
(a) $A = \{a + a^{-1} : a \in \mathbb{Q}, a > 0\}$.
(b) $B = \{a + (2a)^{-1} : a \in \mathbb{Q}, 0.1 \le a \le 5\}$.
(c) $C = \{xe^{-x} : x \in \mathbb{R}\}$.

E. Show that the decimal expansion for the L in the proof of the Least Upper Bound Principle does not end in a tail of all 9's.

2.4 Limits

The notion of a limit is *the* basic notion of analysis. Limits are the culmination of an infinite process. It is the concern with limits in particular that separates analysis from algebra. Intuitively, to say that a sequence a_n converges to a limit L means that eventually *all* the terms of the (tail of the) sequence approximate the limit value L to *any* desired accuracy. To make this precise, we introduce a subtle definition.

2.4.1. DEFINITION OF THE LIMIT OF A SEQUENCE. A real number L is the **limit** of a sequence of real numbers $(a_n)_{n=1}^{\infty}$ if for *every* $\varepsilon > 0$, there is an integer $N = N(\varepsilon) > 0$ such that

$$|a_n - L| < \varepsilon \quad \text{for all} \quad n \ge N.$$

We say that the sequence $(a_n)_{n=1}^{\infty}$ **converges** to L, and we write $\lim_{n \to \infty} a_n = L$.

The important issue in this definition is that for any desired accuracy, there is a point in the sequence such that *every* element after that point approximates the limit L to the desired accuracy. It suffices to consider only values for ε of the form $\frac{1}{2}10^{-k}$. The statement $|a_n - L| < \frac{1}{2}10^{-k}$ means that a_n and L agree to at least k decimal places. Thus a sequence converges to L precisely when for every k, no matter how large, eventually all the terms of the sequence agree with L to at least k decimals of accuracy.

2.4.2. EXAMPLE. Consider the sequence $(a_n) = (n/(n+1))_{n=1}^{\infty}$, which we claim converges to 1. Observe that $\left|\frac{n}{n+1} - 1\right| = \frac{1}{n+1}$. So if $\varepsilon = \frac{1}{2}10^{-k}$, we can choose $N = 2 \cdot 10^k$. Then for all $n \geq N$,

$$\left|\frac{n}{n+1} - 1\right| = \frac{1}{n+1} \leq \frac{1}{2 \cdot 10^k + 1} < \frac{1}{2}10^{-k} = \varepsilon.$$

We could also choose $N = 73 \cdot 10^k$. It is not necessary to find the best choice for N. But in practice, better estimates can lead to better algorithms for computation.

2.4.3. EXAMPLE. Consider the sequence (a_n) with $a_n = (-1)^n$. Since this flips back and forth between two values that are always distance 2 apart, intuition says that it does not converge. To show this using our definition, we need to show that the definition of limit fails for *any* choice of L. However, for each choice of L, we need find *only one* value of ε that violates the definition. Observe that

$$|a_n - a_{n+1}| = |(-1)^n - (-1)^{n+1}| = 2$$

for all n, no matter how large. So let L be any real number. We notice that L cannot be close to both 1 and -1. To avoid cases, we use a trick. For any real number L,

$$|a_n - L| + |a_{n+1} - L| \geq |(a_n - L) - (a_{n+1} - L)| = |a_n - a_{n+1}| = 2.$$

Thus, for every $n \in \mathbb{N}$,

$$\max\{|a_n - L|, |a_{n+1} - L|\} \geq 1. \tag{2.4.4}$$

Now take $\varepsilon = 1$. If this sequence *did* converge, there would be an integer N such that $|a_n - L| < 1$ for all $n \geq N$. In particular, $|a_N - L|$ and $|a_{N+1} - L|$ are both less than 1, contradicting (2.4.4). Consequently, this sequence does not converge.

2.4.5. EXAMPLE. Consider the sequence $((\sin n)/n)_{n=1}^{\infty}$. The numerator oscillates, but it remains bounded between ± 1 while the denominator goes off to infinity. We obtain the estimates

$$-\frac{1}{n} \leq \frac{\sin n}{n} \leq \frac{1}{n}.$$

We know that $\lim_{n \to \infty} 1/n = 0 = \lim_{n \to \infty} -1/n$, since this is exactly like Example 2.4.2. Therefore, the limit can be computed using a familiar principle from calculus:

2.4.6. THE SQUEEZE THEOREM.
Suppose that three sequences (a_n), (b_n), and (c_n) satisfy

$$a_n \leq b_n \leq c_n \quad \text{for all} \quad n \geq 1 \qquad \text{and} \qquad \lim_{n \to \infty} a_n = \lim_{n \to \infty} c_n = L.$$

Then $\lim_{n \to \infty} b_n = L$.

PROOF. Let $\varepsilon > 0$. Since $\lim_{n \to \infty} a_n = L$, there is some N_1 such that

$$|a_n - L| < \varepsilon \quad \text{for all} \quad n \geq N_1,$$

or equivalently, $L - \varepsilon < a_n < L + \varepsilon$ for all $n \geq N_1$. There is also some N_2 such that

$$|c_n - L| < \varepsilon \quad \text{for all} \quad n \geq N_2$$

or $L - \varepsilon < c_n < L + \varepsilon$ for all $n \geq N_2$. Then, if $n \geq \max\{N_1, N_2\}$, we have

$$L - \varepsilon < a_n \leq b_n \leq c_n < L + \varepsilon.$$

Thus $|b_n - L| < \varepsilon$ for $n \geq \max\{N_1, N_2\}$, as required. ∎

Returning to our example $(\sin n / n)_{n=1}^{\infty}$, we have $\lim_{n \to \infty} \frac{1}{n} = \lim_{n \to \infty} \frac{-1}{n} = 0$. By the Squeeze Theorem,

$$\lim_{n \to \infty} \frac{\sin n}{n} = 0.$$

2.4.7. EXAMPLE.
For a more sophisticated example, consider the sequence $\left(n \sin\left(\frac{1}{n}\right)\right)_{n=1}^{\infty}$. To apply the Squeeze Theorem, we need to obtain an estimate for $\sin \theta$ when the angle θ is small. Consider a sector of the circle of radius 1 with angle θ and the two triangles as shown in Figure 2.2.

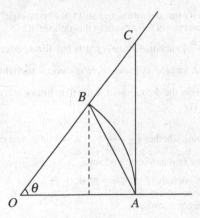

FIG. 2.2 Sector OAB between $\triangle OAB$ and $\triangle OAC$.

Since $\triangle OAB \subset$ sector $OAB \subset \triangle OAC$, we have the same relationship for their areas:

$$\frac{\sin\theta}{2} < \frac{\theta}{2} < \frac{\tan\theta}{2} = \frac{\sin\theta}{2\cos\theta}.$$

A manipulation of these inequalities yields

$$\cos\theta < \frac{\sin\theta}{\theta} < 1.$$

In particular, $\cos\frac{1}{n} < n\sin\frac{1}{n} < 1$. Moreover,

$$\cos\left(\tfrac{1}{n}\right) = \sqrt{1-\sin^2\left(\tfrac{1}{n}\right)} > \sqrt{1-\left(\tfrac{1}{n}\right)^2} > 1 - \frac{1}{n^2}.$$

However,

$$\lim_{n\to\infty} 1 - \frac{1}{n^2} = 1 = \lim_{n\to\infty} 1.$$

Therefore, by the Squeeze Theorem, $\lim_{n\to\infty} n\sin\frac{1}{n} = 1$.

Exercises for Section 2.4

A. In each of the following, compute the limit. Then, using $\varepsilon = 10^{-6}$, find an integer N that satisfies the limit definition.

(a) $\lim_{n\to\infty} \dfrac{\sin n^2}{\sqrt{n}}$ (b) $\lim_{n\to\infty} \dfrac{1}{\log\log n}$ (c) $\lim_{n\to\infty} \dfrac{3^n}{n!}$ (d) $\lim_{n\to\infty} \dfrac{n^2+2n+1}{2n^2-n+2}$ (e) $\lim_{n\to\infty} \sqrt{n^2+n}-$

B. Show that $\lim_{n\to\infty} \sin\frac{n\pi}{2}$ does not exist using the definition of limit.

C. Prove that if $a_n \le b_n$ for $n \ge 1$, $L = \lim_{n\to\infty} a_n$, and $M = \lim_{n\to\infty} b_n$, then $L \le M$.

D. Prove that if $L = \lim_{n\to\infty} a_n$, then $L = \lim_{n\to\infty} a_{2n}$ and $L = \lim_{n\to\infty} a_{n^2}$.

E. Sometimes, a limit is defined informally as follows: "As n goes to infinity, a_n gets closer and closer to L." Find as many faults with this definition as you can.

(a) Can a sequence satisfy this definition and still fail to converge?
(b) Can a sequence converge yet fail to satisfy this definition?

F. Define a sequence $(a_n)_{n=1}^{\infty}$ such that $\lim_{n\to\infty} a_{n^2}$ exists but $\lim_{n\to\infty} a_n$ does not exist.

G. Suppose that $\lim_{n\to\infty} a_n = L$ and $L \ne 0$. Prove there is some N such that $a_n \ne 0$ for all $n \ge N$.

H. Give a careful proof, using the definition of limit, that $\lim_{n\to\infty} a_n = L$ and $\lim_{n\to\infty} b_n = M$ imply that $\lim_{n\to\infty} 2a_n + 3b_n = 2L + 3M$.

I. For each $x \in \mathbb{R}$, determine whether $\left(\dfrac{1}{1+x^n}\right)_{n=1}^{\infty}$ has a limit, and compute it when it exists.

J. Let a_0 and a_1 be positive real numbers, and set $a_{n+2} = \sqrt{a_{n+1}} + \sqrt{a_n}$ for $n \ge 0$.

(a) Show that there is N such that for all $n \ge N$, $a_n \ge 1$.
(b) Let $\varepsilon_n = |a_n - 4|$. Show that $\varepsilon_{n+2} \le (\varepsilon_{n+1} + \varepsilon_n)/3$ for $n \ge N$.
(c) Prove that this sequence converges.

K. Show that the sequence $(\log n)_{n=1}^{\infty}$ does not converge.

2.5 Basic Properties of Limits

2.5.1. PROPOSITION. *If $(a_n)_{n=1}^{\infty}$ is a convergent sequence of real numbers, then the set $\{a_n : n \in \mathbb{N}\}$ is bounded.*

PROOF. Let $L = \lim_{n \to \infty} a_n$. If we set $\varepsilon = 1$, then by the definition of limit, there is some $N > 0$ such that $|a_n - L| < 1$ for all $n \geq N$. In other words,

$$L - 1 < a_n < L + 1 \quad \text{for all} \quad n \geq N.$$

Let $M = \max\{a_1, a_2, \ldots, a_{N-1}, L + 1\}$ and $m = \min\{a_1, a_2, \ldots, a_{N-1}, L - 1\}$. Clearly, for all n, we have $m \leq a_n \leq M$. ∎

It is also crucial that limits respect the arithmetic operations. Proving this is straightforward. The details are left as exercises.

2.5.2. THEOREM. *If $\lim_{n \to \infty} a_n = L$, $\lim_{n \to \infty} b_n = M$, and $\alpha \in \mathbb{R}$, then*

(1) $\lim_{n \to \infty} a_n + b_n = L + M,$

(2) $\lim_{n \to \infty} \alpha a_n = \alpha L,$

(3) $\lim_{n \to \infty} a_n b_n = LM,$ *and*

(4) $\lim_{n \to \infty} \dfrac{a_n}{b_n} = \dfrac{L}{M}$ *if $M \neq 0$.*

In the sequence $(a_n/b_n)_{n=1}^{\infty}$, we ignore terms with $b_n = 0$. There is no problem doing this because $M \neq 0$ implies that $b_n \neq 0$ for all sufficiently large n (see Exercise 2.4.G). (We use "for all sufficiently large n" as shorthand for saying there is some N so that this holds for all $n \geq N$.)

Exercises for Section 2.5

A. Prove Theorem 2.5.2. HINT: For part (4), first bound the denominator away from 0.

B. Compute the following limits.

(a) $\lim_{n \to \infty} \dfrac{\tan \frac{\pi}{n}}{n \sin^2 \frac{2}{n}}$ (b) $\lim_{n \to \infty} \dfrac{2^{100 + 5n}}{e^{4n - 10}}$ (c) $\lim_{n \to \infty} \dfrac{\csc \frac{1}{n}}{n} + \dfrac{2 \arctan n}{\log n}$

C. If $\lim_{n \to \infty} a_n = L > 0$, prove that $\lim_{n \to \infty} \sqrt{a_n} = \sqrt{L}$. Be sure to discuss the issue of when $\sqrt{a_n}$ makes sense. HINT: Express $|\sqrt{a_n} - \sqrt{L}|$ in terms of $|a_n - L|$.

D. Let $(a_n)_{n=1}^{\infty}$ and $(b_n)_{n=1}^{\infty}$ be two sequences of real numbers such that $|a_n - b_n| < \frac{1}{n}$. Suppose that $L = \lim_{n \to \infty} a_n$ exists. Show that $(b_n)_{n=1}^{\infty}$ converges to L also.

E. Find $\lim_{n \to \infty} \dfrac{\log(2 + 3^n)}{2n}$. HINT: $\log(2 + 3^n) = \log 3^n + \log \frac{2 + 3^n}{3^n}$.

F. (a) Let $x_n = \sqrt[n]{n} - 1$. Use the fact that $(1 + x_n)^n = n$ to show that $x_n^2 \leq 2/n$. HINT: Use the Binomial Theorem and throw away most terms.

(b) Hence compute $\lim\limits_{n\to\infty} n^{1/n}$.

G.   Show that the set of rational numbers is **dense** in $\mathbb{R}$, meaning that every real number is a limit of rational numbers.

H.   (a) Show that $\frac{b-1}{b} \le \log b \le b-1$.   HINT: Integrate $1/x$ from 1 to b.
   (b) Apply this to $b = \sqrt[n]{a}$ to show that $\log a \le n(\sqrt[n]{a}-1) \le \sqrt[n]{a}\log a$.
   (c) Hence evaluate $\lim\limits_{n\to\infty} n(\sqrt[n]{a}-1)$.

I.   Suppose that $\lim\limits_{n\to\infty} a_n = L$. Show that $\lim\limits_{n\to\infty} \dfrac{a_1 + a_2 + \cdots + a_n}{n} = L$.

J.   Show that the set $S = \{n + m\sqrt{2} : m, n \in \mathbb{Z}\}$ is dense in $\mathbb{R}$.   HINT: Find infinitely many elements of S in $[0,1]$. Use the Pigeonhole Principle to find two that are close within 10^{-k}.

2.6 Monotone Sequences

We now consider some consequences of the Least Upper Bound Principle (2.3.3).

A sequence (a_n) is **(strictly) monotone increasing** if $a_n \le a_{n+1}$ $(a_n < a_{n+1})$ for all $n \ge 1$. Similarly, we define (strictly) monotone decreasing sequences.

2.6.1. MONOTONE CONVERGENCE THEOREM.
A monotone increasing sequence that is bounded above converges.
A monotone decreasing sequence that is bounded below converges.

PROOF. Suppose $(a_n)_{n=1}^{\infty}$ is an increasing sequence that is bounded above. Then by the Least Upper Bound Principle, there is a number $L = \sup\{a_n : n \in \mathbb{N}\}$. We will show that $\lim\limits_{n\to\infty} a_n = L$.

Let $\varepsilon > 0$ be given. Since $L - \varepsilon$ is not an upper bound for A, there is some integer N such that $a_N > L - \varepsilon$. Then because the sequence is monotone increasing,

$$L - \varepsilon < a_N \le a_n \le L \quad \text{for all} \quad n \ge N.$$

So $|a_n - L| < \varepsilon$ for all $n \ge N$ as required. Therefore, $\lim\limits_{n\to\infty} a_n = L$.

If (a_n) is decreasing and bounded below by B, then the sequence $(-a_n)$ is increasing and bounded above by $-B$. Thus the sequence $(-a_n)_{n=1}^{\infty}$ has a limit $L = \lim\limits_{n\to\infty} -a_n$. Therefore $-L = \lim\limits_{n\to\infty} a_n$ exists.   ∎

2.6.2. EXAMPLE.   Consider the sequence given recursively by

$$a_1 = 1 \quad \text{and} \quad a_{n+1} = \sqrt{2 + \sqrt{a_n}} \quad \text{for all} \quad n \ge 1.$$

Evaluating $a_2, a_3, \ldots, a_9$, we obtain 1.7320508076, 1.8210090645, 1.8301496356, 1.8310735189, 1.831166746, 1.8311761518, 1.8311771007, 1.8311771965. It appears that this sequence increases to some limit.

To prove this, first we show by induction that

$$1 \leq a_n < a_{n+1} < 2 \quad \text{for all} \quad n \geq 1.$$

Since $1 = a_1 < \sqrt{3} = a_2 < 2$, this is valid for $n = 1$. Suppose that it holds for some n. Then

$$a_{n+2} = \sqrt{2 + \sqrt{a_{n+1}}} > \sqrt{2 + \sqrt{a_n}} = a_{n+1} \geq 1,$$

and

$$a_{n+2} = \sqrt{2 + \sqrt{a_{n+1}}} < \sqrt{2 + \sqrt{2}} < 2.$$

This verifies our claim for $n + 1$. Hence by induction, it is valid for each $n \geq 1$.

Therefore, (a_n) is a monotone increasing sequence. So by the Monotone Convergence Theorem (2.6.1), it follows that there is a limit $L = \lim_{n \to \infty} a_n$. It is not clear that there is a nice expression for L. However, once we know that the sequence converges, it is not hard to find a formula for L. Notice that

$$L = \lim_{n \to \infty} a_{n+1} = \lim_{n \to \infty} \sqrt{2 + \sqrt{a_n}} = \sqrt{2 + \sqrt{\lim_{n \to \infty} a_n}} = \sqrt{2 + \sqrt{L}}.$$

We used the fact that the limit of square roots is the square root of the limit (see Exercise 2.5.C). Squaring both sides gives $L^2 - 2 = \sqrt{L}$, and further squaring yields

$$0 = L^4 - 4L^2 - L + 4 = (L - 1)(L^3 + L^2 - 3L - 4).$$

Since $L > 1$, it must be a root of the cubic $p(x) = x^3 + x^2 - 3x - 4$ in the interval $(1, 2)$. There is only one such root. Indeed,

$$p'(x) = 3x^2 + 2x - 3 = 3(x^2 - 1) + 2x$$

is positive on $[1, 2]$. So p is strictly increasing. Since $p(1) = -5$ and $p(2) = 2$, p has exactly one root in between. (See the Intermediate Value Theorem (5.6.1).)

For the amusement of the reader, we give an explicit algebraic formula:

$$L = \frac{1}{3} \left(\sqrt[3]{\frac{79 + \sqrt{2241}}{2}} + \sqrt[3]{\frac{79 - \sqrt{2241}}{2}} - 1 \right).$$

Notice that we proved first that the sequence converged and then evaluated the limit afterward. This is important, for consider the sequence given by $a_1 = 2$ and $a_{n+1} = (a_n^2 + 1)/2$. This is a monotone increasing sequence. Suppose we let L denote the limit and compute

$$L = \lim_{n \to \infty} a_{n+1} = \lim_{n \to \infty} (a_n^2 + 1)/2 = (L^2 + 1)/2.$$

Thus $(L - 1)^2 = 0$, which means that $L = 1$. This is an absurd conclusion because this sequence is monotone increasing and greater than 2. The fault lay in assuming that the limit L actually exists, because instead it diverges to $+\infty$ (see Exercise 2.6.A).

The following easy corollary of the Monotone Convergence Theorem is again a reflection of the completeness of the real numbers. This is just the tool needed to establish the key result of the next section, the Bolzano–Weierstrass Theorem (2.7.2).

Again, the corresponding result for intervals of rational numbers is false. See Example 2.7.6. The result would also be false if we changed closed intervals to open intervals. For example, $\bigcap_{n \geq 1}(0, \frac{1}{n}) = \varnothing$.

2.6.3. NESTED INTERVALS LEMMA.

Suppose that $I_n = [a_n, b_n] = \{x \in \mathbb{R} : a_n \leq x \leq b_n\}$ are nonempty closed intervals such that $I_{n+1} \subseteq I_n$ for each $n \geq 1$. Then the intersection $\bigcap_{n \geq 1} I_n$ is nonempty.

PROOF. Notice that since I_{n+1} is contained in I_n, it follows that

$$a_n \leq a_{n+1} \leq b_{n+1} \leq b_n.$$

Thus (a_n) is a monotone increasing sequence bounded above by b_1; and likewise (b_n) is a monotone decreasing sequence bounded below by a_1. Hence by Theorem 2.6.1, $a = \lim_{n \to \infty} a_n$ exists, as does $b = \lim_{n \to \infty} b_n$. By Exercise 2.4.C, $a \leq b$. Thus

$$a_k \leq a \leq b \leq b_k.$$

Consequently, the point a belongs to I_k for each $k \geq 1$. ∎

Exercises for Section 2.6

A. Say that $\lim_{n \to \infty} a_n = +\infty$ if for every $R \in \mathbb{R}$, there is an integer N such that $a_n > R$ for all $n \geq N$.
Show that a divergent monotone increasing sequence converges to $+\infty$ in this sense.

B. Let $a_1 = 0$ and $a_{n+1} = \sqrt{5 + 2a_n}$ for $n \geq 1$. Show that $\lim_{n \to \infty} a_n$ exists and find the limit.

C. Is $S = \{x \in \mathbb{R} : 0 < \sin(\frac{1}{x}) < \frac{1}{2}\}$ bounded above (below)? If so, find $\sup S$ ($\inf S$).

D. Evaluate $\lim_{n \to \infty} \sqrt[n]{3^n + 5^n}$.

E. Suppose (a_n) is a sequence of positive real numbers such that $a_{n+1} - 2a_n + a_{n-1} > 0$ for all $n \geq 1$. Prove that the sequence either converges or tends to $+\infty$.

F. Let a, b be positive real numbers. Set $x_0 = a$ and $x_{n+1} = (x_n^{-1} + b)^{-1}$ for $n \geq 0$.

(a) Prove that x_n is monotone decreasing.
(b) Prove that the limit exists and find it.

G. Let $a_n = (\sum_{k=1}^{n} 1/k) - \log n$ for $n \geq 1$. **Euler's constant** is defined as $\gamma = \lim_{n \to \infty} a_n$. Show that $(a_n)_{n=1}^{\infty}$ is decreasing and bounded below by zero, and so this limit exists.
HINT: Prove that $1/(n+1) \leq \log(n+1) - \log n \leq 1/n$.

H. Let $x_n = \sqrt{1 + \sqrt{2 + \sqrt{3 + \cdots + \sqrt{n}}}}$.

(a) Show that $x_n < x_{n+1}$.
(b) Show that $x_{n+1}^2 \leq 1 + \sqrt{2} x_n$.   HINT: Square x_{n+1} and factor a 2 out of the square root.
(c) Hence show that x_n is bounded above by 2. Deduce that $\lim_{n \to \infty} x_n$ exists.

I.   (a) Let $(a_n)_{n=1}^\infty$ be a bounded sequence and define a sequence $b_n = \sup\{a_k : k \geq n\}$ for $n \geq 1$. Prove that (b_n) converges. This is the **limit superior** of (a_n), denoted by $\limsup a_n$.

   (b) Without redoing the proof, conclude that the **limit inferior** of a bounded sequence (a_n), defined as $\liminf a_n := \lim_{n \to \infty} (\inf_{k \geq n} a_k)$, always exists.

   (c) Extend the definitions of $\limsup a_n$ and $\liminf a_n$ to unbounded sequences. Provide an example with $\limsup a_n = +\infty$ and $\liminf a_n = -\infty$.

J.   Show that $(a_n)_{n=1}^\infty$ converges to $L \in \mathbb{R}$ if and only if $\limsup a_n = \liminf a_n = L$.

K.   If a sequence (a_n) is not bounded above, show that $\sup\{a_n : n \geq k\} = +\infty$ for all k. What should $\limsup a_n$ be? Formulate and prove a similar statement if (a_n) is not bounded below.

L.   Suppose $(a_n)_{n=1}^\infty$ and $(b_n)_{n=1}^\infty$ are sequences of nonnegative real numbers and $\lim_{n \to \infty} a_n \in \mathbb{R}$ exists. Prove that $\limsup a_n b_n = \lim_{n \to \infty} a_n (\limsup b_n)$.

M.   Suppose that $(a_n)_{n=1}^\infty$ has $a_n > 0$ for all n. Show that $\limsup a_n^{-1} = (\liminf a_n)^{-1}$.

N.   Suppose $(a_n)_{n=1}^\infty$ and $(b_n)_{n=1}^\infty$ are sequences of positive real numbers and $\limsup a_n/b_n < \infty$. Prove that there is a constant M such that $a_n \leq Mb_n$ for all $n \geq 1$.

2.7 Subsequences

Given one sequence, we can build a new sequence, called a subsequence of the original, by picking out some of the entries. Perhaps surprisingly, when the original sequence does not converge, it is often possible to find a subsequence that does.

2.7.1. DEFINITION. A **subsequence** of a sequence $(a_n)_{n=1}^\infty$ is a sequence $(a_{n_k})_{k=1}^\infty = (a_{n_1}, a_{n_2}, a_{n_3}, \ldots)$, where $n_1 < n_2 < n_3 < \cdots$.

For example, $(a_{2k})_{k=1}^\infty$ and $(a_{k^3})_{k=1}^\infty$ are subsequences, where $n_k = 2k$ and $n_k = k^3$, respectively. Notice that if we pick $n_k = k$ for each k, then we get the original sequence; so $(a_n)_{n=1}^\infty$ is a subsequence of itself.

It is easy to verify that if $(a_n)_{n=1}^\infty$ converges to a limit L, then $(a_{n_k})_{k=1}^\infty$ also converges to the same limit. On the other hand, the sequence $(1, 2, 3, \ldots)$ does not have a limit, nor does any subsequence, because any subsequence must diverge to $+\infty$. However, we will show that as long as a sequence remains bounded, it has subsequences that converge.

2.7.2. BOLZANO–WEIERSTRASS THEOREM.
Every bounded sequence of real numbers has a convergent subsequence.

PROOF. Let (a_n) be a sequence bounded by B. Thus the interval $[-B, B]$ contains the whole (infinite) sequence. Now if I is an interval containing infinitely many points of the sequence (a_n), and $I = J_1 \cup J_2$ is the union of two smaller intervals, then at least one of them contains infinitely many points of the sequence, too.

So let $I_1 = [-B, B]$. Split it into two closed intervals of length B, namely $[-B, 0]$ and $[0, B]$. One of these halves contains infinitely many points of (a_n); call it I_2. Similarly, divide I_2 into two closed intervals of length $B/2$. Again pick one, called

I_3, that contains infinitely many points of our sequence. Recursively, we construct a decreasing sequence I_k of closed intervals of length $2^{2-k}B$ such that each contains infinitely many points of our sequence. Figure 2.3 shows the choice of I_3 and I_4, where the terms of the sequence are indicated by vertical lines.

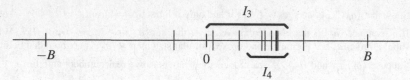

FIG. 2.3 Choice of intervals I_3 and I_4.

By the Nested Interval Lemma (2.6.3), we know that $\bigcap_{k \geq 1} I_k$ contains a number L. Choose an increasing sequence n_k such that a_{n_k} belongs to I_k. This is possible since each I_k contains infinitely many numbers in the sequence, and only finitely many have index less than n_{k-1}. We claim that $\lim_{k \to \infty} a_{n_k} = L$. Indeed, both a_{n_k} and L belong to I_k, and hence

$$|a_{n_k} - L| \leq |I_k| = 2^{-k}(4B).$$

The right-hand side tends to 0, and thus $\lim_{k \to \infty} a_{n_k} = L$. ∎

2.7.3. EXAMPLE. Consider the sequence $(a_n) = (\text{sign}(\sin n))_{n=1}^{\infty}$, where the sign function takes values ± 1 depending on the sign of x except for $\text{sign}\, 0 = 0$. Without knowing anything about the properties of the sine function, we can observe that the sequence (a_n) takes at most three different values. At least one of these values is taken infinitely often. Thus it is possible to deduce the existence of a subsequence that is constant and therefore converges.

Using our knowledge of sine allows us to get somewhat more specific. Now $\sin x = 0$ exactly when x is an integer multiple of π. Since π is irrational, $k\pi$ is never an integer for $k > 0$. Therefore, a_n takes only the values ± 1. Note that $\sin x > 0$ if there is an integer k such that $2k\pi < x < (2k+1)\pi$; and $\sin x < 0$ if there is an integer k such that $(2k-1)\pi < x < 2k\pi$. Observe that n increases by steps of length 1, while the intervals on which $\sin x$ takes positive or negative values has length $\pi \approx 3.14$. Consequently, a_n takes the value $+1$ for three or four terms in a row, followed by three or four terms taking the value -1. Consequently, both 1 and -1 are limits of certain subsequences of (a_n).

2.7.4. EXAMPLE. Consider the sequence $(a_n) = (\sin n)_{n=1}^{\infty}$. As the angles n radians for $n \geq 1$ are marked on a circle, they appear gradually to fill in a dense subset. If this can be demonstrated, we should be able to show that $\sin \theta$ is a limit of a subsequence of our sequence for every θ in $[0, 2\pi]$.

The key is to approximate the angle 0 modulo 2π by integers. Let m be a positive integer and let $\varepsilon > 0$. Choose an integer N so large that $N\varepsilon > 2\pi$. Divide the circle into N arcs of length $2\pi/N$ radians each. Then consider the $N+1$ points $0, m, 2m, \ldots, Nm$ modulo 2π on the circle. Since there are $N+1$ points distributed into only N arcs, the Pigeonhole Principle implies that at least one arc contains two points, say im and jm, where $i < j$. Then $n = jm - im$ represents an angle of absolute value at most $2\pi/N < \varepsilon$ radians up to a multiple of 2π. That is, $n = \psi + 2\pi s$ for some integer s and real number $|\psi| < \varepsilon$. In particular, $|\sin n| < \varepsilon$ and $n \geq m$. Moreover, since π is not rational, n is not an exact multiple of 2π.

So given $\theta \in [0, 2\pi]$, construct a subsequence as follows. Let $n_1 = 1$. Recursively we construct an increasing sequence n_k such that

$$|\sin n_k - \sin \theta| < \frac{1}{k}.$$

Once n_k is defined, take $\varepsilon = \frac{1}{k+1}$ and $m = n_k + 1$. As in the previous paragraph, there is an integer $n > n_k$ such that $n = \psi + 2\pi s$ and $|\psi| < \frac{1}{k+1}$. Thus there is a positive integer t such that $|\theta - t\psi| < \frac{1}{k+1}$. Therefore

$$|\sin(tn) - \sin(\theta)| = |\sin(t\psi) - \sin(\theta)| \leq |t\psi - \theta| < \frac{1}{k+1}. \tag{2.7.5}$$

Set $n_{k+1} = tn$. This completes the induction. The result is a subsequence such that

$$\lim_{k \to \infty} \sin(n_k) = \sin \theta.$$

To verify equation (2.7.5), recall the Mean Value Theorem (6.2.2). There is a point ξ between $t\psi$ and θ such that

$$\left| \frac{\sin(t\psi) - \sin(\theta)}{t\psi - \theta} \right| = |\cos \xi| \leq 1.$$

Rearranging yields $|\sin(t\psi) - \sin(\theta)| \leq |t\psi - \theta|$.

Therefore, we have shown that every value in the interval $[-1, 1]$ is the limit of some subsequence of the sequence $(\sin n)_{n=1}^{\infty}$.

2.7.6. EXAMPLE. Consider the sequence $b_1 = 3$ and $b_{n+1} = (b_n + 8/b_n)/2$. Notice that

$$b_{n+1}^2 - 8 = \frac{b_n^2 + 16 + (64/b_n^2) - 32}{4} = \frac{b_n^2 - 16 + (64/b_n^2)}{4}$$

$$= \frac{(b_n - 8/b_n)^2}{4} = \frac{(b_n^2 - 8)^2}{4b_n^2}.$$

It follows that $b_n^2 > 8$ for all $n \geq 2$, and $b_1^2 - 8 = 1 > 0$ also. Thus

$$0 < b_{n+1}^2 - 8 < \frac{(b_n^2 - 8)^2}{32}.$$

Iterating this, we obtain $b_2^2 - 8 < 32^{-1}$, $b_3^2 - 8 < 32^{-3}$, and $b_4^2 - 8 < 32^{-7}$. In general, we establish by induction that

$$0 < b_n^2 - 8 < 32^{1-2^{n-1}}.$$

Since b_n is positive and $b^2 - 8 = (b - \sqrt{8})(b + \sqrt{8})$, it follows that

$$0 < b_n - \sqrt{8} = \frac{b_n^2 - 8}{b_n + \sqrt{8}} < \frac{32^{1-2^{n-1}}}{2\sqrt{8}} < 6(32^{-2^{n-1}}).$$

Lastly, using the fact that $32^2 = 1024 > 10^3$, we obtain

$$0 < b_n - \sqrt{8} < 10 \cdot 10^{-3 \cdot 2^{n-2}}.$$

In particular, $\lim_{n \to \infty} b_n = \sqrt{8}$. In fact, the convergence is so rapid that b_{10} approximates $\sqrt{8}$ to more than 750 digits of accuracy. See Example 11.2.2 for a more general analysis in terms of Newton's method.

Let $a_n = 8/b_n$. Then a_n is monotone increasing to $\sqrt{8}$. Both a_n and b_n are rational, but $\sqrt{8}$ is irrational. Thus the sets $J_n = \{x \in \mathbb{Q} : a_n \leq x \leq b_n\}$ form a decreasing sequence of nonempty intervals of *rational* numbers with empty intersection.

Exercises for Section 2.7

A. Show that $(a_n) = \left(\dfrac{n \cos^n(n)}{\sqrt{n^2 + 2n}} \right)_{n=1}^{\infty}$ has a convergent subsequence.

B. Does the sequence $(b_n) = \left(n + \cos(n\pi) \sqrt{n^2 + 1} \right)_{n=1}^{\infty}$ have a convergent subsequence?

C. Does the sequence $(a_n) = (\cos \log n)_{n=1}^{\infty}$ converge?

D. Show that every sequence has a monotone subsequence.

E. Use trig identities to show that $|\sin x - \sin y| \leq |x - y|$.
       HINT: Let $a = (x+y)/2$ and $b = (x-y)/2$. Use the addition formula for $\sin(a \pm b)$.

F. Define $x_1 = 2$ and $x_{n+1} = \frac{1}{2}(x_n + 5/x_n)$ for $n \geq 1$.

   (a) Find a formula for $x_{n+1}^2 - 5$ in terms of $x_n^2 - 5$.
   (b) Hence evaluate $\lim_{n \to \infty} x_n$.
   (c) Compute the first ten terms on a computer or a calculator.
   (d) Show that the tenth term approximates the limit to over 600 decimal places.

G. Let $(x_n)_{n=1}^{\infty}$ be a sequence of real numbers. Suppose that there is a real number L such that $L = \lim_{n \to \infty} x_{3n-1} = \lim_{n \to \infty} x_{3n+1} = \lim_{n \to \infty} x_{3n}$. Show that $\lim_{n \to \infty} x_n$ exists and equals L.

H. Let $(x_n)_{n=1}^{\infty}$ be a sequence in $\mathbb{R}$. Suppose there is a number L such that every subsequence $\left(x_{n_k} \right)_{k=1}^{\infty}$ has a subsubsequence $\left(x_{n_{k(l)}} \right)_{l=1}^{\infty}$ with $\lim_{l \to \infty} x_{n_{k(l)}} = L$. Show that the whole sequence converges to L.   HINT: If not, you could find a subsequence bounded away from L.

I. Suppose $(x_n)_{n=1}^{\infty}$ is a sequence in $\mathbb{R}$, and that L_k are real numbers with $\lim_{k \to \infty} L_k = L$. If for each $k \geq 1$, there is a subsequence of $(x_n)_{n=1}^{\infty}$ converging to L_k, show that some subsequence converges to L.   HINT: Find an increasing sequence n_k such that $|x_{n_k} - L| < 1/k$.

J.   (a) Suppose that $(x_n)_{n=1}^{\infty}$ is a sequence of real numbers. If $L = \liminf x_n$, show that there is a subsequence $(x_{n_k})_{k=1}^{\infty}$ such that $\lim_{k \to \infty} x_{n_k} = L$.

      (b) Similarly, prove that there is a subsequence $(x_{n_l})_{l=1}^{\infty}$ such that $\lim_{l \to \infty} x_{n_l} = \limsup x_n$.

K.   Let $(x_n)_{n=1}^{\infty}$ be an arbitrary sequence. Prove that there is a subsequence $(x_{n_k})_{k=1}^{\infty}$ which converges or $\lim_{k \to \infty} x_{n_k} = \infty$ or $\lim_{k \to \infty} x_{n_k} = -\infty$.

L.   Construct a sequence $(x_n)_{n=1}^{\infty}$ such that for every real number L, there is a subsequence $(x_{n_k})_{k=1}^{\infty}$ with $\lim_{k \to \infty} x_{n_k} = L$.

2.8 Cauchy Sequences

Can we decide whether a sequence converges *without* first finding the value of the limit? To do this, we need an intrinsic property of a sequence which is equivalent to convergence that does not make use of the value of the limit. This intrinsic property shows which sequences are 'supposed' to converge. This leads us to the notion of a subset of $\mathbb{R}$ being *complete* if all sequences in the subset that are 'supposed' to converge actually do. As we shall see, this completeness property has been built into the real numbers by our construction of infinite decimals.

To obtain an appropriate condition, notice that if a sequence (a_n) converges to L, then as the terms get close to the limit, they are getting close to each other.

2.8.1. PROPOSITION. *Let $(a_n)_{n=1}^{\infty}$ be a sequence converging to L. For every $\varepsilon > 0$, there is an integer N such that*

$$|a_n - a_m| < \varepsilon \quad for\ all \quad m, n \geq N.$$

PROOF. Fix $\varepsilon > 0$ and use the value $\varepsilon/2$ in the definition of limit. Then there is an integer N such that $|a_n - L| < \varepsilon/2$ for all $n \geq N$. Thus if $m, n \geq N$, we obtain

$$|a_n - a_m| \leq |a_n - L| + |L - a_m| < \frac{\varepsilon}{2} + \frac{\varepsilon}{2} = \varepsilon. \qquad \blacksquare$$

In order for N to work in the conclusion, for every $m \geq N$, a_m must be within ε of a_N. It is not enough to just have a_N and a_{N+1} close (see Exercise 2.8.B).

We make the conclusion of this proposition into a definition. This definition retains the flavour of the definition of a limit, in that it has the same logical structure: *For all $\varepsilon > 0$, there is an integer N ....*

2.8.2. DEFINITION. A sequence $(a_n)_{n=1}^{\infty}$ of real numbers is called a **Cauchy sequence** provided that for *every* $\varepsilon > 0$, there is an integer N such that

$$|a_m - a_n| < \varepsilon \quad for\ all \quad m, n \geq N.$$

2.8.3. PROPOSITION. *Every Cauchy sequence is bounded.*

PROOF. The proof is basically the same as Proposition 2.5.1. Let $(a_n)_{n=1}^{\infty}$ be a Cauchy sequence. Taking $\varepsilon = 1$, find N so large that

$$|a_n - a_N| < 1 \quad \text{for all} \quad n \geq N.$$

It follows that the sequence is bounded by $\max\{|a_1|, \ldots, |a_{N-1}|, |a_N| + 1\}$. ∎

Since the definition of a Cauchy sequence does not require the use of a potential limit L, it permits the following definition.

2.8.4. DEFINITION. A subset S of $\mathbb{R}$ is said to be **complete** if every Cauchy sequence (a_n) in S (that is, $a_n \in S$) converges to a point in S.

This brings us to an important conclusion about the real numbers themselves, another property that distinguishes the real numbers from the rational numbers.

2.8.5. COMPLETENESS THEOREM.
Every Cauchy sequence of real numbers converges. So $\mathbb{R}$ is complete.

PROOF. Suppose that $(a_n)_{n=1}^{\infty}$ is a Cauchy sequence. By Proposition 2.8.3, $\{a_n : n \geq 1\}$ is bounded. By the Bolzano–Weierstrass Theorem (2.7.2), this sequence has a convergent subsequence, say

$$\lim_{k \to \infty} a_{n_k} = L.$$

Let $\varepsilon > 0$. From the definition of Cauchy sequence for $\varepsilon/2$, there is an integer N such that

$$|a_m - a_n| < \frac{\varepsilon}{2} \quad \text{for all} \quad m, n \geq N.$$

And from the definition of limit using $\varepsilon/2$, there is an integer K such that

$$|a_{n_k} - L| < \frac{\varepsilon}{2} \quad \text{for all} \quad k \geq K.$$

Pick any $k \geq K$ such that $n_k \geq N$. Then for every $n \geq N$,

$$|a_n - L| \leq |a_n - a_{n_k}| + |a_{n_k} - L| < \frac{\varepsilon}{2} + \frac{\varepsilon}{2} = \varepsilon.$$

So $\lim_{n \to \infty} a_n = L$. ∎

2.8.6. REMARK. This theorem is not true for the rational numbers. Define the sequence $(a_n)_{n=1}^{\infty}$ by

$$a_1 = 1.4, \quad a_2 = 1.41, \quad a_3 = 1.414, \quad a_4 = 1.4142, \quad a_5 = 1.41421, \ldots$$

and in general, a_n is the first $n+1$ digits in the decimal expansion of $\sqrt{2}$. If n and m are greater than N, then a_n and a_m agree for at least first $N+1$ digits. Thus

$$|a_n - a_m| < 10^{-N} \quad \text{for all} \quad m, n \geq N.$$

This shows that $(a_n)_{n=1}^{\infty}$ is a Cauchy sequence of rational numbers. (Why?)

However, this sequence has no limit *in the rationals*. In our terminology, $\mathbb{Q}$ is not complete. Of course, this sequence does converge to a real number, namely $\sqrt{2}$. This is one way to see the essential difference between $\mathbb{R}$ and $\mathbb{Q}$: the set of real numbers is complete and $\mathbb{Q}$ is not.

2.8.7. EXAMPLE. Let α be an arbitrary real number. Define $a_n = [n\alpha]/n$, where $[x]$ is the nearest integer to x. Then $\big|[n\alpha] - n\alpha\big| \leq 1/2$. So

$$|a_n - \alpha| = \frac{\big|[n\alpha] - n\alpha\big|}{n} \leq \frac{1}{2n}.$$

We claim $\lim_{n\to\infty} a_n = \alpha$. Indeed, given $\varepsilon > 0$, choose N so large that $\frac{1}{N} < \varepsilon$. Then for $n \geq N$, $|a_n - \alpha| < \varepsilon/2$. Moreover, if $m, n \geq N$,

$$|a_n - a_m| \leq |a_n - \alpha| + |\alpha - a_m| < \frac{\varepsilon}{2} + \frac{\varepsilon}{2} = \varepsilon.$$

Thus this sequence is Cauchy.

2.8.8. EXAMPLE. Consider the infinite **continued fraction**

$$\cfrac{1}{2 + \cfrac{1}{2 + \cfrac{1}{2 + \cfrac{1}{2 + \cdots}}}}$$

To make sense of this, it has to be interpreted as the limit of the finite fractions

$$a_1 = \frac{1}{2}, \qquad a_2 = \frac{1}{2 + \frac{1}{2}} = \frac{2}{5}, \qquad a_3 = \frac{1}{2 + \frac{1}{2 + \frac{1}{2}}} = \frac{5}{12}, \qquad \cdots.$$

We need a better way of defining the general term. In this case, there is a recursive formula for obtaining one term from the preceding one:

$$a_1 = \frac{1}{2}, \qquad a_{n+1} = \frac{1}{2 + a_n} \quad \text{for} \quad n \geq 1.$$

In order to establish convergence, we will show that (a_n) is Cauchy. Consider

$$a_{n+1} - a_{n+2} = \frac{1}{2+a_n} - \frac{1}{2+a_{n+1}} = \frac{a_{n+1} - a_n}{(2+a_n)(2+a_{n+1})}.$$

Now $a_1 > 0$, and it is readily follows that $a_n > 0$ for all $n \geq 2$ by induction. Hence the denominator $(2+a_n)(2+a_{n+1})$ is greater than 4. So we obtain

$$|a_{n+1} - a_{n+2}| < \frac{|a_n - a_{n+1}|}{4} \quad \text{for all} \quad n \geq 1.$$

Since $|a_1 - a_2| = 1/10$, we may iterate this inequality to estimate

$$|a_2 - a_3| < \frac{1}{10 \cdot 4}, \quad |a_3 - a_4| < \frac{1}{10 \cdot 4^2}, \quad |a_n - a_{n+1}| < \frac{1}{10 \cdot 4^{n-1}} = \tfrac{2}{5}(4^{-n}).$$

The general formula estimating the difference may be verified by induction.

Now it is straightforward to estimate the difference between arbitrary terms a_m and a_n for $m < n$:

$$\begin{aligned}
|a_m - a_n| &= |(a_m - a_{m+1}) + (a_{m+1} - a_{m+2}) + \cdots + (a_{n-1} - a_n)| \\
&\leq |a_m - a_{m+1}| + |a_{m+1} - a_{m+2}| + \cdots + |a_{n-1} - a_n| \\
&< \tfrac{2}{5}(4^{-m} + 4^{-m-1} + \cdots + 4^{1-n}) < \frac{2 \cdot 4^{-m}}{5(1 - \frac{1}{4})} = \frac{8}{15} 4^{-m} < 4^{-m}.
\end{aligned}$$

This tells us that our sequence is Cauchy. Indeed, if $\varepsilon > 0$, choose N such that $4^{-N} < \varepsilon$. Then

$$|a_m - a_n| < 4^{-m} \leq 4^{-N} < \varepsilon \quad \text{for all} \quad m, n \geq N.$$

Therefore by the Completeness Theorem 2.8.5, it follows that $(a_n)_{n=1}^{\infty}$ converges; say, $\lim_{n \to \infty} a_n = L$. To calculate L, use the recurrence relation

$$L = \lim_{n \to \infty} a_n = \lim_{n \to \infty} a_{n+1} = \lim_{n \to \infty} \frac{1}{2 + a_n} = \frac{1}{2 + L}.$$

It follows that $L^2 + 2L - 1 = 0$. Solving yields $L = \pm\sqrt{2} - 1$. Since $L > 0$, we see that $L = \sqrt{2} - 1$.

We have accumulated five different results for $\mathbb{R}$ that distinguish it from $\mathbb{Q}$.

(1) the Least Upper Bound Principle (2.3.3),
(2) the Monotone Convergence Theorem (2.6.1),
(3) the Nested Intervals Lemma (2.6.3),
(4) the Bolzano–Weierstrass Theorem (2.7.2),
(5) the Completeness Theorem (2.8.5).

It turns out that they are all equivalent. Indeed, each of the proofs of items (2) to (5) relies only on the previous item in our list. To show how the Completeness Theorem implies the Least Upper Bound Principle, go through our proof to obtain an increasing sequence of lower bounds, y_k, and a decreasing sequence of elements $x_k \in S$ with $x_k < y_k + 10^{-k}$. Show that the sequence $x_1, y_1, x_2, y_2, \ldots$ is Cauchy. The limit L will be the greatest lower bound. Fill in the details yourself (Exercise 2.8.G).

Exercises for Section 2.8

A. Let (x_n) be Cauchy with a subsequence (x_{n_k}) such that $\lim_{k \to \infty} x_{n_k} = a$. Show that $\lim_{n \to \infty} x_n = a$.

B. Give a sequence (a_n) such that $\lim_{n \to \infty} |a_n - a_{n+1}| = 0$, but the sequence does not converge.

C. Let (a_n) be a sequence such that $\lim_{N \to \infty} \sum_{n=1}^{N} |a_n - a_{n+1}| < \infty$. Show that (a_n) is Cauchy.

D. If $(x_n)_{n=1}^{\infty}$ is Cauchy, show that it has a subsequence (x_{n_k}) such that $\sum_{k=1}^{\infty} |x_{n_k} - x_{n_{k+1}}| < \infty$.

E. Suppose that (a_n) is a sequence such that $a_{2n} \leq a_{2n+2} \leq a_{2n+3} \leq a_{2n+1}$ for all $n \geq 0$. Show that this sequence is Cauchy if and only if $\lim_{n \to \infty} |a_n - a_{n+1}| = 0$.

F. Give an example of a sequence (a_n) such that $a_{2n} \leq a_{2n+2} \leq a_{2n+3} \leq a_{2n+1}$ for all $n \geq 0$ which does not converge.

G. Fill in the details of how the Completeness Theorem implies the Least Upper Bound Principle.

H. Let $a_0 = 0$ and set $a_{n+1} = \cos(a_n)$ for $n \geq 0$. Try this on your calculator (use radian mode!).

    (a) Show that $a_{2n} \leq a_{2n+2} \leq a_{2n+3} \leq a_{2n+1}$ for all $n \geq 0$.
    (b) Use the Mean Value Theorem to find an explicit number $r < 1$ such that
    $|a_{n+2} - a_{n+1}| \leq r|a_n - a_{n+1}|$ for all $n \geq 0$. Hence show that this sequence is Cauchy.
    (c) Describe the limit geometrically as the intersection point of two curves.

I. Evaluate the continued fraction

$$1 + \cfrac{1}{1 + \cfrac{1}{1 + \cfrac{1}{1 + \cdots}}} \, .$$

J. Let $x_0 = 0$ and $x_{n+1} = \sqrt{5 - 2x_n}$ for $n \geq 0$. Show that this sequence converges and compute the limit. HINT: Show that the even terms increase and the odd terms decrease.

K. Consider an infinite binary expansion $(0.e_1e_2e_3\ldots)_{\text{base } 2}$, where each $e_i \in \{0, 1\}$. Show that $a_n = \sum_{i=1}^{n} 2^{-i} e_i$ is Cauchy for every choice of zeros and ones.

L. One base-independent construction of the real numbers uses Cauchy sequences of rational numbers. This exercise asks for the definitions that go into such a proof.

    (a) Find a way to decide when two Cauchy sequences should determine the same real number without using their limits. HINT: Combine the two sequences into one.
    (b) Your definition in (a) should be an equivalence relation. Is it? (See Appendix 1.3.)
    (c) How are addition and multiplication defined?
    (d) How is the order defined?

2.9 Countable Sets

Cardinality measures the size of a set in the crudest of ways—by counting the numbers of elements. Obviously, the number of elements in a set could be 0, 1, 2, 3, 4, or some other finite number. Or a set can have infinitely many elements. Perhaps

surprisingly, not all infinite sets have the same cardinality. We distinguish only between sets having the smallest infinite cardinality (countably infinite sets) and all
larger cardinalities (uncountable sets). We use the term countable for sets that are
either countably infinite or finite.

2.9.1. DEFINITION. Two sets A and B have the same **cardinality** if there is a
bijection f from A onto B. Write $|A| = |B|$ in this case. We say that the cardinality
of A is at most that of B (write $|A| \leq |B|$) if there is an *injection* f from A into B.

The definition says simply that if all of the elements of A can be paired, one-to-
one, with all of the elements of B, then A and B have the same size. If A fits inside
B in a one-to-one manner, then A is smaller than or equal to B. It is natural to ask
whether $|A| \leq |B|$ and $|B| \leq |A|$ imply $|A| = |B|$. The answer is yes, but this is not
obvious for infinite sets. The Schroeder–Bernstein Theorem establishes this, but we
do not include a proof.

2.9.2. EXAMPLES.

(1) The cardinality of any finite set is the number of elements, and this number
belongs to $\{0, 1, 2, 3, 4, \dots\}$. This property is, essentially, the definition of finite set.

(2) Many sets encountered in analysis are infinite, meaning that they are not fi-
nite. The sets of natural numbers $\mathbb{N}$, integers $\mathbb{Z}$, rational numbers $\mathbb{Q}$, and real numbers $\mathbb{R}$ are all infinite. Moreover, we have the containments $\mathbb{N} \subset \mathbb{Z} \subset \mathbb{Q} \subset \mathbb{R}$.
Therefore $|\mathbb{N}| \leq |\mathbb{Z}| \leq |\mathbb{Q}| \leq |\mathbb{R}|$. Notice that the integers can be written as a
list $0, 1, -1, 2, -2, 3, -3, \dots$. This amounts to defining a bijection $f : \mathbb{N} \to \mathbb{Z}$ by
$f(2n-1) = 1-n$ and $f(2n) = n$ for $n \geq 1$. Therefore, $|\mathbb{N}| = |\mathbb{Z}|$.

2.9.3. DEFINITION. A set A is a **countable set** is it is finite or if $|A| = |\mathbb{N}|$. If
$|A| = |\mathbb{N}|$, we say that A is **countably infinite**. The cardinal $|\mathbb{N}|$ is also denoted by
$\aleph_0$, pronounced **aleph nought**. Aleph is the first letter of the Hebrew alphabet.
An infinite set that is not countable is called an **uncountable set**.

Equivalently, A is countable if the elements of A may be listed as $a_1, a_2, a_3, \dots$.
Indeed, the list itself determines a bijection f from $\mathbb{N}$ to A by $f(k) = a_k$. It is a basic
fact that countable sets are the smallest infinite sets.
Notice that two uncountable sets might have different cardinalities.

2.9.4. LEMMA. *Every infinite subset of $\mathbb{N}$ is countable. Moreover, if A is an
infinite set such that $|A| \leq |\mathbb{N}|$, then $|A| = |\mathbb{N}|$.*

PROOF. Any nonempty subset X of $\mathbb{N}$ has a smallest element. This follows from
induction: if X does not have a smallest element, then $1 \notin X$ and $1, \dots, n$ all not in
X imply $n + 1 \notin X$. By induction, X is empty, a contradiction.

Let B be an infinite subset of $\mathbb{N}$. List the elements of B in increasing order as $b_1 < b_2 < b_3 < \cdots$. This is done by choosing the smallest element b_1, then the smallest of the remaining set $B \setminus \{b_1\}$, then the smallest of $B \setminus \{b_1, b_2\}$, and so on. The result is an infinite list of elements of B in increasing order. It must include every element $b \in B$ because $\{n \in B : n \le b\}$ is finite, containing say k elements. Then $b_k = b$. As noted before the proof, this implies that $|B| = |\mathbb{N}|$.

Now consider an infinite set A with $|A| \le |\mathbb{N}|$. By definition, there is an injection f of A into $\mathbb{N}$. Let $B = f(A)$. Note that f is a bijection of A onto B. Thus B is an infinite subset of $\mathbb{N}$. So $|A| = |B| = |\mathbb{N}|$. ∎

2.9.5. PROPOSITION. *The set* $\mathbb{N} \times \mathbb{N}$ *is countable.*

PROOF. Rather than starting with the formula of a bijection from $\mathbb{N}$ to $\mathbb{N} \times \mathbb{N}$, note that each 'diagonal set' $D_n = \{(i,j) \in \mathbb{N} \times \mathbb{N} : i + j = n + 1\}$, $n \ge 1$, is finite. Thus, if we work through these sets in some methodical way, any pair (i,j) will be reached in finitely many steps. See Figure 2.4.

Noting that $|D_n| = n$ and $1 + 2 + \ldots + n = n(n+1)/2$, we define our bijection for $m \in \mathbb{N}$ by first picking n such that $n(n-1)/2 < m \le n(n+1)/2$. Letting $k = m - n(n-1)/2$, we define $\varphi(m)$ to be $(k, n+1-k)$. It is routine, if tedious, to verify that φ is a bijection, i.e., one-to-one and onto. ∎

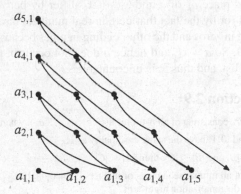

FIG. 2.4 The ordering on $\mathbb{N} \times \mathbb{N}$.

2.9.6. COROLLARY. *The countable union of countable sets is countable.*

PROOF. Let $A_1, A_2, A_3, \ldots$ be countable sets. To avoid repetition, let $B_1 = A_1$ and $B_i = A_i \setminus \cup_{k=1}^{i-1} A_k$. Each B_i is countable, so list its elements as $b_{i,1}, b_{i,2}, b_{i,3}, \ldots$. Map $A = \cup_{i \ge 1} A_i = \cup_{i \ge 1} B_i$ into $\mathbb{N} \times \mathbb{N}$ by $f(b_{ij}) = (i,j)$. This is an injection; therefore $|A| \le |\mathbb{N} \times \mathbb{N}| = |\mathbb{N}|$. Hence the union is countable. ∎

2.9.7. COROLLARY. *The set* $\mathbb{Q}$ *of rational numbers is countable.*

PROOF. Observe that $\mathbb{Z} \times \mathbb{N}$ is countable, since we can take the bijection $f : \mathbb{N} \to \mathbb{Z}$ of Example 2.9.2 (2) and use it to define $g : \mathbb{N} \times \mathbb{N} \to \mathbb{Z} \times \mathbb{N}$ by $g(n,m) = (f(n),m)$, which you can check is a bijection.

Define a map from $\mathbb{Q}$ into $\mathbb{Z} \times \mathbb{N}$ by $h(r) = (a,b)$ if $r = a/b$, where a and b are integers with no common factor and $b > 0$. These conditions uniquely determine the pair (a,b) for each rational r, and so h is a function. Clearly, h is injective since r is recovered from (a,b) by division. Therefore, h is an injection of $\mathbb{Q}$ into a countable set. Hence $\mathbb{Q}$ is an infinite set with $|\mathbb{Q}| \leq |\mathbb{N}|$. So $\mathbb{Q}$ is countable by Lemma 2.9.4. ∎

There are infinite sets that are not countable. The proof uses a **diagonalization** argument due to Cantor.

2.9.8. THEOREM. *The set $\mathbb{R}$ of real numbers is uncountable.*

PROOF. Suppose to the contrary that $\mathbb{R}$ is countable. Then all real numbers may be written as a list $x_1, x_2, x_3, \dots$. Express each x_i as an infinite decimal, which we write as $x_i = x_{i0}.x_{i1}x_{i2}x_{i3} \dots$, where x_{i0} is an integer and x_{ik} is an integer from 0 to 9 for each $k \geq 1$. Our goal is to write down another real number that does not appear in this (supposedly exhaustive) list. Let $a_0 = 0$ and define $a_k = 7$ if $x_{kk} \in \{0,1,2,3,4\}$ and $a_k = 2$ if $x_{kk} \in \{5,6,7,8,9\}$. Define a real number $a = a_0.a_1a_2a_3 \dots$.

Since a is a real number, it must appear somewhere in this list, say $a = x_k$. However, the kth decimal place a_k of a and x_{kk} of x_k differ by between 3 and 7. This cannot be accounted for by the fact that certain real numbers have two decimal expansions, one ending in zeros and the other ending in nines because this changes any digit by either 1 or 9. So $a \neq x_k$, and hence a does not occur in this list. It follows that there is no such list, and thus $\mathbb{R}$ is uncountable. ∎

Exercises for Section 2.9

A. Prove that the set $\mathbb{Z}^n$, consisting of all n-tuples $\mathbf{a} = (a_1, a_2, \dots, a_n)$, where $a_i \in \mathbb{Z}$, is countable.

B. Show that $(0,1)$ and $[0,1]$ have the same cardinality as $\mathbb{R}$.

C. Show that if $|A| \leq |B|$ and $|B| \leq |C|$, then $|A| \leq |C|$.

D. Prove that the set of all infinite sequences of integers is uncountable.
HINT: Modify the diagonalization argument.

E. A real number α is called an **algebraic number** if there is a polynomial with integer coefficients with α as a root. Prove that the set of all algebraic numbers is countable.
HINT: First count the set of all polynomials with integer coefficients.

F. A real number that is not algebraic is called a **transcendental number**. Prove that the set of transcendental numbers has the same cardinality as $\mathbb{R}$.

G. Show that the set of all *finite* subsets of $\mathbb{N}$ is countable.

H. Prove **Cantor's Theorem**: that for any set X, the power set $P(X)$ of all subsets of X satisfies $|X| \neq |P(X)|$.   HINT: If f is an injection from X into $P(X)$, consider $A = \{x \in X : x \notin f(x)\}$.

I. If A is an infinite set, show that A has a countable infinite subset.
HINT: Use recursion to choose a sequence a_n of distinct points in A.

J. Show that A is infinite if and only if there is a proper subset B of A such that $|B| = |A|$.
HINT: Use the previous exercise and let $B = A \setminus \{a_1\}$.

Chapter 3
Series

3.1 Convergent Series

We turn now to the problem of adding up an infinite series of numbers. As we shall quickly see, this is really no different from dealing with the sequence of partial sums of the series. However, there are tests for convergence that are more conveniently expressed for series than for sequences.

3.1.1. DEFINITION. If $(a_n)_{n=1}^{\infty}$ is a sequence of numbers, the **infinite series** with terms a_n is the formal expression $\sum_{n=1}^{\infty} a_n$. Define a sequence of partial sums $(s_n)_{n=1}^{\infty}$ by $s_n = \sum_{k=1}^{n} a_k$. This series **converges**, or equivalently is **summable**, if the sequence of partial sums converges. If $L = \lim_{n \to \infty} s_n$, then we write $L = \sum_{n=1}^{\infty} a_n$. If the series does not converge, then it is said to **diverge**.

It can be fairly difficult or even impossible to find the sum of a series. However, it is not nearly as hard to determine whether a series converges. We devote this chapter to examples of series and to tests for convergence of series. While these tests may be familiar to you from calculus, the proofs may not be.

3.1.2. EXAMPLE. Consider $\sum_{k=1}^{\infty} \frac{1}{k}$, which is known as the **harmonic series**. We will show that this series diverges. The idea is to group the terms cleverly. Suppose that n satisfies $2^k \le n < 2^{k+1}$. Then

$$s_n = s_{2^k} = 1 + \frac{1}{2} + \left(\frac{1}{3} + \frac{1}{4} \right) + \cdots + \left(\frac{1}{2^{k-1}+1} + \cdots + \frac{1}{2^k} \right)$$
$$\ge 1 + \frac{1}{2} + 2\frac{1}{4} + \cdots + 2^{k-1}\frac{1}{2^k} = 1 + \frac{k}{2}.$$

Thus $\lim_{n \to \infty} s_n = +\infty$.

K.R. Davidson and A.P. Donsig, *Real Analysis and Applications: Theory in Practice*,
Undergraduate Texts in Mathematics, DOI 10.1007/978-0-387-98098-0_3,
© Springer Science + Business Media, LLC 2010

There is another way to estimate the terms s_n that gives a more precise idea of the rate of divergence of the harmonic series. Consider the graph of $y = 1/x$, as given in Figure 3.1.

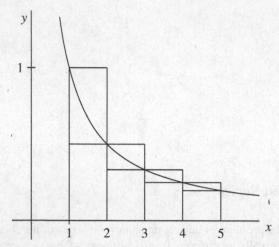

FIG. 3.1 The graph of $1/x$ with bounding rectangles.

It is clear that

$$\frac{1}{k+1} = \int_k^{k+1} \frac{1}{k+1}\,dx < \int_k^{k+1} \frac{1}{x}\,dx < \int_k^{k+1} \frac{1}{k}\,dx = \frac{1}{k}.$$

Notice that s_n is the upper Riemann sum estimate for the integral of $1/x$ from 1 to $n+1$ using the integer partition

$$s_n = \sum_{k=1}^n \frac{1}{k} > \sum_{k=1}^n \int_k^{k+1} \frac{1}{x}\,dx = \int_1^{n+1} \frac{1}{x}\,dx = \log x \Big|_1^{n+1} = \log(n+1).$$

Similarly, $s_n - 1$ is the lower Riemann sum estimate for the integral of $1/x$ from 1 to n using the integer partition

$$s_n - 1 = \sum_{k=2}^n \frac{1}{k} < \sum_{k=2}^n \int_{k-1}^k \frac{1}{x}\,dx = \int_1^n \frac{1}{x}\,dx = \log x \Big|_1^n = \log n.$$

Therefore, $\log(n+1) < s_n < 1 + \log n$ for all $n \geq 1$. Hence s_n diverges to infinity roughly at the same rate as the log function.

3.1.3. EXAMPLE. On the other hand, consider $\sum\limits_{n=1}^{\infty} \dfrac{1}{n(n+3)}$. First observe that

$$\frac{3}{n(n+3)} = \frac{1}{n} - \frac{1}{n+3},$$

and so we have an example of a **telescoping sum** (so named because of the convenient cancellation in the following sum):

$$3s_n = \frac{3}{4} + \frac{3}{10} + \cdots + \frac{3}{n(n+3)}$$

$$= \left(1 - \frac{1}{4}\right) + \left(\frac{1}{2} - \frac{1}{5}\right) + \cdots + \left(\frac{1}{n} - \frac{1}{n+3}\right)$$

$$= \left(1 + \frac{1}{2} + \cdots + \frac{1}{n}\right) - \left(\frac{1}{4} + \frac{1}{5} + \cdots + \frac{1}{n+3}\right)$$

$$= 1 + \frac{1}{2} + \frac{1}{3} - \frac{1}{n+1} - \frac{1}{n+2} - \frac{1}{n+3}.$$

Thus,

$$\sum_{n=1}^{\infty} \frac{1}{n(n+3)} = \lim_{n \to \infty} s_n = \frac{1 + 1/2 + 1/3}{3} = \frac{11}{18}.$$

The harmonic series shows that a series $\sum_{n=1}^{\infty} a_n$ can diverge even if the a_n go to zero. However, if a series $\sum_{n=1}^{\infty} a_n$ does converge, then $\lim_{n \to \infty} a_n$ must be zero.

3.1.4. THEOREM. *If the series $\sum_{n=1}^{\infty} a_n$ is convergent, then $\lim_{n \to \infty} a_n = 0$.*

PROOF. If $(s_n)_{n=1}^{\infty}$ is the sequence of partial sums, then $a_n = s_n - s_{n-1}$ for $n \geq 2$. Using the properties of limits, we have $\lim_{n \to \infty} s_n = \lim_{n \to \infty} s_{n-1}$, and thus

$$\lim_{n \to \infty} a_n = \lim_{n \to \infty} s_n - s_{n-1} = \lim_{n \to \infty} s_n - \lim_{n \to \infty} s_{n-1} = 0. \qquad \blacksquare$$

The rigorous ε–N definition of convergence and the Cauchy criterion have a nice form for series.

3.1.5. CAUCHY CRITERION FOR SERIES.

The following are equivalent for a series $\sum_{n=1}^{\infty} a_n$.

(1) *The series converges.*

(2) *For every $\varepsilon > 0$, there is an $N \in \mathbb{N}$ such that for all $n \geq N$, $\left| \sum_{k=n+1}^{\infty} a_k \right| < \varepsilon$.*

(3) *For every $\varepsilon > 0$, there is an $N \in \mathbb{N}$ such that if $n, m \geq N$, $\left| \sum_{k=n+1}^{m} a_k \right| < \varepsilon$.*

PROOF. Let s_n be the sequence of partial sums of the series. If the series converges to a limit L, then for every $\varepsilon > 0$ there is an integer N such that

$$|L - s_n| < \varepsilon \quad \text{for all} \quad n \geq N.$$

Since

$$L - s_n = \lim_{m \to \infty} s_m - s_n = \lim_{m \to \infty} \sum_{k=n+1}^{m} a_k = \sum_{k=n+1}^{\infty} a_k,$$

this shows that (1) implies (2).

If (2) holds, then there is N such that for all $n \geq N$, $\left| \sum_{k=n+1}^{\infty} a_k \right| < \varepsilon$. If $m, n \geq N$, then the reverse triangle inequality shows that

$$\left| \sum_{k=n+1}^{m} a_k \right| \leq \left| \left| \sum_{k=n+1}^{\infty} a_k \right| - \left| \sum_{k=m+1}^{\infty} a_k \right| \right| \leq \max\left\{ \left| \sum_{k=n+1}^{\infty} a_k \right|, \left| \sum_{k=m+1}^{\infty} a_k \right| \right\} < \varepsilon.$$

So (3) holds.

Finally, if (3) holds, since $|s_m - s_n| = \left| \sum_{k=n+1}^{m} a_k \right|$, then (s_n) is a Cauchy sequence. Therefore the series converges, by the completeness of the real numbers. ∎

Exercises for Section 3.1

A. Sum the series $\sum\limits_{n=1}^{\infty} \dfrac{1}{n(n+2)}$.

B. Sum the series $\sum\limits_{n=1}^{\infty} \dfrac{1}{n(n+1)(n+3)(n+4)}$.

HINT: Show that $\dfrac{12}{n(n+1)(n+3)(n+4)} = \dfrac{1}{n} - \dfrac{2}{n+1} + \dfrac{2}{n+3} - \dfrac{1}{n+4}$.

C. Prove that if $p > 1$ and $\sum\limits_{k=1}^{\infty} t_k$ is a convergent series of nonnegative numbers, $\sum\limits_{k=1}^{\infty} t_k^p$ converges.

D. Let $(a_n)_{n=1}^{\infty}$ be a sequence such that $\lim\limits_{n \to \infty} |a_n| = 0$. Prove that there is a subsequence (a_{n_k}) such that $\sum\limits_{k=1}^{\infty} a_{n_k}$ converges.

E. Compute $\sum\limits_{n=1}^{\infty} \dfrac{1}{(n+1)\sqrt{n} + n\sqrt{n+1}}$. HINT: Multiply the nth term by $1 = \dfrac{\sqrt{n+1} - \sqrt{n}}{\sqrt{n+1} - \sqrt{n}}$.

F. Let $|a| < 1$ and set $S_n = \sum\limits_{k=0}^{n} a^k$ and $T_n = \sum\limits_{k=0}^{n} (k+1)a^k$.

(a) Show that $S_n^2 = \sum\limits_{k=0}^{n} (k+1)a^k + \sum\limits_{k=1}^{n} (n+1-k)a^{n+k}$.

(b) Hence show that $|T_n - S_n^2| \leq \frac{n(n+1)}{2}|a|^{n+1}$.

(c) Show that $\lim\limits_{n \to \infty} T_n = \left(\lim\limits_{n \to \infty} S_n \right)^2$. Hence obtain a formula for this sum.

(d) Evaluate $\sum\limits_{k=0}^{\infty} \dfrac{n+1}{3^n}$.

G. Let $x_0 = 1$ and $x_{n+1} = x_n + 1/x_n$.

(a) Find $\lim\limits_{n \to \infty} x_n$.

(b) Let $y_n = x_n^2 - 2n$. Find a recurrence formula for y_{n+1} in terms of y_n only.

(c) Show that y_n is monotone increasing and $y_n < 2 + \log n$.

(d) Hence show that $\lim\limits_{n \to \infty} x_n - \sqrt{2n} = 0$.

3.2 Convergence Tests for Series

We start by considering infinite series with positive terms. If each $a_n \geq 0$, then $s_{n+1} = s_n + a_{n+1} \geq s_n$, so the sequence of partial sums is monotone increasing. So the Monotone Convergence Theorem (2.6.1) shows that (s_n) converges if and only if it is bounded above. We have established the following proposition.

3.2.1. PROPOSITION. *If $a_k \geq 0$ for $k \geq 1$ and $s_n = \sum\limits_{k=1}^{n} a_k$, then either*

(1) $(s_n)_{n=1}^{\infty}$ *is bounded above, in which case* $\sum\limits_{n=1}^{\infty} a_n$ *converges,*

or

(2) $(s_n)_{n=1}^{\infty}$ *is unbounded, in which case* $\sum\limits_{n=1}^{\infty} a_n$ *diverges.*

A sequence $(a_n)_{n=0}^{\infty}$ is a **geometric sequence** with ratio r if $a_{n+1} = ra_n$ for all $n \geq 0$ or, equivalently, $a_n = a_0 r^n$ for all $n \geq 0$. Finding the sum of a geometric sequence is a standard result from calculus, so we leave the proof as an exercise.

3.2.2. GEOMETRIC SERIES.
A geometric series converges if $|r| < 1$. Moreover, $\sum\limits_{n=0}^{\infty} ar^n = \dfrac{a}{1-r}$.

Of course, if $a \neq 0$ and $|r| > 1$, then the terms ar^n do not converge to 0. In this case, the geometric sequence $(a_n)_{n=0}^{\infty}$ is not summable.

Another test often used in calculus is the Comparison Test.

3.2.3. THE COMPARISON TEST.
Consider two sequences of real numbers (a_n) and (b_n) with $|a_n| \leq b_n$ for all $n \geq 1$. If (b_n) is summable, then (a_n) is summable and

$$\left| \sum_{n=1}^{\infty} a_n \right| \leq \sum_{n=1}^{\infty} b_n.$$

If (a_n) is not summable, then (b_n) is not summable.

PROOF. Let $\varepsilon > 0$ be given. Since (b_n) is summable, Lemma 3.1.5 yields an integer N such that

$$\sum_{k=n+1}^{m} b_k < \varepsilon \quad \text{for all} \quad N \leq n \leq m.$$

Therefore,

$$\left| \sum_{k=n+1}^{m} a_k \right| \leq \sum_{k=n+1}^{m} |a_k| \leq \sum_{k=n+1}^{m} b_k < \varepsilon.$$

Applying Lemma 3.1.5 again shows that $\sum_{n=1}^{\infty} a_n$ converges.

If (a_n) is not summable, then neither is (b_n), by the contrapositive. (The summability of (b_n) would imply that (a_n) was also summable.) ∎

The Root Test can decide the summability of sequences that are dominated by a geometric sequence "at infinity."

3.2.4. THE ROOT TEST.

Suppose $a_n \geq 0$ for all n, and let $\ell = \limsup \sqrt[n]{a_n}$. If $\ell < 1$, then $\sum\limits_{n=1}^{\infty} a_n$ converges; and if $\ell > 1$, then $\sum\limits_{n=1}^{\infty} a_n$ diverges.

NOTE: If $\limsup \sqrt[n]{a_n} = 1$, the series may or may not converge (see Exercise 3.2.L).

PROOF. Suppose that $\limsup \sqrt[n]{a_n} = \ell < 1$. To show that the series converges, we need to show that the sequence of partial sums is bounded above. Pick a number r with $\ell < r < 1$ and let $\varepsilon = r - \ell$. Since $\varepsilon > 0$, we can find an integer $N > 0$ such that

$$a_n^{1/n} < \ell + \varepsilon = r \quad \text{for all} \quad n \geq N.$$

Therefore, $a_n < r^n$ for all $n \geq N$. Consider the sequence $(b_n)_{n=1}^{\infty}$ given by

$$b_n = a_n, \quad 1 \leq n < N, \quad \text{and} \quad b_n = r^n \quad \text{for} \quad n \geq N.$$

This sequence is summable by Theorem 3.2.2. Indeed,

$$\sum_{n=1}^{\infty} b_n = \sum_{n=1}^{N-1} b_n + \sum_{n=N}^{\infty} b_n = \sum_{n=1}^{N-1} b_n + \frac{r^N}{1-r}.$$

Since $|a_n| \leq b_n$ for $n \geq 1$, the Comparison Test (3.2.3) shows that $(a_n)_{n=1}^{\infty}$ is summable.

Conversely, if $\limsup\limits_{n \to \infty} \sqrt[n]{a_n} = \ell > 1$, then let $\varepsilon = \ell - 1$. From the definition of limsup, there is a subsequence $n_1 < n_2 < \cdots$ such that

$$a_{n_k}^{1/n_k} > \ell - \varepsilon = 1 \quad \text{for all} \quad k \geq 1.$$

Therefore, the terms a_n do not converge to 0 and thus the series diverges. ∎

3.2.5. DEFINITION.

A sequence is **alternating** if it has the form $((-1)^n a_n)$ or $((-1)^{n+1} a_n)$, where $a_n \geq 0$ for all $n \geq 1$.

3.2.6. LEIBNIZ ALTERNATING SERIES TEST.

Suppose that $(a_n)_{n=1}^{\infty}$ is a monotone decreasing sequence $a_1 \geq a_2 \geq a_3 \geq \cdots \geq 0$, and $\lim\limits_{n \to \infty} a_n = 0$. Then the alternating series $\sum\limits_{n=1}^{\infty} (-1)^n a_n$ converges.

PROOF. Let $s_n = \sum\limits_{k=1}^{n} (-1)^k a_k$. Intuitively, (s_n) behaves as in Figure 3.2.

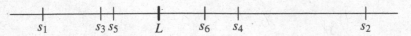

FIG. 3.2 Behaviour of partial sums.

Making this formal, we claim that

(1) $s_2 \geq s_4 \geq s_6 \geq \cdots$,
(2) $s_1 \leq s_3 \leq s_5 \leq \cdots$, and
(3) $s_{2m-1} \leq s_{2n}$ for all $m, n \geq 1$.

To prove (1), notice that $s_{2n} - s_{2n-2} = a_{2n} - a_{2n-1} \leq 0$, since $a_{2n} \leq a_{2n-1}$. For (2), $s_{2n+1} - s_{2n-1} = a_{2n} - a_{2n+1} \geq 0$. For (3), note that if m and n are integers, then for $N = \max\{m, n\}$, we have

$$s_{2m-1} \leq s_{2N-1} \leq s_{2N} \leq s_{2n}.$$

Since the decreasing sequence $(s_2, s_4, \ldots)$ is bounded below by s_1, it converges to some number L by the Monotone Convergence Theorem (2.6.1). Similarly, since $(s_1, s_3, \ldots)$ is increasing and bounded above by s_2, it converges to some number M. Finally,

$$L - M = \lim_{n \to \infty} s_{2n} - \lim_{n \to \infty} s_{2n-1} = \lim_{n \to \infty} s_{2n} - s_{2n-1} = \lim_{n \to \infty} a_{2n} = 0. \qquad \blacksquare$$

3.2.7. EXAMPLE. Consider the alternating harmonic series

$$\sum_{n=1}^{\infty} \frac{(-1)^{n-1}}{n} = 1 - \frac{1}{2} + \frac{1}{3} - \frac{1}{4} + \cdots.$$

This series is alternating and $\frac{1}{n}$ is monotone decreasing to 0, so the series must converge. Note that the harmonic series has the same terms without the sign changes.

It is possible to sum this series in several ways. All rely on calculus in some way. Notice that

$$s_{2n} = \sum_{k=1}^{2n} \frac{(-1)^{k-1}}{k} = 1 - \frac{1}{2} + \cdots + \frac{1}{2n-1} - \frac{1}{2n}$$

$$= \left(1 + \frac{1}{2} + \cdots + \frac{1}{2n-1} + \frac{1}{2n}\right) - 2\left(\frac{1}{2} + \frac{1}{4} + \cdots + \frac{1}{2n}\right)$$

$$= \sum_{k=1}^{2n} \frac{1}{k} - 2\sum_{k=1}^{n} \frac{1}{2k} = \sum_{k=1}^{2n} \frac{1}{k} - \sum_{k=1}^{n} \frac{1}{k} = \sum_{k=1}^{n} \frac{1}{n+k}.$$

We can recognize this as a Riemann sum approximating an integral. Indeed, consider the integral $\int_1^2 \frac{1}{x}\,dx$. Partition the interval $[1,2]$ into n equal pieces. Then from

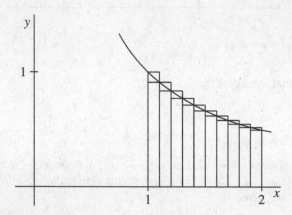

FIG. 3.3 Riemann sum for $\int_1^2 \frac{1}{x}\,dx$.

Figure 3.3, we see that the Riemann (lower) sum for integrating $f(x) = \dfrac{1}{x}$ is

$$\frac{1}{n} \sum_{k=1}^n f\left(1 + \frac{k}{n}\right) = \sum_{k=1}^n \frac{1}{n+k} = s_{2n}.$$

From the calculus, we obtain

$$\sum_{n=1}^\infty \frac{(-1)^{n-1}}{n} = \lim_{n\to\infty} s_{2n} = \int_1^2 \frac{1}{x}\,dx = \log x \Big|_1^2 = \log 2.$$

Exercises for Section 3.2

A. Prove Theorem 3.2.2.

B. Show that if $(|a_n|)_{n=1}^\infty$ is summable, then so is $(a_n)_{n=1}^\infty$.

C. Euler proposed that $1 - 2 + 4 - 8 + \cdots = \sum_{n=0}^\infty (-2)^n = \frac{1}{1-(-2)} = \frac{1}{3}$. What is wrong with this?

D. Let $(a_n)_{n=1}^\infty$ be a monotone decreasing sequence of positive real numbers. Show that the series $\sum_{n=1}^\infty a_n$ converges if and only if the series $\sum_{k=0}^\infty 2^k a_{2^k}$ converges.

E. Apply Exercise D to the series $\sum_{n=1}^\infty \frac{1}{n^p}$ for $p > 0$. For which values of p does this converge?

F. If $\sum_{k=1}^\infty a_k^2$ and $\sum_{k=1}^\infty b_k^2$ both converge, prove that $\sum_{k=1}^\infty a_k b_k$ converges.

G. Find two convergent series $\sum_{k=1}^\infty a_k$ and $\sum_{k=1}^\infty b_k$ such that $\sum_{k=1}^\infty a_k b_k$ diverges.

H. (THE LIMIT COMPARISON TEST) Show that if $\sum\limits_{n=1}^{\infty} a_n$ and $\sum\limits_{n=1}^{\infty} b_n$ are series with $b_n \geq 0$ such

that $\limsup\limits_{n \to \infty} \dfrac{|a_n|}{b_n} < \infty$ and $\sum\limits_{n=1}^{\infty} b_n < \infty$, then the series $\sum\limits_{n=1}^{\infty} a_n$ converges.

I. (THE RATIO TEST) Suppose that $(a_n)_{n=1}^{\infty}$ is a sequence of positive terms. Show that if

$\limsup\limits_{n \to \infty} \dfrac{a_{n+1}}{a_n} < 1$, then $\sum\limits_{n=1}^{\infty} a_n$ converges. Conversely, show that if $\liminf\limits_{n \to \infty} \dfrac{a_{n+1}}{a_n} > 1$, then $\sum\limits_{n=1}^{\infty} a_n$

diverges.   HINT: Imitate the proof of the Root Test (i.e., find a suitable r and integer $N > 0$
and compare a_n with $a_N r^{n-N}$ for all $n \geq N$).

J. Show that the Root Test implies the Ratio Test by proving that if $\lim\limits_{n \to \infty} \dfrac{a_{n+1}}{a_n} = r$, then

$\lim\limits_{n \to \infty} (a_n)^{1/n} = r$.

K. Construct a convergent series of positive terms with $\limsup\limits_{n \to \infty} \dfrac{a_{n+1}}{a_n} = \infty$.

L. (a) Find a convergent series $\sum\limits_{n=1}^{\infty} a_n$, with positive entries, such that $\lim\limits_{n \to \infty} \sqrt[n]{a_n} = 1$.
   (b) Find a divergent series with the same property.

M. If $a_n \geq 0$ for all n, prove that $\sum\limits_{n=1}^{\infty} a_n$ converges if and only if $\sum\limits_{n=1}^{\infty} \dfrac{a_n}{1+a_n}$ converges.

N. (THE INTEGRAL TEST) Let $f(x)$ be a positive, monotone decreasing function on $[1,\infty)$.
Show that the sequence $(f(n))$ is summable if and only if $\int_1^{\infty} f(x)\,dx < \infty$.

HINT: Show that $\sum\limits_{n=2}^{k+1} f(n) < \int_1^{k+1} f(x)\,dx < \sum\limits_{n=1}^{k} f(n)$.

O. Apply the previous exercise to the series $\sum\limits_{n=2}^{\infty} \dfrac{1}{n(\log n)^p}$ for $p > 0$.

P. Determine whether the following series converge or diverge.

(a) $\sum\limits_{n=2}^{\infty} \dfrac{3n}{n^3 + 1}$

(b) $\sum\limits_{n=1}^{\infty} \dfrac{n}{2^n}$

(c) $\sum\limits_{n=2}^{\infty} \dfrac{(-1)^n \log n}{n}$

(d) $\sum\limits_{n=1}^{\infty} \sqrt{n+1} - \sqrt{n}$

(e) $\sum\limits_{n=1}^{\infty} e^{-n^2}$

(f) $\sum\limits_{n=1}^{\infty} \sin(n\pi/4)$

(g) $\sum\limits_{n=1}^{\infty} (-1)^n \sin(1/n)$

(h) $\sum\limits_{n=1}^{\infty} \dfrac{1}{\sqrt{n^3+4}}$

(i) $\sum\limits_{n=1}^{\infty} (\sqrt[n]{n} - 1)^n$

(j) $\sum\limits_{n=2}^{\infty} \dfrac{\sqrt{n+1} - \sqrt{n}}{n}$

(k) $\sum\limits_{n=2}^{\infty} \dfrac{(-1)^n}{\sqrt{n}\log n}$

(l) $\sum\limits_{n=2}^{\infty} \dfrac{(-1)^n}{\sqrt[n]{n}}$

(m) $\sum\limits_{n=2}^{\infty} \dfrac{1}{(\log n)^k}$

(n) $\sum\limits_{n=1}^{\infty} \dfrac{n!}{n^n}$

(o) $\sum\limits_{n=1}^{\infty} \dfrac{(-1)^n \arctan(n)}{n}$

(p) $\sum\limits_{n=2}^{\infty} \dfrac{(-1)^n}{\sqrt{n} + (-1)^n}$

(q) $\sum\limits_{n=1}^{\infty} (-1)^n (e^{1/n} - 1)$

(r) $\sum\limits_{n=1}^{\infty} (-1)^n \dfrac{n^{42}}{(n+1)!}$

(s) $\sum\limits_{n=1}^{\infty} \dfrac{1}{1+n^2}$

(t) $\sum\limits_{n=1}^{\infty} \dfrac{1}{\log(e^n + e^{-n})}$

(u) $\sum\limits_{n=1}^{\infty} \dfrac{\sin(\pi n/3)}{n}$

(v) $\sum\limits_{n=1}^{\infty} \dfrac{n^{10}}{10^n}$

(w) $\sum\limits_{n=2}^{\infty} \dfrac{1}{(\log n)^n}$

(x) $\sum\limits_{n=2}^{\infty} \dfrac{1}{n \log n}$

3.3 Absolute and Conditional Convergence

In this section, we show that the sums of certain series, and even their convergence, depend on the order of terms. The Alternating Series Test shows that badly behaved series such as the harmonic series become more tractable when we introduce appropriate signs to the terms to keep the partial sums close together. However, the following variant on Example 3.2.7 shows that considerable care must be taken when adding this type of series.

3.3.1. EXAMPLE. Consider the series

$$1 - \frac{1}{2} - \frac{1}{4} + \frac{1}{3} - \frac{1}{6} - \frac{1}{8} + \cdots,$$

where $a_{3n-2} = \frac{1}{2n-1}$, $a_{3n-1} = -\frac{1}{4n-2}$ and $a_{3n} = -\frac{1}{4n}$. This has exactly the same terms as the alternating harmonic series except that the negative terms are coming twice as fast as the positive ones.

First let's convince ourselves that this series converges. Notice that

$$a_{3n-2} + a_{3n-1} + a_{3n} = \frac{1}{2n-1} - \frac{1}{4n-2} - \frac{1}{4n} = \frac{1}{4n(2n-1)}.$$

Therefore, $s_{3n} = \sum_{k=1}^{n} \frac{1}{4k(2k-1)}$. The terms of this series are dominated by the series $\sum_{k=1}^{\infty} \frac{1}{4k^2}$. This latter series converges by the Integral Test (Exercise 3.2.N), since

$$\int_1^\infty \frac{1}{4x^2}\,dx = -\frac{1}{4x}\Big|_1^\infty = \frac{1}{4} < \infty.$$

Therefore, $\lim_{n\to\infty} s_{3n} = \sum_{k=1}^{\infty} \frac{1}{4k(2k-1)}$ converges by the Comparison Test (3.2.3). However, $|s_{3n} - s_{3n\pm1}| < \frac{1}{2n}$. Hence

$$\lim_{n\to\infty} s_{3n-1} = \lim_{n\to\infty} s_{3n+1} = \lim_{n\to\infty} s_{3n} = \lim_{n\to\infty} s_n.$$

We can actually sum this series exactly because

$$\frac{1}{4k(2k-1)} = \frac{1}{2}\left(\frac{1}{2k-1} - \frac{1}{2k}\right).$$

Therefore,

$$s_{3n} = \frac{1}{2} \sum_{k=1}^{2n} \frac{(-1)^{n+1}}{k}.$$

By Example 3.2.7, we conclude that the series converges to $\frac{1}{2}\log 2$. Since the alternating harmonic series has the limit $\log 2$, these two series have *different* sums even though they have the same terms.

3.3.2. DEFINITION. A series $\sum\limits_{n=1}^{\infty} a_n$ is called **absolutely convergent** if the series $\sum\limits_{n=1}^{\infty} |a_n|$ converges. A series that converges but is not absolutely convergent is called **conditionally convergent**.

Example 3.2.7 shows that a convergent series need not be absolutely convergent. The next simple fact is that absolute convergence is a stronger notion than convergence (Exercise 3.2.B).

3.3.3. PROPOSITION. *An absolutely convergent series is convergent.*

3.3.4. DEFINITION. A **rearrangement** of a series $\sum\limits_{n=1}^{\infty} a_n$ is another series with the same terms in a different order. This can be described by a permutation π of the natural numbers $\mathbb{N}$ determining the series $\sum\limits_{n=1}^{\infty} a_{\pi(n)}$.

For absolutely convergent series, we get the best possible behaviour under a rearrangement. Example 3.3.1 shows that this fails for conditionally convergent series.

3.3.5. THEOREM. *Every rearrangement of an absolutely convergent series converges to the same limit.*

PROOF. Let $\sum\limits_{n=1}^{\infty} a_n$ be an absolutely convergent series that converges to L. Suppose that π is a permutation of $\mathbb{N}$ and that $\varepsilon > 0$ is given. By the Cauchy Criterion (3.1.5), there is an integer N such that $\sum\limits_{k=N+1}^{\infty} |a_k| < \varepsilon/2$. Since the rearrangement contains exactly the same terms in a different order, the first N terms $a_1, \ldots, a_N$ eventually occur in the rearranged series. Thus there is an integer M such that all of these terms occur in the first M terms of the rearrangement. Hence for $m \geq M$,

$$\left| \sum_{k=1}^{m} a_{\pi(k)} - L \right| \leq \left| \sum_{k=1}^{m} a_{\pi(k)} - \sum_{k=1}^{N} a_k \right| + \left| \sum_{k=1}^{N} a_k - L \right| \leq 2 \sum_{k=N+1}^{\infty} |a_k| < \varepsilon.$$

Therefore, $\sum\limits_{k=1}^{\infty} a_{\pi(k)} = L$. ∎

3.3.6. EXAMPLE. Consider the series $\sum\limits_{n=1}^{\infty} \dfrac{(-1)^{n+1}}{n^4}$. This series is absolutely

convergent since $\sum\limits_{n=1}^{\infty} \dfrac{1}{n^4}$ converges by Exercise 3.2.E. Hence we may manipulate the terms freely. Therefore,

$$\sum_{n=1}^{\infty} \frac{(-1)^{n+1}}{n^4} = \sum_{n=1}^{\infty} \frac{1}{n^4} - 2\sum_{n=1}^{\infty} \frac{1}{(2n)^4} = \frac{7}{8}\sum_{n=1}^{\infty} \frac{1}{n^4}.$$

Using techniques from Fourier series (see Chapter 14), we will be able to show that $\sum\limits_{n=1}^{\infty} 1/n^4 = \pi^4/90$. It follows that the preceding summation equals $7\pi^4/720$.

On the other hand, the worst possible scenario holds for the rearrangements of conditionally convergent series. First, we need the following dichotomy.

3.3.7. REARRANGEMENT THEOREM.
If $\sum\limits_{n=1}^{\infty} a_n$ is a conditionally convergent series, then for every real number L, there is a rearrangement that converges to L.

PROOF. Write the positive terms of this series as $b_1, b_2, \ldots$ and the negative terms as $c_1, c_2, \ldots$. By Theorem 3.1.4, $\lim\limits_{n\to\infty} a_n = 0$; so $\lim\limits_{n\to\infty} b_n = 0$ and $\lim\limits_{n\to\infty} c_n = 0$. We claim that

$$\sum_{k=1}^{\infty} b_k = +\infty \qquad \text{and} \qquad \sum_{k=1}^{\infty} |c_k| = +\infty.$$

Indeed, if both series converged, then $\sum\limits_{n=1}^{\infty} a_n$ would converge absolutely. Suppose that the first series diverges, but $\sum\limits_{k=1}^{\infty} |c_k| = L < \infty$. Then for any $R > 0$, there is an N such that $\sum\limits_{k=1}^{N} b_k > R + L$. Therefore once M is so large that $b_1, \ldots, b_N$ are contained in $a_1, \ldots, a_M$, we have

$$\sum_{i=1}^{M} a_i \geq \sum_{k=1}^{N} b_k - \sum_{k=1}^{\infty} |c_k| > (R+L) - L = R.$$

Since R is arbitrary, the series diverges, contrary to our hypothesis. A similar contradiction occurs if the first series converges. Hence both series must diverge.

Fix $L \in \mathbb{R}$. Choose the least integer m_1 such that $u_1 = b_1 + \cdots + b_{m_1} > L$. Then choose the least integer n_1 such that $v_1 = u_1 + c_1 + \cdots + c_{n_1} < L$. Then choose the least $m_2 > m_1$ such that $u_2 = u_1 + v_1 + b_{m_1+1} + \cdots + b_{m_2} > L$. We continue in this way, adding just enough of the positive terms to make the total greater than L and then switching to negative terms until the total is less than L. In this way, we define increasing sequences m_k and n_k to be the least positive integers greater than m_{k-1}

and n_{k-1}, respectively, such that

$$u_k = \sum_{i=1}^{m_k} b_i + \sum_{j=1}^{n_{k-1}} c_j > L > \sum_{i=1}^{m_k} b_i + \sum_{j=1}^{n_k} c_j = v_k.$$

We will show that this rearranged series,

$$b_1 + \cdots + b_{m_1} + c_1 + \cdots + c_{n_1} + b_{m_1+1} + \cdots + b_{m_2} + c_{n_1+1} + \cdots + c_{n_2} + \cdots,$$

converges to L. By the construction, $u_i - b_{m_i} \leq L < u_i$ and $v_j < L \leq v_j - c_{n_j}$. Therefore,

$$L + c_{n_j} \leq v_j < L < u_i \leq L + b_{m_i}.$$

Since $\lim_{n \to \infty} L + b_n = \lim_{n \to \infty} L + c_n = L$, the Squeeze Theorem (2.4.6) shows that the sequences of the u_i and of the v_j both converge to L. Finally, if s_k is the kth partial sum of the new series, then for k between m_1 and $m_1 + n_1$, $u_1 \geq s_k \geq v_1$, while for k between $m_1 + n_1$ and $m_2 + n_1$, $v_1 \leq s_k \leq u_2$. In general, we have

$$v_{i-1} \leq s_k \leq u_i \quad \text{for} \quad m_{i-1} + n_{i-i} \leq k \leq m_i + n_{i-1},$$

and

$$v_i \leq s_k \leq u_i \quad \text{for} \quad m_i + n_{i-1} \leq k \leq m_i + n_i.$$

Using the Squeeze Theorem shows that the rearranged series converges to L. ∎

Exercises for Section 3.3

A. Find the series in Exercise 3.2.P that converge conditionally but not absolutely.

B. Decide which of the following series converge absolutely, conditionally, or not at all.

(a) $\sum_{n=1}^{\infty} \dfrac{(-1)^n}{n \log(n+1)}$ (b) $\sum_{n=1}^{\infty} \dfrac{(-1)^n}{(2 + (-1)^n)n}$ (c) $\sum_{n=1}^{\infty} \dfrac{(-1)^n \sin(\frac{1}{n})}{n}$

C. Compute the sum of the series $\sum_{n=1}^{\infty} \dfrac{1}{n^2(2n-1)}$ given that $\sum_{n=1}^{\infty} \dfrac{1}{n^2} = \dfrac{\pi^2}{6}$.

HINT: $\dfrac{1}{n^2(2n-1)} = \dfrac{4}{2n(2n-1)} - \dfrac{1}{n^2}$.

D. Show that $\sum_{n=1}^{\infty} \dfrac{\cos(\frac{2n\pi}{3})}{n^2}$ converges absolutely. Find the sum, given that $\sum_{n=1}^{\infty} \dfrac{1}{n^2} = \dfrac{\pi^2}{6}$.
(See Example 13.5.5.)

E. Show that a conditionally convergent series has a rearrangement converging to $+\infty$.

F. Let $a_n = \dfrac{(-1)^k}{n}$ for $(k-1)^2 < n \leq k^2$ and $k \geq 1$. Decide whether the series $\sum_{n=1}^{\infty} a_n$ converges.

Chapter 4
Topology of $\mathbb{R}^n$

The space $\mathbb{R}^n$ is the right setting for many problems in real analysis. For example, in many situations, functions of interest depend on several variables. This puts us into the realm of multivariable calculus, which is naturally set in $\mathbb{R}^n$. We will study normed vector spaces further in Chapter 7, building on the properties and concepts we study here. The space $\mathbb{R}^n$ is the most important normed vector space, after the real numbers themselves.

4.1 n-Dimensional Space

The space $\mathbb{R}^n$ is the set of n-vectors $\mathbf{x} = (x_1, x_2, \ldots, x_n)$ with arbitrary real coefficients x_i for $1 \le i \le n$. Generally, vectors in $\mathbb{R}^n$ will be referred to as **points**. This space has a lot of structure, most of which should be familiar from advanced calculus or linear algebra courses. In particular, we should mention that the **zero vector** is $(0, 0, \ldots, 0)$, which we denote by $\mathbf{0}$.

First, it is a *vector space*. Recall the basic property that vectors may be added and also multiplied by (real) scalars. Indeed, for any $\mathbf{x}$ and $\mathbf{y}$ in $\mathbb{R}^n$ and scalars $t \in \mathbb{R}$,

$$\mathbf{x} + \mathbf{y} = (x_1, x_2, \ldots, x_n) + (y_1, y_2, \ldots, y_n) = (x_1 + y_1, x_2 + y_2, \ldots, x_n + y_n)$$

and

$$t\mathbf{x} = t(x_1, x_2, \ldots, x_n) = (tx_1, tx_2, \ldots, tx_n).$$

We assume that you know the basics of linear algebra. Instead, we concentrate on the properties of $\mathbb{R}^n$ that build on the ideas of distance and convergence.

There is the notion of length of a vector, given by

$$\|\mathbf{x}\| = \|(x_1, x_2, \ldots, x_n)\| = \Big(\sum_{i=1}^{n} |x_i|^2 \Big)^{1/2}.$$

This is called the **Euclidean norm** on $\mathbb{R}^n$, and $\|\mathbf{x}\|$ is the **norm** of $\mathbf{x}$. This conforms to our usual notion of distance in the plane and in space. Moreover, it is the natural

K.R. Davidson and A.P. Donsig, *Real Analysis and Applications: Theory in Practice*, Undergraduate Texts in Mathematics, DOI 10.1007/978-0-387-98098-0_4, © Springer Science + Business Media, LLC 2010

consequence of the Euclidean distance in the plane using the Pythagorean formula
and induction on the number of variables. (See Exercise 4.1.A.) The distance between two points **x** and **y** is then determined by

$$\|\mathbf{x} - \mathbf{y}\| = \Big(\sum_{i=1}^{n} |x_i - y_i|^2 \Big)^{1/2}.$$

An important property of distance is the **triangle inequality**: The distance from
point A to point B and then on to a point C is at least as great as the direct distance
from A to C. This is interpreted geometrically as saying that the sum of the lengths
of two sides of a triangle is greater than the length of the third side (Figure 4.1).
(Equality can occur if the triangle has no area.)

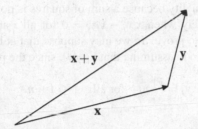

FIG. 4.1 The triangle inequality.

To verify this algebraically, we need an inequality involving the dot product,
which is useful in its own right. Recall that the **dot product** or **inner product** of
two vectors **x** and **y** is given by

$$\langle \mathbf{x}, \mathbf{y} \rangle = \langle (x_1, \ldots, x_n), (y_1, \ldots, y_n) \rangle = \sum_{i=1}^{n} x_i y_i.$$

There is a close connection between the inner product and the Euclidean norm be-
cause of the evident identity $\langle \mathbf{x}, \mathbf{x} \rangle = \|\mathbf{x}\|^2$. The inner product is linear in both vari-
ables:

$$\langle r\mathbf{x} + s\mathbf{y}, \mathbf{z} \rangle = r\langle \mathbf{x}, \mathbf{z} \rangle + s\langle \mathbf{y}, \mathbf{z} \rangle \quad \text{for all} \quad \mathbf{x}, \mathbf{y}, \mathbf{z} \in \mathbb{R}^n \text{ and } r, s \in \mathbb{R}$$

and

$$\langle \mathbf{x}, s\mathbf{y} + t\mathbf{z} \rangle = s\langle \mathbf{x}, \mathbf{y} \rangle + t\langle \mathbf{x}, \mathbf{z} \rangle \quad \text{for all} \quad \mathbf{x}, \mathbf{y}, \mathbf{z} \in \mathbb{R}^n \text{ and } s, t \in \mathbb{R}.$$

4.1.1. SCHWARZ INEQUALITY.
For all **x** *and* **y** *in* $\mathbb{R}^n$,
$$|\langle \mathbf{x}, \mathbf{y} \rangle| \leq \|\mathbf{x}\| \, \|\mathbf{y}\|.$$

Equality holds if and only if **x** *and* **y** *are collinear.*

PROOF. Let $\mathbf{x} = (x_1, \ldots, x_n)$ and $\mathbf{y} = (y_1, \ldots, y_n)$. Then

$$
\begin{aligned}
2\|\mathbf{x}\|^2 \|\mathbf{y}\|^2 - 2|\langle \mathbf{x}, \mathbf{y} \rangle|^2 &= 2\sum_{i=1}^{n}\sum_{j=1}^{n} x_i^2 y_j^2 - 2\Big(\sum_{i=1}^{n} x_i y_i\Big)^2 \\
&= \sum_{i=1}^{n}\sum_{j=1}^{n} x_i^2 y_j^2 + x_j^2 y_i^2 - \sum_{i=1}^{n}\sum_{j=1}^{n} 2x_i y_i x_j y_j \\
&= \sum_{i=1}^{n}\sum_{j=1}^{n} x_i^2 y_j^2 - 2x_i y_j x_j y_i + x_j^2 y_i^2 \\
&= \sum_{i=1}^{n}\sum_{j=1}^{n} (x_i y_j - x_j y_i)^2 \geq 0.
\end{aligned}
$$

This establishes the inequality because a sum of squares is positive.

Equality holds precisely when $x_i y_j - x_j y_i = 0$ for all i and j. If both $\mathbf{x}$ and $\mathbf{y}$ equal $\mathbf{0}$, there is nothing to prove. So we may suppose that at least one coefficient is nonzero. There is no harm in assuming that $x_1 \neq 0$, since the proof is the same in all other cases. Then

$$
y_j = \frac{y_1}{x_1} x_j \quad \text{for all} \quad 1 \leq j \leq n.
$$

Hence $\mathbf{y} = \frac{y_1}{x_1}\mathbf{x}$.   ∎

4.1.2. TRIANGLE INEQUALITY.

The triangle inequality holds for the Euclidean norm on $\mathbb{R}^n$:

$$
\|\mathbf{x} + \mathbf{y}\| \leq \|\mathbf{x}\| + \|\mathbf{y}\| \quad \text{for all} \quad \mathbf{x}, \mathbf{y} \in \mathbb{R}^n.
$$

Moreover, equality holds if and only if either $\mathbf{x} = \mathbf{0}$ or $\mathbf{y} = c\mathbf{x}$ with $c \geq 0$.

PROOF. Use the relationship between the inner product and norm to compute

$$
\begin{aligned}
\|\mathbf{x} + \mathbf{y}\|^2 &= \langle \mathbf{x} + \mathbf{y}, \mathbf{x} + \mathbf{y} \rangle = \langle \mathbf{x}, \mathbf{x} \rangle + \langle \mathbf{x}, \mathbf{y} \rangle + \langle \mathbf{y}, \mathbf{x} \rangle + \langle \mathbf{y}, \mathbf{y} \rangle \\
&\leq \langle \mathbf{x}, \mathbf{x} \rangle + |\langle \mathbf{x}, \mathbf{y} \rangle| + |\langle \mathbf{y}, \mathbf{x} \rangle| + \langle \mathbf{y}, \mathbf{y} \rangle \\
&\leq \|\mathbf{x}\|^2 + \|\mathbf{x}\|\,\|\mathbf{y}\| + \|\mathbf{x}\|\,\|\mathbf{y}\| + \|\mathbf{y}\|^2 = \big(\|\mathbf{x}\| + \|\mathbf{y}\|\big)^2.
\end{aligned}
$$

If equality holds, then we must have $\langle \mathbf{x}, \mathbf{y} \rangle = \|\mathbf{x}\|\,\|\mathbf{y}\|$. In particular, the Schwarz inequality holds. So either $\mathbf{x} = \mathbf{0}$ or $\mathbf{y} = c\mathbf{x}$. Substituting $\mathbf{y} = c\mathbf{x}$ into $\langle \mathbf{x}, \mathbf{y} \rangle = \|\mathbf{x}\|\,\|\mathbf{y}\|$ gives $c = \|\mathbf{y}\|/\|\mathbf{x}\| \geq 0$.   ∎

Collinearity does not imply equality for the triangle inequality in all cases because $\langle x, y \rangle$ could be negative. For example, $\mathbf{x}$ and $-\mathbf{x}$ are collinear for any nonzero vector $\mathbf{x}$, but

$$
0 = \|\mathbf{x} + (-\mathbf{x})\| < \|\mathbf{x}\| + \|-\mathbf{x}\| = 2\|\mathbf{x}\|.
$$

When we write elements of $\mathbb{R}^n$ in vector notation, we are implicitly using the standard basis $\{e_i : 1 \leq i \leq n\}$, where e_i is the vector with a single 1 in the ith position and zeros in the other coordinates. A set $\{v_1, \ldots, v_m\}$ in $\mathbb{R}^n$ is **orthonormal** if $\langle v_i, v_j \rangle = \delta_{ij}$ for $1 \leq i, j \leq m$, where $\delta_{ij} = 0$ when $i \neq j$ and $\delta_{ii} = 1$. If, in addition, $\{v_1, \ldots, v_m\}$ spans $\mathbb{R}^n$, it is called an **orthonormal basis**. In particular, $\{e_1, \ldots, e_n\}$ is an orthonormal basis for $\mathbb{R}^n$.

4.1.3. LEMMA. *Let $\{v_1, \ldots, v_m\}$ be an orthonormal set in $\mathbb{R}^n$. Then*

$$\left\| \sum_{i=1}^m a_i v_i \right\| = \left(\sum_{i=1}^m |a_i|^2 \right)^{1/2}.$$

An orthonormal set in $\mathbb{R}^n$ is linearly independent. So an orthonormal basis for $\mathbb{R}^n$ is a basis and has exactly n elements.

PROOF. Use the inner product to compute

$$\left\| \sum_{i=1}^m a_i v_i \right\|^2 = \sum_{i=1}^m \sum_{j=1}^m \langle a_i v_i, a_j v_j \rangle = \sum_{i=1}^m \sum_{j=1}^m a_i a_j \delta_{ij} = \sum_{i=1}^m |a_i|^2.$$

In particular, if $\sum_{i=1}^m a_i v_i = \mathbf{0}$, we find that $\sum_{i=1}^m |a_i|^2 = 0$ and thus $a_i = 0$ for $1 \leq i \leq m$. This shows that $\{v_1, \ldots, v_m\}$ is linearly independent. Finally, a basis for $\mathbb{R}^n$ is a linearly independent set of vectors that spans $\mathbb{R}^n$. An orthonormal basis spans by definition and is independent, as shown. A basic result of linear algebra shows that every basis for $\mathbb{R}^n$ has exactly n elements. ∎

Exercises for Section 4.1

A. Establish the **Pythagorean formula**: If x and y are orthogonal vectors, prove that
$\|x + y\| = \left(\|x\|^2 + \|y\|^2 \right)^{1/2}$.

B. (a) Suppose $x = \sum_{i=1}^j x_i e_i$ is a vector in $\mathbb{R}^n$ with nonzero coefficients only in the first j positions. Apply the Pythagorean formula to the orthogonal vectors x and $y = x_{j+1} e_{j+1}$.
(b) Show by induction that the Pythagorean formula yields the norm in all dimensions.

C. Show that $\|x + y\|^2 + \|x - y\|^2 = 2\|x\|^2 + 2\|y\|^2$ for all vectors x and y in $\mathbb{R}^n$. This is called the **parallelogram law**. What does it mean geometrically?

D. Prove that if x and y are vectors in $\mathbb{R}^n$, then $\big| \|x\| - \|y\| \big| \leq \|x - y\|$.

E. Prove by induction that $\|x_1 + \cdots + x_k\| \leq \|x_1\| + \cdots + \|x_k\|$ for vectors x_i in $\mathbb{R}^n$.

F. Suppose that x and y are unit vectors in $\mathbb{R}^n$. Show that if $\left\| \frac{x+y}{2} \right\| = 1$, then $x = y$.

G. Let x and y be two nonzero vectors in $\mathbb{R}^2$, and let the angle between them be θ. Prove that $\langle x, y \rangle = \|x\| \|y\| \cos \theta$.   HINT: If x makes the angle α with the positive x-axis, then $x = (\|x\| \cos \alpha, \|x\| \sin \alpha)$.

H. For nonzero vectors x and y in $\mathbb{R}^n$, define θ by $\|x\| \|y\| \cos \theta = \langle x, y \rangle$, and call this the angle between them.

(a) Prove the **cosine law**: If $\mathbf{x}$ and $\mathbf{y}$ are vectors and θ is the angle between them, then
$\|\mathbf{x} + \mathbf{y}\|^2 = \|\mathbf{x}\|^2 + 2\|\mathbf{x}\|\,\|\mathbf{y}\|\cos\theta + \|\mathbf{y}\|^2$.

(b) Prove that $\langle\mathbf{x},\mathbf{y}\rangle$ can be defined using only the norms of related vectors.

I.   Suppose that U is a linear transformation from $\mathbb{R}^n$ into $\mathbb{R}^m$ that is **isometric**, meaning that $\|U\mathbf{x}\| = \|\mathbf{x}\|$ for all $\mathbf{x} \in \mathbb{R}^n$.

(a) Prove that $\langle U\mathbf{x}, U\mathbf{y}\rangle = \langle\mathbf{x},\mathbf{y}\rangle$ for all $\mathbf{x},\mathbf{y} \in \mathbb{R}^n$.

(b) If $\{\mathbf{v}_1,\ldots,\mathbf{v}_k\}$ is an orthonormal set in $\mathbb{R}^n$, show that $\{U\mathbf{v}_1,\ldots,U\mathbf{v}_k\}$ is also orthonormal.

J.   (a) Let U be an isometric linear transformation of $\mathbb{R}^n$ onto itself. Show that the n columns of the matrix of U form an orthonormal basis for $\mathbb{R}^n$.

(b) Conversely, if $\{\mathbf{v}_1,\ldots,\mathbf{v}_n\}$ is an orthonormal basis for $\mathbb{R}^n$, show that the linear transformation $U\mathbf{x} = \sum_{i=1}^{n} x_i\mathbf{v}_i$ is isometric.

K.   Let M be a subspace of $\mathbb{R}^n$ with an orthonormal basis $\{\mathbf{v}_1,\ldots,\mathbf{v}_k\}$. Define a linear transformation on $\mathbb{R}^n$ by $P\mathbf{x} = \sum_{i=1}^{k}\langle\mathbf{x},\mathbf{v}_i\rangle\mathbf{v}_i$.

(a) Show that $P\mathbf{x}$ belongs to M, and $P\mathbf{y} = \mathbf{y}$ for all $\mathbf{y} \in M$. Hence show that $P^2 = P$.

(b) Show that $\langle P\mathbf{x}, \mathbf{x} - P\mathbf{x}\rangle = 0$.

(c) Hence show that $\|\mathbf{x}\|^2 = \|P\mathbf{x}\|^2 + \|\mathbf{x} - P\mathbf{x}\|^2$.

(d) If $\mathbf{y}$ belongs to M, show that $\|\mathbf{x} - \mathbf{y}\|^2 = \|\mathbf{y} - P\mathbf{x}\|^2 + \|\mathbf{x} - P\mathbf{x}\|^2$.

(e) Hence show that $P\mathbf{x}$ is the closest point in M to $\mathbf{x}$.

4.2 Convergence and Completeness in $\mathbb{R}^n$

The notion of norm for points in $\mathbb{R}^n$ immediately allows us to discuss convergence of sequences in this context. The definition of limit of a sequence of points $\mathbf{x}_k$ in $\mathbb{R}^n$ is virtually identical to the definition of convergence in $\mathbb{R}$. The only change is to replace absolute value, which is the measure of distance in the reals, with the Euclidean norm in n-space.

4.2.1. DEFINITION. A sequence of points $(\mathbf{x}_k)$ in $\mathbb{R}^n$ **converges** to a point $\mathbf{a}$ if for *every* $\varepsilon > 0$, there is an integer $N = N(\varepsilon)$ such that

$$\|\mathbf{x}_k - \mathbf{a}\| < \varepsilon \quad \text{for all} \quad k \geq N.$$

In this case, we write $\lim_{k\to\infty}\mathbf{x}_k = \mathbf{a}$.

The parallel between the two definitions of convergence allows us to reformulate the definition of limit of a sequence of points in n-space to the consideration of a sequence of real numbers, namely the Euclidean norms of the points.

4.2.2. LEMMA. *Let $(\mathbf{x}_k)$ be a sequence in $\mathbb{R}^n$. Then $\lim_{k\to\infty}\mathbf{x}_k = \mathbf{a}$ if and only if $\lim_{k\to\infty}\|\mathbf{x}_k - \mathbf{a}\| = 0$.*

The second limit is a sequence of real numbers, and thus it may be understood using only ideas from Chapter 2.

Just as important is the relation between convergence in $\mathbb{R}^n$ and convergence of the coefficients. The following lemma is conceptually quite easy (draw a picture in $\mathbb{R}^2$), but the proof requires careful bookkeeping.

4.2.3. LEMMA. *A sequence $\mathbf{x}_k = (x_{k,1},\ldots,x_{k,n})$ in $\mathbb{R}^n$ converges to a point $\mathbf{a} = (a_1,\ldots,a_n)$ if and only if each coefficient converges:*

$$\lim_{k\to\infty} \mathbf{x}_k = \mathbf{a} \quad \text{if and only if} \quad \lim_{k\to\infty} x_{k,i} = a_i \quad \text{for} \quad 1 \le i \le n.$$

PROOF. First suppose that $\lim_{k\to\infty} \mathbf{x}_k = \mathbf{a}$. Then given $\varepsilon > 0$, we obtain an integer N such that $\|\mathbf{x}_k - \mathbf{a}\| < \varepsilon$ for all $k \ge N$. Then for each $1 \le i \le n$ and all $k \ge N$,

$$|x_{k,i} - a_i| \le \Big(\sum_{j=1}^{n} |x_{k,j} - a_j|^2\Big)^{1/2} = \|\mathbf{x}_k - \mathbf{a}\| < \varepsilon.$$

Therefore, $\lim_{k\to\infty} x_{k,i} = a_i$ for all $1 \le i \le n$.

Conversely, suppose that each coordinate sequence $x_{k,i}$ converges to a real number a_i for $1 \le i \le n$. Then given $\varepsilon > 0$, use ε/n in the definition of limit and choose N_i so large that

$$|x_{k,i} - a_i| < \frac{\varepsilon}{n} \quad \text{for all} \quad k \ge N_i.$$

Then using $N = \max\{N_i : 1 \le i \le n\}$, all n of these inequalities are valid for $k \ge N$. Hence

$$\|\mathbf{x}_k - \mathbf{a}\| = \Big(\sum_{i=1}^{n} |x_{k,i} - a_i|^2\Big)^{1/2} < \Big(\sum_{i=1}^{n} \big(\frac{\varepsilon}{n}\big)^2\Big)^{1/2} < \varepsilon.$$

Therefore, $\lim_{k\to\infty} \mathbf{x}_k = \mathbf{a}$. ∎

Following the same route as for the line, we will define Cauchy sequences and completeness in the higher-dimensional context. For the real line, it was necessary to build the completeness of $\mathbb{R}$ into its construction. However, the completeness of $\mathbb{R}^n$ will be a consequence of the completeness of $\mathbb{R}$.

4.2.4. DEFINITION. A sequence $\mathbf{x}_k$ in $\mathbb{R}^n$ is **Cauchy** if for every $\varepsilon > 0$, there is an integer N such that

$$\|\mathbf{x}_k - \mathbf{x}_l\| < \varepsilon \quad \text{for all} \quad k, l \ge N.$$

A set $S \subset \mathbb{R}^n$ is **complete** if every Cauchy sequence of points in S converges to a point in S.

As in Proposition 2.8.1, it is easy to show that a convergent sequence is Cauchy. Indeed, suppose that $\lim_{k\to\infty} \mathbf{x}_k = \mathbf{a}$ and $\varepsilon > 0$. Then, using $\varepsilon/2$ in the definition of limit, choose an integer N such that $\|\mathbf{x}_k - \mathbf{a}\| < \varepsilon/2$ for all $k \ge N$. Then for all $k, l \ge N$, use the triangle inequality to obtain

$$\|\mathbf{x}_k - \mathbf{x}_l\| \le \|\mathbf{x}_k - \mathbf{a}\| + \|\mathbf{a} - \mathbf{x}_l\| < \frac{\varepsilon}{2} + \frac{\varepsilon}{2} = \varepsilon.$$

The converse lies deeper and is more important.

4.2.5. COMPLETENESS THEOREM FOR $\mathbb{R}^n$.
Every Cauchy sequence in $\mathbb{R}^n$ converges. Thus, $\mathbb{R}^n$ is complete.

PROOF. Let $\mathbf{x}_k$ be a Cauchy sequence in $\mathbb{R}^n$. The proof is accomplished by reducing the problem to each coordinate. Let us write the elements of the sequence as $\mathbf{x}_k = (x_{k,1}, \dots, x_{k,n})$. We will show that the sequences $\left(x_{k,i}\right)_{k=1}^{\infty}$ are Cauchy for each $1 \le i \le n$. Indeed, if $\varepsilon > 0$, choose N so large that

$$\|\mathbf{x}_k - \mathbf{x}_l\| < \varepsilon \quad \text{for all} \quad k,l \ge N.$$

Then

$$|x_{k,i} - x_{l,i}| \le \|\mathbf{x}_k - \mathbf{x}_l\| < \varepsilon \quad \text{for all} \quad k,l \ge N.$$

Thus $\left(x_{k,i}\right)_{k=1}^{\infty}$ are Cauchy for $1 \le i \le n$.

By the completeness of $\mathbb{R}$, Theorem 2.8.5, each of these sequences has a limit, say

$$\lim_{k \to \infty} x_{k,i} = a_i \quad \text{for} \quad 1 \le i \le n.$$

Define a vector $\mathbf{a} \in \mathbb{R}^n$ by $\mathbf{a} = (a_1, \dots, a_n)$. By Lemma 4.2.3, $\lim_{k \to \infty} \mathbf{x}_k = \mathbf{a}$, and hence $\mathbb{R}^n$ is complete. ∎

4.2.6. EXAMPLE.
Let $\mathbf{v}_0 = (0,0)$, and define a sequence $\mathbf{v}_n = (x_n, y_n)$ in $\mathbb{R}^2$ recursively by

$$x_{n+1} = \frac{x_n + y_n + 1}{2}, \qquad y_{n+1} = \frac{x_n - y_n + 1}{2}.$$

The first few terms are

$$(0,0),\ \left(\tfrac{1}{2}, \tfrac{1}{2}\right),\ \left(1, \tfrac{1}{2}\right),\ \left(\tfrac{5}{4}, \tfrac{3}{4}\right),\ \left(\tfrac{3}{2}, \tfrac{3}{4}\right),\ \left(\tfrac{13}{8}, \tfrac{7}{8}\right),\ \left(\tfrac{7}{4}, \tfrac{7}{8}\right),\ \dots.$$

To get an idea of what the limit might be (if it exists), look for fixed points of the map

$$T(x,y) = \left(\frac{x+y+1}{2}, \frac{x-y+1}{2}\right).$$

In other words, solve the equation $T\mathbf{u} = \mathbf{u}$. This is a linear system:

$$x = \tfrac{1}{2}x + \tfrac{1}{2}y + \tfrac{1}{2},$$
$$y = \tfrac{1}{2}x - \tfrac{1}{2}y + \tfrac{1}{2}.$$

Solving, we find the solution $\mathbf{u} = (2,1)$.

This leads us to consider the distance of $\mathbf{v}_n$ to $\mathbf{u}$:

$$\|\mathbf{v}_{n+1} - \mathbf{u}\|^2 = \|(x_{n+1} - 2, y_{n+1} - 1)\|^2 = \left\|\left(\frac{x_n + y_n - 3}{2}, \frac{x_n - y_n - 1}{2}\right)\right\|^2$$

$$= \frac{(x_n + y_n - 3)^2 + (x_n - y_n - 1)^2}{4} = \frac{2x_n^2 + 2y_n^2 - 8x_n - 4y_n + 10}{4}$$

$$= \frac{(x_n - 2)^2 + (y_n - 1)^2}{2} = \frac{1}{2}\|\mathbf{v}_n - \mathbf{u}\|^2.$$

By induction, it follows that

$$\|\mathbf{v}_n - \mathbf{u}\| = 2^{-n/2}\|\mathbf{v}_0 - \mathbf{u}\| = 2^{-n/2}\sqrt{5}.$$

Hence $\lim_{n\to\infty} \|\mathbf{v}_n - \mathbf{u}\| = 0$, which means that $\lim_{n\to\infty} \mathbf{v}_n = (2,1)$.

Exercises for Section 4.2

A. (a) If $(\mathbf{x}_n)_{n-1}^{\infty}$ is a sequence in $\mathbb{R}^n$ with $\lim_{n\to\infty} \mathbf{x}_n = \mathbf{a}$, show that $\lim_{n\to\infty} \|\mathbf{x}_n\| = \|\mathbf{a}\|$.

(b) Show by example that the converse is false.

B. If a sequence $(\mathbf{x}_n)_{n=1}^{\infty}$ in $\mathbb{R}^n$ satisfies $\sum_{n\geq 1} \|\mathbf{x}_n - \mathbf{x}_{n+1}\| < \infty$, show that it is a Cauchy sequence.

C. (a) Give an example of a Cauchy sequence for which the condition of Exercise B fails.

(b) However, show that every Cauchy sequence $(\mathbf{x}_n)_{n=1}^{\infty}$ has a subsequence $(\mathbf{x}_{n_i})_{i=1}^{\infty}$ such that $\sum_{i\geq 1} \|\mathbf{x}_{n_i} - \mathbf{x}_{n_{i+1}}\| < \infty$.

D. Let $\mathbf{x}_0 \in \mathbb{R}^n$ and $R > 0$. Prove that $\{\mathbf{x} \in \mathbb{R}^n : \|\mathbf{x} - \mathbf{x}_0\| \leq R\}$ is complete.

E. (a) Let M be a subspace of $\mathbb{R}^n$, and let $\{\mathbf{v}_1, \ldots, \mathbf{v}_m\}$ be an orthonormal basis for M. Formulate an analogue of Lemma 4.2.3 for M and prove it.

(b) Prove that M is complete.

F. Let $\mathbf{v}_0 = (x_0, y_0)$ with $0 < x_0 < y_0$. Define $\mathbf{v}_{n+1} = (x_{n+1}, y_{n+1}) = \left(\sqrt{x_n y_n}, \frac{x_n + y_n}{2}\right)$ for all $n \geq 0$.

(a) Show by induction that $0 < x_n < x_{n+1} < y_{n+1} < y_n$.

(b) Then estimate $y_{n+1} - x_{n+1}$ in terms of $y_n - x_n$.

(c) Thereby show that there is a number c such that $\lim_{n\to\infty} \mathbf{v}_n = (c,c)$. This value c is known as the **arithmetic–geometric mean** of x_0 and y_0.

G. Let $\mathbf{v}_0 = (x_0, y_0) = (0,0)$, and for $n \geq 0$ define

$$\mathbf{v}_{n+1} = (x_{n+1}, y_{n+1}) = \left(\sqrt{\frac{x_n^2 + 2y_n^2}{4}}, \frac{x_n + y_n + 1}{3}\right).$$

(a) Show that x_n and y_n are increasing sequences that are bounded above.

(b) Prove that $\lim_{n\to\infty} \mathbf{v}_n$ exists, and find the limit.

H. Let $T = \begin{bmatrix} 5/4 & -1/4 \\ 3/4 & 1/4 \end{bmatrix}$. Set $\mathbf{x}_n = T^n(1,0)$ for $n \geq 1$.

(a) Prove that $(\mathbf{x}_n)$ converges and find the limit $\mathbf{y}$.

(b) Find an explicit N such that $\|\mathbf{x}_n - \mathbf{y}\| < \frac{1}{2}10^{-100}$ for all $n \geq N$.

HINT: Show by induction that $\mathbf{x}_n = \left(\frac{3 - 2^{-n}}{2}, \frac{3(1 - 2^{-n})}{2}\right)$.

4.3 Closed and Open Subsets of $\mathbb{R}^n$

Two classes of subsets play a crucial role in analysis: the closed sets, which contain all of their limit points, and the open sets, which contain small balls around each point. These notions will be made precise in this section. From the point of view thus far, closed sets seem more natural because they are directly connected to limiting procedures. Later, however, we shall see that open sets play at least as important a role when we consider continuous functions. These two types of sets are intimately related in any case. We develop closed sets first.

4.3.1. DEFINITION. A point $\mathbf{x}$ is a **limit point** of a subset A of $\mathbb{R}^n$ if there is a sequence $(\mathbf{a}_n)_{n=1}^{\infty}$ with $\mathbf{a}_n \in A$ such that $\mathbf{x} = \lim_{n \to \infty} \mathbf{a}_n$. A set $A \subset \mathbb{R}^n$ is **closed** if it contains all of its limit points.

NOTE: Be warned that some other books define limit point to be a slightly more complicated concept, which we call a cluster point (see Exercise 4.3.N).

4.3.2. EXAMPLES.

(1) $[a,b] = \{x \in \mathbb{R} : a \le x \le b\}$ is closed.

(2) $\varnothing$ and $\mathbb{R}^n$ are both closed.

(3) $[0,+\infty)$ is closed in $\mathbb{R}$.

(4) $(0,1]$ and $(0,1)$ are not closed.

(5) $\{\mathbf{x} \in \mathbb{R}^n : \|\mathbf{x}\| \le 1\}$ is closed.

(6) $\{\mathbf{x} \in \mathbb{R}^n : \|\mathbf{x}\| < 1\}$ is not closed.

(7) $\{(x,y) \in \mathbb{R}^2 : xy \ge 1\}$ is closed.

(8) Finite sets of real numbers are closed in $\mathbb{R}$.

In the following proposition, I denotes an arbitrary index set. This may be an infinite set of very large cardinality (such as the real line) or a countably infinite set (like $\mathbb{N}$) or even a finite set.

4.3.3. PROPOSITION. *If $A, B \subset \mathbb{R}^n$ are closed, then $A \cup B$ is closed. If $\{A_i : i \in I\}$ is a family of closed subsets of $\mathbb{R}^n$, then $\bigcap_{i \in I} A_i$ is closed.*

PROOF. Suppose that $(\mathbf{x}_n)_{n=1}^{\infty}$ is a sequence in $A \cup B$ with limit $\mathbf{x}$. Clearly, either infinitely many of the $\mathbf{x}_n$'s belong to A or infinitely many belong to B. Without loss of generality, we may suppose that A has this property. Hence there is a subsequence $(\mathbf{x}_{n_i})_{i=1}^{\infty}$ of $(\mathbf{x}_n)_{n=1}^{\infty}$ such that each $\mathbf{x}_{n_i}$ belongs to A. But this subsequence has limit $\mathbf{x}$. Since A is closed, we deduce that $\mathbf{x}$ belongs to A, and thus belongs to $A \cup B$. So $A \cup B$ is closed.

Now suppose that $(\mathbf{x}_n)_{n=1}^{\infty}$ is a sequence in $\bigcap_{i \in I} A_i$ with limit $\mathbf{x}$. For each $i \in I$, the sequence $(\mathbf{x}_n)_{n=1}^{\infty}$ belongs to A_i, which is closed. Therefore, the limit $\mathbf{x}$ also belongs to A_i. Since this holds for every $i \in I$, it follows that $\mathbf{x}$ also belongs to the intersection. Hence the intersection is closed.   ∎

Since a closed set has the very useful property of containing all of its limit points, it is natural to want to construct closed sets from other, less well behaved sets. This is the motivation for the following definition.

4.3.4. DEFINITION. If A is a subset of $\mathbb{R}^n$, the **closure** of A is the set $\overline{A}$ consisting of all limit points of A.

To justify the name, we establish some basic properties of the closure operation.

4.3.5. PROPOSITION. *Let A be a subset of $\mathbb{R}^n$. Then $\overline{A}$ is the smallest closed set containing A. In particular, $\overline{\overline{A}} = \overline{A}$.*

PROOF. First notice that for each $\mathbf{a}$ in A, we may consider the sequence $\mathbf{x}_n = \mathbf{a}$ for all $n \geq 1$. This has limit $\mathbf{a}$, and thus A is contained in $\overline{A}$.

To show that $\overline{A}$ is closed, consider a sequence $(\mathbf{x}_n)_{n=1}^{\infty}$, where each $\mathbf{x}_n$ belongs to $\overline{A}$ with limit $\mathbf{x} = \lim\limits_{n \to \infty} \mathbf{x}_n$. For each n, there is a sequence of points in A converging to $\mathbf{x}_n$. Hence we may choose an element $\mathbf{a}_n \in A$ from this sequence such that $\|\mathbf{x}_n - \mathbf{a}_n\| < \frac{1}{n}$. Then

$$\lim_{n \to \infty} \mathbf{a}_n = \lim_{n \to \infty} \mathbf{x}_n + (\mathbf{a}_n - \mathbf{x}_n) = \lim_{n \to \infty} \mathbf{x}_n + \lim_{n \to \infty} \mathbf{a}_n - \mathbf{x}_n = \mathbf{x} + \mathbf{0} = \mathbf{x}.$$

Thus $\mathbf{x}$ is also a limit of points in A, whence it belongs to $\overline{A}$. So $\overline{A}$ is closed.

If C is a closed set containing A, then it also contains all limits of sequences in A and therefore contains $\overline{A}$. So $\overline{A}$ is the smallest closed set containing A. Now $\overline{\overline{A}}$ is the smallest closed set containing $\overline{A}$. Since $\overline{A}$ is already closed, $\overline{\overline{A}} = \overline{A}$.   ∎

4.3.6. DEFINITION. The **ball** about $\mathbf{a}$ in $\mathbb{R}^n$ of radius r is the set

$$B_r(\mathbf{a}) = \{\mathbf{x} \in \mathbb{R}^n : \|\mathbf{x} - \mathbf{a}\| < r\}.$$

A subset U of $\mathbb{R}^n$ is **open** if for every $\mathbf{a} \in U$, there is some $r = r(\mathbf{a}) > 0$ such that the ball $B_r(\mathbf{a})$ is contained in U.

4.3.7. EXAMPLES.

(1) $(a,b) = \{x \in \mathbb{R} : a < x < b\}$ is open.

(2) $\varnothing$ and $\mathbb{R}^n$ are both open.

(3) $(0, +\infty)$ is open in $\mathbb{R}$.

(4) $(0,1]$ and $[0,1]$ are not open.

(5) $B_r(a)$ is open.

(6) $\overline{B_r(a)} = \{x \in \mathbb{R}^n : \|x - a\| \le r\}$ is not open.

(7) $\{(x,y) \in \mathbb{R}^2 : xy < 1\}$ is open.

(8) $\{(x,0) \in \mathbb{R}^2 : 0 < x < 1\}$ is not open.

Figure 4.2 illustrates the definitions of open and closed sets.

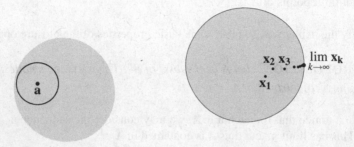

FIG. 4.2 Open and closed sets

It is important to remember that while a door must be either open or closed, a set can be neither. The following connection between open and closed sets makes the relation clear.

4.3.8. THEOREM. *A set $A \subset \mathbb{R}^n$ is open if and only if the complement of A, $A' = \{\mathbf{x} \in \mathbb{R}^n : \mathbf{x} \notin A\}$, is closed.*

PROOF. Let A be open. Let $(\mathbf{x}_n)_{n=1}^{\infty}$ be a sequence in A' with limit $\mathbf{x}$. If $\mathbf{a}$ is any point in A, there is a positive number $r > 0$ such that $B_r(\mathbf{a})$ is contained in A. Hence $\|\mathbf{a} - \mathbf{x}_n\| \ge r$ for all $n \ge 1$. Therefore,

$$\|\mathbf{a} - \mathbf{x}\| = \lim_{n \to \infty} \|\mathbf{a} - \mathbf{x}_n\| \ge r.$$

In particular, $\mathbf{x} \ne \mathbf{a}$. This is true for every point in A, and hence $\mathbf{x}$ belongs to A'. That is, A' is closed.

Conversely, suppose that A is not open. Then there is some $\mathbf{a} \in A$ such that for every $r > 0$, the ball $B_r(\mathbf{a})$ is not contained in A. In particular, if we let $r = \frac{1}{n}$, we can find $\mathbf{x}_n \in A'$ such that $\|\mathbf{a} - \mathbf{x}_n\| < \frac{1}{n}$. Then $\mathbf{a} = \lim_{n \to \infty} \mathbf{x}_n$ is a limit point of A' belonging to A. Hence A' is not closed.   ∎

We have the following proposition, which is dual to Proposition 4.3.3. The proof is left as an exercise.

4.3.9. PROPOSITION. *If U and V are open subsets of $\mathbb{R}^n$, then $U \cap V$ is open. If $\{U_i : i \in I\}$ is a family of open subsets of $\mathbb{R}^n$, then $\bigcup_{i \in I} U_i$ is open.*

There is also a notion for open sets that is dual to the closure. The **interior** of a set X, denoted by $\text{int} X$, is the largest open set contained inside X (see the exercises). If the interior of X is the empty set, then we say that X has **empty interior**.

4.3.10. EXAMPLE. Let $A = \{(x,y) : x \in \mathbb{Q}, y > x^3\}$. This set is neither open nor closed. Indeed, $(0,0) = \lim_{n \to \infty}(0, \frac{1}{n})$ is a limit point of A not contained in A; so A is not closed. The point $(0,1) = \lim_{n \to \infty}(\sqrt{2}/n, 1)$ belongs to A, yet it is the limit of points in A'. So A is not open either.

The closure of A is the set $\overline{A} = \{(x,y) : y \geq x^3\}$. To see this, let (x,y) be given such that $y \geq x^3$. Let x_n be an increasing sequence of rationals converging to x (such as the finite decimal approximations of x). Set $y_n = y + \frac{1}{n}$. Then it is clear that $y_n > x^3 \geq x_n^3$ and thus $\mathbf{a}_n = (x_n, y_n)$ belongs to A. Now

$$\lim_{n \to \infty} x_n = x \qquad \text{and} \qquad \lim_{n \to \infty} y_n = \lim_{n \to \infty} y + \frac{1}{n} = y.$$

Hence $\lim_{n \to \infty} \mathbf{a}_n = (x,y)$. Therefore, $\overline{A}$ contains $\{(x,y) : y \geq x^3\}$. Conversely, if $(x,y) = \lim_{n \to \infty} \mathbf{a}_n$ for any sequence $a_n = (x_n, y_n)$ in A, it follows that

$$y = \lim_{n \to \infty} y_n \geq \lim_{n \to \infty} x_n^3 = x^3.$$

Thus $\{(x,y) : y \geq x^3\}$ contains $\overline{A}$.

In this case, A has empty interior. The reason is that every open ball contains points with irrational coordinates, and A does not.

The interior of $\overline{A}$ is the set $U = \{(x,y) : y > x^3\}$. First we show that U is open. If $\mathbf{a} = (x,y)$ belongs to U, then $s = y - x^3 > 0$. We need to determine a value for r such that $B_r(\mathbf{a})$ is contained in U. Some calculation is needed to determine the proper choice. Suppose that a point (u,v) satisfies $\|(u,v) - (x,y)\| < r$. Then in particular, $|u - x| < r$ and $|v - y| < r$. Hence

$$v - u^3 > (y - r) - (x + r)^3 = y - r - (x^3 + 3rx^2 + 3r^2x + r^3)$$
$$= y - x^3 - r - 3rx^2 - 3r^2x - r^3 = s - (r + 3rx^2 + 3r^2x + r^3).$$

To make the right-hand side positive, which we require, a choice must be made for r so that $r + 3rx^2 + 3r^2x + r^3 \leq s$. Let us decide first that we will choose $r \leq 1$, so that $r^n \leq r$. Then

$$r + 3rx^2 + 3r^2x + r^3 \leq r(2 + 3x^2 + 3|x|).$$

Define $r = \min\{1, s/(2 + 3x^2 + 3|x|)\}$, which is consistent with our choice that $r \leq 1$. Then it follows that

$$v - u^3 > s - (r + 3rx^2 + 3r^2x + r^3) > 0.$$

This shows that $B_r(\mathbf{a})$ is contained in U. Thus U is open.

Now suppose that $\mathbf{a} = (x, y)$ belongs to $\overline{A}$ but is not in U. Then $y \geq x^3$ but $y \not>$ x^3, whence $y = x^3$. To see that $\mathbf{a}$ is not in the interior of $\overline{A}$, it must be shown that whenever $r > 0$, the ball $B_r(\mathbf{a})$ intersects $\overline{A}'$. This is easy, since the point $(x, x^3 - r/2)$ belongs to this ball and does not belong to $\overline{A}$. So $\mathrm{int}\,\overline{A} = U$.

Exercises for Section 4.3

A. Find the closure of the following sets:
   (a) $\mathbb{Q}$ (b) $\{(x, y) \in \mathbb{R}^2 : xy < 1\}$ (c) $\{(x, \sin(\frac{1}{x})) : x > 0\}$ (d) $\{(x, y) \in \mathbb{Q}^2 : x^2 + y^2 < 1\}$.

B. Let $(\mathbf{a}_n)_{n=1}^{\infty}$ be a sequence in $\mathbb{R}^k$ with $\lim_{n \to \infty} \mathbf{a}_n = \mathbf{a}$. Show that $\{\mathbf{a}_n : n \geq 1\} \cup \{\mathbf{a}\}$ is closed.

C. Show that $U = \{(x, y) \in \mathbb{R}^2 : x^2 + 4y^2 < 4\}$ is open by explicitly finding a ball around each point that is contained in U.

D. If A is a bounded subset of $\mathbb{R}$, show that $\sup A$ and $\inf A$ belong to $\overline{A}$.

E. Show that the interior satisfies $\mathrm{int}A = (\overline{A'})'$.

F. Find the interior of $A \cup B$, where $A = \{(x, y) : x \in \mathbb{Q}, y^2 \geq x\}$ and $B = \{(x, y) : x \notin \mathbb{Q}, y \geq x^2\}$.

G. If a subset A of $\mathbb{R}^n$ has no interior, must it be closed?

H. Show that a subset of $\mathbb{R}^n$ is complete if and only if it is closed.

I. Prove Proposition 4.3.9 using Theorem 4.3.8 and Proposition 4.3.3.

J. Show that if U is open and A is closed, then $U \setminus A = \{\mathbf{x} \in U : \mathbf{x} \notin A\}$ is open. What can be said about $A \setminus U$?

K. Suppose that A and B are closed subsets of $\mathbb{R}$.

   (a) Show that the product set $A \times B = \{(x, y) \in \mathbb{R}^2 : x \in A \text{ and } y \in B\}$ is closed.
   (b) Likewise show that if both A and B are open, then $A \times B$ is open.

L. A set A is **dense** in B if B is contained in $\overline{A}$.

   (a) Show that the set of irrational numbers is dense in $\mathbb{R}$.
   (b) Hence show that $\mathbb{Q}$ has empty interior.

M. Suppose that A is a dense subset of $\mathbb{R}^n$.

   (a) Show that if U is open in $\mathbb{R}^n$, then $A \cap U$ is dense in U.
   (b) Show by example that this may fail for sets that are not open.

N. A point $\mathbf{x}$ is a **cluster point** of a subset A of $\mathbb{R}^n$ if there is a sequence $(\mathbf{a}_n)_{n=1}^{\infty}$ with $\mathbf{a}_n \in A \setminus \{\mathbf{x}\}$ such that $\mathbf{x} = \lim_{n \to \infty} \mathbf{a}_n$. Thus, every cluster point is a limit point but not conversely.

   (a) Show that if $\mathbf{x}$ is a limit point of A, then either $\mathbf{x}$ is a cluster point of A or $\mathbf{x} \in A$.
   (b) Hence show that a set is closed if it contains all of its cluster points.
   (c) Find all cluster points of (i) $\mathbb{Q}$, (ii) $\mathbb{Z}$, (iii) $(0, 1)$.

O. Starting with a subset A of $\mathbb{R}^n$, form all the possible sets that may be obtained by repeated use of the operations of closure and complement. Up to 14 different sets can be obtained in this way. Find such a subset of $\mathbb{R}$.

4.4 Compact Sets and the Heine–Borel Theorem

Now we turn to the notion of compactness. At this stage, compactness seems like a convenience and may not appear to be much more useful than completeness. However, when we study continuous functions, compactness will be very useful and then its full power will become apparent.

4.4.1. DEFINITION. A subset A of $\mathbb{R}^n$ is **compact** if every sequence $(\mathbf{a}_k)_{k=1}^\infty$ of points in A has a convergent subsequence $(\mathbf{a}_{k_i})_{i=1}^\infty$ with limit $\mathbf{a} = \lim_{i \to \infty} \mathbf{a}_{k_i}$ in A.

Recall that the Bolzano–Weierstrass Theorem (2.7.2) states that every bounded sequence has a convergent subsequence. Using this new language, we may deduce that every subset of $\mathbb{R}$ that is both closed and bounded is compact. This rephrasing naturally suggests the question, which subsets of $\mathbb{R}^n$ are compact? Before answering this question, we consider a few examples.

4.4.2. EXAMPLES. Consider the set $(0, 1]$. The sequence $1, 1/2, 1/3, \dots$ is in this set but converges to 0, which is not in the set. Since any subsequence will also converge to zero, there is no subsequence of $1, 1/2, 1/3, \dots$ that converges to a number in $(0, 1]$. So this set is not compact.

Next, consider the set $\mathbb{N}$. The sequence $1, 2, 3, \dots$ is in $\mathbb{N}$. However, no subsequence converges (because each subsequence is unbounded and being bounded is a necessary condition for convergence by Proposition 2.5.1). So $\mathbb{N}$ is not compact.

A subset S of $\mathbb{R}^n$ is called **bounded** if there is a real number R such that S is contained in the ball $B_R(0)$. Equivalently, S is bounded if $\sup_{x \in S} \|x\| < \infty$. Notice that when $n = 1$, this definition of bounded agrees with our old definition of bounded subsets of $\mathbb{R}$.

The previous examples suggest that sets that are not closed or not bounded cannot be compact. This is true, and the proofs are an abstraction of the arguments for these examples.

4.4.3. LEMMA. *A compact subset of $\mathbb{R}^n$ is closed and bounded.*

PROOF. Let C be a compact subset of $\mathbb{R}^n$. Suppose that $\mathbf{x}$ is a limit point of C, say $\mathbf{x} = \lim_{n \to \infty} \mathbf{c}_n$ for a sequence $(\mathbf{c}_n)$ in C. Then this sequence has a subsequence $(\mathbf{c}_{n_i})$ converging to a point $\mathbf{c}$ in C. Therefore,

$$\mathbf{x} = \lim_{n \to \infty} \mathbf{c}_n = \lim_{i \to \infty} \mathbf{c}_{n_i} = \mathbf{c} \in C.$$

Thus C is closed.

To show that C is bounded, suppose that it were unbounded. That means that there is a sequence $\mathbf{c}_n \in C$ such that $\|\mathbf{c}_n\| > n$ for each $n \geq 1$. Consider the sequence

$(\mathbf{c}_n)$. If there were a convergent subsequence $(\mathbf{c}_{n_i})$ with limit $\mathbf{c}$, it would follow that

$$\|\mathbf{c}\| = \lim_{i \to \infty} \|\mathbf{c}_{n_i}\| \geq \lim_{i \to \infty} n_i = +\infty.$$

This is an absurd conclusion, and thus C must be bounded. ∎

To establish the converse, we build up a couple of partial results.

4.4.4. LEMMA. *If C is a closed subset of a compact subset of $\mathbb{R}^n$, then C is compact.*

PROOF. Let K be the compact set containing C. Suppose $(\mathbf{x}_n)_{n=1}^{\infty}$ is a sequence in C. To show that C is compact, we must find a subsequence that converges to an element of C.

However, $(\mathbf{x}_n)_{n=1}^{\infty}$ is contained in the compact set K. So it has a subsequence that converges to a number $\mathbf{x}$ in K, say $\mathbf{x} = \lim_{k \to \infty} \mathbf{x}_{n_k}$. Since $(\mathbf{x}_{n_k})_{k=1}^{\infty}$ is contained in C and C is closed, it follows that $\mathbf{x}$ belongs to C as required. ∎

4.4.5. LEMMA. *The cube $[a,b]^n$ is a compact subset of $\mathbb{R}^n$.*

PROOF. Let $\mathbf{x}_k = (x_{k,1}, \ldots, x_{k,n})$ for $k \geq 1$ be a sequence in $\mathbb{R}^n$ such that the co-efficients satisfy $a \leq x_{k,i} \leq b$ for all $k \geq 1$ and $1 \leq i \leq n$. Consider the sequence $(x_{k,1})_{k=1}^{\infty}$ of first coordinates. By the Bolzano–Weierstrass Theorem (2.7.2), there is a subsequence $(x_{k_j,1})_{j=1}^{\infty}$ converging to a point z_1 in $[a,b]$,

$$\lim_{j \to \infty} x_{k_j,1} = z_1.$$

Next consider the sequence $y_j = x_{k_j,2}$ for $j \geq 1$. This sequence is contained in the closed interval $[a,b]$. Thus a second application of the Bolzano–Weierstrass Theorem yields a subsequence $(y_{j_l}) = (x_{k_{j_l},2})$ such that $\lim_{l \to \infty} x_{k_{j_l},2} = z_2$. We still have $\lim_{l \to \infty} x_{k_{j_l},1} = z_1$, since every subsequence of a convergent sequence has the same limit.

Proceeding in this way, finding n consecutive subsubsequences, we obtain a subsequence $p_1 < p_2 < \cdots$ and z_i in $[a,b]$ such that

$$\lim_{j \to \infty} x_{p_j,i} = z_i \quad \text{for} \quad 1 \leq i \leq n.$$

Thus $\lim_{j \to \infty} \mathbf{x}_{p_j} = \mathbf{z} := (z_1, \ldots, z_n)$ by Lemma 4.2.3. ∎

4.4.6. THE HEINE–BOREL THEOREM.
A subset of $\mathbb{R}^n$ is compact if and only if it is closed and bounded.

PROOF. The easy direction is given by Lemma 4.4.3. For the other direction, suppose that C is a closed and bounded subset of $\mathbb{R}^n$. Since it is bounded, there is some $M > 0$ such that $\|x\| \leq M$ for all $x \in C$. In particular, C is contained in the cube $[-M, M]^n$. Now $[-M, M]^n$ is compact by Lemma 4.4.5. So C is a closed subset of a compact set and therefore is compact by Lemma 4.4.4. ∎

This leads to an important generalization of the Nested Interval Theorem.

4.4.7. CANTOR'S INTERSECTION THEOREM.

If $A_1 \supset A_2 \supset A_3 \supset \cdots$ is a decreasing sequence of nonempty *compact subsets of* $\mathbb{R}^n$, *then* $\bigcap_{k \geq 1} A_k$ *is not empty.*

PROOF. Since A_n is not empty, we may choose a point $\mathbf{a}_n$ in A_n for each $n \geq 1$. Then the sequence $(\mathbf{a}_n)_{n=1}^\infty$ belongs to the compact set A_1. By compactness, there is a subsequence $(\mathbf{a}_{n_k})_{k=1}^\infty$ that converges to a limit point $\mathbf{x}$. For each i, the terms a_{n_k} belong to A_i for all $k \geq i$. Thus $\mathbf{x}$ is the limit of points in A_i, whence $\mathbf{x}$ belongs to A_i for all $i > 1$. Therefore, $\mathbf{x}$ belongs to their intersection. ∎

4.4.8. THE CANTOR SET.
We now give a more subtle example of a compact set in $\mathbb{R}$. The **Cantor set** is a fractal subset of the real line. Let $S_0 = [0, 1]$, and construct S_{i+1} from S_i recursively by removing the *middle third* from each interval in S_i. For example, the first three terms are

$$S_1 = [0, 1/3] \cup [2/3, 1],$$
$$S_2 = [0, 1/9] \cup [2/9, 1/3] \cup [2/3, 7/9] \cup [8/9, 1],$$
$$S_3 = [0, 1/27] \cup [2/27, 1/9] \cup [2/9, 7/27] \cup [8/27, 1/3]$$
$$\cup [2/3, 19/27] \cup [20/27, 7/9] \cup [8/9, 25/27] \cup [26/27, 1].$$

By Proposition 4.3.3, the intersection $C = \bigcap_{i \geq 1} S_i$ is a closed set. It is bounded and hence compact. By Cantor's Intersection Theorem, this intersection is not empty. Figure 4.3 shows an approximation to the Cantor set, namely S_5.

0   1

FIG. 4.3 The set S_5 in the construction of the Cantor set.

Every endpoint of an interval in one of the sets S_n belongs to C. But in fact, C contains many other points. Each point in C is determined by a binary decision tree. At the first stage, pick one of the two intervals of S_1 of length $1/3$, which we label 0 and 2. This interval is split into two in S_2 by removing the middle third. Choose either the left (label 0) or right (label 2) to obtain an interval labeled 00, 02, 20, or 22. Continuing in this way, we choose a decreasing sequence of intervals determined by an infinite sequence of 0's and 2's. By Cantor's Intersection Theorem 4.4.7, every choice determines a point of intersection. There is only one point in each of these intersections because the length of the intervals tends to 0. Describe the subset of decision trees that correspond to left or right endpoints. Since this is a proper subset, there are points in C that are not endpoints.

The Cantor set has empty interior. For if C contained an open interval (a,b) with $a < b$, it would also be contained in each S_n. This forces $b - a \leq 3^{-n}$ for every n, whence $a = b$. So the interior of C is empty. A set whose closure has no interior is **nowhere dense**.

Yet C has no isolated points. A point x of a set A is **isolated** if there is an $\varepsilon > 0$ such that the ball $B_\varepsilon(x)$ intersects A only in the singleton $\{x\}$. In fact, C is a **perfect set,** meaning that every point of C is the limit of a sequence of other points in C. In other words, every point is a cluster point. To see this, suppose first that x is not the right endpoint of one of the intervals of some S_n. For each n, let x_n be the right endpoint of the interval of S_n containing x. Then $x_n \neq x$ and $|x_n - x| \leq 3^{-n}$. So $x = \lim_{n \to \infty} x_n$. If x is the right endpoint of one of these intervals, use the left endpoints instead to define the sequence x_n.

The set C is very large, the same size as $[0,1]$, in the sense of cardinality from Section 2.9. Consider the numbers in $[0,1]$ expanded as infinite "decimals" in base 3 (the **ternary expansion**). That is, each number may be expressed as

$$x = (x_0.x_1x_2x_3\dots)_{\text{base 3}} = \sum_{k \geq 0} 3^{-k}x_k,$$

where x_i belongs to $\{0,1,2\}$ for $i \geq 1$. Note that S_1 consists of all numbers in $[0,1]$ that have an expansion with the first digit equal to 0 or 2. In particular,

$$\tfrac{1}{3} = (.1)_{\text{base 3}} = (.02222222\dots)_{\text{base 3}} \quad \text{and} \quad 1 = (.22222222\dots)_{\text{base 3}}.$$

Likewise, S_i consists of all numbers in $[0,1]$ such that the first i terms of some ternary expansion are all 0's and 2's. Since C is the intersection of all the S_i, it consists of precisely all numbers in $[0,1]$ that have a ternary expansion using only 0's and 2's.

Since C is a subset of $[0,1]$, it is clear that C can have no more points than $[0,1]$. To see that $[0,1]$ can have no more points than C, we construct a one-to-one map from $[0,1]$ into C. Think of the points in $[0,1]$ in terms of their binary expansion (base 2). These are all the "decimal" expansions

$$y = (y_0.y_1y_2y_3\dots)_{\text{base 2}} = \sum_{k \geq 0} 2^{-k}y_k,$$

where $y_k \in \{0,1\}$. For each point, pick one binary expansion. (Some numbers such as $\tfrac{1}{2}$ have two possible expansions, one ending in an infinite string of 0's and the other ending in an infinite string of 1's. In this case, pick the expansion ending in 0's.) Send it to the corresponding point in base 3 using 0's and 2's by changing each 1 to 2. Since this corresponding point is in C, we have a map of $[0,1]$ into C. This map is one-to-one because the only duplication of ternary expansions comes from a sequence ending in all 0s corresponding to another ending with all 2s. But we do not send any number to a ternary expansion ending in all 2s.

Warning: This map is not onto because of numbers with two expansions in base 2 such as $\frac{1}{2}$, which in base two equals both $(.1)_{\text{base }2}$ and $(.01111...)_{\text{base }2}$. We used only the first one, which we send to $(.2)_{\text{base }3}$, namely $\frac{2}{3}$. But the other expansion would go to $(.02222...)_{\text{base }3}$, which equals $(.1)_{\text{base }3} = \frac{1}{3}$.

At this point, it seems obvious that there are as many points in C as in $[0,1]$—it can't have any more, and it can't have any less. However, with infinite sets, proving this is a subtle business. The Schroeder–Bernstein Theorem *could* be invoked to obtain a bijection between C and $[0,1]$. However, the special nature of our setup allows a bijection between C and $[0,1]$ to be constructed fairly easily; see Exercise 4.4.K. Therefore, C and $[0,1]$ have the same cardinality, which is uncountable by Theorem 2.9.8.

On the other hand, using a different notion of "size," the Cantor set is very small. We can measure how much of the interval has been removed at each step. The set S_n contains 2^n intervals of length 3^{-n}. The middle third of length 3^{-n-1} is removed from each of these 2^n intervals to obtain S_{n+1}. The total length of the pieces removed is computed by adding an infinite geometric series

$$\sum_{n=0}^{\infty} \frac{2^n}{3^{n+1}} = \frac{1/3}{1-(2/3)} = 1.$$

Thus the Cantor set has *measure zero*, meaning that it can be covered by a countable collection open intervals with arbitrarily small total length. In some sense, C squeezes its very large number of points into a very small space.

Exercises for Section 4.4

A.   Which of the following sets are compact?

(a) $\{(x,y) \in \mathbb{R}^2 : 2x^2 - y^2 \le 1\}$
(b) $\{\mathbf{x} \in \mathbb{R}^n : 2 \le \|\mathbf{x}\| \le 4\}$
(c) $\{(e^{-x}\cos x, e^{-x}\sin x) : x \ge 0\} \cup \{(x,0) : 0 \le x \le 1\}$
(d) $\{(e^{-x}\cos\theta, e^{-x}\sin\theta) : x \ge 0, 0 \le \theta \le 2\pi\}$

B.   Give an example to show that Cantor's Intersection Theorem would not be true if compact sets were replaced by closed sets.

C.   Show that the union of finitely many compact sets is compact.

D.   Show that the intersection of any family of compact sets is compact.

E.   (a) Show that the sum of a closed subset and a compact subset of $\mathbb{R}^n$ is closed. Recall that $A+B = \{\mathbf{a}+\mathbf{b} : \mathbf{a} \in A \text{ and } \mathbf{b} \in B\}$.
(b) Is this true for the sum of two compact sets and a closed set?
(c) Is this true for the sum of two closed sets?

F.   Let $(\mathbf{x}_n)_{n=1}^{\infty}$ be a sequence in a compact set $K \subset \mathbb{R}^n$ that is *not* convergent. Show that there are two subsequences of this sequence that are convergent to *different* limit points.

G.   Prove that a set $S \subset \mathbb{R}$ has no cluster points if and only if $S \cap [-n,n]$ is a finite set for each $n \ge 1$.

H.   Describe all subsets of $\mathbb{R}^n$ that have no cluster points at all.

I.   Let A and B be *disjoint* closed subsets of $\mathbb{R}^n$. Define

$$d(A,B) = \inf\{\|\mathbf{a}-\mathbf{b}\| : \mathbf{a} \in A, \mathbf{b} \in B\}.$$

   (a) If $A = \{\mathbf{a}\}$ is a singleton, show that $d(A,B) > 0$.
   (b) If A is compact, show that $d(A,B) > 0$.
   (c) Find an example of two disjoint closed sets in $\mathbb{R}^2$ with $d(A,B) = 0$.

J.   The **Sierpiński triangle** is constructed in the plane as follows. Start with a solid equilateral
   triangle. Remove the open middle triangle with vertices at the midpoint of each side of the
   larger triangle, leaving three solid triangles with half the side length of the original. From
   each of these three, remove the open middle triangle, leaving nine triangles of one-fourth the
   original side lengths. Proceed in this process ad infinitum. Let S denote the intersection of all
   the finite stages.

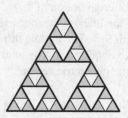

FIG. 4.4 The third stage in constructing the Sierpiński triangle.

   (a) Show that S is a nonempty compact set.
   (b) Show that S has no interior.
   (c) Show that the boundaries of the triangles at the nth stage belong to S. Hence show that
       there is a path in S from the top vertex of the original triangle that gets as close as desired
       (within ε) to any point in S.
   (d) Compute the area of the material removed from the triangle to leave S behind.
   (e) Construct a decision tree for S. Does each decision tree correspond to exactly one point in
       the set? Show that S is uncountable.

K.   Show that there is a bijection from $[0,1]$ onto the Cantor set.
       HINT: Adjust the map constructed in the text by redefining it on a countable sequence to
       include the missing points in the range.

L.   Prove that a countable compact set cannot be perfect.
       HINT: For the case $X = \{x_n : n \geq 1\}$ in $\mathbb{R}^n$, find a decreasing family X_n of closed nonempty
       subsets of X with $x_n \notin X_n$.

Chapter 5
Functions

The main purpose of this chapter is to introduce the notion of a continuous function. Continuity is a basic notion of analysis. It is only with continuous functions that there can be any reasonable approximation or estimation of values at specific points. Most physical phenomena are continuous over most of their domain. The ideas we study in this chapter are sufficiently powerful that even discrete phenomena are sometimes best understood using continuous approximations where these ideas can be used.

5.1 Limits and Continuity

In this chapter, and indeed in most of this book, functions will be defined on some subset S of $\mathbb{R}^n$ with range contained in $\mathbb{R}^m$. Everything is based on the notion of limit, which is a natural variant of the definition for limit of a sequence.

5.1.1. DEFINITION OF THE LIMIT OF A FUNCTION. Let $S \subset \mathbb{R}^n$ and let f be a function from S into $\mathbb{R}^m$. If $\mathbf{a}$ is a limit point of $S \setminus \{\mathbf{a}\}$, then a point $\mathbf{v} \in \mathbb{R}^m$ is the **limit** of f at $\mathbf{a}$ if for every $\varepsilon > 0$, there is an $r > 0$ such that

$$\|f(\mathbf{x}) - \mathbf{v}\| < \varepsilon \quad \text{whenever} \quad 0 < \|\mathbf{x} - \mathbf{a}\| < r \text{ and } \mathbf{x} \in S.$$

We write $\lim\limits_{\mathbf{x} \to \mathbf{a}} f(\mathbf{x}) = \mathbf{v}$.

Geometrically, we have a picture like Figure 5.1. Note that $f(\mathbf{a})$ itself need not be defined. Certainly, saying that $\lim\limits_{\mathbf{x} \to \mathbf{a}} f(\mathbf{x}) = \mathbf{v}$ does not tell us anything about $f(\mathbf{a})$.

Consider the case of a function f defined on an interval (a, b), and let $c \in (a, b)$. Then $\lim\limits_{x \to c} f(x) = L$ means that for every $\varepsilon > 0$, there is an $r > 0$ such that

$$|f(x) - L| < \varepsilon \quad \text{for all} \quad 0 < |x - c| < r.$$

K.R. Davidson and A.P. Donsig, *Real Analysis and Applications: Theory in Practice,*
Undergraduate Texts in Mathematics, DOI 10.1007/978-0-387-98098-0_5,
© Springer Science + Business Media, LLC 2010

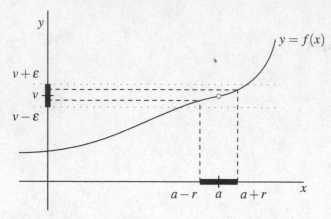

FIG. 5.1 Limit for a function $f : \mathbb{R} \to \mathbb{R}$.

5.1.2. DEFINITION. Let $S \subset \mathbb{R}^n$ and let f be a function from S into $\mathbb{R}^m$. We say that f is **continuous at a** $\in S$ if for every $\varepsilon > 0$, there is an $r > 0$ such that for all $\mathbf{x} \in S$ with $\|\mathbf{x} - \mathbf{a}\| < r$, we have $\|f(\mathbf{x}) - f(\mathbf{a})\| < \varepsilon$. Moreover, f is **continuous on** S if it is continuous at each point $\mathbf{a} \in S$.

If f is not continuous at $\mathbf{a}$, we say that f is **discontinuous** at $\mathbf{a}$.

Continuity can sometimes be described using a limit. If $\mathbf{a}$ is not an isolated point, i.e., $\mathbf{a}$ is a limit point of $S \setminus \{\mathbf{a}\}$, then $\lim_{\mathbf{x} \to \mathbf{a}} f(\mathbf{x})$ makes sense and f is continuous at $\mathbf{a}$ if and only if $\lim_{\mathbf{x} \to \mathbf{a}} f(\mathbf{x}) = f(\mathbf{a})$. Note that if $\mathbf{a}$ is an isolated point of S, then f is always continuous at $\mathbf{a}$.

5.1.3. EXAMPLE. Consider $f : \mathbb{R}^n \setminus \{\mathbf{0}\} \to \mathbb{R}$ given by $f(\mathbf{x}) = 1/\|\mathbf{x}\|$. Let us show that this is continuous on its domain. Fix a point $\mathbf{a} \in \mathbb{R}^n \setminus \{\mathbf{0}\}$. Then

$$|f(\mathbf{x}) - f(\mathbf{a})| = \left| \frac{1}{\|\mathbf{x}\|} - \frac{1}{\|\mathbf{a}\|} \right| = \frac{\big|\|\mathbf{a}\| - \|\mathbf{x}\|\big|}{\|\mathbf{x}\| \, \|\mathbf{a}\|}.$$

Our goal is to make this difference small by controlling the distance $\|\mathbf{x} - \mathbf{a}\|$.

To estimate the numerator, use the triangle inequality.

$$\|\mathbf{x}\| \le \|\mathbf{a}\| + \|\mathbf{x} - \mathbf{a}\| \quad \text{and} \quad \|\mathbf{a}\| \le \|\mathbf{x}\| + \|\mathbf{x} - \mathbf{a}\|.$$

Manipulating these inequalities yields

$$\big|\|\mathbf{x}\| - \|\mathbf{a}\|\big| \le \|\mathbf{x} - \mathbf{a}\|.$$

Now consider the denominator $\|\mathbf{x}\| \, \|\mathbf{a}\|$. Since $\|\mathbf{a}\|$ is a positive constant, it creates no problems. However, $\|\mathbf{x}\|$ must be kept away from 0 to keep the quotient in control. The previous paragraph shows that if $\|\mathbf{a} - \mathbf{x}\| < \|\mathbf{a}\|/2$, then

$$\|\mathbf{x}\| \geq \|\mathbf{a}\| - \|\mathbf{a} - \mathbf{x}\| > \|\mathbf{a}\|/2.$$

Putting this together, choose $r \leq \|\mathbf{a}\|/2$ and consider any $\mathbf{x}$ such that $\|\mathbf{x} - \mathbf{a}\| < r$. Then

$$|f(\mathbf{x}) - f(\mathbf{a})| = \frac{|\|\mathbf{a}\| - \|\mathbf{x}\||}{\|\mathbf{x}\| \|\mathbf{a}\|} \leq \frac{\|\mathbf{x} - \mathbf{a}\|}{\|\mathbf{a}\|^2/2} < \frac{2r}{\|\mathbf{a}\|^2}.$$

To make this less than ε, we need $r \leq \varepsilon \|\mathbf{a}\|^2/2$. Hence $\|f(\mathbf{x}) - f(\mathbf{a})\| < \varepsilon$, provided

$$\|\mathbf{x} - \mathbf{a}\| < r = \min\{\|\mathbf{a}\|/2, \varepsilon \|\mathbf{a}\|^2/2\}.$$

This shows that f is a continuous function. Notice that the function goes to infinity as $\mathbf{x}$ approaches $\mathbf{0}$. So the limit at $\mathbf{0}$ does not exist.

5.1.4. EXAMPLE. A function does not need to have a simple analytic expression to be continuous. However, extra care needs to be taken at the interface. Consider the function graphed in Figure 5.2, given by

$$f(x) = \begin{cases} 0 & \text{if } x \leq 0 \\ e^{-1/x} & \text{if } x > 0. \end{cases}$$

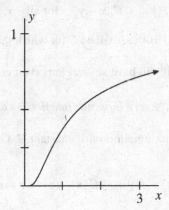

FIG. 5.2 Graph of $y = e^{-1/x}$ for $x > 0$.

When $a < 0$ and $\varepsilon > 0$, we may take $r = |a|$. For if $|x - a| < |a|$, then $x < 0$ and

$$|f(x) - f(a)| = |0 - 0| = 0 < \varepsilon.$$

Hence f is continuous at a.

If $a > 0$, finding an appropriate r for each ε is similar to the previous example. $e^{-1/x}$ is the composition of the simpler functions $g(x) = e^x$ and $h(x) = -1/x$, so that $f(x) = g(h(x))$. We shall see shortly that the composition of continuous functions is continuous. This simplifies the exercise to showing that both g and h are continuous. We leave the details to the reader.

Finally, we must consider $a = 0$ separately because the function f has different definitions on each side of 0. Fix $\varepsilon > 0$. Recall that e^x is an increasing function such that $\lim_{x \to -\infty} e^x = 0$. Hence there is a large negative number $-N$ such that $e^{-N} < \varepsilon$. Therefore, if $0 < x < 1/N$, it follows that $-1/x < -N$ and thus

$$0 < f(x) = e^{-1/x} < e^{-N} < \varepsilon.$$

So take $r = 1/N$. We obtain

$$|f(x) - f(0)| = \begin{cases} 0 & \text{if } -r < x \leq 0 \\ e^{-1/x} & \text{if } 0 < x < r. \end{cases}$$

Since both 0 and $e^{-1/x}$ are less than ε, f is continuous at 0.

Now we treat an important class of functions that are automatically continuous.

5.1.5. DEFINITION. A function f from $S \subset \mathbb{R}^n$ into $\mathbb{R}^m$ is called a **Lipschitz function** if there is a constant C such that

$$\|f(\mathbf{x}) - f(\mathbf{y})\| \leq C\|\mathbf{x} - \mathbf{y}\| \quad \text{for all} \quad \mathbf{x}, \mathbf{y} \in S.$$

The **Lipschitz constant** of f is the smallest C for which this condition holds.

The following easy result will have several important consequences.

5.1.6. PROPOSITION. *Every Lipschitz function is continuous.*

PROOF. Let f be a Lipschitz function with constant C. Given $\varepsilon > 0$, let $r = \varepsilon/C$. Then if $\|\mathbf{x} - \mathbf{y}\| < r$,

$$\|f(\mathbf{x}) - f(\mathbf{y})\| \leq C\|\mathbf{x} - \mathbf{y}\| < Cr = \varepsilon.$$

Therefore, f is continuous. ∎

5.1.7. COROLLARY. *Every linear map A from $\mathbb{R}^n$ to $\mathbb{R}^m$ is Lipschitz, and therefore is continuous.*

PROOF. Recall that every linear transformation is given by an $m \times n$ matrix $[a_{ij}]$, which we also call A, so that

$$A(x_1, x_2, \ldots, x_n) = \left(\sum_{j=1}^{n} a_{1j} x_j, \ldots, \sum_{j=1}^{n} a_{mj} x_j \right).$$

Let $\mathbf{x} = (x_1, x_2, \ldots, x_n)$ and $\mathbf{y} = (y_1, y_2, \ldots, y_m)$. Compute

$$\|A\mathbf{x} - A\mathbf{y}\| = \|A(\mathbf{x} - \mathbf{y})\| = \left(\sum_{i=1}^{m} \left(\sum_{j=1}^{n} a_{ij}(x_j - y_j) \right)^2 \right)^{1/2}.$$

Apply the Schwarz inequality (4.1.1) to obtain that

$$\left| \sum_{j=1}^{n} a_{ij}(x_j - y_j) \right|^2 \le \sum_{j=1}^{n} |a_{ij}|^2 \sum_{j=1}^{n} |x_j - y_j|^2 = \sum_{j=1}^{n} |a_{ij}|^2 \, \|\mathbf{x} - \mathbf{y}\|^2.$$

Setting $C = (\sum_{i=1}^{m} \sum_{j=1}^{n} |a_{ij}|^2)^{1/2}$, we get

$$\|A\mathbf{x} - A\mathbf{y}\| \le \left(\sum_{i=1}^{m} \sum_{j=1}^{n} |a_{ij}|^2 \right)^{1/2} \|\mathbf{x} - \mathbf{y}\| = C\|\mathbf{x} - \mathbf{y}\|.$$

Therefore, linear maps are Lipschitz, and hence are continuous. ∎

There are two basic linear functions, called **coordinate functions**, which we will use regularly. The map $\pi_j(x_1, \ldots, x_n) = x_j$ of $\mathbb{R}^n$ into $\mathbb{R}$ reads off the jth coordinate. And $\varepsilon_i(t) = t e_i$ maps $\mathbb{R}$ into $\mathbb{R}^m$ by sending $\mathbb{R}$ onto the ith coordinate axis. By Exercise 5.1.L, every linear map is a linear combination of the maps $\varepsilon_i \pi_j$.

Exercises for Section 5.1

A. Use the ε–r definition of the limit of a function to show that $\lim_{x \to 2} x^2 = 4$.

B. Let $f(x) = x/\sin x$ for $0 < |x| < \pi/2$ and $f(0) = 1$. Show that f is continuous at 0. Find an $r > 0$ such that $|f(x) - 1| < 10^{-6}$ for all $|x| < r$.   HINT: Use inequalities from Example 2.4.7.

C. Show that the sawtooth function f is continuous, where f is given by
$$f(x) = \begin{cases} x - 2n & \text{if } \ 2n \le x \le 2n+1, \, n \in \mathbb{Z}, \\ 2n - x & \text{if } \ 2n - 1 \le x \le 2n, \, n \in \mathbb{Z}. \end{cases}$$

D. Prove that f is continuous at $(0, y_0)$, where f is defined on $\mathbb{R}^2$ by
$$f(x,y) = \begin{cases} (1 + xy)^{1/x} & \text{if } \ x \ne 0, \\ e^y & \text{if } \ x = 0. \end{cases}$$

E. Consider a function defined on $\mathbb{R}^2$ by
$$f(x,y) = \begin{cases} 0 & \text{if } \ y \le 0 \ \text{or if } \ y \ge x^2, \\ \sin\left(\frac{\pi y}{x^2}\right) & \text{if } \ 0 < y < x^2. \end{cases}$$

(a) Show that f is not continuous at the origin.
(b) Show that the restriction of f to any straight line through the origin is continuous.

F. (a) Show that the definition of limit can be reformulated using open balls instead of norms as follows: A function f mapping a subset $S \subset \mathbb{R}^n$ into $\mathbb{R}^m$ has limit $\mathbf{v}$ as $\mathbf{x}$ approaches $\mathbf{a}$ provided that for every $\varepsilon > 0$, there is an $r > 0$ such that $f(B_r(\mathbf{a}) \cap S \setminus \{\mathbf{a}\}) \subset B_\varepsilon(\mathbf{v})$.

(b) Provide a similar reformulation of the statement that f is continuous at $\mathbf{a}$.

G. Suppose that $f : \mathbb{R}^n \to \mathbb{R}$ is continuous. If there are $\mathbf{x} \in \mathbb{R}^n$ and $C \in \mathbb{R}$ such that $f(\mathbf{x}) < C$, then prove that there is $r > 0$ such that for all $\mathbf{y} \in B_r(\mathbf{x})$, $f(\mathbf{y}) < C$.

H. Suppose that functions f, g, h mapping $S \subset \mathbb{R}^n$ into $\mathbb{R}$ satisfy $f(\mathbf{x}) \le g(\mathbf{x}) \le h(\mathbf{x})$ for $\mathbf{x} \in S$. Suppose that $\mathbf{c}$ is a limit point of S and $\lim_{\mathbf{x} \to \mathbf{c}} f(\mathbf{x}) = \lim_{\mathbf{x} \to \mathbf{c}} h(\mathbf{x}) = L$. Show that $\lim_{\mathbf{x} \to \mathbf{c}} g(\mathbf{x}) = L$.

I. Define a function on the set $S = \{0\} \cup \{\frac{1}{n} : n \ge 1\}$ by $f(\frac{1}{n}) = a_n$ and $f(0) = L$. Prove that f is continuous on S if and only if $\lim_{n \to \infty} a_n = L$.

J. Show that if $f : [a,b] \to \mathbb{R}$ is a differentiable function such that $|f'(x)| \le M$ on $[a,b]$, then f is Lipschitz.   HINT: Mean Value Theorem.

K. Find a bounded continuous function on $\mathbb{R}$ that is not Lipschitz.   HINT: The derivative should blow up somewhere.

L. (a) Show that a linear map $A : \mathbb{R}^n \to \mathbb{R}^m$ with matrix $[a_{ij}]$ can be written as $A = \sum_{i=1}^{m} \sum_{j=1}^{n} a_{ij} \varepsilon_i \pi_j$.

(b) Show that $\varepsilon_i \pi_j$ is Lipschitz with constant 1.
(c) Hence deduce that A is Lipschitz with constant $\sum_{i=1}^{n} \sum_{j=1}^{m} |a_{ij}|$.

M. Consider the linear transformation A on $\mathbb{R}^4$ given by the matrix $A = \frac{1}{2} \begin{bmatrix} 1 & 1 & 1 & 1 \\ 1 & -1 & 1 & -1 \\ 1 & 1 & -1 & -1 \\ 1 & -1 & -1 & 1 \end{bmatrix}$.

(a) Compute the Lipschitz constant obtained in Corollary 5.1.7.
(b) Show that $\|A\mathbf{x}\| = \|\mathbf{x}\|$ for all $\mathbf{x} \in \mathbb{R}^4$. Deduce that the optimal Lipschitz constant is 1. HINT: The columns of A form an orthonormal basis for $\mathbb{R}^4$.

N. At some point, you may have been told that a continuous function is one that can be drawn without lifting your pencil off the paper. Is this actually true?

5.2 Discontinuous Functions

The purpose of this section is to show, through a variety of examples, some of the pathologies that can occur in discontinuous functions. We make no serious attempt to classify discontinuities, although we give names to some of the simpler kinds.

5.2.1. EXAMPLE. Define a function f on $\mathbb{R}$ by $f(0) = 1$ and $f(x) = 0$ for all $x \ne 0$. This function is discontinuous at 0 because $\lim_{x \to 0} f(x) = 0 \ne 1 = f(0)$. This is the simplest kind of discontinuity, known as a **removable singularity** because the function may be altered at the point of discontinuity in order to fix the problem. Changing $f(0)$ to 0 makes the function continuous.

5.2.2. EXAMPLE. Consider the **Heaviside function**, which is much used in engineering. Define H on $\mathbb{R}$ by $H(x) = 0$ for all $x < 0$ and $H(x) = 1$ for all $x \ge 0$.

We claim that $\lim_{x \to 0} H(x)$ does not exist. Suppose that $\lim_{x \to 0} H(x) = L$. Then for any $\varepsilon > 0$, there would be some $r > 0$ such that $|H(x) - L| < \varepsilon$ whenever $0 < |x - 0| < r$. Take $\varepsilon = 1/2$ and let r be any positive real number. The values $\pm r/2$ both satisfy $|\pm r/2 - 0| = r/2 < r$. But for any choice of L, the triangle inequality yields

$$|H(r/2) - L| + |H(-r/2) - L| \geq |H(r/2) - H(-r/2)| = 1.$$

Therefore, $\max\{|H(r/2) - L|, |H(-r/2) - L|\} \geq \frac{1}{2} = \varepsilon$. So no limit exists.

There is no way to "remove" this discontinuity by redefining H at the origin. However, it is not difficult to understand this function's behaviour.

5.2.3. DEFINITION. The **limit** of f as x approaches a **from the right** exists and equals L if for every $\varepsilon > 0$, there is an $r > 0$ such that

$$|f(x) - L| < \varepsilon \quad \text{for all} \quad a < x < a + r.$$

We write $\lim_{x \to a^+} f(x) = L$. Define limits from the left similarly, writing $\lim_{x \to a^-} f(x) = L$.

When a function f on $\mathbb{R}$ has different limits from the left and right at a, we say that f has a **jump discontinuity** at a. A function on an interval is called **piecewise continuous** if on every finite subinterval, it has only a finite number of discontinuities, all of which are jump discontinuities.

The restrictions of H to $(-\infty, 0)$ and $(0, \infty)$ are constant and therefore continuous. What happens at $a = 0$ is that $\lim_{x \to 0^+} H(x) = 1 = H(0)$, and $\lim_{x \to 0^-} H(x) = 0$. Thus H is piecewise continuous with a jump discontinuity at 0.

Piecewise continuity allows a function to have infinitely many jump discontinuities provided the set of jump discontinuities does not have a cluster point (see Exercise 4.3.N). For example, the **ceiling function** on $\mathbb{R}$, defined by letting $\lceil x \rceil$ be the least integer greater than or equal to x, is piecewise continuous on $\mathbb{R}$.

We also consider what it means to have an infinite limit.

5.2.4. DEFINITION. The limit of a function $f(x)$ as x approaches a is $+\infty$ if for every positive integer N, there is an $r > 0$ such that

$$f(x) > N \quad \text{for all} \quad 0 < |x - a| < r.$$

We write $\lim_{x \to a} f(x) = +\infty$. We define the limit $\lim_{x \to a} f(x) = -\infty$ similarly.

5.2.5. EXAMPLE. Recall, from Example 5.1.3, the function f on $\mathbb{R}^n \setminus \{\mathbf{0}\}$ defined by $f(\mathbf{x}) = 1/\|\mathbf{x}\|$. Define $f(\mathbf{0})$ to be 0. Example 5.1.3 showed that f is continuous on $\mathbb{R}^n \setminus \{\mathbf{0}\}$. However,

$$\lim_{\mathbf{x} \to \mathbf{0}} f(\mathbf{x}) = +\infty.$$

Indeed, for each positive integer N, take $r = 1/N$. Then for $0 < \|\mathbf{x}\| < 1/N$, we have $f(\mathbf{x}) > N$ as desired. No redefinition of $f(\mathbf{0})$ can make f continuous.

Nevertheless, this is straightforward behaviour compared to the next examples.

5.2.6. EXAMPLE. Consider the function defined on $\mathbb{R}^2$ by

$$f(x,y) = \begin{cases} \dfrac{x^2}{x^2+y^2} & \text{if} \quad (x,y) \neq (0,0), \\ 0 & \text{if} \quad (x,y) = (0,0). \end{cases}$$

It is easy to verify that f is continuous on $\mathbb{R}^2 \setminus \{(0,0)\}$. However, at the origin, f behaves in a nasty fashion. To understand it, we convert to polar coordinates.

Recall that a vector $(x,y) \neq (0,0)$ is determined by its length $r = \sqrt{x^2+y^2}$ and the angle θ that the vector makes with the positive real axis determined up to a multiple of 2π by $x = r\cos\theta$ and $y = r\sin\theta$. With this notation, we may compute

$$f(x,y) = \frac{x^2}{x^2+y^2} = \frac{r^2\cos^2\theta}{r^2} = \cos^2\theta.$$

Now it is clear that this function is constant on rays from the origin (those points with a fixed angle in polar coordinates). Even though the function remains bounded, its value, $f(x,y)$, oscillates between 0 and 1 as (x,y) progresses around the circle. Thus, for each $r > 0$ and every number $L \in [0,1]$, there is $(x,y) \in B_r((0,0))$ with $f(x,y) = L$. Shrinking r does not change this, so there is no limit at $(0,0)$.

5.2.7. EXAMPLE. A similar phenomenon can be seen in functions on the real line. Consider the function given by

$$f(x) = \begin{cases} \sin\frac{1}{x} & \text{if} \quad x \neq 0, \\ 0 & \text{if} \quad x = 0. \end{cases}$$

First think about the problem of graphing this function. Evidently, $f(-x) = -f(x)$, so it suffices to consider $x > 0$ and then rotate about the origin to obtain the graph of $f(x)$ for $x < 0$. Now as x tends to $+\infty$, the reciprocal $1/x$ tends monotonically to zero. Since $\sin\theta \approx \theta$ for small values of θ, our function is **asymptotic to the curve** $y = 1/x$ as x tends to $+\infty$, meaning $\lim\limits_{x\to\infty} |f(x) - 1/x| = 0$.

On the other hand, as x tends to 0^+, $1/x$ goes off to $+\infty$. Indeed, it passes through values from $2k\pi$ to $2(k+1)\pi$ as x passes from $\frac{1}{2k\pi}$ to $\frac{1}{2(k+1)\pi}$. Hence $\sin$ takes values running from 0 up to 1, down to -1, and back up to 0. This happens infinitely often as x approaches the origin. So the curve oscillates rapidly up and down between 1 and -1. No limit is possible. There is a partial graph in Figure 5.3, not including the graph in the interval $(-0.01, 0.01)$.

General arguments show that this function is continuous on $\mathbb{R} \setminus \{0\}$. However, it has a nasty discontinuity at 0. Let us show that every value in $[-1,1]$ is a limit value along some subsequence. Consider a number $t = \sin\theta$. Notice that

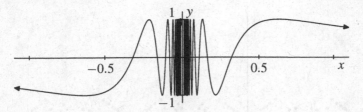

FIG. 5.3 A partial graph of $\sin(1/x)$.

$$f(x) = t \quad \text{if and only if} \quad \sin\frac{1}{x} = \sin\theta$$

$$\text{if and only if} \quad \tfrac{1}{x} = \theta + 2k\pi \text{ or } (\pi - \theta) + 2k\pi, \quad k \in \mathbb{Z}$$

$$\text{if and only if} \quad x = \tfrac{1}{\theta + 2k\pi} \text{ or } \tfrac{1}{(\pi - \theta) + 2k\pi}, \quad k \in \mathbb{Z}.$$

In particular, $\lim\limits_{k\to\infty} f(\frac{1}{\theta + 2k\pi}) = \sin\theta = t$. This shows that every point $(0, t)$ for $|t| \le 1$ lies in the closure of the graph of f. It is not difficult to see that the closure of the graph is precisely this line segment together with the graph itself.

Finally, we look at a couple of bizarre examples.

5.2.8. EXAMPLE. For any subset A of $\mathbb{R}^n$, the **characteristic function** of A is

$$\chi_A(\mathbf{x}) = \begin{cases} 1 & \text{if } \mathbf{x} \in A \\ 0 & \text{if } \mathbf{x} \notin A. \end{cases}$$

The behaviour of χ_A depends on the character of the set A. See Exercise 5.2.A.

Let us take A to be the set $\mathbb{Q}$ of rationals in $\mathbb{R}$. The function $\chi_{\mathbb{Q}}$ takes the values 0 and 1 on every open interval, no matter how small, because the sets of the rational and the irrational numbers are both dense in the line. Thus for every $a \in \mathbb{R}$ and any $r > 0$ there is a point x with $|x - a| < r$ such that $|f(x) - f(a)| = 1$. This function is not continuous at any point!

5.2.9. EXAMPLE. This last example is perhaps the strangest of all. Let

$$f(x) = \begin{cases} 0 & \text{if } x \notin \mathbb{Q}, \\ \tfrac{1}{q} & \text{if } x = \tfrac{p}{q} \text{ in lowest terms and } q > 0, \end{cases}$$

meaning that p, q are integers with no common factor. Figure 5.4 shows part of the graph of this function. We will show that this function is continuous at every irrational point, and discontinuous at every rational point.

First, we show that $\lim_{x \to a} f(x) = 0$ for all $a \in \mathbb{R}$. Let $\varepsilon > 0$ and fix an integer $M > |a|$. There is an integer N sufficiently large that $1/N < \varepsilon$. The set

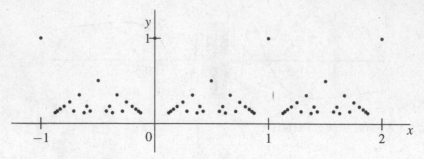

FIG. 5.4 Partial graph of function $f(p/q) = 1/q$.

$$S = \left\{ \frac{p}{q} : 1 \leq q \leq N, \ -Mq \leq p \leq Mq, \right\} \backslash \{a\}$$

is finite and thus is closed. Since S' is open and $a \in S'$, there is a real number $r > 0$ such that $B_r(a) \subset S' \cap (-M, M)$. Now if $x \in (-M, M)$ is not in S, then either it is irrational, whence $f(x) = 0$, or it is a rational p/q with $q > N$, whence $f(x) < \frac{1}{N} < \varepsilon$. Hence

$$|f(x) - 0| = |f(x)| < \varepsilon \quad \text{for all} \quad |x - a| < r.$$

This shows that $\lim_{x \to a} f(x) = 0$.

Thus, for a an irrational number, $\lim_{x \to a} f(x) = 0 = f(a)$ and so f is continuous at a. For a a rational number, say $a = p/q$ in lowest terms, $\lim_{x \to a} f(x) = 0 \neq 1/q = f(a)$ and so f has a removable singularity at a. The surprising fact is that f has a limit at every point in $\mathbb{R}$, yet is discontinuous on a dense set.

Exercises for Section 5.2

A. Let A be a subset of $\mathbb{R}^n$. Show that the characteristic function χ_A is continuous on the interior of A and of its complement A', but is discontinuous on the boundary $\partial A = \overline{A} \cap \overline{A'}$.

B. Show that $f(x) = x \log x^2$ for $x \in \mathbb{R} \backslash \{0\}$ has a removable singularity at $x = 0$.

C. Give an example of a bounded function $f : [-1, 1] \to \mathbb{R}$ that has only jump discontinuities but is not piecewise continuous.

D. What is the nature of the singularity at $x = 1$ of the function $f(x) = x^{\frac{1}{1-x}}$ defined on the set $[0, \infty) \backslash \{1\}$? HINT: $\lim_{x \to 1} \frac{\log x}{x-1}$ is a derivative.

E. Let $f(x) = \arcsin(\sin x)$, where $\arcsin(y)$ is the unique value $\theta \in [-\frac{\pi}{2}, \frac{\pi}{2}]$ such that $\sin \theta = y$.

   (a) Show that f' has limits from the left and the right at every point.
   (b) Where is f' discontinuous?

F. Prove that $L = \lim_{x \to a} f(x)$ if and only if both $\lim_{x \to a-} f(x) = L$ and $\lim_{x \to a+} f(x) = L$.

G. (A monotone convergence test for functions.) Suppose that f is an increasing function on (a, b) that is bounded above. Prove that the one-sided limit $\lim_{x \to b-} f(x)$ exists.

H. Define f on $\mathbb{R}$ by $f(x) = x \chi_{\mathbb{Q}}(x)$. Show that f is continuous at 0 and that this is the *only* point where f is continuous.

5.3 Properties of Continuous Functions

We start with several properties equivalent to continuity. Then we will record various consequences of continuity, most of which are easy to verify. Since the domain of a function is often a proper subset of $\mathbb{R}^n$, we introduce another topological notion. A subset $V \subset S \subset \mathbb{R}^n$ is **open in** S or **relatively open** (with respect to S) if there is an open set U in $\mathbb{R}^n$ such that $U \cap S = V$. In other words, V is open in S if for every $\mathbf{v} \in V$, there is an $\varepsilon > 0$ such that $B_\varepsilon(\mathbf{v}) \cap S \subset V$.

5.3.1. THEOREM. *For a function f mapping $S \subset \mathbb{R}^n$ into $\mathbb{R}^m$, the following are equivalent:*

(1) *f is continuous on S.*

(2) *For every convergent sequence $(\mathbf{x}_k)_{k=1}^\infty$ with $\lim\limits_{k \to \infty} \mathbf{x}_k = \mathbf{a}$ in S, $\lim\limits_{k \to \infty} f(\mathbf{x}_k) = f(\mathbf{a})$.*

(3) *For every open set U in $\mathbb{R}^m$, the set $f^{-1}(U) = \{\mathbf{x} \in S : f(\mathbf{x}) \in U\}$ is open in S.*

PROOF. If we assume (1), that is, f is continuous on S, then combining this with the definition of the limit of sequence gives (2). We leave the details as an exercise. Conversely, assume that (1) is false, and f is not continuous at some point $\mathbf{a} \in S$. Then reversing the definition of continuity, we can find some positive number $\varepsilon > 0$ for which the definition fails, meaning that there is no value of $r > 0$ that works. That is, fixing this ε, for *every* $r > 0$ there is some point $\mathbf{x} \in S$ (depending on r) such that

$$\|\mathbf{x} - \mathbf{a}\| < r \quad \text{and} \quad \|f(\mathbf{x}) - f(\mathbf{a})\| \geq \varepsilon.$$

So take $r = 1/k$ and find an $\mathbf{x}_k \in S$ with

$$\|\mathbf{x}_k - \mathbf{a}\| < \tfrac{1}{k} \quad \text{and} \quad \|f(\mathbf{x}_k) - f(\mathbf{a})\| \geq \varepsilon.$$

It follows that $\lim_{k \to \infty} \mathbf{x}_k = \mathbf{a}$ and $f(\mathbf{x}_k)$ does not converge to $f(\mathbf{a})$. This shows that if (1) fails, then (2) is false also. Therefore, (1) and (2) are equivalent.

Suppose that f is continuous and U is an open subset of $\mathbb{R}^m$. Pick any point $\mathbf{a}$ in $f^{-1}(U)$. Since U is open and contains $\mathbf{u} = f(\mathbf{a})$, there is an $\varepsilon > 0$ such that $B_\varepsilon(\mathbf{u})$ is contained in U. From the continuity of f, there is a real number $r > 0$ such that

$$\|f(\mathbf{x}) - \mathbf{u}\| < \varepsilon \quad \text{for all} \quad \mathbf{x} \in S, \|\mathbf{x} - \mathbf{a}\| < r.$$

This means that $f(B_r(\mathbf{a}) \cap S)$ is contained in $B_\varepsilon(\mathbf{u})$ and thus in U. Hence $f^{-1}(U)$ contains $B_r(\mathbf{a}) \cap S$. Consequently, $f^{-1}(U)$ is open in S.

Conversely, suppose that (3) holds. Fix $\mathbf{a}$ in S and $\varepsilon > 0$. Using the open set $U = B_\varepsilon(f(\mathbf{a}))$, we obtain an open set $f^{-1}(U)$ in S containing $\mathbf{a}$. Therefore, there is a real number $r > 0$ such that

$$B_r(\mathbf{a}) \cap S \subset f^{-1}(U).$$

Thus, $\|f(\mathbf{x}) - f(\mathbf{a})\| < \varepsilon$ for all $\mathbf{x} \in S$ such that $\|\mathbf{x} - \mathbf{a}\| < r$, so (1) holds. ∎

Property (2) could be called the **sequential characterization of continuity**. It will often be more convenient to work with a sequence and this property rather than finding some r for each ε, as in the original definition. Property (3) could be called the **topological characterization of continuity**. This is a formulation that readily generalizes to settings in which there is no appropriate distance function. In certain ways, this version is more powerful than the others. However, it is valid only for continuity on a set, not continuity at a point.

The next two results show that limits and continuity repect the usual arithmetic operations. The proofs are straightforward adaptations of the proof of Theorem 2.5.2. An alternative is to use Exercise 5.3.D and Theorem 2.5.2.

5.3.2. THEOREM. *If f, g are functions from a common domain $S \subset \mathbb{R}^n$ into $\mathbb{R}^m$ and $\mathbf{a} \in S$ such that $\lim_{\mathbf{x} \to \mathbf{a}} f(\mathbf{x}) = \mathbf{u}$ and $\lim_{\mathbf{x} \to \mathbf{a}} g(\mathbf{x}) = \mathbf{v}$, then*

(1) $\lim_{\mathbf{x} \to \mathbf{a}} f(\mathbf{x}) + g(\mathbf{x}) = \mathbf{u} + \mathbf{v}$,

(2) $\lim_{\mathbf{x} \to \mathbf{a}} \alpha f(\mathbf{x}) = \alpha \mathbf{u}$ *for any $\alpha \in \mathbb{R}$.*

When the range is contained in $\mathbb{R}$, say $\lim_{\mathbf{x} \to \mathbf{a}} f(\mathbf{x}) = u$ and $\lim_{\mathbf{x} \to \mathbf{a}} g(\mathbf{x}) = v$, then

(3) $\lim_{\mathbf{x} \to \mathbf{a}} f(\mathbf{x}) g(\mathbf{x}) = uv$, *and*

(4) $\lim_{\mathbf{x} \to \mathbf{a}} \dfrac{f(\mathbf{x})}{g(\mathbf{x})} = \dfrac{u}{v}$ *provided that $v \neq 0$.*

5.3.3. THEOREM. *If f, g are functions from a common domain S into $\mathbb{R}^m$ that are continuous at $\mathbf{a} \in S$, and $\alpha \in \mathbb{R}$, then*

(1) $f + g$ *is continuous at $\mathbf{a}$,*

(2) αf *is continuous at $\mathbf{a}$,*

and when the range is contained in $\mathbb{R}$,

(3) fg *is continuous at $\mathbf{a}$, and*

(4) f/g *is continuous at $\mathbf{a}$ provided that $g(\mathbf{a}) \neq 0$.*

5.3.4. EXAMPLE. Observe that the function $f(x) = x$ is continuous at every $a \in \mathbb{R}$, since $\lim_{x \to a} f(x) = \lim_{x \to a} x = a$. By Theorem 5.3.3 (2), products of this function are continuous, so $g(x) = x^2$, $h(x) = x^3$, and in general $k(x) = x^n$ for every positive integer n are all continuous functions. By Theorem 5.3.3 (1) and (3), linear combinations of these functions are continuous, and so we conclude that every polynomial is continuous on $\mathbb{R}$.

If f is a **rational function**—that is, $f(x) = p(x)/q(x)$, where p and q are polynomials—then f is continuous at all $a \in \mathbb{R}$, where $q(a) \neq 0$. This follows from the previous paragraph and Theorem 5.3.3 (4).

Recall that if f maps a domain $S \subset \mathbb{R}^n$ into a set $T \subset \mathbb{R}^m$, and g maps T into $\mathbb{R}^l$, then the **composition** of g and f, denoted by $g \circ f$, is the function that sends

x to $g(f(x))$. For example, if $f(x,y) = x^2 + y^2$ is defined on $\mathbb{R}^2$ and $g(x) = \sqrt{x}$ for $x \in [0, \infty)$, then $g \circ f(x,y) = \sqrt{x^2 + y^2}$.

5.3.5. THEOREM. *Suppose that f maps a domain S contained in $\mathbb{R}^n$ into a subset T of $\mathbb{R}^m$, and g maps T into $\mathbb{R}^l$. If f is continuous at $\mathbf{a} \in S$ and g is continuous at $f(\mathbf{a}) \in T$, then the function $g \circ f$ is continuous at $\mathbf{a}$. Thus if f and g are continuous, then so is $g \circ f$.*

PROOF. We will use the sequential characterization of continuity. Let $(\mathbf{x}_k)_{k=1}^{\infty}$ be any sequence of points in S with $\lim_{k \to \infty} \mathbf{x}_k = \mathbf{a}$. Since f is continuous at $\mathbf{a}$, we know that $\lim_{k \to \infty} f(\mathbf{x}_k) = f(\mathbf{a})$. Thus $(f(\mathbf{x}_k))_{k=1}^{\infty}$ is a sequence in T with limit $f(\mathbf{a})$, and since g is continuous at $f(\mathbf{a})$, we conclude that

$$\lim_{k \to \infty} g(f(\mathbf{x}_k)) = g(f(\mathbf{a})).$$

Therefore, by Theorem 5.3.1, $g \circ f$ is continuous at $\mathbf{a}$. ∎

5.3.6. EXAMPLE. If f maps $S \subset \mathbb{R}^n$ into $\mathbb{R}^m$, then f_i, the ith coordinate of $f(\mathbf{x})$, is a real-valued function on S. Using this notation, we may write

$$f(\mathbf{x}) = (f_1(\mathbf{x}), \dots, f_m(\mathbf{x})).$$

We claim that f is continuous if and only if each f_i is continuous for $1 \le i \le m$.

One way to see this is to argue exactly as in Lemma 4.2.3. Instead, we will use Corollary 5.1.7 and the (continuous) coordinate functions π_i and ε_i given by $\pi_i(x_1, \dots, x_m) = x_i$ and $\varepsilon_i(t) = t\mathbf{e}_i$. Notice that $f_i(\mathbf{x}) = \pi_i \circ f(\mathbf{x})$. Thus if f is continuous, each f_i is continuous. Conversely, $f(\mathbf{x}) = \sum_{i=1}^{m} \varepsilon_i \circ f_i(\mathbf{x})$. Hence if each f_i is continuous, then each $\varepsilon_i \circ f_i$ is continuous by Theorem 5.3.5; and their sum is continuous by Theorem 5.3.3 (1).

Exercises for Section 5.3

A. Show that the function defined on $\mathbb{R}^2 \setminus \{(0,0)\}$ by $f(x,y) = \dfrac{\sin(\log(x^2 + y^2))}{\cos^2 y + y^2 e^x}$ is continuous.

B. Prove that (1) implies (2) in Theorem 5.3.1.

C. Use Lemma 4.2.3 and the sequential characterization of continuity to give a second proof for Example 5.3.6.

D. Consider f mapping $S \subset \mathbb{R}^n$ into $\mathbb{R}^m$ and two points: $\mathbf{a}$ a limit point of S and $\mathbf{v} \in \mathbb{R}^m$. Show that $\lim_{\mathbf{x} \to \mathbf{a}} f(\mathbf{x}) = \mathbf{v}$ iff for each sequence $(\mathbf{x}_k) \in S \setminus \{\mathbf{a}\}$ with $\lim_{k \to \infty} \mathbf{x}_k = \mathbf{a}$, we have $\lim_{k \to \infty} f(\mathbf{x}_k) = \mathbf{v}$.

E. Suppose that f mapping $S \subset \mathbb{R}^n$ into $\mathbb{R}^m$ is given by $f(\mathbf{x}) = (f_1(\mathbf{x}), \dots, f_m(\mathbf{x}))$ with $f_i : S \to \mathbb{R}$ for each i. Show that $\lim_{\mathbf{x} \to \mathbf{a}} f(\mathbf{x}) = (u_1, \dots, u_m)$ if and only if $\lim_{\mathbf{x} \to \mathbf{a}} f_i(\mathbf{x}) = u_i$ for each i.

F. Let f and g be continuous maps of $S \subset \mathbb{R}^n$ into $\mathbb{R}^m$. Show that the inner product $h(\mathbf{x}) = \langle f(\mathbf{x}), g(\mathbf{x}) \rangle$ is continuous.

G. Finish the proof of Theorem 5.3.2.

H. Suppose f is continuous on $[a,b]$ and g is continuous on $[b,c]$ with $f(b) = g(b)$. Show that
$$h(x) = \begin{cases} f(x) & \text{if} \quad a \le x \le b, \\ g(x) & \text{if} \quad b \le x \le c, \end{cases} \quad \text{is continuous on } [a,c].$$

I. Let f be a continuous real-valued function defined on an open subset U of $\mathbb{R}^n$. Show that $\{(\mathbf{x}, y) : \mathbf{x} \in U, \, y > f(\mathbf{x})\}$ is an open subset of $\mathbb{R}^{n+1}$.

J. (a) Show that $m(x,y) = \max\{x,y\}$ is continuous on $\mathbb{R}^2$.
    (b) Hence show that if f and g are continuous real-valued functions on a set $S \subset \mathbb{R}^n$, then $h(x) = \max\{f(x), g(x)\}$ is continuous on S.
    (c) Use induction to show that if f_i are continuous real-valued functions on S for $1 \le i \le k$, then $h(x) = \max_{1 \le i \le k} f_i(x)$ is continuous.

K. Show that $f : \mathbb{R}^n \to \mathbb{R}^m$ is continuous iff $f^{-1}(C)$ is closed for every closed set $C \subset \mathbb{R}^m$.

L. Suppose that A and B are subsets of $\mathbb{R}^n$. Find necessary and sufficient conditions for there to be a continuous function f on $\mathbb{R}^n$ with $f|_A = 1$ and $f|_B = 0$.
    HINT: Consider $g(x) = \text{dist}(x,A)$ and $h(x) = \text{dist}(x,B)$.

M. Give an example of a continuous function f and an open set U such that $f(U)$ is not open.

N. Suppose that $f : \mathbb{R} \to \mathbb{R}$ satisfies the functional equation
$$f(u+v) = f(u) + f(v) \quad \text{for all} \quad u, v \in \mathbb{R}.$$
    (a) Prove that $f(mx) = mf(x)$ for all $x \in \mathbb{R}$ and $m \in \mathbb{Z}$.   HINT: Use induction for $m \ge 1$.
    (b) Prove that $f(x) = cx$ for all $x \in \mathbb{Q}$, where $c = f(1)$.   HINT: Use (a) to solve for $f(p/q)$.
    (c) Use (b) to prove that if f is continuous on $\mathbb{R}$, then $f(x) = cx$ for all $x \in \mathbb{R}$.

5.4 Compactness and Extreme Values

In every calculus course, a lot of effort is spent finding the maximum or minimum of various functions. Sometimes there were physical reasons why such a point should exist. However, generally it was taken on blind faith and the student dutifully differentiated the function to find critical points. Even when the function is not differentiable, the function may attain its maximum value. On the other hand, many very nice functions do not attain maxima. In this section, we will see how our new topological tools can explain this phenomenon.

First consider a couple of easy examples in which there are no maxima.

5.4.1. EXAMPLE. Consider $f(\mathbf{x}) = -1/(1 + \|\mathbf{x}\|^2)$ for $\mathbf{x} \in \mathbb{R}^n$. This function is bounded above, yet the supremum $0 = \lim_{\|\mathbf{x}\| \to \infty} f(\mathbf{x})$ is never attained. The function $g(\mathbf{x}) = \|\mathbf{x}\|$ is unbounded and thus also does not attain its supremum. These problems can occur whenever the domain of the function is unbounded.

5.4.2. EXAMPLE. Consider $f(x) = -x$ for $x \in (0,1]$. This function is bounded above yet does not attain its supremum $0 = \lim_{x \to 0^+} f(x)$ because the limit point 0 is missing from the domain. Similarly, the function $f(x) = \frac{1}{x}$ for $x \in (0,1]$ is unbounded and thus does not attain its supremum. A modification of this example would show that the same problem results whenever the domain is not closed.

Both of these difficulties may be avoided if the domain is compact. It turns out that this is exactly what we need. As in most proofs using compactness, the aim is to find an appropriate sequence in the compact set C so that a convergent subsequence can be obtained with good properties.

5.4.3. THEOREM. *Let C be a compact subset of $\mathbb{R}^n$, and let f be a continuous function from C into $\mathbb{R}^m$. Then the image set $f(C)$ is compact.*

PROOF. Let $(\mathbf{y}_k)_{k=1}^{\infty}$ be a sequence in $f(C)$. We must find a subsequence converging to a point in the image. First choose points $\mathbf{x}_k$ in C such that $\mathbf{y}_k = f(\mathbf{x}_k)$. Now $(\mathbf{x}_k)_{k=1}^{\infty}$ is a sequence in the compact set C. Therefore, there is a subsequence $(\mathbf{x}_{k_i})$ that converges to some $\mathbf{c}$ in C. By the continuity of f,

$$\lim_{i \to \infty} \mathbf{y}_{k_i} = \lim_{i \to \infty} f(\mathbf{x}_{k_i}) = f\left(\lim_{i \to \infty} \mathbf{x}_{k_i}\right) = f(\mathbf{c}).$$

Thus $(\mathbf{y}_{k_i})$ converges to $f(\mathbf{c}) \in f(C)$, showing that $f(C)$ is compact. ∎

This immediately yields a result often used (without proof) in calculus.

5.4.4. EXTREME VALUE THEOREM.
Let C be a compact subset of $\mathbb{R}^n$, and let f be a continuous function from C into $\mathbb{R}$. Then there are points $\mathbf{a}$ and $\mathbf{b}$ in C attaining the minimum and maximum values of f on C. That is,

$$f(\mathbf{a}) \le f(\mathbf{x}) \le f(\mathbf{b}) \quad \text{for all} \quad \mathbf{x} \in C.$$

PROOF. Since C is compact, Theorem 5.4.3 shows that $f(C)$ is compact. Hence it is closed and bounded in $\mathbb{R}$. Boundedness shows that

$$m = \inf_{\mathbf{x} \in C} f(\mathbf{x}) \quad \text{and} \quad M = \sup_{\mathbf{x} \in C} f(\mathbf{x})$$

are both finite. From the definition of supremum, M is a limit of values in $f(C)$. Thus since $f(C)$ is closed, $M \in f(C)$. This means that there is a point $\mathbf{b} \in C$ such that $f(\mathbf{b}) = M$. Similarly, the infimum is attained at some point $\mathbf{a} \in C$. ∎

Exercises for Section 5.4

A. If A is a noncompact subset of $\mathbb{R}^n$, show that there is a bounded continuous real-valued function on A that does not attain its maximum.

B. Find a *discontinuous* function on $[0,1]$ that is bounded but does not achieve its supremum.

C. Suppose that f is a continuous function on $[a,b]$ with no local maximum or local minimum. Prove that f is monotone.

D. Find a linear transformation T on $\mathbb{R}^2$ and a closed subset C of $\mathbb{R}^2$ such that $T(C)$ is not closed.

E. Show that f mapping a *compact* set $S \subset \mathbb{R}^n$ into $\mathbb{R}^m$ is continuous iff its graph $G(f) = \{(\mathbf{x}, f(\mathbf{x})) : \mathbf{x} \in S\}$ is compact.   HINT: ($\Rightarrow$) use Theorem 5.4.3. ($\Leftarrow$) use Theorem 5.3.1 (2).

F. Give a function defined on $[0,1]$ that has a closed graph but is not continuous.

G. Suppose that f is a *positive* continuous function on $\mathbb{R}^n$ such that $\lim_{\|\mathbf{x}\| \to \infty} f(\mathbf{x}) = 0$ Show that f attains its maximum.

H. Let f be a **periodic function** on $\mathbb{R}$, i.e., there is a $d > 0$ with $f(x+d) = f(x)$ for all $x \in \mathbb{R}$. Show that if f is continuous, then f attains its maximum and minimum on $\mathbb{R}$.

I. (a) Give an example of a continuous function on $\mathbb{R}^2$ satisfying $f(x+1,y) = f(x,y)$ for all $x, y \in \mathbb{R}$ that does not attain its maximum.
    (b) Find and prove a variant of the previous exercise that is valid for functions on $\mathbb{R}^2$.

J. Let A be a compact subset of $\mathbb{R}^n$. Show that for any point $\mathbf{x} \in \mathbb{R}^n$, there is a closest point $\mathbf{a}$ in A to $\mathbf{x}$; i.e., $\mathbf{a} \in A$ satisfies $\|\mathbf{x} - \mathbf{a}\| \le \|\mathbf{x} - \mathbf{b}\|$ for all $\mathbf{b} \in A$.   HINT: Define a useful continuous function on A.

K. For a function f on $[0,\infty)$, we say that $\lim_{x \to \infty} f(x) = L$ if for every $\varepsilon > 0$, there is some $N > 0$ such that $|f(x) - L| < \varepsilon$ for all $x > N$. Suppose that f is continuous. and $\lim_{x \to \infty} f(x) = f(0)$. Prove that f attains its maximum and minimum values.

L. Suppose that C is a compact subset of $\mathbb{R}^n$ and that f is a continuous, one-to-one function of C onto $D \subset \mathbb{R}^m$. Prove that the inverse function f^{-1} is continuous.   HINT: For $d_0 \in D$, let $c_0 = f^{-1}(d_0)$. If $\varepsilon > 0$, find $r > 0$ such that $B_r(d_0)$ is disjoint from $f(C \setminus B_\varepsilon(c_0))$.

M. *A space-filling curve.* Let T be a right triangle with side lengths 3, 4, and 5. Drop a perpendicular line from the right angle to the opposite side, splitting the triangle into two similar pieces. Label the smaller triangle $T(0)$ and the larger one $T(1)$. Then divide each $T(\varepsilon)$ into two pieces in the same way, labeling the smaller $T(\varepsilon 0)$ and the larger $T(\varepsilon 1)$. Recursively divide each triangle $T(\varepsilon_1 \ldots \varepsilon_n)$ into two smaller similar triangles labeled $T(\varepsilon_1 \ldots \varepsilon_n 0)$ and $T(\varepsilon_1 \ldots \varepsilon_n 1)$. Now consider each point $x \in [0,1]$ in its base-2 (binary) expansion $x = 0.\varepsilon_1 \varepsilon_2 \varepsilon_3 \ldots$, where ε_i is 0 or 1. Define a function $f : [0,1] \to T$ by defining $f(x)$ to be the point in $\bigcap_{n \ge 1} T(\varepsilon_1 \ldots \varepsilon_n)$.

  (a) Prove that $T(\varepsilon_1 \ldots \varepsilon_n)$ has diameter at most $5(0.8)^n$.
  (b) If $x = 0.\varepsilon_1 \ldots \varepsilon_{n-1} 100000 \ldots$ has a finite binary expansion, then it has a second binary representation $x = 0.\varepsilon_1 \ldots \varepsilon_{n-1} 011111 \ldots$ ending in ones. Prove that both expansions yield the same value for $f(x)$.
  (c) Hence prove that $f(x)$ is well defined for each $x \in [0,1]$.
  (d) Prove that f is continuous.   HINT: If x and y agree to the nth decimal, what do $f(x)$ and $f(y)$ have in common?
  (e) Prove that f maps $[0,1]$ onto T.

N. *A space-filling curve II.* Adapt the triangle-filling function of the previous exercise to construct a continuous function on $\mathbb{R}$ that maps onto the entire plane.   HINT: Cover $\mathbb{R}^2$ by triangles.

5.5 Uniform Continuity

Mathematical terminology is not always consistent, but the adjective *uniform* is (almost) always used the same way. A property is uniform on a set if that property holds at every point in the set with common estimates. Uniform estimates often lead to more powerful conclusions.

5.5.1. DEFINITION. A function f from $S \subset \mathbb{R}^n$ into $\mathbb{R}^m$ is **uniformly contin-uous** if for every $\varepsilon > 0$, there is a positive real number $r > 0$ such that

$$\|f(\mathbf{x}) - f(\mathbf{a})\| < \varepsilon \quad \text{whenever} \quad \|\mathbf{x} - \mathbf{a}\| < r, \ \mathbf{x}, \mathbf{a} \in S.$$

Read the definition carefully to note where it differs from continuity at each point $\mathbf{a} \in S$. For f to be continuous, we fix both $\varepsilon > 0$ and $\mathbf{a} \in S$ before obtaining the value of r. So the choice of r might depend on $\mathbf{a}$ as well as on ε. Uniform continuity means that for each $\varepsilon > 0$, the value $r > 0$ that we obtain can be chosen independently of the point $\mathbf{a}$. This is a subtle difference, so we look at some examples.

5.5.2. EXAMPLE. Consider the function $f(x) = x^2$ defined on the bounded in-terval $[c, d]$. Let us try to obtain a common estimate for r for each $\varepsilon > 0$. Remember that we are trying to control the difference $|f(x) - f(a)|$ *only* by controlling $|x - a|$. Hence we always look for a method of getting a factor close to $|x - a|$ into our estimate while gaining some (perhaps crude) control over the rest. Compute

$$|f(x) - f(a)| = |x^2 - a^2| = |x + a| \, |x - a|.$$

In this case, the factor of $|x - a|$ comes out naturally. A bound must be found for $|x + a|$. For any $x \in [c, d]$, we have $|x| \le \max\{|c|, |d|\}$ so let $M = \max\{|c|, |d|\}$. Hence

$$|x + a| \le |x| + |a| \le 2M.$$

If we choose $r > 0$ and consider $|x - a| < r$, then

$$|f(x) - f(a)| = |x + a| \, |x - a| < 2Mr.$$

To make this at most ε, it suffices to choose $r = \varepsilon/(2M)$, whence

$$|f(x) - f(a)| < 2Mr = \varepsilon \quad \text{for all} \quad |x - a| < r, \ x, a \in [c, d].$$

Hence f is uniformly continuous.

On the other hand, consider $f(x) = x^2$ defined on the whole real line. The pre-ceding argument doesn't work because M would be infinite. As Figure 5.5 suggests, as a goes to infinity, the interval between $f(a)$ and $f(a + r)$ becomes huge. Indeed, $f(k + 1/k) - f(k) = 2 + k^{-2} > 2$. Therefore there is no $r > 0$ such that $|x - y| < r$ implies $|f(x) - f(y)| < 1$. Therefore, f is not uniformly continuous on $\mathbb{R}$.

5.5.3. EXAMPLE. Consider the function $f(x) = 1/x$ on $(0, 1]$. Notice that the graph blows up at the origin and becomes very steep. This is the same property that we just exploited for x^2 as x goes off to infinity. Very close values in the domain are mapped to points that are far apart. Let $x_k = 1/k$. Then

$$|f(x_{k+1}) - f(x_k)| = (k + 1) - k = 1.$$

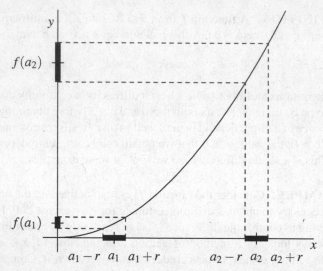

FIG. 5.5 The function $f(x) = x^2$ on $\mathbb{R}$.

However, $|x_{k+1} - x_k| = 1/(k + k^2)$ tends to 0. So let $\varepsilon = 1$ and consider any $r > 0$. For k large enough, $|x_{k+1} - x_k| < r$, but $|f(x_{k+1}) - f(x_k)| = 1 = \varepsilon$. So f is not uniformly continuous.

A number of properties of functions imply uniform continuity.

5.5.4. PROPOSITION. *Every Lipschitz function is uniformly continuous.*

PROOF. Recall that f is Lipschitz on S means that there is a constant C such that $\|f(\mathbf{x}) - f(\mathbf{y})\| \leq C\|\mathbf{x} - \mathbf{y}\|$. Given $\varepsilon > 0$, choose $r = \varepsilon/C$. Then if $\mathbf{x}, \mathbf{a} \in S$ and $\|\mathbf{x} - \mathbf{a}\| < r$,

$$\|f(\mathbf{x}) - f(\mathbf{a})\| \leq C\|\mathbf{x} - \mathbf{a}\| < C\frac{\varepsilon}{C} = \varepsilon.$$

Thus uniform continuity is established (almost by definition). ∎

Corollary 5.1.7 shows that every linear transformation is a Lipschitz function and Exercise 5.1.J shows that every function $f : [a, b] \to \mathbb{R}$ with a bounded derivative is a Lipschitz function. Hence we obtain the following:

5.5.5. COROLLARY. *Every linear transformation from $\mathbb{R}^n$ to $\mathbb{R}^m$ is uniformly continuous.*

5.5.6. COROLLARY. *Let f be a differentiable real-valued function on $[a, b]$ with a bounded derivative. Then f is uniformly continuous on $[a, b]$.*

Before getting to our main result, we look at two more examples. The first is to show that a function does not need to be unbounded in order to fail uniform continuity. However, the previous theorem shows us that the function should be very steep frequently. This suggests returning to one of our favourite functions.

5.5.7. EXAMPLE. Let $f(x) = \sin\frac{1}{x}$ on $(0,1]$. A computation of the derivative is not necessary, since the qualitative features of this function have been considered before in Example 5.2.7. The function oscillates wildly between $+1$ and -1 as x approaches 0. In particular,

$$\lim_{k\to\infty} f\left(\frac{1}{(2k+\frac{1}{2})\pi}\right) = 1 \quad\text{and}\quad \lim_{k\to\infty} f\left(\frac{1}{(2k-\frac{1}{2})\pi}\right) = -1.$$

Letting $x_k = 1/((2k+\frac{1}{2})\pi)$ and $a_k = 1/((2k-\frac{1}{2})\pi)$, then $f(x_k) - f(a_k) = 2$, while

$$\lim_{k\to\infty} |x_k - a_k| = \lim_{k\to\infty} \frac{1}{(4k^2-\frac{1}{4})\pi} = 0.$$

As before, this means that f is not uniformly continuous.

Finally, let us look at an example in which the derivative is unbounded yet the function is still uniformly continuous.

5.5.8. EXAMPLE. Let $f(x) = x\sin\frac{1}{x}$ on $(0,\infty)$, which we graph in Figure 5.6. What makes this different from the previous examples is the behaviour at the end-points. At 0, the Squeeze Theorem shows that

$$\lim_{x\to 0} x\sin\frac{1}{x} = 0.$$

Thus we may define $f(0) = 0$ and obtain a continuous function on $[0,\infty)$. At infinity, the substitution $y = \frac{1}{x}$ yields

$$\lim_{x\to\infty} x\sin\frac{1}{x} = \lim_{y\to 0} \frac{\sin y}{y} = 1.$$

This latter limit is established in Example 2.4.7, along with the estimates

$$1 - \frac{1}{x^2} < x\sin\frac{1}{x} < 1 \quad\text{for}\quad x \geq 1.$$

We will show that f is uniformly continuous. The two limits will be used to deal with points near 0 and near infinity (sufficiently large). Let $0 < \varepsilon < 1$ be given. First consider values near the origin. If $|x| < \varepsilon/2$ and $|y| < \varepsilon/2$, then

$$|f(x) - f(y)| \leq |x||\sin\frac{1}{x}| + |y||\sin\frac{1}{y}| \leq |x| + |y| < \varepsilon.$$

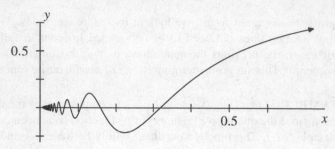

FIG. 5.6 Partial graph of $x \sin \frac{1}{x}$.

Thus if $x \in [0, \varepsilon/4]$ and $|x - y| < \varepsilon/4$, then this estimate holds.

Now do the same thing near infinity. Pick an integer $N > \varepsilon^{-1/2}$. If x and y are greater than N, then

$$1 - \varepsilon < 1 - \frac{1}{N^2} < f(x) < 1.$$

The same is true for y, and thus if $x \geq N + 1$ and $|x - y| < 1$, then

$$|f(y) - f(x)| < \varepsilon.$$

These two estimates show that $|f(y) - f(x)| < \varepsilon$ if $|x - y| < \varepsilon/4$ and either x or y lies in either $[0, \varepsilon/4]$ or $[N + 1, \infty)$.

Now consider the case in which both x and y lie in the interval $[\varepsilon/4, N + 1]$. On this interval, the function f has a continuous derivative

$$f'(x) = \sin \frac{1}{x} - \frac{1}{x} \cos \frac{1}{x}.$$

An easy estimate shows that

$$|f'(x)| \leq 1 + \frac{1}{x} \leq 1 + \frac{4}{\varepsilon},$$

so we let $M = 1 + 4/\varepsilon$. Hence, if x and y are in $[\varepsilon/4, N + 1]$ and $|x - y| < \varepsilon/M$, then the Mean Value Theorem (6.2.2) implies that $|f(x) - f(y)| \leq M|x - y| < \varepsilon$.

Finally, since $\varepsilon/M = \varepsilon^2/(4 + \varepsilon)$, we can choose $r = \min\{\varepsilon/4, \varepsilon^2/(4 + \varepsilon)\}$. Then if $|x - y| < r$, one of the preceding estimates applies to show that $|f(y) - f(x)| < \varepsilon$. Therefore, f is uniformly continuous.

Having emphasized the differences between continuity and uniform continuity up to this point, we conclude this section by admitting that in some important situations, the two notions coincide.

5.5.9. THEOREM. *Suppose that $C \subset \mathbb{R}^n$ is compact and $f : C \to \mathbb{R}^m$ is continuous. Then f is uniformly continuous on C.*

PROOF. Suppose that f were not uniformly continuous. Then there would be some $\varepsilon > 0$ for which no $r > 0$ satisfies the definition. That is, for each $r = 1/k$, there are points $\mathbf{a}_k$ and $\mathbf{x}_k$ in C such that $\|\mathbf{x}_k - \mathbf{a}_k\| < 1/k$ but $\|f(\mathbf{x}_k) - f(\mathbf{a}_k)\| \geq \varepsilon$.

Since C is compact and $(\mathbf{a}_k)_{k=1}^{\infty}$ is a sequence in C, there is a convergent subsequence $(\mathbf{a}_{k_i})$ with

$$\lim_{i \to \infty} \mathbf{a}_{k_i} = \mathbf{a} \in C.$$

Thus we also have

$$\lim_{i \to \infty} \mathbf{x}_{k_i} = \lim_{i \to \infty} \mathbf{a}_{k_i} + (\mathbf{x}_{k_i} - \mathbf{a}_{k_i}) = \mathbf{a} + 0 = \mathbf{a}.$$

By the continuity of f, we have

$$\lim_{i \to \infty} f(\mathbf{a}_{k_i}) = f(\mathbf{a}) \quad \text{and} \quad \lim_{i \to \infty} f(\mathbf{x}_{k_i}) = f(\mathbf{a}).$$

Consequently,

$$\lim_{i \to \infty} \|f(\mathbf{a}_{k_i}) - f(\mathbf{x}_{k_i})\| = 0.$$

This contradicts $\|f(\mathbf{a}_k) - f(\mathbf{x}_k)\| \geq \varepsilon > 0$ for all n. Therefore, the function must be uniformly continuous. ∎

Compare how the sequences $(\mathbf{a}_k)$ and $(\mathbf{x}_k)$ are used in this proof to how similar sequences are used to show that specific functions are not uniformly continuous in Examples 5.5.3, 5.5.2, and 5.5.7.

Exercises for Section 5.5

A.   Show that $g(x) = \sqrt{x}$ is uniformly continuous on $[0, +\infty)$.
HINT: Show that $\sqrt{a-b} \geq \sqrt{a} - \sqrt{b}$ and $\sqrt{a+b} \leq \sqrt{a} + \sqrt{b}$.

B.   Given a polynomial $p(x,y) = \sum_{m,n=0}^{N} a_{mn} x^m y^n$ in variables x and y and an $\varepsilon > 0$, find an explicit $\delta > 0$ establishing uniform continuity on the square $[-R, R]^2$.
HINT: Try $\delta = \varepsilon/C$, where $C = \sum_{m,n=0}^{N} |a_{mn}| (m+n) R^{m+n-1}$.

C.   (a) Show that $f(x) = \frac{1}{x} \sin x$ for $x \neq 0$ can be extended to a continuous function on $\mathbb{R}$.
(b) Prove that it is uniformly continuous on $\mathbb{R}$.

D.   Show that $f(x) = x^p$ is not uniformly continuous on $\mathbb{R}$ if $p > 1$.

E.   If f is continuous on $(0, 1)$ and $\lim_{x \to 0+} f(x) = +\infty$, show that f is not uniformly continuous.

F.   Show that a periodic continuous function on $\mathbb{R}$ is bounded and uniformly continuous.

G.   Consider a continuous function $f = (f_1, \ldots, f_m)$ from an open subset S of $\mathbb{R}^n$ into $\mathbb{R}^m$ with bounded partial derivatives $\left| \frac{\partial f_i}{\partial x_j}(\mathbf{x}) \right| \leq M$ for all $\mathbf{x} \in S$. Prove that f is uniformly continuous.

H.   Suppose that f is continuous on (a, c) and that $a < b < c$. Show that if f is uniformly continuous on both $(a, b]$ and $[b, c)$, then f is uniformly continuous on (a, c).

I.   Let $f(x)$ be continuous on $(0, 1]$. Show that f is uniformly continuous iff $\lim_{x \to 0+} f(x)$ exists.

J.   For which real values of α is the function $g_\alpha(x) = x^\alpha \log(x)$ uniformly continuous on $(0, \infty)$?
HINT: Use Exercises D, E, and H. For $0 < \alpha < 1$, consider $[0, 1]$ and $[1, \infty)$ separately.

K. A function $f : [a,b] \to \mathbb{R}$ satisfies a **Lipschitz condition of order** $\alpha > 0$ if there is some positive constant M such that $|f(x_1) - f(x_2)| \leq M|x_1 - x_2|^\alpha$. Let Lip α denote the set of all functions satisfying a Lipschitz condition of order α.

(a) Prove that if $f \in \mathrm{Lip}\,\alpha$, then f is uniformly continuous.

(b) Prove that if $f \in \mathrm{Lip}\,\alpha$ and $\alpha > 1$, then f is constant.

(c) For $\alpha \in (0,1)$, show that $f(x) = x^\alpha$ belongs to Lip α.

5.6 The Intermediate Value Theorem

Here is another fundamental result often used in calculus without proof.

5.6.1. INTERMEDIATE VALUE THEOREM.

If f is a continuous real-valued function on $[a,b]$ and $z \in \mathbb{R}$ satisifies $f(a) < z < f(b)$, then there exists a point $c \in (a,b)$ such that $f(c) = z$.

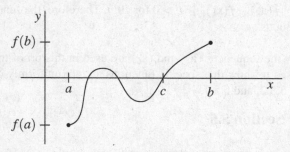

FIG. 5.7 Applying the Intermediate Value Theorem to a function $f : [a,b] \to \mathbb{R}$ with $z = 0$.

Figure 5.7 shows the conclusion graphically. Like the Extreme Value Theorem, this result seems intuitively clear, but its proof depends on the completeness of the real numbers. For example, the function $f : \mathbb{Q} \to \mathbb{R}$ given by $f(x) = x^3 - 2$ for $x \in \mathbb{Q}$ satisfies $f(1) = -1 < 0$ and $f(2) = 6 > 0$ but there is no *rational* x for which $f(x) = 0$.

PROOF. Define $A = \{x \in [a,b] : f(x) < z\}$. Since $a \in A$, A is not empty. And since b is an upper bound for A, the Least Upper Bound Principle allows us to define $c = \sup A$; and it belongs to $[a,b]$. We claim that $f(c) = z$.

First, since c is the least upper bound for A, there is a sequence (a_n) in A such that $c - \frac{1}{n} < a_n \leq c$. So $c = \lim_{n\to\infty} a_n$. Therefore,

$$f(c) = \lim_{n\to\infty} f(a_n) \leq z.$$

In particular, $c \neq b$, and thus $c < b$. Choose any sequence $c < b_n \leq b$ such that $c = \lim_{n\to\infty} b_n$. Since c is the upper bound for A, it follows that $b_n \notin A$ and so $f(b_n) \geq z$.

Consequently,

$$f(c) = \lim_{n \to \infty} f(b_n) \geq z.$$

Putting the two inequalities together yields $f(c) = z$. ∎

5.6.2. COROLLARY. *If f is a continuous real-valued function on $[a,b]$, then $f([a,b])$ is a closed interval.*

PROOF. The Extreme Value Theorem (5.4.4) shows that the range of f is bounded, and the extrema are attained. Thus there are points c and d in $[a,b]$ such that

$$f(c) = m := \inf_{x \in [a,b]} f(x) \quad \text{and} \quad f(d) = M := \sup_{x \in [a,b]} f(x).$$

Suppose that $c \leq d$. (The case $c > d$ is similar.) Let z be any value in (m,M). Then $f(c) = m < z < M = f(d)$. Since f is continuous on $[c,d]$, there is $x \in (c,d)$ such that $f(x) = z$. Therefore, $f([a,b]) = [m,M]$. ∎

A **path** in $S \subset \mathbb{R}^n$ from $\mathbf{a}$ to $\mathbf{b}$, both points in S, is the image of a continuous function γ from $[0,1]$ into S such that $\gamma(0) = \mathbf{a}$ and $\gamma(1) = \mathbf{b}$.

5.6.3. COROLLARY. *Suppose that $S \subset \mathbb{R}^n$ and f is a continuous real-valued function on S. If there is a path from $\mathbf{a}$ to $\mathbf{b}$ in S and $z \in \mathbb{R}$ with $f(\mathbf{a}) < z < f(\mathbf{b})$, then there is a point $\mathbf{c}$ on the path such that $f(\mathbf{c}) = z$.*

PROOF. Let $\gamma : [0,1] \to S$ define the path from $\mathbf{a}$ to $\mathbf{b}$. Consider the continuous function $g = f \circ \gamma$. Then $g(0) < z < g(1)$. By the Intermediate Value Theorem, there is a point x in $(0,1)$ such that $g(x) = z$. Then $\mathbf{c} = \gamma(x)$ is the desired point. ∎

We look at the following concept in several of the exercises.

5.6.4. DEFINITION. A subset A of $\mathbb{R}^n$ is **not connected** if there are *disjoint* open sets U and V such that $A \subset U \cup V$ and $A \cap U \neq \varnothing \neq A \cap V$. Otherwise, the set A is said to be **connected**.

Exercises for Section 5.6

A. (a) Show that there is some $x \in (0, \pi/2)$ such that $\cos x = x$.

    (b) Prove that this is the only real solution.

B. How many solutions are there to $\tan x = x$ in $[0, 11]$?

C. Show that $2 \sin x + 3 \cos x = x$ has three solutions.

D. Show that a polynomial of odd degree has at least one real root.

E.   The temperature $T(\mathbf{x})$ at each point $\mathbf{x}$ on the surface of Mars (a sphere) is a continuous func-
     tion. Show that there is a point $\mathbf{x}$ on the surface such that $T(\mathbf{x}) = T(-\mathbf{x})$.   HINT: Represent
     the surface of Mars as $\{\mathbf{x} \in \mathbb{R}^3 : \|\mathbf{x}\| = 1\}$. Consider the function $f(\mathbf{x}) = T(\mathbf{x}) - T(-\mathbf{x})$.

F.   Let f be a continuous function from a circle into $\mathbb{R}$. Show that f cannot be one-to-one.

G.   If f is a continuous real-valued function on $(0,1)$, what are the possibilities for the range of
     f? Give examples for each possibility, and prove that your list is complete.

H.   (a) Show that a continuous function on $(-\infty, +\infty)$ cannot take *every* real value *exactly twice*.
     (b) Find a continuous function on $(-\infty, +\infty)$ that takes *every* real value *exactly three times*.

I.   Show that $\mathbb{Q}$ is not connected.

J.   Show that $[a,b]$ is connected.   HINT: if $[a,b]$ were contained in the disjoint union of open
     sets U and V, then the characteristic function of $U \cap [a,b]$ would be continuous on $[a,b]$.

5.7 Monotone Functions

Most functions we encounter are increasing or decreasing on intervals contained in
their domain. So we explore some of special properties of monotone functions.

5.7.1. DEFINITION. A function f is called **increasing** on an interval (a,b) if
$f(x) \le f(y)$ whenever $a < x < y < b$. It is **strictly increasing** on (a,b) if $f(x) < f(y)$
whenever $a < x < y < b$. Similarly, we define **decreasing** and **strictly decreasing**
functions. All of these functions are called **monotone**.

Sometimes, **monotone increasing** and **monotone decreasing** are used as syn-
onyms for increasing and decreasing.

5.7.2. PROPOSITION. *If f is an increasing function on the interval (a,b),
then the one-sided limits of f exist at each point $c \in (a,b)$, and*

$$\lim_{x \to c^-} f(x) \le f(c) \le \lim_{x \to c^+} f(x).$$

For decreasing functions, the inequalities are reversed.

PROOF. Consider the case of f increasing. The set $F = \{f(x) : a < x < c\}$ is a
nonempty set of real numbers bounded above by $f(c)$. Hence $L = \sup_{a<x<c} f(x)$
is defined by the Least Upper Bound Principle (2.3.3), and $L \le f(c)$. Let $\varepsilon > 0$.
Since $L - \varepsilon$ is not an upper bound for F, there is a point x_0 with $a < x_0 < c$ such
that $f(x_0) > L - \varepsilon$. Hence for all $x_0 < x < c$, we have $L - \varepsilon < f(x_0) \le f(x) \le L$. It
follows that $\lim_{x \to c^-} f(x) = L \le f(c)$.

The limit from the right and the decreasing case are handled similarly.   ∎

For brevity, we use $f(x^+)$ for $\lim_{x \to c^+} f(x)$ and $f(x^-)$ for $\lim_{x \to c^-} f(x)$ in this section.

5.7.3. COROLLARY. *The only type of discontinuity that a monotone function
on an interval can have is a jump discontinuity.*

PROOF. If $f(x^-) = f(x^+)$, then $\lim_{x \to c} f(x)$ exists and equals $f(c)$, so f is continuous at c. If $f(x^-) < f(x^+)$, f has a jump disconinuity at c, by the definition. ∎

5.7.4. COROLLARY. *If f is a monotone function on $[a,b]$ and the range of f intersects every nonempty open interval in $[f(a), f(b)]$, then f is continuous.*

PROOF. Suppose that f is increasing and has a jump discontinuity at c, i.e., $f(c^-) < f(c^+)$. Then the range of f intersects the interval $(f(c^-), f(c^+))$ in a single point, $f(c)$. Thus either $(f(c^-), f(c))$ or $(f(c), f(c^+))$ is a nonempty interval in $[f(a), f(b)]$ that is disjoint from the range of f. Consequently, if the range of f meets every open interval in $[f(a), f(b)]$, then f must be continuous. ∎

Here is a stronger conclusion.

5.7.5. THEOREM. *A monotone function on $[a,b]$ has at most countably many discontinuities.*

PROOF. Assume f is increasing, and let $D = f(b) - f(a)$. Define the *jump* of f at x to be $J(x) = f(x^+) - f(x^-)$. When $J(x) > 0$, f is discontinuous at x and $J(x)$ is the length of the gap $(f(x^-), f(x^+))$ in the range of f. These intervals are disjoint; hence the sum of all the jumps is at most D.

Let N_k be the number of discontinuities with $J(x) \geq 2^{-k}$. Then $2^{-k} N_k \leq D$ or $N_k \leq 2^k D < \infty$. Therefore

$$\{x : J(x) > 0\} = \bigcup_{k \geq 0} \{x : J(x) \geq 2^{-k}\}$$

is a countable union of finite sets, and hence is countable by Corollary 2.9.6. ∎

A strictly monotone function is one-to-one, and so has an inverse function.

5.7.6. THEOREM. *Let f be a continuous strictly increasing function on $[a,b]$. Then f maps $[a,b]$ one-to-one and onto $[f(a), f(b)]$. The inverse function f^{-1} is also continuous and strictly increasing.*

PROOF. Since f is strictly increasing, it is clearly one-to-one. Since f is monotone, $f(a) < f(x) < f(b)$ on (a,b). By the Intermediate Value Theorem (5.6.1), the range of f contains $[f(a), f(b)]$. So f maps $[a,b]$ one-to-one and onto $[f(a), f(b)]$.

Let g be the inverse function. Suppose that $f(a) \leq s < t \leq f(b)$, and let $x = g(s)$ and $y = g(t)$. Then $x < y$ because $f(x) = s < t = f(y)$. Hence g is strictly increasing. The range of g is $[a,b]$. Thus by Corollary 5.7.4, g is continuous. ∎

5.7.7. EXAMPLE. Consider the function $\tan x$. This function has period π, so clearly it is not monotone. However, it is monotone on subintervals of its range. A natural choice is the interval $(-\pi/2, \pi/2)$. Note that $\tan x$ is strictly increasing on this interval and that $\lim_{x\to-\pi/2^+}\tan x = -\infty$ and $\lim_{x\to\pi/2^-}\tan x = +\infty$. So $\tan x$ maps $(-\pi/2,\pi/2)$ one-to-one and onto $\mathbb{R}$. The inverse function arctan is the unique function that assigns to each real y the value $x \in (-\pi/2,\pi/2)$, satisfying $\tan x = y$. In particular, $\lim_{y\to+\infty}\arctan(y) = \pi/2$ and $\lim_{y\to-\infty}\arctan(y) = -\pi/2$.

5.7.8. EXAMPLE. We construct a function associated to the Cantor set known as the **Cantor function**. Recall from Example 4.4.8 that the Cantor set is constructed by successively removing the middle thirds from the unit interval and obtaining the set as the intersection of the sets S_k obtained from this procedure. We define a function on $[0,1]$ as follows. Set $f(0) = 0$ and $f(1) = 1$; next set $f(x) = 1/2$ on $[1/3,2/3]$; then $f(x) = 1/4$ on $[1/9,2/9]$ and $3/4$ on $[7/9,8/9]$; and so on.

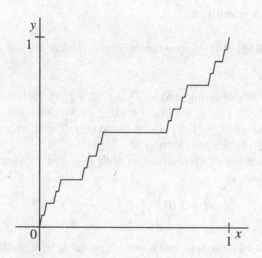

FIG. 5.8 An approximation to the Cantor function.

We can understand this function using ternary (base 3) expansions. Write a number in $[0,1]$ in base 3 as $x = (0.x_1x_2x_3\ldots)_{\text{base 3}} = \sum_{k\geq 1}3^{-k}x_k$, where each x_i belongs to $\{0,1,2\}$. If $k \geq 1$ is the smallest integer with $x_k = 1$, the point x belongs to one of the $2^k - 1$ closed intervals on which f is assigned a value of the form $m/2^k$. More precisely, if $x = (0.x_1\ldots x_{k-1}1x_{k+1}\ldots)_{\text{base 3}}$, where $x_i \in \{0,2\}$ for $1 \leq i < k$, we set $f(x) = (0.\frac{x_1}{2}\ldots\frac{x_{k-1}}{2}1000\ldots)_{\text{base 2}}$. For the remaining points in C that have a ternary expansion using only 0's and 2's, we have $f(x) = (0.\frac{x_1}{2}\ldots\frac{x_k}{2}\ldots)_{\text{base 2}}$. The restriction of f to C was considered in Example 4.4.8 in order to show that C was very large. Some numbers have two ternary expansions. Rather than verify that both expansions lead to the same function value (which they do), we merely choose the expansion that contains a 1. This leads to a well-defined function.

Clearly f is increasing. The range of f contains all numbers of the form $m/2^k$ for $0 \leq m \leq 2^k$. Since these are dense in $[0,1] = [f(0),f(1)]$, the range of f has no gaps. So f is continuous by Corollary 5.7.4.

Note that f is constant on each interval removed from C. So the function has a rather flat appearance. It follows that $f(C) = [0,1]$. This function f provides another way to show that the cardinality of C is the same as that of the real line.

Exercises for Section 5.7

A. Let f and g be decreasing functions defined on $\mathbb{R}$.

    (a) Is the composition $c(x) = f(g(x))$ monotone?
    (b) Is the sum $s(x) = f(x) + g(x)$ monotone?
    (c) Is the product $p(x) = f(x)g(x)$ monotone?

B. Show that if f is continuous on $[0,1]$ and one-to-one, then it is monotone.

C. What is the inverse function of $f(x) = x^2$ on $(-\infty, 0]$?

D. (a) Show that the restriction f of $\cos x$ to $[0, \pi]$ has an inverse function, and graph them both.
    (b) Why do we choose the interval $[0, \pi]$?
    (c) Let g be the restriction of $\cos x$ to $[3\pi, 4\pi]$. What is the relationship between f^{-1} and g^{-1}?

E. Show that the cubic $f(x) = ax^3 + bx^2 + cx + d$, with $a \neq 0$, is one-to-one and thus has an inverse function if and only if $3ac \geq b^2$.   HINT: Compute the derivative and its discriminant.

F. Verify that the formula for the Cantor function in terms of the ternary expansion yields the same answer for both expansions of a point x when two expansions exist.

G. Define f on $S = [0,1] \cup (2,3]$ by $f(x) = \begin{cases} x & \text{for } 0 \leq x \leq 1, \\ x-1 & \text{for } 2 < x \leq 3. \end{cases}$

    (a) Show that f is continuous and strictly increasing on S.
    (b) Show that f maps S one-to-one and onto $[0,2]$.
    (c) Show that f^{-1} is not continuous.
    (d) Why is this not a contradiction to Theorem 5.7.6?

H. For $x \in [0,1]$, express it as a decimal $x = x_0.x_1 x_2 x_3 \ldots$. Use a finite decimal expansion without repeating 9's when there is a choice. Then define a function f by $f(x) = x_0.0x_1 0x_2 0x_3 \ldots$.

    (a) Show that f is strictly increasing.
    (b) Compute $\lim_{x \to 1^-} f(x)$.
    (c) Show that $\lim_{x \to a^+} f(x) = f(a)$ for $0 \leq a < 1$.
    (d) Find all discontinuities of f.

Chapter 6
Differentiation and Integration

In this chapter, we examine the mathematical foundations of differentiation and integration. The theorems of this chapter are useful not only to make calculus work but also for studying functions in many other contexts. We do not spend any time on the important applications that typically appear in courses devoted to calculus, such as optimization problems. Rather we will highlight those aspects that either depend on or apply to results in real analysis.

Since we assume that the reader has already seen calculus, we dive right in with key definitions, assuming that the motivating examples are familiar.

6.1 Differentiable Functions

6.1.1. DEFINITION. A real-valued function $f : (a,b) \to \mathbb{R}$ is **differentiable at a point** $x_0 \in (a,b)$ if

$$\lim_{h \to 0} \frac{f(x_0 + h) - f(x_0)}{h} = \lim_{x \to x_0} \frac{f(x) - f(x_0)}{x - x_0}$$

exists. In this case, we write $f'(x_0)$ for this limit.

If a function is defined on a closed interval $[a,b]$, then we say it is differentiable at a or b if the appropriate one-sided limit exists. The function f is **differentiable on an interval** $[a,b]$ if it is differentiable at each point x_0 in the interval.

When f is differentiable at x_0, we define the **tangent line** to f at x_0 to be the linear function $T(x) = f(x_0) + f'(x_0)(x - x_0)$. See Figure 6.1.

The phrase *linear function* has two different meanings. In linear algebra, a linear function is one satisfying $f(\alpha \mathbf{x} + \beta \mathbf{y}) = \alpha f(\mathbf{x}) + \beta f(\mathbf{y})$; but in calculus, a linear function is one whose graph is a line $f(x) = mx + b$.

An immediate consequence of differentiability is continuity.

K.R. Davidson and A.P. Donsig, *Real Analysis and Applications: Theory in Practice*,
Undergraduate Texts in Mathematics, DOI 10.1007/978-0-387-98098-0_6,
© Springer Science + Business Media, LLC 2010

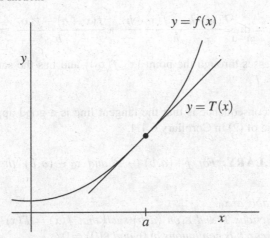

FIG. 6.1 The tangent line T to f at a.

6.1.2. PROPOSITION. *If $f : (a,b) \to \mathbb{R}$ is differentiable at $x_0 \in (a,b)$, then it is continuous at x_0. So every differentiable function is continuous.*

PROOF. We compute

$$\lim_{x \to x_0} f(x) = \lim_{x \to x_0} f(x_0) + (x - x_0)\frac{f(x) - f(x_0)}{x - x_0} = f(x_0) + 0f'(x_0) = f(x_0).$$ ∎

When f is differentiable at x_0, the tangent line T passes through the point $(x_0, f(x_0))$ with slope $f'(x_0)$. In fact, $T(x)$ is the best linear approximation to $f(x)$ for x very close to x_0.

6.1.3. LEMMA. *Let f be a function on (a,b) that is differentiable at x_0. Let T be the tangent line to f at x_0. Then T is the unique linear function with the property that*

$$\lim_{x \to x_0} \frac{f(x) - T(x)}{x - x_0} = 0.$$

PROOF. First we compute the limit by substituting $h = x - x_0$,

$$\lim_{x \to x_0} \frac{f(x) - T(x)}{x - x_0} = \lim_{h \to 0} \frac{(f(x_0 + h) - f(x_0)) - f'(x_0)h}{h} = f'(x_0) - f'(x_0) = 0.$$

If another linear function $L(x)$ satisfies this limit, then by continuity,

$$f(x_0) - L(x_0) = \lim_{x \to x_0} f(x) - L(x) = \lim_{x \to x_0} (x - x_0)\frac{f(x) - L(x)}{x - x_0} = 0.$$

So $L(x) = f(x_0) + m(x - x_0)$, where m is its slope, and

$$m = L'(x_0) = \lim_{h \to 0} \frac{L(x_0+h) - f(x_0+h)}{h} + \frac{f(x_0+h) - f(x_0)}{h} = f'(x_0).$$

Thus the line L passes through the point $(x_0, f(x_0))$ and has the same slope as T. Consequently, $L = T$. ∎

An immediate consequence is that the tangent line is a good approximant to f near x_0 in the sense of (2) in Corollary 6.1.4.

6.1.4. COROLLARY. *For $f : (a,b) \to \mathbb{R}$ and $x_0 \in (a,b)$, the following are equivalent:*

(1) *f is differentiable at x_0.*
(2) *There are functions T and ε on (a,b) such that $f(x) = T(x) + \varepsilon(x)(x - x_0)$, where T is linear, ε is continuous at 0, and $\varepsilon(0) = 0$.*
(3) *There is a function φ on (a,b) such that $f(x) = f(x_0) + \varphi(x)(x - x_0)$, where φ is continuous at x_0.*

If these hold, then in (2), T is the tangent line, and in (3), $\varphi(x_0) = f'(x_0)$.
   Moreover, if f is continuous on (a,b), then so are ε and φ.

PROOF. Clearly, we must define $\varepsilon(x) = \frac{f(x) - T(x)}{x - x_0}$ and $\varphi(x) = \frac{f(x) - f(x_0)}{x - x_0}$ for $x \neq x_0$. Set $\varepsilon(x_0) = 0$ and $\varphi(x_0) = f'(x_0)$. Lemma 6.1.3 shows that (1) and (2) are equivalent and that T is the tangent line. Thus if (2) holds, then $\varphi(x) = f'(x_0) + \varepsilon(x)$ satisfies $f(x) = f(x_0) + \varphi(x)(x - x_0)$ and $\lim_{x \to x_0} \varphi(x) = f'(x_0)$. So (3) holds. Conversely, if (3) holds, then $T(x) = f(x_0) + \varphi(x_0)(x - x_0)$, $\varepsilon(x) = \varphi(x) - \varphi(x_0)$ satisfy (2). ∎

6.1.5. EXAMPLES. The prototypical example of a continuous function that is not differentiable everywhere is $f(x) = |x|$. It is differentiable except at $x = 0$. Here it has left and right derivatives ± 1. Since this function comes to a point at the origin, it is intuitively clear that no straight line is a good approximant near $x = 0$.
   A more subtle example is $g(x) = \sqrt[3]{x}$; see Figure 6.2 (a). This has derivative $g'(x) = x^{-2/3}/3$ for $x \neq 0$. And at $x = 0$, $\lim_{h \to 0} \frac{g(h) - g(0)}{h} = \lim_{h \to 0} h^{-2/3} = +\infty$. This function is not differentiable because it has a vertical tangent at the origin.
   A related example is the function $h(x) = \sqrt{|x|}$; see Figure 6.2 (b). This function has a cusp, with right derivative $+\infty$ and left derivative $-\infty$.

The familiar differentiation rules like $(\alpha f + \beta g)'(x) = \alpha f'(x) + \beta g'(x)$ and the product rule, $(fg)'(x) = f'(x)g(x) + f(x)g'(x)$, are left as exercises. The chain rule is more subtle, so we provide a proof.

6.1.6. THE CHAIN RULE.
Suppose that $f : [a,b] \to [c,d]$ is differentiable at $x_0 \in [a,b]$ and $g : [c,d] \to \mathbb{R}$ is differentiable at $f(x_0)$. Then the composition $h = g \circ f$ is differentiable at x_0, and $h'(x_0) = g'(f(x_0))f'(x_0)$.

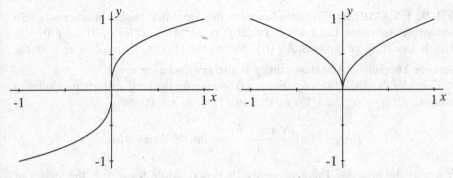

FIG. 6.2 (a) The graph of $y = \sqrt[3]{x}$.   (b) The graph of $y = \sqrt{|x|}$.

PROOF. Using Corollary 6.1.4 part (3), write $f(x) = f(x_0) + \varphi(x)(x - x_0)$, where $\varphi(x_0) = f'(x_0)$. Similarly, we can write $g(y) = g(f(x_0)) + \psi(y)(y - f(x_0))$, where $\psi(f(x_0)) = g'(f(x_0))$. Then

$$h(x) = g(f(x_0)) + \psi(f(x))(f(x) - f(x_0)) = g(f(x_0)) + \psi(f(x))\varphi(x)(x - x_0).$$

Since $x \mapsto \psi(f(x))\varphi(x)$ is continuous at x_0 (being a composition and product of maps continuous at x_0), Corollary 6.1.4 implies that h is differentiable at x_0 and $h'(x_0) = \psi(f(x_0))\varphi(x_0) = g'(f(x_0))f'(x_0)$.   ■

We use the chain rule to compute the derivatives of inverse functions. One issue is that a zero derivative results in a vertical tangent in the inverse function. For example, $g(x) = x^3$ has $g'(0) = 0$. The inverse $g^{-1}(y) = \sqrt[3]{y}$ has a vertical tangent line at $y = 0$. See Figure 6.2 (a).

6.1.7. THEOREM. *Suppose $f : (a,b) \to \mathbb{R}$ is continuous and one-to-one. If f is differentiable at x_0 and $f'(x_0) \neq 0$, then f^{-1} is differentiable at $y_0 = f(x_0)$ and*

$$(f^{-1})'(y_0) = \frac{1}{f'(f^{-1}(y_0))} = \frac{1}{f'(x_0)}.$$

PROOF. Since f is differentiable at x_0, use Corollary 6.1.4 to obtain a continuous function φ on (a,b) such that $\varphi(x_0) = f'(x_0) \neq 0$ and $f(x) = f(x_0) + \varphi(x)(x - x_0)$. For $x \in (a,b)$, setting $y = f(x)$ yields $x = f^{-1}(y)$. In particular, $x_0 = f^{-1}(y_0)$. Therefore $y = y_0 + \varphi(f^{-1}(y))(f^{-1}(y) - f^{-1}(y_0))$. Solving for $f^{-1}(y)$ gives

$$f^{-1}(y) = f^{-1}(y_0) + \frac{1}{\varphi(f^{-1}(y))}(y - y_0), \tag{6.1.8}$$

provided $\varphi(f^{-1}(y)) \neq 0$. Now $\varphi(f^{-1}(y_0)) = f'(f^{-1}(y_0)) = f'(x_0) \neq 0$ and f^{-1} is continuous. So there is an interval around y_0 where $\varphi(f^{-1}(y))$ is nonzero, which follows from Exercise 5.1.G. Thus, (6.1.8) makes sense in this interval. Invoking Corollary 6.1.4 again yields the conclusion.   ■

6.1.9. EXAMPLE. There are functions that are differentiable but do not have a continuous derivative. Let $a > 0$. Consider $f(x) = x^a \sin(1/x)$ for $x > 0$ and $f(0) = 0$. This is evidently continuous $\mathbb{R} \setminus \{0\}$. Since $|\sin(1/x)| \leq 1$ and $\lim\limits_{x \to 0^+} x^a = 0$, the Squeeze Theorem (2.4.6) shows that f is also continuous at $x = 0$.

Consider the derivative of f. For $x > 0$, we invoke the product and chain rules to compute $f'(x) = ax^{a-1} \sin(1/x) - x^{a-2} \cos(1/x)$. At $x = 0$,

$$f'(0) = \lim_{h \to 0^+} \frac{h^a \sin(1/h)}{h} = \lim_{h \to 0^+} h^{a-1} \sin(1/h).$$

If $a > 1$, the Squeeze Theorem yields $f'(0) = 0$, while for $a \leq 1$, the values of $h^{a-1} \sin(1/h)$ oscillate wildly, and the limit does not exist.

So for $0 < a \leq 1$, f is continuous but not differentiable at $x = 0$. Our interest is in the values $1 < a \leq 2$. Here, $f(x)$ is differentiable at every point but the derivative is wildly discontinuous at $x = 0$. So a function may be differentiable but not C^1.

Exercises for Section 6.1

A. If functions f, g are differentiable at x_0, prove that $(\alpha f + \beta g)'(x_0) = \alpha f'(x_0) + \beta g'(x_0)$.

B. If f is differentiable at x_0 and $a \in \mathbb{R}$, show that $\lim\limits_{h \to 0} \frac{f(x_0 + ah) - f(x_0)}{h} = af'(x_0)$.

C. Explain what goes wrong with the proof of Proposition 6.1.2 if f is not differentiable.

D. Let f and g be differentiable functions on (a, b). Suppose there is a point x_0 in (a, b) with $f(x_0) = g(x_0)$ and $f(x) \leq g(x)$ for $a < x < b$. Prove that $f'(x_0) = g'(x_0)$.

E. Show that the derivative of an even function is odd, and the derivative of an odd function is even. Recall that a function f is **even** if $f(-x) = f(x)$ and is **odd** if $f(-x) = -f(x)$.

F. Prove the product rule for functions f and g on $[a, b]$ that are differentiable at x_0.
HINT: $f(x_0+h)g(x_0+h) - f(x_0)g(x_0) = (f(x_0+h) - f(x_0))g(x_0+h) + f(x_0)(g(x_0+h) - g(x_0))$.

G. For each positive integer n, give a function that is C^n but not C^{n+1}.   HINT: Example 6.1.9.

H. Prove the quotient rule: $(f/g)'(x_0) = \frac{f'(x_0)g(x_0) - f(x_0)g'(x_0)}{g(x_0)^2}$ for f and g differentiable at x_0 provided $g(x_0) \neq 0$.   HINT: Let $h = f/g$ and use the product rule on $f = gh$.

I. Say that f is **right-differentiable** at x_0 if $f'(x_0+) := \lim\limits_{h \to 0^+} [f(x_0 + h) - f(x_0)]/h$ exists.

   (a) Define **left-differentiable** at x_0 and assign a meaning to $f'(x_0-)$.
   (b) Show that f is differentiable at x_0 if and only if it is both left-differentiable and right-differentiable at x_0 and $f'(x_0+) = f'(x_0-)$.

J. Find left and right derivatives of $f(x) = \sqrt{1 - \sin x}$ at every point. Where does f fail to be differentiable?

K. Suppose $f : [a, b] \to \mathbb{R}$ is differentiable on (a, b) and continuous on $[a, b]$. Does it follow that f is right-differentiable at a and left-differentiable at b?

L. Find several things wrong with the formula $(f^{-1})' = 1/f'$. Which is the most egregious?

M. The function $\sin x$ is 2π-periodic and consequently is definitely not one-to-one.
   (a) How do we define the inverse function $\arcsin y$?
   (b) How does the choice you make in part (a) affect the graph of $\arcsin y$? What is the effect on the derivative?

N. If f is periodic with period T, show that f' is also T-periodic.

O. A function $f(x)$ is asymptotic to a curve $c(x)$ as $x \to +\infty$ if $\lim\limits_{x \to +\infty} |f(x) - c(x)| = 0$.

    (a) Show that if $f(x)$ is asymptotic to a line $L(x) = ax + b$ as $x \to +\infty$, then $a = \lim\limits_{x \to +\infty} \frac{f(x)}{x}$ and $b = \lim\limits_{x \to +\infty} f(x) - ax$. (As usual, this includes showing that the limits exist.)

    (b) Find all of the asymptotes (including horizontal and vertical ones) for $f(x) = \frac{(x-2)^3}{(x+1)^2}$. Sketch the graph.

P. Sketch the curve $f(x) = xe^{-\frac{5}{x} - \frac{2}{x^2}}$ for $x \neq 0$. Pay attention to the (a) asymptotic behaviour at $\pm\infty$, (b) all critical points, (c) limits of f and f' at 0, and (d) points of inflection.

Q. Sketch the curve $f(x) = \frac{(\log x)^2 + 4\log x}{(1 + \log x)^2}$ for $x > 0$. Pay attention to (a) points where f is zero or undefined, (b) limits of f and f' at 0^+, (c) all asymptotes and local extrema, and (d) indicate all points of inflection on the graph, but do not compute the second derivative.

R. Establish the Leibniz formula that the nth derivative of a product $f(x)g(x)$ is given by $\sum_{k=0}^{n} \binom{n}{k} f^{(k)}(x) g^{(n-k)}(x)$. HINT: Use induction.

S. (a) Suppose that g is continuous at $x = 0$. Prove that $f(x) = xg(x)$ is differentiable at $x = 0$.

    (b) Conversely, suppose that $f(0) = 0$ and f is differentiable at $x = 0$. Prove that there is a function g that is continuous at $x = 0$ and satisfies $f(x) = xg(x)$.

6.2 The Mean Value Theorem

The Mean Value Theorem is a fundamental approximation result of differential calculus. The proof depends on the Extreme Value Theorem, putting a valid proof beyond many calculus courses. The starting point is the following basic result on which all of calculus relies—the location of possible extrema. Simply stated, extrema of a continuous function occur at the endpoints or at critical points, where the derivative is either undefined or equal to 0. This result has been credited to Fermat.

6.2.1. FERMAT'S THEOREM.

Let $f : [a,b] \to \mathbb{R}$ be a continuous function that takes its maximum or minimum value at a point $x_0 \in (a,b)$. If f is differentiable at x_0, then $f'(x_0) = 0$.

PROOF. For convenience, assume that x_0 is a maximum. Since $f(x_0 + h) - f(x_0) \leq 0$, the limits from the left and right yield

$$f'(x_0) = \lim_{h \to 0^+} \frac{f(x_0 + h) - f(x_0)}{h} \leq 0 \leq \lim_{h \to 0^-} \frac{f(x_0 + h) - f(x_0)}{h} = f'(x_0). \quad \blacksquare$$

6.2.2. MEAN VALUE THEOREM.

Suppose that f is a function that is continuous on $[a,b]$ and differentiable on (a,b). Then there is a point $c \in (a,b)$ such that

$$f'(c) = \frac{f(b) - f(a)}{b - a}.$$

This theorem provides a connection between the derivative of the function and its average slope. Figure 6.3 shows the theorem graphically.

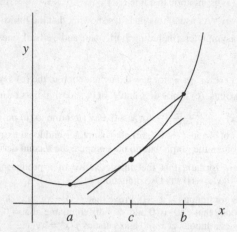

FIG. 6.3 Graph for the Mean Value Theorem.

PROOF. First we prove a special case known as **Rolle's Theorem**. Assume that $f(a) = f(b) = 0$. If the maximum value and minimum value of f are both 0, then f is constant and $f'(c) = 0$ for every $c \in (a,b)$. Otherwise, either the maximum is greater than 0 or the minimum is less. For convenience, assume the former.

By the Extreme Value Theorem (5.4.4), there is a point c at which f attains its maximum value. Evidently, c is an interior point. So by Fermat's Theorem, we have $f'(c) = 0$, which happens to equal $(f(b) - f(a))/(b-a)$.

Now a rescaling trick yields the general result. Let $L(x)$ be the linear function through $(a, f(a))$ and $(b, f(b))$, i.e., $L(x) = f(a) + ((f(b) - f(a))/(b-a))(x - a)$. Consider the function $g(x) = f(x) - L(x)$. Then g is continuous on $[a,b]$ and differentiable on $[a,b]$, since f and L are. Moreover, $g(a) = g(b) = 0$. So by Rolle's Theorem, there is a point $c \in (a,b)$ at which

$$0 = g'(c) = f'(c) - \frac{f(b) - f(a)}{b - a}.$$

This is the desired point. ∎

6.2.3. EXAMPLE. Differentiability is necessary at every interior point for the Mean Value Theorem to be valid. The continuous function $f(x) = 1 - |x|$ on $[-1,1]$ satisfies $f(1) = f(-1) = 0$, but it does not have any point at which $f'(x) = 0$. Of course, the only point where f is not differentiable is $x = 0$, but this is where the maximum occurs, and so Fermat's Theorem cannot be applied.

The first important consequence is used repeatedly in calculus arguments.

6.2.4. COROLLARY. *Let f be a differentiable function on* $[a,b]$.

(1) *If* f' *is* (*strictly*) *positive, then* f *is* (*strictly*) *increasing.*
(2) *If* f' *is* (*strictly*) *negative, then* f *is* (*strictly*) *decreasing.*
(3) *If* $f'(x) = 0$ *at every* $x \in (a,b)$, *then* f *is constant.*

PROOF. We prove only (1). For any $a \le x < y \le b$, apply the Mean Value Theorem on $[x,y]$ to obtain a point c in between such that $f(y) - f(x) = f'(c)(y - x)$. Since $f'(c) \ge 0$, we deduce that $f(y) - f(x) \ge 0$ and so f is increasing. If f' is strictly positive, then the same argument yields a strict inequality.   ∎

6.2.5. EXAMPLE. We show how the Mean Value Theorem may be used to obtain useful approximations. Consider $f(x) = \sin x$ on $[0, \pi/2]$. For any $x \in (0, \pi/2]$, we may apply the theorem on $[0,x]$ to find a point c with $0 < c < x$ such that

$$\frac{f(x) - f(0)}{x - 0} = \frac{\sin x}{x} = f'(c) = \cos c < 1.$$

Thus we obtain the well-known inequality $\sin x < x$ for $0 < x \le \pi/2$. It is evidently valid for $x > \pi/2$ as well. Now consider $g(x) = 1 - x^2/2 - \cos x$. Applying the Mean Value Theorem again, we obtain a (different) point c such that

$$\frac{g(x) - g(0)}{x - 0} = \frac{1 - x^2/2 - \cos x}{x} = g'(c) = \sin c - c < 0.$$

So $\cos x > 1 - x^2/2$ on $0 < x \le \pi/2$. Next consider $h(x) = \sin x - x + x^3/6$. Once again, apply the Mean Value Theorem on $[0,x]$ with $0 < x \le \pi/2$:

$$\frac{\sin x - x + x^3/6}{x} = h'(c) = \cos c - 1 + \frac{c^2}{2} > 0.$$

Hence $x - x^3/6 < \sin x < x$ on $(0, \pi/2]$.

Many functions that occur in practice can be differentiated several times. We define the higher-order derivatives recursively by $f^{(n+1)}(x) = \left(f^{(n)}\right)'(x)$, assuming that $f^{(n)}$ turns out to be differentiable.

It stands to reason that information about higher-order derivatives should add to our information about the behaviour of the function f. The simplest example is the sign of the second derivative. If $f''(x) \ge 0$ on $[a,b]$, then the derivative $f'(x)$ is increasing. Thus the graph of f curves upward (even if $f'(x) < 0$) and f is said to be **convex** or **concave up**. See Exercise 6.2.L. Likewise, if $f''(x) \le 0$, then f' is decreasing and f curves downward. Such functions are called **concave** or **concave down**. Points at which $f''(x)$ changes sign are called **inflection points** to indicate that the curvature of the graph changes direction. These points are easily identified by eye, since these are points where the tangent line "crosses" the graph of the function. Changes in higher derivatives are more subtle, and not so easily recognized.

Exercises for Section 6.2

A. If f and g are differentiable on $[a,b]$ and $f'(x) = g'(x)$ for all $a < x < b$, show that $g(x) = f(x) + c$ for some constant c.

B. Suppose that $f : (a,b) \to \mathbb{R}$ has a continuous derivative on (a,b). If $f'(x_0) \neq 0$, prove that there is an interval $(c,d) \ni x_0$ such that f is one-to-one on (c,d).

C. If f is strictly increasing on $[a,b]$, is $f'(x) > 0$ for all $x \in (a,b)$?

D. Suppose that f is C^3 on (a,b), and f has four zeros in (a,b). Show that $f^{(3)}$ has a zero.

E. Extend Example 6.2.5 to show that $1 - x^2/2 < \cos x < 1 - x^2/2 + x^4/24$ on $(0, \pi/2]$.

F. (a) Show that $\tan x > x + \frac{x^3}{3} + \frac{2x^5}{15}$ for $0 < x < \frac{\pi}{2}$.
   (b) Show that $\tan x < x + \frac{x^3}{3} + \frac{2x^5}{5}$ for $0 < x < 1$.

G. Suppose that f is continuous on an interval $[a,b]$ and is differentiable at all points of (a,b) except possibly at a single point $x_0 \in (a,b)$. If $\lim_{x \to x_0} f'(x)$ exists, show that $f'(x_0)$ exists and
$$f'(x_0) = \lim_{x \to x_0} f'(x). \quad \text{HINT: Consider the intervals } [x_0, x_0 + h] \text{ and } [x_0 - h, x_0].$$

H. (a) Show that the error between a differentiable function $f(x)$ on $[a,b]$ and its tangent line $T(x)$ at a satisfies $|f(x) - T(x)| \leq C|x - a|$, where $C = \sup_{a \leq y \leq x} |f'(y) - f'(a)|$.
   (b) If f is C^2, refine this estimate to $|f(x) - T(x)| \leq D|x - a|^2$, where $D = \sup_{a \leq y \leq x} |f''(y)|$.

I. (a) Let $f(x) = x^2 \sin(1/x)$ for $x \neq 0$ and $f(0) = 0$ as in Example 6.1.9. Show that 0 is a critical point of f that is not a local maximum nor a local minimum nor an inflection point.
   (b) Let $g(x) = 2x^2 + f(x)$. Show that g does have a global minimum at 0, but $g'(x)$ changes sign infinitely often on both $(0, \varepsilon)$ and $(-\varepsilon, 0)$ for any $\varepsilon > 0$.
   (c) Let $h(x) = x + 2f(x)$. Show that $h'(0) > 0$ but h is not monotone increasing on any interval including 0.

J. Suppose $g \in C^1[a,b]$. Prove that for all $\varepsilon > 0$, there is $\delta > 0$ such that $\left| g'(c) - \frac{g(d) - g(c)}{d - c} \right| < \varepsilon$ for all points $c, d \in [a,b]$ with $0 < |d - c| < \delta$.   HINT: Use the uniform continuity of g'.

K. Suppose that f is differentiable on $[a,b]$ and $f'(a) < 0 < f'(b)$.

   (a) Show that there are points $a < c < d < b$ such that $f(c) < f(a)$ and $f(d) < f(b)$.
   (b) Show that the minimum on $[a,b]$ occurs at an interior point.
   (c) Hence show that there is a point x_0 in (a,b) such that $f'(x_0) = 0$.
   (d) Prove **Darboux's Theorem**: If f is differentiable on $[a,b]$ and $f'(a) < L < f'(b)$, then there is a point x_0 in (a,b) at which $f'(x_0) = L$.

L. A function f is **convex** on $[a,b]$ if $f(tx + (1-t)y) \leq tf(x) + (1-t)f(y)$ for all $x, y \in [a,b]$ and all $t \in [0,1]$; i.e., the graph of f lies below the line segment joining $(x, f(x))$ and $(y, f(y))$.

   (a) If f is differentiable on $[a,b]$ and f' is increasing, then f is convex on $[a,b]$.
      HINT: If $x < y$ and $z = tx + (1-t)y$, apply the Mean Value Theorem to $[x,z]$ and $[z,y]$.
   (b) If $f \in C^2[a,b]$ and $x_0 \in [a,b]$ such that $f''(x_0) > 0$, prove that f is convex in an interval about x_0.
   (c) If $f \in C^2[a,b]$ and $f''(x) \geq 0$ for all $x \in (a,b)$, show that f is convex on $[a,b]$.

M. Suppose that f is differentiable on $[0, \infty)$ and f' is strictly increasing.

   (a) Show that $f'(x)$ is continuous.
   (b) Suppose that $f(0) = 0$. Set $g(0) = f'(0)$ and $g(x) = f(x)/x$ for $x > 0$. Show that g is continuous and strictly increasing.

N. (a) Suppose that f is a continuous function on $\mathbb{R}$ such that $\lim_{h \to 0} \frac{f(x+h) - f(x-h)}{h} = 0$ for every $x \in \mathbb{R}$. Prove that f is constant.
   (b) Find a discontinuous function f on $\mathbb{R}$ such that $\lim_{h \to 0} \frac{f(x+h) - f(x-h)}{h} = 0$ for every $x \in \mathbb{R}$.

6.3 Riemann Integration

We turn now to integration. A crucial point is that the notion of *derivative* is not used. An integral is *defined* as a limit related to area, not as an antiderivative. In the next section, we establish the Fundamental Theorem of Calculus, which shows that integration and differentiation are, in some sense, inverse operations.

Riemann's idea is simple and geometric: partition the interval into a number of smaller subintervals, approximate f above and below by functions that are constant on each subinterval, and the region bounded by f is approximated above and below by the union of a number of rectangles. The areas of these upper and lower approximations are called upper and lower sums of the partition. For a 'reasonable' function, these upper and lower estimates will converge to a common value as the partition is made finer and finer, called the integral of f from a to b.

6.3.1. DEFINITION. Let f be a *bounded* function defined on an interval $[a,b]$. A **partition** of $[a,b]$ is a finite set $P = \{a = x_0 < x_1 < \cdots < x_{n-1} < x_n = b\}$. Define $\Delta_j = x_j - x_{j-1}$. The **mesh** of P is defined as $\mathrm{mesh}(P) = \max\{\Delta_j : 1 \leq j \leq n\}$. On each interval $[x_{j-1}, x_j]$ of P, define the maximum and minimum of f by

$$M_j(f,P) = \sup\{f(x) : x_{j-1} \leq x \leq x_j\}, \quad m_j(f,P) = \inf\{f(x) : x_{j-1} \leq x \leq x_j\}.$$

Then define the **upper and lower sums** of f with respect to the partition P by

$$U(f,P) = \sum_{j=1}^{n} M_j(f,P)\Delta_j \quad \text{and} \quad L(f,P) = \sum_{j=1}^{n} m_j(f,P)\Delta_j.$$

For a partition P, we call $X = (x'_1, x'_2, \ldots, x'_n)$, with $x'_j \in [x_{j-1}, x_j]$, an **evaluation sequence** for P. The associated **Riemann sum** is $I(f,P,X) = \sum_{j=1}^{n} f(x'_j)\Delta_j$.

A partition R is a **refinement** of a partition P if $P \subset R$. If P and Q are two partitions, then R is a **common refinement** of P and Q if $P \cup Q \subset R$.

Figure 6.4 illustrates upper and lower sums. We always have

$$L(f,P) \leq I(f,P,X) \leq U(f,P).$$

6.3.2. LEMMA. *If P and Q are partitions of $[a,b]$, then $L(f,P) \leq U(f,Q)$.*

PROOF. First suppose that R is a refinement of P. Each interval $[x_{j-1}, x_j]$ of P may be subdivided further by R into $x_{j-1} = t_k < \cdots < t_l = x_j$. Hence for $k+1 \leq i \leq l$,

$$m_j(f,P) = \inf_{x_{j-1} \leq x \leq x_j} f(x) \leq \inf_{t_{i-1} \leq t \leq t_i} f(t) = m_i(f,R).$$

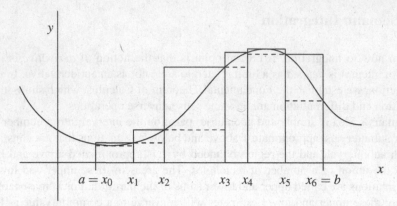

FIG. 6.4 Example of upper and lower sums.

Thus $m_j(f,P)(x_j - x_{j-1}) \leq \sum\limits_{i=k+1}^{l} m_i(f,R)(t_i - t_{i-1})$. Summing over all the intervals of P yields $L(f,P) \leq L(f,R)$. Similarly, $U(f,R) \leq U(f,P)$.

Now consider two arbitrary partitions P and Q. Let $R = P \cup Q$, which refines both P and Q. Then $L(f,P) \leq L(f,R) \leq U(f,R) \leq U(f,Q)$. ∎

In particular, we see that the set of numbers $\{L(f,P)\}$ is bounded above by any $U(f,Q)$. Hence by the completeness of $\mathbb{R}$, $\sup_P L(f,P)$ is defined. Moreover, $\sup_P L(f,P) \leq U(f,Q)$ for every partition Q. Therefore, $\inf_P U(f,P)$ is defined and $\sup_P L(f,P) \leq \inf_P U(f,P)$.

6.3.3. DEFINITION. Define $L(f) = \sup_P L(f,P)$ and $U(f) = \inf_P U(f,P)$. As noted, $L(f) \leq U(f)$. A bounded function f on $[a,b]$ is called **Riemann integrable** if $L(f) = U(f)$. In this case, we write $\int_a^b f(x)\,dx$ for the common value.

We establish Riemann's Condition for integrability, which follows easily from our definition.

6.3.4. RIEMANN'S CONDITION.
Suppose $f : [a,b] \to \mathbb{R}$ is bounded. Then f is Riemann integrable if and only if for each $\varepsilon > 0$, there is a partition P of $[a,b]$ such that $U(f,P) - L(f,P) < \varepsilon$.

PROOF. Assume that the ε condition holds. For an $\varepsilon > 0$, there is P such that

$$L(f,P) \leq L(f) \leq U(f) \leq U(f,P).$$

Hence $0 \leq U(f) - L(f) \leq U(f,P) - L(f,P) < \varepsilon$ for each $\varepsilon > 0$. So $U(f) = L(f)$.

If f is Riemann integrable, let $L = L(f) = U(f)$. Let $\varepsilon > 0$. We can find two partitions P_1 and P_2 such that $U(f,P_1) < L + \varepsilon/2$ and $L(f,P_2) > L - \varepsilon/2$. Let P be their common refinement, $P_1 \cup P_2$. By Lemma 6.3.2,

$$L - \frac{\varepsilon}{2} < L(f,P) \leq U(f,P) < L + \frac{\varepsilon}{2},$$

and so $U(f,P) - L(f,P) < \varepsilon$, as required. ∎

A typical choice is the uniform partition $P = \{a + j(b-a)/n : 0 \leq j \leq n\}$. But it is sometimes convenient to choose a partition better suited to the function.

6.3.5. EXAMPLE. Consider the function $f(x) = x^p$ on $[a,b]$, where $p \neq -1$ and $0 < a < b$. Take the partition $P_n = \{a = x_0 < x_1 < \cdots < x_n = b\}$, where $x_j = a(b/a)^{j/n}$ for $0 \leq j \leq n$. To keep the notation under control, let $R = (b/a)^{1/n}$. So, $x_j = aR^j$ and $\Delta_j = x_j - x_{j-1} = aR^{j-1}(R-1)$. Since f is monotone increasing when $p \geq 0$, we easily compute

$$m_j(f,P_n) = x_{j-1}^p = a^p R^{p(j-1)} \quad \text{and} \quad M_j(f,P_n) = x_j^p = a^p R^{pj} = R^p m_j(f,P_n).$$

When $p < 0$, m_j and M_j are reversed. The details of this case are left to the reader. So for $p > 0$, we have $U(f,P_n) = R^p L(f,P_n)$ and

$$L(f,P_n) = \sum_{j=1}^n m_j(f,P_n)\Delta_j = \sum_{j=1}^n a^p R^{p(j-1)} aR^{j-1}(R-1) = a^{p+1}(R-1)\sum_{j=0}^{n-1} R^{(p+1)j}.$$

Summing the geometric series and rearranging, we have

$$L(f,P_n) = (R-1)\frac{(aR^n)^{p+1} - a^{p+1}}{R^{p+1} - 1} = (b^{p+1} - a^{p+1})\frac{R-1}{R^{p+1} - 1}.$$

To show the role of n clearly, we set $r = b/a$ and $h = 1/n$, so that $R = r^{1/n} = r^h$. The key is to recognize the limit as the quotient of two derivatives:

$$\lim_{n \to \infty} \frac{r^{1/n} - 1}{r^{(p+1)/n} - 1} = \lim_{h \to 0} \frac{r^h - 1}{h} \frac{h}{r^{(p+1)h} - 1} = \frac{\frac{d}{dx}(r^x)(0)}{\frac{d}{dx}(r^{(p+1)x})(0)} = \frac{\log r}{(p+1)\log r} = \frac{1}{p+1}.$$

Hence $\lim_{n \to \infty} L(f,P_n) = (b^{p+1} - a^{p+1})/(p+1)$.

Since $U(f,P_n) = (a/b)^{p/n} L(f,P_n)$ has the same limit, we conclude that

$$\frac{b^{p+1} - a^{p+1}}{p+1} \leq L(f) \leq U(f) \leq \frac{b^{p+1} - a^{p+1}}{p+1}.$$

So this function is Riemann integrable with $\int_a^b x^p \, dx = (b^{p+1} - a^{p+1})/(p+1)$.

Next, we give two variations on Riemann's Condition. Condition (3) is a "δ–ε" formulation of integrablility. Condition (4) below shows that for suitable partitions we can use the Riemann sum for arbitrarily chosen points, instead of finding maximum and minimum values. So once f is known to be integrable, there are many choices of partition or evaluation sequence that 'work'.

6.3.6. THEOREM. *Let $f(x)$ be bounded on $[a,b]$. The following are equivalent:*

(1) *f is Riemann integrable.*
(2) *For each $\varepsilon > 0$, there is a partition P such that $U(f,P) - L(f,P) < \varepsilon$.*
(3) *For every $\varepsilon > 0$, there is a $\delta > 0$ such that every partition Q with $\mathrm{mesh}(Q) < \delta$ satisfies $U(f,Q) - L(f,Q) < \varepsilon$.*
(4) *For every $\varepsilon > 0$, there is a $\delta > 0$ such that every partition Q with $\mathrm{mesh}(Q) < \delta$ and every evaluation sequence for Q, X satisfies $\left| I(f,Q,X) - \int_a^b f(x)\,dx \right| < \varepsilon$.*

PROOF. We have verified that (1) and (2) are equivalent. Clearly, (3) implies (2).

Let us prove that (1) implies (3). If f is Riemann integrable, let $L = L(f) = U(f)$. Let $\varepsilon > 0$. We can find two partitions P_1 and P_2 such that $U(f,P_1) < L + \varepsilon/4$ and $L(f,P_2) > L - \varepsilon/4$. Let $P = P_1 \cup P_2$ be their common refinement, say with n points. By Lemma 6.3.2,

$$L - \tfrac{\varepsilon}{4} < L(f,P) \le U(f,P) < L + \tfrac{\varepsilon}{4}.$$

Let $\delta = \varepsilon/(8n\|f\|_\infty)$, where $\|f\|_\infty = \sup\{|f(x)| : x \in [a,b]\}$.

Now suppose that Q is any partition with $\mathrm{mesh}(Q) < \delta$. Define $R = P \cup Q$ to be the common refinement of P and Q. By Lemma 6.3.2 again, we obtain

$$L - \tfrac{\varepsilon}{4} < L(f,R) \le U(f,R) < L + \tfrac{\varepsilon}{4}.$$

The intervals of R coincide with the intervals of Q except for at most $n-1$ intervals of Q, which are split in two by points of P. Thus in the sums determining $L(f,R)$ and $L(f,Q)$, all terms are the same except for terms from these $n-1$ intervals. On each such interval, say $[x_{j-1}, x_j] = [t_{l-1}, t_l] \cup [t_l, t_{l+1}]$, we have $-\|f\|_\infty \le f \le \|f\|_\infty$. So $m_l(f,R) - m_j(f,Q) \le 2\|f\|_\infty$. Adding up the differences over the $n-1$ such intervals of Q, the total can be no more than

$$L(f,R) - L(f,Q) \le (n-1)2\|f\|_\infty \mathrm{mesh}(Q) < \tfrac{2n\|f\|_\infty \varepsilon}{8n\|f\|_\infty} = \tfrac{\varepsilon}{4}.$$

Hence $L(f,Q) > L - \tfrac{\varepsilon}{2}$. Likewise, $U(f,Q) < L + \tfrac{\varepsilon}{2}$. So $U(f,Q) - L(f,Q) < \varepsilon$.

To see that (3) implies (4), fix a partition Q satisfying (3) and an evaluation sequence X. Then $L(f,Q) \le I(f,Q,X) \le U(f,Q) < L(f,Q) + \varepsilon$. Now we also know that $L(f,Q) \le \int_a^b f(x)\,dx \le U(f,Q)$. Hence $\left| I(f,Q,X) - \int_a^b f(x)\,dx \right| < \varepsilon$.

Conversely, if (4) holds, then *every* choice of $X = (x_1', \ldots, x_n')$ satisfies this inequality for $\varepsilon/3$. If x_j' satisfy $f(x_j') = \inf_{x_{j-1} \le x \le x_j} f(x)$, then $I(f,Q,X) = L(f,Q)$. Hence $\left| L(f,Q) - \int_a^b f(x)\,dx \right| < \varepsilon/3$. If the infimum is not attained, then we can choose X such that $f(x_j')$ are sufficiently close to this infimum to obtain the inequality $\left| L(f,Q) - \int_a^b f(x)\,dx \right| < \varepsilon/2$. The details are left as an exercise. Similarly, $\left| U(f,Q) - \int_a^b f(x)\,dx \right| < \varepsilon/2$. Hence $U(f,Q) - L(f,Q) < \varepsilon$. So (3) holds. ∎

Using Riemann's condition, we can now show that many functions are integrable.

6.3.7. THEOREM. *Every monotone function on $[a,b]$ is Riemann integrable.*

PROOF. We may assume that f is monotone increasing. Consider the uniform partition P given by $x_j = a + \frac{j(b-a)}{n}$ for $0 \leq j \leq n$. Notice that $m_j(f,P) = f(x_{j-1})$ and $M_j(f,P) = f(x_j)$. Thus we obtain a telescoping sum

$$U(f,P) - L(f,P) = \sum_{j=1}^{n} f(x_j) \frac{b-a}{n} - \sum_{j=1}^{n} f(x_{j-1}) \frac{b-a}{n} = \frac{(f(b) - f(a))(b-a)}{n}.$$

For $\varepsilon > 0$, choose $n > (f(b) - f(a))(b-a)\varepsilon^{-1}$, so that $U(f,P) - L(f,P) < \varepsilon$. This verifies Riemann's condition, and therefore f is integrable. ∎

6.3.8. THEOREM. *Every continuous function on $[a,b]$ is Riemann integrable.*

PROOF. This result is deeper than the result for monotone functions because we must use Theorem 5.5.9 to deduce that a continuous function f on $[a,b]$ is uniformly continuous. Let $\varepsilon > 0$. By uniform continuity, there is a $\delta > 0$ such that for $x, y \in [a,b]$ with $|x - y| < \delta$, we have $|f(x) - f(y)| < \varepsilon/(b-a)$. Let P be any partition with mesh$(P) < \delta$. Then for any points x, y in a common interval $[x_{j-1}, x_j]$, we have $|f(x) - f(y)| < \varepsilon/(b-a)$. Hence $M_j(f,P) - m_j(f,P) \leq \varepsilon/(b-a)$. Thus,

$$U(f,P) - L(f,P) = \sum_{j=1}^{n} (M_j(f,P) - m_j(f,P))\Delta_j \leq \frac{\varepsilon}{b-a} \sum_{j=1}^{n} \Delta_j = \varepsilon.$$

Thus Riemann's condition is satisfied, and f is Riemann integrable. ∎

6.3.9. EXAMPLE. There do exist functions that are not Riemann integrable. For example, consider $f : [0,1] \to \mathbb{R}$ defined by $f(x) = 1$ if $x \in \mathbb{Q}$ and $f(x) = 0$ if $x \notin \mathbb{Q}$. Let P be any partition. Notice that $M_j(f,P) = 1$ and $m_j(f,P) = 0$ for all j. Thus we see that $U(f,P) = \sum_{j=1}^{n} x_j - x_{j-1} = 1$ and $L(f,P) = 0$. This holds for all P. Thus $L(f) = 0$ and $U(f) = 1$; so f is not Riemann integrable. The reason for this failure is that f is discontinuous at every point in $[0,1]$.

6.3.10. EXAMPLE. On the other hand, there are discontinuous functions that are Riemann integrable. The characteristic function $\chi_{(.5,1]}$ is Riemann integrable on $[0,1]$ because it is monotone. However, the discontinuity is rather banal.

Consider $f(x) = \sin(1/x)$ on $(0,1]$, and set $f(0) = 0$. This function has a nasty discontinuity at the origin, as discussed in Example 5.2.7. See Figure 5.3 on page 75. Provided the function remains bounded, even a bad discontinuity like this one does not prevent integrability.

Let $\varepsilon > 0$ be given. We will choose a partition P with $x_1 = \varepsilon/4$. Notice that since f is continuous on $[\varepsilon/4, 1]$, it is integrable there. Thus there is a partition $Q = \{x_1 = \varepsilon/4 < \cdots < x_n = 1\}$ of $[\varepsilon/4, 1]$ with $U(f|_{[x_1,1]}, Q) - L(f|_{[x_1,1]}, Q) < \frac{\varepsilon}{2}$. Now take $P = \{0\} \cup Q$ as a partition of $[0,1]$. Then since $\sin(1/x)$ oscillates wildly between ± 1 near $x = 0$, it follows that $M_1(f,P) = 1$ and $m_1(f,P) = -1$. So

$$U(f,P) = \tfrac{\varepsilon}{4} + U(f|_{[x_1,1]},Q) \quad \text{and} \quad L(f,P) = -\tfrac{\varepsilon}{4} + L(f|_{[x_1,1]},Q).$$

Therefore, $U(f,P) - L(f,P) = \tfrac{\varepsilon}{2} + U(f|_{[x_1,1]},Q) - L(f|_{[x_1,1]},Q) < \varepsilon$. So f is integrable.

In fact, even functions with many bad discontinuities can be integrated. Consider the function $g(x) = \sin(\csc(1/x))$. This function behaves badly at those x where $\csc(1/x)$ is undefined, namely where $\sin(1/x) = 0$. Since $\sin(t) = 0$ for $t = n\pi$, these discontinuities occur at $x = 1/(n\pi)$ for $n \geq 1$ and at the endpoint $x = 0$, where g is undefined. Moreover, on an interval around one of these discontinuities, say $[1/(x\pi), 1/((x+1)\pi)]$, where $x = n + 1/2$, g has, qualitatively, the same behaviour as $\sin(1/x)$ has around the origin. So we have a $\sin(1/x)$-type discontinuity at a sequence of points converging to 0. Nonetheless, the function is integrable on $[0,1]$.

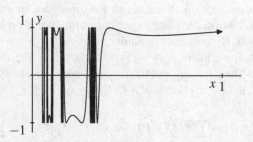

FIG. 6.5 A partial graph of $y = \sin(\csc(1/x))$ from 0 to π.

Exercises for Section 6.3

A. (a) Compute the upper Riemann sum for $\int_a^b \tfrac{1}{x}\,dx$ using $P_n = \{x_j = a(b/a)^{j/n} : 0 \leq j \leq n\}$.

  (b) Evaluate the integral $\int_a^b \tfrac{1}{x}\,dx$. HINT: Recognize $\lim_{n\to\infty} U(f,P_n)$ as a derivative.

B. (a) Compute the upper Riemann sum for $f(x) = x^2$ on $[a,b]$ using the uniform partition $P_n = \{x_j = a + j(b-a)/n : 0 \leq j \leq n\}$. HINT: $\sum_{j=1}^{n} j^2 = n(n+1)(2n+1)/6$.

  (b) Hence evaluate the integral $\int_a^b x^2\,dx$.

C. Show that if a function $f : [a,b] \to \mathbb{R}$ is Lipschitz with constant C, then for any partition P of $[a,b]$, we have $U(f,P) - L(f,P) \leq C(b-a)\,\mathrm{mesh}(P)$.

D. Show that if f and g are Riemann integrable on $[a,b]$, then so is $\alpha f + \beta g$; and

$$\int_a^b \alpha f(x) + \beta g(x)\,dx = \alpha \int_a^b f(x)\,dx + \beta \int_a^b g(x)\,dx.$$

E. Show that every piecewise continuous function is Riemann integrable.

F. Show that if f is Riemann integrable on $[a,b]$, then so is $|f|$.
  HINT: Show that $M_i(|f|,P) - m_i(|f|,P) \leq M_i(f,P) - m_i(f,P)$.

G. Show that f is Riemann integrable if and only if for each $\varepsilon > 0$, there are step functions f_1 and f_2 on $[a,b]$ with $f_1(x) \leq f(x) \leq f_2(x)$ such that $\int_a^b f_2(x) - f_1(x)\,dx < \varepsilon$.

H. (a) Show that if $f \geq 0$ is Riemann integrable on $[a,b]$, then $\int_a^b f(x)\,dx \geq 0$.

(b) If f and g are integrable on $[a,b]$ and $f(x) \le g(x)$, prove that $\int_a^b f(x)\,dx \le \int_a^b g(x)\,dx$.

(c) Show that $\left| \int_a^b f(x)\,dx \right| \le \int_a^b |f(x)|\,dx$.

I. **Translation invariance.** Suppose f is integrable on $[a,b]$ and $c \in \mathbb{R}$. Define g on $[a+c,b+c]$ by $g(x) = f(x-c)$. Show that g is integrable and $\int_{a+c}^{b+c} g(x)\,dx = \int_a^b f(x)\,dx$.

J. If f is Riemann integrable on $[a,b]$, show that $F(x) = c + \int_a^x f(t)\,dt$ is Lipschitz.

K. Show that if f is integrable on $[a,b]$, then it is integrable on each interval $[c,d] \subset [a,b]$ as well. Moreover, for $a < c < b$, $\int_a^b f(x)\,dx = \int_a^c f(x)\,dx + \int_c^b f(x)\,dx$.

L. If $b < a$, define $\int_a^b f(x)\,dx = -\int_b^a f(x)\,dx$. Show that the formula of the previous exercise also holds for c outside $[a,b]$. Where must f be integrable for this to make sense?

M. Show that $\sin(\csc(1/x))$ is integrable on $[0,1]$. HINT: mimic Example 6.3.10 near every point where $\csc(1/x)$ blows up.

N. If f and g are both Riemann integrable on $[a,b]$, show that fg is also integrable. HINT: Use the identity $f(x)g(x) - f(t)g(t) = f(x)\big(g(x)-g(t)\big) + \big(f(x)-f(t)\big)g(t)$ to show that $M_i(fg,P) - m_i(fg,P)$ is bounded by $\|f\|_\infty \big(M_i(g,P) - m_i(g,P)\big) + \|g\|_\infty \big(M_i(f,P) - m_i(f,P)\big)$.

O. Show that the function of Example 5.2.9 is Riemann integrable, even though it is discontinuous at every rational number. HINT: For $\varepsilon > 0$, there are only finitely many points taking values greater than ε. Choose a partition that includes those points in very small intervals.

6.4 The Fundamental Theorem of Calculus

Calculating integrals via Riemann sums is of theoretical importance, but it is not a practical method. Fortunately, there is a crucial connection between integrals and derivatives that makes evaluating integrals by hand effective and efficient. We stress that this is *not* the definition of integral. The word *fundamental* is used to emphasize that this is the central result connecting differential and integral calculus.

We split the main theorem into two parts. First, a simple estimate.

6.4.1. LEMMA. *Suppose that f is integrable on $[a,b]$ and bounded by M. Then*

$$\left| \int_a^b f(t)\,dt \right| \le M(b-a).$$

PROOF. For the partition $\{a,b\}$, clearly $U(f,\{a,b\}) \le M(b-a)$. Hence $\int_a^b f(t)\,dt = \inf_P U(f,P) \le M(b-a)$. Similarly, we obtain $\int_a^b f(t)\,dt \ge -M(b-a)$. ∎

6.4.2. FUNDAMENTAL THEOREM OF CALCULUS, PART 1.

Let f be integrable on $[a,b]$. Define $F(x) = \int_a^x f(t)\,dt$ for $a \le x \le b$. Then F is continuous. When f is continuous at x_0, F is differentiable at x_0 and $F'(x_0) = f(x_0)$.

PROOF. Let f be bounded by M. For x,y in $[a,b]$, compute

$$|F(x) - F(y)| = \left| \int_a^x f(t)\,dt - \int_a^y f(t)\,dt \right| = \left| \int_y^x f(t)\,dt \right| \le M|x-y|.$$

Thus F is Lipschitz with constant M, and so is continuous by Proposition 5.1.6.

Suppose f is continuous at x_0. Given $\varepsilon > 0$, choose $\delta > 0$ such that $|y - x_0| < \delta$ implies that $|f(y) - f(x_0)| < \varepsilon$. Then for $|h| < \delta$, compute

$$\left| \frac{F(x_0 + h) - F(x_0)}{h} - f(x_0) \right| = \left| \frac{1}{h} \int_{x_0}^{x_0+h} f(t)\,dt - \frac{1}{h} \int_{x_0}^{x_0+h} f(x_0)\,dt \right|$$

$$\leq \frac{1}{h} \int_{x_0}^{x_0+h} |f(t) - f(x_0)|\,dt \leq \varepsilon.$$

Thus $F'(x_0) = \lim\limits_{h \to 0} \dfrac{F(x_0 + h) - F(x_0)}{h} = f(x_0)$.   ∎

6.4.3. FUNDAMENTAL THEOREM OF CALCULUS, PART 2.

Let f be integrable on $[a,b]$. If there is a continuous function g on $[a,b]$ that is differentiable on (a,b) such that $g'(x) = f(x)$ for $a < x < b$, then

$$\int_a^b f(x)\,dx = g(b) - g(a).$$

PROOF. Let $P = \{a = x_0 < x_1 < \cdots < x_n = b\}$ be a partition. On $[x_{j-1}, x_j]$, the Mean Value Theorem supplies a point $x_j' \in [x_{j-1}, x_j]$ such that

$$g(x_j) - g(x_{j-1}) = g'(x_j')(x_j - x_{j-1}) = f(x_j')\Delta_j.$$

Let $X = (x_1', \ldots, x_n')$. Then we obtain a telescoping sum

$$I(f, P, X) = \sum_{j=1}^n f(x_j')\Delta_j = \sum_{j=1}^n g(x_j) - g(x_{j-1}) = g(b) - g(a).$$

Since P is arbitrary, Theorem 6.3.6 part (4) shows that $\int_a^b f(x)\,dx = g(b) - g(a)$. ∎

6.4.4. REMARK.
A jump discontinuity in the integrand f can result in a point where the integral is not differentiable. For example, take $f(x) = 1$ for $0 \leq x \leq 1$ and 2 for $1 < x \leq 2$. Then $F(x) = \int_0^x f(t)\,dt = x$ on $[0,1]$ and $F(x) = 2x - 1$ on $[1,2]$. This function is continuous on $[0,2]$, but is not differentiable at $x = 1$. It has a left derivative of 1 and a right derivative of 2.

Nor is it the case that every differentiable function is an integral. An easy way for this to fail is when the derivative is unbounded. Recall from Example 6.1.9 the function $F(x) = x^a \sin(1/x)$ on $[0,1]$ for some constant a in $(1,2)$. Then $F'(x) = ax^{a-1}\sin(1/x) - x^{a-2}\cos(1/x)$ for $x > 0$ and $F'(0) = 0$. Thus F is differentiable, but the derivative is an unbounded function and thus is not Riemann integrable.

We can take various formulae for differentiation and, using the Fundamental Theorem, integrate them to obtain useful integration techniques. We are not concerned here with the all tricks of the trade, but just a glance at the major methods.

The product rule translates into **integration by parts**. We assume that F and G are C^1 to avoid pathology. Since $(FG)'(x) = F'(x)G(x) + F(x)G'(x)$, we obtain

$$\int_a^b F'(x)G(x) + F(x)G'(x)\,dx = F(b)G(b) - F(a)G(a). \qquad (6.4.5)$$

Using $F(x)G(x)\big|_a^b$ for the right-hand side, we obtain the usual formulation

$$\int_a^b F'(x)G(x)\,dx = F(x)G(x)\Big|_a^b - \int_a^b F(x)G'(x)\,dx.$$

The chain rule (6.1.6) corresponds to the substitution rule, also called the **change of variable formula**. Let u be a C^1 function on $[a,b]$, and let F be C^1 on an interval $[c,d]$ containing the range of u. Then if $G(x) = F(u(x))$, the chain rule states that $G'(x) = F'(u(x))u'(x)$. Thus if we set $f = F'$,

$$\int_a^b f(u(x))u'(x)\,dx = G(b) - G(a) = F(u(b)) - F(u(a)) = \int_{u(a)}^{u(b)} f(t)\,dt. \quad (6.4.6)$$

We interpret this as making the substitution $t = u(x)$ and think of dt as $u'(x)\,dx$.

This change of variables is sometimes formulated somewhat differently. Suppose that the function u satisfies $u'(x) \neq 0$ for all $x \in [a,b]$. If we set $c = u(a)$ and $d = u(b)$, we obtain

$$\int_c^d f(x)\,dx = \int_{u^{-1}(c)}^{u^{-1}(d)} f(u(t))u'(t)\,dt. \qquad (6.4.7)$$

This corresponds to the substitution $x = u(t)$.

Without any attempt to be complete, we give a couple of examples of integration technique to refresh the reader's memory.

6.4.8. EXAMPLE. Consider $\int_0^1 \arctan(x)\,dx$. The derivative of arctan is well known to be $1/(1+x^2)$, but the integral is probably not memorized. This suggests that an integration by parts approach might help. Of course, you need something to integrate, and so we put in a factor of 1, which integrates to x. Thus

$$\int \arctan(x)\,dx = x\arctan(x) - \int \frac{x}{1+x^2}\,dx.$$

Now substitute $u = 1 + x^2$, which has derivative $du = 2x\,dx$, to obtain

$$= x\arctan(x) - \int \frac{du}{2u} = x\arctan(x) - \frac{1}{2}\log u = x\arctan(x) - \frac{1}{2}\log(1+x^2).$$

Thus

$$\int_0^1 \arctan(x)\,dx = x\arctan(x) - \frac{1}{2}\log(1+x^2)\Big|_0^1 = \frac{\pi}{4} - \frac{1}{2}\log 2.$$

6.4.9. EXAMPLE. Now consider the integral $\int_0^8 e^{\sqrt[3]{x}} dx$. This integrand has a complicated exponent, $\sqrt[3]{x}$, which can be simplified by substituting $x = u^3$. Then $dx = 3u^2 du$ and $u = \sqrt[3]{x}$ runs from 0 to 2 as x runs from 0 to 8. This now can be integrated by parts twice by integrating e^u and differentiating $3u^2$:

$$\int_0^8 e^{\sqrt[3]{x}} dx = \int_0^2 e^u 3u^2 \, du = 3u^2 e^u \Big|_0^2 - \int_0^2 6u e^u \, du$$

$$= (3u^2 - 6u)e^u \Big|_0^2 + \int_0^2 6e^u \, du = (3u^2 - 6u + 6)e^u \Big|_0^2 = 6(e^2 - 1).$$

Exercises for Section 6.4

A. Evaluate the following by recognizing them as Riemann sums.

(a) $\lim\limits_{n \to \infty} \sum\limits_{j=1}^{n} \dfrac{1}{n + jc}$ for $c > 1$. (b) $\lim\limits_{n \to \infty} \dfrac{1}{n^{a+1}} + \dfrac{2^a}{n^{a+1}} + \cdots + \dfrac{(n-1)^a}{n^{a+1}}$ for $a > -1$.

B. For $x > 0$, define $L(x) = \int_1^x 1/t \, dt$. Manipulate integrals to prove $L(ab) = L(a) + L(b)$.

C. (a) Prove the **Mean Value Theorem for Integrals**: If f is a continuous function on $[a, b]$, then there is a point $c \in (a, b)$ such that $\frac{1}{b-a} \int_a^b f(x) \, dx = f(c)$.
(b) Show by example that this may fail for a discontinuous but integrable function.

D. Let $f(x) = \text{sign}(x)$, and $F(x) = |x|$. Show that f is Riemann integrable on $[a, b]$ and that $\int_a^b f(x) \, dx = F(b) - F(a)$ for any $a < b$. Why is F not an antiderivative of f?

E. Let f be a continuous function on $\mathbb{R}$, and suppose that $b(x)$ is a C^1 function. Define $G(x) = \int_a^{b(x)} f(t) \, dt$. Compute $G'(x)$. HINT: Let $F(x) = \int_a^x f(t) \, dt$ and note that $G(x) = F(b(x))$.

F. Compute the following integrals:

(a) $\displaystyle\int_1^e (\log x)^2 \, dx$ (b) $\displaystyle\int_0^{\pi/2} \frac{\sin^3 x}{\sqrt{\cos x}} \, dx$ (c) $\displaystyle\int_1^{125} \frac{dt}{\sqrt{t} + \sqrt[3]{t}}$

G. Let f be a continuous function on $\mathbb{R}$, and fix $\varepsilon > 0$. Define a function $G(x) = \frac{1}{\varepsilon} \int_x^{x+\varepsilon} f(t) \, dt$. Show that G is C^1 and compute G'.

H. Let u be a strictly increasing C^1 function on $[a, b]$.

(a) By considering the area under the graph plus the area between the graph and the y-axis, establish the formula $\int_a^b u(x) \, dx + \int_{u(a)}^{u(b)} u^{-1}(t) \, dt = bu(b) - au(a)$.
(b) Use the substitution formula (6.4.6) using $f(x) = u^{-1}(x)$ and integrate the second expression by parts to derive the same formula as in part (a).

I. Suppose f is twice differentiable on $\mathbb{R}$, $\|f\|_\infty = A$ and $\|f''\|_\infty = C$. Prove that $\|f'\|_\infty \le \sqrt{2AC}$. HINT: If $f'(x_0) = b > 0$, show that $f'(x_0 + t) \ge b - C|t|$. Integrate from $x_0 - b/C$ to $x_0 + b/C$.

J. **Improper Integrals.** Say that f is integrable on $[a, \infty)$ if $\lim\limits_{b \to \infty} \int_a^b f(x) \, dx =: \int_a^\infty f(x) \, dx$ exists.

(a) For which real values of p does $\int_e^\infty \frac{(\log x)^p}{x} \, dx$ exist?
(b) Show that $\int_0^\infty \frac{\sin x}{x} \, dx$ exists. HINT: Consider the alternating series $\sum_{n \ge 0} \int_{n\pi}^{(n+1)\pi} \frac{\sin x}{x} \, dx$.

K. If f is unbounded as it approaches a, define an improper integral by $\lim\limits_{\varepsilon \to 0} \int_{a+\varepsilon}^b f(x) \, dx$, when the limit exists. Of course, $\int_a^b f(x) \, dx$ is used to denote this limit, so be careful.

(a) For which real values of p does $\int_0^1 x^p \, dx$ exist?
(b) Show that $\lim\limits_{\varepsilon \to 0} \int_{-b}^{a-\varepsilon} f(x) \, dx + \int_{a+\varepsilon}^b f(x) \, dx$ can exist even though $\lim\limits_{\varepsilon \to 0} \int_{a+\varepsilon}^b f(x) \, dx$ does not.

Chapter 7
Norms and Inner Products

In this chapter, we generalize to more abstract settings two key properties of $\mathbb{R}^n$: the Euclidean norm of a vector and the dot product of two vectors. The generalizations, norms and inner products, respectively, are set in a general vector space. Many of our applications will be set in this framework.

The basic notions of topology go through with almost no change in the definitions. However, some theorems can be quite different. For example, being closed and bounded is not sufficient to imply compactness in an infinite-dimensional normed vector space, such as $C[a,b]$.

7.1 Normed Vector Spaces

The Euclidean norm on $\mathbb{R}^n$, the focus of Chapter 4, is crucial to analysis on $\mathbb{R}^n$. To apply these ideas to other vector spaces, we need a general definition of a norm, one that captures the essential properties of the Euclidean norm. Vector spaces over $\mathbb{R}$ are discussed in more detail in Section 1.2; We continue to use boldface letters, such as $\mathbf{x}$, for elements of $\mathbb{R}^n$, but not for elements of a general vector space. An exception is the zero vector, which we always write as $\mathbf{0}$, to distinguish it from the number 0.

7.1.1. DEFINITION. Let V be a vector space over $\mathbb{R}$. A **norm** on V is a function $\|\cdot\|$ on V taking values in $[0, +\infty)$ with the following properties:

(1) (positive definite) $\|x\| = 0$ if and only if $x = \mathbf{0}$,

(2) (homogeneous) $\|\alpha x\| = |\alpha|\|x\|$ for all $x \in V$ and $\alpha \in \mathbb{R}$, and

(3) (triangle inequality) $\|x+y\| \leq \|x\| + \|y\|$ for all $x, y \in V$.

We call the pair $(V, \|\cdot\|)$ a **normed vector space**.

The first two properties are usually easy to verify. The positive definite property just says that nonzero vectors have nonzero length. And the homogeneous property

K.R. Davidson and A.P. Donsig, *Real Analysis and Applications: Theory in Practice*,
Undergraduate Texts in Mathematics, DOI 10.1007/978-0-387-98098-0_7,
© Springer Science + Business Media, LLC 2010

says that the norm is scalable. The important property, which often requires some cleverness to verify, is the triangle inequality. It says that the path from point A to B and on to C is at least as long as the direct route from A to C. As we indicated in Figure 4.1 in Chapter 4, this algebraic inequality is equivalent to the geometric statement that the length of one side of a triangle is at most sum of the lengths of the other two sides.

7.1.2. EXAMPLE. Consider the vector space $\mathbb{R}^n$. In Chapter 4, we showed that the Euclidean norm

$$\|\mathbf{x}\| = \|(x_1,\ldots,x_n)\|_2 = \Big(\sum_{i=1}^n |x_i|^2\Big)^{1/2},$$

is a norm. Indeed, properties (1) and (2) are evident, and the triangle inequality was a consequence of Schwarz's inequality (4.1.1).

Consider two other functions:

$$\|\mathbf{x}\|_1 = \|(x_1,\ldots,x_n)\|_1 = \sum_{i=1}^n |x_i|,$$

$$\|\mathbf{x}\|_\infty = \|(x_1,\ldots,x_n)\|_\infty = \max_{1\le i\le n} |x_i|.$$

Again it is easy to see that they are positive definite and homogeneous. The key is the triangle inequality. But for these functions, even that is straightforward. Compute

$$\|\mathbf{x}+\mathbf{y}\|_1 = \sum_{i=1}^n |x_i+y_i| \le \sum_{i=1}^n |x_i| + |y_i| = \|\mathbf{x}\|_1 + \|\mathbf{y}\|_1,$$

$$\|\mathbf{x}+\mathbf{y}\|_\infty = \max_{1\le i\le n} |x_i+y_i| \le \max_{1\le i\le n} |x_i| + \max_{1\le i\le n} |y_i| = \|\mathbf{x}\|_\infty + \|\mathbf{y}\|_\infty.$$

To illustrate the differences between these norms, consider the vectors of norm at most 1 in $\mathbb{R}^2$ for these three norms, as given in Figure 7.1. This set is sufficiently useful that we give it a name. For any normed vector space $(V, \|\cdot\|)$, the **unit ball** of V is the set $\{x \in V : \|x\| \le 1\}$.

The next example is very important for our applications, since it allows us to apply vector space methods to collections of functions, that is, to think of functions as vectors.

7.1.3. EXAMPLE. Let K be a compact subset of $\mathbb{R}^n$, and let $C(K)$ denote the vector space of all continuous real-valued functions on K. If $f,g \in C(K)$ and $\alpha \in \mathbb{R}$, then $f + g$ and αf are the functions given by $(f+g)(x) := f(x)+g(x)$ and $(\alpha f)(x) := \alpha f(x)$. There are several different possible norms on $C(K)$. The most natural and most important is the **uniform norm**, given by

$$\|f\|_\infty = \sup_{x\in K} |f(x)|.$$

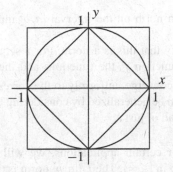

FIG. 7.1 The unit balls $\{(x,y) : \|(x,y)\|_p \leq 1\}$ for $p = 1, 2, \infty$.

By the Extreme Value Theorem (5.4.4), $|f|$ achieves its maximum at some point $x_0 \in K$. So using this point x_0, we have $\|f\|_\infty = |f(x_0)| < \infty$.

Clearly, the map $f \mapsto \|f\|_\infty$ is nonnegative. To see that it is really a norm, observe first that if $\|f\|_\infty = 0$, then $|f(x)| = 0$ for all $x \in K$ and so f is the zero function. For homogeneity, we have

$$\|\alpha f\|_\infty = \sup_{x \in K} |\alpha f(x)| = |\alpha| \sup_{x \in K} |f(x)| = |\alpha| \|f\|_\infty.$$

Finally, the triangle inequality is proved as follows:

$$\|f + g\|_\infty = \sup_{x \in K} |f(x) + g(x)| \leq \sup_{x \in K} |f(x)| + |g(x)|$$

$$\leq \sup_{x \in K} |f(x)| + \sup_{x \in K} |g(x)| = \|f\|_\infty + \|g\|_\infty.$$

We shall see in Chapter 8 that a sequence of functions f_n converges to a function f in $\big(C(K), \|\cdot\|_\infty\big)$ if and only if the sequence converges uniformly. Since uniform convergence is often the "right" notion of convergence for many applications, we will use this normed vector space often.

Controlling the derivatives of a function, as well as the function itself, is often important. Fortunately, this is easy to do, and we will regularly use normed vector spaces like the one in the following example.

7.1.4. EXAMPLE. For simplicity, we restrict our attention to an interval of $\mathbb{R}$, but the same idea readily generalizes. Let $C^3[a,b]$ denote the vector space of all functions $f : [a,b] \to \mathbb{R}$ such that f and its first three derivatives f', f'', f''' are all defined and continuous. Using $f^{(j)}$ for the jth derivative (and $f^{(0)}$ for f), we can define a new norm $\|\cdot\|_{C^3}$ by

$$\|f\|_{C^3} = \max_{0 \leq j \leq 3} \|f^{(j)}\|_\infty,$$

where $\| \cdot \|_\infty$ is the uniform norm on the interval $[a, b]$ introduced in the previous example.

It is an exercise to verify that this is a norm. For a sequence of functions f_n in $C^3[a, b]$ to converge to a function f, the functions and their first three derivatives $f_n^{(j)}$ for $0 \leq j \leq 3$ must all converge uniformly to the corresponding derivative $f^{(j)}$.

Clearly, this example can be generalized by considering the first k derivatives, for any positive integer k instead of just $k = 3$.

7.1.5. EXAMPLE. For certain applications, we will need the L^p norms on $C[a, b]$. Fix a real number p in $[1, \infty)$. The $L^p[a, b]$ norm is defined on $C[a, b]$ by

$$\|f\|_p = \left(\int_a^b |f(x)|^p \, dx \right)^{1/p}.$$

First notice that $\|f\|_p \geq 0$. Moreover, if $f \neq 0$, then there is a point $x_0 \in [a, b]$ such that $f(x_0) \neq 0$. Take $\varepsilon = |f(x_0)|/2$ and use the continuity of f to find an $r > 0$ such that

$$|f(x) - f(x_0)| < \frac{|f(x_0)|}{2} \quad \text{for} \quad x_0 - r < x < x_0 + r.$$

Hence

$$|f(x)| \geq |f(x_0)| - |f(x) - f(x_0)| > \frac{|f(x_0)|}{2}$$

for $x_0 - r < x < x_0 + r$. We may suppose that r is small enough that $a \leq x_0 - r$ and $x_0 + r \leq b$. (If $x_0 = a$ or b, the simple modification is left to the reader.) Consequently,

$$\|f\|_p \geq \left(\int_{x_0 - r}^{x_0 + r} \left(\frac{|f(x_0)|}{2} \right)^p dx \right)^{1/p} \geq \frac{(2r)^{1/p} |f(x_0)|}{2} > 0.$$

So the p-norms are positive definite.

Homogeneity is easy to verify from the definition. However, the triangle inequality is tricky and we prove it only for p equal to 1 or 2. To prove the triangle inequality for $p = 1$, suppose f and g are in $C[a, b]$ and compute

$$\|f + g\|_1 = \int_a^b |f(x) + g(x)| \, dx \leq \int_a^b |f(x)| + |g(x)| \, dx$$

$$= \int_a^b |f(x)| \, dx + \int_a^b |g(x)| \, dx = \|f\|_1 + \|g\|_1.$$

The case $p = 2$, which follows from results in Section 7.4, is easier because it arises from an inner product; see Example 7.4.2.

Exercises for Section 7.1

A.   Show that $\|(x, y, z)\| = |x| + 2\sqrt{y^2 + z^2}$ is a norm on $\mathbb{R}^3$. Sketch the unit ball.

B.   Is $\|(x, y)\| = \left(|x|^{1/2} + |y|^{1/2} \right)^2$ a norm on $\mathbb{R}^2$?

C. For f in $C^1[a,b]$, define $\rho(f) = \|f'\|_\infty$. Show that ρ is nonnegative, homogeneous, and satisfies the triangle inequality. Why is it not a norm?

D. If $(V, \|\cdot\|)$ is a normed vector space, show that $\big|\|x\| - \|y\|\big| \leq \|x - y\|$ for all $x, y \in V$.

E. Show that the unit ball of a normed vector space, $(V, \|\cdot\|)$, is convex, meaning that if $\|x\| \leq 1$ and $\|y\| \leq 1$, then every point on the line segment between x and y has norm at most 1. HINT: Describe the line segment algebraically in terms of x and y and a parameter t.

F. Let K be a compact subset of $\mathbb{R}^n$, and let $C(K, \mathbb{R}^m)$ denote the vector space of all continuous functions from K into $\mathbb{R}^m$. Show that for f in $C(K, \mathbb{R}^m)$, the quantity $\|f\|_\infty = \sup_{x \in K} \|f(x)\|_2$ is finite and $\|\cdot\|_\infty$ is a norm on $C(K, \mathbb{R}^m)$.

G. Define $C^p[a,b]$ for $p \in \mathbb{N}$ and verify that the $C^p[a,b]$ norm is indeed a norm. Is there a reasonable definition for $C^0[a,b]$?

H. (a) Show that if $\|\cdot\|$ and $\|\!|\cdot\|\!|$ are norms on V, then $\|v\|_m := \max\{\|v\|, \|\!|v\|\!|\}$ is a norm on V.
(b) Take $V = \mathbb{R}^2$ and $\|(x,y)\| = \sqrt{x^2 + y^2}$ and $\|\!|(x,y)\|\!| = \frac{3}{2}|x| + |y|$. Then define $\|(x,y)\|_m$ as in part (a). Draw a sketch of the unit balls for these three norms.

I. Let S be any subset of $\mathbb{R}^n$. Let $C_b(S)$ denote the vector space of all *bounded* continuous functions on S. For $f \in C(S)$, define $\|f\|_\infty = \sup_{x \in S} |f(x)|$.

(a) Show that this is a norm on $C_b(S)$.
(b) When is this a norm on the vector space of *all* continuous functions on S?

J. (a) Consider $C[a,b]$ and let $x_1, \dots, x_{n+1}$ be distinct points in $[a,b]$. Show that there are polynomials p_i of degree n such that $p_i(x_j) = \delta_{ij}$ for $1 \leq i, j \leq n+1$.
(b) Deduce that the polynomials of degree at most n form an $(n+1)$-dimensional subspace.
(c) Deduce that $C[a,b]$ is infinite-dimensional.

K. (a) Let $a = x_0 < \cdots < x_n = b$ be distinct points in a compact subset K of $\mathbb{R}$. For $0 \leq k \leq n$, let $h_k : \mathbb{R} \to \mathbb{R}$ be the piecewise linear function that is 1 at x_k and 0 at all other x_l, and zero off $[a,b]$. That is,

$$h_0(x) = \begin{cases} \frac{x - x_1}{a - x_1}, & a \leq x \leq x_1, \\ 0, & x_1 \leq x \leq b, \end{cases} \qquad h_n(x) = \begin{cases} 0, & a \leq x \leq x_{n-1}, \\ \frac{x - x_{n-1}}{b - x_{n-1}}, & x_{n-1} \leq x \leq b, \end{cases}$$

and for $1 \leq k \leq n-1$, let

$$h_k(x) = \begin{cases} 0, & a \leq x \leq x_{k-1} \text{ and } x_{k+1} \leq x \leq b, \\ \frac{x - x_{k-1}}{x_k - x_{k-1}}, & x_{k-1} \leq x \leq x_k, \\ \frac{x_{k+1} - x}{x_{k+1} - x_k}, & x_k \leq x \leq x_{k+1}. \end{cases}$$

Describe the linear span of the restrictions to K of $h_0, h_1, \dots, h_n$ as a subspace of $C(K)$.
(b) Hence show that $C(K)$ is infinite-dimensional if K is an infinite set.

7.2 Topology in Normed Spaces

The point of this section is to show how the notions of convergence and topology, which we developed in $\mathbb{R}$ and in $\mathbb{R}^n$, can be generalized to any normed vector space. While the definitions and some simple properties correspond exactly to the situation in $\mathbb{R}^n$, there are significant differences. The most important difference is that the Heine–Borel Theorem (4.4.6) does not hold in general. For such properties, we will have to study specific kinds of normed vector spaces individually using their special properties. When in doubt, you can test your intuition by looking at such spaces.

The definitions of convergence and Cauchy sequences are almost exactly the same as our definitions in $\mathbb{R}^n$.

7.2.1. DEFINITION. In a normed vector space $(V, \|\cdot\|)$, we say that a sequence $(v_n)_{n=1}^{\infty}$ **converges** to $v \in V$ if $\lim_{n \to \infty} \|v_n - v\| = 0$. Equivalently, for every $\varepsilon > 0$, there is an integer $N > 0$ such that $\|v_n - v\| < \varepsilon$ for all $n \geq N$. This is written $\lim_{n \to \infty} v_n = v$.

Call $(v_n)_{n=1}^{\infty}$ a **Cauchy sequence** if for every $\varepsilon > 0$, there is an integer $N > 0$ such that $\|v_n - v_m\| < \varepsilon$ for all $n, m \geq N$.

This leads to the notion of completeness in this context.

7.2.2. DEFINITION. Say that $(V, \|\cdot\|)$ is **complete** if every Cauchy sequence in V converges to some vector in V. A complete normed space is a **Banach space**.

Completeness is the fundamental property that distinguishes the real numbers from the rational numbers, and several crucial theorems depend on completeness. So it should not surprise the reader to find out that this is also a fundamental property of bigger normed spaces such as $(C(K), \|\cdot\|_{\infty})$. That $C(K)$ is complete will be established in the next chapter; see Theorem 8.2.2.

We can reformulate convergence using open and closed sets, exactly as for $\mathbb{R}^n$.

7.2.3. DEFINITION. For a normed vector space $(V, \|\cdot\|)$, we define the **open ball** with centre $a \in V$ and radius $r > 0$ to be $B_r(a) = \{v \in V : \|v - a\| < r\}$.

A subset U of V is **open** if for every $a \in U$, there is some $r > 0$ such that $B_r(a) \subset U$.

A subset C of V is **closed** if it contains all of its limit points. That is, whenever (x_n) is a sequence in C and $x = \lim_{n \to \infty} x_n$, then x belongs to C.

Proposition 4.3.8 works just as well for any normed space. So the open sets are precisely the complements of closed sets. Here is a sample result in showing the relationship between convergence and topology.

7.2.4. PROPOSITION. *A sequence (x_n) in a normed vector space V converges to a vector x if and only if for each open set U containing x, there is an integer N such that $x_n \in U$ for all $n \geq N$.*

PROOF. Suppose that $x = \lim_{n \to \infty} x_n$ and U is an open set containing x. There is an $r > 0$ such that $B_r(x) \subset U$. From the definition of limit, there is an integer N such that

$$\|x - x_n\| < r \quad \text{for all} \quad n \geq N.$$

This just says that $x_n \in B_r(x) \subset U$ for all $n \geq N$.

Conversely, suppose that the latter condition holds. In order to establish that $x = \lim_{n \to \infty} x_n$, let $r > 0$ be given. Take the open set $U = B_r(x)$. By hypothesis, there is an integer N such that $x_n \in U$ for all $n \geq N$. As before, this just means that

$$\|x - x_n\| < r \quad \text{for all} \quad n \geq N.$$

Hence $\lim_{n \to \infty} x_n = x$.   ∎

There is one more fundamental property that we can define in this general context, compactness. However, the reader should be warned that the main theorem about compactness in $\mathbb{R}^n$, the Heine–Borel Theorem (4.4.6), is *not valid* in infinite-dimensional spaces. The correct characterization of compact sets is given by the Borel–Lebesgue Theorem (9.2.3), which we will prove in the context of metric spaces.

7.2.5. DEFINITION. A subset K of a normed vector space V is **compact** if every sequence (x_n) in K has a subsequence (x_{n_i}) that converges to a point in K.

Exercises for Section 7.2

A. If $x = \lim_{n \to \infty} x_n$ and $y = \lim_{n \to \infty} y_n$ in a normed space V and $\alpha = \lim_{n \to \infty} \alpha_n$, show that $x + y = \lim_{n \to \infty} x_n + y_n$ and $\alpha x = \lim_{n \to \infty} \alpha_n x_n$.

B. Show that every convergent sequence in a normed space is a Cauchy sequence.

C. If A is a subset of $(V, \| \cdot \|)$, let $\bar{A}$ denote its closure. Show that if $x \in V$ and $\alpha \in \mathbb{R}$, then $x + \bar{A} = \overline{x + A}$ and $\alpha \bar{A} = \overline{\alpha A}$.

D. Show that if A is an arbitrary subset of a normed space V and U is an open subset, then $A + U = \{a + u : a \in A, u \in U\}$ is open.

E. Prove that if two norms on V have the same unit ball, then the norms are equal.

F. Which of the following sets are open in $C^2[0,1]$? Explain.

   (a) $A = \{f \in C^2[0,1] : f(x) > 0, \|f'\|_\infty < 1, |f''(0)| > 2\}$.
   (b) $B = \{f \in C^2[0,1] : f(1) < 0, f'(1) = 0, f''(1) > 0\}$.
   (c) $C = \{f \in C^2[0,1] : f(x)f'(x) > 0 \text{ for } 0 \leq x \leq 1\}$.
       HINT: Extreme Value Theorem and Intermediate Value Theorem
   (d) $D = \{f \in C^2[0,1] : f(x)f'(x) > 0 \text{ for } 0 < x < 1\}$.
       HINT: Why is this different from the previous example?

G. Prove that a compact subset of a normed vector space is closed and bounded.

H. (a) Prove that a compact subset of a normed vector space is complete.
    (b) Prove that a closed subset of a complete normed vector space is complete.

I. Consider the piecewise linear functions in $C[-1,1]$ given by $f_n(x) = 0$ for $-1 \leq x \leq 0$, $f_n(x) = nx$ for $0 \leq x \leq 1/n$, and $f_n(x) = 1$ for $1/n \leq x \leq 1$.

   (a) Show that $\|f_n - f_m\|_\infty \geq \frac{1}{2}$ if $m \geq 2n$.
   (b) Hence show that no subsequence of $(f_n)_{n=1}^{\infty}$ converges.
   (c) Conclude that the unit ball of $C[-1,1]$ is not compact.
   (d) Show that the unit ball of $C[-1,1]$ is closed and bounded and complete.

J.   Prove that the following are equivalent for a normed vector space $(V, \| \cdot \|)$.

    (1) $(V, \| \cdot \|)$ is complete.

    (2)   Every decreasing sequence of closed balls has a nonempty intersection. Note that the balls need not be concentric.

    (3)   Every decreasing sequence of closed balls with radii $r_i \to 0$ has nonempty intersection.

    HINT: For (1) $\implies$ (2), show that the centres of the balls form a Cauchy sequence.

K.   (a)   Show that if A is a closed subset of a normed vector space and C is a compact subset, then $A + C = \{a + c : a \in A, c \in C\}$ is closed.

    (b)   Is it enough for C to be only closed, or is compactness necessary?

    (c)   If A and C are both compact, show that $A + C$ is compact.

L.   Let $X_n = \{f \in C[0,1] : f(0) = 0, \|f\|_\infty \le 1, \text{ and } f(x) \ge \frac{1}{2} \text{ for } x \ge \frac{1}{n}\}$.

    (a)   Show that X_n is a closed bounded subset of $C[0,1]$.

    (b)   Show that X_{n+1} is a proper subset of X_n for $n \ge 1$, and compute $\bigcap_{n \ge 1} X_n$.

    (c)   Compare this with Cantor's Intersection Theorem (4.4.7). Why does the theorem fail in this context?

M.   Let c_0 be the vector space of all sequences $\mathbf{x} = (x_n)_{n=1}^\infty$ such that $\lim\limits_{n \to \infty} x_n = 0$. Define a norm on c_0 by $\|\mathbf{x}\|_\infty = \sup\limits_{n \ge 1} |x_n|$. Prove that c_0 is complete.   HINT: Let $\mathbf{x}_k = (x_{k,n})_{n=1}^\infty$ be Cauchy.

    (a)   Show that $(x_{k,n})_{k=1}^\infty$ is Cauchy for each $n \ge 1$. Hence define $\mathbf{y} = (y_n)$ by $y_n = \lim\limits_{k \to \infty} x_{k,n}$.

    (b)   Given $\varepsilon > 0$, apply the Cauchy criterion. Then show that there is an integer K such that $|y_n - x_{k,n}| \le \varepsilon$ for all $n \ge 1$ and all $k \ge K$.

    (c)   Conclude that $\mathbf{y}$ belongs to c_0 and that $\lim\limits_{k \to \infty} \mathbf{x}_k = \mathbf{y}$.

N.   Consider the sequence f_n in $C[-1,1]$ from Exercise 7.2.I, but use the $L^1[-1,1]$ norm.

    (a)   Show that f_n is Cauchy in the L^1 norm.

    (b)   Show that f_n converges to $\chi_{(0,1]}$, the characteristic function of $(0,1]$, in the L^1 norm.

    (c)   Show that $\|\chi_{(0,1]} - h\|_1 > 0$ for every h in $C[-1,1]$.

    (d)   Conclude that $C[-1,1]$ is not complete in the L^1 norm.

7.3 Finite-Dimensional Normed Spaces

The key point of the last section is that many normed vector spaces have topologies quite different from $\mathbb{R}^n$ with the Euclidean norm. On the other hand, the point of this section is that a finite-dimensional normed space has the same topology as $\mathbb{R}^n$. A particularly important result is Theorem 7.3.5, which will be very useful in the chapters on approximation.

7.3.1. PROPOSITION. *If $\{v_1, v_2, \ldots, v_n\}$ is a linearly independent set in a normed vector space $(V, \| \cdot \|)$, then there exist positive constants $0 < c < C$ such that for all $\mathbf{a} = (a_1, \ldots, a_n) \in \mathbb{R}^n$, we have*

$$c\|\mathbf{a}\|_2 \le \left\| \sum_{i=1}^n a_i v_i \right\| \le C\|\mathbf{a}\|_2.$$

PROOF. By the triangle inequality and the Schwarz inequality (4.1.1),

$$\Big\| \sum_{i=1}^{n} a_i v_i \Big\| \le \sum_{i=1}^{n} |a_i| \|v_i\| \le \Big(\sum_{i=1}^{n} a_i^2 \Big)^{1/2} \Big(\sum_{i=1}^{n} \|v_i\|^2 \Big)^{1/2} = C\|\mathbf{a}\|_2,$$

where $C = \Big(\sum_{i=1}^{n} \|v_i\|^2 \Big)^{1/2}$.

Define a function N on $\mathbb{R}^n$ by $N(\mathbf{a}) = \big\| \sum_{i=1}^{n} a_i v_i \big\|$. If $\alpha \in \mathbb{R}$, then $N(\alpha \mathbf{a}) = |\alpha| N(\mathbf{a})$ by the homogenity of the norm. Also, N is Lipschitz, and so a continuous function, since

$$|N(\mathbf{x}) - N(\mathbf{y})| = \Big| \Big\| \sum_{i=1}^{n} x_i v_i \Big\| - \Big\| \sum_{i=1}^{n} y_i v_i \Big\| \Big|$$

$$\le \Big\| \sum_{i=1}^{n} (x_i - y_i) v_i \Big\| \le C\|\mathbf{x} - \mathbf{y}\|_2.$$

Let S be $\{\mathbf{a} \in \mathbb{R}^n : \|\mathbf{a}\|_2 = 1\}$, the unit sphere of $\mathbb{R}^n$. Since the set $\{v_1, \dots, v_n\}$ is linearly independent, it follows that $N(\mathbf{x}) > 0$ when $\mathbf{x} \ne 0$. So N never vanishes on the compact set S. By the Extreme Value Theorem (5.4.4), N must achieve its minimum value c at some point on S, whence $c > 0$. So if $\mathbf{a}$ is an arbitrary vector in $\mathbb{R}^n$, we obtain

$$\Big\| \sum_{i=1}^{n} a_i v_i \Big\| = \|\mathbf{a}\|_2 N\Big(\frac{\mathbf{a}}{\|\mathbf{a}\|_2} \Big) \ge c\|\mathbf{a}\|_2. \qquad \blacksquare$$

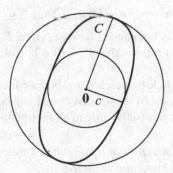

FIG. 7.2 Euclidean balls inside and outside the unit ball of V.

The effect of this result is that every finite-dimensional normed space has the same topology as $(\mathbb{R}^n, \|\cdot\|_2)$, in the sense that they have the "same" convergent sequences, the "same" open sets, and so on. Let us make this more precise. If $\{v_1, v_2, \dots, v_n\}$ is a basis for an n-dimensional normed space V, define a linear transformation from $\mathbb{R}^n$ into V by $T\mathbf{a} = \sum_{i=1}^{n} a_i v_i$. The map T carries $\mathbb{R}^n$ one-to-one and onto V by the definition of a basis. Since every element of V is a unique linear combination of $\{v_1, \dots, v_n\}$, we can define the inverse map by $T^{-1}\big(\sum_{i=1}^{n} a_i v_i \big) = \mathbf{a}$.

7.3.2. COROLLARY. *Suppose that V is an n-dimensional normed space with basis $\{v_1, v_2, \ldots, v_n\}$. Then the maps T and T^{-1} defined previously are both Lipschitz. Further, a set A in V is closed, bounded, open, or compact if and only if $T^{-1}(A)$ is closed, bounded, open, or compact in $\mathbb{R}^n$.*

PROOF. We have for vectors $\mathbf{x}$ and $\mathbf{y}$ in $\mathbb{R}^n$,

$$\|T\mathbf{x} - T\mathbf{y}\| = N(\mathbf{x} - \mathbf{y}) \leq C\|\mathbf{x} - \mathbf{y}\|_2,$$

where C is the constant from Proposition 7.3.1. So T is Lipschitz with constant C. Similarly, if $u = T\mathbf{x}$ and $v = T\mathbf{y}$ are typical vectors in V, then

$$\|T^{-1}u - T^{-1}v\|_2 = \|T^{-1}(T\mathbf{x}) - T^{-1}(T\mathbf{y})\|_2 = \|\mathbf{x} - \mathbf{y}\|_2$$
$$\leq \frac{1}{c}\|T\mathbf{x} - T\mathbf{y}\| = \frac{1}{c}\|u - v\|,$$

where c comes from Proposition 7.3.1. So T^{-1} is Lipschitz with constant $1/c$.

This means that if $\mathbf{x}_k$ is a sequence of vectors in $\mathbb{R}^n$ converging to a point $\mathbf{x}$, then $T\mathbf{x}_k$ converges to $T\mathbf{x}$ because of the continuity of T. And conversely, if v_k is a sequence of vectors in V converging to a vector v, then $T^{-1}v_k$ converges to $T^{-1}v$ in $\mathbb{R}^n$. This just says that there is a direct correspondence between convergent sequences in $\mathbb{R}^n$ and V. Since closed and compact sets are defined in terms of convergent sequences, these sets correspond as well. Open sets are the complements of closed sets, so open sets correspond. If $A \subset \mathbb{R}^n$ is bounded by L, the Lipschitz condition shows that

$$\|T\mathbf{a}\| = \|T\mathbf{a} - T\mathbf{0}\| \leq C\|\mathbf{a}\|_2 \leq CL$$

for every $\mathbf{a}$ in A. So $T(A)$ is bounded. Likewise, since T^{-1} is Lipschitz, if B is a subset of V bounded by L, it follows that $T^{-1}(B)$ is bounded by L/c in $\mathbb{R}^n$. ∎

Notice that we may conclude that closed and bounded sets also correspond. Since the Heine–Borel Theorem (4.4.6) shows that closed and bounded sets are compact in $\mathbb{R}^n$, we can conclude that this is also true in all finite-dimensional normed spaces.

7.3.3. COROLLARY. *A subset of a finite-dimensional normed vector space is compact if and only if it is closed and bounded.*

Another immediate consequence refers to the way a finite-dimensional subspace sits inside an arbitrary normed space. Arbitrary normed spaces are in general not complete. However, finite-dimensional subspaces are, because in $\mathbb{R}^n$ we have the Heine–Borel Theorem (4.4.6).

7.3.4. COROLLARY. *A finite-dimensional subspace of a normed vector space is complete, and in particular, it is closed.*

PROOF. Let V be a normed vector space and let W be an n-dimensional subspace. Let T be a linear invertible map from $\mathbb{R}^n$ onto W as just constructed. Suppose that $(w_k)_{k=1}^\infty$ is a Cauchy sequence in W. Then since T^{-1} is Lipschitz, the sequence $x_k = T^{-1} w_k$ for $k \geq 1$ is Cauchy in $\mathbb{R}^n$. (Check this yourself!) Since $\mathbb{R}^n$ is complete (Theorem 4.2.5), the sequence x_k converges to a vector x. Again by Corollary 7.3.2, we see that w_k must converge to $w = Tx$. Thus W is complete. In particular, all of the limit points of W lie in W, so W is closed. ∎

As an application of these corollaries, we prove the following result, which is fundamental to approximation theory.

7.3.5. THEOREM. *Let* $(V, \|\cdot\|)$ *be a normed vector space, and let* W *be a finite-dimensional subspace of* V. *Then for any* $v \in V$, *there is at least one* closest *point* $w^* \in W$ *to* v. *That is, there is* $w^* \in W$ *such that* $\|v - w^*\| = \inf\{\|v - w\| : w \in W\}$.

PROOF. Notice that the zero vector is in W, and so

$$\inf\{\|v - w\| : w \in W\} \leq \|v - 0\| = \|v\|.$$

Let $M = \|v\|$. If w satisfies $\|v - w\| \leq \|v\|$, then

$$\|w\| \leq \|w - v\| + \|v\| \leq M + M = 2M.$$

Thus if we define $K := \{w \in W : \|w\| \leq 2M\}$, then

$$\inf\{\|v - w\| : w \subset K\} = \inf\{\|v - w\| : w \in W\}.$$

We will show that K is compact. Clearly K is bounded by $2M$. The norm function is Lipschitz, and hence continuous. Thus any convergent sequence of vectors in K will converge to a vector of norm at most $2M$. And since W is complete, this limit also lies in W, whence the limit lies in K. This shows that K is closed and bounded. By our corollary, it follows that K is compact.

Now define a function on K by $f(w) = \|v - w\|$. This function has Lipschitz constant 1, since

$$|f(w) - f(x)| = \big| \|v - w\| - \|v - x\| \big| \leq \|w - x\|.$$

By the Extreme Value Theorem (5.4.4), f achieves its minimum at some point w^* in K. This is a closest point to v in W. ∎

It is not true that w^* is unique; for example, see Exercise 7.3.F.

Exercises for Section 7.3

A.   Let V be a finite-dimensional vector space with two norms $\|\cdot\|$ and $\|\|\cdot\|\|$. Show that there are constants $0 < a < A$ such that $a\|v\| \leq \|\|v\|\| \leq A\|v\|$ for all $v \in V$.

B.   Let T be the invertible linear map from Corollary 7.3.2.

(a) Use the Lipschitz property of T and T^{-1} to show that $T(B_r(x))$ contains a ball about Tx in V, and that $T^{-1}(B_r(Tx))$ contains a ball about x in $\mathbb{R}^n$.

(b) Hence show directly that U is open if and only if $T(U)$ is open.

C. Write out a careful proof of Corollary 7.3.3.

D. Suppose that $(w_k)_{k=1}^\infty$ is a Cauchy sequence in a normed space W. If $T : W \to V$ is a Lipschitz map into another normed space V, show that the sequence $v_k = Tw_k$ for $k \geq 1$ is Cauchy in V.

E. Show that for each integer n and each function f in $C[a,b]$, there is a polynomial of degree at most n that is closest to f in the max norm on $C[a,b]$.

F. Let $\mathbb{R}^n$ have the max norm $\|\mathbf{x}\|_\infty = \max\{|x_i| : 1 \leq i \leq n\}$. Let K be the unit ball of V and let $v = (2, 0, \ldots, 0)$. Find all closest points to v in K.

7.4 Inner Product Spaces

In studying $\mathbb{R}^n$, we constructed the Euclidean norm using the dot product. An inner product on a vector space is a generalization of the dot product. It is one of the most important sources of norms, and the norms obtained from inner products are particularly tractable. For example, the L^2 norm on $C[a,b]$ arises in this way.

7.4.1. DEFINITION. An **inner product** on a vector space V is a function $\langle x, y \rangle$ on pairs (x, y) of vectors in $V \times V$ taking values in $\mathbb{R}$ satisfying the following properties:

(1) (positive definiteness) $\langle x, x \rangle \geq 0$ for all $x \in V$ and $\langle x, x \rangle = 0$ only if $x = 0$.

(2) (symmetry) $\langle x, y \rangle = \langle y, x \rangle$ for all $x, y \in V$.

(3) (bilinearity) For all $x, y, z \in V$ and scalars $\alpha, \beta \in \mathbb{R}$,

$$\langle \alpha x + \beta y, z \rangle = \alpha \langle x, z \rangle + \beta \langle y, z \rangle.$$

An **inner product space** is a vector space with an inner product.

Given an inner product space, the following definition provides a norm:

$$\|x\| = \langle x, x \rangle^{1/2}.$$

The first two properties of a norm are easy to verify; the triangle inequality is proved as Corollary 7.4.6.

The dot product on $\mathbb{R}^n$ is an inner product and the norm obtained from it is the usual Euclidean norm.

The bilinearity condition is just given as linearity in the first variable. But as the term suggests, it really means a twofold linearity because combining it with symmetry yields linearity in the second variable as well. For x, y in V and scalars α, β in $\mathbb{R}$,

$$\langle z, \alpha x + \beta y \rangle = \langle \alpha x + \beta y, z \rangle = \alpha \langle x, z \rangle + \beta \langle y, z \rangle = \alpha \langle z, x \rangle + \beta \langle z, y \rangle.$$

7.4.2. EXAMPLE. The space $C[a,b]$ can be given an inner product

$$\langle f, g \rangle = \int_a^b f(x)g(x)\,dx.$$

This gives rise to the L^2 norm, which we defined in Example 7.1.5. Positive definiteness of this norm was established in Example 7.1.5 for arbitrary p, including $p = 2$. The other two properties follow from the linearity of the integral and are left as an exercise for the reader.

7.4.3. EXAMPLE. The space $\mathbb{R}^n$ can be given other inner product structures by weighting the vectors by a matrix $A = \left[a_{ij}\right]$ by

$$\langle \mathbf{x}, \mathbf{y} \rangle_A = \langle A\mathbf{x}, \mathbf{y} \rangle = \sum_{i=1}^n \sum_{j=1}^n a_{ij} x_i y_j.$$

This is easily seen to be bilinear because A is linear and the standard inner product is bilinear. To be symmetric, we must require that $a_{ij} = a_{ji}$, so A is a symmetric matrix. Moreover, we need an additional condition to ensure that the inner product is positive definite. It turns out that the necessary condition is that the eigenvalues of A all be strictly positive. The proof of this is an important result from linear algebra known as the Spectral Theorem for Symmetric Matrices or as the Principal Axis Theorem.

For the purposes of this example, consider the 2×2 matrix $A = \left[\begin{smallmatrix} 3 & 1 \\ 1 & 2 \end{smallmatrix}\right]$. We noted that since A is symmetric, the inner product $\langle \cdot, \cdot \rangle_A$ is symmetric and bilinear. Let us establish directly that it is positive definite. Take a vector $\mathbf{x} = (x, y)$:

$$\langle \mathbf{x}, \mathbf{x} \rangle_A = 3x^2 + xy + yx + 2y^2 = 2x^2 + (x+y)^2 + y^2$$

From this identity, it is clear that $\langle \mathbf{x}, \mathbf{x} \rangle_A \geq 0$. Moreover, equality requires that x, y and $x+y$ all be 0, whence $\mathbf{x} = \mathbf{0}$. So it is positive definite.

It is a fundamental fact that every inner product space satisfies the Schwarz inequality. Our proof for $\mathbb{R}^n$ was special, using the specific formula for the dot product. We now show that it follows just from the basic properties of an inner product.

7.4.4. CAUCHY–SCHWARZ INEQUALITY.
For all vectors x, y in an inner product space V,

$$|\langle x, y \rangle| \leq \|x\|\,\|y\|.$$

Equality holds if and only if x and y are collinear.

PROOF. If either x or y is $\mathbf{0}$, both sides of the inequality are 0. Equality holds here, and these vectors are collinear. So we may assume that x and y are nonzero.
Apply the positive definite property to the vector $x - ty$ for $t \in \mathbb{R}$:

$$0 \le \langle x - ty, x - ty \rangle = \langle x, x - ty \rangle - t \langle y, x - ty \rangle$$
$$= \langle x, x \rangle - t \langle x, y \rangle - t \langle x, y \rangle + t^2 \langle y, y \rangle = \|x\|^2 - 2t \langle x, y \rangle + t^2 \|y\|^2.$$

Substitute $t = \langle x, y \rangle / \|y\|^2$ to obtain

$$0 \le \|x\|^2 - \frac{\langle x, y \rangle^2}{\|y\|^2}.$$

Hence $\langle x, y \rangle^2 \le \|x\|^2 \|y\|^2$, establishing the inequality.

For equality to hold, the vector $x - ty$ must have norm 0. By the positive definite property, this means that $x = ty$, and so they are collinear. Conversely, if $x = ty$, then $\|x\|^2 = \langle ty, ty \rangle = t^2 \|y\|^2$ and

$$|\langle x, y \rangle| = |t| \langle y, y \rangle = \sqrt{t^2 \langle y, y \rangle} \sqrt{\langle y, y \rangle} = \|x\| \, \|y\|.$$

∎

7.4.5. COROLLARY. *For $f, g \in C[a, b]$, we have*

$$\left| \int_a^b f(x) g(x) \, dx \right| \le \left(\int_a^b f(x)^2 \, dx \right)^{1/2} \left(\int_a^b g(x)^2 \, dx \right)^{1/2}.$$

As for $\mathbb{R}^n$, the triangle inequality in an immediate consequence. In particular, the L^2 norms on $C[a, b]$ are indeed norms.

7.4.6. COROLLARY. *An inner product space V satisfies the triangle inequality*

$$\|x + y\| \le \|x\| + \|y\| \quad \text{for all} \quad x, y \in V.$$

Moreover, if equality occurs, then x and y are collinear.

PROOF. This proof is same as for $\mathbb{R}^n$. Using the Cauchy–Schwarz inequality,

$$\|x + y\|^2 = \langle x + y, x + y \rangle = \langle x, x \rangle + 2 \langle x, y \rangle + \langle y, y \rangle$$
$$\le \|x\|^2 + 2\|x\| \, \|y\| + \|y\|^2 = (\|x\| + \|y\|)^2.$$

Moreover, equality occurs only if $\langle x, y \rangle = \|x\| \, \|y\|$, which by the Cauchy–Schwarz inequality can happen only when x and y are collinear. ∎

Another fundamental consequence of the Cauchy–Schwarz inequality is that the inner product is continuous with respect to the induced norm. We leave the proof as an exercise.

7.4.7. COROLLARY. *Let V be an inner product space with induced norm $\| \cdot \|$. Then the inner product is continuous (i.e., if x_n converges to x and y_n converges to y, then $\langle x_n, y_n \rangle$ converges to $\langle x, y \rangle$).*

Exercises for Section 7.4

A. Let $A = \begin{bmatrix} 3 & 1 & 2 \\ 1 & 2 & 1 \\ 2 & 1 & 4 \end{bmatrix}$. Show that the form $\langle \cdot, \cdot \rangle_A$ is positive definite on $\mathbb{R}^3$.

B. Show that every inner product space satisfies the parallelogram law:

$$\|x+y\|^2 + \|x-y\|^2 = 2\|x\|^2 + 2\|y\|^2 \quad \text{for all} \quad x, y \in V.$$

C. Minimize the quantity $\|x\|^2 - 2t\langle x, y \rangle + t^2\|y\|^2$ over $t \in \mathbb{R}$. You will see why we chose t as we did in the proof of the Cauchy–Schwarz inequality.

D. Show that x and y can be collinear yet the triangle inequality is still a strict inequality.

E. Prove Corollary 7.4.7.

F. Let $w(x)$ be a strictly positive continuous function on $[a,b]$. Define a form on $C[a,b]$ by the formula $\langle f, g \rangle_w = \int_a^b f(x)g(x)w(x)\,dx$ for $f, g \in C[a,b]$. Show that this is an inner product.

G. A normed vector space V is **strictly convex** if $\|u\| = \|v\| = \|(u+v)/2\| = 1$ for vectors $u, v \in V$ implies that $u - v$.

    (a) Show that an inner product space is always strictly convex.
    (b) Show that $\mathbb{R}^2$ with the norm $\|(x,y)\|_\infty = \max\{|x|, |y|\}$ is not strictly convex.

H. For strictly convex normed vector spaces, we have better approximation results.

    (a) Show that if W is a finite-dimensional subspace of a strictly convex normed vector space V, then each point $v \in V$ has a *unique* closest point in W.
    (b) Show that $\mathbb{R}^n$ with the standard Euclidean norm is strictly convex.
    (c) Show that $\mathbb{R}^2$ with the norm $\|(x,y)\|_1 = |x| + |y|$ is not strictly convex.
    (d) Find a subspace W of $V = (\mathbb{R}^2, \|\cdot\|_1)$ such that every point in V that is not in W has more than one closest point in W.

I. Let T be an $n \times n$ matrix. Define a form on $\mathbb{R}^n$ by $\langle \mathbf{x}, \mathbf{y} \rangle_T = \langle T\mathbf{x}, T\mathbf{y} \rangle$ for $\mathbf{x}, \mathbf{y} \in \mathbb{R}^n$. Show that this is an inner product if and only if T is invertible.

J. Let $\text{Tr}(A) = \sum_{i=1}^n a_{ii}$ denote the trace on the space $\mathscr{M}_n$ of all $n \times n$ matrices. Show that there is an inner product on $\mathscr{M}_n$ given by $\langle A, B \rangle = \text{Tr}(AB^t) = \sum_{i=1}^n \sum_{j=1}^n a_{ij}b_{ij}$. The norm $\|A\|_2 = \langle A, A \rangle^{1/2}$ is called the **Hilbert–Schmidt norm**.

K. Let $A = [a_{ij}]$ be an $n \times n$ matrix, and let $\mathbf{x} = (x_1, \ldots, x_n)$ be a vector in $\mathbb{R}^n$. Using the notation of Exercise J, show that $\|A\mathbf{x}\| \leq \|A\|_2 \|\mathbf{x}\|$.   HINT: Compute $\|A\mathbf{x}\|^2$ using coordinates, and apply the Cauchy–Schwarz inequality.

L. Let T be an invertible $n \times n$ matrix. Prove that a sequence of vectors $\mathbf{x}_k$ in $\mathbb{R}^n$ converges to a vector $\mathbf{x}$ in the usual Euclidean norm if and only if it converges to $\mathbf{x}$ in the norm of Exercise I, namely $\|\mathbf{x}\|_T := \|T\mathbf{x}\|$.
HINT: First use Exercise K to show that $\|\mathbf{x}\|/\|T^{-1}\|_2 \leq \|T\mathbf{x}\| \leq \|T\|_2 \|\mathbf{x}\|$.

M. Given an inner product space V, define a function on $V \setminus \{0\}$ by $Rx = x/\|x\|^2$. This R is called *inversion* with respect to the unit sphere $\{x \in V : \|x\| = 1\}$.

    (a) Prove that $\|Rx - Ry\| = \|x - y\|/(\|x\|\,\|y\|)$.
    (b) Hence show that the inversion R is continuous.
    (c) For all $w, x, y, z \in V$, show that $\|w - y\|\,\|x - z\| \leq \|w - x\|\,\|y - z\| + \|w - z\|\,\|x - y\|$.
    HINT: Reduce to the case $w = 0$, and reinterpret the inequality using inversion.

7.5 Finite Orthonormal Sets

7.5.1. DEFINITION. Two vectors x and y are called **orthogonal** if $\langle x, y \rangle = 0$. A collection of vectors $\{e_n : n \in S\}$ in V is called **orthonormal** if $\|e_n\| = 1$ for all $n \in S$ and $\langle e_n, e_m \rangle = 0$ for $n \neq m \in S$.

The space $\mathbb{R}^n$ has many orthonormal bases, including the canonical one given by $\mathbf{e}_1 = (1, 0, \ldots, 0), \ldots, \mathbf{e}_n = (0, \ldots, 0, 1)$. In the next section, we look carefully at an infinite orthonormal set based on trigonometric functions.

We have a pair of simple observations that prove to be useful.

7.5.2. LEMMA. *Let $\{e_1, e_2, \ldots, e_n\}$ be a finite orthonormal set and let $\alpha_1, \ldots, \alpha_n$ be real numbers. If w is $\sum\limits_{i=1}^{n} \alpha_i e_i$, then for each j, $\langle w, e_j \rangle = \alpha_j$.*

For any vector v, $v - \sum\limits_{i=1}^{n} \langle v, e_i \rangle e_i$ is orthogonal to each e_j.

PROOF. For j between 1 and n, $\langle e_j, w \rangle = \sum\limits_{i=1}^{n} \alpha_i \langle e_j, e_i \rangle$. Observe that $\langle e_j, e_i \rangle$ is always zero unless $i = j$, when it is 1. Thus, the summation reduces to α_j, as required.

Letting $x = v - \sum_{i=1}^{n} \langle v, e_i \rangle e_i$, observe that $\langle x, e_j \rangle = \langle v, e_j \rangle - \langle v, e_j \rangle = 0$, by linearity and the previous paragraph. ∎

7.5.3. COROLLARY. *Let $\{e_1, e_2, \ldots, e_n\}$ be a finite orthonormal set. If x is in $\mathrm{span}\{e_1, e_2, \ldots, e_n\}$, then x can be written uniquely as $\sum\limits_{i=1}^{n} \langle x, e_i \rangle e_i$.*

Further, all orthonormal sets are linearly independent.

PROOF. If $x = \alpha_1 e_1 + \cdots + \alpha_n e_n$ then by Lemma 7.5.2, $\langle x, e_i \rangle = \alpha_i$. So the coefficients of x are uniquely determined.

If an orthonormal set is not linearly independent, then there is some finite linear combination that is zero, say $\alpha_1 e_1 + \cdots + \alpha_n e_n = \mathbf{0}$ with not all α_i zero. By the previous paragraph with $x = \mathbf{0}$, all the α_i must be zero, a contradiction. ∎

7.5.4. COROLLARY. *In a finite-dimensional vector space, an orthonormal set is a basis if and only if it is maximal with respect to being an orthonormal set.*

PROOF. Let V be the vector space. Recall that a linearly independent set has at most as many elements as a basis. In particular, an orthonormal set in V must be finite.

Suppose $B = \{e_1, \ldots, e_n\}$ is an orthonormal set and a basis. For any nonzero vector $v \in V$, there are real numbers $\alpha_1, \ldots, \alpha_n$, not all zero, such that $v = \alpha_1 e_1 + \cdots + \alpha_n e_n$. Applying Lemma 7.5.2, v is not orthogonal to at least one e_j. So B is maximal as an orthonormal set.

Conversely, if v is a vector not in the span of an orthonormal set $B = \{e_1, \ldots, e_n\}$, then by Lemma 7.5.2,

$$w = v - \sum_{i=1}^{n} \langle v, e_i \rangle e_i$$

is a nonzero vector orthogonal to each element of B. Thus, we can enlarge B to the orthonormal set $B \cup \{w/\|w\|\}$, i.e., B is not maximal. ∎

7.5.5. THE GRAM–SCHMIDT PROCESS.

The idea used in the proof of Lemma 7.5.2 can be turned into an algorithm for producing orthonormal sets. We outline this algorithm, leaving the proofs of the various claims as an exercise.

We start with a linearly independent set $x_1, \ldots, x_n$ and inductively define an orthonormal basis $f_1, \ldots, f_n$. Note that $x_1 \neq \mathbf{0}$ and let $f_1 = x_1/\|x_1\|$.

Having computed $f_1, \ldots, f_k$ where $k < n$, next compute

$$y_{k+1} = x_{k+1} - \sum_{i=1}^{l} \langle x_{k+1}, f_i \rangle f_i.$$

Linear independence of $x_1, \ldots, x_{k+1}$ ensures that $y_{k+1} \neq \mathbf{0}$. We define $f_{k+1} = y_{k+1}/\|y_{k+1}\|$. Then $\{f_1, \ldots, f_n\}$ is an orthonormal basis for span$\{x_1, \ldots, x_n\}$.

Applying this algorithm to any basis for a finite-dimensional inner product space produces an orthonormal basis of the same size. So we have the following corollary.

7.5.6. COROLLARY.
An inner product space of dimension n has an orthonormal basis with n elements.

The major theorem of this section, the Projection Theorem, deals with finite orthonormal sets in general inner product spaces. We start with a lemma about computing inner products for such a set.

7.5.7. LEMMA.
Let $\{e_1, \ldots, e_n\}$ be an orthonormal set in an inner product space V. If $x = \sum_{j=1}^{n} \alpha_j e_j$ and $y \in V$ satisfies $\langle y, e_j \rangle = \beta_j$ for each e_j, then

$$\langle x, y \rangle = \sum_{j=1}^{n} \alpha_j \beta_j.$$

In particular, $\|x\|^2 = \sum_{j=1}^{n} \alpha_j^2.$

PROOF. Compute

$$\langle x, y \rangle = \left\langle \sum_{j=1}^{n} \alpha_j e_j, y \right\rangle = \sum_{j=1}^{n} \alpha_j \langle e_j, y \rangle = \sum_{j=1}^{n} \alpha_j \beta_j. \qquad \blacksquare$$

This lemma suffices for us to understand finite-dimensional inner product spaces. In particular, we see that every inner product space of dimension n behaves exactly like $\mathbb{R}^n$ with the dot product, once we coordinatize it using an orthonormal basis.

7.5.8. COROLLARY. *If V is an inner product space of finite dimension n, then it has an orthonormal basis $\{e_i : 1 \le i \le n\}$ and the inner product is given by*

$$\left\langle \sum_{i=1}^{n} \alpha_i e_i, \sum_{j=1}^{n} \beta_j e_j \right\rangle = \sum_{i=1}^{n} \alpha_i \beta_i.$$

PROOF. By definition of dimension, V has a basis consisting of n linearly independent vectors. Apply the Gram–Schmidt process (7.5.5) to this basis to obtain an orthonormal basis spanning V. Now Lemma 7.5.7 provides the formula for inner product. ∎

7.5.9. DEFINITION. A **projection** on an inner product space V is a linear map $P : V \to V$ such that $P^2 = P$. In addition, we say that P is an **orthogonal projection** if $\ker P = \{v \in V : Pv = 0\}$ is orthogonal to $\operatorname{ran} P = PV$.

The identity map and the zero map are projections in any vector space, called the **trivial projections**. A nontrivial projection, P, in $\mathbb{R}^2$ is determined by two lines through the origin, call them R and K. Given a point p in $\mathbb{R}^2$, there is a line K_p parallel to K through p. The image of p under the projection is the intersection of K_p and R. Clearly, $\operatorname{ran} P$ is the line R and if $p \in R$, then $Pp = p$, so $P^2 = P$. Since $\ker P$ is the line K, P is an orthogonal projection exactly when the lines R and K are orthogonal. This description, using subspaces instead of lines, can be extended to all projections in $\mathbb{R}^n$.

7.5.10. PROPOSITION. *If P is a projection on a normed vector space V, then*

(1) $\ker P = \operatorname{ran}(I - P)$.

If, in addition, V is an inner product space and P is an orthogonal projection, then

(2) *For all $x \in V$, $\|x\|^2 = \|Px\|^2 + \|(I - P)x\|^2$.*
(3) *P is uniquely determined by its range.*

PROOF. That $\ker P \subset \operatorname{ran}(I - P)$ follows from $Px = 0$ if and only if $(I - P)x = x$. Conversely, $x = (I - P)y$ implies that $Px = (P - P^2)y = 0$.

If P is an orthogonal projection, the vectors Px and $(I - P)x$ are orthogonal. This gives (2) above. We leave the proof of (3) as an exercise. ∎

7.5.11. PROJECTION THEOREM.

Let M be a finite-dimensional subspace of an inner product space V and P the orthogonal projection with $\operatorname{ran} P = M$. For all $y \in V$ and all $x \in M$,

$$\|y - x\|^2 = \|y - Py\|^2 + \|Py - x\|^2. \tag{7.5.12}$$

In particular, Py is the closest vector in M to y.

If $\{e_1, \ldots, e_n\}$ is an orthonormal basis for M, then $Py = \sum_{j=1}^{n} \langle y, e_j \rangle e_j$ for each $y \in V$. Further,

$$\|y\|^2 \geq \sum_{j=1}^{n} \langle y, e_j \rangle^2. \tag{7.5.13}$$

The first part of the theorem shows that the orthogonal projection gives the best approximation to a vector in a finite-dimensional subspace. The second shows that we can use finite orthonormal sets to compute orthogonal projections. Figure 7.3 illustrates the theorem for M a plane in $\mathbb{R}^3$.

PROOF. We handle the second part first. Let $Qy = \langle y, e_1 \rangle e_1 + \cdots + \langle y, e_n \rangle e_n$. We leave it to the reader to verify that Q is linear. By definition, Q maps V into M. A typical vector in M is expressed as $x = \sum_{j=1}^{n} \alpha_j e_j$. By Lemma 7.5.7,

$$Qx = \sum_{j=1}^{n} \langle x, e_j \rangle e_j = \sum_{j=1}^{n} \alpha_j e_j = x.$$

Hence for any $y \in V$, $Q^2 y = Q(Qy) = Qy$, since $Qy \in M$. Therefore, Q is a projection of V onto M. Since Qy belongs to M, Lemma 7.5.7 shows that $\|Qy\|^2 = \sum_{j=1}^{n} \beta_j^2$, where $\beta_j = \langle y, e_j \rangle$. By Proposition 7.5.10 (2), $\|y\|^2 \geq \|Qy\|^2$; so (7.5.13) follows.

To see that Q is the orthogonal projection onto $M = \operatorname{ran} Q$, suppose that $x \in M$ and $y \in \ker Q$. From the definition of Q, we see that $\langle y, e_j \rangle = 0$ for each j. By Lemma 7.5.7,

$$\langle x, y \rangle = \sum_{j=1}^{n} \langle x, e_j \rangle \langle y, e_j \rangle = 0.$$

So $\ker Q$ is orthogonal to $\operatorname{ran} Q$, showing that $Q = P$.

To establish (7.5.12), consider $x \in M$ and $y \in V$ as Lemma 7.5.7 and compute

$$\|x - y\|^2 = \langle x - y, x - y \rangle = \langle x, x \rangle - 2\langle x, y \rangle + \langle y, y \rangle$$

$$= \sum_{j=1}^{n} \alpha_j^2 - 2 \sum_{j=1}^{n} \alpha_j \beta_j + \left(\sum_{j=1}^{n} \beta_j^2 - \sum_{j=1}^{n} \beta_j^2 \right) + \|y\|^2$$

$$= \sum_{j=1}^{n} (\alpha_j - \beta_j)^2 - \|Py\|^2 + \|y\|^2 = \|x - Py\|^2 + \|y\|^2 - \|Py\|^2.$$

If $x = Py$, then $\|x - Py\|^2 = 0$ and the equation becomes $\|Py - y\|^2 = \|y\|^2 - \|Py\|^2$.
Substituting this into the preceding equation yields (7.5.12). ∎

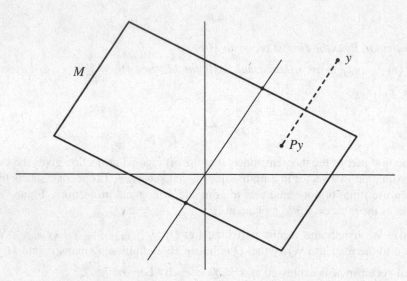

FIG. 7.3 The projection of a point.

Exercises for Section 7.5

A.   **The Gram–Schmidt process.** We use the notation of 7.5.5.

(a) Show, by induction on k, that $f_1, \dots, f_k$ has the same span as $x_1, \dots, x_k$ and that $y_{k+1} \neq \mathbf{0}$.

(b) Show that $f_1, \dots, f_n$ is an orthonormal set.

(c) Hence conclude that the Gram–Schmidt process produces an orthonormal basis for span$\{x_1, \dots, x_n\}$.

B.   Modify the Gram–Schmidt process if the initial list of vectors is not linearly independent. Make sure your process works correctly if $x_1 = \mathbf{0}$.

C.   Show that if $\{x_k : k \geq 1\}$ is a countable set of vectors in an inner product space, then the Gram–Schmidt process produces an orthonormal set $\{f_k : k \in S\}$ for $S \subset \mathbb{N}$ such that $k \notin S$ if and only if $x_k \in \text{span}\{x_1, \dots, x_{k-1}\}$, and for each $n \in \mathbb{N}$, span$\{f_k : k \in S, k \leq n\} = \text{span}\{x_k | 1 \leq k \leq n\}$.

D.   Find an orthonormal basis for the $n \times n$ matrices using the inner product of Exercise 7.4.J.

7.6 Fourier Series

To motivate infinite orthonormal sets, it helps to have an interesting example. We use Fourier series, which have many applications. Some of these applications appear in Chapters 13 and 14. To do these applications, we need to work with piecewise continuous functions.

Let $PC[-\pi,\pi]$ be the vector space of piecewise continuous functions f on $[-\pi,\pi]$ with the usual operations. From Definition 5.2.3, there is a partition of $[-\pi,\pi]$ such that f is continuous on each subinterval and has one-sided limits at the nodes. The Extreme Value Theorem (5.4.4) shows that f is bounded on each subinterval and hence on $[-\pi,\pi]$. Thus, f and $|f|$ are Riemann integrable. The product of two piecewise continuous functions is also piecewise continuous and thus is integrable. Thus, we have a well-defined inner product on $PC[-\pi,\pi]$ given by

$$\langle f,g \rangle = \frac{1}{2\pi} \int_{-\pi}^{\pi} f(\theta)g(\theta)\,d\theta.$$

Dividing by 2π gives the constant function 1 unit length. The norm is given by

$$\|f\|_2 := \langle f,f \rangle^{1/2} = \left(\frac{1}{2\pi} \int_{-\pi}^{\pi} |f(\theta)|^2\,d\theta \right)^{1/2}.$$

This inner product and norm are the same, except for the constants, as the L^2 inner product and norm defined in Examples 7.1.5 and 7.4.2 for $C[a,b]$. We will always use these normalizations for $C[-\pi,\pi]$ or $PC[-\pi,\pi]$.

7.6.1. PROPOSITION. *The functions* $\{1, \sqrt{2}\cos n\theta, \sqrt{2}\sin n\theta : n \geq 1\}$ *form an orthonormal set in* $PC[-\pi,\pi]$ *with this inner product.*

PROOF. Starting with the cosines for $n \geq m \geq 1$, we have

$$\langle \sqrt{2}\cos n\theta, \sqrt{2}\cos m\theta \rangle - \frac{1}{\pi} \int_{-\pi}^{\pi} \cos n\theta \cos m\theta\,dt$$

$$= \frac{1}{2\pi} \int_{-\pi}^{\pi} \cos(m+n)\theta + \cos(m-n)\theta\,dt,$$

where we have used the identity $2\cos A \cos B = \cos(A+B) + \cos(A-B)$. If $n > m$, then both $m+n$ and $m-n$ are not zero, and the integral is

$$\langle \sqrt{2}\cos n\theta, \sqrt{2}\cos m\theta \rangle = \frac{1}{2\pi} \left(\frac{\sin(m+n)\theta}{m+n} + \frac{\sin(m-n)\theta}{m-n} \right) \Big|_{-\pi}^{\pi} = 0.$$

If $n = m$, then

$$\langle \sqrt{2}\cos n\theta, \sqrt{2}\cos n\theta \rangle = \frac{1}{2\pi} \left(\frac{\sin 2n\theta}{2n} + \theta \right) \Big|_{-\pi}^{\pi} = 1.$$

The other cases are similar and are left to the reader. ∎

A **trigonometric polynomial** is a finite sum

$$f(\theta) = A_0 + \sum_{k=1}^{N} A_k \cos k\theta + B_k \sin k\theta.$$

We say that f as above has degree N if A_N or B_N is nonzero. We use $\mathbb{TP}$ for the set of all such functions and $\mathbb{TP}_N$ for the set of trigonometric polynomials of degree at most N. Applying Lemma 7.5.2 and adjusting for the constants, we obtain

$$\langle f, 1 \rangle = A_0, \qquad \langle f, \cos n\theta \rangle = \frac{A_n}{2}, \qquad \langle f, \sin n\theta \rangle = \frac{B_n}{2}.$$

7.6.2. DEFINITION. Denote the **Fourier series** of a piecewise continuous function $f : [-\pi, \pi] \to \mathbb{R}$ to be

$$f \sim A_0 + \sum_{n=1}^{\infty} A_n \cos n\theta + B_n \sin n\theta,$$

where $A_0 = \dfrac{1}{2\pi} \displaystyle\int_{-\pi}^{\pi} f(t)\,dt$, and for $n \geq 1$,

$$A_n = \frac{1}{\pi} \int_{-\pi}^{\pi} f(t) \cos nt\,dt \qquad \text{and} \qquad B_n = \frac{1}{\pi} \int_{-\pi}^{\pi} f(t) \sin nt\,dt.$$

The sequences $(A_n)_{n\geq 0}$ and $(B_n)_{n\geq 1}$ are the **Fourier coefficients** of f.

This definition makes sense even if f is only Riemann integrable.

By construction, the Fourier series of a trigonometric polynomial is itself. It is often possible to compute Fourier series exactly.

7.6.3. EXAMPLE. Consider the Fourier series of the function $f(\theta) = |\theta|$ for $-\pi \leq \theta \leq \pi$. First note that f is even, so $B_n = 0$ for all n (Exercise 7.6.I). For A_n, we compute

$$A_0 = \frac{1}{2\pi} \int_{-\pi}^{\pi} |t|\,dt = \frac{1}{\pi} \int_0^{\pi} t\,dt = \frac{\pi}{2},$$

and, using integration by parts,

$$A_n = \frac{1}{\pi} \int_{-\pi}^{\pi} |t| \cos nt\,dt = \frac{2}{\pi} \int_0^{\pi} t \cos nt\,dt$$

$$= \frac{2t}{\pi} \frac{\sin nt}{n} \bigg|_0^{\pi} - \frac{2}{\pi} \int_0^{\pi} 1 \frac{\sin nt}{n}\,dt = 0 + \frac{2\cos nt}{\pi n^2} \bigg|_0^{\pi}$$

$$= \frac{2}{\pi n^2}\left((-1)^n - 1 \right) = \begin{cases} 0 & \text{if } n \text{ is even,} \\ -\dfrac{4}{\pi n^2} & \text{if } n \text{ is odd.} \end{cases}$$

Thus the Fourier series is

$$|\theta| \sim \frac{\pi}{2} - \frac{4}{\pi} \sum_{k=0}^{\infty} \frac{\cos(2k+1)\theta}{(2k+1)^2}.$$

Even if a function f is continuous, it is not immediately evident that the Fourier series of f will converge to f. This is a serious problem, which required a lot of careful thought and helped force mathematicians to adopt the careful definitions and concern with proofs that drive this book. For one thing, there is more than one natural notion of convergence in $C[-\pi, \pi]$. In the next section, we look at complete inner product spaces, where the norm arises from an inner product. We will return to this question after discussing more notions of convergence in Chapter 8.

To finish this section, we show that the sequence of Fourier coefficients is bounded. Since piecewise continuous functions on a finite interval must be bounded,

$$\|f\|_1 := \frac{1}{2\pi} \int_{-\pi}^{\pi} |f(\theta)| d\theta < \infty.$$

This is the L^1 norm of Example 7.1.5 divided by 2π. Notice that $\|f\|_1 \leq \|f\|_\infty$.

If a function is not bounded, then we can still compute an improper integral, taking limits around each infinite discontinuity. All functions with $\|f\|_1 < \infty$ using either the Riemann integral or improper Riemann integrals are called **absolutely integrable functions**.

7.6.4. PROPOSITION. *If f is absolutely integrable on $[-\pi, \pi]$, then*

$$|A_0| \leq \|f\|_1, \qquad |A_n| \leq 2\|f\|_1, \quad and \quad |B_n| \leq 2\|f\|_1 \quad for \quad n \geq 1.$$

In particular, if $f \in PC[-\pi, \pi]$, then its Fourier coefficients are bounded.

PROOF. This is a routine integration. For example,

$$|B_n| \leq \frac{1}{\pi} \int_{-\pi}^{\pi} |f(\theta) \sin n\theta| d\theta \leq \frac{1}{\pi} \int_{-\pi}^{\pi} |f(\theta)| d\theta = 2\|f\|_1.$$

Moreover, it is evident that if f is bounded,

$$\|f\|_1 = \frac{1}{2\pi} \int_{-\pi}^{\pi} |f(\theta)| d\theta \leq \frac{1}{2\pi} \int_{-\pi}^{\pi} \|f\|_\infty d\theta = \|f\|_\infty.$$

So continuous functions are absolutely integrable, and thus the Fourier coefficients are bounded by $2\|f\|_\infty$. ∎

Exercises for Section 7.6

A. Complete the proof of Proposition 7.6.1.

B. Show that $f_n = \sin(n\pi x)$ for $n \geq 1$ forms an orthonormal set in $C[0,1]$ with respect to the $L^2[0,1]$ norm.

C. (a) Find the Fourier series for $\cos^3(\theta)$, which will be a trigonometric polynomial.
(b) Use trig identities to verify that $\cos^3(\theta)$ equals your answer to part (a).

D. If $f \in C[-\pi, \pi]$ has the Fourier series $f \sim A_0 + \sum_{n=1}^{\infty} A_n \cos n\theta + B_n \sin n\theta$, show that
$$A_0^2 + \frac{1}{2} \sum_{n=1}^{\infty} |A_n|^2 + |B_n|^2 \leq \frac{1}{2\pi} \int_{-\pi}^{\pi} |f(x)|^2 dx. \qquad \text{HINT: Consider the finite sums.}$$
NOTE: We will show in Example 13.5.5 that this is an equality.

E. Find the Fourier series of the following functions:

(a) $f(\theta) = |\sin\theta|$

(b) $f(\theta) = \theta$ for $-\pi \le \theta \le \pi$

F. (a) Suppose that $f(\theta)$ is a 2π-periodic function with known Fourier series. Let α be a real number, and let $g(\theta) = f(\theta - \alpha)$ for $\theta \in \mathbb{R}$. Find the Fourier series of g.

(b) Combine part (a) with Exercise E(a) to find the Fourier series of $|\cos\theta|$.

G. (a) Let $f(\theta)$ be a 2π-periodic function with a given Fourier series. Let $g(\theta) = f(-\theta)$ for $\theta \in \mathbb{R}$. Find the Fourier series of g.

(b) Suppose that h is a 2π-periodic function such that $h(\pi - \theta) = h(\theta)$. What does this imply about its Fourier series?

H. (a) Compute the Fourier series of $f(\theta) = \begin{cases} 1 - |\theta| & \text{for } -1 \le \theta \le 1, \\ 0 & \text{otherwise.} \end{cases}$

(b) Compute the Fourier series of $g(\theta) = \begin{cases} 1 & \text{for } -1 \le \theta \le 0, \\ -1 & \text{for } \quad 0 < \theta \le 1, \\ 0 & \text{otherwise.} \end{cases}$

(c) What relationship do you see between these two functions and series?

I. Show that if $f \in C[-\pi, \pi]$ is an odd function, then the Fourier series of f involves only functions of the form $\sin k\theta$. Similarly, if $f \in C[-\pi, \pi]$ is an even function, then the Fourier series of f involves only the constant term and the cosine functions.

J. For $f \in C[-\pi, \pi]$, define $f_e(\theta) = \frac{1}{2}\big(f(\theta) + f(-\theta)\big)$ and $f_o(\theta) = \frac{1}{2}\big(f(\theta) - f(-\theta)\big)$. Compute the Fourier series of f_e and f_o in terms of the series for f.

K. Show that $f_0(\theta) = 1$ and $f_n(\theta) = \sqrt{2}\cos n\theta$ on $0 \le \theta \le \pi$ for $n \ge 1$ is an orthonormal set in $C[0, \pi]$ for the inner product $\langle f, g \rangle = \frac{1}{\pi}\int_0^\pi f(\theta)g(\theta)\,d\theta$.

L. Find an inner product on $C[0, 1]$ such that $\{\sqrt{2}\sin n\pi x : n \ge 1\}$ is an orthonormal set, and verify that this set is orthonormal for your choice of inner product.

M. Let $f(x) = x$ for $-\pi \le x \le \pi$. Compute the inner product $\langle f, \sin nx \rangle$ for $n \ge 1$. Hence show that $\sum_{n=1}^\infty \frac{1}{n^2} \le \frac{\pi^2}{6}$.   HINT: Integrate by parts. (See Example 13.5.5.)

7.7 Orthogonal Expansions and Hilbert Spaces

Given our success in Section 7.5 in understanding inner product spaces using finite orthonormal sets, it is natural to look at infinite orthonormal sets. Our goal is to extend the Projection Theorem (7.5.11). We start by generalizing the inequality (7.5.13) to a countable orthonormal set in an inner product space. Further results in this section depend on making Cauchy series of vectors converge, and so require inner product spaces that are complete.

7.7.1. BESSEL'S INEQUALITY.
Let $S \subseteq \mathbb{N}$ and let $\{e_n : n \in S\}$ be an orthonormal set in an inner product space V. For $x \in V$,

$$\sum_{n \in S} |\langle x, e_n \rangle|^2 \le \|x\|^2.$$

PROOF. Let us write $\alpha_n = \langle x, e_n \rangle$. If S is a finite set, then the Projection Theorem applies. In particular, if Px denotes the projection onto the span of the e_n, then

$$\sum_{n \in S} \left| \langle x, e_n \rangle \right|^2 = \|Px\|^2 \leq \|x\|^2.$$

So suppose that $S = \mathbb{N}$ is infinite. Using limits and the preceding argument for the finite set $\{e_n : 1 \leq n \leq N\}$ gives

$$\sum_{n=1}^{\infty} \left| \langle x, e_n \rangle \right|^2 = \lim_{N \to \infty} \sum_{n=1}^{N} \left| \langle x, e_n \rangle \right|^2 \leq \lim_{N \to \infty} \|x\|^2 = \|x\|^2. \qquad \blacksquare$$

To extend other parts of the Projection Theorem, we need to have infinite series of vectors converge. This can be a delicate issue. However, the problem has an accessible solution if the inner product space is complete.

7.7.2. DEFINITION. A **Hilbert space** is a complete inner product space.

We give a Hilbert space in the next example, but it is important to know that not all "natural" inner product spaces are complete. For example, the space $C[-\pi, \pi]$ with the L^2 norm, examined in the previous section, is not complete (the argument is outlined in Exercise 7.7.G). It is possible to complete $C[-\pi, \pi]$ in the L^2 norm by an abstract completion process to obtain a Hilbert space, $L^2(-\pi, \pi)$. Another way to do this is to develop the more powerful theory of integration known as the Lebesgue integral, which is a central topic in a course on measure theory. This larger class of square integrable functions turns out to be complete.

7.7.3. EXAMPLE. The space ℓ^2 consists of all sequences $\mathbf{x} = (x_n)_{n=1}^{\infty}$ such that
$\|\mathbf{x}\|_2 := \left(\sum_{n=1}^{\infty} x_n^2 \right)^{1/2}$ is finite. The inner product on ℓ^2 is given by $\langle \mathbf{x}, \mathbf{y} \rangle = \sum_{n=1}^{\infty} x_n y_n$.
In order for this inner product to be well defined, we need to know that this series always converges. In fact, it always converges absolutely (see Exercise 7.7.B).

7.7.4. THEOREM. *The inner product space ℓ^2 is complete.*

PROOF. We must show that if a sequence $\mathbf{x}_k = (x_{k,n})_{n=1}^{\infty}$ is Cauchy, then it converges to a vector $\mathbf{x}$ in ℓ^2. We know that for every $\varepsilon > 0$, there is a number K so large that $\|\mathbf{x}_k - \mathbf{x}_l\| < \varepsilon$ for all $k, l \geq K$. In particular,

$$|x_{k,n} - x_{l,n}| \leq \|\mathbf{x}_k - \mathbf{x}_l\| < \varepsilon \quad \text{for all} \quad k, l \geq K.$$

So for each coordinate n, the sequence $(x_{k,n})_{k=1}^{\infty}$ is a Cauchy sequence of real numbers. By the completeness of $\mathbb{R}$ (Theorem 2.8.5), there exists $y_n \in \mathbb{R}$ such that

$$y_n = \lim_{k \to \infty} x_{k,n} \quad \text{for each } n \geq 1.$$

Let $\mathbf{y} = (y_n)_{n=1}^{\infty}$. We need to show two things: first, that $\mathbf{y}$ is in ℓ^2, and second, that $\mathbf{x}_k$ converges in ℓ^2 to $\mathbf{y}$.

It also follows from the triangle inequality that

$$\left| \|\mathbf{x}_k\| - \|\mathbf{x}_l\| \right| \le \|\mathbf{x}_k - \mathbf{x}_l\| < \varepsilon \quad \text{for all} \quad k, l \ge K.$$

Hence the sequence $(\|\mathbf{x}_k\|)_{k=1}^{\infty}$ is Cauchy. Let $L = \lim_{k \to \infty} \|\mathbf{x}_k\|$.

Fix an integer N. Then compute

$$\sum_{n=1}^{N} |y_n|^2 = \lim_{k \to \infty} \sum_{n=1}^{N} |x_{k,n}|^2 \le \lim_{k \to \infty} \|\mathbf{x}_k\|^2 = L^2.$$

To show that $\mathbf{y}$ belongs to ℓ^2, take a limit as N tends to infinity to obtain

$$\|\mathbf{y}\|^2 = \lim_{N \to \infty} \sum_{n=1}^{N} |y_n|^2 \le L^2.$$

A similar argument shows that $\mathbf{x}_k$ converges to $\mathbf{y}$. Indeed, fix $\varepsilon > 0$ and choose K as before using the Cauchy criterion. Then fix N and compute

$$\sum_{n=1}^{N} |y_n - x_{k,n}|^2 = \lim_{l \to \infty} \sum_{n=1}^{N} |x_{l,n} - x_{k,n}|^2 \le \lim_{l \to \infty} \|\mathbf{x}_l - \mathbf{x}_k\|^2 \le \varepsilon^2.$$

The right-hand side is now independent of N, so letting N tend to infinity yields

$$\|\mathbf{y} - \mathbf{x}_k\|^2 = \lim_{N \to \infty} \sum_{n=1}^{N} |y_n - x_{k,n}|^2 \le \varepsilon^2$$

for all $k \ge K$. Since $\varepsilon > 0$ is arbitrary, this establishes convergence. ∎

In a Hilbert space, the **closed span** of a set of vectors S, denoted by $\overline{\mathrm{span}\,S}$, is the closure of the linear subspace spanned by S. This is still a subspace. Since it is a closed subset of a complete space, it is also complete. Thus closed subspaces of Hilbert spaces are themselves Hilbert spaces in the given inner product.

We call an orthonormal set S an **orthonormal basis** for a Hilbert space H if it is maximal as an orthonormal set. For H a finite-dimensional space, this is equivalent to the usual definition of a basis, by Corollary 7.5.4. In infinite-dimensional spaces, it is only a basis in a topological sense, where we allow convergent infinite linear combinations of the orthonormal set.

We can now sharpen Bessel's inequality (7.7.1), telling us precisely when it is an equality. Solely to avoid the technicalities of uncountable bases (see Appendix 2.9), we assume that our Hilbert spaces are **separable**, meaning that every orthonormal set is countable. In other words, we assume that an orthonormal set can be indexed by either by a finite set or $\mathbb{N}$.

7.7.5. PARSEVAL'S THEOREM.

Let $S \subset \mathbb{N}$ and $E = \{e_n : n \in S\}$ be an orthonormal set in a Hilbert space H. Then the subspace $M = \overline{\operatorname{span} E}$ consists of all vectors $x = \sum\limits_{n \in S} \alpha_n e_n$, where the coefficient sequence $(\alpha_n)_{n=1}^{\infty}$ belongs to ℓ^2. Further, for $x \in H$, then $x \in M$ if and only if

$$\sum_{n \in S} |\langle x, e_n \rangle|^2 = \|x\|^2.$$

PROOF. When S is a finite set, this theorem follows from the Projection Theorem (7.5.11). So suppose that S is infinite, say $S = \{j_1, j_2, j_3, \dots\}$.

Suppose that $(\alpha_n)_{n=1}^{\infty} \in \ell^2$. Define $x_k = \sum_{n=1}^{k} \alpha_n e_{j_n}$. We will show that this is a Cauchy sequence. Indeed, if $\varepsilon > 0$, then the convergence of $\sum_{n \geq 1} |\alpha_n|^2$ shows that there is an integer K such that $\sum_{n=K+1}^{\infty} |\alpha_n|^2 < \varepsilon^2$. Thus if $l \geq k \geq K$,

$$\|x_l - x_k\|^2 = \left\| \sum_{n=k+1}^{l} \alpha_n e_{j_n} \right\| = \sum_{n=k+1}^{l} |\alpha_n|^2 < \varepsilon^2.$$

Since H is complete, this sequence converges to a vector x.

Since M is closed and each x_k lies in M, it follows that x belongs to M. Moreover, using Corollary 7.4.7,

$$\langle x, e_{j_n} \rangle = \lim_{k \to \infty} \langle x_k, e_{j_n} \rangle = \alpha_n \quad \text{for all} \quad n \geq 1.$$

So we can write without confusion $x = \sum\limits_{n=1}^{\infty} \alpha_n e_{j_n}$. Thus M contains all of the ℓ^2 linear combinations of the basis vectors.

Now let x be an arbitrary vector in H and set $\alpha_n = \langle x, e_{j_n} \rangle$. By Bessel's inequality (7.7.1), the sequence $(\alpha_n)_{n=1}^{\infty}$ belongs to ℓ^2 and $\sum\limits_{n \geq 1} |\alpha_n|^2 \leq \|x\|^2$. Let $y = \sum\limits_{n=1}^{\infty} \alpha_n e_{j_n}$. Compute

$$\|x - y\|^2 = \|x\|^2 - 2\langle x, y \rangle + \|y\|^2 = \|x\|^2 - 2 \sum_{n=1}^{\infty} \langle x, \alpha_n e_{j_n} \rangle + \sum_{n=1}^{\infty} |\alpha_n|^2$$

$$= \|x\|^2 - 2 \sum_{n=1}^{\infty} |\alpha_n|^2 + \sum_{n=1}^{\infty} |\alpha_n|^2 = \|x\|^2 - \sum_{n=1}^{\infty} |\alpha_n|^2.$$

Thus if Bessel's inequality is an equality, then $x = y$ and thus it belongs to M.

Conversely, if x belongs to M, we must show that the series $\sum_{n=1}^{\infty} \alpha_n e_n$ actually converges to x itself. Since x belongs to M, it is the limit of vectors in the algebraic span of the basis vectors. So given any $\varepsilon > 0$, there are an integer N and a vector z in $\operatorname{span}\{e_{j_n} : 1 \leq n \leq N\}$ such that $\|x - z\| < \varepsilon$. By the Projection Theorem (7.5.11), the vector $x_N = \sum_{n=1}^{N} \alpha_n e_{j_n}$ is closer to x, whence $\|x - x_N\| \leq \|x - z\| < \varepsilon$.

Since this holds for all $\varepsilon > 0$, we deduce that a subsequence of the sequence $(x_k)_{n=1}^{\infty}$ converges to x. But this whole sequence converges (as shown in the second paragraph), so that $x = \sum_{n=1}^{\infty} \alpha_n e_{j_n}$. ∎

7.7.6. COROLLARY. *Let $E = \{e_n : n \in S\}$ be an orthonormal set in a Hilbert space H. Then there is a continuous linear orthogonal projection P_M of H onto $M = \overline{\operatorname{span} E}$ given by $P_M x = \sum_{n \in S} \langle x, e_n \rangle e_n$.*

PROOF. The preceding proof established that $y = P_M x = \sum_{n \in S} \langle x, e_n \rangle e_n$ is defined and that

$$\|x\|^2 = \|y\|^2 + \|x - y\|^2 = \|P_M x\|^2 + \|x - P_M x\|^2.$$

Since the coefficients are determined in a linear fashion,

$$\langle \alpha x + \beta y, e_n \rangle = \alpha \langle x, e_n \rangle + \beta \langle y, e_n \rangle,$$

it follows that $P_M(\alpha x + \beta y) = \alpha P_M x + \beta P_M y$ for all $x, y \in H$ and scalars $\alpha, \beta \in \mathbb{R}$. So P_M is linear.

Since $\|P_M x\| \le \|x\|$, we have $\|P_M x - P_M y\| = \|P_M(x - y)\| \le \|x - y\|$. So P_M is Lipschitz and thus (uniformly) continuous. Parseval's Theorem also established that $P_M x = x$ if and only if $x \in M$. In particular, the range of P_M is precisely M.

Finally, $P_M y = 0$ if and only if $\langle y, e_n \rangle = 0$ for all $n \in S$. Thus if $x = P_M x = \sum_{n \in S} \alpha_n e_n$ and $P_M y = 0$, it follows that

$$\langle x, y \rangle = \sum_{n \in S} \alpha_n \langle e_n, y \rangle = 0.$$

Hence P_M is an orthogonal projection. ∎

Recall that an orthonormal basis for a Hilbert space was defined as a maximal orthonormal set. Every Hilbert space has an orthonormal basis, but a proof of this fact requires assumptions from set theory, including the Axiom of Choice.

7.7.7. COROLLARY. *If $E = \{e_i : i \ge 1\}$ is an orthonormal basis for a Hilbert space H, every vector $x \in H$ may be uniquely expressed as $x = \sum_{i=1}^{\infty} \langle x, e_i \rangle e_i$.*

PROOF. We need to show that the closed span of a basis is H. If $M = \overline{\operatorname{span} E}$ is a proper subspace, there is a vector $x \in H$ that is not in M. Invoking the previous corollary, $y = x - P_M x \ne 0$. So $e = y/\|y\|$ is a unit vector such that

$$P_M e = \frac{P_M(x - P_M x)}{\|y\|} = 0.$$

Thus e is orthogonal to M. In particular, $\{e, e_i : i \ge 1\}$ is orthonormal, which contradicts the maximality of E. Since E is maximal, it follows that $M = H$.

Parseval's Theorem shows that the closed span of an orthonormal set consists of all ℓ^2 combinations of this orthonormal set. So every vector $x \in H$ may be expressed as $\sum_{i=1}^{\infty} \alpha_i e_i$. The coefficients are unique because this expression for x implies that

$$\langle x, e_j \rangle = \lim_{n \to \infty} \left\langle \sum_{i=1}^{n} \alpha_i e_i, e_j \right\rangle = \sum_{i=1}^{n} \alpha_i \langle e_i, e_j \rangle = \alpha_j.$$

■

Exercises for Section 7.7

A. When does equality hold in equation (7.5.13)?

B. Let $\mathbf{x} = (x_n)_{n=1}^{\infty}$ and $\mathbf{y} = (y_n)_{n=1}^{\infty}$ be elements of ℓ^2.

    (a) Show that $\sum_{n=1}^{N} |x_n y_n| \leq \|\mathbf{x}\| \|\mathbf{y}\|$.   HINT: Schwarz inequality.

    (b) Hence prove that $\sum_{n=1}^{\infty} x_n y_n$ converges absolutely.

C. If M is a closed subspace of a Hilbert space H, define the **orthogonal complement** of M to be $M^{\perp} = \{x : \langle x, m \rangle = 0 \text{ for all } m \in M\}$.

    (a) Show that every vector in H can be written uniquely as $x = m + y$, where $m \in M$ and $y \in M^{\perp}$. Moreover, $\|x\|^2 = \|m\|^2 + \|y\|^2$.

    (b) Show that $M = (M^{\perp})^{\perp}$.

D. Let P be a projection on an inner product space V. Prove that the following are equivalent:

    (a) P is an orthogonal projection.

    (b) $\|v\|^2 = \|Pv\|^2 + \|v - Pv\|^2$ for all $v \in V$.

    (c) $\|Pv\| \leq \|v\|$ for all $v \in V$.

    (d) $\langle Pv, w \rangle = \langle v, Pw \rangle$ for all $v, w \in V$.

    HINT: For (c) $\Longrightarrow$ (d), show that *not* (d) implies there are vectors $v = Pv$ and $w = (I - P)w$ such that $\langle v, w \rangle > 0$. Compute $\|v - tw\|^2 - \|v\|^2$ for small $t > 0$. For (d) $\Longrightarrow$ (a), show that $\langle Pv, (I - P)w \rangle = 0$.

E. Formulate and prove a precise version of the following statement: "A separable infinite-dimensional Hilbert space with an orthonormal basis $\{e_n : n \geq 1\}$ behaves like ℓ^2."

    HINT: Look at the finite-dimensional statement, Corollary 7.5.8.

F. (a) If M and N are closed subspaces of a Hilbert space, show that $(M \cap N)^{\perp} = \overline{M^{\perp} + N^{\perp}}$.

    (b) Let $\{e_n : n \geq 1\}$ be an orthonormal basis for ℓ^2. Let $M = \overline{\text{span}}\{e_{2n} : n \geq 1\}$ and $N = \overline{\text{span}}\{e_{2n-1} + n e_{2n} : n \geq 1\}$. Show that $M + N$ is not closed.

    (c) Use (b) to show that closure is needed in part (a).

G. Consider the functions f_n in $C[-\pi, \pi]$ given by $f_n(x) = \begin{cases} 0, & -\pi \leq x \leq 0, \\ nx, & 0 \leq x \leq \frac{1}{n}, \\ 1, & \frac{1}{n} \leq x \leq \pi. \end{cases}$

    (a) Show that f_n converges in the L^2 norm to the characteristic function χ of $(0, \pi]$. In particular, $(f_n)_{n=1}^{\infty}$ is an L^2 Cauchy sequence.

    (b) Show that $\|\chi - h\|_2 > 0$ for every function h in $C[-\pi, \pi]$.

    (c) Hence conclude that $C[-\pi, \pi]$ is not complete in the L^2 norm.

Chapter 8
Limits of Functions

8.1 Limits of Functions

There are several reasonable definitions for the limit of a sequence of functions. Clearly the entries of the sequence should approximate the limit function f to greater and greater accuracy in some sense. But there are different ways of measuring the accuracy of an approximation, depending on the problem. Different approximation schemes generally correspond to different norms, although not all convergence criteria come from a norm. In this section, we consider two natural choices and see why the stronger notion is better for many purposes.

8.1.1. DEFINITION. Let (f_k) be a sequence of functions from $S \subset \mathbb{R}^n$ into $\mathbb{R}^m$. This sequence **converges pointwise** to a function f if

$$\lim_{k \to \infty} f_k(x) = f(x) \quad \text{for all} \quad x \in S.$$

This is the most obvious and perhaps simplest notion of convergence. It is also a rather weak concept fraught with difficulties.

8.1.2. EXAMPLE. Define piecewise linear continuous functions f_k on $[0,1]$ by connecting the points $(0,0)$, $(\frac{1}{k}, k)$, $(\frac{2}{k}, 0)$, and $(1,0)$ by straight lines, namely

$$f_k(x) = \begin{cases} k^2 x & \text{for } 0 \le x \le \frac{1}{k}, \\ k^2(\frac{2}{k} - x) & \text{for } \frac{1}{k} \le x \le \frac{2}{k}, \\ 0 & \text{for } \frac{2}{k} \le x \le 1. \end{cases}$$

See Figure 8.1. This sequence converges pointwise to the zero function; that is,

$$\lim_{k \to \infty} f_k(x) = 0 \quad \text{for all} \quad 0 \le x \le 1.$$

K.R. Davidson and A.P. Donsig, *Real Analysis and Applications: Theory in Practice*, Undergraduate Texts in Mathematics, DOI 10.1007/978-0-387-98098-0_8, © Springer Science + Business Media, LLC 2010

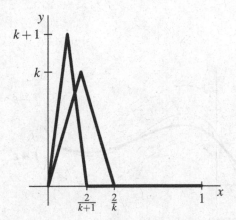

FIG. 8.1 Graphs of f_k and f_{k+1}.

Indeed, at $x = 0$, we have $f_k(x) = 0$ for all $k \geq 1$; and if $x > 0$, then there is an integer N such that $x \geq 2/N$. Thus once $k \geq N$, we have $f_k(x) = 0$. So at every point, the functions are eventually constant. Notice, however, that the closer x is to zero, the larger the choice of N must be.

The limit is a continuous function. However, the limit of the integrals is not the integral of the limit. The area between the graph of f_k and the x-axis forms a triangle with base $2/k$ and height k and thus has area 1 for all k. Therefore,

$$\lim_{k \to \infty} \int_0^1 f_k(x)\, dx = 1 \neq 0 = \int_0^1 0\, dx.$$

Easy modifications of this example yield sequences of functions converging pointwise to 0 with integrals tending to infinity or any finite value or oscillating wildly.

The other notion of convergence that we study, uniform convergence, will demand that convergence occur at a uniform rate on the whole space S. To formulate this, we first consider the ε–N version of pointwise limit. A sequence (f_k) converges pointwise to f if for every $x \in S$ and $\varepsilon > 0$, there is an integer N such that

$$\|f_k(x) - f(x)\| < \varepsilon \quad \text{for all} \quad k \geq N.$$

In this case, N depends on both ε and on the point x. (Look at how we chose different values of N for different $x \in [0,1]$ in Example 8.1.2.) Uniform convergence demands that this choice depend only on ε, providing a common N that works for all x in S.

8.1.3. DEFINITION. Let (f_k) be a sequence of functions from $S \subset \mathbb{R}^n$ into $\mathbb{R}^m$. This sequence **converges uniformly** to a function f if for every $\varepsilon > 0$, there is an integer N such that

$$\|f_k(x) - f(x)\| < \varepsilon \quad \text{for all} \quad x \in S \text{ and } k \geq N.$$

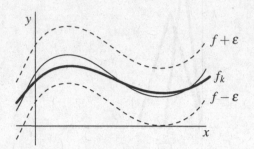

FIG. 8.2 Graph of neighbourhood of f and a sample f_k.

To understand this definition, look at Figure 8.2. The point is that the graph of f_k must lie between the graphs of $f + \varepsilon$ and $f - \varepsilon$.

Clearly, if (f_k) converges uniformly to f, then (f_k) also converges pointwise to f. But this is not reversible.

As we have seen in Example 7.1.3, when K is a compact subset of $\mathbb{R}^n$, we may define a norm on the space $C(K)$ of all real-valued continuous functions on K by

$$\|f\|_\infty = \sup_{x \in K} |f(x)|.$$

This is defined because the Extreme Value Theorem (5.4.4) guarantees that the supremum is finite.

When S is a subset of $\mathbb{R}^n$ that is not compact, there are unbounded continuous functions on S. Nevertheless, we may restrict ourselves to the subspace $C_b(S)$, consisting of all bounded continuous functions from S to $\mathbb{R}$. Then the supremum becomes a norm in the same manner. Similarly, we may consider bounded continuous functions with values in $\mathbb{R}^m$. This space is denoted by $C_b(S, \mathbb{R}^m)$ and has the norm

$$\|f\|_\infty = \sup_{x \in S} \|f(x)\|_2,$$

where $\| \cdot \|_2$ is the usual Euclidean norm in $\mathbb{R}^m$. We have the following theorem.

8.1.4. THEOREM. *Given $S \subset \mathbb{R}^n$ and a sequence of functions (f_k) in $C(S, \mathbb{R}^m)$, (f_k) converges uniformly to f if and only if $f_k - f \in C_b(S, \mathbb{R}^m)$ for all k sufficiently large and*

$$\lim_{k \to \infty} \|f_k - f\|_\infty = 0.$$

After the preceding discussion, the proof is immediate. Indeed, the statement $\|f_k(x) - f(x)\| \le \varepsilon$ for all $x \in S$ is equivalent to saying that $f_k - f$ is bounded and

$$\|f_k - f\|_\infty \le \varepsilon.$$

Returning to Example 8.1.2, the maximum of f_k occurs at $\frac{1}{k}$ with $f_k(\frac{1}{k}) = k$, and so

$$\|f_k - 0\|_\infty = k.$$

This does not converge to 0. So f_k does not converge uniformly to the zero function (or any other bounded function, for that matter).

8.1.5. EXAMPLE. Consider $f_k(x) = x^k$ for $x \in [0,1]$. It is easy to check that

$$\lim_{k \to \infty} f_k(x) = \lim_{k \to \infty} x^k = \begin{cases} 0 & \text{for } 0 \le x < 1, \\ 1 & \text{for } x = 1. \end{cases}$$

Thus the pointwise limit is the function $\chi_{\{1\}}$, the characteristic function of $\{1\}$. The functions f_k are polynomials, and hence not only continuous but even smooth, while the limit function has a discontinuity at the point 1.

For each $k \ge 1$, we have $f_k(1) = 1$, and so

$$\|f_k - \chi_{\{1\}}\|_\infty = \sup_{0 \le x < 1} |x^k - 0| = 1.$$

So f_k does not converge in the uniform norm. Indeed, to contradict the definition, take $\varepsilon = 1/2$. For each k, let $x_k = 2^{-1/2k}$. Then

$$|f_k(x_k) - \chi_{\{1\}}(x_k)| = \frac{1}{\sqrt{2}} > \varepsilon.$$

Hence there is no integer N satisfying the definition.

8.1.6. EXAMPLE. Consider the functions f_n on $[0, \pi]$ given by

$$f_n(x) = \frac{1}{n} \sin nx.$$

Several of the f_n are graphed in Figure 8.3. By the Squeeze Theorem (2.4.6),

$$\lim_{n \to \infty} \frac{1}{n} \sin nx = 0 \quad \text{for all} \quad 0 \le x \le \pi.$$

Moreover, $\|f_n\|_\infty = \sup_{0 \le x \le \pi} \frac{1}{n} |\sin nx| = \frac{1}{n}$. Thus this sequence converges uniformly to 0. If $\varepsilon > 0$, we may choose N so large that $\frac{1}{N} < \varepsilon$. Then for any $n \ge N$,

$$|f_n(x) - 0| = \left| \frac{1}{n} \sin nx \right| \le \frac{1}{N} < \varepsilon \quad \text{for all} \quad 0 \le x \le \pi.$$

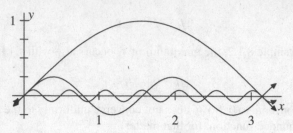

FIG. 8.3 The graphs of $y = f_n(x)$ for $n = 1, 4$, and 13.

This sequence does not behave well with respect to derivatives—a typical feature of uniform approximation. Compute

$$f_n'(x) = \cos nx.$$

Hence $\lim_{n \to \infty} f_n'(0) = \lim_{n \to \infty} 1 = 1 \neq 0 = f'(0)$, while $\lim_{n \to \infty} f_n'(\pi) = \lim_{n \to \infty} (-1)^n$ does not even exist. Indeed, this limit does not exist at any point of $[0, \pi]$ except 0.

The intuition is that for any nice smooth function, there are functions that oscillate up and down very rapidly and yet remain close to the nice function, such as our previous functions. The sequence $g_n(x) = \frac{1}{n} \sin n^2 x$ converges uniformly to 0 as well, yet has derivatives, $g_n'(x) = n \cos n^2 x$, that do not converge anywhere. So uniform convergence does not give control of derivatives.

Exercises for Section 8.1

A. Let $f_n(x) = xne^{-nx}$ for all $x \geq 0$ and $n \geq 1$. Show that (f_n) converges to zero on $[0, \infty)$ pointwise but not uniformly.

B. Let $f_n(x) = nx(1 - x^2)^n$ on $[0, 1]$ for $n \geq 1$. Find $\lim_{n \to \infty} f_n(x)$. Is the convergence uniform?
HINT: Recall that $\lim_{n \to \infty} (1 - \frac{h}{n})^n = e^{-h}$.

C. For the sequence of functions in the previous exercise, compare the limit of the integrals (from 0 to 1) with the integral of the limit.

D. Does the sequence $f_n(x) = \dfrac{x}{1 + nx^2}$ converge uniformly on $\mathbb{R}$?

E. Show that $f_n(x) = \dfrac{\arctan(nx)}{n}$, $n \geq 1$ converges uniformly on $\mathbb{R}$.

F. Show that $f_n(x) = n \sin(x/n)$ converges uniformly on $[-R, R]$ for any finite R but does not converge uniformly on $\mathbb{R}$.

G. Find all intervals on which the sequence $f_n(x) = \dfrac{x^{2n}}{n + x^{2n}}$, $n \geq 1$, converges uniformly.

H. Suppose that $f_n : [0, 1] \to \mathbb{R}$ is a sequence of C^1 functions (i.e., functions with continuous derivatives) that converges pointwise to a function f. If there is a constant M such that $\|f_n'\|_\infty \leq M$ for all n, then prove that (f_n) converges to f uniformly.

I. Prove **Dini's Theorem**: if f and f_n are continuous functions on $[a, b]$ such that $f_n \leq f_{n+1}$ for all $n \geq 1$ and (f_n) converges to f pointwise, then (f_n) converges to f uniformly.

HINT: Work with $g_n = f - f_n$, which decrease to 0. Show that for any point x_0 and $\varepsilon > 0$, there are an integer N and a positive $r > 0$ such that $g_N(x) \leq \varepsilon$ on $(x_0 - r, x_0 + r)$. If convergence is not uniform, say $\lim \|g_n\|_\infty = d > 0$, find x_n such that $\lim g_n(x_n) = d$. Obtain a contradiction.

J. Find an example which shows that Dini's Theorem is false if $[a,b]$ is replaced with a non-compact subset of $\mathbb{R}$.

K. (a) Suppose that $f : \mathbb{R} \to \mathbb{R}$ is uniformly continuous. Let $f_n(x) = f(x + 1/n)$. Prove that f_n converges uniformly to f on $\mathbb{R}$.
   (b) Does this remain true if f is just continuous? Prove it or provide a counterexample.

L. For which values of $x \geq 1$ does the expression $x^{x^{x^{x^{\cdots}}}}$ make sense?
   HINT: Define $f_1(x) = x$ and $f_{n+1}(x) = x^{f_n(x)}$ for $n \geq 1$. Then

   (a) Show that $f_{n+1}(x) \geq f_n(x)$ for all $n \geq 1$.
   (b) When $L(x) = \lim\limits_{n \to \infty} f_n(x)$ exists, find optimal upper bounds for x and L.
   (c) For these values of x, show by induction that $f_n(x)$ is bounded above by e for all $n \geq 1$. What can you conclude?
   (d) What happens for larger x?

M. The behaviour of $x^{x^{x^{\cdots}}}$ when $0 < x < 1$ is more complicated and so more interesting. To get started, compute $f_n(1/16)$ for small values of n, using the functions f_n from the previous exercise, and see what occurs.

8.2 Uniform Convergence and Continuity

Our first positive result is that uniform convergence preserves continuity and so is (almost) always the right notion of convergence for continuous functions.

8.2.1. THEOREM. *Let (f_k) be a sequence of continuous functions mapping a subset S of $\mathbb{R}^n$ into $\mathbb{R}^m$ that converges uniformly to a function f. Then f is continuous.*

PROOF. Fix a point $a \in S$ and an $\varepsilon > 0$. We must control $\|f(x) - f(a)\|$ only by controlling the bound on $\|x - a\|$. To this end, we make use of the proximity of one of the continuous functions f_k and compute

$$\|f(x) - f(a)\| = \|f(x) - f_k(x) + f_k(x) - f_k(a) + f_k(a) - f(a)\|$$
$$\leq \|f(x) - f_k(x)\| + \|f_k(x) - f_k(a)\| + \|f_k(a) - f(a)\|.$$

Note that the first and last terms may be controlled by

$$\|f(x) - f_k(x)\| \leq \|f_k - f\|_\infty \quad \text{for all} \quad x \in S,$$

including $x = a$. The middle term may be controlled by the continuity of f_k.

To be precise, first choose N so large that

$$\|f_N - f\|_\infty < \frac{\varepsilon}{3}.$$

Then using the continuity of f_N at a, choose a positive number $r > 0$ such that

$$\|f_N(x) - f_N(a)\| < \frac{\varepsilon}{3} \quad \text{for all} \quad \|x - a\| < r.$$

Then for all $x \in S$ with $\|x - a\| < r$, we obtain

$$\|f(x) - f(a)\| \le \|f(x) - f_N(x)\| + \|f_N(x) - f_N(a)\| + \|f_N(a) - f(a)\|$$
$$< \frac{\varepsilon}{3} + \frac{\varepsilon}{3} + \frac{\varepsilon}{3} = \varepsilon.$$

Thus f is continuous.   ∎

This method of proof, often called an '$\varepsilon/3$ argument', is one that we shall use often in this chapter. Also note that in this proof, smaller values of ε require using a larger value of N, i.e., a closer approximant f_N, in order to achieve the desired estimate.

Now we will use the compactness of K and Theorem 8.2.1 to show that $C(K, \mathbb{R}^m)$ is complete. Just as we used completeness to understand the real line, we can use it to understand other spaces. First, of course, we have to prove that the space in question is complete.

8.2.2. COMPLETENESS THEOREM FOR $C(K, \mathbb{R}^m)$.

If $K \subset \mathbb{R}^n$ is a compact set, the space $C(K, \mathbb{R}^m)$ of all continuous $\mathbb{R}^m$-valued functions on K with the sup norm is complete.

PROOF. A sequence (f_k) in $C(K, \mathbb{R}^m)$ is a Cauchy sequence for the sup norm if for every $\varepsilon > 0$, there is an integer N such that

$$\|f_k - f_l\|_\infty < \varepsilon \quad \text{for all} \quad k, l \ge N.$$

We must show that every Cauchy sequence has a (uniform) limit in $C(K, \mathbb{R}^m)$.

First consider an arbitrary point $x \in K$. Using $\| \cdot \|$ for the Euclidean norm in $\mathbb{R}^m$, we have

$$\|f_k(x) - f_l(x)\| \le \|f_k - f_l\|_\infty < \varepsilon \quad \text{for all} \quad k, l \ge N.$$

Hence the sequence $(f_k(x))_{k=1}^\infty$ is a Cauchy sequence of real numbers. Since $\mathbb{R}^m$ is complete (Theorem 4.2.5), this has a pointwise limit

$$f(x) := \lim_{n \to \infty} f_n(x).$$

This must be shown to converge uniformly. With ε and N as before, we obtain the estimate

$$\|f(x) - f_m(x)\| = \lim_{n \to \infty} \|f_n(x) - f_m(x)\| \le \varepsilon \quad \text{for all} \quad m \ge N.$$

Since this holds for *all* $x \in K$, it follows that $\|f - f_n\|_\infty \le \varepsilon$. Therefore, the limit is uniform.

By the previous theorem, the uniform limit of continuous functions is continuous. Thus f is continuous and hence belongs to $C(K, \mathbb{R}^m)$. This establishes that $C(K, \mathbb{R}^m)$ is complete. $\blacksquare$

That K is compact is used only implicitly in the proof, when we assume that $\|f_m - f_n\|_\infty$ is well defined. So the same proof shows that for any $S \subseteq \mathbb{R}^m$, $C_b(S, \mathbb{R}^m)$ is complete.

We have used this method of proof before and will use it again. For example, Theorem 4.2.5, showing the completeness of $\mathbb{R}^n$, and the Weierstrass M-test (8.4.7), later in this chapter, both follow a similar strategy.

Exercises for Section 8.2

A. Find the limits of the following functions. Find an interval on which convergence is uniform and another on which it is not. Explain.

   (a) $f_n(x) = \left(\dfrac{x}{2}\right)^n + \left(\dfrac{1}{x}\right)^n$ (b) $g_n(x) = \dfrac{nx}{2 + 5nx}$

B. Show that $h_n(x) = \dfrac{n+x}{4n+x}$ converges uniformly on $[0, N]$ for any $N < \infty$ but not uniformly on $[0, \infty)$.

C. Consider a sequence of continuous functions $f_n : (0,1) \to \mathbb{R}$. Suppose there is a function $f : (0,1) \to \mathbb{R}$ such that whenever $0 < a < b < 1$, f_n converges uniformly on $[a, b]$ to f. Prove that f is continuous on $(0,1)$.

D. Let (f_n) and (g_n) be sequences of continuous functions on $[a, b]$. Suppose that (f_n) converges uniformly to f and (g_n) converges uniformly to g on $[a, b]$. Prove that $(f_n g_n)$ converges uniformly to fg on $[a, b]$.

E. Suppose that (f_k) converges uniformly to f on a compact subset K of $\mathbb{R}^n$ and that (g_k) converges uniformly on K to a continuous function g such that $g(x) \neq 0$ for all $x \in K$. Prove that $f_k(x)/g_k(x)$ is everywhere defined for large k and that this quotient converges uniformly to $f(x)/g(x)$ on K.

F. Let $f_n(x) = \arctan(nx)/\sqrt{n}$.

   (a) Find $f(x) = \lim\limits_{n \to \infty} f_n(x)$, and show that (f_n) converges uniformly to f on $\mathbb{R}$.

   (b) Compute $\lim\limits_{n \to \infty} f_n'(x)$, and compare this with $f'(x)$.

   (c) Where is the convergence of f_n' uniform? Prove your answer.

G. Suppose that functions f_k defined on $\mathbb{R}^n$ converge uniformly to a function f. Suppose that each f_n is bounded, say by A_k. Prove that f is bounded.

H. Suppose that f_n in $C[0, 1]$ all have Lipschitz constant L. Show that if (f_n) converges pointwise to f, then the convergence is uniform and f is Lipschitz with constant L.

I. Give an example of a sequence of discontinuous functions f_n that converges uniformly to a continuous function.

J. Let V be a complete normed vector space.

   (a) Let (f_n) be a Cauchy sequence in $C([a, b], V)$. Show that for each $x \in [a, b]$, $(f_n(x))$ is Cauchy, and so define the pointwise limit $f(x) = \lim\limits_{n \to \infty} f_n(x)$.

   (b) Prove that (f_n) converges uniformly. HINT: Use the Cauchy criterion to obtain an estimate for $\|f_n(x) - f(x)\|$ that is independent of the point x.

   (c) Prove that f is continuous, and deduce that $C([a, b], V)$ is complete.

8.3 Uniform Convergence and Integration

A useful feature of uniform convergence is its good behaviour with respect to limits. We now show that integration over a compact set respects uniform limits.

8.3.1. INTEGRAL CONVERGENCE THEOREM.
Let (f_n) be a sequence of continuous functions on the closed interval $[a,b]$ converging uniformly to $f(x)$ and fix $c \in [a,b]$. Then the functions

$$F_n(x) = \int_c^x f_n(t)\,dt \quad for \quad n \geq 1$$

converge uniformly on $[a,b]$ to the function $F(x) = \int_c^x f(t)\,dt$.

PROOF. The proof is straightforward:

$$|F_n(x) - F(x)| = \left| \int_c^x f_n(t) - f(t)\,dt \right|$$

$$\leq \int_c^x |f_n(t) - f(t)|\,dt \leq \int_c^x \|f_n - f\|_\infty\,dt$$

$$\leq |x - c|\,\|f_n - f\|_\infty \leq (b-a)\,\|f_n - f\|_\infty.$$

The upper bound does not depend on x. Hence

$$\|F_n - F\|_\infty \leq (b-a)\|f_n - f\|_\infty.$$

Since (f_n) converges uniformly to f,

$$\lim_{n \to \infty} \|F_n - F\|_\infty \leq (b-a) \lim_{n \to \infty} \|f_n - f\|_\infty = 0.$$

That is, (F_n) converges uniformly to F.   ∎

This can be reformulated in terms of derivatives as follows.

8.3.2. COROLLARY. *Suppose that (f_n) is a sequence of continuously differentiable functions on $[a,b]$ such that (f_n') converges uniformly to a function g and there is a point $c \in [a,b]$ such that $\lim_{n \to \infty} f_n(c) = \gamma$ exists. Then (f_n) converges uniformly to a differentiable function f with $f(c) = \gamma$ and $f' = g$.*

PROOF. By the Fundamental Theorem of Calculus, part 2 (6.4.3), f_n is the unique antiderivative of f_n' whose value at c is $f_n(c)$. That is,

$$f_n(x) = f_n(c) + \int_c^x f_n'(t)\,dt.$$

By the previous theorem, the sequence of functions $F_n(x) = \int_c^x f_n'(t)\,dt$ for $n \geq 1$ converges uniformly to $F(x) = \int_c^x g(t)\,dt$. Since $\lim_{n \to \infty} f_n(c) = \gamma$, it follows that

$$\lim_{n \to \infty} \|f_n - (\gamma + F)\|_\infty \leq \lim_{n \to \infty} |f_n(c) - \gamma| + \|F_n - F\|_\infty = 0.$$

Therefore, (f_n) converges uniformly to

$$f(x) = \gamma + \int_c^x g(x)\,dx.$$

Finally, the Fundamental Theorem of Calculus, part 1 (6.4.2) shows that f is differentiable and $f' = g$. ∎

Consider a function of two variables $f(x,t)$. Notice that

$$F(x) = \int_c^d f(x,t)\,dt$$

is a function of x. The previous theorem can be seen as a special case of this situation, where x is in $\mathbb{N}$ and $f(t,n)$ is written as $f_n(t)$. It turns out that $F'(x)$ equals the integral of $\partial f / \partial x$, but proving it requires some careful estimates. We begin with a continuous parameter version of the Integral Convergence Theorem.

8.3.3. PROPOSITION. *Let $f(x,t)$ be a continuous function on $[a,b] \times [c,d]$. Define $F(x) = \int_c^d f(x,t)\,dt$. Then F is continuous on $[a,b]$.*

PROOF. Since f is continuous on a compact set, it is uniformly continuous. Therefore, given $\varepsilon > 0$, there is a $\delta > 0$ such that $|f(x,t) - f(y,t)| < \varepsilon/(d-c)$ whenever $|x - y| < \delta$. Therefore,

$$|F(x) - F(y)| = \left| \int_c^d f(x,t) - f(y,t)\,dt \right| \leq \int_c^d |f(x,t) - f(y,t)|\,dt$$

$$\leq \int_c^d \frac{\varepsilon}{d-c}\,dt = \varepsilon.$$

Thus F is uniformly continuous. ∎

8.3.4. LEIBNIZ'S RULE.
Suppose that $f(x,t)$ and $\frac{\partial}{\partial x} f(x,t)$ are continuous functions on $[a,b] \times [c,d]$. Then the function $F(x)$ on $[a,b]$ given by $F(x) = \int_c^d f(x,t)\,dt$ is differentiable and

$$F'(x) = \int_c^d \frac{\partial}{\partial x} f(x,t)\,dt.$$

PROOF. For brevity, we use f_1 for the partial derivative of f with respect to its first variable, that is, $f_1(x,t) = \frac{\partial}{\partial x}f(x,t)$. Fix $x_0 \in [a,b]$ and let $h \neq 0$. Observe that

$$\frac{F(x_0+h) - F(x_0)}{h} = \int_c^d \frac{f(x_0+h,t) - f(x_0,t)}{h}\,dt.$$

Since $f(x,t)$ is a differentiable function of x for fixed t, we may apply the Mean Value Theorem (6.2.2) to obtain a point $x(t)$ depending on t such that $|x(t) - x| < h$ and

$$\frac{f(x_0+h,t) - f(x_0,t)}{h} = f_1(x(t),t).$$

The Mean Value Theorem does not show that the function $x(t)$ is continuous. However, in this situation, the left-hand side of this identity is evidently a continuous function of t, and hence so is the right-hand side, $f_1(x(t),t)$.

Since $f_1(x,t)$ is continuous on a compact set, it is uniformly continuous. So for $\varepsilon > 0$, we can choose $\delta > 0$ such that

$$\left|f_1(x,t) - f_1(y,t)\right| < \frac{\varepsilon}{d-c}$$

whenever $|x - y| < \delta$. Therefore, if $|h| < \delta$, then $|x(t) - x| < \delta$; so

$$\left|\frac{F(x_0+h) - F(x_0)}{h} - \int_c^d f_1(x,t)\,dt\right| = \left|\int_c^d \frac{f(x_0+h,t) - f(x_0,t)}{h} - f_1(x,t)\,dt\right|$$

$$\leq \int_c^d \left|f_1(x(t),t) - f_1(x,t)\right|\,dt$$

$$\leq \int_c^d \frac{\varepsilon}{d-c}\,dt = \varepsilon.$$

Since $\varepsilon > 0$ was arbitrary, we obtain

$$F'(x_0) = \lim_{h \to 0} \frac{F(x_0+h) - F(x_0)}{h} = \int_c^d f_1(x,t)\,dt = \int_c^d \frac{\partial}{\partial x}f(x,t)\,dt. \qquad \blacksquare$$

8.3.5. EXAMPLE. We will establish the improper integral $\int_0^\infty e^{-x^2}\,dx = \frac{\sqrt{\pi}}{2}$.

It is known that the integral $g(u) = \int_0^u e^{-x^2}\,dx$ cannot be expressed in closed form in terms of the standard elementary functions. However, the definite integral can be evaluated in a number of ways. Here we exploit Leibniz's rule to accomplish this. The auxiliary function that we introduce is unmotivated, but the rest of the proof is straightforward.

Before we begin computing, observe that e^{-x^2} is positive and thus $g(u)$ is monotone increasing. Thus to prove that a limit exists as u tends to $+\infty$, it suffices to show that g is bounded. However, $e^{-x^2} \leq 1$ for all x and $e^{-x^2} \leq e^{-x}$ when $x \geq 1$. So

$$g(u) \leq \int_0^1 1\, ds + \int_1^u e^{-s}\, ds = 1 + (e - e^{-u}) \leq 1 + e.$$

Consequently, $\int_0^\infty e^{-x^2}\, dx$ is defined and finite.

Consider

$$F(x) = \int_0^1 \frac{e^{-x(1+t^2)}}{1+t^2}\, dt.$$

Observe that

$$F(0) = \int_0^1 \frac{1}{1+t^2}\, dt = \arctan t \Big|_0^1 = \frac{\pi}{4}.$$

The integrand $f(x,t) = \dfrac{e^{-x(1+t^2)}}{1+t^2}$ is continuous on $[0,\infty) \times [0,1]$. We define $f_x(t) = f(x,t)$ and observe that $0 \leq f_x(t) \leq e^{-x}$. Hence f_x converges uniformly to 0 on $[0,1]$ as $x \to +\infty$. By the Integral Convergence Theorem (8.3.1), we conclude that

$$\lim_{x\to\infty} f(x) = \int_0^1 \lim_{x\to\infty} f_x(t)\, dt = \int_0^1 0\, dt = 0.$$

Now apply the Leibniz rule to compute

$$F'(x) = \int_0^1 \frac{\partial}{\partial x}\left(\frac{e^{-x(1+t^2)}}{1+t^2}\right)\, dt$$

$$= \int_0^1 \frac{e^{-x(1+t^2)}(-(1+t^2))}{1+t^2}\, dt = -e^{-x}\int_0^1 e^{-xt^2}\, dt.$$

Make the change of variables $s = \sqrt{x}\,t$ (where x is held constant), to obtain

$$F'(x) = -e^{-x}\int_0^{\sqrt{x}} \frac{e^{-s^2}}{\sqrt{x}}\, ds = -\frac{e^{-x}}{\sqrt{x}}g(\sqrt{x}).$$

Next, we relate $F(0)$ to the improper integral we want to evaluate:

$$\frac{\pi}{4} = \lim_{n\to\infty} F(0) - F(n) = \lim_{n\to\infty} -\int_0^n F'(x)\, dx = \lim_{n\to\infty}\int_0^n \frac{e^{-x}}{\sqrt{x}}g(\sqrt{x})\, dx.$$

Substitute $s = \sqrt{x}$. By the Fundamental Theorem of Calculus, part 1, $g'(s) = e^{-s^2}$. So

$$\frac{\pi}{4} = \lim_{n\to\infty}\int_0^{\sqrt{n}} 2e^{-s^2}g(s)\, ds$$

$$= \lim_{n\to\infty}\int_0^{\sqrt{n}} 2g'(s)g(s)\, ds = \lim_{n\to\infty} g^2(s)\Big|_0^n = \left(\int_0^\infty e^{-x^2}\, dx\right)^2.$$

Taking square roots gives the result.

Exercises for Section 8.3

A. For $x \in [-1,1]$, let $F(x) = \int_0^1 x(1-x^2y^2)^{-1/2}\,dy$. Show that $F'(x) = (1-x^2)^{-1/2}$ and deduce that $F(x) = \arcsin(x)$.

B. For $n \geq 1$, define functions f_n on $[0,\infty)$ by

$$f_n(x) = \begin{cases} e^{-x} & \text{for } 0 \leq x \leq n, \\ e^{-2n}(e^n + n - x) & \text{for } n \leq x \leq n + e^n, \\ 0 & \text{for } x \geq n + e^n. \end{cases}$$

(a) Find the pointwise limit f of (f_n). Show that the convergence is uniform on $[0,\infty)$.

(b) Compute $\int_0^\infty f(x)\,dx$ and $\lim_{n\to\infty} \int_0^\infty f_n(x)\,dx$.

(c) Why does this not contradict Theorem 8.3.1?

C. Suppose that $g \in C[0,1]$ and (f_n) is a sequence in $C[0,1]$ that converges uniformly to f. Prove that

$$\lim_{n\to\infty} \int_0^1 f_n(x)g(x)\,dx = \int_0^1 f(x)g(x)\,dx.$$

D. Find $\lim_{n\to\infty} \int_0^\pi \frac{\sin nx}{nx}\,dx$.   HINT: Find the limit of the integral over $[\varepsilon, \pi]$ and estimate the rest.

E. Define $f(x) = \int_0^\pi \frac{\sin xt}{t}\,dt$.

(a) Prove that this integral is defined.
(b) Compute $f'(x)$ explicitly.
(c) Prove that f' is continuous at 0.

F. Define the **Bessel function** J_0 by $J_0(x) = \frac{1}{\pi} \int_{-1}^1 \frac{\cos(xt)}{\sqrt{1-t^2}}\,dt$. Prove that J_0 satisfies the differential equation $y'' + y'/x + y = 0$, that is, $J_0'' + J_0'/x + J_0$ is identically zero.

G. With the setup for the Leibniz rule, let b be a variable and set $F(x,b) = \int_a^b f(x,t)\,dt$. Let $b(x)$ be a differentiable function, and define $G(x) = F(x,b(x)) = \int_a^{b(x)} f(x,t)\,dt$. Show that $G'(x) = \int_a^{b(x)} \frac{\partial f}{\partial x}(x,t)\,dt + f(x,b(x))b'(x)$. HINT: $G'(x) = \frac{\partial F}{\partial x}(x,b(x)) + \frac{\partial F}{\partial y}(x,b(x))b'(x)$.

H. Suppose that $f \in C^2[0,1]$ such that $f''(x) + bf'(x) + cf(x) = 0$, $f(0) = 0$, and $f'(0) = 1$. Let $d(x)$ be continuous on $[0,1]$ and define $g(x) = \int_0^x f(x-t)d(t)\,dt$. Prove that $g(0) = g'(0) = 0$ and $g''(x) + bg'(x) + cg(x) = d(x)$.

8.4 Series of Functions

By analogy with series of numbers, we define a series of functions, $\sum_{n=1}^\infty f_n(x)$, as the limit of the sequence of partial sums $\sum_{n=1}^k f_n(x)$. Thus, we say that $\sum_{n=1}^\infty f_n(x)$ converges pointwise (or uniformly) if the partial sums converge pointwise (or uniformly).

8.4.1. EXAMPLE. Consider the series of functions $\sum\limits_{n=1}^{\infty} \dfrac{\sin(nx)}{n^2}$. To see that the partial sums converge, first observe that if $k \geq l$, then

$$\left| \sum_{n=1}^{k} f_n(x) - \sum_{n=1}^{l} f_n(x) \right| \leq \sum_{n=l+1}^{k} |f_n(x)| \leq \sum_{n=l+1}^{k} \frac{1}{n^2}.$$

Since $\sum\limits_{n=1}^{\infty} 1/n^2$ is a convergent series, the Cauchy Criterion (3.1.5) shows that for any $\varepsilon > 0$, there is an integer N such that if $l, k \geq N$, then $\sum\limits_{n=l+1}^{k} 1/n^2 < \varepsilon$. Thus, for $l, k \geq N$,

$$\left| \sum_{n=1}^{k} f_n(x) - \sum_{n=1}^{l} f_n(x) \right| < \varepsilon,$$

proving that for each x, the partial sums are Cauchy and so converge.

8.4.2. EXAMPLE. On the other hand, consider the sequence of functions f_n on $[0,1]$ given by $f_n = \chi_{(0,1/n)}$. For any x in $[1/(n+1), 1/n)$, the values $f_{n+1}(x)$, $f_{n+2}(x), \ldots$ are all zero and the values $f_1(x), \ldots, f_n(x)$ are all one. Hence

$$\sum_{k=0}^{\infty} f_k(x) = n \quad \text{for} \quad \frac{1}{n+1} \leq x < \frac{1}{n}.$$

Thus, the series $\sum_{k=0}^{\infty} f_k(x)$ converges at each point of $[0,1]$. It does not converge uniformly, since for all $k > l$, we have

$$\left| \sum_{n=1}^{k} f_n(x) - \sum_{n=1}^{l} f_n(x) \right| \geq f_{l+1}(x) = 1 \quad \text{for all} \quad x \in (0, 1/(l+1)).$$

8.4.3. EXAMPLE. One of the most important types of series of functions is a **power series**. This is a series of the form

$$\sum_{n=1}^{\infty} a_n x^n = a_0 + a_1 x + a_2 x^2 + a_3 x^3 + \cdots.$$

We will consider these series in detail in the next section. As a starter, consider the series $\sum\limits_{n=0}^{\infty} x^n/n!$. For each $x \in \mathbb{R}$, we can apply the Ratio Test, to obtain

$$\lim_{n \to \infty} \frac{x^{n+1}/(n+1)!}{x^n/n!} = \lim_{n \to \infty} \frac{x}{n+1} = 0.$$

So this series converges pointwise for on $\mathbb{R}$. After Theorem 8.4.7, we will show that it converges uniformly on each interval $[-A, A]$; we find the sum in Example 8.5.4.

Using the partial sums, we can translate all of the results of the previous two sections about sequences of functions into results about series of functions. Here is an example. We leave the reformulation of the other theorems as exercises.

8.4.4. THEOREM. *Let (f_k) be a sequence of continuous functions from a subset S of $\mathbb{R}^n$ into $\mathbb{R}^m$. If $\sum\limits_{k=1}^{\infty} f_k(x)$ converges uniformly, then it is continuous.*

8.4.5. DEFINITION. Let $S \subset \mathbb{R}^n$. We say that a series of functions (f_k) from S to $\mathbb{R}^m$ is **uniformly Cauchy** on S if for every $\varepsilon > 0$, there is an N such that

$$\left\| \sum_{i=k+1}^{l} f_i(x) \right\|_{\infty} \leq \varepsilon \quad \text{whenever} \quad x \in S \text{ and } l > k \geq N.$$

The proof that a sequence of real numbers converges if and only if it is Cauchy (Theorem 2.8.5) can be modified in a straightforward way to show the following.

8.4.6. THEOREM. *A series of functions converges uniformly if and only if it is uniformly Cauchy.*

PROOF. Let f_k be the kth partial sum. If f_k converges uniformly to f, then for each $\varepsilon > 0$, there is $N \in \mathbb{N}$ such that $\|f_k - f\| < \varepsilon/2$ for all $k \geq N$. If $k, l \geq N$,

$$\|f_k - f_l\| \leq \|f_k - f\| + \|f - f_l\| < \frac{\varepsilon}{2} + \frac{\varepsilon}{2} = \varepsilon.$$

Conversely, if (f_k) is uniformly Cauchy, then $(f_k(x))$ is Cauchy for every x, and thus $f(x) = \lim\limits_{k \to \infty} f_k(x)$ exists as a pointwise limit. Moreover, if $\varepsilon > 0$ and $\|f_k - f_l\| < \varepsilon$ for all $k, l \geq N$, then

$$\|f - f_k\| = \lim_{l \to \infty} \|f_k - f_l\| \leq \varepsilon.$$

Thus this convergence is uniform.   ∎

There is a useful test for uniform convergence of a series of functions. The proof is easy, and the test comes up often in practice.

8.4.7. WEIERSTRASS M-TEST.
Suppose that $a_k(x)$ is a sequence of functions on $S \subset \mathbb{R}^n$ into $\mathbb{R}^m$, (M_k) is a sequence of real numbers, and there is N such that for all $k \geq N$ and all $x \in S$,

$$\|a_k\|_{\infty} = \sup_{x \in S} \|a_k(x)\| \leq M_k.$$

If $\sum\limits_{k=1}^{\infty} M_k$ converges, then the series $\sum\limits_{k=1}^{\infty} a_k(x)$ converges uniformly on S.

PROOF. For each $x \in S$, the sequence $(a_k(x))$ is an absolutely convergent sequence of real numbers, since

$$\sum_{k=1}^{\infty} \|a_k(x)\| < \sum_{k=1}^{\infty} \|a_k\|_\infty \le \sum_{k=1}^{N} \|a_k\|_\infty + \sum_{k=N+1}^{\infty} M_k < \infty.$$

Thus the sum exists. Define $f(x) = \sum_{k=1}^{\infty} a_k(x)$. Then for each $x \in S$,

$$\left\| f(x) - \sum_{k=1}^{l} a_k(x) \right\| = \left\| \sum_{k=l+1}^{\infty} a_k(x) \right\| \le \sum_{k=l+1}^{\infty} \|a_k(x)\| \le \sum_{k=l+1}^{\infty} \|a_k\|_\infty \le \sum_{k=l+1}^{\infty} M_k,$$

for all l with $l \ge N$. This estimate does not depend on x. Thus

$$\lim_{l \to \infty} \left\| f - \sum_{k=1}^{l} a_k \right\|_\infty \le \lim_{l \to \infty} \sum_{k=l+1}^{\infty} M_k = 0.$$

Therefore, this series converges uniformly to f. ∎

As an application, we return to the series $\sum_{n=0}^{\infty} \frac{x^n}{n!}$ considered in Example 8.4.3. On any interval $[-A, A]$ with $A \ge 0$, $|x^n/n!| \le A^n/n! =: M_n$. Applying the Ratio Test to M_n shows that $\sum_{n=0}^{\infty} M_n$ converges. Hence by the M-test, the series converges uniformly on $[-A, A]$. The series does not converge uniformly on the whole real line, but since it converges uniformly on every bounded interval, we may conclude that the limit is continuous on the whole line.

8.4.8. EXAMPLE. Consider the geometric series $\sum_{n=0}^{\infty} (-x^2)^n$. The ratio of successive terms of this series at the point x is $-x^2$. Thus for $|x| < 1$, this series converges, while it diverges for $|x| > 1$. By inspection, it also diverges at $x = \pm 1$. For each x in $(-1, 1)$, we readily obtain that

$$\sum_{n=0}^{\infty} (-x^2)^n = \frac{1}{1 - (-x^2)} = \frac{1}{1 + x^2}.$$

On the interval of convergence $(-1, 1)$, the convergence is not uniform. Indeed, for any integer N, take $a = 2^{-1/2N}$ and note that the Nth term $(-a^2)^N = \frac{1}{2}$ is large. However, on the interval $[-r, r]$ for any $r < 1$, we have

$$\sup_{|x| \le r} |(-x^2)^n| = r^{2n}.$$

Since $\sum\limits_{n=0}^{\infty} r^{2n} = \dfrac{1}{1-r^2} < \infty$, the Weierstrass M-test shows that the series converges

uniformly to $f(x) = \dfrac{1}{1+x^2}$ on $[-r, r]$.

Consider the functions

$$F_n(x) := \int_0^x \sum_{k=0}^{n} (-t^2)^k \, dt = \sum_{k=0}^{n} \frac{(-1)^k}{2k+1} x^{2k+1}.$$

Apply Theorem 8.3.1 to see that the F_n converge uniformly on $[-r, r]$ to the function

$$F(x) = \int_0^x \frac{1}{1+t^2} \, dt = \arctan(x).$$

This yields the Taylor series for arctan about 0; see also Example 10.1.5,

$$\arctan(x) = \sum_{n=0}^{\infty} \frac{(-1)^n}{2n+1} x^{2n+1}.$$

The radius of convergence of this series is still 1. This converges at $x = \pm 1$ as well because it is an alternating series in which the terms are monotone decreasing to zero.

Figure 8.4.8 gives the graphs of $y = \arctan(x)$ and the degree-5 and -11 Taylor series. Notice that the degree-11 series is closer to arctan than the degree-5 series on $[-1, 1]$ but is further away outside that interval.

It happens that this convergence is uniform on the whole interval $[-1, 1]$. Note that the series is an alternating series for every $x \in \mathbb{R}$. The terms converge monotonely to 0 precisely when $|x| \leq 1$. To see this, we need the error estimate for alternating series. Since the terms decrease in absolute value, the error is never greater than the next term in the series. So the error between the nth partial sum and the limit is no greater than

$$\sup_{|x| \leq 1} \left| \frac{(-1)^n}{2n+1} x^{2n+1} \right| \leq \frac{1}{2n+1}.$$

Since this tends to 0, the series converges uniformly on $[-1, 1]$ to $\arctan(x)$.

8.4.9. EXAMPLE. The reason we have said nothing so far about the derivatives of uniformly convergent sequences is that in general, there is nothing good to say. Indeed, there are continuous functions that are not differentiable at any point. These are called **nowhere differentiable functions**. The first example was constructed by Bolzano sometime before 1830 but was not published. Weierstrass independently discovered such functions in 1861 and he published his construction in 1872.

To construct a continuous nowhere differentiable function, let

$$f(x) = \sum_{k \geq 1} 2^{-k} \cos(10^k \pi x).$$

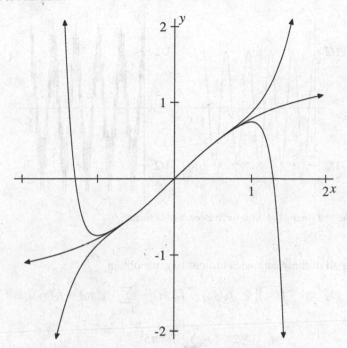

FIG. 8.4 Graphs for $y = \arctan(x)$ and its degree-5 and -11 Taylor series.

Set $f_k(x) = 2^{-k}\cos(10^k\pi x)$. Then $\sum_{k\geq 1}\|f_k\|_\infty = \sum_{k\geq 1}2^{-k} = 1$ converges. Thus by the Weierstrass M-test, this series converges uniformly on the whole real line to a continuous function.

Figure 8.5 gives the graphs of the first two partial sums, namely $y = \cos(10\pi x)/2$ and $y = \cos(10\pi x)/2 + \cos(100\pi x)/4$. The rapid oscillation in the second graph suggests how the limit function could fail to be differentiable. Bear in mind that each partial sum is infinitely differentiable, being a finite linear combination of infinitely differentiable functions.

Consider an arbitrary point x in $\mathbb{R}$, say $x = x_0.x_1x_2x_3\ldots$. We will show that f is not differentiable at x by constructing a sequence z_n converging to x such that the difference quotient $|f(z_n) - f(x)|/|z_n - x|$ goes to $+\infty$.

Fix $n \geq 1$. Let $y_0 = x_0.x_1x_2\ldots x_n$ and $y_1 = y_0 + 10^{-n}$. So $y_0 \leq x \leq y_1$. Let us estimate $|f(y_0) - f(y_1)|$. Since $10^n\pi y_0$ and $10^n\pi y_1$ are integer multiples of π, we have
$$f_n(y_0) = (-1)^{x_n}2^{-n} \quad \text{and} \quad f_n(y_1) = (-1)^{x_n+1}2^{-n}.$$
Hence $|f_n(y_0) - f_n(y_1)| = 2^{1-n}$. For $k > n$, $10^k y_i\pi$ is an integer multiple of 2π. So $f_k(y_0) = f_k(y_1) = 2^{-k}$. For $1 \leq k < n$, the Mean Value Theorem (6.2.2) yields
$$|f_k(y_0) - f_k(y_1)| \leq \|f_k'\|_\infty|y_0 - y_1| = (2^{-k}10^k\pi)10^{-n} = 2^{-n}\pi 5^{k-n}.$$

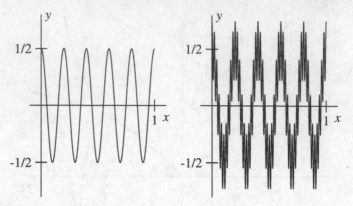

FIG. 8.5 The first two partial sums for Weierstrass's function.

Combining all of these estimates judiciously, we obtain

$$|f(y_0) - f(y_1)| \geq |f_n(y_0) - f_n(y_1)| - \sum_{k \neq n} |f_k(y_0) - f_k(y_1)|$$

$$\geq 2^{1-n} - \sum_{k=1}^{n-1} 2^{-n} \pi 5^{k-n}$$

$$> 2^{-n} \left(2 - \frac{\pi}{4} \right) > 2^{-n}.$$

One of these values is far from $f(x)$, since

$$|f(y_0) - f(x)| + |f(x) - f(y_1)| \geq |f(y_0) - f(y_1)| > 2^{-n}.$$

Choose $i = 0$ or 1 such that $|f(y_i) - f(x)| > 2^{-n-1}$, and set $z_n = y_i$. Clearly $|z_n - x| \leq |y_1 - y_0| = 10^{-n}$. Therefore,

$$\left| \frac{f(z_n) - f(x)}{z_n - x} \right| \geq \left| \frac{2^{-n-1}}{10^{-n}} \right| = 5^n/2.$$

As n tends to infinity, it is clear that the sequence (z_n) converges to x while the differential quotient blows up. Therefore, f is not differentiable at x.

Exercises for Section 8.4

A.   Reformulate Theorem 8.3.1 and Corollary 8.3.2 in terms of series of functions.

B.   Prove Theorem 8.4.6.

C.   (a) Show that $\sum_{n=1}^{\infty} x^n e^{-nx}$ converges uniformly on $[0, A]$ for each $A > 0$.

   (b) Does it converge uniformly on $[0, \infty)$?

D.   Does $\sum_{n=1}^{\infty} \frac{1}{x^2 + n^2}$ converge uniformly on the whole real line?

E. Show that if $\sum\limits_{n=1}^{\infty} |a_n| < \infty$, then $\sum\limits_{n=1}^{\infty} a_n \cos nx$ converges uniformly on $\mathbb{R}$.

F. (a) Let $f_n(x) = \dfrac{x^2}{(1+x^2)^n}$ for $x \in \mathbb{R}$. Evaluate the sum $S(x) = \sum\limits_{n=0}^{\infty} f_n(x)$.

   (b) Is this convergence uniform? For which values $a < b$ does this series converge uniformly on $[a,b]$?

G. Find the sum of $\sum\limits_{n=0}^{\infty} \left(\dfrac{x-7}{x+1}\right)^n$ for $x \neq -1$. Where is the convergence uniform?

H. Suppose that $a_k(x)$ are continuous functions on $[0,1]$, and define $s_n(x) = \sum\limits_{k=1}^{n} a_k(x)$. Show that if (s_n) converges uniformly on $[0,1]$, then (a_n) converges uniformly to 0.

I. Prove the series version of Dini's Theorem (Exercise 8.1.I): If g_n are nonnegative continuous functions on $[a,b]$ and $\sum\limits_{n=1}^{\infty} g_n$ converges pointwise to a *continuous* function on $[a,b]$, then it converges uniformly.

J. Let (f_n) be a sequence of functions defined on $\mathbb{N}$ such that $\lim\limits_{k\to\infty} f_n(k) = L_n$ exists for each $n \geq 0$. Suppose that $\|f_n\|_{\infty} \leq M_n$, where $\sum\limits_{n=0}^{\infty} M_n < \infty$. Define a function $F(k) = \sum\limits_{n=0}^{\infty} f_n(k)$.

Prove that $\lim\limits_{k\to\infty} F(k) = \sum\limits_{n=0}^{\infty} L_n$.

HINT: Think of f_n as a function g_n on $\{\frac{1}{k} : k \geq 1\} \cup \{0\}$. How will you define $g_n(0)$?

K. Apply the previous exercise to the functions $f_n(k) = \dbinom{k}{n}\left(\dfrac{x}{k}\right)^n$ for $n \geq 0$ and $k \geq 1$. Hence show that $\lim\limits_{k\to\infty} \left(1 + \dfrac{x}{k}\right)^k = e^x$.

L. In Example 8.4.9, we could use $f(x) = \sum\limits_{k\geq 1} b^k \cos(a^k \pi x)$, where $b < 1$ and a is an even integer. Prove that if $ab > 1 + \pi/2$, then f is nowhere differentiable.

M. Let $d(x) = \mathrm{dist}(x, \mathbb{N})$ and $f_k(x) = 2^{-k}\left(d(2^k x) - 2d(2^{k-1}x)\right)$ for $k \geq 1$.

   (a) Compute $g_k(x) = d(x) + \sum\limits_{k=1}^{n} f_k(x)$.

   (b) Where does g_k fail to be differentiable? This is an increasing sequence of sets with union dense in $\mathbb{R}$.

   (c) Find the limit of g_k. How can it turn out to be differentiable?

8.5 Power Series

As mentioned in the previous section, a power series is a series of functions of the form

$$\sum_{n=0}^{\infty} a_n x^n = a_0 + a_1 x + a_2 x^2 + a_3 x^3 + \cdots .$$

Formally, this is a power series in x and we could also consider a power series in $x - x_0$, namely

$$\sum_{n=0}^{\infty} a_n (x - x_0)^n = a_0 + a_1(x - x_0) + a_2(x - x_0)^2 + a_3(x - x_0)^3 + \cdots .$$

This increase in generality is only apparent, since we can set $y = x - x_0$ and work with a power series in y.

Clearly, a power series converges when $x = 0$. This may be the only value of x for which the series converges. For example, apply the Ratio Test (Exercise 3.2.I) to $\sum_{n=1}^{\infty} n! x^n$. The following theorem provides a full answer to the general question of when a power series converges.

8.5.1. HADAMARD'S THEOREM.

Given a power series $\sum_{n=0}^{\infty} a_n x^n$, there is R in $[0, +\infty) \cup \{+\infty\}$ so that the series converges for all x with $|x| < R$ and diverges for all x with $|x| > R$. Moreover, the series converges uniformly on each closed interval $[a,b]$ contained in $(-R,R)$.

Finally, if $\alpha = \limsup_{n \to \infty} |a_n|^{1/n}$, then

$$
R = \begin{cases} +\infty & \text{if} \quad \alpha = 0, \\ 0 & \text{if} \quad \alpha = +\infty, \\ \frac{1}{\alpha} & \text{if} \quad \alpha \in (0, +\infty). \end{cases}
$$

We call R the **radius of convergence** of the power series.

PROOF. Fixing $x \in \mathbb{R}$ and applying the Root Test (3.2.4) to $\sum_{n=0}^{\infty} a_n x^n$ gives

$$
\limsup |a_n x^n|^{1/n} = |x| \limsup |a_n|^{1/n} = |x| \alpha.
$$

So if $\alpha = 0$, then $|x|\alpha < 1$ for all choices of x, and so the series always converges, as claimed. If $\alpha = +\infty$, then $|x|\alpha > 1$ for all $x \neq 0$, and so the series diverges for nonzero x, again as claimed. Otherwise, $|x|\alpha < 1$ if and only if $|x| < R$ and $|x|\alpha > 1$ if and only if $|x| > R$. By the Root Test again, it follows that we have convergence and divergence on the required intervals.

It remains only to show uniform convergence on each interval $[a,b]$ contained in $(-R,R)$. There is some $c < R$ such that $[a,b] \subset [-c,c]$ for some $c < R$. Observe that for $x \in [-c,c]$, $|a_n x^n| \leq |a_n| c^n$. Since $c < R$, the previous paragraph shows that $\sum_{n=0}^{\infty} |a_n| c^n$ converges. By the M-test (8.4.7), it follows that $\sum_{n=0}^{\infty} a_n x^n$ converges uniformly on $[-c,c]$ and hence on $[a,b]$. ∎

By Exercise 3.2.J, if $\lim_{n \to \infty} \left| \frac{a_{n+1}}{a_n} \right|$ is defined, then $\lim_{n \to \infty} |a_n|^{1/n} = \lim_{n \to \infty} \left| \frac{a_{n+1}}{a_n} \right|$. Thus, we can often use ratios instead of roots to compute radii of convergence. See Exercise 8.5.C.

8.5.2. EXAMPLE.
Hadamard's Theorem contains no information about what happens if $|x| = R$. The answer, as with the Ratio and Root Tests for series of numbers, is that the series may converge or diverge at these points. We consider three

series,

$$\sum_{n=1}^{\infty} \frac{x^n}{2^n n^2}, \qquad \sum_{n=1}^{\infty} \frac{x^n}{2^n n}, \qquad \sum_{n=1}^{\infty} \frac{x^{2n}}{2^n n},$$

to illustrate this. For the first series, the limit ratio of successive coefficients is

$$\lim_{n \to \infty} \frac{2^n n^2}{2^{n+1}(n+1)^2} = \lim_{n \to \infty} \frac{n^2}{2(n+1)^2} = \frac{1}{2}.$$

Therefore, the radius of convergence is 2. Similarly, the second series also has radius of convergence 2.

Consider the first series at $x = \pm 2$. We have $\sum_{n=1}^{\infty} 1/n^2$ and $\sum_{n=1}^{\infty} (-1)^n/n^2$, both of which converge. For the second series at $x = \pm 2$, we have $\sum_{n=1}^{\infty} 1/n$ and $\sum_{n=1}^{\infty} (-1)^n/n$, which diverge and converge, respectively. The first series has interval of convergence $[-2,2]$ while the second has $[-2,2)$.

The third series shows that some care is needed in using Hadamard's Theorem. To write this series in the form $\sum_{n=1}^{\infty} a_n x^n$, we cannot define a_n to be $1/(2^n n)$, but rather $a_{2k+1} = 0$ and $a_{2k} = 2^{-k}/k$ for $k \geq 0$. Using this formula for a_n, only the even terms matter:

$$\limsup_{n \to \infty} |a_n|^{1/n} = \lim_{k \to \infty} \left| \frac{1}{2^k k} \right|^{1/2k} = \frac{1}{\sqrt{2}} \lim_{k \to \infty} \left(\frac{1}{k} \right)^{1/2k} = \frac{1}{\sqrt{2}}.$$

So the series has radius of convergence $\sqrt{2}$. At $x = \pm\sqrt{2}$, this series is $\sum_{n=1}^{\infty} \frac{1}{n}$, which diverges. Therefore, the interval of convergence is $(-\sqrt{2}, \sqrt{2})$.

It seems natural to differentiate and integrate a power series term by term. That is, the derivative and the indefinite integral of $f(x) = \sum_{n=0}^{\infty} a_n x^n$ should be the sums of the terms $n a_n x^{n-1}$ and $a_n x^{n+1}/(n+1)$, respectively. The badly behaved examples of previous sections show that such hopes are misplaced for arbitrary series of functions.

The next theorem shows that power series fulfil these hopes; such properties make power series particularly useful. For example, this theorem implies that if a power series has radius of convergence $R > 0$, then it is infinitely differentiable on $(-R, R)$.

8.5.3. TERM-BY-TERM OPERATIONS ON SERIES.

If $f(x) = \sum_{n=0}^{\infty} a_n x^n$ has radius of convergence $R > 0$, then $\sum_{n=1}^{\infty} n a_n x^{n-1}$ has radius of convergence R, f is differentiable on $(-R, R)$, and for $x \in (-R, R)$,

$$f'(x) = \sum_{n=1}^{\infty} n a_n x^{n-1}.$$

Furthermore, $\sum\limits_{n=0}^{\infty} \dfrac{a_n}{n+1} x^{n+1}$ *has radius of convergence R, and for* $x \in (-R,R)$,

$$\int_0^x f(t)\,dt = \sum_{n=0}^{\infty} \frac{a_n}{n+1} x^{n+1}.$$

PROOF. Observe that $\sum\limits_{n=0}^{\infty} na_n x^{n-1}$ and $\sum\limits_{n=0}^{\infty} na_n x^n$ have the same radius of convergence. We have

$$\limsup_{n\to\infty} |na_n|^{1/n} = \lim_{n\to\infty} n^{1/n} \limsup_{n\to\infty} |a_n|^{1/n} = \frac{1}{R}.$$

Thus $\sum\limits_{n=0}^{\infty} na_n x^{n-1}$ has radius of convergence R. Since the partial sums $\sum\limits_{n=0}^{k} na_n x^{n-1}$ converge uniformly on each interval $[-a,a] \subset (-R,R)$, we can apply Corollary 8.3.2 (with $c=0$) to show that f is differentiable and $f'(x) = \sum\limits_{n=0}^{\infty} na_n x^{n-1}$.

Similarly, $\sum\limits_{n=0}^{\infty} \dfrac{a_n}{n+1} x^{n+1}$ and $\sum\limits_{n=0}^{\infty} \dfrac{a_n}{n+1} x^n$ have the same radius of convergence and

$$\limsup_{n\to\infty} \left| \frac{a_n}{n+1} \right|^{1/n} = \lim_{n\to\infty} \frac{1}{(n+1)^{1/n}} \limsup_{n\to\infty} |a_n|^{1/n} = \frac{1}{R}.$$

So $\sum\limits_{n=0}^{\infty} \dfrac{a_n}{n+1} x^{n+1}$ has radius of convergence R. If $x \in (-R,R)$, then since $\sum\limits_{k=0}^{n} a_k t^k$ converges uniformly to $f(t)$ on the interval $[0,x]$, Theorem 8.3.1 implies that the sequence of integrals

$$F_n(x) = \int_0^x \sum_{k=0}^{n} a_k t^k\,dt = \sum_{k=0}^{n} \frac{a_k}{k+1} x^{k+1}$$

converges uniformly to $F(x) = \displaystyle\int_0^x f(t)\,dt$ on each interval $[-a,a] \subset (-R,R)$. Thus,

$$\sum_{n=0}^{\infty} \frac{a_n}{n+1} x^{n+1} = \int_0^x f(t)\,dt. \qquad \blacksquare$$

8.5.4. EXAMPLE. We return to $f(x) = \sum\limits_{n\geq 0} \dfrac{x^n}{n!}$ from Example 8.4.3. After proving the M-test (8.4.7), we showed that this series has infinite radius of convergence, and converges uniformly on $[-A,A]$ for all finite A. Using term-by-term differentiation,

$$f'(x) = \sum_{n=1}^{\infty} \frac{x^{n-1}}{(n-1)!} = \sum_{k=0}^{\infty} \frac{x^k}{k!} = f(x).$$

The differential equation $f'(x) = f(x)$ may be rewritten as

$$1 = \frac{f'(x)}{f(x)} = (\log f)'(x).$$

Integrating from 0 to t, we obtain

$$t = \int_0^t 1\,dx = \int_0^t (\log f)'(x)\,dx = \log f(t) - \log f(0).$$

It is evident that $f(0) = 1$ and therefore $\log f(t) = t$, whence $f(t) = e^t$.

8.5.5. EXAMPLE. Consider the power series $\sum_{n\geq 1} n^2 x^n$. Since $\lim_{n\to\infty} \frac{(n+1)^2}{n^2} = 1$, the Ratio Test tells us that the radius of convergence is 1. When $|x| = 1$, the terms do not tend to 0, and thus the series diverges. So $\sum_{n\geq 1} n^2 x^n$ is a well-defined function for $x \in (-1, 1)$.

Turn now to the function $g(x) = \sum_{n\geq 0} x^n$, which also has radius of convergence 1. Since g is defined a geometric series, we have $g(x) = 1/(1-x)$ for $|x| < 1$. Applying Theorem 8.5.3 yields

$$\sum_{n\geq 1} nx^{n-1} = g'(x) = \frac{1}{(1-x)^2}.$$

This series has the same radius of convergence, 1, as does

$$\sum_{n\geq 1} nx^n = \frac{x}{(1-x)^2}.$$

A second application of Theorem 8.5.3 yields

$$\sum_{n\geq 1} n^2 x^{n-1} = \left(\frac{x}{(1-x)^2}\right)' = \frac{1+x}{(1-x)^3}.$$

Multiplying by x gives $f(x) = \sum_{n\geq 1} n^2 x^n = \frac{x(1+x)}{(1-x)^3}$. In particular, $\sum_{n\geq 1} \frac{n^2}{2^n} = f(\frac{1}{2}) = 6$.

8.5.6. EXAMPLE. In this example, we obtain the Binomial Theorem for fractional powers. That is, we derive the power series expansion of $(1+x)^\alpha$ for $\alpha \in \mathbb{R}$. If $g(x) = (1+x)^\alpha$, then $g'(x) = \alpha(1+x)^{\alpha-1}$, and so g satisfies the differential equation (DE)

$$(1+x)g'(x) = \alpha g(x), \quad g(0) = 1.$$

Suppose there is a power series $f(x) = \sum_{n=0}^\infty a_n x^n$ that satisfies this DE. Then we have

$$(1+x)\sum_{n=1}^\infty na_n x^{n-1} = \alpha \sum_{n=0}^\infty a_n x^n.$$

Collecting terms, we have

$$\sum_{n=0}^{\infty} (na_n + (n+1)a_{n+1})x^n = \sum_{n=0}^{\infty} \alpha a_n x^n,$$

and so $na_n + (n+1)a_{n+1} = \alpha a_n$, giving $a_{n+1} = \dfrac{\alpha - n}{n+1} a_n$. Since $a_0 = f(0) = 1$, we

have $a_1 = \alpha$, $a_2 = \dfrac{\alpha(\alpha - 1)}{2}$, $a_3 = \dfrac{\alpha(\alpha - 1)(\alpha - 2)}{6}$, and so on. In general, we

obtain the **fractional binomial coefficients**

$$a_n = \frac{\alpha(\alpha - 1)\cdots(\alpha - n + 2)(\alpha - n + 1)}{n!} = \binom{\alpha}{n}.$$

It remains to show that this series has a positive radius of convergence and that it actually converges to $(1+x)^\alpha$.

If α is a nonnegative integer, then the a_n are eventually zero, and so the series reduces to the usual Binomial Theorem. In this case, the radius of convergence is infinite. Otherwise, $a_n \neq 0$ for all n, and we can apply the Ratio Test to obtain

$$\lim_{n \to \infty} \left| \frac{a_{n+1}}{a_n} \right| = \lim_{n \to \infty} \left| \frac{\alpha - n}{n+1} \right| = 1.$$

Hence the series has radius of convergence 1.

To show that $f(x) = (1+x)^\alpha$, consider the ratio $f(x)/(1+x)^\alpha$. Differentiating the ratio with respect to x gives

$$\frac{(1+x)^\alpha f'(x) - \alpha(1+x)^{\alpha-1}f(x)}{(1+x)^{2\alpha}},$$

and since we have shown that $(1+x)f'(x) = \alpha f(x)$, it follows that the derivative is zero. However, setting $x = 0$ in $f(x)/(1+x)^\alpha$ gives $1/1 = 1$, and so the ratio is constantly equal to 1, showing that $f(x) = (1+x)^\alpha$.

Thus, for $|x| < 1$ and any real α,

$$(1+x)^\alpha = \sum_{n=0}^{\infty} \binom{\alpha}{n} x^n.$$

Exercises for Section 8.5

A. Determine the interval of convergence of the following power series:

(a) $\displaystyle\sum_{n=0}^{\infty} n^3 x^n$

(b) $\displaystyle\sum_{n=1}^{\infty} \frac{(-1)^n}{n^2} x^n$

(c) $\displaystyle\sum_{n=0}^{\infty} \frac{n^2}{2^n} x^n$

(d) $\displaystyle\sum_{n=0}^{\infty} \sqrt{n} x^n$

(e) $\displaystyle\sum_{n=0}^{\infty} (-1)^n \frac{x^{2n}}{(2n)!}$

(f) $\displaystyle\sum_{n=0}^{\infty} x^{n!}$

$$\text{(g)} \sum_{n=1}^{\infty} \frac{n!}{n^n} x^n \qquad\qquad \text{(h)} \sum_{n=0}^{\infty} \frac{(n!)^2}{(2n)!} x^n \qquad\qquad \text{(i)} \sum_{n=0}^{\infty} \frac{1}{n} x^n$$

B. Find a power series $\sum_{n=0}^{\infty} a_n x^n$ that has a different *interval* of convergence than $\sum_{n=0}^{\infty} n a_n x^{n-1}$.

C. Suppose that $\lim_{n \to \infty} \dfrac{a_n}{a_{n+1}} = L$ exists. Find the radius of convergence of the power series $\sum_{n=1}^{\infty} a_n x^n$. (See Exercise 3.2.J.)

D. Using the method of Example 8.5.6, show that if $f(x) = \sum_{n=0}^{\infty} a_n x^n$ satisfies the DE $f'(x) = f(x)$ and $f(0) = 1$, then $f(x) = \sum_{n=0}^{\infty} x^n / n!$.

E. Repeat the previous exercise with the conditions $f''(x) = -f(x)$, f is an odd function, and $f(0) = 0$.

F. Prove that if infinitely many of the a_n are nonzero integers, then the radius of convergence of $\sum_{n=1}^{\infty} a_n x^n$ is at most 1.

G. (a) Compute $f(x) = \sum_{n=1}^{\infty} x^n / n$.

  (b) Compute $\sum_{n=1}^{\infty} 2^n / (n 5^n)$. Justify your method.

H. (a) Compute $f(x) = \sum_{n=0}^{\infty} (n+1) x^n$.

  (b) Compute $\sum_{n=0}^{\infty} n / 3^n$. Justify your method.

  (c) Is the substitution of $x = -1$ justified?

I. (a) Compute $g(x) = \sum_{n=0}^{\infty} (n^2 + n) x^n$.

  (b) Compute $\sum_{n=0}^{\infty} (n^2 + n) / 2^n$. Justify your method.

J. Using the binomial series for $(1-x)^{-1/2}$ and the formula

$$\int_0^{\pi/2} \sin^{2n} t \, dt = \frac{1 \cdot 3 \cdot 5 \cdots (2n-1)}{2 \cdot 4 \cdots (2n)} \frac{\pi}{2},$$

show that for $\kappa \in (0,1)$, the integral $\int_0^{\pi/2} (1 - \kappa^2 \sin^2 t)^{-1/2} dt$ equals

$$\frac{\pi}{2} \left(1 + \left(\frac{1}{2}\right)^2 \kappa^2 + \left(\frac{1 \cdot 3}{2 \cdot 4}\right)^2 \kappa^4 + \left(\frac{1 \cdot 3 \cdot 5}{2 \cdot 4 \cdot 6}\right)^2 \kappa^6 + \cdots \right).$$

K. Recall that the Fibonacci sequence is defined by $F(0) = F(1) = 1$ and the recurrence formula $F(n+2) = F(n) + F(n+1)$ for $n \geq 0$. Set $f(x) = \sum_{n=0}^{\infty} F(n) x^n$.

  (a) Show that $F(n) \leq 2^n$ and hence deduce a positive lower bound for the radius of convergence of this power series.

  (b) Compute $\lim_{n \to \infty} F(n+1)/F(n)$ and hence find the radius of convergence R.

  HINT: Let $r_n = F(n+1)/F(n)$. Show by induction that $r_{n+1} - r_n = -\frac{r_n - r_{n-1}}{r_n r_{n-1}}$ and that $r_n r_{n+1} \geq 2$. Hence deduce that the limit R exists. Show that R satisfies a quadratic equation.

  (c) Compute $(1 - x - x^2) f(x)$ for $|x| < R$, and justify your steps. Hence compute $f(x)$.

  (d) Show that $f(x) = \sum_{n=0}^{\infty} (x + x^2)^n$, and that this converges for $|x| < R$.

8.6 Compactness and Subsets of $C(K)$

We saw in Chapter 4 that compactness is a very powerful property. This showed up particularly in Chapter 5 in the proofs of the Extreme Value Theorem (5.4.4) and of uniform continuity for a continuous function on a compact set, Theorem 5.5.9. In this section, we characterize the subsets of $C(K)$ that are themselves compact. We will need this characterization to prove Peano's Theorem (12.8.1), which shows that a wide range of differential equations have solutions.

We restrict our attention to sets of functions in $C(K)$ when K is a compact subset of $\mathbb{R}^n$. Recall, from Definition 7.2.5, that $\mathscr{F} \subseteq C(K)$ is called compact if every sequence (f_n) of functions in $\mathscr{F}$ has a subsequence (f_{n_i}) that converges uniformly to a function f in $\mathscr{F}$. The Heine–Borel Theorem showed that a subset of $\mathbb{R}^n$ is compact if and only if it is closed and bounded. However, $\mathbb{R}^n$ is a finite-dimensional space; and this is a critical fact. $C(K)$ is infinite-dimensional, so some of those arguments are invalid. But then, so is the conclusion—see Exercise 7.2.I, for example, or Example 8.6.1 below.

The arguments of Lemma 4.4.3 are still valid. If $\mathscr{F}$ is not closed, there is a sequence (f_n) in $\mathscr{F}$ that has a uniform limit $f = \lim_{n \to \infty} f_n$ that is not in $\mathscr{F}$. Every subsequence (f_{n_i}) also has limit f, which is not in $\mathscr{F}$. So $\mathscr{F}$ is not compact.

Likewise, if $\mathscr{F}$ is unbounded, it contains a sequence (f_n) such that $\|f_n\| > n$ for $n \geq 1$. Any subsequence (f_{n_i}) satisfies $\lim_{n \to \infty} \|f_{n_i}\| = \infty$. Consequently, it cannot converge uniformly to any function.

For $C(K)$, there are other ways in which a subset can fail to be compact; we work out an example in detail.

8.6.1. EXAMPLE. Look again at Example 8.1.5. We will show that the set $\mathscr{F} = \{f_n(x) = x^n : n \geq 1\}$ is closed and bounded but not compact. The functions f_n on $[0,1]$ are all bounded by 1. Suppose that (f_{n_i}) is any subsequence of (f_n). By Example 8.1.5,

$$\lim_{i \to \infty} f_{n_i}(x) = \lim_{n \to \infty} f_n(x) = \begin{cases} 0 & \text{for } 0 \leq x < 1, \\ 1 & \text{for } x = 1. \end{cases}$$

However, this limit $\chi_{\{1\}}$ is not continuous, and the convergence is not uniform. Thus *no* subsequence converges. It follows that the only limit points of $\mathscr{F}$ are the points in $\mathscr{F}$ themselves. In particular, $\mathscr{F}$ contains all of its limit points, and therefore it is closed. On the other hand, $\mathscr{F}$ is not compact because the sequence (f_n) has no convergent subsequence.

8.6.2. EXAMPLE. Consider a sequence (g_n) of continuous functions on K such that g_n converges uniformly to a function g. Then the set

$$\mathscr{G} = \{g_n : n \geq 1\} \cup \{g\}$$

is compact. Indeed, suppose that (f_n) is a sequence in $\mathcal{G}$. Either some element of $\mathcal{G}$ is repeated infinitely often, or infinitely many g_k's are represented in this sequence. In the first case, there is a constant subsequence that evidently converges in $\mathcal{G}$. Otherwise, there is a subsequence (f_{n_i}) such that $f_{n_i} = g_{k_i}$ and $\lim_{i \to \infty} k_i = \infty$. In this case, the subsequence converges uniformly to g.

Now consider a point a in K and an $\varepsilon > 0$. Since g is continuous, there is an $r_0 > 0$ such that

$$\|g(x) - g(a)\| < \frac{\varepsilon}{3} \quad \text{whenever} \quad \|x - a\| < r_0.$$

Since g_n converges uniformly to g, there is an integer N such that

$$\|g - g_n\|_\infty < \frac{\varepsilon}{3} \quad \text{for all} \quad n \geq N.$$

Combining these estimates, we can show that for $n \geq N$ and $\|x - a\| < r_0$,

$$\|g_n(x) - g_n(a)\| \leq \|g_n(x) - g(x)\| + \|g(x) - g(a)\| + \|g(a) - g_n(a)\|$$
$$\leq \frac{\varepsilon}{3} + \frac{\varepsilon}{3} + \frac{\varepsilon}{3} = \varepsilon.$$

Now we can modify this to obtain a statement for all functions in $\mathcal{G}$. Each g_n for $n < N$ is continuous. So there are positive real numbers $r_n > 0$ such that

$$\|g_n(x) - g_n(a)\| < \varepsilon \quad \text{whenever} \quad \|x - a\| < r_n.$$

Set $r = \min\{r_n : 0 \leq n \leq N\}$. We have shown that

$$\|f(x) - f(a)\| < \varepsilon \quad \text{whenever} \quad \|x - a\| < r \text{ and } f \in \mathcal{G}.$$

This property suggests a new variant of continuity in which a whole family of functions satisfy the same inequalities.

8.6.3. DEFINITION. A family of functions $\mathcal{F}$ mapping a set $S \subset \mathbb{R}^n$ into $\mathbb{R}^m$ is **equicontinuous at a point** $a \in S$ if for every $\varepsilon > 0$, there is an $r > 0$ such that

$$\|f(x) - f(a)\| < \varepsilon \quad \text{whenever} \quad \|x - a\| < r \text{ and } f \in \mathcal{F}.$$

The family $\mathcal{F}$ is **equicontinuous on a set** S if it is equicontinuous at every point in S. The family $\mathcal{F}$ is **uniformly equicontinuous** on S if for each $\varepsilon > 0$, there is an $r > 0$ such that

$$\|f(x) - f(y)\| < \varepsilon \quad \text{whenever} \quad \|x - y\| < r, \ x, y \in S \text{ and } f \in \mathcal{F}.$$

Reconsider the previous two examples. In the second example, we established that $\mathcal{G}$ is equicontinuous. However, in the first example, the set $\mathcal{F} = \{x^n : n \geq 1\}$ is *not* equicontinuous if $x = 1$. Indeed, take $\varepsilon = 1/10$ and let $0 < r < 1$ be an arbitrary positive number. Take $x = 1 - r/2$. Since $\lim_{n\to\infty} x^n = 0$, there is an integer N sufficiently large that

$$|1 - x^n| > 0.5 \quad \text{for all} \quad n \geq N.$$

Hence this r does not work in the definition of equicontinuity, since $|1 - x| < r$ and $|1 - x^n| > 0.5$. Since r is arbitrary, there is no choice of r that will work, and so $\mathscr{F}$ is not equicontinuous at 1.

8.6.4. LEMMA. *Let K be a compact subset of $\mathbb{R}^n$. A compact subset $\mathscr{F}$ of $C(K, \mathbb{R}^m)$ is equicontinuous.*

PROOF. Suppose to the contrary that for a certain point a in K and $\varepsilon > 0$, the definition of equicontinuity is not satisfied for $\mathscr{F}$. This means that for each choice of $r = 1/n$, there are a function $f_n \in \mathscr{F}$ and a point $x_n \in K$ such that

$$\|x_n - a\| < \frac{1}{n} \quad \text{and} \quad \|f_n(x_n) - f_n(a)\| \geq \varepsilon.$$

It is evident that no subsequence of (f_n) can be equicontinuous either.

Now if $\mathscr{F}$ were compact, there would be a subsequence (f_{n_i}) that converged uniformly to some function f. By Example 8.6.2, this subsequence would be equicontinuous. This contradiction shows that $\mathscr{F}$ must be equicontinuous. ∎

8.6.5. PROPOSITION. *If $\mathscr{F}$ is an equicontinuous family of functions on a compact set, then it is uniformly equicontinuous.*

PROOF. This is a modification of Theorem 5.5.9. If the result is false, there is an $\varepsilon > 0$ for which the definition of equicontinuity fails. This means that for each $r = 1/n$, there are points x_n and y_n in K and a function f_n in $\mathscr{F}$ such that

$$\|x_n - y_n\| < \frac{1}{n} \quad \text{and} \quad \|f_n(x_n) - f_n(y_n)\| \geq \varepsilon.$$

Since K is compact, the sequence (x_n) has a convergent subsequence with $\lim_{i\to\infty} x_{n_i} = a$. Hence

$$\lim_{i\to\infty} y_{n_i} = \lim_{i\to\infty} x_{n_i} + \lim_{i\to\infty} y_{n_i} - x_{n_i} = a + 0 = a.$$

By the equicontinuity of $\mathscr{F}$ at a, there is an $r > 0$ such that

$$\|f(x) - f(a)\| < \frac{\varepsilon}{2} \quad \text{for all} \quad f \in \mathscr{F}, \ \|x - a\| < r.$$

There is an integer I such that for all $i \geq I$,

$$\|x_{n_i} - a\| < r \quad \text{and} \quad \|y_{n_i} - a\| < r.$$

Combining these estimates, for all $i \geq I$,

$$\|f_{n_i}(x_{n_i}) - f_{n_i}(y_{n_i})\| \leq \|f_{n_i}(x_{n_i}) - f_{n_i}(a)\| + \|f_{n_i}(a) - f_{n_i}(y_{n_i})\|$$
$$< \frac{\varepsilon}{2} + \frac{\varepsilon}{2} = \varepsilon.$$

This contradicts the hypothesis that uniform equicontinuity fails, so it must hold. ∎

To prove our main result, characterizating compact subsets of $C(K, \mathbb{R}^m)$, we need a new concept, total boundedness.

8.6.6. DEFINITION. A subset S of $K \subseteq \mathbb{R}^m$ is called an ε-net of K if

$$K \subset \bigcup_{a \in S} B_\varepsilon(a).$$

A set K is **totally bounded** if it has a *finite* ε-net for every $\varepsilon > 0$.

In this section, we consider total boundedness only for subsets of $\mathbb{R}^m$. In $\mathbb{R}^m$, total boundedness is equivalent to boundedness (Lemma 8.6.7 proves the nontrivial direction). In fact, Corollary 7.3.2 implies that proving the result for $\mathbb{R}^m$ with the Euclidean norm will show that it is true for all finite-dimensional normed vector spaces. In more general settings, compactness (not boundedness) implies total boundedness, and *complete* totally bounded sets are compact. We prove this in the next chapter, as part of the Borel–Lebesgue Theorem (9.2.3).

8.6.7. LEMMA. *Let K be a bounded subset of $\mathbb{R}^m$. Then K is totally bounded.*

PROOF. As usual, we use the Euclidean norm on $\mathbb{R}^m$. Fix $\varepsilon > 0$. Choose N such that $1/N < \min\{\varepsilon, m^{-1/2}\}$. Since K is bounded, there is $L > 0$ such that $K \subseteq \{x \in \mathbb{R}^m : |x_i| \leq L, 1 \leq i \leq m\}$. Let

$$F = \left\{ \frac{k_i}{2N^2} \in [-L, L] : k_i \in \mathbb{Z} \right\}$$

and $A = \{(a_1, a_2, \ldots, a_m) \in \mathbb{R}^m : a_i \in F, i = 1, \ldots, m\}$. It is easy to see that F is a finite $1/(2N^2)$-net for $[-L, L]$. Let $\tilde{A} = \{a \in A : B_{\varepsilon/2}(a) \cap K \neq \varnothing\}$, and for each $a \in \tilde{A}$, pick $x_a \in B_{\varepsilon/2}(a) \cap K$.

We claim that $\{x_a : a \in \tilde{A}\}$ is a finite ε-net for K. If $x = (x_1, \ldots, x_n) \in K$, then for $i = 1, \ldots, n$, there is $a_i \in F$ with $|x_i - a_i| < 1/(2N^2)$. Letting $a = (a_1, a_2, \ldots, a_m)$, a short calculation shows that $\|x - a\| \leq 1/(2N) < \varepsilon/2$, since $\sqrt{m} \leq N$. Since $B_{\varepsilon/2}(a) \cap K$ is nonempty, $a \in \tilde{A}$, and hence x_a is defined, and further,

$$\|x - x_a\| \leq \|x - a\| + \|a - x_a\| < \frac{\varepsilon}{2} + \frac{\varepsilon}{2} = \varepsilon.$$

∎

8.6.8. COROLLARY. *Let K be a bounded subset of $\mathbb{R}^m$. Then K contains a sequence $\{x_i : i \geq 1\}$ that is dense in K. Moreover, for any $\varepsilon > 0$, there is an integer N such that $\{x_i : 1 \leq i \leq N\}$ forms an ε-net for K.*

PROOF. For each integer $k \geq 1$, let B_k be a finite $1/k$-net for K. Form a sequence (x_i) by listing all of the elements of each B_k in turn as the points $x_{N_{k-1}+1}, \dots, x_{N_k}$. If $x \in K$, then for each k there is an element x_{n_k} that is a point in B_k such that $\|x - x_{n_k}\| < 1/k$. Thus $(x_{n_k})_{k=1}^{\infty}$ is a subsequence converging to x. Since x is an arbitrary point in K, the sequence (x_i) is dense. By construction, if $\varepsilon > 1/k$, the set $\{x_i : 1 \leq i \leq N_k\}$ forms an ε-net for K. ∎

8.6.9. ARZELÀ–ASCOLI THEOREM.

Let K be a compact subset of $\mathbb{R}^n$. A subset $\mathscr{F}$ of $C(K, \mathbb{R}^m)$ is compact if and only if it is closed, bounded, and equicontinuous.

PROOF. The easy direction has been established: If $\mathscr{F}$ is compact, then it is closed, bounded, and equicontinuous.

So assume that $\mathscr{F}$ has these three properties and let (f_n) be a sequence in $\mathscr{F}$. We will construct a convergent subsequence. By Corollary 8.6.8, there is a sequence (x_i) such that for each $r > 0$, there is an integer N such that $\{x_1, \dots, x_N\}$ forms an r-net for K.

We claim that there is a subsequence of (f_n), call it (f_{n_k}), such that

$$\lim_{k \to \infty} f_{n_k}(x_i) = L_i \quad \text{exists for all } i \geq 1.$$

To prove this, let Λ_0 denote the set of positive integers. Since $(f_n(x_1))$ is a bounded sequence, the Bolzano–Weierstrass Theorem (2.7.2) provides a convergent subsequence. That is, there is an infinite subset $\Lambda_1 \subset \Lambda_0$ such that

$$\lim_{n \in \Lambda_1} f_n(x_1) = L_1 \quad \text{exists}.$$

Next, $(f_n(x_2))_{n \in \Lambda_1}$ is bounded sequence, so there is an infinite subset $\Lambda_2 \subset \Lambda_1$ such that

$$\lim_{n \in \Lambda_2} f_n(x_2) = L_2 \quad \text{exists}.$$

Continuing in this way, we obtain a decreasing sequence $\Lambda_0 \supset \Lambda_1 \supset \Lambda_2 \supset \cdots$ of infinite sets such that

$$\lim_{n \in \Lambda_i} f_n(x_i) = L_i \quad \text{converges for each } i \geq 1.$$

We now use a diagonalization method, similar to the proof that $\mathbb{R}$ is uncountable, Theorem 2.9.8. Let n_k be the kth entry of Λ_k; and let $\Lambda = \{n_k : k \in \mathbb{N}\}$. Then for each i, there are at most $i - 1$ entries of Λ that are not in Λ_i. Thus,

$$\lim_{k \to \infty} f_{n_k}(x_i) = \lim_{n \in \Lambda_i} f_n(x_i) = L_i \quad \text{for all} \quad i \geq 1,$$

proving the claim.

For simplicity of notation, we use g_k for f_{n_k} in the remainder of the proof. Now fix $\varepsilon > 0$. By uniform equicontinuity, there is $r > 0$ such that

$$\|f(x) - f(y)\| < \frac{\varepsilon}{3} \quad \text{for all} \quad f \in \mathscr{F} \text{ and } \|x - y\| < r.$$

Choose N such that $\{x_1, \ldots, x_N\}$ is an r-net for K. Since the g_k converge at each of these N points, there is some integer M such that

$$\|g_k(x_i) - g_l(x_i)\| \leq \frac{\varepsilon}{3} \quad \text{for all} \quad k, l \geq M \text{ and } 1 \leq i \leq N.$$

Let $k, l \geq M$ and pick $x \in K$. Since $\{x_1, \ldots, x_N\}$ is an r-net for K, there is some $i \leq N$ such that $\|x - x_i\| < r$. We need an $\varepsilon/3$-argument to finish the proof:

$$\|g_k(x) - g_l(x)\| \leq \|g_k(x) - g_k(x_i)\| + \|g_k(x_i) - g_l(x_i)\| + \|g_l(x_i) - g_l(x)\|$$
$$\leq \frac{\varepsilon}{3} + \frac{\varepsilon}{3} + \frac{\varepsilon}{3} = \varepsilon.$$

Thus g_k is uniformly Cauchy, and so converges uniformly by Theorem 8.2.2. The limit g belongs to $\mathscr{F}$ because $\mathscr{F}$ is closed. Finally, since every sequence in $\mathscr{F}$ has a convergent subsequence, it follows that $\mathscr{F}$ is compact. ∎

In particular, this theorem shows that if a sequence of functions (f_n) in $C[a, b]$ forms a bounded equicontinuous subset, then (f_n) has a subsequence that converges uniformly to some function in $C[a, b]$.

8.6.10. EXAMPLE. Consider the subset K of $C[0, 1]$ consisting of all functions $f \in C[0, 1]$ such that $|f(0)| \leq 5$ and f has Lipschitz constant at most 47.

Notice that K is closed. For if $f_n \in K$ converge uniformly to a function f, then

$$|f(0)| = \lim_{n \to \infty} |f_n(0)| \leq 5,$$

and

$$|f(x) - f(y)| = \lim_{n \to \infty} |f_n(x) - f_n(y)| \leq \lim_{n \to \infty} 47|x - y| = 47|x - y|.$$

In particular,

$$|f(x)| \leq |f(x) - f(0)| + |f(0)| \leq 47 + 5 = 52,$$

so K is bounded.

Finally, observe that K is equicontinuous. For if $\varepsilon > 0$, take $r = \varepsilon/47$. Then if $|x - y| < r$,

$$|f(x) - f(y)| \leq 47|x - y| < 47r = \varepsilon.$$

Therefore, all of the hypotheses of the Arzelà–Ascoli Theorem (8.6.9) are satisfied. Hence this is a compact subset of $C[0,1]$.

Exercises for Section 8.6

A. Use $f_n(x) = x^n$ on $[0,1]$ to show that $B = \{f \in C[0,1] : \|f\| \leq 1\}$ is not compact.

B. Show that $\{f \in C[0,1] : f(x) > 0 \text{ for all } x \in [0,1]\}$ is open.   HINT: Extreme Value Theorem (5.4.4).

C. What is the interior of $\{f \in C_b(\mathbb{R}) : f(x) > 0 \text{ for all } x \in \mathbb{R}\}$?
HINT: Compare with the previous exercise.

D. Prove that the family $\{\sin(nx) : n \geq 1\}$ is not an equicontinuous subset of $C[0,\pi]$.

E. (a) Show that
$$\mathscr{F} = \left\{ F(x) = \int_0^x f(t)\,dt : f \in C[0,1],\ \|f\|_\infty \leq 1 \right\}$$
   is a bounded and equicontinuous subset of $C[0,1]$.
   (b) Why is $\mathscr{F}$ not closed?
   (c) Show that the closure of $\mathscr{F}$ is all functions f with Lipschitz constant 1 such that $f(0) = 0$.
   HINT: Construct F_n in $\mathscr{F}$ such that $F_n(2^{-n}k) = \frac{n}{n+1} f(2^{-n}k)$ for $0 \leq k \leq 2^n$.

F. (a) Let $\mathscr{F}$ be a subset of $C[0,1]$ that is closed, bounded, and equicontinuous. Prove that there is a function $g \in \mathscr{F}$ such that
$$\int_0^1 g(x)\,dx \geq \int_0^1 f(x)\,dx \quad \text{for all} \quad f \in \mathscr{F}.$$
   (b) Construct a closed bounded subset $\mathscr{F}$ of $C[0,1]$ for which the conclusion of the previous problem is false.

G. Let $K \subset \mathbb{R}^n$ be compact. Show that a subset S of $C(K)$ is compact if and only if it is closed and totally bounded.   HINT: Show that totally bounded sets are bounded and equicontinuous.

H. Let $\mathscr{F}$ be an equicontinuous family of functions in $C(K)$, where K is a compact subset of $\mathbb{R}^n$. Prove that if for each $x \in K$, $\sup\{f(x) : f \in \mathscr{F}\} = M_x < \infty$, then $\mathscr{F}$ is bounded.
HINT: Use equicontinuity to bound $\mathscr{F}$ by $M_x + 1$ on a ball about x. Suppose that $|f_n(x_n)|$ tends to $+\infty$. Extract a convergent subsequence of $\{x_n\}$.

I. Let $\mathscr{F}$ be a family of continuous functions defined on $\mathbb{R}$ that is (i) equicontinuous and satisfies (ii) $\sup\{f(x) : f \in \mathscr{F}\} = M_x < \infty$ for every x. Show that every sequence $(f_n)_{n=1}^\infty$ has a subsequence that converges uniformly on $[-k,k]$ for every $k > 0$.
HINT: Find a subsequence $(f_{1,n})_{n=1}^\infty$ that converges uniformly on $[-1,1]$. Then extract a subsequence $(f_{2,n})_{n=1}^\infty$ that converges uniformly on $[-2,2]$, and so on. Now use a diagonal argument.

Chapter 9
Metric Spaces

This text focuses on subsets of a normed space, since this is the natural setting for most of our applications. In this chapter, we introduce an apparently more general framework, metric spaces, and some new ideas that are somewhat more advanced. They play an occasional role in the advanced sections of the applications.

9.1 Definitions and Examples

In a normed vector space, the distance between elements is found using the norm of the difference. However, a distance function can be defined abstractly on any set using the idea of a metric. Most of the arguments we used in the normed context will also work for metric spaces, with only minimal changes. The crucial difference is that in a metric space, we do not work in a vector space, so we cannot use the addition or scalar multiplication.

9.1.1. DEFINITION. Let X be a set. A **metric** on a set X is a function ρ defined on $X \times X$ taking values in $[0, \infty)$ with the following properties:

(1) (positive definiteness) $\quad \rho(x, y) = 0$ if and only if $x = y$,

(2) (symmetry) $\qquad\qquad \rho(x, y) = \rho(y, x)$ for all $x, y \in X$,

(3) (triangle inequality) $\quad \rho(x, z) \le \rho(x, y) + \rho(y, z)$ for all $x, y, z \in X$.

A **metric space** is a set X with a metric ρ, denoted by (X, ρ). If the metric is understood, we use X alone.

9.1.2. EXAMPLES.

(1) If X is a subset of a normed space V, define $\rho(x, y) = \|x - y\|$. This is our standard example.

(2) Put a metric on the surface of the sphere by setting $\rho(x, y)$ to be the length of the shortest path from x to y (known as a **geodesic**). This is the length of the shorter

K.R. Davidson and A.P. Donsig, *Real Analysis and Applications: Theory in Practice*, Undergraduate Texts in Mathematics, DOI 10.1007/978-0-387-98098-0_9, © Springer Science + Business Media, LLC 2010

arc of the great circle passing through x and y. More generally, we can define such a metric on any smooth surface.

(3) The **discrete metric** on a set X is given by

$$d(x,y) = \begin{cases} 0 & \text{if } x = y, \\ 1 & \text{if } x \neq y. \end{cases}$$

(4) Define a metric on $\mathbb{Z}$ by $\rho_2(n,n) = 0$ and $\rho_2(m,n) = 2^{-d}$, where d is the largest power of 2 dividing $m - n \neq 0$. It is trivial to verify properties (1) and (2). If $\rho_2(l,m) = 2^{-d}$ and $\rho_2(m,n) = 2^{-e}$, then $2^{\min\{d,e\}}$ divides $l - n$, and so

$$\rho(l,n) \leq 2^{-\min\{d,e\}} = \max\{\rho_2(l,m), \rho_2(m,n)\}.$$

This metric is known as the **2-adic metric**. Replacing 2 with another prime p, we can define similarly the **p-adic metric**.

(5) If X is a closed subset of $\mathbb{R}^n$, let $K(X)$ denote the collection of all nonempty compact subsets of X. If A is a compact subset of X and $x \in X$, we define

$$\text{dist}(x,A) = \inf_{a \in A} \|x - a\|.$$

Then we define the **Hausdorff metric** on $K(X)$ by

$$d_H(A,B) = \max\left\{ \sup_{q \in A} \text{dist}(a,B), \sup_{b \in B} \text{dist}(b,A) \right\}$$

$$= \max\left\{ \sup_{a \in A} \inf_{b \in B} \|a - b\|, \sup_{b \in B} \inf_{a \in A} \|a - b\| \right\}.$$

Since A is closed, $\text{dist}(x,A) = 0$ if and only if $x \in A$. In particular, we see that $d_H(A,B) = 0$ if and only if $A = B$. So d_H is positive definite and is evidently symmetric. For the triangle inequality, let A, B, C be three compact subsets of X. For each $a \in A$, the Extreme Value Theorem (5.4.4) yields the existence of a closest point $b \in B$, so $\|a - b\| = \text{dist}(a,B)$. Then there is a closest point $c \in C$ to b with $\|b - c\| = \text{dist}(b,C)$. Therefore,

$$\text{dist}(a,C) \leq \|a - c\| \leq \|a - b\| + \|b - c\|$$
$$= \text{dist}(a,B) + \text{dist}(b,C)$$
$$\leq d_H(A,B) + d_H(B,C).$$

Therefore, $\sup_{a \in A} \text{dist}(a,C) \leq d_H(A,B) + d_H(B,C)$. Reversing the roles of A and C and combining the two inequalities, we obtain $d_H(A,C) \leq d_H(A,B) + d_H(B,C)$.

Let $A_\varepsilon := \{x \in \mathbb{R}^n : \text{dist}(x,A) \leq \varepsilon\}$. Note that $d_H(A,B) \leq \varepsilon$ if and only if $A \subset B_\varepsilon$ and $B \subset A_\varepsilon$.

This example will be used in Section 11.7, when we study fractal sets.

The notions of convergence and open set can be carried over to metric spaces by replacing $\|x - y\|$ with $\rho(x,y)$. Once this is done, most of the other definitions do not need to be changed at all.

9.1.3. DEFINITION. The **ball** $B_r(x)$ of radius $r > 0$ about a point x is defined as $\{y \in X : \rho(x,y) < r\}$. We write $B_r^\rho(x)$ if the metric is ambiguous. A subset U is **open** if for every $x \in U$, there is an $r > 0$ so that $B_r(x)$ is contained in U and the interior of a set A, $\int A$, is the largest open set contained in A,

A sequence (x_n) is said to **converge** to x if $\lim_{n \to \infty} \rho(x,x_n) = 0$. A set C is **closed** if it contains all limit points of sequences of points in C and the **closure** of a set A, $\overline{A}$, is the set of all limit points of A.

If X is a subset of a normed space, then these definitions agree with our old definitions; so the language is consistent. It is easy to see that a set is open precisely when the complement is closed (adapt the proof of Theorem 4.3.8 for normed vector spaces). Because metric spaces do not have a vector space structure, the topology is not always like that of $\mathbb{R}^n$ or a normed vector space. For example, $\overline{B_r(a)}$ can be a proper subset of $\{x : \rho(x,a) \le r\}$, as it is for the discrete metric.

9.1.4. DEFINITION. A sequence $(x_n)_{n=1}^\infty$ in a metric space (X,ρ) is a **Cauchy sequence** if for every $\varepsilon > 0$, there is an integer N such that $\rho(x_i,x_j) < \varepsilon$ for all $i,j \ge N$.

A metric space X is **complete** if every Cauchy sequence converges (in X).

9.1.5. EXAMPLES.

(1) Every convergent sequence is Cauchy; the proof of Proposition 2.8.1 carries over.

(2) It is easy to show that a subset of a complete metric space is complete if and only if it is closed. So one useful way to construct many complete metric spaces is to take a closed subset of a complete normed space. The purpose of Exercise 9.1.M is to show that every metric space arises in this manner, although it is not always a natural context.

(3) If X has the discrete metric, the only way a sequence may converge to x is if the sequence is eventually constant (i.e., $x_n = x$ for all $n \ge N$). This is because the ball $B_{1/2}(x)$ equals $\{x\}$. So every subset of X is both open and closed. Also X is complete.

(4) Consider the 2-adic metric of Example 9.1.2 (4) again. The balls have the form $B_{2^{-d}}(n) = \{m \in \mathbb{Z} : m \equiv n \pmod{2^d}\}$. The sequence $(2^n)_{n=1}^\infty$ converges to 0 because $\rho_2(2^n,0) = 2^{-n} \to 0$. Observe that $1 - (-2)^n$ is an odd multiple of 3 for all $n \ge 1$. The sequence $a_n = (1 - (-2)^n)/3$ is Cauchy because if $n > m \ge N$, then $a_n - a_m = (-2)^m a_{n-m}$ and therefore $\rho_2(a_m,a_n) = 2^{-m} \le 2^{-N}$. This sequence does not converge.

(5) If $X \subseteq \mathbb{R}^n$ is closed, the metric space $(K(X), d_H)$ is complete; see Theorem 11.7.2.

Continuous functions are defined by analogy with the norm case.

9.1.6. DEFINITION. A function f from a metric space (X, ρ) into a metric space (Y, σ) is **continuous** if for every $x_0 \in X$ and $\varepsilon > 0$, there is a $\delta > 0$ such that $\sigma(f(x), f(x_0)) < \varepsilon$ whenever $\rho(x, x_0) < \delta$.

The proof of Theorem 5.3.1 goes through without change.

9.1.7. THEOREM. *Let f map a metric space (X, ρ) into a metric space (Y, σ). The following are equivalent:*

(1) *f is continuous on X;*
(2) *for every sequence (x_n) with $\lim_{n\to\infty} x_n = a \in X$, we have $\lim_{n\to\infty} f(x_n) = f(a)$; and*
(3) *$f^{-1}(U) = \{x \in X : f(x) \in U\}$ is open in X for every open set U in Y.*

As in the norm case, if the domain of a continuous function is not compact, the function need not be bounded. However, we have the same solution: consider the normed vector space $C_b(X, \mathbb{R}^m)$ of all *bounded* continuous functions $f : X \to \mathbb{R}^m$ with the sup norm on functions, i.e., $\|f\|_\infty = \sup_{x\in X} \|f(x)\|$, where $\|\cdot\|$ is the Euclidean norm on $\mathbb{R}^m$.

9.1.8. THEOREM. *The space $C_b(X, \mathbb{R}^m)$ of all bounded continuous functions on a metric space X with the sup norm $\|f\| = \sup\{\|f(x)\| : x \in X\}$ is complete.*

PROOF. The proofs of Theorems 8.2.1 and 8.2.2 work with $C_b(X, \mathbb{R}^m)$. As we noted after the latter theorem, compactness is used in its proof only to ensure that the sup norm is finite. Therefore, $C_b(X, \mathbb{R}^m)$ is a complete normed vector space. The details are left as an exercise. ∎

Exercises for Section 9.1

A. Show that $\rho(x, y) = |e^x - e^y|$ is a metric on $\mathbb{R}$.

B. Show that every subset of a discrete metric space is both open and closed.

C. Prove that U is open in (X, ρ) if and only if $X \setminus U$ is closed.

D. Prove Theorem 9.1.7.

E. Given a metric space (X, ρ), define a new metric on X by $\sigma(x, y) = \min\{\rho(x, y), 1\}$.

   (a) Show that σ is a metric on X. Observe that X has finite diameter in the σ metric.
   (b) Show that $\lim_{n\to\infty} x_n = x$ in (X, ρ) if and only if $\lim_{n\to\infty} x_n = x$ in (X, σ).
   (c) Show that (x_n) is Cauchy in (X, ρ) if and only if it is Cauchy in (X, σ). Hence completeness is the same for these two metrics.

F.   Suppose that V is a vector space with norm $\|\cdot\|$. If (X,ρ) is a metric space, observe that the space $C_b(X,V)$ of all bounded continuous functions from X to V is a vector space. Show that $\|f\|_\infty = \sup_{x\in X} \|f(x)\|$ is a norm on $C_b(X,V)$.

G.   Two metrics ρ and σ on a set X are **topologically equivalent** if for each $x \in X$ and $r > 0$, there is an $s = s(r,x) > 0$ such that $B_s^\rho(x) \subset B_r^\sigma(x)$ and $B_s^\sigma(x) \subset B_r^\rho(x)$.

    (a) Prove that topologically equivalent metrics have the same open and closed sets.
    (b) Prove that topologically equivalent metrics have the same convergent sequences.
    (c) Give examples of topologically equivalent metrics with different Cauchy sequences.

H.   Two metrics ρ and σ on a set X are **equivalent** if there are constants $0 < c < C$ such that $c\rho(x,y) \le \sigma(x,y) \le C\rho(x,y)$ for all $x,y \in X$.

    (a) Prove that equivalent metrics are topologically equivalent.
    (b) Prove that equivalent metrics have the same Cauchy sequences.
    (c) Give examples of topologically equivalent metrics that are not equivalent.

I.   Define a function on $\mathcal{M}_n \times \mathcal{M}_n$ by $\rho(A,B) = \text{rank}(A-B)$. Prove that ρ is a metric that is topologically equivalent to the discrete metric.

J.   Put a metric ρ on all the words in a dictionary by defining the distance between two distinct words to be 2^{-n} if the words agree for the first n letters and are different at the $(n+1)$st letter. Here we agree that a space is distinct from a letter. For example, $\rho(\text{car}, \text{cart}) = 2^{-3}$ and $\rho(\text{car}, \text{call}) = 2^{-2}$.

    (a) Verify that this is a metric.
    (b) Suppose that words w_1, w_2, and w_3 are listed in alphabetical order. Show that $\rho(w_1,w_2) \le \rho(w_1,w_3)$.
    (c) Suppose that words w_1, w_2, and w_3 are listed in alphabetical order. Find a formula for $\rho(w_1,w_3)$ in terms of $\rho(w_1,w_2)$ and $\rho(w_2,w_3)$.

K.   Recall the 2-adic metric of Examples 9.1.2 (4) and 9.1.5 (4). Extend it to $\mathbb{Q}$ by setting $\rho_2(a/b,a/b) = 0$ and, if $a/b \ne c/d$, then $\rho_2(a/b,c/d) = 2^{-e}$, where e is the unique integer such that $a/b - c/d = 2^e(f/g)$ and both f and g are odd integers.

    (a) Prove that ρ_2 is a metric on $\mathbb{Q}$.
    (b) Show that the sequence of integers $a_n = (1-(-2)^n)/3$ converges in $(\mathbb{Q},\rho_2)$.
    (c) Find the limit of $\dfrac{n!}{n!+1}$ in this metric.

L.   Complete the details of Theorem 9.1.8 as follows:

    (a) Prove that Theorem 8.2.1 is valid when S is replaced by a metric space X.
    (b) Prove that $C_b(X,\mathbb{R}^m)$ is a complete normed vector space.   HINT: Theorem 8.2.2

M.   Suppose that (X,ρ) is a nonempty metric space. Let $C_b(X)$ be the normed space of all bounded continuous functions on X with the sup norm $\|f\|_\infty = \sup\{|f(x)| : x \in X\}$.

    (a) Fix x_0 in X. For each $x \in X$, define $f_x(y) = \rho(x,y) - \rho(x_0,y)$ for $y \in X$. Show that f_x is a bounded continuous function on X.
    (b) Show that $\|f_x - f_y\|_\infty = \rho(x,y)$.
    (c) Hence deduce that the map that takes $x \in X$ to the function f_x identifies X with a subset F of $C_b(X)$ that induces the same metric.

N.   (a) Give an example of a decreasing sequence of closed balls in a complete metric space with empty intersection. Compare with Exercise 7.2.J.
       HINT: Use a metric on $\mathbb{N}$ topologically equivalent to the discrete metric so that $\{n \ge k\}$ are closed balls.
    (b) Show that a metric space (M,d) is complete if and only if every decreasing sequence of closed balls with radii going to zero has a nonempty intersection.

9.2 Compact Metric Spaces

As we have mentioned in the last two chapters, in general, a closed and bounded set need not be compact. There are several characterizations of compactness in metric spaces, every bit as useful and important as the Heine–Borel Theorem (4.4.6). In this section, we make the needed definitions and then prove the characterizations.

We have to change our language slightly. Our old definition of compactness will be renamed **sequential compactness**. Although we introduce a new definition and call it compactness, we will prove that the two notions are equivalent in normed spaces and in metric spaces. There is a more general setting, topological spaces, where they differ, but we never deal with it in this book. The new definition also makes sense in normed spaces and could have been introduced there.

9.2.1. DEFINITION. A collection of open sets $\{U_\alpha : \alpha \in A\}$ in X is called an **open cover** of $Y \subseteq X$ if $Y \subseteq \bigcup_{\alpha \in A} U_\alpha$. A **subcover** of Y in $\{U_\alpha : \alpha \in A\}$ is just a subcollection $\{U_\alpha : \alpha \in B\}$ for some $B \subseteq A$ that is still a cover of Y. In particular, it is a **finite subcover** if B is finite, that is, a finite collection of the U_α that covers Y.

A collection of closed sets $\{C_\alpha : \alpha \in A\}$ has the **finite intersection property** if every *finite* subcollection has nonempty intersection.

For example, consider the cover of $[a,b]$ indicated in Figure 9.1, where there are three sets, each on its own horizontal line.

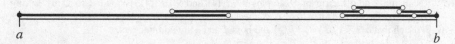

a b

FIG. 9.1 An open cover for $X = [a,b]$.

9.2.2. DEFINITION. A metric space X is **compact** if every open cover of X has a finite subcover. A metric space X is **sequentially compact** if every sequence of points in X has a convergent subsequence.

A metric space X is **totally bounded** if for every $\varepsilon > 0$, there are finitely many points $x_1, \ldots, x_k \in X$ such that $\{B_\varepsilon(x_i) : 1 \le i \le k\}$ is an open cover. This is an immediate generalization of the definition for subsets of $\mathbb{R}^m$, Definition 8.6.6.

9.2.3. BOREL–LEBESGUE THEOREM.
For a metric space X, the following are equivalent.

(1) *X is compact.*
(2) *Every collection of closed subsets of X with the finite intersection property has nonempty intersection.*
(3) *X is sequentially compact.*
(4) *X is complete and totally bounded.*

PROOF. We assume (1), X is compact, and prove (2). Suppose that $\{C_\alpha : \alpha \in A\}$ is a collection of closed sets such that $\bigcap_\alpha C_\alpha = \varnothing$. Consider the open sets $U_\alpha = C_\alpha'$, the complements of C_α. Then $\bigcup_\alpha U_\alpha = (\bigcap_\alpha C_\alpha)' = X$. Thus there is a finite subcover $X = U_{\alpha_1} \cup \cdots \cup U_{\alpha_k}$. Consequently,

$$C_{\alpha_1} \cap \cdots \cap C_{\alpha_k} = (U_{\alpha_1} \cup \cdots \cup U_{\alpha_k})' = \varnothing.$$

So no collection of closed sets with empty intersection has the finite intersection property, the contrapositive of (2).

Now assume (2) and let (x_i) be a sequence in X. Define $C_n = \overline{\{x_i : i \geq n\}}$. This is a decreasing sequence of closed sets. The collection $\{C_n : n \geq 1\}$ has the finite intersection property, since the intersection of $C_{n_1}, \ldots, C_{n_k}$ contains the point x_n, where $n = \max\{n_1, \ldots, n_k\}$. By hypothesis, there is a point x in $\bigcap_{n \geq 1} C_n$. Recursively define a sequence as follows. Let $n_0 = 1$. If n_{k-1} is defined, use the fact that $x \in \overline{\{x_i : i > n_{k-1}\}}$ to choose $n_k > n_{k-1}$ with $\rho(x, x_{n_k}) < 1/k$. By construction, $\lim_{k \to \infty} x_{n_k} = x$.

Assume sequential compactness, (3), and consider (4). If (x_i) is a Cauchy sequence, select a convergent subsequence, say $\lim_{k \to \infty} x_{n_k} = x$. Given $\varepsilon > 0$, use the Cauchy property to find N such that $i, j \geq N$ implies that $\rho(x_i, x_j) < \varepsilon/2$. Then use the convergence to find $n_k > N$ such that $\rho(x, x_{n_k}) < \varepsilon/2$. Then if $i \geq N$,

$$\rho(x, x_i) \leq \rho(x, x_{n_k}) + \rho(x_{n_k}, x_i) < \tfrac{\varepsilon}{2} + \tfrac{\varepsilon}{2} = \varepsilon.$$

So $\lim_{i \to \infty} x_i = x$. Thus X is complete.

If X were not totally bounded, then there would be some $\varepsilon > 0$ such that no finite collection of ε-balls could cover X. Recursively select points $x_k \in X$ such that $x_k \notin B_\varepsilon(x_1) \cup \cdots \cup B_\varepsilon(x_{k-1})$ for all $k \geq 2$. Consider the sequence (x_i). We will obtain a contradiction by showing that there is no convergent subsequence. Indeed, if (x_{n_i}) were a convergent subsequence, then it would be Cauchy. So for some N large, all $i > N$ satisfy $\rho(x_i, x_N) < \varepsilon$, contrary to the fact that $x_i \notin B_\varepsilon(x_N)$. Therefore, X must be totally bounded.

Finally, we show that (4) implies (1). For each $k \geq 1$, choose a finite set $X_k \subset X$ such that $\bigcup_{x \in X_k} B_{1/k}(x)$ covers X. Suppose (1) is false, that is, there is an open cover $\mathscr{U}$ with no finite subcover. This leads to a contradiction.

We will recursively define a sequence of points $y_k \in X_k$ such that $\bigcap_{j=1}^k \overline{B_{1/j}(y_j)}$ does not have a finite subcover from $\mathscr{U}$. For $k = 1$, suppose, for each $x \in X_1$, that $\overline{B_1(x)}$ had a finite subcover, call it $\mathscr{U}_x \subseteq \mathscr{U}$. Then $\bigcup_{x \in X_1} \mathscr{U}_x$ covers X and, being a finite union of finite sets, is finite. This contradicts $\mathscr{U}$ having no finite subcovers of X, and so there is some $y_1 \in X_1$ such that $\overline{B_1(y_1)}$ has no finite subcover from $\mathscr{U}$.

Suppose that we have defined $y_1, \ldots, y_{k-1}$. Let $S = \bigcap_{j=1}^{k-1} \overline{B_{1/j}(y_j)}$. For each $x \in X_k$, assume $S \cap \overline{B_{1/k}(x)}$ had a finite subcover from $\mathscr{U}$, call it $\mathscr{U}_x$. Then S would have a finite subcover from $\mathscr{U}$, namely $\bigcup_{x \in X_k} \mathscr{U}_x$, contradicting our choice of $y_1, \ldots, y_{k-1}$. Thus, there is $y_k \in X_k$ such that $\bigcap_{j=1}^k \overline{B_{1/j}(y_j)}$ has no finite subcover from $\mathscr{U}$.

We claim that (y_k) is Cauchy. Suppose $k, j \in \mathbb{N}$ with $k \geq j$. Now $\cap_{i=1}^{k} \overline{B_{1/i}(y_i)}$ is nonempty because the empty set has a finite subcover, the one with no elements of $\mathcal{U}$. Picking x in this set, we have $x \in \overline{B_{1/j}(y_j)} \cap \overline{B_{1/k}(y_k)}$. Hence

$$\rho(y_j, y_k) \leq \rho(y_j, x) + \rho(x, y_k) \leq \tfrac{1}{j} + \tfrac{1}{k}.$$

Given $\varepsilon > 0$, choose N so large that $N > 2/\varepsilon$. Then for $j, k \geq N$, it follows that $\rho(y_j, y_k) \leq 2/N < \varepsilon$.

But (4) states that X is complete, and thus $y = \lim\limits_{k \to \infty} y_k$ exists. There is some $V \in \mathcal{U}$ such that $y \in V$. Since V is open, there is $\varepsilon > 0$ such that $B_\varepsilon(y) \subseteq V$. Choose $k > 2/\varepsilon$ with $\rho(y_k, y) < \varepsilon/2$. Then since $1/k < \varepsilon/2$,

$$\cap_{i=1}^{k} \overline{B_{1/i}(y_i)} \subset \overline{B_{1/k}(y_k)} \subset B_\varepsilon(y) \subset V.$$

Thus $\cap_{i=1}^{k} \overline{B_{1/i}(y_i)}$ has a finite subcover, namely $\{Y\}$. But $\cap_{i=1}^{k} \overline{B_{1/i}(y_i)}$ does not a finite subcover, a contradiction, and so (1) holds. ∎

In general, being complete and bounded is not sufficient to imply that a metric space is compact. For example, by Exercise 9.1.E the real line has a bounded metric topologically equivalent to the usual one. In this new metric it is bounded and complete but is not compact.

Perhaps a more natural example is the closed unit ball B of $C[0, 1]$ in the max norm. This is complete because it is a closed subset of a complete space. However, it is not totally bounded. To see this, let f_n be the piecewise linear functions that take the value 0 on $[0, 1/(n+1)] \cup [1/(n-1), 1]$ and 1 at $1/n$. Since $\|f_n\|_\infty = 1$, they all lie in B. But $\|f_n - f_m\|_\infty = 1$ if $n \neq m$. Thus no $1/2$-ball can contain more than one of them. Hence no finite family of $1/2$-balls covers B. We have a complete description of the compact subsets of $C[0, 1]$, by the Arzelà–Ascoli Theorem (8.6.9).

All of our basic theorems on continuous functions go through for continuous functions on metric spaces. In particular, Theorem 5.4.3 can be established with the same proof. We will give a proof based on open covers. In Exercise 9.2.H, this will yield the metric space version of the Extreme Value Theorem (5.4.4).

9.2.4. THEOREM. *Let $f : X \to Y$ be a continuous map between metric spaces. If $C \subseteq X$ is compact, then the image set $f(C)$ is compact.*

PROOF. Let $\mathcal{U} = \{U_\alpha : \alpha \in A\}$ be an open cover of $f(C)$ in Y. Since f is continuous, $V_\alpha := f^{-1}(U_\alpha)$ are open sets in X. The collection $\mathcal{V} = \{V_\alpha : \alpha \in A\}$ is an open cover of C. Indeed, for each $x \in C$, $f(x) \in f(C)$ and thus $f(x) \in U_\alpha$ for some α. Hence x belongs to V_α. By the compactness of C, select a finite subcover $V_{\alpha_1}, \ldots, V_{\alpha_k}$. Then

$$f(C) \subset \bigcup_{i=1}^{k} f(V_{\alpha_i}) \subset \bigcup_{i=1}^{k} U_{\alpha_i}.$$

So $U_{\alpha_1}, \ldots, U_{\alpha_k}$ is a finite subcover of $f(C)$. Therefore, $f(C)$ is compact. ∎

Exercises for Section 9.2

A. Let (X, ρ) be a metric space and consider $Y \subset X$ as a metric space with the same metric.

(a) Show that every open set U in Y has the form $V \cap Y$ for some open set V in X.
HINT: This is easy for balls.
(b) Show that Y is compact if and only if every collection $\{V_\alpha : \alpha \in A\}$ of open sets in X that covers Y has a finite subcover.

B. Show that if Y is a subset of a complete metric space X, then Y is compact if and only if it is closed and totally bounded.

C. Show that a closed subset of a compact metric space is compact.

D. Show that every compact metric space is **separable** (i.e., it has a countable dense subset).

E. (a) Prove that every open subset U of $\mathbb{R}^n$ is the countable union of compact subsets.
HINT: Use the distance to U^c and the norm to define the sets.
(b) Show that every open cover of an open subset of $\mathbb{R}^n$ has a countable subcover.

F. Prove **Cantor's Intersection Theorem**: A decreasing sequence of nonempty compact subsets $A_1 \supset A_2 \supset \cdots$ of a metric space (X, ρ) has nonempty intersection.

G. Show that a continuous function from a compact metric space (X, ρ) into a metric space (Y, d) is uniformly continuous.   HINT: Fix $\varepsilon > 0$. For $x \in X$, choose $\delta_x > 0$ such that $\rho(x, t) < 2\delta_x$ implies $d(f(x), f(t)) < \varepsilon/2$. Then $\{B_{\delta_x}(x) : x \in X\}$ covers X.

H. If f is a continuous function from a compact metric space (X, ρ) into $\mathbb{R}$, prove that there is a point $x_0 \in X$ such that $|f(x_0)| = \sup\{|f(x)| : x \in X\}$.
HINT: Compare to the proof of the Extreme Value Theorem (5.4.4).

I. Let S_n for $n \geq 1$ be a finite union of disjoint closed balls in $\mathbb{R}^k$ of radius at most 2^{-n} such that $S_{n+1} \subset S_n$ and S_{n+1} has at least two balls inside each ball of S_n. Prove that $C = \bigcap_{n \geq 1} S_n$ is a perfect, nowhere dense compact subset of $\mathbb{R}^k$.
HINT: Compare with Example 4.4.8.

J. If f is a continuous one-to-one function of a compact metric space X onto Y, show that f^{-1} is continuous.   HINT: Theorem 9.2.4.

K. Show that the previous exercise is false if X is not compact.
HINT: Map $(0, 1]$ onto a circle with a tail (i.e., the figure 6).

L. We say that (X, ρ) is a **second countable metric space** if there is a countable collection $\mathscr{U}$ of open balls in X such that for every $x \in X$ and $r > 0$, there is a ball $U \in \mathscr{U}$ with $x \in U \subset B_r(x)$.
Prove that (X, ρ) is second countable if and only if it is separable.
HINT: For $\Rightarrow$, take the centres of balls in $\mathscr{U}$. For $\Leftarrow$, take all balls of radius $1/k$, $k \geq 1$, about each point in a countable dense set.

9.3 Complete Metric Spaces

Now we turn to an important consequence of completeness.

9.3.1. DEFINITION. A subset A of a metric space (X, ρ) is **nowhere dense** if $\text{int}(\overline{A}) = \varnothing$ (i.e., the closure of A has no interior). A subset B is said to be **first category** if it is the *countable* union of nowhere dense sets. A subset Y of a *complete* metric space is a **residual set** if the complement Y' is first category.

Nowhere dense sets are small in a certain sense, and thus sets of first category are also considered to be small. For example, the Cantor set is nowhere dense in $\mathbb{R}$. Our main result, which has surprisingly powerful consequences, is that complete metric spaces are never first category.

9.3.2. BAIRE CATEGORY THEOREM.

Let (X,ρ) be a complete metric space. Then the union of countably many nowhere dense subsets of X has no interior, and in particular is a proper subset of X. Equivalently, the intersection of countably many dense open subsets of X is dense in X.

PROOF. Consider a sequence $(A_n)_{n=1}^{\infty}$ of nowhere dense subsets of X. To show that the complement of $\bigcup_{n\geq 1} A_n$ is dense in X, take any ball $\overline{B_{r_0}(x_0)}$. We will construct a point in this ball that is not in any $\overline{A_n}$. It will then follow that this union has no interior, whence the complement contains points arbitrarily close to each point $x \in X$, and so is dense.

Since $\overline{A_1}$ has no interior, it does not contain $B_{r_0/2}(x_0)$. Pick x_1 in $B_{r_0/2}(x_0) \setminus \overline{A_1}$. Since $\overline{A_1}$ is closed, $\mathrm{dist}(x_1,\overline{A_1}) > 0$. So we may choose an $0 < r_1 < r_0/2$ such that $\overline{B_{r_1}(x_1)}$ is disjoint from $\overline{A_1}$. Note that $\overline{B_{r_1}(x_1)} \subset \overline{B_{r_0}(x_0)}$. Proceed recursively choosing a point $x_{n+1} \in B_{r_n/2}(x_n)$ and an $r_{n+1} \in (0,r_n/2)$ such that $\overline{B_{r_{n+1}}(x_{n+1})}$ is disjoint from $\overline{A_{n+1}}$. Clearly, $\overline{B_{r_{n+1}}(x_{n+1})} \subset \overline{B_{r_n}(x_n)}$.

The sequence $(x_n)_{n=1}^{\infty}$ is Cauchy. Indeed, given $\varepsilon > 0$, choose N such that $2^{-N}r_0 < \varepsilon/2$. Then $r_N < \varepsilon/2$. However, all x_n for $n \geq N$ lie in $\overline{B_{r_N}(x_N)}$. Thus for $n,m \geq N$,

$$\rho(x_n,x_m) \leq \rho(x_n,x_N) + \rho(x_N,x_m) < \frac{\varepsilon}{2} + \frac{\varepsilon}{2} = \varepsilon.$$

Because X is a complete space, there is a limit $x_\infty = \lim_{n \to \infty} x_n$ in X. The point x_∞ belongs to $\bigcap_{n \geq 1} \overline{B_{r_n}(x_n)}$. Hence it is disjoint from every $\overline{A_n}$ for $n \geq 1$.

If U_n are dense open subsets of X, their complements A_n are closed and nowhere dense. By the previous paragraphs, $\bigcup_{n \geq 1} A_n$ has dense complement. This complement is exactly $\bigcap_{n \geq 1} U_n$. ∎

Here is one interesting and unexpected consequence. It says that *most* continuous functions are nowhere differentiable. This is rather nonintuitive, since it was hard work to construct even one explicit example of such a function in Example 8.4.9.

9.3.3. PROPOSITION. *The set of continuous, nowhere differentiable functions on an interval $[a,b]$ is a residual set and in particular is dense in $C[a,b]$.*

PROOF. Say that a function f is Lipschitz at x_0 if there is a constant L such that $|f(x) - f(x_0)| \leq L|x - x_0|$ for all $x \in [a,b]$. Our first observation is that if f is differentiable at x_0, then it is also Lipschitz at x_0. From the definition of derivative, $\lim_{x \to x_0} \frac{f(x) - f(x_0)}{x - x_0} = f'(x_0)$. Choose $\delta > 0$ such that for $|x - x_0| < \delta$,

$$\frac{|f(x) - f(x_0)|}{|x - x_0|} \le |f'(x_0)| + 1.$$

Then for $|x - x_0| < \delta$, $|f(x) - f(x_0)| \le (|f'(x_0)| + 1)|x - x_0|$. If $|x - x_0| \ge \delta$,

$$|f(x) - f(x_0)| \le 2\|f\|_\infty \le \frac{2\|f\|_\infty}{\delta}|x - x_0|.$$

So $L = \max\{|f'(x_0)| + 1, 2\|f\|_\infty/\delta\}$ is a Lipschitz constant at x_0.

Let A_n consist of all functions $f \in C[a,b]$ such that f has Lipschitz constant $L \le n$ at some point $x_0 \in [a,b]$. Let us show that A_n is closed. Suppose that $(f_k)_{k=1}^\infty$ is a sequence of functions in A_n converging uniformly to a function f. Each f_k has Lipschitz constant n at some point $x_k \in [a,b]$. Since $[a,b]$ is compact, there is a subsequence x_{k_i} converging to a point $x_0 \in [a,b]$. Then

$$\begin{aligned}
|f(x) - f(x_0)| &\le |f(x) - f_{k_i}(x)| + |f_{k_i}(x) - f_{k_i}(x_{k_i})| \\
&\quad + |f_{k_i}(x_{k_i}) - f_{k_i}(x_0)| + |f_{k_i}(x_0) - f(x_0)| \\
&\le \|f - f_{k_i}\|_\infty + n|x - x_{k_i}| + n|x_{k_i} - x_0| + \|f - f_{k_i}\|_\infty \\
&= 2\|f - f_{k_i}\|_\infty + n(|x - x_{k_i}| + |x_{k_i} - x_0|).
\end{aligned}$$

Take a limit as $i \to \infty$ to obtain $|f(x) - f(x_0)| \le n|x - x_0|$. Thus A_n is closed.

Next we show that A_n has no interior and hence is nowhere dense. Fix a function $f \in A_n$ and an $\varepsilon > 0$. Let us look for a function g in $B_\varepsilon(f)$ that is not in A_n. By Theorem 5.5.9, f is uniformly continuous. Choose $\delta > 0$ such that $|x - y| < \delta$ implies that $|f(x) - f(y)| < \varepsilon/4$. Construct a piecewise linear continuous function h that agrees with f on a sequence $a = x_0 < x_1 < \cdots < x_N = b$, where $x_k - x_{k-1} < \delta$ for $1 \le k \le N$. A simple estimate shows that $\|f - h\|_\infty < \varepsilon/2$. The function h is Lipschitz with constant L, say. Choose $M > 4\pi(L+n)/\varepsilon$ and set $g = h + \frac{\varepsilon}{2}\sin Mx$. Then

$$\|f - g\|_\infty \le \|f - h\|_\infty + \|h - g\|_\infty < \frac{\varepsilon}{2} + \frac{\varepsilon}{2} < \varepsilon.$$

To see that $g \notin A_n$, take any point x_0. We can always choose $x \in [a,b]$ with $|x - x_0| < 2\pi/M$ such that $\sin Mx = \pm 1$ has sign opposite to the sign of $\sin Mx_0$. Thus,

$$\begin{aligned}
|g(x) - g(x_0)| &\ge \frac{\varepsilon}{2}|\sin Mx - \sin Mx_0| - |h(x) - h(x_0)| \\
&\ge \frac{\varepsilon}{2} - L|x - x_0| \ge \left(\frac{M\varepsilon}{4\pi} - L\right)|x - x_0| > n|x - x_0|.
\end{aligned}$$

So g does not have Lipschitz constant n at any point $x_0 \in [a,b]$.

Recall from Theorem 8.2.2 that $C[a,b]$ is complete. By the Baire Category Theorem, it is not the union of countably many nowhere dense sets. The first category set $\bigcup_{n \ge 1} A_n$ contains all functions that are differentiable at any single point. Thus the complement, consisting entirely of nowhere differentiable functions, is a residual set. Therefore, nowhere differentiable functions are dense in $C[a,b]$. ∎

Exercises for Section 9.3

A. Show that A is nowhere dense in X if and only if $X \setminus \overline{A}$ is dense in X.

B. Show that $\mathbb{R}^2$ is not the union of countably many lines.

C. Show that a countable complete metric space has isolated points, i.e., points that are open (and closed).
HINT: Write the space as a union of points and apply Baire Category.

D. Show that a complete metric space containing more than one point that has no isolated points is uncountable.   HINT: Use the previous exercise.

E. A **G_δ set** is the intersection of a countable family of open sets.

(a) Show that if $A \subset \mathbb{R}$ is closed, then A is a G_δ set.
(b) Show that $\mathbb{Q}$ is not a G_δ subset of $\mathbb{R}$.   HINT: Show that $\mathbb{R} \setminus \mathbb{Q}$ is not first category.

F. (a) If f is a real-valued function on a metric space X, show that the set of points at which f is continuous is a G_δ set.   HINT: Show that the set U_k of points x for which there are $i \in \mathbb{N}$ and $\delta > 0$ such that $|f(y) - \frac{i}{k}| < \frac{1}{k}$ for $y \in B_\delta(x)$ is open.
(b) Show that no function on $[0,1]$ can be continuous just on $\mathbb{Q}$. Compare Example 5.2.9.

G. Suppose that (f_n) is a sequence of continuous real-valued functions on a complete metric space X that converges pointwise to a function f.

(a) Prove that there are $M > 0$ and an open set $U \subset X$ such that $\sup\{|f_n(x)| : n \geq 1\} \leq M$ for all $x \in U$.   HINT: Let $A_k = \{x \in X : \sup_{n \geq 1} |f_n(x)| \leq k\}$.
(b) If f is continuous and $\varepsilon > 0$, show that there are an open set U and an integer N such that $|f(x) - f_n(x)| < \varepsilon$ for all $x \in U$ and $n \geq N$.
HINT: Let $B_k = \{x \in X : \sup_{n \geq k} |f(x) - f_n(x)| \leq \varepsilon/2\}$.

H. (a) Show that the set of compact nowhere dense subsets of $\mathbb{R}^n$ is dense in $K(X)$, where $K(X)$ is equipped with the Hausdorff metric of Example 9.1.2 (5).
HINT: If C is compact and $\varepsilon > 0$, use a finite ε-net. Finite sets are nowhere dense.
(b) Show that the set A_n of those compact sets in $K(X)$ that contain a ball of radius $1/n$ is a closed set with no interior.

I. **Banach–Steinhaus Theorem.** Suppose that X and Y are complete normed vector spaces, and $\{T_\alpha : \alpha \in A\}$ is a family of continuous linear maps of X into Y such that for each $x \in X$, $K_x = \sup_{\alpha \in A} \|T_\alpha x\| < \infty$.

(a) Let $A_n = \{x \in X : K_x \leq n\}$. Show that A_n is closed.
(b) Prove that there is some n_0 such that A_{n_0} has interior, say containing $B_\varepsilon(x_0)$.
(c) Show that there is a finite constant L such that every T_α has Lipschitz constant L.
HINT: If $\|x\| < 1$, then $x_0 + \varepsilon x \in A_{n_0}$. Estimate $\|T_\alpha x\|$.

J. Show that $[0,1]$ is not the *disjoint* union of a countable family $\{A_n : n \geq 1\}$ of nonempty closed sets.   HINT: If $U_n = \mathrm{int}A_n$, observe that $X := [0,1] \setminus \bigcup_{n \geq 1} U_n = \bigcup_{n \geq 1}(A_n \setminus U_n)$ is complete. Find $n_0 \geq 1$ and (a,b) such that $\varnothing \neq X \cap (a,b) \subset A_{n_0}$. Show that $U_n \cap (a,b) = \varnothing$ for $n \neq n_0$.

K. A function on $[0,1]$ that is not monotonic on any interval is called a **nowhere monotonic function**. Show that these functions are a residual subset of $C[0,1]$.
HINT: Let $A_n = \{\pm f : \text{there is } x \in [0,1], (f(y) - f(x))(y - x) \geq 0 \text{ for } |y - x| \leq \frac{1}{n}\}$.

Part B
Applications

Chapter 10
Approximation by Polynomials

This chapter introduces some of the essentials of approximation theory, in particular approximating functions by "nice" ones such as polynomials. In general, the intention of approximation theory is to replace some complicated function with a new function, one that is easier to work with, at the price of some (hopefully small) difference between the two functions. The new function is called an approximation. There are two crucial issues in using an approximation: first, how much simpler is the approximation? and second, how close is the approximation to the original function? Deciding which approximation to use requires an analysis of the trade-off between these two issues.

Of course, the answers to these two questions depend on the exact meanings of *simpler* and *close*, which vary according to the context. In this chapter, we study approximations by polynomials. *Close* is measured by some norm. We concentrate on the uniform norm, so that a polynomial approximation should be close to the function everywhere on a given interval.

Approximations are closely tied to the notions of limit and convergence, since a sequence of functions approximating a function f to greater and greater accuracy converges to f in the norm used. Different approximation schemes correspond to different norms.

10.1 Taylor Series

The first approximation taught in calculus and the most often used approximation is the tangent line approximation: If $f : \mathbb{R} \to \mathbb{R}$ is differentiable at $a \in \mathbb{R}$, then for x near a, we have

$$f(x) \approx f(a) + f'(a)(x - a).$$

However, you should learn early that an approximation is only as good as the error estimate that can be verified. Unless we can estimate the error, the difference between $f(x)$ and its approximation $f(a) + f'(a)(x - a)$, it is impossible to say whether

K.R. Davidson and A.P. Donsig, *Real Analysis and Applications: Theory in Practice*,
Undergraduate Texts in Mathematics, DOI 10.1007/978-0-387-98098-0_10,
© Springer Science + Business Media, LLC 2010

the approximation is worth the trouble. For example, why not just approximate $f(x)$ with the constant $f(a)$, since x is 'near' a and f is continuous?

We start with the error estimate for the constant approximation $f(a)$. This error estimate comes from the Mean Value Theorem (6.2.2), which gives us the estimate $|f(x) - f(a)| = |f'(c)(x-a)| \leq C|x-a|$, where c is some point between a and x and $C = \sup\{|f'(c)| : c \text{ between } x \text{ and } a\}$. When C is finite, we obtain a useful error estimate for this constant approximation. Notice that this estimate does not require us to find $f(x)$ exactly. If we could easily find $f(x)$, we wouldn't bother with the approximation.

A more sophisticated use of the Mean Value Theorem shows that the tangent line has an error of the form $M(x-a)^2$ for a constant M that depends on f''. For x very close to a, this is a considerable improvement on $C|x-a|$. See Exercise 10.1.B.

In this section, we generalize these two approximations and their error estimates to take account of higher derivatives—in other words, we generalize the Mean Value Theorem. Since this method requires many derivatives, and because it uses information at only one point, it will not be an ideal method for uniform approximation over an interval. Nevertheless, it works very well in certain instances of great importance.

The role of the tangent line to f is replaced by a polynomial $P_n(x)$ of degree at most n that has the same derivatives at a as f up to the nth degree. This is all the agreement that the parameters of a polynomial of degree at most n permit.

10.1.1. DEFINITION. If f has n derivatives at a point $a \in [A, B]$, the **Taylor polynomial** of order n for f at a is

$$P_n(x) = f(a) + f'(a)(x-a) + \frac{f''(a)}{2}(x-a)^2 + \cdots + \frac{f^{(n)}(a)}{n!}(x-a)^n$$

$$= \sum_{k=0}^{n} \frac{f^{(k)}(a)}{k!}(x-a)^k.$$

10.1.2. LEMMA. *Let $f(x)$ belong to $C^n[A, B]$ (i.e., f has n continuous derivatives), and let $a \in [A, B]$. The Taylor polynomial $P_n(x)$ of order n for f at a is the unique polynomial $p(x)$ of degree at most n such that $p^{(k)}(a) = f^{(k)}(a)$ for $0 \leq k \leq n$.*

PROOF. Every polynomial of degree at most n has the form $p(x) = \sum_{j=0}^{n} a_j(x-a)^j$. We may differentiate this k times to obtain

$$p^{(k)}(x) = \sum_{j=k}^{n} j(j-1) \cdots (j+1-k)(x-a)^{j-k}.$$

Substituting $x = a$ yields $p^{(k)}(a) = k!a_k$. Therefore, we must choose the coefficients $a_k = f^{(k)}(a)/k!$, which yields the Taylor polynomial $P_n(x)$. ∎

The preceding simple lemma established that the Taylor polynomial is the appropriate analogue of the tangent line for higher-order polynomials. The hard work, and indeed the total content of this approximation, comes from the error estimate. In this case, the estimate is good only for points sufficiently close to a when there is reasonable control on the size of the $(n+1)$st derivative. The case $n = 0$ is a direct consequence of the Mean Value Theorem.

10.1.3. TAYLOR'S THEOREM.

Let $f(x)$ belong to $C^n[A,B]$, and furthermore assume that $f^{(n+1)}$ is defined and $|f^{(n+1)}(x)| \le M$ for $x \in [A,B]$. Let $a \in [A,B]$, and let $P_n(x)$ be the Taylor polynomial of order n for f at a. Then for each $x \in [A,B]$, the error of approximation $R_n(x) = f(x) - P_n(x)$ satisfies

$$|R_n(x)| \le \frac{M|x-a|^{n+1}}{(n+1)!}.$$

PROOF. Notice that for $0 \le k \le n$,

$$R_n^{(k)}(a) = f^{(k)}(a) - P_n^{(k)}(a) = 0.$$

Because P_n is a polynomial of degree at most n,

$$R_n^{(n+1)}(x) = f^{(n+1)}(x) - P_n^{(n+1)}(x) = f^{(n+1)}(x).$$

Applying the Mean Value Theorem to $R_n^{(n)}$ gives

$$|R_n^{(n)}(x)| = |R_n^{(n)}(x) - R_n^{(n)}(a)| \le M|x-a|.$$

Suppose that for some k with $0 \le k < n$, we have shown that

$$|R_n^{(n-k)}(x)| \le \frac{M|x-a|^{k+1}}{(k+1)!}.$$

Then we integrate to obtain

$$|R_n^{(n-k-1)}(x)| = \left| R_n^{(n-k-1)}(a) + \int_a^x R_n^{(n-k)}(t)\,dt \right|$$
$$\le \left| 0 + \int_a^x \frac{M|t-a|^{k+1}}{(k+1)!}\,dt \right| = \frac{M|x-a|^{k+2}}{(k+2)!}.$$

We established the formula for $k = 0$, and have now completed the induction step. Eventually we obtain the desired formula when $k = n$,

$$|R_n(x)| \le \frac{M|x-a|^{n+1}}{(n+1)!}. \qquad \blacksquare$$

When f is C^∞, the **Taylor series** of f about a is $\sum\limits_{k=0}^{\infty} \dfrac{f^{(k)}(a)}{k!}(x-a)^k$. This is a power series, and so we must watch out for problems with convergence.

10.1.4. EXAMPLE. Consider $f(x) = e^x$. This function has the very nice property that $f' = f$. Thus $f^{(n)}(x) = e^x$ for all $n \geq 0$. Expanding around $a = 0$, we obtain Taylor polynomials $P_n(x) = \sum\limits_{k=0}^{n} \dfrac{x^k}{k!}$. Note that $|f^{(n+1)}(t)| = e^t \leq \max\{1, e^x\}$ if t lies between 0 and x. Thus Taylor's Theorem for the interval $[0,x]$ or $[x,0]$ says that the error is at most

$$\left| e^x - \sum_{k=0}^{n} \frac{x^k}{k!} \right| \leq \max\{1, e^x\} \frac{|x|^{n+1}}{(n+1)!}.$$

Even for small values of n, $P_n(x)$ is close to e^x around the origin; see Figure 10.1.

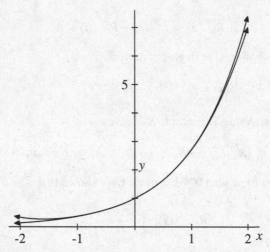

FIG. 10.1 The graphs of $y = e^x$ and $y = P_4(x)$.

The Ratio Test shows that $\lim\limits_{n \to \infty} \dfrac{|x|^{n+1}}{(n+1)!} = 0$ for every $x \in \mathbb{R}$. Thus the Taylor series converges to e^x on the whole real line. Moreover, this series converges uniformly on any interval $[-A,A]$. To see this, notice that the error estimate at any point $x \in [-A,A]$ is greatest for $x = A$. Hence

$$\sup_{|x| \leq A} \left| e^x - \sum_{k=0}^{n} \frac{x^k}{k!} \right| \leq \frac{e^A A^{n+1}}{(n+1)!}.$$

To compute e, we could use this formula to compute

$$\left| e - \sum_{k=0}^{n} \frac{1}{k!} \right| \leq \frac{e}{(n+1)!} \leq \frac{3}{(n+1)!}.$$

To obtain e to 10 decimal places, we need $\frac{3}{(n+1)!} < 0.5(10)^{-10}$ or $(n+1)! > 6(10)^{10}$. A calculation shows that we need $n = 13$.

This is not too bad, yet we can significantly increase the rate of convergence by using a smaller value of x. For example, suppose that we use $x = 1/16$ to compute $e^{1/16}$. We can then square this number four times to obtain e. If we use just the first ten terms, we have

$$\left| e^{1/16} - \sum_{k=0}^{10} \frac{1}{(16)^k k!} \right| \leq \frac{e^{1/16}}{(16)^{11}(11)!} < 1.6(10)^{-21}.$$

Then we take the number $a = \sum_{k=0}^{10} \frac{1}{(16)^k k!}$ and square it four times to obtain a^{16} as an approximation to e. Since we know that $e^{1/16} - \varepsilon < a < e^{1/16}$, where $\varepsilon = 1.6(10)^{-21}$, we have
$$e > a^{16} > (e^{1/16} - \varepsilon)^{16} > e - 16e^{15/16}\varepsilon > e - 7(10)^{-20}.$$

So roughly the same number of calculations yields almost double the number of digits of accuracy.

Consider the power series $\sum_{n=0}^{\infty} \frac{x^n}{n!}$. The Ratio Test shows that

$$\lim_{n \to \infty} \frac{|x|^{n+1}/(n+1)!}{|x|^n/n!} = \lim_{n \to \infty} \frac{|x|}{n+1} = 0$$

for every real x. Thus this power series has an infinite radius of convergence. Moreover, as we have shown, this series converges to the function e^x; and this convergence is uniform on each bounded interval.

A similar situation occurs for $\sin x$ and $\cos x$.

Many functions in common use are C^∞, meaning that they have continuous derivatives of all orders. For such functions, the Taylor polynomials of all orders are defined. Thus it is natural to consider the convergence of the Taylor series of f around $x = a$. Recall from Section 8.5 that every power series has a radius of convergence. In the previous example, the best possible result occurred—the power series of e^x has an infinite radius of convergence, and the limit of the series is the function itself. Unfortunately, things are not always so good.

10.1.5. EXAMPLE. Consider the function $f(x) = \dfrac{1}{1+x^2}$. The formulae for the derivatives are a bit complicated, so we use a trick. Consider the polynomials

$$P_{2n}(x) = \sum_{k=0}^{n} (-x^2)^k = 1 - x^2 + x^4 + \cdots + (-x^2)^n = \frac{1 - (-x^2)^{n+1}}{1+x^2}.$$

We deduce the estimate

$$\left| \frac{1}{1+x^2} - P_{2n}(x) \right| = \frac{x^{2n+2}}{1+x^2}.$$

It follows that $\lim_{n \to \infty} P_{2n}(x) = f(x)$ provided that $|x| < 1$, and this convergence is uniform on any interval $[-r, r]$ for $0 < r < 1$. Moreover, this estimate shows that

$$\lim_{x \to 0} \frac{|f(x) - P_{2n}(x)|}{x^{2n+1}} = \lim_{x \to 0} \frac{|x|}{1+x^2} = 0.$$

By Exercise 10.1.C, it follows that this is indeed the Taylor polynomial for f not only of order $2n$, but also of order $2n + 1$.

So the Taylor series for f about 0 is $\sum_{k=0}^{\infty} (-x^2)^k$. The radius of convergence is readily seen to be 1, since for $|x| \geq 1$ the terms to do not go to 0, while for $|x| < 1$, the geometric series does converge. Moreover, the limit is our function $f(x)$. Notice that even though f is defined and C^∞ on the whole real line, the Taylor series converges only on a finite interval.

10.1.6. EXAMPLE. In the last example, we showed that for any $r < 1$, the series for $1/(1 + x^2)$ converges uniformly on $[-r, r]$. Thus we may integrate this series term by term by Theorem 8.3.1. For $|x| = r < 1$,

$$\arctan(x) = \int_0^x \frac{1}{1+t^2}\, dt = \int_0^x \lim_{n \to \infty} \sum_{k=0}^{n} (-t^2)^k\, dt$$

$$= \lim_{n \to \infty} \int_0^x \sum_{k=0}^{n} (-t^2)^k\, dt = \lim_{n \to \infty} \sum_{k=0}^{n} \frac{(-1)^k}{2k+1} x^{2k+1} = \sum_{k=0}^{\infty} \frac{(-1)^k}{2k+1} x^{2k+1}.$$

This is the Taylor series for $\arctan(x)$. It also has radius of convergence 1 and converges uniformly on $[-r, r]$ for any $r < 1$. This series also converges at $x = \pm 1$ by the alternating series test.

The next thing to notice is that this series does converge to $\arctan(x)$ uniformly on $[-1, 1]$. To see this, we use the alternating series test at each point $x \in [-1, 1]$. We have $P_{2n}(x) = \sum_{k=0}^{n-1} \frac{(-1)^k}{2k+1} x^{2k+1}$. This is a sequence of polynomials of degree $2n - 1$ that converges to $\arctan(x)$ at each point in $[-1, 1]$. The corresponding series is alternating in sign with terms of modulus $|x|^{2k+1}/(2k+1)$ tending monotonically to 0. Thus the error is no greater than the modulus of the next term.

$$\left| \arctan(x) - P_{2n}(x) \right| < \frac{|x|^{2n+1}}{2n+1} \leq \frac{1}{2n+1}.$$

Therefore,

$$\sup_{|x| \leq 1} \left| \arctan(x) - P_{2n}(x) \right| \leq \frac{1}{2n+1}.$$

So P_{2n} converges uniformly to $\arctan(x)$ on $[-1, 1]$.

However, this sequence converges very slowly. Indeed, by the triangle inequality for the max norm in $C[-1,1]$,

$$\frac{1}{2n+1} = \|P_{2n} - P_{2n+2}\|_\infty \leq \|P_{2n} - f\|_\infty + \|f - P_{2n+2}\|_\infty.$$

So $\max\left\{\|P_{2n} - f\|_\infty, \|f - P_{2n+2}\|_\infty\right\} \geq \dfrac{1}{4n+2}$, a rather slow rate of convergence.

On the other hand, this estimate shows that the error on $[-r,r]$ is no more than $r^{2n+1}/(2n+1)$, which goes to zero quite quickly as $n \to \infty$, if r is small. So Taylor series can sometimes be a good approximation in a limited range. See Exercise 10.1.E for a method of rapidly computing π using these polynomials.

10.1.7. EXAMPLE. Even if f has derivatives of all orders that we can evaluate accurately *and* the Taylor series converges uniformly, the Taylor polynomials may not converge to the right function! The classic example of this is the function

$$f(x) = \begin{cases} e^{-1/x^2} & \text{if } x \neq 0, \\ 0 & \text{if } x = 0. \end{cases}$$

Figure 10.2 shows just how flat this function is near zero.

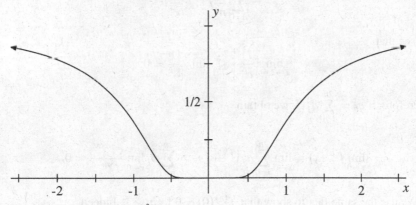

FIG. 10.2 The graph of $y = e^{-1/x^2}$.

We will show that f is C^∞ on all of $\mathbb{R}$ and $f^{(n)}(0) = 0$ for all n. We claim that there is a polynomial $q_n(x)$ of degree at most $2n$ such that

$$f^{(n)}(x) = \frac{q_n(x)}{x^{3n}} e^{-1/x^2} \quad \text{for } x \neq 0.$$

Indeed, this is obvious for $n = 0$, where $q_0 = 1$. We proceed by induction. If it is true for n, then by the product rule,

$$f^{(n+1)}(x) = f^{(n)\prime}(x) = e^{-1/x^2}\left(\frac{q_n'(x)}{x^{3n}} - \frac{3nq_n(x)}{x^{3n+1}} + \frac{q_n(x)}{x^{3n}}\left(\frac{-2}{x^3}\right)\right)$$

$$= \frac{x^3 q_n'(x) - 3nx^2 q_n(x) - 2q_n(x)}{x^{3n+3}}e^{-1/x^2}.$$

To check the degree of

$$q_{n+1}(x) = x^3 q_n'(x) - 3nx^2 q_n(x) - 2q_n(x),$$

note that

$$\deg(x^3 q_n') \le 3 + \deg q_n' \le 2 + \deg q_n \le 2n+2$$

and

$$\deg(3nx^2 q_n) \le 2 + \deg q_n \le 2n+2.$$

Clearly this nice algebraic formula for the derivatives is continuous on both $(-\infty,0)$ and $(0,\infty)$. So f is C^∞ everywhere except possibly at $x = 0$. Also,

$$\lim_{x\to 0}\frac{e^{-1/x^2}}{x^k} = \lim_{t\to\pm\infty}\frac{t^k}{e^{t^2}},$$

where we substitute $t = 1/x$. Since

$$e^{t^2} = \sum_{n=0}^{\infty}\frac{1}{n!}t^{2n} \ge \frac{1}{k!}t^{2k},$$

we see that

$$\lim_{t\to\pm\infty}\left|\frac{t^k}{e^{t^2}}\right| \le \lim_{t\to\pm\infty}\frac{k!}{|t|^k} = 0.$$

Therefore, if $q_n = \sum_{j=0}^{2n} a_j x^j$, we obtain

$$\lim_{x\to 0}f^{(n)}(x) = \lim_{x\to 0}\sum_{j=0}^{2n}a_j x^j\frac{e^{-1/x^2}}{x^{3n}} = \sum_{j=0}^{2n}a_j\lim_{x\to 0}\frac{e^{-1/x^2}}{x^{3n-j}} = 0.$$

We use the same fact to show that $f^{(n)}(0) = 0$ for $n \ge 1$. Indeed,

$$f^{(n+1)}(0) = \lim_{h\to 0}\frac{f^{(n)}(h) - f^{(n)}(0)}{h} = \lim_{h\to 0}\frac{q_n(h)e^{-1/h^2}}{h^{3n+1}} = 0.$$

So $f^{(n)}$ is defined on the whole line and is continuous for each n. Therefore, f is C^∞.

Because all of the derivatives of f vanish at $x = 0$, *all* of the Taylor polynomials are $P_n(x) = 0$. While this certainly converges rapidly on the whole real line, it converges to the wrong function! The Taylor polynomials completely fail to approximate f anywhere except at the one point $x = 0$.

Exercises for Section 10.1

A.   Find the Taylor polynomials of order 3 for each of the following functions at the given point a, and estimate the error at the point b.

(a) $f(x) = \tan x$ about $a = \frac{\pi}{4}$ and $b = 0.75$
(b) $g(x) = \sqrt{1+x^2}$ about $a = 0$ and $b = 0.1$
(c) $h(x) = x^4$ about $a = 1$ and $b = 0.99$
(d) $k(x) = \sinh x$ about $a = 0$ and $b = 0.003$

B.   Let $a \in [A,B]$, $f \in C^2[A,B]$, and let $P_1(x) = f(a) + f'(a)(x-a)$ be the first-order Taylor polynomial. Fix a point x_0 in $[A,B]$.

(a) Define $h(t) = f(t) + f'(t)(x_0 - t) + D(x_0 - t)^2$. Find D such that $h(a) = h(x_0)$.
(b) Find c between a and x_0 such that $f(x_0) - P_1(x_0) = \frac{1}{2}f''(c)(x_0 - a)^2$.
(c) Find a constant M such that $|f(x) - f(a)| \le M(x-a)^2$ for all $x \in [A,B]$.

C.   Let f satisfy the hypotheses of Taylor's Theorem at $x = a$.

(a) Show that $\lim\limits_{x \to a} \dfrac{f(x) - P_n(x)}{(x-a)^n} = 0$.

(b) If $Q(x) \in \mathbb{P}_n$ and $\lim\limits_{x \to a} \dfrac{f(x) - Q(x)}{(x-a)^n} = 0$, prove that $Q = P_n$.

D.   (a) Find the Taylor series for $\sin x$ about $x = 0$, and prove that it converges to $\sin x$ uniformly on any bounded interval $[-N,N]$.
(b) Find the Taylor expansion of $\sin x$ about $x = \pi/6$. Hence show how to approximate $\sin(31°)$ to 10 decimal places. Do careful estimates.

E.   (a) Verify that $4\arctan(\frac{1}{5}) - \arctan(\frac{1}{239}) = \frac{\pi}{4}$.   HINT: Take the tan of both sides.
(b) Using the estimates for $\arctan(x)$ derived in Example 10.1.6, compute how many terms are needed to approximate π to 1000 decimal places of accuracy using this formula.
(c) Calculate π to 6 decimal places of accuracy using this method.

F.   Let $f(x) = \log x$.

(a) Find the Taylor series of f about $x = 1$.
(b) What is the radius of convergence of this series?
(c) What happens at the two endpoints of the interval of convergence? Hence find a series converging to $\log 2$.
(d) By observing that $\log 2 = \log 4/3 - \log 2/3$, find another series converging to $\log 2$. Why is this series more useful?
(e) Show that $\log 3 = 3\log 0.96 + 5\log \frac{81}{80} - 11\log 0.9$. Find a finite expression that does not involve logs which estimates $\log 3$ to 50 decimal places.

G.   Suppose that $f, g \in C^{n+1}[a-\delta, a+\delta]$ and $f^{(k)}(a) = g^{(k)}(a) = 0$ for $0 \le k < n$ and $g^{(n)}(a) \ne 0$. Use Taylor polynomials to show that $\lim\limits_{x \to a} \dfrac{f(x)}{g(x)} = \dfrac{f^{(n)}(a)}{g^{(n)}(a)}$.

H.   Show that $\displaystyle\int_{-\infty}^{+\infty} \log\left|\frac{1+x}{1-x}\right| \frac{dx}{x} = \pi^2$ as follows:

(a) Reduce the integral to $4\displaystyle\int_0^1 \log\left|\frac{1+x}{1-x}\right| \frac{dx}{x}$.
(b) Use the Taylor series for $\log x$ about $x = 1$ found in Exercise 10.1.F. Use your knowledge of convergence and integration to evaluate $\displaystyle\int_0^r \log\left|\frac{x+1}{x-1}\right| \frac{dx}{x}$ as a series when $r < 1$.
(c) Justify the improper integral obtained by letting r go to 1.
(d) Use the famous identity $\sum\limits_{n=1}^{\infty} \frac{1}{n^2} = \frac{\pi^2}{6}$ to complete the argument.

I.   Let $f(x) = (1+x)^{-1/2}$.

(a) Find a formula for $f^{(k)}(x)$. Hence show that

$$f^{(k)}(0) = \binom{-\frac{1}{2}}{k} := \frac{-\frac{1}{2}(-\frac{1}{2}-1)\cdots(-\frac{1}{2}+1-k)}{k!} = \frac{(-1)^k(2k)!}{2^{2k}(k!)^2} = \left(\frac{-1}{4}\right)^k\binom{2k}{k}.$$

(b) Show that the Taylor series for f about $x = 0$ is $\displaystyle\sum_{k=0}^{\infty}\binom{2k}{k}\left(\frac{-x}{4}\right)^k$, and compute the radius of convergence.

(c) Show that $\sqrt{2} = 1.4f(-0.02)$. Hence compute $\sqrt{2}$ to 8 decimal places.

(d) Express $\sqrt{2} = 1.415f(\varepsilon)$, where ε is expressed as a fraction in lowest terms. Use this to obtain an alternating series for $\sqrt{2}$. How many terms are needed to estimate $\sqrt{2}$ to 100 decimal places?

J.   Let a be the number with 198 ones (i.e., $a = \underbrace{11\ldots11}_{198 \text{ ones}}$). Find $\sqrt{a}$ to 500 decimal places.

HINT: $a = \left(\frac{10^{99}}{3}\right)^2(1 - 10^{-198})$. Your decimal expansion should end in 97916.

10.2 How Not to Approximate a Function

Given a continuous function $f : [a,b] \to \mathbb{R}$, can we approximate f by a polynomial? For example, suppose you need to write a computer program to evaluate f. Since some round-off errors are inevitable, why not replace f with a polynomial that is close to f, since the polynomial will be easy to evaluate?

What precisely do we mean by *close*? This depends on the context, but in the context of the preceding programming example, we mean a polynomial p such that

$$\|f - p\|_{\infty} = \max_{x \in [a,b]} |f(x) - p(x)|$$

is small. This is known as **uniform approximation**. Such approximations are important both in practical work and in theory. Later in this chapter, there are several methods for computing such approximations, including methods well adapted to programming.

We start by looking at several plausible methods that do not work.

It might seem that the Taylor polynomials of the previous section are a good answer to this problem. However, there are several serious flaws. First, and most important, the function f may not be differentiable at all and f must have at least n derivatives in order to compute P_n. Moreover, the bound on the $(n+1)$st derivative is the crucial factor in the error estimate, so f must have $n+1$ derivatives. The bounds may be so large that this estimate is useless. Second, even for very nice functions, the Taylor series may converge only on a small interval about the point a. Recall Example 10.1.6, which shows that the Taylor series for $\arctan(x)$ converges only on the small interval $[-1,1]$ even though the function is C^{∞} on all of $\mathbb{R}$. Moreover, the convergence is very slow unless we further restrict the interval to something like $[-0.5,0.5]$. Worse yet was $f(x) = e^{-1/x^2}$ of Example 10.1.7, which is C^{∞} and for which the Taylor series about $x = 0$ converges everywhere, but to the wrong function. Because the Taylor series uses information at only one point, it cannot be expected to always do a good job over an entire interval.

Differentiation is a very unstable process when the function is known only approximately in the uniform norm—a small error in evaluating the function can result in a huge error in the derivative (see Example 10.2.1). So even when the Taylor series does converge, it can be difficult to compute the coefficients numerically.

10.2.1. EXAMPLE. One reason that Taylor polynomials fail is that they use information available at only one point. Another failing is that they try too hard to approximate derivatives at the same time. Here is an example where convergence for the function is good, but not for the derivative. Let

$$f_n(x) = x + \frac{1}{\sqrt{n}} \sin nx \quad \text{for} \quad -\pi \le x \le \pi.$$

It is easy to verify that f_n converges uniformly on $[-\pi, \pi]$ to the function $f(x) = x$. Indeed,

$$\|f - f_n\|_\infty = \max_{-\pi \le x \le \pi} \frac{1}{\sqrt{n}} |\sin nx| = \frac{1}{\sqrt{n}} \quad \text{for} \quad n \ge 1.$$

However $f'(x) = 1$ everywhere, while $f_n'(x) = 1 + \sqrt{n} \cos nx$. Therefore,

$$\|f' - f_n'\|_\infty = \max_{-\pi \le x \le \pi} \sqrt{n} |\cos nx| = \sqrt{n} \quad \text{for} \quad n > 1.$$

As the approximants f_n get closer to f, they oscillate more and more dramatically. So the derivatives of the f_n are very far from the derivative of f.

Another possible method for finding good approximants is to use polynomial interpolation. Pick $n + 1$ points distributed over $[a, b]$, for example,

$$x_i = a + \frac{i(b-a)}{n} \quad \text{for} \quad i = 0, 1, \ldots, n.$$

There is a unique polynomial p_n of degree at most n that goes through the $n + 1$ points $(x_i, f(x_i))$, $i = 0, 1, \ldots, n$.

It seems reasonable to suspect that as n increases, p_n will converge uniformly to f. However, this is not true. In 1901, Runge showed that the polynomial interpolants on $[-5, 5]$ to the function

$$f(x) = \frac{1}{1 + x^2}$$

do not converge to f. In fact, $\lim_{n \to \infty} \|f - p_n\|_\infty = \infty$. Proving this is more than a little tricky. If instead of choosing $n + 1$ equally spaced points, we cleverly choose the points $x_i = 5 \cos(i\pi/n)$, $i = 0, 1, \ldots, n$, then the interpolating polynomials $p_n(x)$ *will* converge uniformly to this particular function.

However, if you specify the points in advance, then no matter which points you choose at each stage, there is some continuous function $f \in C[a, b]$ such that the interpolating polynomials of degree n do not converge uniformly to f. This was proved by Bernstein and Faber independently in 1914.

There are algorithms using interpolation that involve varying the points of interpolation strategically. There are also ways of making interpolation into a practical method by using splines instead of polynomials. We will return to the latter idea in the last two sections of this chapter.

After all of these negative results, we might wonder whether it is possible to approximate an arbitrary continuous function by a polynomial. It is a remarkable and important theorem, proved by Weierstrass in 1885, that this is possible.

10.2.2. WEIERSTRASS APPROXIMATION THEOREM.
Let f be any continuous real-valued function on $[a,b]$. Then there is a sequence of polynomials p_n that converges uniformly to f on $[a,b]$.

In the language of normed vector spaces, this theorem says that the polynomials are dense in $C[a,b]$ in the max norm.

In fact, this theorem is sufficiently important that many different proofs have been found. The proof we give was found in 1912 by Bernstein, a Russian mathematician. It explicitly constructs the approximating polynomial. This algorithm is not the most efficient, but the problem of finding efficient algorithms can wait until we have proved that the theorem is true.

Exercises for Section 10.2

A. Assuming that the Weierstrass Theorem is true for $C[0,1]$, prove that it is true for $C[a,b]$, for an arbitrary interval $[a,b]$. HINT: For $f \in C[a,b]$, use $g(t) := f(a+(b-a)t)$ in $C[0,1]$.

B. Let $\alpha > 0$. Using the Weierstrass Theorem, prove that every continuous function $f(x)$ on $[0,+\infty]$ with $\lim_{x\to\infty} f(x) = 0$ can be uniformly approximated as closely as we like by a function of the form $q(x) = \sum_{n=1}^{N} C_n e^{-n\alpha x}$. HINT: Consider $g(y) = f(-\log(y)/\alpha)$ on $(0,1]$.

C. (a) Show that every continuous function f on $[a,b]$ is the uniform limit of polynomials of the form $p_n(x^3)$.
    (b) Describe the subspace of $C[-1,1]$ consisting of functions that are uniform limits of polynomials of the form $p_n(x^2)$.

D. Suppose that f is a continuous function on $[0,1]$ such that $\int_0^1 f(x)x^n \, dx = 0$ for all $n \geq 0$. Prove that $f = 0$. HINT: Use the Weierstrass Theorem to show that $\int_0^1 |f(x)|^2 \, dx = 0$.

E. Let X be a compact subset of $[-N,N]$.

(a) Show that every continuous function f on X may be extended to a continuous function g defined on $[-N,N]$ with $\|g\|_\infty = \|f\|_\infty$.
(b) Show that every continuous function on X is the uniform limit of polynomials.

F. If f is continuously differentiable on $[0,1]$, show that there is a sequence of polynomials p_n converging uniformly to f such that p_n' converge uniformly to f'. HINT: Start with f'.

G. Show that if f is in $C^\infty[0,1]$, then there is a sequence p_n of polynomials such that the kth derivatives $p_n^{(k)}$ converge uniformly to $f^{(k)}$ for every $k \geq 0$.
    HINT: Adapt Exercise 10.2.F to find p_n with $\|f^{(k)} - p_n^{(k)}\|_\infty < \frac{1}{n}$ for $0 \leq k \leq n$.

H. Prove that e^x is not a polynomial. HINT: Consider behaviour at $\pm\infty$.

I.   (a) If $0 \notin [a,b]$, show that every continuous function f on $[a,b]$ is the uniform limit of a
         sequence of polynomials (q_n), where $q_n(x) = x^n p_n(x)$ for polynomials p_n.
         (b) If $0 \in [a,b]$, show that a continuous function f on $[a,b]$ is the uniform limit of a sequence
         of polynomials (q_n), where $q_n(x) = x^n p_n(x)$ for polynomials p_n, if and only if $f(0) = 0$.

J.   (a) If $x_0, \ldots, x_n$ are points in $[a,b]$ and $\mathbf{a} = (a_0, \ldots, a_n) \in \mathbb{R}^{n+1}$, show that there is a unique
         polynomial $p_\mathbf{a}$ in $\mathbb{P}_n$ such that $p(x_i) = a_i$ for $0 \le i \le n$.
         HINT: Find polynomials q_j such that $q_j(x_i) = \delta_{ij}$ is 1 if $i = j$ and is 0 if $0 \le i \ne j \le n$.
         (b) Show that there is a constant M (depending on n) such that $\|p_\mathbf{a}\|_\infty \le M\|\mathbf{a}\|_2$.

K.   Suppose that $f \in C[a,b]$, $\varepsilon > 0$ and $x_1, \ldots, x_n$ are points in $[a,b]$. Prove that there is a polyno-
         mial p such that $p(x_i) = f(x_i)$ for $1 \le i \le n$ and $\|f - p\|_\infty < \varepsilon$.   HINT: First approximate f
         closely by some polynomial. Then use the previous exercise to adjust the difference.

10.3 Bernstein's Proof of the Weierstrass Theorem

Recall the binomial formula, $(a+b)^n = \sum\limits_{k=0}^{n} \binom{n}{k} a^k b^{n-k}$. If we set $a = x$ and $b = 1 - x$,
then we obtain

$$1 = \sum_{k=0}^{n} \binom{n}{k} x^k (1-x)^{n-k}.$$

Bernstein started by considering the functions

$$P_k^n(x) = \binom{n}{k} x^k (1-x)^{n-k} \quad \text{for} \quad k = 0, 1, \ldots, n,$$

now called **Bernstein polynomials**. They have several virtues. They are polynomials of degree n. They take only nonnegative values on $[0,1]$. And they add up to 1.
Moreover, P_k^n is a "bump" function with a maximum at k/n, as a routine calculus
calculation shows. For example, the four functions P_k^3 for $0 \le k \le 3$ are given in
Figure 10.3.

Given a continuous function f on $[0,1]$, define a polynomial $B_n f$ by

$$(B_n f)(x) = \sum_{k=0}^{n} f\!\left(\tfrac{k}{n}\right) P_k^n(x) = \sum_{k=0}^{n} f\!\left(\tfrac{k}{n}\right) \binom{n}{k} x^k (1-x)^{n-k}.$$

This is a linear combination of the polynomials P_k^n; and so $B_n f$ is a polynomial of
degree at most n. We think of B_n as a function from the vector space $C[0,1]$ into
itself. This map has several easy but important properties. If $f, g \in C[0,1]$, we say
that $f \ge g$ if $f(x) \ge g(x)$ for all $0 \le x \le 1$.

10.3.1. PROPOSITION. *The map B_n is linear and monotone. That is, for all*
$f, g \in C[0,1]$ and $\alpha, \beta \in \mathbb{R}$,

(1) $B_n(\alpha f + \beta g) = \alpha B_n f + \beta B_n g$,
(2) $B_n f \ge B_n g$ if $f \ge g$,
(3) $|B_n f| \le B_n g$ if $|f| \le g$.

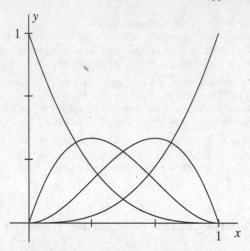

FIG. 10.3 The Bernstein polynomials of degree 3.

The only part that requires any cleverness is the monotonicity. However, since each $P_k^n \geq 0$, it follows that when $f \geq 0$, then $B_n f$ is also positive. So if $f \geq g$, then $B_n f - B_n g = B_n(f - g) \geq 0$. In particular, $|f| \leq g$ means that $-g \leq f \leq g$; and hence $-B_n g \leq B_n f \leq B_n g$. The details are left to the reader.

Next let us compute $B_n f$ for three basic polynomials: 1, x, and x^2.

10.3.2. LEMMA. $B_n 1 = 1$, $\quad B_n x = x$, $\quad$ and

$$B_n x^2 = \frac{n-1}{n} x^2 + \frac{1}{n} x = x^2 + \frac{x - x^2}{n}.$$

PROOF. For the first equation, observe that $B_n 1$ is the sum of P_0^n through P_n^n, which we have already noted is identically 1.

Differentiating the binomial identity $\sum_{k=0}^n \binom{n}{k} a^k b^{n-k} = (a+b)^n$ with respect to a gives

$$\sum_{k=0}^n k \binom{n}{k} a^{k-1} b^{n-k} = n(a+b)^{n-1}. \tag{10.3.3}$$

Substitute $a = x$ and $b = 1 - x$ and multiply by x/n to get

$$B_n x = \sum_{k=0}^n \frac{k}{n} \binom{n}{k} x^k (1-x)^{n-k} = \frac{x}{n} n \big(x + (1-x)\big)^{n-1} = x.$$

Multiplying (10.3.3) by a and differentiating again with respect to a yields

$$\sum_{k=0}^n k^2 \binom{n}{k} a^{k-1} b^{n-k} = n(a+b)^{n-1} + n(n-1)a(a+b)^{n-2}.$$

Substitute $a = x$ and $b = 1 - x$ and multiply by x/n^2 to obtain

$$B_n x^2 = \sum_{k=1}^{n} \frac{k^2}{n^2} \binom{n}{k} x^k (1-x)^{n-k} = \frac{x}{n^2}(n+n(n-1)x) = \frac{x+(n-1)x^2}{n}. \quad \blacksquare$$

PROOF OF WEIERSTRASS'S THEOREM. By Exercise 10.2.A, it suffices to prove the theorem for the interval $[0,1]$. Fix a continuous function f in $C[0,1]$. We will prove that for each $\varepsilon > 0$, there is some $N > 0$ such that

$$\|f(x) - B_n f(x)\| < \varepsilon \quad \text{for all} \quad n \geq N.$$

Since $[0,1]$ is compact, f is uniformly continuous on $[0,1]$ by Theorem 5.5.9. Thus for our given $\varepsilon > 0$, there is some $\delta > 0$ such that

$$|f(x) - f(y)| \leq \frac{\varepsilon}{2} \quad \text{for all} \quad |x-y| \leq \delta, \ x, y \in [0,1].$$

Also, f is bounded on $[0,1]$ by the Extreme Value Theorem (5.4.4). So let

$$M = \|f\|_\infty = \sup_{x \in [0,1]} |f(x)|.$$

Fix any point $a \in [0,1]$. We claim that $|f(x) - f(a)| \leq \frac{\varepsilon}{2} + \frac{2M}{\delta^2}(x-a)^2$. Indeed, if $|x-a| \leq \delta$, then

$$|f(x) - f(a)| \leq \frac{\varepsilon}{2} \leq \frac{\varepsilon}{2} + \frac{2M}{\delta^2}(x-a)^2$$

by our estimate of uniform continuity. And if $|x-a| \geq \delta$, then

$$|f(x) - f(a)| \leq 2M \leq 2M\left(\frac{x-a}{\delta}\right)^2 \leq \frac{\varepsilon}{2} + \frac{2M}{\delta^2}(x-a)^2.$$

By linearity of B_n and $B_n 1 = 1$, we obtain $B_n(f - f(a))(x) = B_n f(x) - f(a)$. Now use the positivity of our map B_n to obtain

$$|B_n f(x) - f(a)| \leq B_n\left(\frac{\varepsilon}{2} + \frac{2M}{\delta^2}(x-a)^2\right)$$

$$= \frac{\varepsilon}{2} + \frac{2M}{\delta^2}\left(x^2 + \frac{x-x^2}{n} - 2ax + a^2\right)$$

$$= \frac{\varepsilon}{2} + \frac{2M}{\delta^2}(x-a)^2 + \frac{2M}{\delta^2}\frac{x-x^2}{n}.$$

Evaluate this at $x = a$ to obtain

$$|B_n f(a) - f(a)| \leq \frac{\varepsilon}{2} + \frac{2M}{\delta^2}\frac{a-a^2}{n} \leq \frac{\varepsilon}{2} + \frac{M}{2\delta^2 n}.$$

We use the fact that $\max\{a - a^2 : 0 \leq a \leq 1\} = \frac{1}{4}$.

This estimate does not depend on the point a. So we have found that

$$\|B_n f - f\|_\infty \leq \frac{\varepsilon}{2} + \frac{M}{2\delta^2 n}.$$

So now choose $N \geq \dfrac{M}{\delta^2 \varepsilon}$ such that $\dfrac{M}{2\delta^2 N} < \dfrac{\varepsilon}{2}$. Then for all $n \geq N$,

$$\|B_n f - f\|_\infty \leq \frac{\varepsilon}{2} + \frac{\varepsilon}{2} = \varepsilon. \qquad \blacksquare$$

As was already mentioned, using Bernstein polynomials is not an efficient way of finding polynomial approximations. However, Bernstein polynomials have other advantages, which are developed in the exercises.

Exercises for Section 10.3

A.   Show that $P_k^n(x) = \binom{n}{k} x^k (1-x)^{n-k}$ attains its maximum at $\frac{k}{n}$.

B.   Show that $\|B_n f\|_\infty \leq \|f\|_\infty$.   HINT: Use monotonicity.

C.   Prove that $B_n(f)^2 \leq B_n(f^2)$.   HINT: Expand $B_n((f-a)^2)$.

D.   (a) Compute $B_n x^3$.
   (b) Compute $\lim\limits_{n\to\infty} n(B_n x^3 - x^3)$.

E.   Work through our proof of the Weierstrass theorem with the function $f(x) = |x - \frac{1}{2}|$ on $[0,1]$ to obtain an estimate for the degree of a polynomial p needed to ensure that $\|f - p\|_\infty < 0.0005$.

F.   (a) Show that $B_n(e^x) = (1 + (e^{1/n} - 1)x)^n$.
   (b) Show that this may be rewritten as $(1 + \frac{x}{n} + x\frac{c_n}{n^2})^n$, where $0 \leq c_n \leq 1$.
   (c) Hence prove directly that $B_n(e^x)$ converges uniformly to e^x on $[0,1]$.

G.   (a) Show that the derivative of $B_{n+1}f$ is

$$(B_{n+1}f)'(x) = \sum_{k=0}^n \frac{f\left(\frac{k+1}{n+1}\right) - f\left(\frac{k}{n+1}\right)}{\frac{1}{n+1}} \binom{n}{k} x^k (1-x)^{n-k}.$$

   (b) If f has a continuous first derivative, use the Mean Value Theorem and the uniform continuity of f to show that $\lim\limits_{n\to\infty} \|(B_n f)' - f'\|_\infty = 0$.

H.   (a) Set $f_{nm}(x) = x\left(x - \frac{1}{n}\right)\left(x - \frac{2}{n}\right) \cdots \left(x - \frac{m-1}{n}\right)$ for $m \geq 0, n \geq 1$. Show that $B_n f_{nm} = f_{nm}(1)x^m$.
   (b) Hence show that both sequences (f_{nm}) and $(B_n f_{nm})$ converge uniformly to x^m.
   (c) Show $B_n x^m$ converges to x^m using $\|B_n x^m - x^m\|_\infty \leq \|B_n(x^m - f_{nm})\|_\infty + \|B_n f_{nm} - x^m\|_\infty$.
   (d) Use this to give another proof that $B_n p$ converges uniformly to p for every polynomial p.

10.4 Accuracy of Approximation

In this section, we measure the rate of convergence of polynomial approximations. We define the optimal error. The aim is to get a reasonable idea of what it is for a given function, and how well a given approximation compares with it.

Let $\mathbb{P}_n$ denote the vector space of polynomials of degree at most n. We will write $\mathbb{P}_n[a,b]$ to mean that $\mathbb{P}_n$ is considered as a subspace of $C[a,b]$ with norm given by the maximum modulus over the interval $[a,b]$.

10.4.1. DEFINITION. If $f \in C[a,b]$, define the **error function** $E_n(f)$ by

$$E_n(f) = \inf\{\|f - q\|_\infty : q \in \mathbb{P}_n\}.$$

Likewise, if $\mathscr{F}$ is a set of functions, we let

$$E_n(\mathscr{F}) = \sup_{f \in \mathscr{F}} E_n(f).$$

We can determine how good a polynomial approximation $p \in \mathbb{P}_n$ to f is by how close $\|f - p\|_\infty$ is to $E_n(f)$.

A little thought reveals that wildly oscillating functions will not be well approximated by polynomials of low degree. For example, the function $f(x) = \cos(n\pi x)$ in $C[0,1]$ alternately takes the extreme values ± 1 at $\frac{k}{n}$ for $0 \le k \le n$. Any function close to f (within 1) will have to switch signs between these points. This suggests that in order to get a reasonable estimate, we must measure how quickly f varies.

10.4.2. DEFINITION. The **modulus of continuity** of $f \in C[a,b]$ is defined for each $\delta > 0$ by

$$\omega(f;\delta) = \sup\{|f(x_1) - f(x_2)| : |x_1 - x_2| < \delta, \ x_1, x_2 \in [a,b]\}.$$

In other words, $\omega(f,\delta)$ is the smallest choice of ε for which δ "works" in the definition of uniform continuity.

By Theorem 5.5.9, every continuous function on the compact set $[0,1]$ is uniformly continuous. Therefore, for each $\varepsilon > 0$, there is a $\delta > 0$ such that

$$|f(x) - f(y)| < \varepsilon \quad \text{for all} \quad |x - y| < \delta, \ x, y \in [0,1].$$

Restating this with our new terminology, we see that for every $\varepsilon > 0$, there is a $\delta > 0$ such that $\omega(f;\delta) < \varepsilon$. Thus the uniform continuity of f is equivalent to

$$\lim_{\delta \to 0^+} \omega(f;\delta) = 0.$$

10.4.3. EXAMPLE. Consider $f(x) = \sqrt{x}$ on $[0,1]$. Fix $\delta \ge 0$ and look at

$$\sup_{0 \le t \le \delta} f(x+t) - f(x) = \sqrt{x+\delta} - \sqrt{x} = \frac{\delta}{\sqrt{x+\delta} + \sqrt{x}} \le \sqrt{\delta}.$$

This inequality is sharp at $x = 0$. Thus, $\omega(f;\delta) = \sqrt{\delta}$.

The class of functions f with $\omega(f;\delta) \le \delta$ for all $\delta > 0$ are precisely the functions satisfying

$$|f(x) - f(y)| \le |x - y|,$$

namely the functions with Lipschitz constant 1. Denote by $\mathscr{S}$ the class of functions in $C[0,1]$ with Lipschitz constant 1. We will prove our results first for the class $\mathscr{S}$.

A good lower bound for the error is obtained using an idea due to Chebyshev. We will use the idea behind the next proof repeatedly, so examine it carefully.

10.4.4. PROPOSITION. $E_n(\mathscr{S}) \geq \dfrac{1}{2n+2}$ *for* $n \geq 0$.

PROOF. Fix $n \geq 0$. Consider the sawtoothed function f that takes the values

$$f\left(\tfrac{k}{n+1}\right) = \frac{(-1)^k}{2n+2} \quad \text{for} \quad 0 \leq k \leq n+1$$

and is linear in between with slope ± 1. Clearly, f belongs to $\mathscr{S}$.

We will show that the closest polynomial to f in $\mathbb{P}_n$ is the zero polynomial, which is clearly distance $1/(2n+2)$ from f. To this end, suppose that p is a polynomial with $\|p - f\|_\infty < \frac{1}{2n+2}$. Then

$$\left| p\left(\tfrac{k}{n+1}\right) - \frac{(-1)^k}{2n+2} \right| < \frac{1}{2n+2}.$$

It follows that $\operatorname{sign} p\left(\tfrac{k}{n+1}\right) = (-1)^k$. Consequently, p changes sign between $\tfrac{k}{n+1}$ and $\tfrac{k+1}{n+1}$ for each $0 \leq k \leq n$. By the Intermediate Value Theorem (5.6.1), p has a root in the open interval $\left(\tfrac{k}{n+1}, \tfrac{k+1}{n+1}\right)$. So p is a nonconstant polynomial with at least $n+1$ roots, and thus is not in $\mathbb{P}_n$. Consequently,

$$E_n(\mathscr{S}) \geq E_n(f) = \|f\|_\infty = \frac{1}{2n+2}. \qquad \blacksquare$$

To obtain an upper bound, let us look carefully at the estimate that comes out of Bernstein's proof of the Weierstrass Theorem.

10.4.5. PROPOSITION. $E_n(\mathscr{S}) \leq 1/\sqrt{n}$ *for* $n \geq 1$.

PROOF. Fix $f \in \mathscr{S}$. We recall the details of the proof of the Weierstrass Approximation Theorem in our context. Let ε be any positive number. We claim that the Lipschitz condition gives the strong inequality

$$|f(x) - f(a)| \leq |x - a| \leq \varepsilon + \frac{(x-a)^2}{\varepsilon}.$$

To check this, consider the cases $|x - a| \leq \varepsilon$ and $|x - a| > \varepsilon$ separately.

Now apply the Bernstein map B_n, which by monotonicity yields

$$|B_n f(x) - f(a)| \leq \varepsilon + \frac{B_n\big((x-a)^2\big)}{\varepsilon} = \varepsilon + \frac{(x-a)^2}{\varepsilon} + \frac{x - x^2}{n\varepsilon}.$$

Substituting $x = a$ and maximizing over $[0,1]$, we obtain

$$\|B_n f - f\|_\infty \le \varepsilon + \frac{1}{n\varepsilon}\|x - x^2\|_\infty = \varepsilon + \frac{1}{4n\varepsilon}.$$

Minimizing this leads to the choice of $\varepsilon = \frac{1}{2\sqrt{n}}$. Thus $\|B_n f - f\|_\infty \le \frac{1}{\sqrt{n}}$. ∎

There is quite a gap between our upper and lower bounds when n is large. In fact, the lower bound has the correct order of growth. In order to obtain superior upper bounds, we need to replace Bernstein approximations $B_n f$ with a better method of polynomial approximation. We do this in Section 14.9 using Fourier series.

Exercises for Section 10.4

A. Show that $\omega(f; \delta_1) \le \omega(f; \delta_2)$ if $\delta_1 \le \delta_2$.

B. If f is C^1 on $[a,b]$, show that $\omega(f; \delta) \le \|f'\|_\infty \delta$.

C. Show that a function f on $\mathbb{R}$ is uniformly continuous if and only if $\lim_{\delta \to 0^+} \omega(f; \delta) = 0$.

D. Show that f is Lipschitz with constant L if and only if f satisfies $\omega(f, \delta) \le L\delta$.

E. If f is Lipschitz with constant L, prove that $\|B_n f - f\| \le \dfrac{L}{\sqrt{n}}$.

F. For $f, g \in C[a,b]$ and $\alpha, \beta \in \mathbb{R}$,

    (a) Show that $E_n(\alpha f + \beta g) \le |\alpha| E_n(f) + |\beta| E_n(g)$.
    (b) Show that $E_{m+n}(fg) \le \|f\|_\infty E_n(g) + \|g\|_\infty E_m(f)$.

G. Show that if $\lim_{\delta \to 0^+} \dfrac{\omega(f; \delta)}{\delta} = 0$, then f is constant.

H. (a) In $C[0,1]$, show that $E_n(\cos m\pi x) = 1$ for $n < m$.
    (b) Use the Taylor series about $a = 1/2$ to show that $E_{10n}(\cos n\pi x) < 10^{-3n}$.

I. Let $f(x) = |2x - 1|$ on $[0,1]$.

    (a) Show that $B_n f(\frac{1}{2}) = 2^{-2n}\binom{2n}{n}$.
    (b) Compute $\lim_{n \to \infty} \sqrt{n} B_n f(\frac{1}{2})$.   HINT: Use Stirling's formula to approximate the factorials.
    (c) Hence show that Proposition 10.4.5 is the right order of magnitude.

10.5 Existence of Best Approximations

Suppose that f is a continuous function on $[a,b]$. We may search for the optimal polynomial approximation of given degree. The analysis tools that we have developed will allow us to show that such an optimal approximation always exists. Moreover, in the next section, the best approximation will be shown to be unique.

A polynomial $p(x) = a_0 + a_1 x + \cdots + a_n x^n$ of degree at most n is determined by the $n+1$ coefficients $a_0, \ldots, a_n$. Moreover, a nonzero polynomial of this form has at most n zeros and thus is not equal to the zero function on $[a,b]$. Hence $\mathbb{P}_n[a,b]$ is an $(n+1)$-dimensional vector subspace of $C[a,b]$ with basis $1, x, \ldots, x^n$. It may be identified with $\mathbb{R}^{n+1}$ by associating p to the vector $(a_0, \ldots, a_n)$. The norm is quite different from the Euclidean norm. However, the results of Section 7.3 are precisely what we need to solve our problem.

First, Lemma 7.3.1 shows that there are constants $0 < c < C$ (depending on a, b and n) such that every polynomial $p(x) = a_0 + a_1 x + \cdots + a_n x^n$ in $\mathbb{P}_n[a,b]$ satisfies

$$c\Big(\sum_{k=0}^{n} |a_k|^2\Big)^{1/2} \le \|p\|_\infty = \sup_{a \le x \le b} |p(x)| \le C\Big(\sum_{k=0}^{n} |a_k|^2\Big)^{1/2}.$$

This lemma allows us to transfer our convergence results for $\mathbb{R}^{n+1}$ over to $\mathbb{P}_n$. It is important to note that these results depend on having a fixed bound on the degree of the polynomials. They are false for polynomials of unbounded degree.

10.5.1. COROLLARY. *$\mathbb{P}_n[a,b]$ has the same convergent sequences as $\mathbb{R}^{n+1}$ in the sense that the sequence $p_i = \sum_{k=0}^{n} a_{ik} x^k$ in $\mathbb{P}_n[a,b]$ converges uniformly on $[a,b]$ to a polynomial $p = \sum_{k=0}^{n} a_k x^k$ if and only if $\lim_{i \to \infty} a_{ik} = a_k$ for $0 \le k \le n$.*

PROOF. Let p_i correspond to the vector $\mathbf{a}_i = (a_{i0}, \ldots, a_{in})$ for $i \ge 1$, and let p correspond to the vector $\mathbf{a}$. If $\mathbf{a}_i$ converges to $\mathbf{a}$, then

$$\lim_{i \to \infty} \|p - p_i\|_\infty \le \lim_{i \to \infty} C \|\mathbf{a} - \mathbf{a}_i\|_2 = 0.$$

Hence p_i converges uniformly to p on $[a,b]$. Conversely, if p_i converges uniformly to p on $[a,b]$, then

$$\lim_{i \to \infty} \|\mathbf{a} - \mathbf{a}_i\|_2 \le \lim_{i \to \infty} \frac{1}{c} \|p - p_i\|_\infty = 0.$$

So $\mathbf{a}_i$ converges to $\mathbf{a}$ in the Euclidean norm.

The second statement follows from Lemma 4.2.3, which shows that a sequence converges in $\mathbb{R}^{n+1}$ if and only if each coefficient converges. ∎

Second, Corollary 7.3.3 applies directly. Again this is false if the degree of the polynomials is not bounded.

10.5.2. COROLLARY. *A subset of $\mathbb{P}_n[a,b]$ is compact if and only if it is closed and bounded.*

An immediate consequence of Theorem 7.3.5 is the result we are looking for.

10.5.3. THEOREM. *Let f be a continuous function on $[a,b]$. For each $n \ge 0$, there exists a closest polynomial to f of degree at most n in the max norm on $C[a,b]$.*

We consider an example that shows that a best approximation in certain more general circumstances may not exist; and when it does exist, it may not be unique.

10.5.4. EXAMPLE. Consider the subspace

$$S = \{h \in C[0,1] : h(0) = 0\}$$

of $C[0,1]$. Note that if f is any function in $C[0,1]$, then $f - f(0)$ belongs to S. This shows that the linear span of S and the constant function 1 is all of $C[0,1]$. We therefore say that S is a subspace of **codimension one**. In particular, it is infinite-dimensional, since $C[0,1]$ is infinite-dimensional. So the type of arguments we used for $\mathbb{P}_n$ do not apply.

Consider the function $f = 1$. What are the best approximations to f in S? Clearly, for any $h \in S$,

$$\|f - h\|_\infty \geq |f(0) - h(0)| = 1.$$

On the other hand, $\|f - h\|_\infty = 1$ is equivalent to the inequalities

$$0 \leq h(x) \leq 2 \quad \text{for all} \quad 0 \leq x \leq 1.$$

There are many functions $h \in S$ within these constraints. For example, $h_0 = 0$, $h_1(x) = x/2$, $h_2(x) = 2x$, and $h_3(x) = \frac{\pi}{2} \sin^2(6\pi x)$. We see that there are (infinitely) many closest points.

Now consider the subspace

$$T = \left\{ h \in S : \int_0^1 h(x)\,dx - 0 \right\}.$$

This is a subspace because if g and h belong to T and α and β are in $\mathbb{R}$, then $(\alpha g + \beta h)(0) = 0$ and

$$\int_0^1 (\alpha g + \beta h)(x)\,dx = \alpha \int_0^1 g(x)\,dx + \beta \int_0^1 h(x)\,dx = 0.$$

The subspace T has codimension 2 because $C[0,1]$ is spanned by T, 1, and x (see Exercise 10.5.E). Moreover, T is closed. For if $h_n \in T$ converge uniformly to a function h, then $h(0) = \lim_{n\to\infty} h_n(0) = 0$, and by Theorem 8.3.1,

$$\int_0^1 h(x)\,dx = \lim_{n\to\infty} \int_0^1 h_n(x)\,dx = 0.$$

So h belongs to T.

Let $g(x) = x$ and consider the distance of g to T. Note that $g(0) = 0$ but

$$\int_0^1 g(x)\,dx = \frac{1}{2}.$$

Suppose that $h \in T$, and compute that

$$\frac{1}{2} = \int_0^1 g(x) - h(x)\,dx \le \int_0^1 \|g - h\|_\infty\,dx = \|g - h\|_\infty.$$

If $\|g - h\|_\infty = \frac{1}{2}$, then this inequality must be an equality. This can occur only if

$$g(x) - h(x) = \|g - h\|_\infty = \frac{1}{2} \quad \text{for all} \quad 0 \le x \le 1.$$

This implies that $h(x) = x - \frac{1}{2}$. Note that h does not lie in T because $h(0) \ne 0$. So the distance $1/2$ is not attained.

However, we can easily come arbitrarily close to this distance. Indeed, we will show that for any integer n, there will be a continuous function h_n in T such that $\|g - h_n\|_\infty = \frac{1}{2} + \frac{1}{n}$. The idea is to make $h_n(x) = x - \frac{1}{2} - \frac{1}{n}$ on $[a_n, 1]$, $h_n(0) = 0$ and linear in between, with a_n chosen so that the integral is zero. It is easy to check that the function with these properties does the job. A calculation shows that $a_n = \frac{4}{n+2}$. We find that

$$h_n(x) = \begin{cases} -\dfrac{(n-2)^2}{8n}x & \text{for} \quad 0 \le x \le \dfrac{4}{n+2}, \\[2mm] x - \dfrac{1}{2} - \dfrac{1}{n} & \text{for} \quad \dfrac{4}{n+2} \le x \le 1. \end{cases}$$

This shows that when infinite-dimensional subspaces are involved, there need not be a closest point.

Exercises for Section 10.5

A. Suppose that $f \in C[0,1]$ satisfies $f(0) = f(1) = 0$.

   (a) Show that f is a limit of polynomials such that $p(0) = p(1) = 0$.
   (b) Show that there is a closest polynomial of degree at most n with this property.

B. Let $f \in C^1[0,1]$. Show that there is a closest polynomial of degree at most n to f in the $C^1[0,1]$ norm, analogous to the $C^3[a,b]$ norm defined in Example 7.1.4.

C. Find *all* closest lines $p(x) = ax + b$ to $f(x) = x^2$ in the $C^1[0,1]$ norm. Note that the best approximation is not unique.

D. Find the closest polynomial to $\sin x$ on $\mathbb{R}$.

E. For the subspace T of Example 10.5.4, show that $\mathrm{span}\{T, 1, x\} = C[0,1]$.
   HINT: for $f \in C[0,1]$, find a such that $f(x) - f(0) - ax \in T$.

F. (a) Show that for every bounded function on $[a,b]$, there is a closest polynomial $p \in \mathbb{P}_n$ in the max norm.
   (b) Show by example that a closest polynomial need not be unique.

G. Recall that a norm is strictly convex if $\|x\| = \|y\| = \|(x+y)/2\|$ implies that $x = y$.

   (a) Suppose that V is a vector space with a strictly convex norm and M is a finite-dimensional subspace of V. Prove that each $v \in V$ has a unique closest point in M.
   (b) Prove that an inner product norm is strictly convex.
   (c) Show by example that $C[0,1]$ is not strictly convex.

10.6 Characterizing Best Approximations

Perhaps in view of the previous examples, it is surprising that the best polynomial approximant of degree n to any continuous function f is uniquely determined. However, it is unique. We are able to show this because there is an interesting condition that characterizes this best approximation. This result was established by Borel in 1905, building on work of Chebyshev.

10.6.1. EXAMPLE. Consider any continuous function f in $C[0, 1]$. What is the best approximation by a polynomial of degree 0 (i.e., a constant)? We want to make $\|f - c\|_\infty$ as small as possible. By the Extreme Value Theorem, there are two points $x_{\min}$ and $x_{\max}$ in $[0, 1]$ such that

$$f(x_{\min}) \le f(x) \le f(x_{\max}) \quad \text{for all} \quad 0 \le x \le 1.$$

Clearly, $\|f - c\|_\infty$ is the maximum of $|f(x_{\min}) - c|$ and $|f(x_{\max}) - c|$. To make both as small as possible, we must take

$$c = \frac{f(x_{\min}) + f(x_{\max})}{2}.$$

With this choice, the error $r(x) = f(x) - c$ satisfies

$$r(x_{\max}) = \|r\|_\infty = \frac{f(x_{\max}) - f(x_{\min})}{2}$$

and

$$r(x_{\min}) = -\|r\|_\infty = -\frac{f(x_{\max}) - f(x_{\min})}{2}.$$

10.6.2. EXAMPLE. Consider the continuous function $f(x) = x^2$ in $C[0, 1]$. What is the best linear approximation? One approach is to find the maximum modulus of $x^2 - ax - b$ and then minimize over choices of a and b. This is a calculus problem that is not too difficult. However, our approach will be to "guess" the answer and to verify it by geometric means.

First subtract x from f to get $g(x) = x^2 - x$. This function is symmetric about the line $x = \frac{1}{2}$. It takes its maximum value 0 at both 0 and 1, while its minimum is $-\frac{1}{4}$ at $x = \frac{1}{2}$. From the previous example, we know that to minimize $\|g(x) - b\|_\infty$ we should set the constant b equal to $-\frac{1}{8}$ so that the maximum and minima have the same absolute value, $\frac{1}{8}$. This intuitive approach yields a guess that the best linear approximation is $x - \frac{1}{8}$. The error is $r(x) = x^2 - x + \frac{1}{8}$. We know that

$$r(0) = \tfrac{1}{8}, \qquad r(\tfrac{1}{2}) = -\tfrac{1}{8}, \quad \text{and} \quad r(1) = \tfrac{1}{8}.$$

Now we will show that $y = x - \frac{1}{8}$ is indeed the closest line to x^2 on $[0, 1]$. Equivalently, it suffices to show that $y = 0$ is the closest line to $y = r(x)$ on $[0, 1]$. Suppose that some linear function g satisfies $\|r - g\| < \frac{1}{8}$. Then

$$g(0) \in (0, \tfrac{1}{4}), \quad g(\tfrac{1}{2}) \in (-\tfrac{1}{4}, 0), \quad \text{and} \quad g(1) \in (0, \tfrac{1}{4}).$$

Therefore,

$$g(0) > 0 > g(\tfrac{1}{2}) < 0 < g(1).$$

By the Intermediate Value Theorem, g has a zero between 0 and $\frac{1}{2}$ and another zero between $\frac{1}{2}$ and 1. But g is linear, and thus it has at most one root. This contradiction shows that no better linear approximation exists.

Notice that the strategy we used in this example is essentially the same as that used in proving Proposition 10.4.4.

In the first example, the best approximation yields an error function r that achieves the values $\pm\|r\|_\infty$. In the case of our linear approximation, we found three points at which r alternately achieved the values $\pm\|r\|_\infty$. This notion generalizes to give a condition that is sufficient to be the best approximation.

10.6.3. DEFINITION. A function $g \in C[a, b]$ satisfies the **equioscillation condition** of degree n if there are $n + 2$ points $x_1 < x_2 < \cdots < x_{n+2}$ in $[a, b]$ such that

$$g(x_i) = (-1)^i \|g\|_\infty \quad \text{or} \quad g(x_i) = (-1)^{i+1} \|g\|_\infty \quad \text{for} \quad 1 \leq i \leq n + 2.$$

In other words, g attains its maximum absolute value at $n + 2$ points and it alternates in sign between these points.

Figure 10.4 shows a function that satisfies the equioscillation condition.

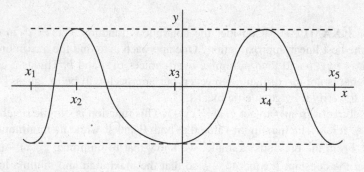

FIG. 10.4 A function satisfying the equioscillation condition for $n = 3$

10.6.4. THEOREM. *Suppose $f \in C[a, b]$ and $p \in \mathbb{P}_n$. If $r = f - p$ satisfies the equioscillation condition of degree n, then*

$$\|f - p\|_\infty = \inf\{\|f - q\|_\infty : q \in \mathbb{P}_n\}.$$

PROOF. If the equality were not true, then there would be some nonzero $q \in \mathbb{P}_n$ such that $p + q$ is a better approximation to f; that is,

$$\|f - (p+q)\|_\infty < \|f - p\|_\infty,$$

or equivalently, $\|r - q\|_\infty < \|r\|_\infty$. In particular, if $x_1, \ldots, x_{n+2}$ are the points from the equioscillation condition for r, then

$$|r(x_i) - q(x_i)| < \|r\|_\infty = |r(x_i)| \quad \text{for} \quad 1 \le i \le n+2.$$

It follows that $q(x_i)$ is nonzero and has the same sign as $r(x_i)$ for $1 \le i \le n+2$.

Therefore, q changes sign between x_i and x_{i+1} for $1 \le i \le n+1$. By the Intermediate Value Theorem, q has a root between x_i and x_{i+1} for $1 \le i \le n+1$. Hence it has at least $n+1$ zeros. But q is a polynomial of degree at most n, so the only way it can have $n+1$ zeros is if q is the zero polynomial. This is false, and thus no better approximation exists. ∎

The important insight of Chebyshev and Borel is that this condition is not only sufficient, but also necessary. The argument is more subtle. When the function r fails the equioscillation condition, a better approximation of degree n must be found.

10.6.5. THEOREM. *If $f \in C[a,b]$ and $p \in \mathbb{P}_n$ satisfy*

$$\|f - p\|_\infty = \inf\{\|f - q\|_\infty : q \in \mathbb{P}_n\},$$

then $f - p$ satisfies the equioscillation condition of degree n.

PROOF. Let $r = f - p \in C[a,b]$ and set $R = \|r\|_\infty$. By Theorem 5.5.9, r is uniformly continuous on $[a,b]$. Thus there is a $\delta > 0$ such that

$$|r(x) - r(y)| < \frac{R}{2} \quad \text{for all} \quad |x - y| < \delta, \ x, y \in [a,b].$$

Partition $[a,b]$ into disjoint intervals of length less than δ. Let $I_1, I_2, \ldots, I_l$ denote those intervals I of the partition such that $|r(x)| = R$ for some $x \in \overline{I}$. Notice that since each I_j has length less than δ, if $r(x) = R$ for some $x \in \overline{I_j}$, then for all $y \in I_j$ we have

$$r(y) \ge r(x) - |r(x) - r(y)| \ge R/2.$$

Similarly, if $r(x) = -R$ for some $x \in \overline{I_j}$, then $r(y) \le -R/2$ for all $y \in I_j$. Let ε_j be $+1$ or -1 according to whether $r(x)$ is positive or negative on I_j.

CLAIM: The sequence $(\varepsilon_1, \ldots, \varepsilon_l)$ has at least $n+1$ changes of sign.

Accepting this claim for a moment, we produce the points required in the definition of the equioscillation condition. Group together adjacent intervals with the same sign, and label these new intervals $J_1, J_2, \ldots, J_k$, where $k \ge n+2$. For each i between 1 and $n+2$, pick a point $x_i \in J_i$ such that $|r(x_i)| = R$. By the choice of J_i, the signs alternate and so the equioscillation condition holds.

Thus, it remains only to prove the claim. Suppose the claim is false; that is, there are at most n changes of sign in $(\varepsilon_1, \ldots, \varepsilon_l)$. We will construct a better approximating polynomial, contradicting the choice of p.

Again, group together adjacent intervals with the same sign, and label these new intervals $J_1, J_2, \ldots, J_k$, where $k \leq n+1$. Because r changes sign between J_i and J_{i+1}, it is possible to pick a point $a_i \in \mathbb{R}$ that lies between them. Define a polynomial q of degree $k-1 \leq n$ by

$$q(x) = \prod_{i=1}^{k-1} x - a_i.$$

Since $q \in \mathbb{P}_n$ and q changes sign at each a_i, either q or $-q$ agrees in sign with $r(x)$ on each set J_i, $i = 1, \ldots, k$. If necessary, replace q with $-q$ such that this agreement holds.

Let $L_0 = \overline{\bigcup_{j=1}^{l} I_j}$ and $L_1 = \overline{[a,b] \setminus L_0}$. Since L_0 is compact and q is never zero on L_0, the minimum,

$$m = \min\{|q(x)| : x \in L_0\},$$

is strictly positive by the Extreme Value Theorem. Let $M = \|q\|_\infty$.

Since L_1 is the union of finitely many closed intervals on which the maximum of $|r(x)|$ is not attained, $|r(x)|$ does not attain the value R on L_1. Again by the Extreme Value Theorem, there is some $d > 0$ such that

$$\max\{|r(x)| : x \in L_1\} = R - d < R.$$

We will show that the polynomial

$$s(x) = p(x) + \frac{d}{2M} q(x)$$

is a better approximation to f than $p(x)$, contradicting the choice of p. Notice that

$$f(x) - s(x) = r(x) - \frac{d}{2M} q(x).$$

Because r and q have the same sign on each I_j,

$$\max_{x \in L_0} |f(x) - s(x)| \leq R - \frac{dm}{2M}.$$

On the remainder, we have

$$\max_{x \in L_1} |f(x) - s(x)| \leq \max_{x \in L_1} |f(x) - p(x)| + \max_{x \in L_1} |p(x) - s(x)|$$

$$\leq R - d + \frac{d}{2M} \|q\| = R - \frac{d}{2}.$$

Now $[a,b] = L_0 \cup L_1$. Thus

$$\|f - s\| = \max_{a \le x \le b} |f(x) - s(x)| \le \max\left\{R - \frac{dm}{2M}, R - \frac{d}{2}\right\} < R.$$

This contradicts the minimality of p and so proves the claim. ∎

Let us put these results together with one more idea to complete the main result.

10.6.6. CHEBYSHEV APPROXIMATION THEOREM.

For each continuous function f in $C[a,b]$, there is a unique polynomial p of degree at most n such that

$$\|f - p\|_\infty = \inf\{\|f - q\|_\infty : q \in \mathbb{P}_n[a,b]\}.$$

This best approximant is characterized by the fact that $f - p$ either is 0 or satisfies the equioscillation condition of degree n.

PROOF. By Theorem 10.5.3, there is at least one closest polynomial to f in $\mathbb{P}_n$. Suppose $p, q \in \mathbb{P}_n$ are both closest polynomials in $\mathbb{P}_n$, and let

$$R = \|f - p\|_\infty = \|f - q\|_\infty.$$

Then the midpoint $(p+q)/2$ is also a polynomial in $\mathbb{P}_n$ that is closest to f because of the triangle inequality:

$$R \le \left\|f - \frac{p+q}{2}\right\|_\infty \le \tfrac{1}{2}\|f \quad p\|_\infty + \tfrac{1}{2}\|f - q\|_\infty = R.$$

Thus by Theorem 10.6.5, $r = f - \frac{1}{2}(p+q)$ satisfies the equioscillation condition. Let $x_1 < x_2 < \cdots < x_{n+2}$ be the required points such that

$$|r(x_i)| = R \quad \text{for} \quad 1 \le i \le n+2.$$

Another use of the triangle inequality yields

$$R = \left|f(x_i) - \frac{p(x_i) + q(x_i)}{2}\right| \le \frac{1}{2}|f(x_i) - p(x_i)| + \frac{1}{2}|f(x_i) - q(x_i)| \le R.$$

Consequently, $f(x_i) - p(x_i)$ and $f(x_i) - q(x_i)$ both have absolute value R. Since there is no cancellation when they are added, they must have the same sign. Therefore, $f(x_i) - p(x_i) = f(x_i) - q(x_i)$, and hence

$$p(x_i) = q(x_i) \quad \text{for} \quad 1 \le i \le n+2.$$

Therefore, $p - q$ is a polynomial of degree at most n with $n + 2$ roots, and so is identically equal to zero. In other words, the closest point is unique. ∎

10.6.7. EXAMPLE. Chebyshev's characterization sometimes allows exact calculation of the best polynomial. Consider the following problem: Find the closest cubic to $f(x) = \cos x$ on $[-\frac{\pi}{2}, \frac{\pi}{2}]$.

Since f is an even function, we expect that the best approximation will be even. This is indeed the case. Let $p \in \mathbb{P}_3$ be the closest cubic, and let $\tilde{p}(x) = p(-x)$. Then

$$\|f - \tilde{p}\|_\infty = \max_{-1 \le x \le 1} |f(x) - p(-x)|$$
$$= \max_{-1 \le x \le 1} |f(-x) - p(-x)| = \|f - p\|_\infty.$$

Since the closest polynomial is unique, it follows that $p = \tilde{p}$, namely $p(-x) = p(x)$. So we are looking for $p(x) = ax^2 + b$.

From Chebyshev's Theorem, we are looking for a polynomial that differs from f by $\pm d$ at five points with alternating signs, where $d = \|f - p\|_\infty$. Consider the derivatives of $r(x) = \cos x - ax^2 - b$:

$$r'(x) = -\sin x - 2ax \quad \text{and} \quad r''(x) = -\cos x - 2a.$$

Since $-\cos x$ is concave on $[-\frac{\pi}{2}, \frac{\pi}{2}]$, the second derivative r'' has at most two zeros. This happens only if $-\frac{1}{2} < a < 0$; and these zeros are at $\pm z = \pm \arccos(2a)$. So r' is decreasing on $[-1, -z]$, then increasing on $[-z, z]$, and then decreasing again on $[z, 1]$. Thus r' can have at most three zeros. But r achieves its extreme values five times. So r' has exactly three zeros corresponding to extrema of r; and the other two extrema must be at the endpoints. Because of the symmetry of even functions, one extremum is at 0, and the other two critical points will be called $\pm x_0$. Moreover, $r'(0) = 0$ and $r''(0) = -1 - 2a < 0$; so this is a maximum.

So far, we have

$$d = r(0) = 1 - b,$$
$$d = r(\pm \frac{\pi}{2}) = -a\left(\frac{\pi}{2}\right)^2 - b,$$
$$-d = r(\pm x_0) = \cos x_0 - ax_0^2 - b,$$
$$0 = r'(x_0) = -\sin x_0 - 2ax_0.$$

Solving the first two equations for a yields $a = -4/\pi^2$. Plugging this into the fourth yields $\sin x_0 = 8x_0/\pi^2$.

Since $\sin x$ is concave on $[0, \frac{\pi}{2}]$, this equation has a unique positive solution. It may be found numerically to be approximately $x_0 := 1.0988243$. From the third equation (and the first), we obtain

$$b = \frac{1}{2}\left(1 + \cos x_0 + \frac{4x_0^2}{\pi^2}\right) := 0.9719952.$$

So the closest cubic to $\cos x$ on $[-\frac{\pi}{2}, \frac{\pi}{2}]$ is $p(x) = -\frac{4}{\pi^2}x^2 + 0.9719952$ and the error is $d = 0.0280048$.

Exercises for Section 10.6

A. Find the closest line to e^x on $[0,1]$.

B. Find the cubic polynomial that best approximates $|x|$ on the interval $[-1,1]$. HINT: Use symmetry first.

C. Suppose that $f \in C[a,b]$ is a twice continuously differentiable function with $f''(x) > 0$ on $[a,b]$. Show that the best linear approximation to f has slope $\dfrac{f(b)-f(a)}{b-a}$.

D. Apply the previous exercise to find the closest line to $f(x) = \sqrt{1+3x^2}$ on $[0,1]$, and compute the error.

E. If f in $C[-1,1]$ is an even (odd) function, show that the best approximation of degree n is also even (odd).

F. Let p be the best polynomial approximation of degree n to $\sqrt{x}$ on $[0,1]$. Show that $q(x) = p(x^2)$ is the best polynomial approximation of degree $2n+1$ to $|x|$ on $[-1,1]$. HINT: How does the equioscillation condition on $\sqrt{x}-p(x)$ translate to the approximation of $|x|$?

10.7 Expansions Using Chebyshev Polynomials

Ideally, we would like to find the polynomial that is exactly the best approximation to a given continuous function f. There is an algorithm that constructs a sequence of polynomials converging uniformly to the best approximating polynomial of degree n known as **Remes's algorithm**. Roughly, it works as follows: Pick $n+2$ points $x_1 < x_2 < \cdots < x_{n+2}$ in $[a,b]$. These points might be equally spaced, but foreknowledge of the function could lead you to pick points clustered in regions where f behaves more wildly. Then solve the linear equations for $a_0, a_1, \ldots, a_n$ and d:

$$
\begin{aligned}
a_0 + x_1 a_1 + x_1^2 a_2 + \cdots + x_1^n a_n - d &= f(x_1), \\
a_0 + x_2 a_2 + x_2^2 a_2 + \cdots + x_2^n a_n + d &= f(x_2), \\
&\ \ \vdots \\
a_0 + x_{n+2} a_1 + x_{n+2}^2 a_2 + \cdots + x_{n+2}^n a_n + (-1)^{n+2} d &= f(x_{n+2}).
\end{aligned}
$$

This method attempts to find a polynomial that satisfies Chebyshev's Theorem. However, the function may well take its extrema on other points. So the algorithm proceeds to choose new points $x_1', \ldots, x_{n+2}'$ by selecting points where the error is largest (or close to it) near each point. Eventually, this procedure converges to the nearest polynomial of degree n.

However, each step of this process involves many calculations, so convergence to the optimal polynomial is slow. In practice, it is better to find quickly an approximating polynomial that is not quite the best. Solutions that are less than optimal, but still quite good, can be found very efficiently. We develop such an algorithm in this section using Chebyshev polynomials. Chebyshev polynomials are useful in numerical analysis, algebra, and other areas.

10.7.1. DEFINITION. For $n \geq 0$, define the **Chebyshev polynomial** of degree n in $\mathbb{P}_n[-1,1]$ by

$$T_n(x) = \cos(n \arccos x).$$

It is not immediately obvious that T_n is a polynomial, much less a polynomial of degree n. The T in T_n comes from the continental transliterations, such as Tchebycheff, of the original Russian. The graphs of T_1 through T_8 in Figure 10.5 suggests some of the many nice properties of these polynomials; for instance, T_n is an even or odd function, according to n being even or odd.

10.7.2. LEMMA. $T_0(x) = 1$, $T_1(x) = x$ and

$$T_n(x) = 2xT_{n-1}(x) - T_{n-2}(x) \quad for \quad n \geq 2.$$

For each $n \geq 1$, $T_n(x)$ is a polynomial of degree n with leading coefficient 2^{n-1}. Also, $\|T_n\|_\infty = 1$, and

$$T_n\big(\cos\big(\tfrac{k}{n}\pi\big)\big) = (-1)^k \quad for \quad 0 \leq k \leq n.$$

PROOF. Recall the sum and difference of angles formulas for cosine:

$$\cos(A \pm B) = \cos A \cos B \mp \sin A \sin B.$$

Let $A = n\theta$ and $B = \theta$, and add these two formulas to get

$$\cos(n+1)\theta + \cos(n-1)\theta = 2\cos n\theta \cos\theta.$$

Substituting $\theta = \arccos x$ gives

$$T_{n+1}(x) + T_{n-1}(x) = 2xT_n(x) \quad \text{for all} \quad n \geq 1.$$

Evidently, $T_0 = 1$ and $T_1(x) = x$. The next few terms are

$$T_2(x) = 2xT_1(x) - T_0(x) = 2x^2 - 1,$$
$$T_3(x) = 2xT_2(x) - T_1(x) = 4x^3 - 3x,$$
$$T_4(x) = 2xT_3(x) - T_2(x) = 8x^4 - 8x^2 + 1.$$

By induction, it follows that $T_n(x)$ is a polynomial of degree n with leading coefficient 2^{n-1}.

Since $|\cos\theta| \leq 1$ for all values of θ, it follows that $\|T_n\|_\infty \leq 1$. Now $|\cos(\theta)| = 1$ only when θ is an integer multiple of π. It follows that T_n attains its maximum modulus when $n \arccos x = k\pi$ for some integer k. Solving, we obtain

$$x_k = \cos\big(\tfrac{k}{n}\pi\big) \quad \text{for} \quad 0 \leq k \leq n.$$

Other choices of k just repeat these values. Finally,

$$T_n(x_k) = (-1)^k \quad \text{for} \quad 0 \leq k \leq n. \qquad \blacksquare$$

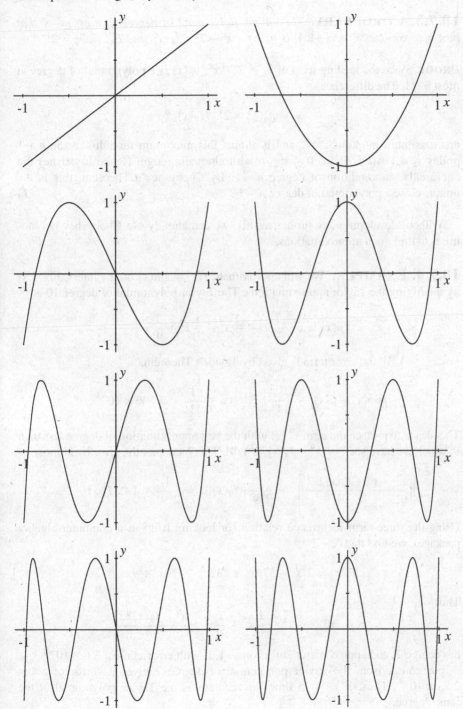

FIG. 10.5 The Chebyshev polynomials T_1 through T_8.

10.7.3. COROLLARY. *The unique polynomial of degree at most $n-1$ that best approximates x^n on $[-1,1]$ is $p_n(x) = x^n - 2^{1-n}T_n(x)$, and $E_{n-1}(x^n) = 2^{1-n}$.*

PROOF. Since the leading term of T_n is $2^{n-1}x^n$, $p_n(x)$ is a polynomial of degree at most $n-1$. The difference

$$x^n - p_n(x) = 2^{1-n}T_n(x)$$

has maximum modulus 2^{1-n}, and it attains this maximum modulus at the $n+1$ points $x_k = \cos(k\pi/n)$ for $0 \le k \le n$ with alternating sign. Hence it satisfies the equioscillation condition of degree $n-1$. By Chebyshev's Theorem, this is the unique closest polynomial of degree $n-1$. ∎

Without developing any further results, we can already use Chebyshev polynomials to find good approximations.

10.7.4. EXAMPLE. We will approximate $f(x) = \sin(x)$ on the interval $[-1,1]$ by modifying the Taylor approximations. The Taylor polynomial of degree 10 is

$$p(x) = x - \frac{1}{3!}x^3 + \frac{1}{5!}x^5 - \frac{1}{7!}x^7 + \frac{1}{9!}x^9.$$

For $x \in [-1,1]$, the error term is given by Taylor's Theorem,

$$|\sin(x) - p(x)| \le \frac{|x|^{11}}{11!}\|f^{(11)}\|_\infty \le \frac{1}{11!} < 2.506 \times 10^{-8}.$$

The idea is to replace the term $x^9/9!$ with the best approximation of degree less than 9, which we have seen is $(x^9 - T_9(x)/2^8)/9!$. This increases the error by at most

$$\left\|\frac{x^9}{9!} - \frac{x^9 - T_9(x)/2^8}{9!}\right\|_\infty = \frac{1}{2^8 9!}\|T_9(x)\|_\infty = \frac{1}{2^8 9!} \le 1.077 \times 10^{-8}.$$

Using the three-term recurrence relation (or looking it up in a computer algebra package), we find that

$$T_9(x) = 2^8 x^9 - 576 x^7 + 432 x^5 - 120 x^3 + 9x.$$

Thus

$$p(x) = x - \frac{1}{3!}x^3 - \frac{1}{5!}x^5 + \frac{1}{7!}x^7 + \frac{x^9 - T_9(x)/2^8}{9!}$$

has degree 7, and approximates $\sin(x)$ on $[-1,1]$ with error at most 3.6×10^{-8}.

For comparison, the Taylor polynomial of degree 7 gives an error of about 2.73×10^{-6}. Thus, $p(x)$ is 75 times as accurate as the Taylor polynomial of the same degree.

In practice, we want to do away with ad hoc methods and find an algorithm that yields reasonably good approximations quickly. We need the following inner product for $f, g \in C[-1, 1]$:

$$\langle f, g \rangle_T = \frac{1}{\pi} \int_{-1}^{1} f(x) g(x) \frac{dx}{\sqrt{1 - x^2}}.$$

It is easy to verify that this is an inner product on $C[-1, 1]$ (i.e., it is linear in both variables, positive definite, and symmetric). The crucial property we need is that the Chebyshev polynomials are orthogonal with respect to this inner product. The constant $1/\pi$ makes the constant function 1 have norm 1, which is computationally convenient.

10.7.5. LEMMA.

$$\langle T_n, T_m \rangle_T = \begin{cases} 0 & \text{if } m \neq n, \\ \frac{1}{2} & \text{if } m = n \neq 0, \\ 1 & \text{if } m = n = 0. \end{cases}$$

PROOF. Make the substitution $\cos \theta = x$ in the integral, so that $-\sin \theta \, d\theta = dx$, whence $d\theta = -dx/\sqrt{1 - x^2}$. We have

$$\langle T_n, T_m \rangle_T = \frac{1}{\pi} \int_{-1}^{1} T_n(x) T_m(x) \frac{dx}{\sqrt{1 - x^2}}$$

$$= \frac{1}{\pi} \int_{0}^{\pi} \cos n\theta \cos m\theta \, d\theta$$

$$= \frac{1}{2\pi} \int_{0}^{\pi} \cos(m+n)\theta + \cos(m-n)\theta \, d\theta,$$

where again we have used the identity $2 \cos A \cos B = \cos(A + B) + \cos(A - B)$.

There are three different integrals, depending on the values of m and n. If $m \neq n$, then both $m + n$ and $m - n$ are not zero and the integral is

$$\frac{1}{2\pi} \left(\frac{\sin(m+n)\theta}{m+n} + \frac{\sin(m-n)\theta}{m-n} \right) \Big|_{0}^{\pi} = 0.$$

If $m = n \neq 0$, then $m - n$ is zero and $m + n$ is not, so the integral is

$$\frac{1}{2\pi} \left(\frac{\sin(m+n)\theta}{m+n} + x \right) \Big|_{0}^{\pi} = \frac{1}{2}.$$

Finally, if $m = n = 0$, then the integral is 1. ∎

Suppose that a function $f \in C[-1, 1]$ can be expressed as an infinite sum of Chebyshev polynomials, say $f(x) = \sum_{n=1}^{\infty} a_n T_n(x)$ for all $x \in [-1, 1]$. Then, willfully ignoring the issue of convergence, we can write

$$\langle f, T_k \rangle_T = \left\langle \sum_{n=1}^{\infty} a_n T_n, T_k \right\rangle_T = \sum_{n=1}^{\infty} \langle a_n T_n, T_k \rangle_T = a_k \langle T_k, T_k \rangle_T.$$

Solving for a_k in the preceding equation and using the definition of the inner product, we have a possible formula for the coefficients:

$$a_k = 2\langle f, T_k \rangle_T = \frac{2}{\pi} \int_{-1}^{1} f(x) T_k(x) \frac{dx}{\sqrt{1-x^2}} \quad \text{for} \quad k \geq 1$$

and

$$a_0 = \langle f, T_0 \rangle_T = \frac{1}{\pi} \int_{-1}^{1} f(x) \frac{dx}{\sqrt{1-x^2}}.$$

10.7.6. DEFINITION. We define the **Chebyshev series** for f in $C[-1,1]$ to be $\sum_{n=1}^{\infty} a_n T_n(x)$, where the sequence (a_n) is given by the preceding formulas.

There are a host of questions about this series. For which x does this infinite series converge? Is the resulting function continuous? Does it equal f or not?

In Chapter 14, we connect Chebyshev series to Fourier series and obtain Theorem 14.8.2, which shows that the Chebyshev series converges uniformly for all Lipschitz functions. For now, we state a weaker result and leave its proof as an exercise.

10.7.7. THEOREM. *If $f \in C[-1,1]$ has a continuous second derivative, then the Chebyshev series of f converges uniformly to f.*

Exercises for Section 10.7

A. Verify the following properties of the Chebyshev polynomials $T_n(x)$.

(a) If m is even, then T_m is an even function; and if m is odd, then T_m is odd.
(b) Show that every polynomial p of degree n has a *unique* representation using Chebyshev polynomials: $p(x) = a_0 T_0(x) + a_1 T_1(x) + \cdots + a_n T_n(x)$.
(c) $T_m(T_n(x)) = T_{mn}(x)$.
(d) $(1-x^2)T_n''(x) - xT_n'(x) + n^2 T_n(x) = 0$.

B. Show by induction that $T_n(x) = \dfrac{\left(x + \sqrt{x^2-1}\right)^n + \left(x - \sqrt{x^2-1}\right)^n}{2}$.

C. Find a sequence of polynomials converging uniformly to $f(x) = |x|^3$ on $[-1,1]$. HINT: $f \in C^2$.

D. Prove Theorem 10.7.7 using the following outline.

(a) Show that there is $M > 0$ such that the Chebyshev coefficients of f satisfy $a_n \leq M/n^2$. HINT: Make the change of variable $x = \cos\theta$ and integrate by parts twice.
(b) Deduce that the Chebyshev series converges uniformly to some function F.
(c) Show that $F = f$. HINT: compute $\langle f - F, 2T_n \rangle_T$ and find a way to use Exercise 10.2.D.

E.   Suppose that $f \in C[-1,1]$ has a Chebyshev series $\sum\limits_{n=0}^{\infty} a_n T_n$. If $\sum\limits_{n=0}^{\infty} |a_n| < \infty$, show that the Chebyshev series converges uniformly to f.   HINT: Study the proof of Theorem 10.7.7.

F.   Verify the following expansions in Chebyshev polynomials:

(a) $|x| = \dfrac{2}{\pi} - \dfrac{4}{\pi} \sum\limits_{j=1}^{\infty} \dfrac{(-1)^j}{4j^2 - 1} T_{2j}(x)$.

(b) $\sqrt{1 - x^2} = \dfrac{2}{\pi} - \dfrac{4}{\pi} \sum\limits_{j=1}^{\infty} \dfrac{1}{4j^2 - 1} T_{2j}(x)$.   HINT: Substitute $x = \cos\theta$. Apply Exercise E.

G.   Suppose that $f \in C[-1,1]$ has a Chebyshev series $\sum\limits_{n=0}^{\infty} a_n T_n$.

(a) Show that $E_n(f) \leq \sum\limits_{k=n+1}^{\infty} |a_k|$.

(b) Show that $E_n(T_{n+1}) = 1$.   HINT: Theorem 10.6.4

(c) Show that $\left| E_n(f) - |a_{n+1}| \right| \leq \sum\limits_{k=n+2}^{\infty} |a_k|$.   HINT: Show $E_n(f) \geq E_n\big(|a_{n+1}|T_{n+1}\big) - \sum\limits_{k=n+2}^{\infty} |a_k|$.

(d) Show that if $\lim\limits_{n \to \infty} \dfrac{\sum_{k=n+1}^{\infty} |a_k|}{|a_n|} = 0$, then $\lim\limits_{n \to \infty} \dfrac{E_n(f)}{|a_{n+1}|} = 1$.

H.   Let a_n be a sequence of real numbers monotone decreasing to 0. Define the sequence of polynomials $p_n(x) = \sum\limits_{k=1}^{n} (a_k - a_{k+1}) T_{3^k}(x)$.

(a) Show that this sequence converges uniformly on $[-1,1]$ to a continuous function $f(x)$. HINT: Weierstrass M-test.

(b) Evaluate $(f - p_n)\big(\cos(3^{-n-1}k\pi)\big)$ for $0 \leq k \leq 3^{n+1}$.

(c) Show that $E_{3^n}(f) = a_{n+1}$. Conclude that there are continuous functions for which the optimal sequence of polynomials converges exceedingly slowly.

10.8 Splines

Splines are smooth piecewise polynomials. They are well adapted to use on computers, and are often used in practice. Because they are closely related to polynomials, we give a brief treatment of splines here, concentrating on issues related to real analysis. For algorithmic and implementation issues, we refer the reader to [14].

To motivate the idea behind splines, observe that approximation by polynomials can be improved either by increasing the degree of the polynomial or by decreasing the size of the interval on which the approximation is used. Splines take the latter approach, successively chopping the interval into small pieces and approximating the function on each piece by a polynomial of fixed small degree, such as a cubic.

We search for a relatively smooth function that is piecewise a polynomial of low degree but is not globally a polynomial at all. This turns out to be worth the additional theoretical complications. Why? First, evaluating the approximation will be easier on each subinterval because it is a polynomial of small degree. Instead of having to do some multiplications, we have several comparisons to decide which interval we are in. Comparison is much simpler than multiplication, so evaluation can be much faster, even if we have to do many comparisons.

Second, since the degree is small, we can use simple methods like interpolation to find the polynomial on each subinterval. Interpolation is both easy to implement on the computer and (mostly) easy to understand.

Third, local irregularities of the function affect the approximation only locally, in contrast to polynomial approximation. For example, a sharp spike in f affects the polynomial approximation globally; but for splines, this affects the approximation on only a single subinterval.

The discussion so far has supposed that we are free to choose completely different polynomials on each subinterval. In fact, we would like the polynomials to fit together smoothly. To start, we begin with the revealing special case of approximation by a piecewise linear continuous function.

Choose a partition Δ of the interval $[a,b]$ into k subintervals with endpoints $a = x_0 < x_1 < \cdots < x_k = b$. We define $\mathbb{S}_1(\Delta)$ to be the subspace of $C[a,b]$ given by

$$\mathbb{S}_1(\Delta) = \{g \in C[a,b] : g|_{[x_i,x_{i+1}]} \text{ is linear for } 0 \leq i < k\}.$$

Clearly, a function $g \in \mathbb{S}_1(\Delta)$ is uniquely determined by its values $g(x_i)$ at the **nodes** x_i for $0 \leq i \leq k$. Indeed, we just construct the line segments between the points $(x_i, g(x_i))$. Thus $\mathbb{S}_1(\Delta)$ is a finite-dimensional subspace of dimension $k+1$. The elements of this space are called **linear splines**. Figure 10.6 shows an element of $\mathbb{S}_1(\Delta)$ approximating a given continuous function.

Now take a continuous function f in $C[a,b]$. By Theorem 7.3.5, there is some g in $\mathbb{S}_1(\Delta)$ such that

$$\|f - g\|_\infty = \inf\{\|f - h\|_\infty : h \in \mathbb{S}_1(\Delta)\}.$$

Instead of trying to find this optimal choice, we choose the function h in $\mathbb{S}_1(\Delta)$ such that $h(x_i) = f(x_i)$ for $0 \leq i \leq k$. Define $J_1 : C[a,b] \to \mathbb{S}_1(\Delta)$ by letting $J_1 f$ be this function h. Notice that our characterization of functions in $\mathbb{S}_1(\Delta)$ shows that $J_1 g = g$ for $g \in \mathbb{S}_1(\Delta)$. Also, J_1 is linear:

$$J_1(af + bg) = aJ_1 f + bJ_1 g \quad \text{for} \quad f,g \in C[a,b] \text{ and } a,b \in \mathbb{R}.$$

The following lemma shows that choosing $J_1 f$ instead of the best approximant does not increase the error too much.

10.8.1. LEMMA. *If $f \in C[a,b]$, then*

$$\|f - J_1 f\|_\infty \leq 2\inf\{\|f - g\|_\infty : g \in \mathbb{S}_1(\Delta)\}.$$

PROOF. Notice that for any $f \in C[a,b]$,

$$\|J_1 f\|_\infty = \max\{|f(x_i)| : 0 \leq i \leq k\} \leq \|f\|_\infty.$$

If $g \in \mathbb{S}_1(\Delta)$ is the closest point to f, we use linearity and $J_1 g = g$ to obtain

$$\|f - J_1 f\|_\infty = \|f - g - J_1(f - g)\|_\infty \leq \|f - g\|_\infty + \|J_1(f - g)\|_\infty \leq 2\|f - g\|_\infty. \blacksquare$$

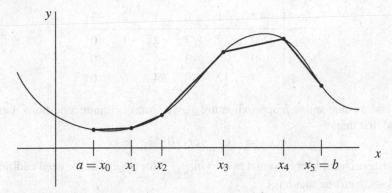

FIG. 10.6 Approximation in $\mathbb{S}_1(\Delta)$.

10.8.2. DEFINITION. A **cubic spline** for a partition Δ of $[a,b]$ is a C^2 function h such that $h|_{[x_i,x_{i+1}]}$ is a polynomial of degree at most 3 for $0 \le i < k$. Let $\mathbb{S}(\Delta)$ denote the vector space of all cubic splines for the partition Δ.

Cubic spline interpolation is popular in practice. It may seem rather surprising that it is possible to fit cubics together and remain twice continuously differentiable and still have the flexibility to approximate functions well.

10.8.3. EXAMPLE. Consider

$$h(x) = \begin{cases} 2x^3 + 12x^2 + 24x + 16 & \text{if } -2 \le x \le -1, \\ -7x^3 - 15x^2 - 3x + 7 & \text{if } -1 \le x \le 0, \\ 9x^3 - 15x^2 - 3x + 7 & \text{if } 0 \le x \le 1, \\ -5x^3 + 27x^2 - 45x + 21 & \text{if } 1 \le x \le 2, \\ x^3 - 9x^2 + 27x - 27 & \text{if } 2 \le x \le 3. \end{cases}$$

We readily compute the following:

$h'(x)$	$h''(x)$	interval
$6x^2 + 24x + 24$	$12x + 12$	$-2 \le x \le -1$
$-21x^2 - 30x - 3$	$-42x - 15$	$-1 \le x \le 0$
$27x^2 - 30x - 3$	$54x - 15$	$0 \le x \le 1$
$-15x^2 + 54x - 45$	$-30x + 27$	$1 \le x \le 2$
$3x^2 - 18x + 27$	$6x - 9$	$2 \le x \le 3$

We can now verify the following table of values. Since the first and second derivatives match up at the endpoints of each interval, h is C^2 and so is a cubic spline:

x_i	-2	-1	0	1	2	3
$h(x_i)$	0	2	7	-2	-1	0
$h'(x_i)$	0	6	-3	-6	3	0
$h''(x_i)$	0	12	-30	24	-6	0

To find a cubic spline h approximating f, we specify certain conditions. Let us demand first that

$$h(x_i) = f(x_i) \quad \text{for} \quad 0 \le i \le k.$$

Let's write each cubic polynomial as $h_i = h|_{[x_i, x_{i+1}]}$ for $0 \le i < k$. We need additional conditions to ensure that h is C^2:

$$h_i'(x_i) = h_{i+1}'(x_i) \quad \text{and} \quad h_i''(x_i) = h_{i+1}''(x_i) \quad \text{for} \quad 1 \le i \le k-1.$$

A cubic has four parameters, and these equations put four conditions on each cubic except for the two on the ends, where there are three constraints. To finish specifying the spline, we add two endpoint conditions:

$$h_1'(x_0) = f'(x_0) \qquad \text{and} \qquad h_k'(x_k) = f'(x_k),$$

assuming that these derivatives exist. (If they do not, we may set them equal to 0.) For convenience, we shall assume that f is C^2, which ensures that these data are defined and allows some interesting theoretical consequences.

We shall see that a cubic spline h in $\mathbb{S}(\Delta)$ is uniquely determined by these equations. There are $k+3$ data conditions determined by f, namely $f(x_0), \ldots, f(x_k)$ and $f'(x_0)$ and $f'(x_k)$. Hence we expect to find that $\mathbb{S}(\Delta)$ is a finite-dimensional subspace of dimension $k+3$. This will allow us to define a map J from $C^2[a,b]$ to $\mathbb{S}(\Delta)$ by setting Jf to be the function h specified by these equations.

10.8.4. LEMMA. *Given $c < d$ and real numbers a_1, a_2, s_1, s_2, there is a unique cubic polynomial p satisfying*

$$p(c) = a_1, \qquad p(d) = a_2, \qquad p'(c) = s_1, \quad \text{and} \quad p'(d) = s_2.$$

Setting $\Delta = d - c$, we obtain

$$p''(c) = \frac{6(a_2 - a_1)}{\Delta^2} - \frac{4s_1 + 2s_2}{\Delta} \quad \text{and} \quad p''(d) = -\frac{6(a_2 - a_1)}{\Delta^2} + \frac{2s_1 + 4s_2}{\Delta}.$$

PROOF. Consider the cubics

$$p_1(x) = \frac{(x-d)^2}{(d-c)^2}\left(1 + 2\frac{x-c}{d-c}\right), \qquad p_2(x) = \frac{(x-c)^2}{(d-c)^2}\left(1 - 2\frac{x-d}{d-c}\right),$$

$$q_1(x) = \frac{(x-c)(x-d)^2}{(d-c)^2}, \qquad q_2(x) = \frac{(x-c)^2(x-d)}{(d-c)^2}.$$

For example, $p_1(c) = 1$ and $p_1'(c) = p_1'(d) = p_1(d) = 0$. The reader can verify that $p(x) = a_1 p_1(x) + a_2 p_2(x) + s_1 q_1(x) + s_2 q_2(x)$ is the desired cubic.

For uniqueness, we can note that the difference of two such cubics is a cubic q such that $q(c) = q(d) = q'(c) = q'(d) = 0$. The first two conditions show that c and d are roots of q; and the second two conditions then imply that they are double roots. So $(x-c)^2(x-d)^2$ divides q. Since q has degree at most 3, this forces $q = 0$.

Finding the value of p'' at c and d is a routine calculation. ∎

10.8.5. THEOREM. *Given a partition $\Delta : a = x_0 < x_1 < \cdots < x_k = b$ of the interval $[a,b]$ and real numbers $a_0, \ldots, a_k$, s_0, and s_k, there is a unique cubic spline $h \in \mathbb{S}(\Delta)$ such that $h(x_i) = a_i$ for $0 \leq i \leq k$ and $h'(a) = s_0$ and $h'(b) = s_k$.*

PROOF. If such a spline exists, we could define $s_i = h'(x_i)$ for $1 \leq i \leq k-1$. We search for such values of s_i that allow a spline. Given the values a_i of h and s_i of h' at the points x_{i-1} and x_i, the previous lemma determines a unique cubic h_i on the interval $[x_{i-1}, x_i]$. So for each choice of $(s_1, \ldots, s_{k-1})$, there is one piecewise cubic function on $[a,b]$ that interpolates the values a_i and derivatives s_i at each point x_i for $0 \leq i \leq k$. However, in general this will not be C^2. There are $k-1$ conditions that must be satisfied:

$$h_i''(x_i) = h_{i+1}''(x_i) \quad \text{for} \quad 1 \leq i < k-1.$$

Our job is to compute a formula for these second derivatives to obtain conditions on the hypothetical data $s_1, \ldots, s_{k-1}$.

Let us write $\Delta_i = x_i - x_{i-1}$ for $1 \leq i \leq k$. By the previous lemma, the second derivative conditions at x_i for $1 \leq i \leq k-1$ are

$$h''(x_i) = -\frac{6(a_i - a_{i-1})}{\Delta_i^2} + \frac{2s_{i-1} + 4s_i}{\Delta_i} = \frac{6(a_{i+1} - a_i)}{\Delta_{i+1}^2} - \frac{4s_i + 2s_{i+1}}{\Delta_{i+1}}. \qquad (10.8.6)$$

Rearranging this yields a linear system of $k-1$ equations in the $k-1$ unknowns $s_1, \ldots, s_{k-1}$. For $1 \leq i \leq k-1$,

$$\Delta_{i+1} s_{i-1} + 2(\Delta_i + \Delta_{i+1}) s_i + \Delta_i s_{i+1} = \frac{3\Delta_i(a_{i+1} - a_i)}{\Delta_{i+1}} + \frac{3\Delta_{i+1}(a_i - a_{i-1})}{\Delta_i}.$$

The terms involving s_0 and s_k may be moved to the right-hand side.

It now remains to show that this system has a unique solution. Let

$$X = \begin{bmatrix} 2(\Delta_1 + \Delta_2) & \Delta_1 & 0 & 0 & \cdots & 0 \\ \Delta_3 & 2(\Delta_2 + \Delta_3) & \Delta_2 & 0 & \cdots & 0 \\ \vdots & \vdots & \vdots & \ddots & \vdots & \vdots \\ 0 & \cdots & 0 & \Delta_{k-1} & 2(\Delta_{k-2} + \Delta_{k-1}) & \Delta_{k-2} \\ 0 & \cdots & 0 & 0 & \Delta_k & 2(\Delta_{k-1} + \Delta_k) \end{bmatrix}.$$

This will follow if we can show that X is invertible. The property of this system that makes it possible is that the matrix is **diagonally dominant**, which means that the diagonal entries are greater than the sum of all other entries in each row.

To see this, suppose that $y = (y_1, \ldots, y_{k-1})$ is in the kernel of X. Choose a co-efficient i_0 such that $|y_{i_0}| \geq |y_j|$ for $1 \leq j \leq k-1$. Then looking only at the i_0th coefficient of $0 = Xy$, we obtain

$$0 = \left| \Delta_{i_0+1} y_{i_0-1} + 2(\Delta_{i_0} + \Delta_{i_0+1}) y_{i_0} + \Delta_{i_0} y_{i_0+1} \right|$$
$$\geq 2(\Delta_{i_0} + \Delta_{i_0+1}) |y_{i_0}| - \Delta_{i_0+1} |y_{i_0-1}| - \Delta_{i_0} |y_{i_0+1}|$$
$$\geq (\Delta_{i_0} + \Delta_{i_0+1}) |y_{i_0}|.$$

Hence $y = 0$ and so X has trivial kernel. Therefore X is invertible. So for each set of data $a_0, \ldots, a_k, s_0$, and s_k, there is a unique choice of points $s_1, \ldots, s_{k-1}$ solving our system. Thus there is a unique spline $h \in \mathbb{S}(\Delta)$ satisfying these data. ∎

Thus if f is a C^2 function on $[a,b]$, there is a unique cubic spline h such that $h(x_i) = a_i := f(x_i)$ for $0 \leq i \leq k$, $h'(a) = s_0 := f'(a)$, and $h'(b) = s_k := f'(b)$. We denote the function h by Jf. Let us show that J is linear. If f_1 and f_2 are functions in $C^2[a,b]$ with $h_i = Jf_i$, then $h = \alpha_1 h_1 + \alpha_2 h_2$ is a spline such that

$$h(x_i) = (\alpha_1 h_1 + \alpha_2 h_2)(x_i) = (\alpha_1 f_1(x_i) + \alpha_2 f_2)(x_i)$$

and

$$h'(a) = (\alpha_1 h'_1 + \alpha_2 h'_2)(a) = \alpha_1 f'_1(a) + \alpha_2 f'_2(a) = (\alpha_1 f_1(x_i) + \alpha_2 f_2)'(a).$$

Similarly, this holds at b. By the uniqueness of the spline, it follows that

$$J(\alpha_1 f_1 + \alpha_2 f_2) = \alpha_1 h_1 + \alpha_2 h_2 = \alpha_1 Jf_1 + \alpha_2 Jf_2.$$

In particular, we may find specific splines c_i satisfying

$$c'_i(a) = c'_i(b) = 0 \quad \text{and} \quad c_i(x_j) = \begin{cases} 1 & \text{if} \quad j = i, \\ 0 & \text{if} \quad j \neq i, \end{cases}$$

for $0 \leq i \leq k$. The linear space $\mathbb{S}(\Delta)$ is spanned by $\{x, x^2, c_i : 0 \leq i \leq k\}$. To see this, let h be the spline with data $a_0, \ldots, a_k, s_0$, and s_k. Let q be the unique quadratic $q(x) = cx + dx^2$ such that $q'(a) = s_1$ and $q'(b) = s_k$. Then $g = h - q$ is a spline with $g'(a) = g'(b) = 0$. Form the spline

$$s(x) = q(x) + \sum_{i=0}^{n} g(x_i) c_i(x).$$

It is easy to check that s is another spline with the same data as h. Since this uniquely determines the spline, $h = s$ has the desired form. This exhibits a specific basis for $\mathbb{S}(\Delta)$. Figure 10.7 shows c_0 and c_2 for a particular partition.

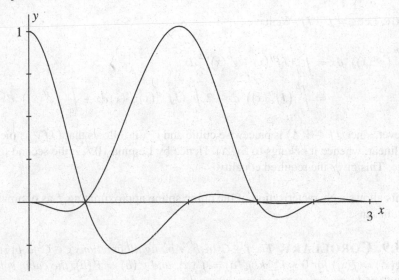

FIG. 10.7 Graphs of c_0 and c_2 for $\Delta = \{0, .5, 1.4, 2, 2.5, 3\}$.

Now that we have shown that cubic splines are plentiful, we investigate how well Jf approximates the original function f. We need the following:

10.8.7. LEMMA. If $f \in C^2[a, b]$, then for all $\varphi \in \mathbb{S}_1(\Delta)$,

$$\int_a^b \varphi(x)(f - Jf)''(x)\, dx = 0.$$

PROOF. We use integration by parts twice. Letting $du = (f - Jf)''(x)\, dx$ and $v = \varphi(x)$, we have

$$\int_a^b \varphi(x)(f - Jf)''(x)\, dx = \varphi(x)(f - Jf)'(x)\Big|_a^b - \int_a^b \varphi'(x)(f - Jf)'(x)\, dx.$$

Observe that $f'(a) = (Jf)'(a)$ and $f'(b) = (Jf)'(b)$, so the first term above is zero. Using integration by parts again with $du = (f - Jf)'(x)\, dx$ and $v = \varphi'(x)$,

$$\int_a^b \varphi(x)(f - Jf)''(x)\, dx = \varphi'(x)(f - Jf)(x)\Big|_a^b - \int_a^b \varphi''(x)(f - Jf)(x)\, dx.$$

Now, $f(a) = Jf(a)$ and $f(b) = Jf(b)$, so the first term is zero. For the second term, we observe that since φ is piecewise linear, φ'' is equal to zero except for the points $x_0, x_1, \dots, x_n$, where it is not defined. Thus the integral is zero. ∎

10.8.8. THEOREM. If $f \in C^2[a, b]$, then

$$\int_a^b \left(f''(x)\right)^2 dx = \int_a^b \left((Jf)''(x)\right)^2 dx + \int_a^b \left((f - Jf)''(x)\right)^2 dx.$$

PROOF. Let $g = f - Jf$. We have

$$\int_a^b \left(f''(x)\right)^2 dx = \int_a^b \left(Jf''(x) + g''(x)\right)^2 dx$$

$$= \int_a^b \left(Jf''(x)\right)^2 dx + 2\int_a^b (Jf)''(x)g''(x)\,dx + \int_a^b \left(g''(x)\right)^2 dx.$$

However, since $Jf \in \mathbb{S}(\Delta)$ is piecewise cubic and C^2, it follows that $(Jf)''$ is piecewise linear, whence it belongs to $\mathbb{S}_1(\Delta)$. Hence by Lemma 10.8.7, the second term is zero. This gives the required equality.   ∎

This allows a characterization of the cubic spline approximating f as optimal in a certain sense.

10.8.9. COROLLARY. *Fix $f \in C^2[a,b]$. Among all functions $g \in C^2[a,b]$ such that $g(x_i) = f(x_i)$ for $0 \le i \le k$, $g'(a) = f'(a)$, and $g'(b) = f'(b)$, the cubic spline interpolant Jf minimizes the energy integral*

$$\int_a^b \left(g''(x)\right)^2 dx.$$

PROOF. For any such function g, we have $Jg = Jf$. So by the previous theorem,

$$\int_a^b \left(g''(x)\right)^2 dx = \int_a^b \left((Jf)''(x)\right)^2 dx + \int_a^b \left((g - Jf)''(x)\right)^2 dx$$

$$\ge \int_a^b \left((Jf)''(x)\right)^2 dx.$$

This inequality becomes an equality only if $g'' = (Jf)''$. Since we also have $g(a) = Jf(a)$ and $g'(a) = (Jf)'(a)$ by hypothesis, this implies that $g = Jf$ by integrating twice.   ∎

This property is called the **smoothest interpolation property** of cubic spline interpolation. Minimizing $\int_a^b \left(g''\right)^2(x)\,dx$ is roughly equivalent to minimizing the strain energy. Historically, flexible thin strips of wood called splines were used in drafting to approximate curves through a set of points. In 1946, when Schoenberg introduced spline curves, he observed that they represent the curves drawn by means of wooden splines, hence the name. Splines appear to be smooth since they avoid discontinuous first derivatives, which people recognize as "spikes," and avoid discontinuous second derivatives, which are recognized as sudden changes in curvature. Discontinuous third derivatives are not visible in any obvious geometric way.

Exercises for Section 10.8

A.   Fill in the details of the proof of Lemma 10.8.4.

B.   Find a nice explicit basis for $\mathbb{S}_1(\Delta)$.

C. Show that if $f \in \mathbb{S}_1(\Delta)$ and f' is continuous on $[a,b]$, then f is a straight line.

D. Prove the Weierstrass Approximation Theorem (10.2.2) using the following outline. Define $\mathrm{abs}_a(x) = |x - a|$ and use abs for abs_0.

(a) Show that abs is a uniform limit of the polynomials (p_n) given by $p_0 = 0$ and $p_{k+1} = p_k + (x - p_k^2)/2$. HINT: First show that $p_{k+1} \geq p_k$ and $0 \leq p_k(x) \leq \sqrt{x}$ by induction on k.
(b) Deduce that for each $a \in \mathbb{R}$, abs_a is a uniform limit of polynomials.
(c) Show that $\mathbb{S}_1(\Delta)$ is spanned by 1 and $\{\mathrm{abs}_a : a \in \Delta\}$.
(d) If $f \in C[a,b]$ and $\varepsilon > 0$, show that there is a partition of $[a,b]$, Δ, and $g \in \mathbb{S}_1(\Delta)$ such that $\|f - g\| < \varepsilon$.

E. Show that if $f \in \mathbb{S}(\Delta)$ and f''' is continuous on $[a,b]$, then f is a cubic polynomial.

F. Show that for $f \in C^2[a,b]$, $\|f - Jf\|_\infty \leq 2\inf\{\|f - g\|_\infty : g \in \mathbb{S}(\Delta)\}$.
HINT: Compare with Lemma 10.8.1.

G. Suppose that Δ has $k > 4$ intervals. Let $1 \leq i \leq k - 4$. If $h \in \mathbb{S}(\Delta)$ is 0 everywhere except on (x_i, x_{i+3}), show that $h = 0$. HINT: What derivative conditions are forced?

H. Let x_+^3 denote the function $\max\{x^3, 0\}$. Show that every cubic spline in $\mathbb{S}(\Delta)$ has the form
$$p(x) + \sum_{i=1}^{k-1} c_i (x - x_i)_+^3,$$
where $p(x)$ is a cubic polynomial and $c_i \in \mathbb{R}$.
HINT: Given $h \in \mathbb{S}(\Delta)$, let $c_i = \delta_i/6$, where δ_i is the change in h''' at x_i.

I. Find a nonzero spline h for the partition $\{-1, 0, 1, 2, 3, 4, 5\}$ such that h is 0 on $[-1, 0] \cup [4, 5]$. HINT: Use the previous exercise.

10.9 Uniform Approximation by Splines

To complete our analysis of cubic splines, we will obtain an estimate for the error of approximation. This is a rather delicate argument that combines a generalized mean value theorem with another system of linear equations. Our goal is to establish the following theorem:

10.9.1. THEOREM. *Let Δ be a partition $a = x_0 < x_1 < \cdots < x_k = b$ of the interval $[a,b]$ and set $\delta = \max\{x_i - x_{i-1} : 1 \leq i \leq k\}$. Let $f \in C^2[a,b]$ and let $h = Jf$ be the cubic spline in $\mathbb{S}(\Delta)$ approximating f. Then*

$$\|f - h\|_\infty \leq \frac{5}{2} \delta^2 \, \omega(f''; \delta),$$
$$\|f' - h'\|_\infty \leq 5\delta \, \omega(f''; \delta),$$
$$\|f'' - h''\|_\infty \leq 5 \omega(f''; \delta).$$

Since the proof is long and computational, we give an overview first. Following the algebra of the last section, we obtain a system of $k + 1$ linear equations satisfied by $h''(x_0), \ldots, h''(x_k)$. The constant terms in this system are estimated using a second-order Mean Value Theorem. Then the equations are used to show that $|h''(x_i) - f''(x_i)| \leq 4\omega(f''; \delta)$. It is then straightforward to bound $\|h'' - f''\|_\infty$, and then integratation gives bounds for $\|f' - h'\|_\infty$ and $\|f - h\|_\infty$.

We begin with a second-order Mean Value Theorem.

10.9.2. LEMMA. *Suppose that $f \in C^2[a,c]$ and $a < b < c$. There is a point ξ in (a,c) such that*

$$f(b) - \left(\frac{c-b}{c-a} f(a) + \frac{b-a}{c-a} f(c) \right) = \frac{-(c-b)(b-a)}{2} f''(\xi).$$

PROOF. Let $L(x)$ be the straight line through $(a, f(a))$ and $(c, f(c))$, namely

$$L(x) = \frac{c-x}{c-a} f(a) + \frac{x-a}{c-a} f(c).$$

Consider the function

$$g(x) = (c-b)(b-a)\big(f(x) - L(x)\big) - (c-x)(x-a)\big(f(b) - L(b)\big).$$

Notice that $g(a) = g(b) = g(c) = 0$. So by Rolle's Theorem, there are points $\xi_1 \in (a,b)$ and $\xi_2 \in (b,c)$ such that $g'(\xi_1) = g'(\xi_2) = 0$. Applying Rolle's Theorem to g' now yields a point ξ in (ξ_1, ξ_2) such that

$$0 = g''(\xi) = (c-b)(b-a)f''(\xi) + 2\big(f(b) - L(b)\big).$$

This is just a rearrangement of the desired formula. ∎

Notice that there is a limiting situation in which b equals a or c. Take $b = a$, for example. Divide both sides by $b - a$ and take the limit, ignoring the important point that ξ depends on b. Then we obtain

$$-\frac{c-a}{2} f''(\xi) = \frac{f(a) - f(c)}{c-a} + \lim_{b \to a} \frac{f(b) - f(a)}{b - a} = \frac{f(a) - f(c)}{c-a} + f'(a).$$

Rearranging, this becomes

$$f(c) = f(a) + f'(a)(c-a) + \frac{(c-a)^2}{2} f''(\xi) \quad \text{for some } \xi \in (a,c).$$

We could make the limit argument correct, but we do not need to do so because this is just a consequence of the order-one Taylor's Theorem (see Exercise 10.1.B).

PROOF OF THEOREM 10.9.1. We need to show that h'' is close to f'' at the points x_i. We rewrite the formula (10.8.6) as

$$h''(x_i) = \frac{6(a_{i+1} - a_i)}{\Delta_{i+1}^2} - \frac{4s_i + 2s_{i+1}}{\Delta_{i+1}} \qquad \text{for } 0 \le i \le k-1,$$

$$h''(x_i) = \frac{-6(a_i - a_{i-1})}{\Delta_i^2} + \frac{2s_{i-1} + 4s_i}{\Delta_i} \qquad \text{for } 1 \le i \le k.$$

The idea is to eliminate the unknown $s_1, \ldots, s_{k-1}$ from these $2k$ equations to yield $k+1$ equations for the $h''(x_i)$, $0 \le i \le k$.

We save some space by presenting the list of equations and ask the interested reader to verify that they are correct:

$$2\Delta_1 h''(x_0) + \Delta_1 h''(x_1) = 6\left(\frac{a_1 - a_0}{\Delta_1} - s_0\right),$$

$$\Delta_i h''(x_{i-1}) + 2(\Delta_i + \Delta_{i+1})h''(x_i) + \Delta_{i+1}h''(x_{i+1}) = 6\left(\frac{a_{i+1} - a_i}{\Delta_{i+1}} - \frac{a_i - a_{i-1}}{\Delta_i}\right)$$

$$\text{for} \quad 1 \le i \le k-1,$$

$$\Delta_k h''(x_{k-1}) + 2\Delta_k h''(x_k) = 6\left(s_k - \frac{a_k - a_{k-1}}{\Delta_k}\right).$$

Let us define the matrix

$$Y = \begin{bmatrix} 2\Delta_1 & \Delta_1 & 0 & 0 & \cdots & 0 & 0 & 0 \\ \Delta_1 & 2(\Delta_1+\Delta_2) & \Delta_2 & 0 & \cdots & 0 & 0 & 0 \\ 0 & \Delta_2 & 2(\Delta_2+\Delta_3) & \Delta_3 & \cdots & 0 & 0 & 0 \\ \vdots & \vdots & \vdots & \vdots & \ddots & \vdots & \vdots & \vdots \\ 0 & 0 & 0 & \cdots & \Delta_{k-2} & 2(\Delta_{k-2}+\Delta_{k-1}) & \Delta_{k-1} & 0 \\ 0 & 0 & 0 & \cdots & 0 & \Delta_{k-1} & 2(\Delta_{k-1}+\Delta_k) & \Delta_k \\ 0 & 0 & 0 & \cdots & 0 & 0 & \Delta_k & 2\Delta_k \end{bmatrix}.$$

So if we set

$$\mathbf{h}'' = \begin{bmatrix} h''(x_0) \\ h''(x_1) \\ \vdots \\ h''(x_{k-1}) \\ h''(x_k) \end{bmatrix}, \quad \mathbf{f}'' = \begin{bmatrix} f''(x_0) \\ f''(x_1) \\ \vdots \\ f''(x_{k-1}) \\ f''(x_k) \end{bmatrix}, \text{ and } \mathbf{z} = \begin{bmatrix} 6\left(\frac{a_1-a_0}{\Delta_1} - s_0\right) \\ 6\left(\frac{a_2-a_1}{\Delta_2} - \frac{a_1-a_0}{\Delta_1}\right) \\ \vdots \\ 6\left(\frac{a_k-a_{k-1}}{\Delta_k} - \frac{a_{k-1}-a_{k-2}}{\Delta_{k-1}}\right) \\ 6\left(s_k - \frac{a_k-a_{k-1}}{\Delta_k}\right) \end{bmatrix},$$

then the equation becomes $Y\mathbf{h}'' = \mathbf{z}$.

Now we apply Lemma 10.9.2 to approximate $\mathbf{z}$. Use x_{i-1}, x_i, x_{i+1} for a, b, c. There is a point ξ_i in $[x_{i-1}, x_{i+1}]$ such that

$$6\left(\frac{a_{i+1} - a_i}{\Delta_{i+1}} - \frac{a_i - a_{i-1}}{\Delta_i}\right)$$

$$= 6\frac{(x_i - x_{i-1})f(x_{i+1}) - (x_{i+1} - x_{i-1})f(x_i) + (x_{i+1} - x_i)f(x_{i-1})}{(x_{i+1} - x_i)(x_i - x_{i-1})}$$

$$= 3(x_{i+1} - x_{i-1})f''(\xi_i) = 3(\Delta_i + \Delta_{i+1})f''(\xi_i)$$

for $1 \leq i \leq k - 1$. The two end terms are approximated using the limit version, namely Taylor's formula of order 2:

$$6\left(\frac{a_1 - a_0}{\Delta_1} - s_0\right) = 6\left(\frac{f(x_1) - f(x_0)}{x_1 - x_0} - f'(x_0)\right)$$
$$= 3(x_1 - x_0)f''(\xi_0) = 3\Delta_1 f''(\xi_0)$$

for some ξ_0 in $[x_0, x_1]$. Similarly, there is a point ξ_k in $[x_{k-1}, x_k]$ such that

$$6\left(s_k - \frac{a_k - a_{k-1}}{\Delta_k}\right) = 6\left(f'(x_k) - \frac{f(x_k) - f(x_{k-1})}{x_k - x_{k-1}}\right)$$
$$= 3(x_k - x_{k-1})f''(\xi_k) = 3\Delta_k f''(\xi_k).$$

Now we approximate $\mathbf{h}'' - \mathbf{f}''$ by evaluating $Y(\mathbf{h}'' - \mathbf{f}'')$. The first coefficient is estimated by

$$\left|3\Delta_1 f''(\xi_0) - 2\Delta_1 f''(x_0) - \Delta_1 f''(x_1)\right| \leq 2\Delta_1 |f''(\xi_0) - f''(x_0)| + \Delta_1 |f''(\xi_0) - f''(x_1)|$$
$$\leq 3\Delta_1\, \omega(f''; \delta).$$

By Lemma 14.9.10, $\omega(f''; 2\delta) \leq 2\omega(f''; \delta)$. For $1 \leq i \leq k - 1$, we obtain

$$\left|3(\Delta_i + \Delta_{i+1})f''(\xi_i) - \Delta_i f''(x_{i-1}) - 2(\Delta_i + \Delta_{i+1})f''(x_i) - \Delta_{i+1}f''(x_{i+1})\right|$$
$$\leq 2(\Delta_i + \Delta_{i+1})\left|f''(\xi_i) - f''(x_i)\right| + \Delta_i\left|f''(\xi_i) - f''(x_{i-1})\right| + \Delta_{i+1}\left|f''(\xi_i) - f''(x_{i+1})\right|$$
$$\leq 2(\Delta_i + \Delta_{i+1})\omega(f''; \delta) + (\Delta_i + \Delta_{i+1})\omega(f''; 2\delta)$$
$$\leq 4(\Delta_i + \Delta_{i+1})\omega(f''; \delta).$$

Finally, the last term is estimated:

$$\left|3\Delta_k f''(\xi_k) - \Delta_k f''(\xi_{k-1}) - 2\Delta_k f''(x_k)\right|$$
$$\leq \Delta_k |f''(\xi_k) - f''(\xi_{k-1})| + 2\Delta_k |f''(\xi_k) - f''(x_k)|$$
$$\leq 3\Delta_k\, \omega(f''; \delta).$$

Let $A = \max\{|h''(x_i) - f''(x_i)| : 0 \leq i \leq k\}$ occur at i_0. Then looking at the i_0th coefficient of $Y(\mathbf{h}'' - \mathbf{f}'')$, we obtain

$$4(\Delta_{i_0} + \Delta_{i_0+1})\omega(f''; \delta)$$
$$\geq \left|\Delta_{i_0}\left(h''(x_{i_0-1}) - f''(x_{i_0-1})\right) + 2\left(\Delta_{i_0} + \Delta_{i_0+1}\right)\left(h''(x_{i_0}) - f''(x_{i_0})\right)\right.$$
$$\left. + \Delta_{i_0+1}\left(h''(x_{i_0+1}) - f''(x_{i_0+1})\right)\right|$$
$$\geq 2\left|\left(\Delta_{i_0} + \Delta_{i_0+1}\right)\left(h''(x_{i_0}) - f''(x_{i_0})\right)\right| - \left|\Delta_{i_0}\left(h''(x_{i_0-1}) - f''(x_{i_0-1})\right)\right|$$
$$- \left|\Delta_{i_0+1}\left(h''(x_{i_0+1}) - f''(x_{i_0+1})\right)\right|$$
$$\geq (\Delta_{i_0} + \Delta_{i_0+1})A.$$

So $A \leq 4\omega(f''; \delta)$. When the $i_0 = 0$ or k, we obtain $3\omega(f''; \delta)$ instead.

We are almost done with the estimate for the second derivative. Notice that on $[x_{i-1}, x_i]$, $h''(x) = h_i''(x)$ is linear. So $h''(x)$ lies between $h''(x_{i-1})$ and $h''(x_i)$. For convenience, suppose that $h''(x_{i-1}) \leq h''(x_i)$. The other case is similar. Then

$$\begin{aligned}
-5\omega(f''; \delta) &\leq \left(h''(x_{i-1}) - f''(x_{i-1})\right) + \left(f''(x_{i-1}) - f''(x)\right) \\
&\leq h''(x) - f''(x) \\
&\leq \left(h''(x_i) - f''(x_i)\right) + \left(f''(x_i) - f''(x)\right) \leq 5\omega(f''; \delta).
\end{aligned}$$

Thus $\|h'' - f''\|_\infty \leq 5\omega(f''; \delta)$.

The rest is easy. Since $f(x_{i-1}) - h(x_{i-1}) = 0 = f(x_i) - h(x_i)$, Rolle's Theorem provides a point ζ_i in $[x_{i-1}, x_i]$ such that $f'(\zeta_i) - h'(\zeta_i) = 0$. Hence for any point x in $[x_{i-1}, x_i]$,

$$|f'(x) - h'(x)| = \left| \int_{\zeta_i}^x f''(t) - h''(t)\, dt \right| \leq 5\omega(f''; \delta)|x - \zeta_i| \leq 5\delta\omega(f''; \delta).$$

Therefore $\|h' - f'\|_\infty \leq 5\delta\omega(f''; \delta)$. Now pick the nearest partition point x_i to x, so that $|x - x_i| \leq \delta/2$. Since $f(x_i) = h(x_i)$,

$$|f(x) - h(x)| = \left| \int_{x_i}^x f'(t) - h'(t)\, dt \right| \leq 5\delta\omega(f''; \delta)|x - x_i| \leq \frac{5}{2}\delta^2\omega(f''; \delta).$$

So $\|h - f\|_\infty \leq \frac{5}{2}\delta^2\omega(f''; \delta)$. ∎

Exercises for Section 10.9

A. Show that $\mathbb{S}(\Delta)$ has dimension $k+3$.

B. The second-order Mean Value Theorem (10.9.2) suggests the possibility of a third-order Mean Value Theorem. Suppose that $f \in C^3[a,d]$ and b,c in (a,d) with $b \neq c$. If P is the unique quadratic polynomial through $(a, f(a))$, $(c, f(c))$, and $(d, f(d))$, show that there is $\xi \in [a,d]$ with
$$f(b) - P(b) = \frac{(b-c)(b-a)(b-d)}{6} f'''(\xi).$$

C. Prove that every continuous function on $[0,1]$ is the uniform limit of the sequence of cubic splines h_k with nodes at $\{j2^{-k} : 0 \leq j \leq 2^k\}$.

10.10 The Stone–Weierstrass Theorem

We conclude this chapter with a very general approximation theorem that has many applications to approximation problems. It provides a very simple, easy to check criterion for when all continuous functions on a compact metric space can be approximated by some element of a subalgebra of functions. In particular, we shall see immediate consequences for approximation by polynomials in several variables and by trigonometric polynomials.

10.10.1. DEFINITION. A subset A of $C(X)$, the space of continuous real-valued functions on a compact metric space X, is an **algebra** if it is a subspace of $C(X)$ that is closed under multiplication (i.e., if $f, g \in A$, then $fg \in A$).

For $f, g \in C(X)$, define new elements of $C(X)$, $f \vee g$ and $f \wedge g$, by

$$(f \vee g)(x) = \max\{f(x), g(x)\} \quad \text{and} \quad (f \wedge g)(x) = \min\{f(x), g(x)\}.$$

A subset L of $C(X)$ is a **vector lattice** if it is a subspace that is closed under these two operations, that is, f, g both in L imply $f \vee g$ and $f \wedge g$ are in L.

It is easy to verify the two identities $f \vee g = 1/2(f+g) + 1/2|f-g|$ and that $f \wedge g = 1/2(f+g) - 1/2|f-g|$. Conversely, $|f| = f \vee (-f)$. It follows that an algebra A is a vector lattice if and only if $|f| \in A$ for each $f \in A$.

10.10.2. DEFINITION. A set S of functions on X **separates points** if for each pair of points $x, y \in X$, there is a function $f \in S$ such that $f(x) \neq f(y)$. Say that S **vanishes** at x_0 if $f(x_0) = 0$ for all $f \in S$.

In order to approximate arbitrary continuous functions on X from elements of A, a moment's thought shows that A must separate points. Moreover, A cannot vanish at any point, for then we could not approximate the constant function 1. These rather modest requirements, combined with the algebraic structure of an algebra, yield the following beautiful result.

10.10.3. STONE–WEIERSTRASS THEOREM.

An algebra A of continuous real-valued functions on a compact metric space X that separates points and does not vanish at any point is dense in $C(X)$.

We break the proof into several parts. Because norms over several domains occur in this proof, let us write $\|f\|_X$ for the uniform norm over X, and write $\|f\|_\infty$ for the uniform norm of a function on a real interval $[a, b]$ if the interval is understood.

10.10.4. LEMMA. *If A is an algebra of real-valued continuous functions on X, then its closure $\overline{A}$ is a closed algebra and a vector lattice.*

PROOF. It is easy to check that the closure of a subspace is still a subspace. The verification is left as an exercise. To see that $\overline{A}$ is closed under multiplication, take $f, g \in \overline{A}$. Choose sequences f_n and g_n in A that converge uniformly to f and g, respectively. Since A is an algebra, $f_n g_n$ belongs to A. Then the sequence $(f_n g_n)$ converges uniformly to fg (see Exercise 8.2.D). So $\overline{A}$ is closed under multiplication, and thus is an algebra.

Fix a function $f \in \overline{A}$. By the Weierstrass Approximation Theorem, the function $h(t) = |t|$ for $t \in [-\|f\|_X, \|f\|_X]$ is the uniform limit of a sequence p_n of polynomials. We can arrange that $p_n(0) = 0$. Indeed, $0 = h(0) = \lim\limits_{n \to \infty} p_n(0)$. Thus $q_n(x) = p_n(x) - p_n(0)$ satisfies $q_n(0) = 0$ and

$$\|h - q_n\|_\infty \le \|h - p_n\|_\infty + |p_n(0)|.$$

The right-hand side converges to 0, and thus (q_n) converges uniformly to h.

We will show that $|f|$ belongs to $\overline{A}$ also. Since $\overline{A}$ is an algebra, all linear combinations of $f, f^2, f^3, \ldots$ belong to $\overline{A}$. So if q is a polynomial with $q(0) = 0$, say $q(x) = a_1 x + a_2 x^2 + \cdots + a_k x^k$, then

$$q(f) = a_1 f + a_2 f^2 + \cdots + a_k f^k$$

belongs to $\overline{A}$. Moreover, if p, q are two such polynomials, then

$$\|p(f) - q(f)\|_X = \sup_{x \in X} |p(f(x)) - q(f(x))|$$
$$\le \sup_{t \in [-\|f\|_X, \|f\|_X]} |p(t) - q(t)| = \|p - q\|_\infty.$$

Since $\|q_n(f) - q_m(f)\|_X \le \|q_n - q_m\|_\infty$ and (q_n) is a Cauchy sequence, we conclude that $(q_n(f))$ is also a Cauchy sequence. Thus the limit g belongs to $\overline{A}$. But

$$g(x) = \lim_{n \to \infty} q_n(f(x)) = h(f(x)) - |f(x)|.$$

So $|f|$ belongs to A for each $f \in \overline{A}$. Therefore, $\overline{A}$ is a vector lattice. ∎

10.10.5. LEMMA. *If A is an algebra on X that separates points and never vanishes, then for any $x, y \in X$ and $\alpha, \beta \in \mathbb{R}$, there is a function $h \in A$ such that $h(x) = \alpha$ and $h(y) = \beta$.*

PROOF. There is a function $f \in A$ such that $f(x) \ne f(y)$. We may assume that $f(y) \ne 0$. If $f(x) \ne 0$ also, try $h = f^2 - tf$. We require

$$f(x)^2 - t f(x) = \alpha,$$
$$f(y)^2 - t f(y) = \beta.$$

Solving yields $t = \dfrac{\alpha - \beta}{f(y) - f(x)} + f(x) + f(y)$. If $f(x) = 0$, there is a function $g \in A$ with $g(x) \ne 0$. In this case,

$$h = \frac{\alpha}{g(x)} g + \frac{\beta g(x) - \alpha g(y)}{g(x) f(y)} f$$

will suffice. ∎

PROOF OF THE STONE–WEIERSTRASS THEOREM. Fix a function $f \in C(X)$ and $\varepsilon > 0$. We will approximate f within ε by functions in $\overline{A}$. For each pair of points $x, y \in X$, use Lemma 10.10.5 to find functions $g_{x,y} \in A$ such that $g_{x,y}(x) = f(x)$ and $g_{x,y}(y) = f(y)$.

Fix y. For each $x \neq y$,

$$U_x = \{z \in X : g_{x,y}(z) > f(z) - \varepsilon\} = (g_{x,y} - f)^{-1}(-\varepsilon, \infty)$$

is an open set containing x and y. Thus $\{U_x : x \in X \setminus \{y\}\}$ is an open cover of X. By the Borel–Lebesgue Theorem (9.2.3), this cover has a finite subcover $U_{x_1}, \ldots, U_{x_k}$. Let

$$g_y = g_{x_1, y} \vee g_{x_2, y} \vee \cdots \vee g_{x_k, y}.$$

By Lemma 10.10.4, g_y belongs to $\overline{A}$. By construction, $g_y(y) = f(y)$ and $g_y(x) > f(x) - \varepsilon$ for all $x \in X$.

Now define $V_y = \{x \in X : g_y(x) < f(x) + \varepsilon\}$, which is an open set containing y. Then $\{V_y : y \in X\}$ is an open cover of X. By the Borel–Lebesgue Theorem, this cover has a finite subcover $V_{y_1}, \ldots, V_{y_l}$. Let $g = g_{y_1} \wedge g_{y_2} \wedge \cdots \wedge g_{y_l}$. By Lemma 10.10.4, g belongs to $\overline{A}$. By construction, $g(x) < f(x) + \varepsilon$ for all $x \in X$. Moreover, since $g_{y_j} > f(x) - \varepsilon$ for $1 \leq j \leq l$,

$$f(x) - \varepsilon < g(x) < f(x) + \varepsilon \quad \text{for all} \quad x \in X.$$

Thus $\|f - g\|_X < \varepsilon$ as desired.   ∎

10.10.6. COROLLARY. *Let X be a compact subset of $\mathbb{R}^n$. The algebra of all polynomials $p(x_1, x_2, \ldots, x_n)$ in the n coordinates is dense in $C(X)$.*

PROOF. It is clear that the set $\mathbb{P}$ of all polynomials in n variables is an algebra. The constant function 1 does not vanish at any point. Finally, any two distinct points are separated by at least one of the coordinate functions x_i. Therefore, $\mathbb{P}$ satisfies the hypotheses of the Stone–Weierstrass Theorem and hence is dense in $C(X)$.   ∎

Another application of the Stone–Weierstrass Theorem yields an abstract proof of the following corollary, which will be given a different and more direct proof in Fejér's Theorem (14.4.5). See also Corollary 13.5.6 for yet another proof. Recall (Section 7.5) that a trigonometric polynomial is a function of the form $f(t) = a_0 + \sum\limits_{i=1}^{n} a_k \cos kt + b_k \sin kt$.

10.10.7. COROLLARY. *The set $\mathbb{TP}$ of all trigonometric polynomials is dense in $C_*[-\pi, \pi]$, the space of 2π-periodic functions on $\mathbb{R}$.*

PROOF. The set $\mathbb{TP}$ is clearly a vector space but is not obviously an algebra. The identities $\sin kt \sin lt = \frac{1}{2} \cos(k-l)t - \frac{1}{2} \cos(k+l)t$, $\sin kt \cos lt = \frac{1}{2} \sin(k+l)t + \frac{1}{2} \sin(k-l)t$, and $\cos kt \cos lt = \frac{1}{2} \cos(k-l)t + \frac{1}{2} \cos(k+l)t$ show that the four possible products of functions of the form $\sin kt$ and $\cos lt$ are each a trigonometric polynomial. By linearity, $\mathbb{TP}$ is an algebra.

To put this problem into the context of the Stone–Weierstrass Theorem, let T be the unit circle. Consider the map $P(t) = (\cos t, \sin t)$ of $\mathbb{R}$ onto T. Observe that $P(s) = P(t)$ if and only if $s - t$ is an integer multiple of 2π. In particular, $[-\pi, \pi]$ is wrapped around the circle with only the endpoints $\pm \pi$ coinciding. A continuous function $g \in C(T)$ may be identified with the function $f(t) = g(P(t))$. It is easy to see that $f(-\pi) = f(\pi)$. If we define the closed subspace

$$C_*[-\pi, \pi] = \{ f \in C[-\pi, \pi] : f(-\pi) = f(\pi) \},$$

then it is also easy to see that every function in $C_*[-\pi, \pi]$ is obtained as $g(P(t))$ for some $g \in C(T)$. Moreover, this correspondence between $C(T)$ and $C_*[-\pi, \pi]$ is isometric, linear, and preserves multiplication.

The two coordinate functions x_1 and x_2 on T send the point $P(t)$ to $\cos t$ and $\sin t$, respectively. So the first paragraph shows that polynomials in x_1 and x_2 are actually trigonometric polynomials of the variable t. So the algebra $\mathbb{TP}$ of all trigonometric polynomials in $C_*[-\pi, \pi]$ is identified with the algebra of all usual polynomials on T. By Corollary 10.10.6, the algebra of polynomials is dense in $C(T)$. Therefore, the algebra of trigonometric polynomials is dense in $C_*[-\pi, \pi]$.   ■

Exercises for Section 10.10

A. Show that the closure of a subspace of $C(X)$ is also a subspace of $C(X)$.

B. Let X and Y be compact metric spaces. Show that the set of all functions of the form
$$\sum_{i=1}^{k} f_i(x) g_i(y) \text{ for } k \geq 1 \text{ and } f_i \in C(X) \text{ and } g_i \in C(Y) \text{ is dense in } C(X \times Y).$$

C. Let $h \in C[0, 1]$. Show that every $f \in C[0, 1]$ is a limit of polynomials in h if and only if h is strictly monotone.

D. Let X be a compact metric space. Suppose A is a subalgebra of $C(X)$ that separates the points of X. If $\overline{A} \neq C(X)$, show that there is a point $x_0 \in X$ such that $\overline{A} = \{ f \in C(X) : f(x_0) = 0 \}$.
HINT: Show that $A + \mathbb{R}1$ is an algebra that does not vanish at any point. Can A vanish at more than one point?

E. Let A be an algebra of complex-valued continuous functions on X, i.e., a vector space over $\mathbb{C}$ that is closed under multiplication. Suppose A separates points, does not vanish at any point, and further, if $f \in A$, then $\overline{f}$ belongs to A. Show that A is dense in $C_{\mathbb{C}}(X)$, the complex-valued continuous functions on X. HINT: Consider the set A_r of real-valued functions in A.

F. Let X be a compact metric space. A subset J of $C(X)$ is an **ideal** if it is a vector space with the property that if $p \in J$ and $f \in C(X)$, then $fp \in J$.

(a) Let $E = \{ x \in X : J \text{ vanishes at } x \}$. Show that E is closed. .
(b) Show that J separates points of $X \setminus E$.
(c) Show that J is dense in the set of continuous functions on X that vanish on E.
   HINT: Fix f and $\varepsilon > 0$, and set $F = \{ x \in X : |f(x)| \geq \varepsilon \}$. Find a finite set of elements $p_i \in J$ such that $p(x) = \sum_{i=1}^{k} p_i(x)^2 \geq 1$ for $x \in F$. Let $q = fnp/(1 + np)$ for n large.

Chapter 11
Discrete Dynamical Systems

Suppose we wish to describe some physical system. The dynamical systems approach considers the space X of all possible states of the system—think of a point x in X as representing physical data. We will assume that X is a subset of some normed vector space, often $\mathbb{R}$. The evolution of the system over time determines a function T of X into itself that takes each state to a new state, one unit of time later.

Suppose that x_0 is the state of the system at time zero and x_n is the state of the system at time n. If $T^{n+1}(x) := T(T^n(x))$ for $n \geq 1$, then we have

$$x_n = Tx_{n-1} = T^2x_{n-2} = \cdots = T^nx_0.$$

In this chapter, T^2x virtually always means $T(T(x))$ and not $(T(x))^2$.

11.0.1. DEFINITION. Suppose that X is a subset of a normed vector space, and T is a continuous map from X into itself. The pair (X,T) is called a **discrete dynamical system**. For each point $x \in X$, the **forward orbit** of x is the sequence $\mathcal{O}(x) := \{T^nx : n \geq 0\}$.

To determine the behaviour of this system, we study the forward orbit $\mathcal{O}(x)$. What is the limit of this orbit for each x, as n goes to $+\infty$? If x is changed a little (in some sense), then how does this limit change? This turns out to be an interesting problem. There are systems in which the limit can change dramatically for tiny changes in the starting point. This leads to the idea of chaos, which we study in Section 11.5.

We concentrate first on fixed points, that is, points $x_0 \in X$ with $x_0 = Tx_0$. These are interesting for their own sake, since analysis problems can sometimes be formulated so that the solution is the fixed point of some dynamical system. Viewing a problem in this way can give important results that are otherwise hard to obtain. In this chapter, we will take this approach to Newton's method for solving equations numerically. In Chapter 12, we will use it to study differential equations.

Later in this chapter, we describe other types of orbits and then look at the notion of chaos for dynamical systems. The final section of the chapter considers iterated function schemes, which provide a means of generating fractals. It requires only Section 11.1 and the first section of Chapter 9.

K.R. Davidson and A.P. Donsig, *Real Analysis and Applications: Theory in Practice*, Undergraduate Texts in Mathematics, DOI 10.1007/978-0-387-98098-0_11, © Springer Science + Business Media, LLC 2010

11.1 Fixed Points and the Contraction Principle

Fixed points are so important in analysis that we will discuss two methods for ensuring their existence. Our first method has the added advantage that it comes equipped with a natural algorithm for computing the limit point. This has both theoretical and computational advantages over purely existential arguments.

As motivation, consider the function $Tx = 1.8(x - x^3)$ graphed in Figure 11.1. Factoring shows that this cubic has roots at -1, 0, and 1. It has a local maximum at the point $(1/\sqrt{3}, 2\sqrt{3}/5)$. Since T is an odd function, it has a local minimum at $(-1/\sqrt{3}, -2\sqrt{3}/5)$. To find the fixed points, we solve the equation

$$x = Tx = 1.8x - 1.8x^3.$$

This has three solutions, $x = -2/3, 0,$ and $2/3$.

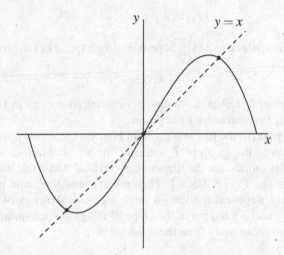

FIG. 11.1 Graph of $Tx = 1.8(x - x^3)$ showing fixed points.

We classify fixed points based on the orbits of nearby points. This behaviour is crucial to understanding how to approximate a fixed point.

11.1.1. DEFINITION. A fixed point x^* is called an **attractive fixed point** or a **sink** is there is an open neighbourhood $U = (a, b)$ containing x^* such that for every point x in (a, b), the orbit $\mathcal{O}(x)$ converges to x^*. A fixed point x^* is called a **repelling fixed point** or **source** if there is a neighbourhood $U = (a, b)$ containing x^* such that for every point x in (a, b) except for x^* itself, the orbit $\mathcal{O}(x)$ leaves the interval U.

Notice that for an attractive fixed point x^*, the interval U around x^* may be quite small. Also, for a repelling fixed point x^*, the orbit $\mathcal{O}(x)$ may return to the interval U after it leaves. However, it must then leave the interval again eventually.

We shall see that if T has a continuous derivative, the difference between attractive and repelling fixed points comes down to the size of the derivative at x^*. A fixed point x^* is attracting if $|T'(x^*)| < 1$ and repelling if $|T'(x^*)| > 1$. The case $|T'(x^*)| = 1$ is ambiguous and might be one, the other, or neither.

In our example,

$$T'(x) = 1.8 - 5.4x^2.$$

So $T'(0) = 1.8 > 1$. The tangent line at the origin is $L(x) = 1.8x$. Since the tangent line is a good approximation to T near $x = 0$, it follows that T roughly multiplies x by the factor 1.8 when x is small. So repeated application of this to a very small nonzero number will eventually move the point far from 0. We will make this precise in Lemma 11.1.2 using the Mean Value Theorem (6.2.2) to show that 0 is a repelling point in the interval $(-1/3, 1/3)$.

On the other hand, at $x = \pm 2/3$, $T'(x) = -0.6$. This has absolute value less than 1. So near $x = 2/3$, the function is approximated by the tangent line

$$L(x) = \tfrac{2}{3} - 0.6(x - \tfrac{2}{3}).$$

This decreases the distance to $2/3$ by approximately a factor of 0.6 on each iteration; i.e.,

$$T^{n+1}x - \tfrac{2}{3} \approx 0.6(T^n x - \tfrac{2}{3}).$$

So $T^n x$ converges to $2/3$. Again, we obtain a precise inequality in Lemma 11.1.2. So the points $\pm 2/3$ are attractive fixed points.

Consider the graph of the function given in Figure 11.2. Fixed points correspond to the intersection of the graph of T with the line $y = x$. Starting with any point x_0, mark the point (x_0, x_0) on the diagonal. A vertical line from this point meets the graph of T at $(x_0, Tx_0) = (x_0, x_1)$. Then a horizontal line from here meets the diagonal at (x_1, x_1). Repeated application yields a graphical picture of the dynamics. Note that starting near a fixed point, the slope of the graph determines whether the points approach or move away from the fixed point.

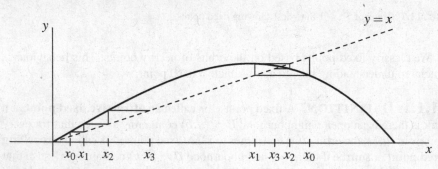

FIG. 11.2 Fixed points for $Tx = 1.8(x - x^3)$.

In our example, another typical behaviour is exhibited by all $|x|$ sufficiently large. For the sake of simplicity, consider $|x| > 2$. Then

$$|Tx| = 1.8(x^2 - 1)|x| \geq 5.4|x|.$$

It is clear then that $\lim_{n\to\infty} |T^n x| = +\infty$. So all of these orbits go off to infinity. Usually we will try to restrict our domain to a bounded region that is mapped back into itself by the transformation T.

We now connect our classification of fixed points with derivatives. Say T is a C^1 **dynamical system** on $X \subset \mathbb{R}$ if the function T is C^1, i.e., has a continuous derivative.

11.1.2. LEMMA. *Suppose that T is a C^1 dynamical system with a fixed point x^*. If $|T'(x^*)| < c < 1$, then x^* is an attractive fixed point. Moreover, there is an interval $U = (x^* - r, x^* + r)$ about x^* such that for every $x_0 \in U$, the sequence $x_n = T^n x_0$ satisfies*

$$|x_n - x^*| \leq c^n |x_0 - x^*| \leq \frac{c^n}{1-c}|x_1 - x_0|.$$

If $|T'(x^*)| > 1$, then x^* is a repelling fixed point.

PROOF. Suppose that $|T'(x^*)| < c < 1$. Since $x \mapsto |T'(x)|$ is continuous, by Exercise 5.1.G, there is $r > 0$ such that for all x in the interval $U = (x^* - r, x^* + r)$, $|T'(x)| < c$.

Let x_0 be an arbitrary point in U, and consider the sequence $x_n = T^n x_0$. Applying the Mean Value Theorem (6.2.2) to T at x_n and x^*, there is a point z between them such that

$$\frac{Tx_n - Tx^*}{x_n - x^*} = T'(z).$$

Rewriting this using $Tx_n = x_{n+1}$ and $Tx^* = x^*$, we obtain

$$|x_{n+1} - x^*| = |T'(z)| \, |x_n - x^*|.$$

Provided that x_n belongs to the interval U, we obtain that

$$|x_{n+1} - x^*| < c|x_n - x^*|.$$

In particular, x_{n+1} is closer to x^* than x_n is; and therefore x_{n+1} also belongs to U. By induction, we obtain (verify this!) that

$$|x_n - x^*| < c^n |x_0 - x^*| \quad \text{for all} \quad n \geq 1.$$

Hence $\lim_{n\to\infty} |x_n - x^*| = 0$ by the Squeeze Theorem (2.4.6). That is, $\lim_{n\to\infty} x_n = x^*$. So x^* is an attractive fixed point.

Similarly, suppose that $|T'(x^*)| > c > 1$. Using Exercise 5.1.G and a few calculations with inequalities, there is $r > 0$ such that for all x in the interval $U = (x^* - r, x^* + r)$, $|T'(x)| > c$.

The Mean Value Theorem argument works the same way as long as x_n belongs to the interval U. However, in this case, repeated iteration eventually moves x_n outside of U, at which point we have almost no information about the dynamics of T. So as long as x_n is in U, we obtain a point z between x_n and x^* such that

$$|x_{n+1} - x^*| = |T'(z)| \, |x_n - x^*| > c|x_n - x^*|.$$

This can be repeated as long as x_n remains inside U to obtain

$$|x_n - x^*| > c^n |x_0 - x^*|.$$

Since this distance to x^* is tending to $+\infty$, repeated iteration eventually will move x_{n+1} outside of U. Therefore, x^* is a repelling fixed point.   ∎

Notice that in this proof, we only used differentiability in order to apply the Mean Value Theorem and obtain a distance estimate. The following definition describes this distance estimate directly, and so allows us to abstract the arguments of the previous proof to other settings, where differentiability may not hold. However, we demand that these estimates hold on the whole domain. While this may seem to be excessively strong, remember that we may be able to restrict our attention to a smaller domain in which these conditions apply. For example, in our previous proof, the interval U would be a suitable domain.

11.1.3. DEFINITION. Let X be a subset of a normed vector space $(V, \|\cdot\|)$. A map $T : X \to X$ is a **contraction** on X if there is a positive constant $c < 1$ such that

$$\|Tx - Ty\| \le c\|x - y\| \quad \text{for all} \quad x, y \in X.$$

That is, T is Lipschitz with constant $c < 1$.

11.1.4. EXAMPLE. Let us look more closely at the map $Tx = 1.8(x - x^3)$ introduced in the first section. The fixed points are $-2/3$, 0, and $2/3$.

Now $|T'(2/3)| = |-0.6| < 1$, so this is an attracting fixed point. We will show that T is a contraction on the interval $[0.5, 0.7]$. In this interval, T has a single critical point at $1/\sqrt{3}$, a local maximum, and

$$T(0.5) = 0.675, \quad T(1/\sqrt{3}) = 0.4\sqrt{3} \approx 0.6928, \quad \text{and} \quad T(0.7) = 0.6426.$$

Since T is increasing on $[0.5, 1/\sqrt{3}]$ and decreasing on $[1/\sqrt{3}, 0.7]$, it follows that T maps $[0.5, 0.7]$ into itself. Moreover,

$$\sup_{0.5 \le x \le 0.7} |T'(x)| = \sup_{0.5 \le x \le 0.7} 1.8|1 - 3x^2| = \max\{|0.45|, |-0.846|\} = 0.846.$$

Therefore, T is a contraction on $[0.5, 0.7]$ with contraction constant $c = 0.846$.

Now consider the fixed point 0. Since $T'(0) = 1.8 > 1$, this is a repelling fixed point. On the interval $[-1/3, 1/3]$, we have

$$\inf_{|x| \le 1/3} T'(x) = \inf_{|x| \le 1/3} 1.8(1 - 3x^2) = 1.2.$$

So the proof of Lemma 11.1.2 shows that

$$|Tx| \ge 1.2|x| \quad \text{for all} \quad x \in [-\tfrac{1}{3}, \tfrac{1}{3}].$$

So the sequence $T^n x$ moves away from 0 until it leaves this interval.

11.1.5. EXAMPLE. Consider the linear function $T(x) = mx + b$ for $x \in \mathbb{R}$. Then

$$|Tx - Ty| = |m||x - y|.$$

Hence T is a contraction on $\mathbb{R}$ provided that $|m| < 1$. This map has a fixed point if there is a solution to $x = Tx = mx + b$. It is easy to compute that $x^* = \frac{b}{1-m}$ is the unique solution provided that $m \ne 1$. What happens when $m = 1$?

We may think of T as the dynamical system on $\mathbb{R}$ that maps each point x to Tx. Consider the forward orbit $\mathcal{O}(x) = \{T^n x : n \ge 0\}$ of a point x. We obtain a sequence defined by the recurrence

$$x_{n+1} = Tx_n \quad \text{for} \quad n \ge 0.$$

A simple calculation shows that

$$x_1 = mx_0 + b,$$
$$x_2 = m^2 x_0 + (1 + m)b,$$
$$x_3 = m^3 x_0 + (1 + m + m^2)b,$$
$$x_4 = m^4 x_0 + (1 + m + m^2 + m^3)b.$$

It appears that there is a general formula

$$x_n = m^n x_0 + (1 + m + m^2 + \cdots + m^{n-1})b.$$

We may verify this by induction. It evidently holds true for $n = 1$. Suppose that it is valid for a given n. Then

$$x_{n+1} = mx_n + b = m\big(m^n x_0 + (1 + m + m^2 + \cdots + m^{n-1})b\big) + b$$
$$= m^{n+1} x_0 + (m + m^2 + \cdots + m^n)b + b$$
$$= m^{n+1} x_0 + (1 + m + m^2 + \cdots + m^n)b.$$

Hence the formula follows for $n + 1$, and so for all positive integers by induction.

When $|m| < 1$, the contraction case, this sequence has a limit that we obtain by summing an infinite geometric series:

$$\lim_{n\to\infty} x_n = \lim_{n\to\infty} m^n x_0 + (1 + m + m^2 + \cdots + m^{n-1})b = 0 + b\sum_{k=0}^{\infty} m^k = \frac{b}{1-m}.$$

Hence this sequence converges to the fixed point x^*. Therefore, x^* is an attractive fixed point that may be located by starting *anywhere* and iterating T. This means that an approximate solution to $Tx = x$ will be **stable**, meaning that the orbit of any point close to x^* will remain close to x^*. In fact, it was not even necessary to start close to x^* to obtain convergence.

On the other hand, when $|m| > 1$, $\lim_{n\to\infty} |x_n| = \infty$, and so this sequence diverges. There is still a fixed point, but it is repelling. In this case, the answer is to invert T. The map T is one-to-one, and thus we may solve for

$$T^{-1}x = (x - b)/m = \tfrac{1}{m}x - \tfrac{b}{m}.$$

This map T^{-1} is a contraction, and we can apply the previous analysis to it. Each point x comes from the point $x_{-1} = T^{-1}x$. Going "back into the past" by setting $x_{-n-1} = T^{-1}x_{-n}$ converges to the fixed point x^*. Since points close to x^* move outward and eventually go off to infinity, x^* is a source.

Finally, when $m = -1$, there is a unique fixed point. But this point cannot be located as the limit of an orbit. The reason is that

$$T^2x = -(Tx) + b = -(-x + b) + b = x.$$

That is, T^2 equals the identity map on $\mathbb{R}$. So with the exception of the fixed point $x^* = b/2$, every point has a period-2 orbit.

11.1.6. THE BANACH CONTRACTION PRINCIPLE.

Let X be a closed subset of a complete normed vector space $(V, \|\cdot\|)$. If T is a contraction map of X into X, then T has a unique fixed point x^*. Furthermore, if x is any vector in X, then $x^* = \lim_{n\to\infty} T^n x$ and

$$\|T^n x - x^*\| \le c^n \|x - x^*\| \le \frac{c^n}{1-c}\|x - Tx\|,$$

where c is the Lipschitz constant for T.

PROOF. The statement of the theorem suggests how the proof should proceed. Pick any point x_0 in X and form the sequence (x_n) given by $x_{n+1} = Tx_n$ for $n \ge 0$.

We claim that this sequence is Cauchy. To see this, first observe that

$$\|x_{n+1} - x_n\| = \|Tx_n - Tx_{n-1}\| \le c\|x_n - x_{n-1}\|$$
$$\le c^2\|x_{n-1} - x_{n-2}\| \le c^n\|x_1 - x_0\| = c^n D,$$

where $D = \|x_1 - x_0\| < \infty$. Using this fact and the triangle inequality, we compute

$$\|x_{n+m} - x_n\| \leq \sum_{i=0}^{m-1} \|x_{n+i+1} - x_{n+i}\| \leq \sum_{i=0}^{m-1} c^{n+i}D < \sum_{i=0}^{\infty} c^{n+i}D = \frac{c^n D}{(1-c)}. \quad (11.1.7)$$

Given $\varepsilon > 0$, choose N so large that $c^N < \varepsilon(1-c)/D$, which is possible since $\lim_{n \to \infty} c^n = 0$. Hence for $n \geq N$ and $m \geq 0$, we have $\|x_{n+m} - x_n\| < \varepsilon$. So the sequence (x_n) is Cauchy, as claimed.

Because V is complete, the sequence (x_n) converges to some vector $x^* \in V$. Since X is closed, this limit point belongs to X. Observe that T is uniformly continuous because it is Lipschitz (Proposition 5.5.4). Using continuity, we have

$$Tx^* = T\left(\lim_{n \to \infty} x_n\right) = \lim_{n \to \infty} Tx_n = \lim_{n \to \infty} x_{n+1} = x^*.$$

Hence x^* is a fixed point.

Suppose that $y \in X$ is also a fixed point: so $Ty = y$. Then

$$\|x^* - y\| = \|Tx^* - Ty\| \leq c\|x^* - y\|.$$

Since $c < 1$, this implies that $\|x^* - y\| = 0$, whence $x^* = y$. Therefore, x^* is the unique fixed point, independent of the choice of x_0.

Starting at any vector x and using the estimate (11.1.7), we obtain

$$\|T^n x - x^*\| = \|T^n x - T^n x^*\| \leq c^n \|x - x^*\|$$
$$= c^n \lim_{m \to \infty} \|x - x_m\| \leq c^n (1-c)^{-1} \|Tx - x\|. \quad \blacksquare$$

11.1.8. EXAMPLE. Let $V = \mathbb{R}$, $X = [-1, 1]$, and $Tx = \cos x$. By the Mean Value Theorem, for any $x, y \in X$, there is a point z between x and y such that

$$|Tx - Ty| = |x - y| \,|\sin z|.$$

In particular, $|z| < 1$. Since $\sin x$ is increasing on $[-1, 1]$,

$$\max_{|z| \leq 1} |\sin z| = |\sin \pm 1| = \sin 1 < 1.$$

Thus T is a contraction. To find the fixed point experimentally, type any value into your calculator and repeatedly push the $\boxed{\cos}$ button. If your calculator is set for radians, the sequence will converge rapidly to $0.73908513321516064\ldots$.

What happens when you do the same for $Tx = \sin x$? It is not a contraction. Nevertheless, there is a unique fixed point $\sin 0 = 0$, and the iterated sequence converges. But it converges at a painfully slow rate. Try it on your calculator. It is slow because the derivative at the fixed point is $\cos 0 = 1$.

Indeed, the inequality $0 < \sin\theta < \theta$ for θ in $(0, \pi/2]$ shows that the only fixed point in $[0, \infty)$ is $x = 0$. Since $\sin x$ is odd, the same is true on $(-\infty, 0]$. The first iteration $x_1 = \sin x_0$ lies in $[-1, 1]$; and thereafter this inequality shows that if $0 < x_1 \leq 1$, then $0 \leq x_{n+1} < x_n < 1$ for $n \geq 1$. Likewise, if $-1 \leq x_1 < 0$, then the sequence x_n is monotone increasing and bounded above by 0. Hence the limit exists and must be the unique fixed point of $\sin x$. We can use the Taylor series of $\sin x$ (see Exercise 10.1.D) to show that

$$|\sin x - x| \leq \frac{|x|^3}{6}.$$

Thus if $x_0 = 0.1$, it follows that

$$|x_n - x_{n+1}| \leq \frac{x_0^3}{6} = 10^{-3}/6.$$

Since each step moves us at most $10^{-3}/6$, it will take over $\dfrac{0.1 - 0.01}{10^{-3}/6} = 540$ iterations before $x_n \leq 0.01$. After that, it will take at least 54,000 iterations to get below 0.001 and 5.4 million more steps to obtain a fourth decimal of accuracy in the approximation of the fixed point.

One moral of this example is that for maps T that are *not* contractions, even if applying T moves a point x very little, x need not be very close to a fixed point x^*. For contractions, it is true that if T moves x little, then x is close to a fixed point x^* and, in addition, we have a numerical estimate for $\|x - x^*\|$ in terms of $\|Tx - x\|$ and the Lipschitz constant.

11.1.9. EXAMPLE. If in the definition of contraction, the condition is weakened to $\|Tx - Ty\| \leq \|x - y\|$, then we cannot conclude that there is a fixed point. For example, take $X = \mathbb{R}$ and $Tx = x + 1$. Clearly, T has no fixed points. But since $\|Tx - Ty\| = \|x - y\|$, distance is preserved.

Even the strict inequality $\|Tx - Ty\| < \|x - y\|$ is not sufficient. Consider $X = [1, \infty)$ and $Sx = x + x^{-1}$. Then

$$|Sx - Sy| = (x + \tfrac{1}{x}) - (y + \tfrac{1}{y}) = |x - y|(1 - \tfrac{1}{xy}) < |x - y|$$

for all $x, y \geq 1$. However, $Sx > x$ for all x, so S has no fixed point.

11.1.10. REMARK. The first part of Lemma 11.1.2 may be proved as a simple consequence of the Banach Contraction Principle. Indeed, suppose that T is a C^1 dynamical system with a fixed point x^* such that $|T'(x^*)| < c < 1$. As before, use the continuity of T' to find an interval $I = [x^* - r, x^* + r]$ such that $|T'(x)| \leq c$ for all $x \in I$. The only difference here is that we are using a closed interval and a $\leq$ sign instead of an open interval and strict inequality. Apply the Mean Value Theorem to any two points x, y in I. For such points, there is a point z between them such that $|Tx - Ty| = |T'(z)| \, |x - y| \leq c|x - y|$.

In particular,

$$|Tx - x^*| = |Tx - Tx^*| \le c|x - x^*| \le cr.$$

Therefore, T maps the interval I into itself, and it is a contraction with contraction constant c. By the Banach Contraction Principle, we see that x^* is the unique fixed point in I. Moreover, we obtain the desired distance estimates.

11.1.11. EXAMPLE. This example deals with a system of linear equations and looks for a condition that guarantees an attractive fixed point. Let $A = \begin{bmatrix} a_{ij} \end{bmatrix}$ be an $n \times n$ matrix, and let $\mathbf{b} = (b_1, \dots, b_n)$ be a (column) vector in $\mathbb{R}^n$. Consider the system of equations

$$\mathbf{x} = A\mathbf{x} + \mathbf{b}. \tag{11.1.12}$$

We will try to analyze this problem by studying the dynamical system given by the map $T : \mathbb{R}^n \to \mathbb{R}^n$ according to the rule

$$T\mathbf{x} = A\mathbf{x} + \mathbf{b}.$$

A solution to (11.1.12) corresponds to a fixed point of T.

There are various norms on $\mathbb{R}^n$, and different norms lead to different criteria for T to be a contraction. In this example, we will use the max norm,

$$\|\mathbf{x}\|_\infty = \max_{1 \le i \le n} |x_i|.$$

If we think of points in $\mathbb{R}^n$ as real-valued functions on $\{1, \dots, n\}$, say $\mathbf{x}(i) = x_i$ for $1 \le i \le n$, then this is the uniform norm on $C(\{1, \dots, n\})$. We will show that T is a contraction on $\mathbb{R}^n$ in the max norm if and only if

$$c := \max_{1 \le i \le n} \sum_{j=1}^n |a_{ij}| < 1.$$

First suppose that $c \ge 1$. There is some integer i_0 such that $c = \sum_{j=1}^n |a_{i_0 j}|$. Set $x_j = \text{sign}(a_{i_0 j})$. Then

$$\|\mathbf{x} - \mathbf{0}\|_\infty = \|\mathbf{x}\|_\infty = \max_{1 \le i \le n} |x_i| = 1,$$

while

$$\|T\mathbf{x} - T\mathbf{0}\|_\infty = \|A\mathbf{x}\|_\infty \ge |(A\mathbf{x})_{i_0}| = \sum_{j=1}^n a_{i_0 j} x_j$$

$$= \sum_{j=1}^n |a_{i_0 j}| = c \ge 1 = \|\mathbf{x} - \mathbf{0}\|_\infty.$$

So T is not contractive.

On the other hand, if $c < 1$, then we compute for $\mathbf{x}, \mathbf{y} \in \mathbb{R}^n$,

$$\|T\mathbf{x} - T\mathbf{y}\|_\infty = \|A(\mathbf{x} - \mathbf{y})\|_\infty$$

$$= \max_{1 \le i \le n} \Big| \sum_{j=1}^{n} a_{ij}(x_j - y_j) \Big| \le \max_{1 \le i \le n} \sum_{j=1}^{n} |a_{ij}| \, |x_j - y_j|$$

$$\le \Big(\max_{1 \le i \le n} \sum_{j=1}^{n} |a_{ij}| \Big) \Big(\max_{1 \le j \le n} |x_j - y_j| \Big) = c\|\mathbf{x} - \mathbf{y}\|_\infty.$$

So T is a contraction.

Thus by the Banach Contraction Principle, we know that there is a unique solution that can be computed iteratively from any starting point. Choose $\mathbf{x}_0 = 0$. Then

$$\mathbf{x}_1 = T\mathbf{x}_0 = \mathbf{b},$$
$$\mathbf{x}_2 = T\mathbf{x}_1 = \mathbf{b} + A\mathbf{b} = (I + A)\mathbf{b},$$
$$\mathbf{x}_3 = T\mathbf{x}_2 = \mathbf{b} + A(\mathbf{b} + A\mathbf{b}) = (I + A + A^2)\mathbf{b}.$$

We will show that $\mathbf{x}_n = (I + A + \cdots + A^{n-1})\mathbf{b}$. This is evident for $n = 0, 1, 2, 3$ by the previous calculations. Assume that it is true for some integer n. Then

$$\mathbf{x}_{n+1} = T\mathbf{x}_n = \mathbf{b} + A(I + A + \cdots + A^{n-1})\mathbf{b} = (I + A + \cdots + A^n)\mathbf{b}.$$

So the formula is established by induction.

The solution to (11.1.12) is the unique fixed point

$$\mathbf{x}^* = \lim_{n \to \infty} \mathbf{x}_n = \lim_{n \to \infty} (I + A + \cdots + A^{n-1})\mathbf{b} = \sum_{k=0}^{\infty} A^k \mathbf{b}.$$

The important factor making this infinite sum convergent is that because T and A are contractions,

$$\|A^k \mathbf{b}\|_\infty = \|A^k \mathbf{b} - A^k \mathbf{0}\| \le c^k \|\mathbf{b} - \mathbf{0}\|_\infty = c^k \|\mathbf{b}\|_\infty.$$

So this series is dominated by a convergent geometric series and thus converges by the comparison test. Indeed, the same argument shows that the series $\sum_{k=0}^{\infty} A^k \mathbf{x}$ converges for *every* vector $\mathbf{x} \in \mathbb{R}^n$. So the sum $C = \sum_{k=0}^{\infty} A^k$ makes sense as a linear transformation. We note that in particular, if $\mathbf{x}$ is any vector in $\mathbb{R}^n$, then $\lim_{n \to \infty} A^n \mathbf{x} = \mathbf{0}$. The solution to our problem is $\mathbf{x}^* = C\mathbf{b}$.

We know from linear algebra how to solve the equation $\mathbf{x} = A\mathbf{x} + \mathbf{b}$. This leads to $(I - A)\mathbf{x} = \mathbf{b}$. When $I - A$ is invertible, there is a unique solution $\mathbf{x} = (I - A)^{-1}\mathbf{b}$. This suggests that our contractive condition (11.1.11) leads to the conclusion that $I - A$ is invertible with inverse $C = \sum_{k=0}^{\infty} A^k$. To see that this is the case, compute

$$(I - A)C\mathbf{x} = \lim_{n \to \infty} (I - A)(I + A + \cdots + A^{n-1})\mathbf{x}$$

$$= \lim_{n \to \infty} (I + A + \cdots + A^{n-1})\mathbf{x} - (A + A^2 + \cdots + A^n)\mathbf{x}$$

$$= \lim_{n \to \infty} \mathbf{x} - A^n \mathbf{x} = \mathbf{x}.$$

Since this holds for all $\mathbf{x} \in \mathbb{R}^n$, $I - A$ has inverse C. This formula $(I - A)^{-1} = \sum_{k=0}^{\infty} A^k$ should be seen as parallel to the power series identity

$$\frac{1}{1-x} = \sum_{k=0}^{\infty} x^k \quad \text{for} \quad |x| < 1.$$

The condition $|x| < 1$ guarantees that the series converges. The contractive condition (11.1.11) plays the same role in the matrix equation.

Exercises for Section 11.1

A. Let $Tx = 1.8(x - x^3)$. Find the smallest R such that $\lim_{n \to \infty} |T^n x| = +\infty$ for all $|x| > R$.

B. Show that $Tx = \sin x$ is not a contraction on $[-1, 1]$.

C. Give an example of a differentiable map T from $\mathbb{R}$ to $\mathbb{R}$ whose fixed points are exactly the set of integers. Find points where $|T'(x)| > 1$.

D. Explain why in the previous example, points with $|T'(x)| > 1$ necessarily exist.

E. Suppose that S and T are contractions with Lipschitz constants s and t, respectively. Prove that the composition ST is a contraction with Lipschitz constant st.

F. Consider the special case of Example 11.1.11 where $A = \left[\begin{smallmatrix} 0.5 & 0.4 \\ 0 & 0.8 \end{smallmatrix}\right]$ and $b = \left[\begin{smallmatrix} 0.1 \\ 0.2 \end{smallmatrix}\right]$. Explicitly compute the infinite sum $\sum_{k=0}^{\infty} A^k$ in order to solve for the fixed point of $Tx = Ax + b$.

G. Redo Example 11.1.11 using the 1-norm $\|\mathbf{x}\|_1 = \sum_{i=1}^{n} |x_i|$ on $\mathbb{R}^n$ in place of $\|\mathbf{x}\|_\infty$.
  HINT: Show that T is a contraction if and only if $\max_{1 \le j \le n} \sum_{i=1}^{n} |a_{ij}| < 1$.

H. Define T on $C[-1, 1]$ by $Tf(x) = f(x) + (x^2 - f(x)^2)/2$. Set $f_0 = 0$ and $f_{n+1} = Tf_n$ for $n \ge 0$. Prove that f_n is a monotone increasing sequence of functions such that

$$0 \le x^2 - f_{n+1}(x)^2 \le (1 - \tfrac{1}{4}x^2)(x^2 - f_n(x)^2).$$

  Hence show that f_n is a sequence of polynomials converging uniformly to $|x|$.

I. Suppose S and T are contractions on X with Lipschitz constant $c < 1$ and fixed points x_s and x_t respectively. Prove that $\|x_s - x_t\| \le (1 - c)^{-1} \|S - T\|_\infty$, where $\|S - T\|_\infty = \sup_{x \in X} \|Sx - Tx\|$.
  HINT: Estimate $\|x_s - x_t\|$ in terms of $\|x_s - Tx_s\|$.

J. Suppose that for $0 \le s \le 1$, T_s is a contraction of a complete normed space X with Lipschitz constant $c < 1$. Moreover, assume that this is a continuous path of contractions. That is, $\lim_{s \to s_0} \|T_s - T_{s_0}\|_\infty = 0$. Prove that the fixed points x_s of T_s form a continuous path.
  HINT: Use the previous exercise.

K. Define a map $\mathscr{D}$ on $C[0, 1]$ as follows:

$$\mathscr{D}f(x) = \begin{cases} \frac{2}{3} + \frac{1}{3}f(3x) & \text{for} \quad 0 \le x \le \frac{1}{3}, \\ (2 + f(1))(\frac{2}{3} - x) & \text{for} \quad \frac{1}{3} \le x \le \frac{2}{3}, \\ x - \frac{2}{3} & \text{for} \quad \frac{2}{3} \le x \le 1. \end{cases}$$

  (a) Sketch the graph of some function f and $\mathscr{D}f$.
  (b) Show that $\mathscr{D}$ is a contraction.
  (c) Describe the fixed point.   HINT: Repeatedly apply $\mathscr{D}$ to the function $f(x) = 1/3$.

11.2 Newton's Method

Newton's method is an iterative procedure to rapidly compute a zero of a differentiable function. The Contraction Principle provides a nice proof that it works.

Suppose we have a function f on $\mathbb{R}$ and a reasonable guess x_0 for the zero (or root) of f. If f is differentiable at x_0, then we can draw the tangent line at x_0. It seems plausible that *if* x_0 is close to the root of the function, the root of the tangent line will be even closer. This uses one of the basic ideas of differential calculus: The tangent line to f at x_0 is a good approximation to f near x_0.

The equation of the tangent line is

$$y = f(x_0) + f'(x_0)(x - x_0).$$

To find the root, we set $y = 0$ and solve for x, to obtain

$$x = x_0 - \frac{f(x_0)}{f'(x_0)}.$$

Call this root x_1. By repeating the same calculation for x_1, we obtain a sequence $(x_n)_{n=1}^{\infty}$, where

$$x_{n+1} = x_n - \frac{f(x_n)}{f'(x_n)} \quad \text{for} \quad n \geq 1.$$

Our hope is that the sequence converges to a point x^* satisfying $f(x^*) = 0$. See Figure 11.3 for an example of (the first few terms) of such a sequence.

To prove this, we reformulate the problem using fixed points. Suppose f is twice differentiable and $f(x^*) = 0$ for some point x^*. Define a dynamical system T by

$$Tx = x - \frac{f(x)}{f'(x)}.$$

It is easy to see that if $f(x^*) = 0$, then $Tx^* = x^*$ and vice versa. So we are looking for a fixed point for the function T. Notice that for this definition to make sense, we require $f'(x) \neq 0$ on our domain. This will show up in our hypotheses as the condition $f'(x^*) \neq 0$.

We compute the derivative

$$T'(x) = 1 - \frac{f'(x)^2 - f(x)f''(x)}{f'(x)^2} = \frac{f(x)f''(x)}{f'(x)^2}.$$

Notice that $T'(x^*) = 0$. So x^* is an attractive fixed point by Lemma 11.1.2. Moreover, the constant c used may be any small positive number. Indeed, the closer we get to x^*, the smaller the value of c that may be used. This will be important in obtaining very rapid convergence.

In addition to verifying that Newton's method works, we can estimate how quickly the error decreases. This kind of error analysis is fundamental to deciding the effectiveness of an algorithm. It is considered minimally acceptable if the

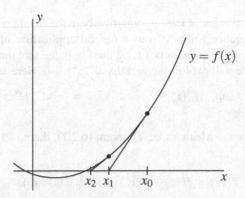

FIG. 11.3 Example of a function f and sequence (x_n).

error tends to zero **geometrically** in the sense that

$$|x_{n+1} - x^*| \le c|x_n - x^*| \quad \text{for} \quad n \ge N$$

for some constant $c < 1$. For if $c^s < 0.1$, we obtain an additional digit of accuracy every s iterations. Compare Example 11.1.8.

However, if great accuracy is desired, this is not all that fast. Some algorithms, such as Newton's method, converge **quadratically**, meaning that there is a constant M such that

$$|x_{n+1} - x^*| \le M|x_n - x^*|^2 \quad \text{for} \quad n \ge N.$$

Once $|x_n - x^*| < 0.1/M$, the number of digits of accuracy doubles every iteration. Thus, once we get sufficiently close to the solution, the sequence approaches x^* very rapidly.

11.2.1. NEWTON'S METHOD.

Suppose f is C^2 and there is $x^* \in \mathbb{R}$ such that $f(x^*) = 0$ and $f'(x^*) \ne 0$. There is an $r > 0$ such that if $|x_0 - x^*| \le r$, then the iterates $x_{n+1} = x_n - f(x_n)/f'(x_n)$ converge to x^*. Moreover, the iterates converge quadratically; i.e., there is a constant M such that

$$|x_{n+1} - x^*| < M|x_n - x^*|^2 \quad \text{for} \quad n \ge 1.$$

PROOF. Let $Tx = x - f(x)/f'(x)$. Then $Tx^* = x^*$. As we just computed,

$$T'(x) = \frac{f(x)f''(x)}{f'(x)^2}.$$

In particular, $T'(x^*) = 0$. Also, $T'(x)$ is defined for x near x^*, since $f'(x) \ne 0$ for x near x^*. Choose $r > 0$ so small that $|T'(x_0)| \le 1/2$ for all $x_0 \in [x^* - r, x^* + r]$. By Lemma 11.1.2, iterates $x_{n+1} = Tx_n$ converge to x^*, and

$$|x_n - x^*| \le 2^{-n}|x_0 - x^*| \le 2^{1-n}|x_0 - x_1|.$$

In particular, (x_n) converges at least geometrically to the fixed point.

Quadratic convergence follows from a careful application of the Mean Value Theorem. The estimate from Lemma 11.1.2 used only the fact that $|T'(x)| \leq c$ near the fixed point x^*. We will exploit the fact that $T'(x^*) = 0$. Here are the details. Let

$$A = \sup_{|x-x^*| \leq r} |f''(x)| \qquad \text{and} \qquad B = \inf_{|x-x^*| \leq r} |f'(x)|,$$

and set $M = A/B$. By the Mean Value Theorem (6.2.2), there is a point a_n between x_n and x^* such that

$$f(x_n) = f(x_n) - f(x^*) = f'(a_n)(x_n - x^*).$$

Solve for $x_n - x^*$ and substitute into the following:

$$x_{n+1} - x^* = (x_n - x^*) + (x_{n+1} - x_n) = \frac{f(x_n)}{f'(a_n)} - \frac{f(x_n)}{f'(x_n)}$$

$$= \frac{f(x_n)}{f'(a_n)f'(x_n)}\left(f'(x_n) - f'(a_n)\right) = \frac{x_n - x^*}{f'(x_n)}\left(f'(x_n) - f'(a_n)\right).$$

Applying the Mean Value Theorem again, this time to f', provides a point b_n between x_n and a_n such that

$$|f'(x_n) - f'(a_n)| = |f''(b_n)(x_n - a_n)| \leq A|x_n - x^*|.$$

Combining this with $|f'(x_n)| \geq B$ and substituting again yields

$$|x_{n+1} - x^*| \leq \frac{|x_n - x^*|}{B}A|x_n - x^*| = M|x_n - x^*|^2.$$

This establishes the quadratic convergence.   ∎

11.2.2. EXAMPLE. COMPUTATION OF SQUARE ROOTS.

At one time, everyone had to compute square roots by hand. Today, a few people have to program calculators to compute them. Newton's method is an excellent way for a person or a computer to find square roots rapidly to high accuracy. To illustrate the method, we will compute a few square roots. Unlike a calculator, we also give an upper bound for the error between our computed value and the true value.

Finding the square root of $a > 0$ means finding the positive root of the function $f(x) = x^2 - a$. Applying Newton's method, a simple computation gives

$$Tx = x - \frac{x^2 - a}{2x} = \frac{1}{2}\left(x + \frac{a}{x}\right).$$

In fact, we will see that any initial positive choice of x_1 will converge to $\sqrt{a}$. Try this on your calculator as a method of computing $\sqrt{2}$ or $\sqrt{71}$.

Suppose that you are asked to compute $\sqrt{149}$ to 15 digits of accuracy. Since $12^2 = 144 < 149 < 169 = 13^2$, let $x_0 = 12$ be our first approximation. Clearly, the error is at most 0.5. Moreover,

$$T'(x) = \frac{1}{2}\left(1 - \frac{149}{x^2}\right).$$

This is evidently monotone. Thus on $[12, 13]$, the derivative is bounded by

$$\max\{|T'(12)|, |T'(13)|\} = \max\left\{\frac{5}{288}, \frac{10}{169}\right\} = \frac{10}{169} < 0.06.$$

Therefore, Newton's method guarantees that the sequence

$$x_0 = 12, \qquad x_{n+1} = \frac{1}{2}\left(x_n + \frac{149}{x_n}\right) \quad \text{for} \quad n \geq 0$$

converges quadratically to $\sqrt{149}$.

Let us compute the other constants involved. Since $f'(x) = 2x$ and $f''(x) = 2$, we obtain

$$A = \sup_{12 \leq x \leq 13} |f''(x)| = 2, \qquad B = \inf_{12 \leq x \leq 13} |f'(x)| = 24, \quad \text{and} \quad M = \frac{A}{B} < 0.084.$$

Using the Mean Value Theorem (6.2.2), note that

$$|f(x_0)| = |f(x_0) - f(x^*)| = |f'(c)| \, |x_0 - x^*| \geq B|x_0 - x^*|.$$

Thus

$$|x_0 - x^*| < |f(x_0)|/24 = 5/24 < 0.21.$$

From the error estimate for Newton's method, we obtain

$$|x_{n+1} - x^*| \leq 0.084|x_n - x^*|^2.$$

Starting with $x_0 = 12$, we have the following table of terms and bounds on the error:

n	x_n	Bound on $|x_n - x^*|$	x_n rounded to accuracy
0	12	0.21	12
1	12.208333333333333	3.704×10^{-3}	12.21
2	12.206555745164960	1.153×10^{-6}	12.20656
3	12.206555615733703	3.572×10^{-13}	12.206555615734
4	12.206555615733702	1.117×10^{-26}	12.2065556157347029...

So we obtain 15 digits of accuracy, in fact, 26 digits, at $n = 4$. To progress further, we need to worry more about the round-off error of our calculations than with Newton's method.

Look at the global aspects of this example. It is easy to see from the graph that $f(x) = x^2 - 149$ is concave, much like Figure 11.3. If x_0 lies in $(0, \sqrt{149})$, then

$x_1 > \sqrt{149}$. However, taking x_0 very small will make x_1 very large. After that, it is also apparent from the graph that x_n decreases monotonely and quickly to $\sqrt{149}$. Similarly, starting with $x_0 < 0$, the same procedure follows from reflection of the whole picture in the y-axis—the sequence converges rapidly to $-\sqrt{149}$. The point $x_0 = 0$ is a nonstarter because $f'(0) = 0$.

Exercises for Section 11.2

A. Show that $f(x) = x^3 + x + 1$ has exactly one real root. Use Newton's method to approximate it to eight decimal places. Show your error estimates.

B. The equation $\sin x = x/2$ has exactly one positive solution. Use Newton's method to approximate it to eight decimal places. Show your error estimates.

C. (a) Set up Newton's method for computing cube roots.
(b) Show by hand that $\sqrt[3]{2} - 1.25 < 0.01$.
(c) Compute $\sqrt[3]{2}$ to eight decimal places.

D. Let $f(x) = (\sqrt{2})^x$ for $x \in \mathbb{R}$. Sketch $y = f(x)$ and $y = x$ on the same graph. Given $x_0 \in \mathbb{R}$, define a sequence by $x_{n+1} = f(x_n)$ for $n \geq 0$.

(a) Find all fixed points of f.
(b) Show that the sequence $(x_n)_{n=1}^{\infty}$ is monotone.
(c) If (x_n) converges to a number x^*, prove that $f(x^*) = x^*$.
(d) For which $x_0 \in \mathbb{R}$ does the sequence (x_n) converge, and what is the limit?

E. Find the largest critical point of $f(x) = x^2 \sin(1/x)$ to four decimal places.

F. Find the minimum value of $f(x) = (\log x)^2 + x$ on $(0, \infty)$ to four decimal places.

G. Apply Newton's method to find the root of $f(x) = (x - r)^{1/3}$. Start with any point $x_0 \neq r$, and compute $|x_n - r|$. Explain what went wrong here.

H. **Modified Newton's method**. With the same setup as for Newton's method, show that the sequence $x_{n+1} = x_n - \dfrac{f(x_n)}{f'(x_0)}$ for $n \geq 0$ converges to x^*.

I. Before computers had high-precision division, the following algorithms were used. Notice that they involve only multiplication.

(a) Let $a > 0$. Show that Newton's formula for solving $1/x = a$ yields the iteration $x_{n+1} = 2x_n - ax_n^2$.
(b) Suppose that $x_0 = (1 - \varepsilon)/a$ for some $|\varepsilon| < 1$. Derive the formula for x_n.
(c) Do the same analysis for the iteration scheme $x_{n+1} = x_n(1 + (1 - ax_n)(1 + (1 - ax_n)))$. Explain why this is a superior algorithm.

J. Let $h(x) = x^{1/3}e^{-x^2}$.

(a) Set up Newton's method for this function.
(b) If $0 < |x_n| < 1/\sqrt{6}$, then $|x_{n+1}| > 2|x_n|$.
(c) Show that if $x_n > 1/\sqrt{6}$, then $x_{n+1} > x_n + \frac{1}{2x_n}$.
(d) Hence show that Newton's method never works unless $x_0 = 0$. However, given $\varepsilon > 0$, there will be an N so large that $|x_{n+1} - x_n| < \varepsilon$ for $n \geq N$.
(e) Sketch h and try to explain this nasty behaviour.

K. Three towns are situated around the shore of a circular lake of radius 1 km. The largest town, Alphaville, claims one-half of the area of the lake as its territory. The town mathematician is charged with computing the radius r of a circle from the town hall (which is right on the shore) that will cut off half the area of the lake. Compute r to seven decimal places.

HINTS: Let T denote the town hall, and O the centre of the lake. The circle of radius r meets the shoreline at X and Y. The area enclosed is the union of two segments of circles cut off by the chord XY. Express r and the area A as functions of the angle $\theta = \angle OTX$.

MORE HINTS:(1) Show that (i) $1 < r < 2$; (ii) $\angle TOX = \pi - 2\theta$; (iii) the area of the segment of a circle of radius ρ cut off by a chord that subtends an angle α is $\rho^2(\alpha - \sin\alpha)/2$. (2) Show that $r = 2\cos\theta$ and $A(\theta) = 2\pi - 4\theta\sin^2\theta - 2\sin(2\theta)$. (3) Solve $A(\theta) = \pi$ using Newton's method. Show error estimates.

11.3 Orbits of a Dynamical System

There are several possibilities for the structure of the orbit of a point x_0. Fixed points, which we have discussed in detail, have the simplest possible orbits, namely $\mathcal{O}(x_0) = \{x_0\}$. Almost as good as fixed points from the point of view of dynamics, and certainly more common, are periodic points. We say that x^* is a **periodic point** if there is a positive integer n such that $T^n x^* = x^*$. The smallest positive n for which this holds is called the **period**. Notice that x^* is a fixed point of T^n. We can therefore call x^* an **attractive periodic point** or a **repelling periodic point** for T if it is an attractive or repelling fixed point of T^n. Because T maps an open set around x^* into an open set around Tx^*, it is easy to check that points in the same periodic orbit are either all attractive or all repelling.

Let us discuss the terminology of dynamical systems by examining a particular map, namely the map $Tx = 1.8(x - x^3)$ that we discussed in Section 11.1. Our example contains an orbit of period 2, namely $\{\pm\sqrt{14}/3\}$. Indeed, it is an easy calculation to see that

$$T(\sqrt{14}/3) = -\sqrt{14}/3 \quad \text{and} \quad T(-\sqrt{14}/3) = \sqrt{14}/3.$$

See Figure 11.4. This is a repelling orbit, since

$$(T^2)'(\sqrt{14}/3) = T'(T\sqrt{14}/3)T'(\sqrt{14}/3) = (1.8 - 5.4(14/3))^2 = 547.56 > 1.$$

A point x is called an **eventually periodic point** if $T^n x$ eventually belongs to a period. Consider our example again. Notice that $T(1) = 0$. So $T^n(1) = 0$ for all $n \geq 1$. The equation $Tx = 1$ has a solution that is the root of the cubic

$$1 = 1.8x - 1.8x^3.$$

This has a solution r, since every real polynomial of odd degree has a root by Exercise 5.6.D and we may calculate $r \approx -1.20822631883$. Then with $x_0 = r$, we obtain $x_1 = Tx_0 = 1$ and $x_2 = T(1) = 0$, and so $x_0 = T^n x_0 = 0$ for all $n \geq 2$. Similarly, we have a sequence (r_n) of points such that $T^n(r_n) = 0$, such as $r_0 = 0$, $r_1 = 1$, $r_2 \approx -1.20823$, $r_3 \approx 1.24128$. Observe that r_n is close to $(-1)^{n-1}\sqrt{14}/3$ for large n, in the sense that $r_n - (-1)^{n-1}\sqrt{14}/3$ converges to zero.

Likewise, there are points that are eventually mapped onto the two attractive fixed points $\pm 2/3$.

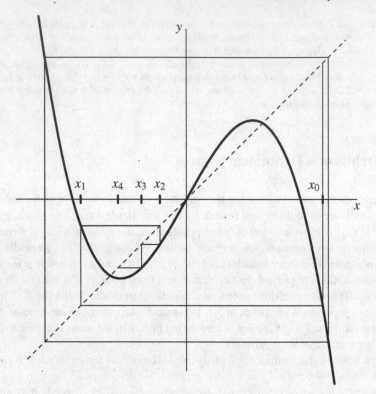

FIG. 11.4 Graph of $Tx = 1.8(x - x^3)$ showing period-2 orbit and a nearby orbit.

Figure 11.5, called a **phase portrait**, attempts to show graphically the behaviour of our mapping T. Rather than using a graph, it represents the space as a line and the function as arrows taking a point x to Tx. In particular, you should observe that except for a sequence of points that eventually map to 0, every other point in $(-\sqrt{14}/3, \sqrt{14}/3)$ has an orbit that converges to one of the two attractive fixed points $\pm 2/3$. Notice that except for the two arrows for the period-2 orbit, arrows above the line show the first few terms of the sequence (r_n) mentioned previously, while those below the line show the "nearby" orbit from Figure 11.4.

Of course, this example does not exhaust the possibilities for the behaviour of an orbit. It can happen that the orbit of a single point is dense in the whole space. A point x is called a **transitive point** if $\overline{\mathscr{O}(x)} = X$. The following examples are more complicated than our first one and exhibit this new possibility as well as having many periodic points.

11.3.1. EXAMPLE. Let the space be the unit circle $\mathbb{T}$. We can describe a typical point on the circle by the angle θ it makes to the positive real axis in radians. If θ is changed to $\theta + 2n\pi$, the point determined remains the same. So this angle is determined "up to a multiple of 2π." We may add two angles, or multiply them

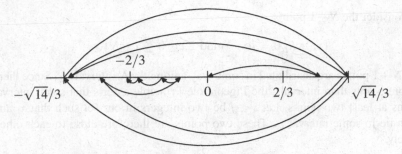

FIG. 11.5 Phase portrait of $Tx = 1.8(x - x^3)$

by an integer. The resulting point does not depend on how we initially measure the angle. (Notice that when multiplying by a fraction such as $1/2$, the answer does depend on which way the angle is represented. Half of 1 does not determine the same point as half of $1 + 2\pi$.) We call this arithmetic **modulo** 2π. We will write

$$\theta \equiv \varphi \quad (\text{mod } 2\pi)$$

to mean that $\theta - \varphi$ is an integer multiple of 2π, which is to say that θ and φ represent the same point on the circle. In this way, we may think of the unit circle $\mathbb{T}$ as the real line wrapping around the circle infinitely often, meeting up at the same point every 2π. This will be convenient for calculation.

Consider the rotation map R_α through an angle α given by

$$R_\alpha \theta \equiv \theta + \alpha \quad (\text{mod } 2\pi).$$

It is rather easy to analyze what happens here. The only way to obtain a periodic point is to have

$$\theta \equiv R_\alpha^n \theta = \theta + n\alpha \quad (\text{mod } 2\pi).$$

This requires $n\alpha$ to be an integer multiple of 2π. When this occurs and n is as small as possible, it follows that *every* point in $\mathbb{T}$ has period n. Indeed, $R_\alpha^n = \text{id}$, the identity map. This is the case for those α that are a rational multiple of 2π. Equivalently, R_α is periodic if and only if $\alpha/2\pi \in \mathbb{Q}$.

On the other hand, for all irrational values of $\alpha/2\pi$, R_α has no periodic points. We will show that the orbit of every point is dense in the circle. Indeed, suppose that $\varepsilon > 0$ and that θ and φ are two points on the circle. We wish to find an integer n such that $|R_\alpha^n \theta - \varphi| < \varepsilon$.

To do this, we first solve a simpler problem. We will find a positive integer m such that $|R_\alpha^m 0| < \varepsilon$. This says that while there are no periodic points, points do come back very close to where they start every once in a while. Our method uses the Pigeonhole Principle.

Pick an integer N so large that $2\pi < N\varepsilon$. Divide the circle into N intervals $I_k = [(k-1)2\pi/N, k2\pi/N)$ for $1 \leq k \leq N$. Note that each interval has length $2\pi/N < \varepsilon$.

Now consider the $N+1$ points

$$\{x_j \equiv R_\alpha^j 0 = j\alpha \quad (\text{mod } 2\pi) : 0 \le j \le N\}.$$

These $N+1$ points are distributed in some way among the N intervals I_k. Since there are more points than intervals, the Pigeonhole Principle asserts that some interval contains at least two points. Let $i < j$ be two integers in our set such that x_i and x_j both lie in some interval I_k. These two points are therefore close to each other. Precisely,

$$|x_j - x_i| < \frac{2\pi}{N} < \varepsilon.$$

Now we use the fact that R_α is isometric, meaning that

$$|R_\alpha \theta - R_\alpha \varphi| = |\theta - \varphi|$$

for any two points $\theta, \varphi \in \mathbb{T}$. This just says that the map R_α is a rigid rotation that does not change the distance between points. Let $m = j - i$. Hence

$$|R_\alpha^m 0| = |x_{j-i} - x_0| = |R_\alpha^i x_{j-i} - R_\alpha^i x_0| = |x_j - x_i| < \varepsilon.$$

It is also important that x_m is not 0. This follows because $\alpha/2\pi$ is irrational. So we observe that

$$R_\alpha^{km} 0 \equiv k x_m \quad (\text{mod } 2\pi) \quad \text{for} \quad k \ge 0$$

forms a sequence of points that move around the circle in steps smaller than ε. So it is possible to choose k such that $k x_m$ is close to $\varphi - \theta$, and thus

$$|\varphi - R_\alpha^{km} \theta| = |(\varphi - \theta) - R_\alpha^{km} 0| = |(\varphi - \theta) - k x_m| < \varepsilon.$$

So every orbit is dense in the whole circle.

11.3.2. EXAMPLE. We will now look at an example that has both interesting periodic points and transitive orbits. Some of the proofs must be left until later (see Examples 11.5.5, 11.5.10, and 11.5.14). We shall see that this example is chaotic. While this word is suggestive of wild behaviour, it actually has a precise mathematical meaning, which we will explore in Section 11.5.

Consider the map T from $\mathbb{T}$ into itself given by $T\theta = 2\theta$. This is called the **doubling map** on the circle. Essentially this map wraps the circle twice around itself. That is, the top semicircle $[0, \pi)$ is mapped one-to-one and onto the whole circle; and the bottom semicircle $[\pi, 2\pi)$ is also mapped one-to-one and onto the whole circle. Thus this map is two-to-one.

A point θ is periodic of period $n \ge 1$ if

$$\theta \equiv T^n \theta = 2^n \theta \quad (\text{mod } 2\pi).$$

This happens if and only if $(2^n - 1)\theta$ is an integer multiple of 2π. The period of $\mathcal{O}(\theta)$ will be the smallest positive integer k such that $(2^k - 1)\theta$ is an integer multiple of 2π. Thus the point $2\pi/(2^n - 1)$ is periodic of period n, and

$$\mathcal{O}\left(\frac{2\pi}{2^n - 1}\right) = \left\{\frac{2^j\pi}{2^n - 1} : 1 \leq j \leq n\right\}.$$

Indeed, every point $2\pi s/(2^n - 1)$ for every $n \geq 1$ and $1 \leq s \leq 2^n - 1$ is a periodic point, although the period will possibly be a proper divisor of n rather than n itself. These points are dense in the whole circle. Because the derivative of T is 2 as a map from $\mathbb{R}$ to $\mathbb{R}$, it follows that every periodic point is repelling.

There are eventually periodic points, namely $2\pi s/(2^p(2^n - 1))$ for $p \geq 1$. After p iterations, these points join the periods identified previously. It is not difficult to see that this is a complete list of all the periodic and eventually periodic points. So every other point has infinite orbit. Unlike our first example, these orbits will not converge to some period, since every periodic point is repelling.

This example also has a dense set of transitive points, although we only outline the argument. Write a point θ as $2\pi t$ for $0 \leq t < 1$. Then write t in binary as $t = (0.\varepsilon_1\varepsilon_2\varepsilon_3\ldots)_{\text{base }2}$. Then $T^k\theta \equiv 2\pi t_k \pmod{2\pi}$, and the binary expansion is $t_k = (0.\varepsilon_{k+1}\varepsilon_{k+2}\varepsilon_{k+3}\ldots)_{\text{base }2}$. The set of possible limit points of this orbit has little to do with the first few (say) billion coefficients. So we may use these to specify θ close to any point in the circle. Now arrange the tail of the binary expansion to include all possible finite sequences of 0's and 1's. Then by applying T repeatedly, each of these finite sequences eventually appears as the initial part of the binary expansion of t_k. This shows that the orbit is dense in the whole circle.

Exercises for Section 11.3

A. Suppose that x^* is a point of period n. Show that if x^* is attracting (or repelling) for T^n, then each $T^i x^*$, $i = 0, \ldots, n-1$, is an attracting (or repelling) periodic point.

B. Draw a phase diagram of the dynamics of $Tx = 0.5(x - x^3)$ for $x \in \mathbb{R}$.

C. Find the periodic points of the **tripling map** on the circle: $T : \mathbb{T} \to \mathbb{T}$ given by $T\theta = 3\theta$.

D. Consider $Tx = a(x - x^3)$ for $x \in \mathbb{R}$ and $a > 0$.

   (a) Find all fixed points. Decide whether they are attracting or repelling.
   (b) Find all points of period-2.   HINT: First look first for solutions of $Tx = -x$. To factor $T^2x - x$, use the fact that each fixed point is a root to factor out a cubic, and factor out a quadratic corresponding to the period-2 cycle already found.
   (c) Decide whether the period-2 points are attracting or repelling.
   (d) Find the three bifurcation points corresponding to the changes in the period-1 and -2 points (i.e., at which values of the parameter a do changes in the dynamics occur?).
   (e) Draw a phase diagram of the dynamics for $a = 2.1$.

E. Consider the **tent map** T of $[0, 1]$ onto itself by $Tx = \begin{cases} 2x & \text{if } 0 \leq x \leq \frac{1}{2}, \\ 2 - 2x & \text{if } \frac{1}{2} \leq x \leq 1. \end{cases}$

   (a) Graph T^n for $n = 1, 2, 3, 4$.
   (b) Using the graphs, show there are exactly 2^n fixed points for T^n. How are they distributed?
   (c) Use (b) to show that the periodic points are dense in X.
   (d) Show that there are two distinct orbits of period 3.
       HINT: Solve $T^3x = x$ for $x \in [\frac{1}{8}, \frac{1}{4}]$ and for $x \in [\frac{1}{4}, \frac{3}{8}]$.

(e) Show that there are points of period n for every positive integer n.
(f) Find all points that are not fixed but are eventually fixed. Show that they are dense in X.

F.   Let $\omega(x) = \bigcap_{n \geq 0} \overline{\mathcal{O}(T^n x)}$ be the **cluster set** of the forward orbit of $T : X \to X$.

(a) Show that T maps $\omega(x)$ into itself.
(b) Show that $\overline{\mathcal{O}(x)} = \mathcal{O}(x) \cup \omega(x)$.
(c) Show by example that $\omega(x)$ can be empty.
(d) If $\mathcal{O}(x)$ is compact, show that $\omega(x)$ is a nonempty subset of $\mathcal{O}(x)$.

G.   Suppose that $\mathcal{O}(x)$ is compact.

(a) Let n_0 be the smallest integer such that $T^{n_0} x \in \omega(x)$. Prove $\mathcal{O}(T^{n_0} x) = \omega(x) = \omega(T^{n_0} x)$.
(b) If $\omega(x)$ is infinite, show that it must be perfect. This contradicts Exercise 4.4.L.
(c) Show that $\mathcal{O}(x)$ is a compact set if and only if there is an n_0 such that $T^{n_0} x$ is periodic.

H.   A dynamical system (X, T) is **minimal** if the only closed sets F with $TF \subset F$ are $\varnothing$ and X.

(a) Show that (X, T) is minimal if and only if every point $x \in X$ is a transitive point, meaning that $\mathcal{O}(x)$ is dense in X.
(b) Show that the rotation R_α on the circle $\mathbb{T}$ is minimal if and only if $\alpha / 2\pi$ is irrational.

I.   Let C be the Cantor set of Example 4.4.8, and represent each point $x \in C$ in its ternary expansion using only 0's and 2's. Define $T : C \to C$ by

$$T\, 0.\overbrace{2 \ldots 2}^{k} 0\varepsilon_{k+2} \varepsilon_{k+3} \ldots = 0.\overbrace{0 \ldots 0}^{k} 2\varepsilon_{k+2} \varepsilon_{k+3} \ldots$$

for $k = 0, 1, \ldots$ and $T\, 0.22222 \ldots = 0.00000 \ldots$.

(a) Prove that T is continuous and bijective.
(b) Show that $\mathcal{O}(0)$ is dense in C.
(c) Prove that (C, T) is minimal.

11.4 Periodic Points

In this section, we use the Intermediate Value Theorem (5.6.1) to establish the *existence* of fixed points and periodic points. This new technique applies more widely than the Banach Contraction Principle (11.1.6), but it is not constructive, that is, it does not generally yield a computational scheme.

First, we look at a couple of situations that imply a fixed point.

11.4.1. LEMMA. *Suppose that T is a continuous function of a closed bounded interval $I = [a, b]$ into itself. Then T has a fixed point.*

PROOF. Consider the function

$$f(x) = Tx - x \quad \text{for} \quad a \leq x \leq b.$$

Notice that

$$f(a) = Ta - a \geq 0 \quad \text{and} \quad f(b) = Tb - b \leq 0.$$

By the Intermediate Value Theorem (5.6.1), there is a point x^* in $[a,b]$ such that

$$0 = f(x^*) = Tx^* - x^*.$$

Thus x^* is the desired fixed point. ∎

The second result is very similar, but instead of mapping an interval *into* itself, we have an interval mapping *onto* itself.

11.4.2. LEMMA. *Let T be a continuous function on a closed bounded interval $I = [a,b]$ such that $T(I)$ contains I. Then T has a fixed point.*

PROOF. Again consider the function $f(x) = Tx - x$ for $a \leq x \leq b$. By hypothesis, there are points c and d such that $Tc = a$ and $Td = b$. Thus

$$f(c) = Tc - c = a - c \leq 0,$$
$$f(d) = Td - d = b - d \geq 0.$$

Again by the Intermediate Value Theorem, there is a point x^* in $[c,d]$ such that

$$0 = f(x^*) = Tx^* - x^*.$$

So x^* is the desired fixed point. ∎

11.4.3. EXAMPLE. Consider the family of quadratic maps $Q_a x = a(x - x^2)$ for $a > 1$ known as the **logistic functions**. These maps are inverted parabolas with zeros at 0 and 1 and a maximum at $(1/2, a/4)$. Each map Q_a takes positive values on $[0,1]$ and negative values elsewhere. The derivative is $Q'_a(x) = a(1 - 2x)$. It is evident that $|Q'_a(x)| > 1$ on $\mathbb{R} \setminus [0,1]$. So it is not difficult to show that if $x < 0$ or $x > 1$, then $Q^n_a x$ diverges to $-\infty$. For this reason, we restrict our domain to the interval $I = [0,1]$.

There are two cases. Suppose that $a \leq 4$. Then Q_a maps I into itself. Thus it has a fixed point by Lemma 11.4.1. On the other hand, if $a \geq 4$, then since $Q_a 0 = 0$ and $Q_a 1/2 = a/4 \geq 1$, it follows from the Intermediate Value Theorem that $Q_a(I)$ contains I. So Q_a has a fixed point by Lemma 11.4.2.

It will be an added convenience, when the image of one interval contains another, to find a smaller interval that exactly maps onto the target interval. While this is intuitively clear, the details need to be checked. The proof is left as an exercise.

11.4.4. LEMMA. *Let $T : [a,b] \to \mathbb{R}$ be a continuous function such that $T([a,b])$ contains an interval $[c,d]$. Then there is a (smaller) interval $[a',b'] \subseteq [a,b]$ such that $T([a',b']) = [c,d]$ and $T(\{a',b'\}) = \{c,d\}$.*

For closed bounded intervals I and J, we will write $I \to J$ to indicate that $T(I)$ contains J. Let us see how this can be used to find periodic points.

11.4.5. EXAMPLE. Consider the family of logistic functions $Q_a x = a(x - x^2)$ acting on $I = [0, 1]$ for $a \geq 4$. Write $I_0 = [0, \frac{1}{2}]$ and $I_1 = [\frac{1}{2}, 1]$. By the argument in the previous example, $Q_a(I_0)$ and $Q_a(I_1)$ both contain I (i.e., $I_0 \to I_0 \cup I_1$ and $I_1 \to I_0 \cup I_1$).

In particular, $I_0 \to I_0 \to I_1 \to I_0$. Use Lemma 11.4.4 repeated as follows. First find an interval $J_2 \subset I_1$ such that $Q_a(J_2) = I_0$. Then find $J_1 \subset I_0$ such that $Q_a(J_1) = J_2$. Finally, find $J_0 \subset I_0$ such that $Q_a(J_0) = J_1$. Then

$$Q_a^3(J_0) = Q_a^2(J_1) = Q_a(J_2) = I_0.$$

By Lemma 11.4.1, Q_a^3 has a fixed point in J_0, say x_0. We will show that this point has period 3. Indeed, let $x_i = Q_a^i x_0$, $i = 1, 2$. By construction, $Q_a x_2 = Q_a^3 x_0 = x_0$. Thus x_0 is either a period-3 point or a fixed point. Now x_0 and x_1 belong to I_0 and x_2 belongs to I_1. If x_0 were a fixed point, it would belong to $I_0 \cap I_1 = \{\frac{1}{2}\}$. But $Q_a \frac{1}{2} = \frac{a}{4} \geq 1$ is not fixed or even periodic. Thus x_2 is different from x_0 and x_1, and consequently Q_a has an orbit of length 3.

This *proof* requires $a \geq 4$ to work. However, it is actually the case that period-3 orbits begin to appear when a is about 3.8284.

11.4.6. EXAMPLE. **Period doubling**. Let f be a map from $I = [0, 1]$ into itself. Define a map $\mathscr{D}f$ as follows:

$$\mathscr{D}f(x) = \begin{cases} \frac{2}{3} + \frac{1}{3}f(3x) & \text{for} \quad 0 \leq x \leq \frac{1}{3} \\ (2 + f(1))(\frac{2}{3} - x) & \text{for} \quad \frac{1}{3} \leq x \leq \frac{2}{3} \\ x - \frac{2}{3} & \text{for} \quad \frac{2}{3} \leq x \leq 1. \end{cases}$$

We claim that with the single exception of a repelling fixed point in the interval $I_2 = [\frac{1}{3}, \frac{2}{3}]$, the periodic orbits of $\mathscr{D}f$ correspond to the periodic orbits of f, the periods are exactly double, and the dynamics (attracting or repelling) of these orbits are preserved.

Indeed, it is clear from the graphs in Figure 11.6 that there is an (easily computed) fixed point in the interval I_2. Since the function $\mathscr{D}f$ has slope $-2 - f(1) \leq -2$, this evidently is a repelling fixed point. Any other point in I_2 is mapped farther and farther from this point until it leaves this middle third.

The interval $I_1 = [0, \frac{1}{3}]$ is mapped into $I_3 = [\frac{2}{3}, 1]$ and I_3 is mapped bijectively onto I_1. These orbits map back and forth between I_1 and I_3, never intersecting I_2. Since no orbit, other than the fixed point, stays in I_2, eventually every orbit (other than the fixed point) alternates between I_1 and I_3.

Now notice that if $x \in [0, 1]$, then

$$(\mathscr{D}f)^2(x/3) = \mathscr{D}f(\tfrac{2}{3} + \tfrac{1}{3}f(x)) = \tfrac{1}{3}f(x).$$

This means that the graph of $(\mathscr{D}f)^2$ on I_1 is identical to the graph of f on I except that it is scaled (in both the x-direction and the y-direction) by a factor of one-third. To make this precise, let $\sigma(x) = x/3$ be a map of I onto I_1. This map is a continuous

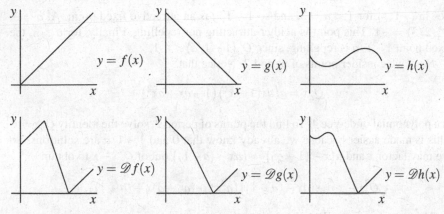

FIG. 11.6 Applying $\mathscr{D}$ to various functions.

bijection with continuous inverse $\sigma^{-1}(x) = 3x$. The relation between f and $(\mathscr{D}f)^2$ can be expressed as

$$\sigma f(x) = (\mathscr{D}f)^2 \sigma \quad \text{or} \quad (\mathscr{D}f)^2 - \sigma f \sigma^{-1}.$$

In Section 11.6, we will see that this relationship is a special case of an important notion, topological conjugacy.

So if x is a periodic point for f of period n, then $x/3$ has period n for $(\mathscr{D}f)^2$ and vice versa. Since $\mathscr{D}f$ flips back and forth between I_1 and I_3, it follows that $x/3$ has period $2n$ for $\mathscr{D}f$. Moreover, it will be attracting or repelling as x is.

Now consider a map f_0 with a unique attracting fixed point such as the constant function $f_0(x) = \frac{1}{3}$. Define a sequence of functions by

$$f_{n+1} = \mathscr{D}f_n \quad \text{for all} \quad n \geq 1.$$

Then f_1 has an attracting orbit of period 2 and has a repelling fixed point in between. The function f_2 will have an attracting orbit of period 4, and in between, there will be a repelling orbit of period 2 and a repelling fixed point. Recursively we find that f_n has an attracting orbit of period 2^n and in between it has repelling orbits of lengths $1, 2, \ldots, 2^{n-1}$.

Exercise 11.4.G outlines how to show that this sequence of functions converges uniformly to a function f_∞ that has one repelling orbit of period 2^n for each $n \geq 0$ and no other periods.

11.4.7. EXAMPLE. The logistic functions of Example 11.4.3 give a dramatic demonstration of how periodic points can change as a parameter is varied. Recall $Q_a(x) = a(x - x^2)$ on $I = [0, 1]$. When $a > 1$, there are two fixed points: a repelling fixed point at 0 and another fixed point at $1 - 1/a$. The derivative

$$Q_a'\left(1 - \tfrac{1}{a}\right) = a\left(1 - 2(1 - \tfrac{1}{a})\right) = 2 - a$$

lies in $(-1, 1)$ for $1 < a < 3$, and so $1 - 1/a$ is an attractive fixed point. At $a = 3$, $Q_3'(2/3) = -1$. This point is neither attracting nor repelling. Finally, for $a > 3$, the fixed point $1 - 1/a$ is repelling, since $Q_a'(1 - 1/a) < -1$.

Next, we consider points of period 2. Notice that

$$Q_a^2 x = a(ax(1-x))(1 - ax + ax^2)$$

is a polynomial of degree 4. To find the points of period 2, solve the identity $Q_a^2 x = x$. This is made easier because we already know that 0 and $1 - 1/a$ are solutions. So we may factor x and $a(x - (1 - \frac{1}{a})) = (ax - (a-1))$ out of $Q_a^2 x - x$ to obtain

$$Q_a^2 x - x = x(ax - (a-1))(a^2 x^2 - (a^2 + a)x + (a+1)).$$

The quadratic factor has discriminant

$$(a^2 + a)^2 - 4a^2(a+1) = a^4 - 2a^3 - 3a^2 = a^2((a-1)^2 - 4).$$

This is positive precisely when $a > 3$. So for $1 < a < 3$, there are no period-2 points. At $a = 3$, $Q_3^2 x - x = x(3x - 2)^3$ also has no period-2 orbits. However, once $a > 3$, a period-2 orbit appears consisting of the points

$$p_\pm = \frac{a^2 + a \pm a\sqrt{(a-1)^2 - 4}}{2a^2}.$$

When a parametric family of maps changes its dynamical behaviour, this is called a **bifurcation**. The associated value of the parameter is called a **bifurcation point**.

To compute the derivative of Q_a^2 at $p_\pm$, we use the chain rule:

$$\begin{aligned}
(Q_a^2)'(p_+) &= Q_a'(Q_a p_+) Q_a'(p_+) = Q_a'(p_-) Q_a'(p_+) \\
&= a(1 - 2p_-)a(1 - 2p_+) \\
&= a^2 - 2a^2(p_- + p_+) + 4a^2(p_- p_+) \\
&= a^2 - 2(a^2 + a) + 4(a+1) = 5 - (a-1)^2.
\end{aligned}$$

It follows that this period is attracting for $3 < a < 1 + \sqrt{6}$. Then another bifurcation occurs at $a = 1 + \sqrt{6}$ when this becomes a repelling orbit. It turns out that at this point, an attracting orbit of period 4 appears.

However, the story does not stop here. An infinite sequence of bifurcations occurs, at each point of which an attractive period of length 2^n appears and the period of length 2^{n-1} becomes repelling. This is sometimes called the 'period-doubling route to chaos'. The limit of this procedure is a point $a_\infty \approx 3.5699$. For every $a \geq a_\infty$, Q_a has repelling orbits of period 2^n for all $n \geq 0$. The story continues and yet more bifurcations happen between a_∞ and 4. For example, period-3 orbits first appear at about 3.8284.

There is an ordering among the possible periods of orbits that was discovered by Sharkovskii. In this ordering, the existence of a period-n orbit implies the existence of all periods that occur later in the ordering:

$$3 \rhd 5 \rhd 7 \rhd 9 \rhd \cdots \quad \rhd 6 \rhd 10 \rhd 14 \rhd \cdots \quad \rhd 12 \rhd 20 \rhd 28 \rhd \cdots$$
$$\rhd 3 \cdot 2^n \rhd 5 \cdot 2^n \rhd 7 \cdot 2^n \rhd \cdots \quad \rhd 3 \cdot 2^{n+1} \rhd 5 \cdot 2^{n+1} \rhd 7 \cdot 2^{n+1} \rhd \cdots$$
$$\rhd \cdots \quad \cdots \rhd 2^n \rhd 2^{n-1} \rhd \cdots \rhd 4 \rhd 2 \rhd 1.$$

We will prove a special case of this result, that period 3 is preeminent among all periods. The general argument follows the same lines but is more complicated.

11.4.8. LEMMA. *Let T be a continuous map from an interval I into itself. Suppose that there are intervals such that $I_1 \to I_2 \to \cdots \to I_n$. Then there are intervals $J_k \subset I_k$ for $1 \le k \le n-1$ such that*

$$T(J_k) = J_{k+1} \quad for \quad 1 \le k \le n-2 \quad and \quad TJ_{n-1} = I_n.$$

PROOF. This is an easy application of Lemma 11.4.4. First, find $J_{n-1} \subset I_{n-1}$ such that $T(J_{n-1}) = I_n$. Then use the lemma again to obtain an interval $J_{n-2} \subset I_{n-2}$ such that $T(J_{n-2}) = J_{n-1}$. Proceed in this way to define the sequence recursively. ∎

11.4.9. SHARKOVSKII'S THEOREM.

Suppose that T is a continuous map of an interval $I = [a, b]$ into itself that has an orbit of period 3. Then T has an orbit of period n for every $n \ge 1$.

PROOF. For $n = 1$, we may invoke Lemma 11.4.1 to obtain a fixed point a_1.

For $n \ge 2$, we need to look at the period-3 orbit, so suppose $x_1 < x_2 < x_3$ is the period-3 orbit. Either $Tx_1 = x_2$ or $Tx_1 = x_3$. But in the second case, we may consider the interval with order reversed, in which case x_3 is the smallest, and it maps to the second point x_2. So the argument for the first case must equally apply in the second. Thus we may assume that

$$x_1 < x_2 = Tx_1 < x_3 = Tx_2 \quad and \quad Tx_3 = x_1.$$

Let $I_0 = [x_1, x_2]$ and $I_1 = [x_2, x_3]$. It is immediate from the Intermediate Value Theorem that TI_0 contains I_1 and TI_1 contains $I_0 \cup I_1$. That is, $I_0 \to I_1$ and $I_1 \to I_0$ and $I_1 \to I_1$. By Lemma 11.4.4, there is an interval J_0 contained in I_0 and there are intervals J_1 and J_2 contained in I_1 such that

$$T(J_0) = I_1, \qquad T(J_1) = I_0, \quad and \quad T(J_2) = I_1.$$

So for $n = 2$, we use the fact that $J_0 \to J_1 \to J_0$. From this it follows that $T^2(J_0)$ contains J_0. So by Lemma 11.4.2, there is a fixed point a_2 of T^2 in J_0. We note that if $J_0 \cap J_1$ is nonempty, it must consist only of the point x_2. Since $T^2 x_2 = x_1$, it is not fixed for T^2. Thus a_2 is some other point in J_0 and so $Ta_2 \neq a_2$. Hence a_2 has period 2.

Now consider $n \geq 4$. We will proceed as in Example 11.4.5. Notice that

$$J_0 \to \underbrace{J_2 \to \cdots \to J_2}_{n-2 \text{ copies}} \to J_1 \to J_0.$$

Apply Lemma 11.4.8 to find intervals $K_0 \subset J_0$, $K_i \subset J_2$ for $1 \leq i \leq n-2$ and $K_{n-1} \subset J_1$ such that

$$T(K_i) = K_{i+1} \quad \text{for} \quad 0 \leq i \leq n-2 \quad \text{and} \quad T(K_{n-1}) = J_0.$$

In particular, $T^n(K_0) = J_0$ contains K_0. Applying Lemma 11.4.2 again yields a fixed point a_n of T^n in K_0.

We must verify that a_n has no smaller period. But this is a consequence of a fact guaranteed by our construction:

$$T^i a_n \in J_2 \quad \text{for} \quad 1 \leq i \leq n-2 \quad \text{and} \quad T^{n-1} a_n \in J_1.$$

CLAIM: None of these points is equal to a_n. Indeed, as in the period-2 case, the only possible intersection of J_0 and $J_1 \cup J_2$ is the point x_2. However, were $a_n = x_2$, it would follow that J_2 would be an interval containing $T^2 a_n = x_1$, which is not possible. Hence the period of a_n is exactly n. ∎

Exercises for Section 11.4

A. Find a continuous function from $(0,1)$ onto itself with no fixed points. Why does this not contradict Lemma 11.4.1 or 11.4.2?

B. Prove Lemma 11.4.4. HINT: Pick a_0 and b_0 in $[a,b]$ such that $Ta_0 = c$ and $Tb_0 = d$. If $a_0 < b_0$, let $a' = \sup\{x \in [a_0,b_0] : Tx = c\}$ and $b' = \inf\{x \in [a',b_0] : Tx = d\}$. Consider the case $a_0 > b_0$ separately.

C. Consider the function T mapping $I = [0,4]$ onto itself by

$$Tx = \begin{cases} 2x+2 & \text{for} \quad 0 \leq x \leq 1, \\ 5-x & \text{for} \quad 1 \leq x \leq 2, \\ 7-2x & \text{for} \quad 2 \leq x \leq 3, \\ 4-x & \text{for} \quad 3 \leq x \leq 4. \end{cases}$$

Figure 11.7 gives the graph of T. Note that $\{0,2,3,1,4\}$ is a period-5 orbit.

(a) Sketch the graphs of T^2 and T^3.
(b) Show that T has one period-2 orbit.
(c) Show that T has no period-3 orbit. HINT: Show that $T^3 x > x$ on $[0,2]$, $T^3 x < x$ on $[3,4]$, and T^3 is monotone decreasing on $[2,3]$.

D. Suppose that T is a continuous map from an interval I into itself. Moreover, suppose that there are points $x_1 < x_2 < x_3 < x_4$ such that $Tx_1 = x_2$, $Tx_2 = x_3$, $Tx_3 = x_4$, and $Tx_4 \leq x_1$. Show that T has an orbit of period 3. HINT: Let $I_k = [x_k, x_{k+1}]$. Show that $I_1 \to I_2 \to I_3 \to I_1$.

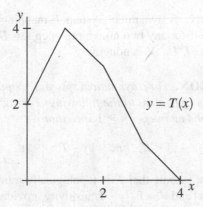

FIG. 11.7 The graph of T.

E.   Give an example of a map with an orbit of period 6, but no odd orbits.

F.   This is a computer experiment for the family of logistic maps Q_a.

(a) Let $a = 3.46$. Use a computer to calculate $x = Q_a^{100}(0.5)$. Then compute $Q_a x$, $Q_a^2 x$, $Q_a^3 x, \ldots, Q_a^{10} x$. What do you observe? Why did this happen?
(b) Try this for $a = 3.55$. What is different now? What bifurcation occurred?
(c) Do the same for $a = 3.83$. What do you observe? Try this for $a = 3.8$ and $a = 3.9$. The two sequences do not behave in the same way, but the reasons are different.

G.   Consider the period-doubling method of Example 11.4.6. Start with the constant function $f_0(x) = \frac{1}{3}$. Define a sequence of functions by $f_{n+1} = \mathscr{D} f_n$ for $n \geq 0$.

(a) Show that $f_{n+1}(x) = f_n(x)$ for all $3^{-n} \leq x \leq 1$ and $1 - 3^{-n} \leq f_n(x) \leq 1$ for all $0 \leq x \leq 3^{-n}$.
(b) Use part (a) to show that f_n converges uniformly to a continuous limit function f_∞.
(c) Calculate the point x_n of intersection between the line $y = 1 - 3^{-n} + x$ and the graph of $f_{n+1}(x)$. Show that this is a point of period 2^n for f_∞.
(d) Show that these are the only periods of the function f_∞.

11.5 Chaotic Systems

In this section, we will define and examine chaotic systems, which are systems of striking complexity with seemingly "wild" behaviour. The surprise is that this complexity arises in seemingly simple situations, as the examples of this section will show. In mathematical physics, it was an important insight that very simple, commonly occurring differential equations exhibit chaotic behaviour. Part of the definition of chaos is that very small perturbations in initial conditions lead to wildly different orbits. For example, this phenomenon makes detailed weather prediction over the long term impossible, even if it is weather produced in a laboratory using an apparently simple model. This is also the reason that water flowing in a river produces complicated eddying that is constantly changing and unpredictable.

The mathematical notion of chaos depends on three things. The first is a dense set of periodic points. The other two items are new, and we study them in turn.

11.5.1. DEFINITION. A dynamical system T mapping a set X into itself is **topologically transitive** if for any two nonempty open sets U and V in X, there is an integer $n \geq 1$ such that $T^n U \cap V$ is nonempty.

11.5.2. PROPOSITION. *For a dynamical system T mapping a set X into itself, topological transitivity is equivalent to the following: For each $x, y \in X$ and $\varepsilon > 0$, there are a point $z \in X$ and an integer $n \geq 1$ such that*

$$\|x - z\| < \varepsilon \quad and \quad \|y - T^n z\| < \varepsilon.$$

PROOF. To see this, first assume that T is topologically transitive. Given $x, y \in X$ and $\varepsilon > 0$, take $U = B_\varepsilon(x)$ and $V = B_\varepsilon(y)$. Transitivity provides n such that $T^n U \cap V$ is nonempty. Pick $z \in U$ such that $T^n z \in V$, and we are done.

Conversely, let nonempty open sets U and V be given. Pick points $x \in U$ and $y \in V$. Since U and V are open, there is an $\varepsilon > 0$ such that

$$B_\varepsilon(x) \subset U \quad and \quad B_\varepsilon(y) \subset V.$$

Let z and $n \geq 1$ be chosen such that $z \in B_\varepsilon(x)$ and $T^n z \in B_\varepsilon(y)$. Then $T^n z$ belongs to $T^n U \cap V$. ∎

If there is a **transitive point** x_0, meaning that $\mathscr{O}(x_0)$ is dense in X, then T is topologically transitive. To see this, suppose $x, y \in X$ and $\varepsilon > 0$ are given, then pick m such that $\|x - T^m x_0\| < \varepsilon$. Notice that the orbit $\mathscr{O}(T^m x_0)$ is the same as $\mathscr{O}(x_0)$ except for the first m points, and hence it is also dense (explain this). So there is another integer $n \geq 1$ such that $\|y - T^{m+n} x_0\| < \varepsilon$. So $z = T^m x_0$ does the job.

It is perhaps a surprising fact that the converse is true. We require that X be infinite just to avoid the trivial case in which X consists of a single finite orbit. The proof depends on the Baire Category Theorem, from Section 9.3.

11.5.3. THE BIRKHOFF TRANSITIVITY THEOREM.
If a mapping T is topologically transitive on an infinite closed subset of $\mathbb{R}^k$, then it has a dense set of transitive points.

PROOF. Let $\{V_n : n \geq 1\}$ be a collection of open sets with the property that every open set V contains one of these V_n. For example, let $\{x_n : n \geq 1\}$ be a dense subset of X in which every point in this set is repeated infinitely often. Then the sets $V_n = B_{1/n}(x_n)$ have this property (verify!).

For each V_n, the set $U_n = \{x \in X : T^k x \in V_n \text{ for some } k \geq 1\}$ is the union of the open sets $T^{-k}(V_n)$ for $k \geq 1$, and thus is open. Since T is topologically transitive, given any open set U, there is some $k \geq 1$ such that $T^k U \cap V_n$ is nonempty. Therefore, $U_n \cap U \neq \varnothing$, and thus U_n is dense.

Consider $R = \bigcap_{n \geq 1} U_n$. Take any point x_0 in R. For each $n \geq 1$, there is an integer k such that $T^k x_0 \in V_n$. Therefore, $\mathscr{O}(x_0)$ intersects every V_n. This shows that $\mathscr{O}(x_0)$ is dense. So R is the set of transitive points of T. Since R is the intersection of countably many dense open sets, the Baire Category Theorem (9.3.2) shows that R is dense in X. ∎

11.5.4. EXAMPLE. In Example 11.3.1, if $\alpha/2\pi$ is not rational, then the irrational rotation R_α of the circle $\mathbb{T}$ has transitive points. Hence it is topologically transitive. Indeed, every point is transitive.

11.5.5. EXAMPLE. In Example 11.3.2, the map $T\theta = 2\theta \pmod{2\pi}$ was shown to have a dense set of repelling periodic points, and we outlined how to show that it has a dense set of transitive points. This would imply that it is topologically transitive. We will verify this again directly from the definition.

Let U and V be nonempty open subsets of the circle. Then U contains an interval I of length $\varepsilon > 0$. It follows that $T^n U$ contains $T^n I$, which is an interval of length $2^n \varepsilon$. Eventually $2^n \varepsilon > 2\pi$, at which point $T^n I$ must contain the whole circle. In particular, the intersection of $T^n U$ with V is V itself.

11.5.6. EXAMPLE. Again we consider the quadratic family of logistic maps $Q_a x = a(x - x^2)$ on the unit interval I for large a. Our arguments will work for $a > 2 + \sqrt{5} \approx 4.2361$. However, more delicate arguments work for any $a > 4$.

The first thing to notice about the case $a > 4$ is that Q_a does not map I into itself. Notice that once $Q_a^k x$ is mapped outside of $[0, 1]$, it remains outside, since Q_a maps $(-\infty, 0) \cup (1, \infty)$ into $(-\infty, 0)$. We recall from Example 11.4.3 that once a point is outside $[0, 1]$, the orbit goes off to $-\infty$.

There is an open interval

$$J_1 = \{x \in [0, 1] : Q_a x > 1\}$$

centred around $x = \frac{1}{2}$. The remainder I_1 consists of two closed intervals, and each is mapped one-to-one and onto $[0, 1]$. In particular, in the middle of each of these closed intervals is an open interval that is mapped onto J_1. Hence

$$J_2 = \{x \in I_1 : Q_a^2 x > 1\}$$

is the union of these two intervals. What remains is the union of four intervals that Q_a^2 maps one-to-one and onto $[0, 1]$.

Proceeding in this way, we may define

$$I_n = \{x \in [0, 1] : Q_a^n x \in [0, 1]\}$$

and

$$J_n = \{x \in I_{n-1} : Q_a^n x > 1\}.$$

See Figure 11.8 for an example. Notice that $I_n = [0, 1] \setminus \bigcup_{k=1}^n J_k$ consists of the union of 2^n disjoint intervals and Q_a^n maps each of these intervals one-to-one and onto $[0, 1]$. We call these 2^n intervals the component intervals of I_n.

We are interested in the set

$$X_a = \{x \in [0,1] : Q_a^n x \in [0,1] \text{ for all } n \geq 1\}.$$

If $x \in X_a$, then it is clear that $Q_a x$ remains in X_a. So this set is mapped into itself, making (X_a, Q_a) a dynamical system.

From our construction, we see that $X_a = \bigcap_{n \geq 1} I_n$. In fact, this looks a lot like the construction of the Cantor set C (Example 4.4.8) and X_a has many of the same properties. By Cantor's Intersection Theorem (4.4.7), it follows that X_a is nonempty and compact. We will show that it is perfect (no point is isolated) and nowhere dense (it contains no intervals). A set with these properties is often called a **generalized Cantor set**, or sometimes just a Cantor set.

To simplify the argument, we will assume that $a > 2 + \sqrt{5} \approx 4.236$.

11.5.7. LEMMA. *If $a > 2 + \sqrt{5}$, then $c := \min_{x \in I_1} |Q_a'(x)| > 1$. Thus each of the 2^n component intervals of I_n has length at most c^{-n}.*

PROOF. The graph of Q_a is symmetric about the line $x = \frac{1}{2}$. The set I_1 consists of two intervals $[0, s]$ and $[1 - s, 1]$, where s is the smaller root of $a(x - x^2) = 1$, namely

$$s = \frac{a - \sqrt{a^2 - 4a}}{2a} = \frac{1}{2} - \frac{\sqrt{a^2 - 4a}}{2a}.$$

Note that s is a decreasing function of a for $a \geq 4$. Also, $|Q_a'(x)| = a|1 - 2x|$ is decreasing on $[0, \frac{1}{2}]$. So the minimum value is taken at s, which is

$$c = Q_a'(s) = \sqrt{a^2 - 4a} = \sqrt{(a-2)^2 - 4}.$$

This is an increasing function of a and takes the value 1 when $a^2 - 4a = 1$. Rearranging, we have $(a - 2)^2 = 5$, so that $a = 2 + \sqrt{5}$. Any larger value of a yields a value of c greater than 1.

We will verify that the intervals in I_n have length at most c^{-n} by induction. For $n = 0$, this is clear. Suppose that the conclusion is valid for $n - 1$. Notice that Q_a maps each component interval $[p, q]$ of I_n onto an interval of I_{n-1}. The Mean Value

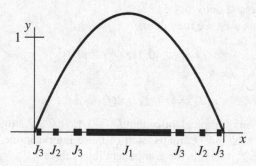

FIG. 11.8 The graph of Q_5, showing J_1, J_2, and J_3.

Theorem implies that there is a point r between p and q such that

$$\left| \frac{Q_a(q) - Q_a(p)}{q - p} \right| = |Q_a'(r)| \geq c.$$

Hence $|q - p| \leq c^{-1} |Q_a(q) - Q_a(p)| \leq c^{-1} c^{1-n} = c^{-n}$. ∎

We can apply this to any interval contained in X_a. Since it would also be an interval contained in I_n for all $n \geq 1$, it must have zero length. So X_a has no interior.

Now let x be a point in X_a. It is clear from the construction of X_a that the endpoints of each component interval of I_n belongs to X_a. (In fact, these are eventually fixed points whose orbits end at 0.) If x is not the left endpoint of one of the intervals in some I_n, let x_n be the left endpoint of the component interval of I_n that contains x. By Lemma 11.5.7, it follows that $|x - x_n| \leq c^{-n}$ and so $x = \lim_{n \to \infty} x_n$. If x happens to be a left endpoint, then use the right endpoints instead. Hence X_a is perfect. This verifies our claim that X_a is a Cantor set.

Now we are ready to establish topological transitivity.

11.5.8. PROPOSITION. *If $a > 2 + \sqrt{5}$, the quadratic map $Q_a = a(x - x^2)$ is topologically transitive on the generalized Cantor set X_a.*

PROOF. Suppose that $x, y \in X_a$ and $\varepsilon > 0$. Choose n so large that $c^{-n} < \varepsilon$, and let J be the component interval of I_n containing x. Then since J has length at most c^{-n}, it is contained in $(x - \varepsilon, x + \varepsilon)$. Now $Q_a^n J$ is the whole interval I. Pick z to be the point in J such that $Q_a^n z = y$. Since y belongs to X_a, it is clear that the orbit of z consists of a few points in $[0, 1]$ together with the orbit of y, which also remains in I. Therefore, z belongs to X_a. We have found a point z in X_a near x that maps precisely onto y via Q_a^n. Therefore, Q_a is topologically transitive on X_a. ∎

The third notion we need is the crucial one of sensitive dependence on initial conditions. Roughly, it says that for every point x we can find a point y, as close as we like to x, such that the orbits of x and y are eventually far apart. This means that no measurement of initial conditions, however accurate, can predict the long-term behaviour of the orbit of a point.

11.5.9. DEFINITION. A map T mapping X into itself exhibits **sensitive dependence on initial conditions** if there is a real number $r > 0$ such that for every point $x \in X$ and any $\varepsilon > 0$, there is a point $y \in X$ and $n \geq 1$ such that

$$\|x - y\| < \varepsilon \quad \text{and} \quad \|T^n x - T^n y\| \geq r.$$

11.5.10. EXAMPLE. Consider the circle-doubling map $T\theta \equiv 2\theta \pmod{2\pi}$ again. It is easy to see that this map has sensitive dependence on initial conditions. Indeed, let $r = 1$. For any $\varepsilon > 0$ and any $\theta \in \mathbb{T}$, pick any other point $\varphi \neq \theta$ with

$|\theta - \varphi| < \varepsilon$. Choose n such that $1 \leq 2^n|\theta - \varphi| \leq 2$. Then it is clear that

$$|T^n\theta - T^n\varphi| = 2^n|\theta - \varphi| \geq 1.$$

11.5.11. EXAMPLE. On the other hand, the rotation map R_α of the circle $\mathbb{T}$ through an angle α is rigid: $|T^n\theta - T^n\varphi| = |\theta - \varphi|$ for all $n \geq 1$. So this map is not sensitive to initial conditions.

11.5.12. PROPOSITION. *When $a > 2 + \sqrt{5}$, then the quadratic logistic map $Q_a x = a(x - x^2)$ exhibits sensitive dependence on initial conditions on the generalized Cantor set X_a.*

PROOF. Set $r = \frac{1}{2}$. Given $x \in X_a$ and $\varepsilon > 0$, we find as before an integer n and a component interval J of I_n that is contained in $(x - \varepsilon, x + \varepsilon)$. Then Q_a^n maps J one-to-one and onto $[0, 1]$. In particular, the two endpoints y and z of J are mapped to 0 and 1. So

$$|Q_a^n z - Q_a^n x| + |Q_a^n x - Q_a^n y| = 1.$$

So $\max\left\{ |Q_a^n z - Q_a^n x|, |Q_a^n x - Q_a^n y| \right\} \geq \frac{1}{2}$ as desired.   ∎

Now we can define chaos.

11.5.13. DEFINITION. We call (X, T) a **chaotic dynamical system** if

(1) The set of periodic points is dense in X.
(2) T is topologically transitive on X.
(3) T exhibits sensitive dependence on initial conditions.

This definition demands lots of wild behaviour. In order for the periodic points to be dense, there need to be infinitely many distinct finite orbits. The existence of transitive points already means that orbits are distributed everywhere throughout X. Sensitive dependence on initial conditions means that orbits that start out nearby can be expected to diverge eventually.

These notions are interrelated. For any infinite metric space, the conditions of dense periodic points and topological transitivity together imply sensitive dependence on initial conditions. The proof is elementary but delicate; see [38]. However, (2) and (3) do not imply (1), nor do (1) and (3) imply (2). But if the space X is an interval in $\mathbb{R}$, then (2) implies both (1) and (3); a simple proof of this result is given in [43]. Some authors drop condition (1), arguing that it is the other two conditions that are paramount.

11.5.14. EXAMPLE. We have shown that the circle-doubling map has a dense set of periodic points in Example 11.3.2. In Example 11.5.5, it was shown to be topologically transitive. And in Example 11.5.10, it was seen to have sensitive dependence on initial conditions. Hence this system is chaotic.

11.5.15. EXAMPLE. The quadratic family $Q_a x = a(x - x^2)$ of logistic maps is chaotic for $a > 2 + \sqrt{5}$. Indeed, Proposition 11.5.8 established topological transitivity and Proposition 11.5.12 established sensitive dependence on initial conditions. In Example 11.4.5, it was established that Q_a has orbits of period 3. Hence by Sharkovskii's Theorem (11.4.9), there are orbits of every possible period. But this does not show that they are dense.

It suffices to show that each component interval J of I_n contains periodic points, since as we have argued before, every interval $(x - \varepsilon, x + \varepsilon)$ contains such an interval. Now Q_a^n maps J onto I, which contains J. Therefore, by Lemma 11.4.2, there is a point $y \in J$ that is a fixed point for Q_a^n. So y is a periodic point (whose period is a divisor of n). Moreover, y must belong to X_a, since the whole orbit of y remains in $[0, 1]$. It follows that periodic points are dense in X_a and that Q_a is chaotic.

In fact, all of this analysis remains valid for $a > 4$. But because the Mean Value Theorem argument based on Lemma 11.5.7 is no longer valid, the proof is different.

For our last example in this section, we will do a complete proof of chaos for a new system that will be useful in the next section for understanding the relationship between the quadratic maps Q_a for large a.

11.5.16. EXAMPLE. Recall from Example 4.4.8 that the middle thirds Cantor set C can be described as the set of all points x in $[0, 1]$ that have a ternary expansion (base 3) using only 0's and 2's. It is a compact set that is nowhere dense (contains no intervals) and perfect (has no isolated points). It was constructed by removing, in succession, the middle third of each interval remaining at each stage. The endpoints of the removed intervals belong to C and consist of those points that have two different ternary expansions. However, only one of these expansions consists of 0's and 2's alone.

Define the shift map on the Cantor set C by

$$Sy = 3y \,(\text{mod } 1) = (0.y_2 y_3 y_4 \ldots)_{\text{base } 3} \quad \text{for} \quad y = (0.y_1 y_2 y_3 \ldots)_{\text{base } 3} \in C.$$

It is easy to see that

$$Sy = \begin{cases} 3y & \text{for } y \in C \cap [0, 1/3], \\ 3y - 2 & \text{for } y \in C \cap [2/3, 1]. \end{cases}$$

It follows that S is a continuous map. Moreover, the range is contained in C since every point in the image has a ternary expansion with only 0's and 2's. Clearly, S maps each of the sets $C \cap [0, 1/3]$ and $C \cap [2/3, 1]$ bijectively onto C.

Let us examine the dynamics of the shift map, starting with periodic points. A moment's reflection shows that $S^n y = y$ if and only if $y_{k+n} = y_k$ for all $k \geq 1$. That is, y has period n exactly when the ternary expansion of y is periodic of period n. There are precisely 2^n points such that $S^n y = y$. Indeed, the first n ternary digits $a_1, \ldots, a_n$ form an arbitrary finite sequence of 0's and 2's, and this forces

$$y = (0.a_1 \ldots a_n a_1 \ldots a_n a_1 \ldots a_n \ldots)_{\text{base3}}$$

$$= \sum_{k=1}^{n} a_k 3^{-k} \left(1 + 3^{-n} + 3^{-2n} + \cdots\right) = \frac{1}{1 - 3^{-n}} \sum_{k=1}^{n} a_k 3^{-k}.$$

From this, it is evident that the set of periodic points is dense in C. Indeed, given $y = (0.y_1 y_2 y_3 \ldots)_{\text{base3}}$ in C and $\varepsilon > 0$, choose N so large that $3^{-N} < \varepsilon$. Then let x be the periodic point determined by the sequence $y_1, \ldots, y_N$. Then x and y both belong to the interval $[(0.y_1 y_2 \ldots y_N)_{\text{base3}}, (0.y_1 y_2 \ldots y_N)_{\text{base3}} + 3^{-N}]$, which has length 3^{-N}. Hence $|x - y| \leq 3^{-N} < \varepsilon$.

The set of aperiodic points that are eventually fixed is also dense. Indeed, the points in C that are eventually mapped to 0 are exactly those with a finite ternary expansion, namely

$$y = (0.y_1 \ldots y_n)_{\text{base3}} = (0.y_1 \ldots y_n 000 \ldots)_{\text{base3}}.$$

Next we will show that the set of transitive points is dense. The hard part is to describe one such point. List all finite sequences of 0's and 2's by first listing all sequences of length 1 in increasing order, then those of length 2, and so on:

0, 2,   00, 02, 20, 22,   000, 002, 020, 022, 200, 202, 220, 222,
0000, 0002, 0020, 0022, 0200, 0202, 0220, 0222, ...

String them all together to give the infinite ternary expansion of a point:

$$a = (0.0200022022000002020022200202220222000000200200022 \ldots)_{\text{base3}}.$$

Suppose y is any point in C and $\varepsilon > 0$ is given. Pick an integer N such that $3^{-N} < \varepsilon$. Somewhere in the expansion of a we can find the first N digits of y in sequence, say starting in the $(p+1)$st place of a. Then $S^p a$ starts with these same N digits. Hence

$$|y - S^p a| \leq 3^{-N} < \varepsilon.$$

To see that the transitive points are dense, notice that if $S^N x = a$, then x is also transitive. So let x be the point beginning with the first N digits of y followed by the digits of a. Then x is transitive. As before, we obtain that $|x - y| < \varepsilon$.

Finally, we need to verify that S has sensitive dependence on initial conditions. This is easy. Let $r = 1/4$. If x and $\varepsilon > 0$ are given, choose $N > 1$ such that $3^{-N} < \varepsilon$. Let y be the point in C obtained by changing the ternary expansion of x only in the Nth digit from a 0 to a 2, or vice versa. Then $|x - y| < \varepsilon$. Also $S^{N-1} x$ and $S^{N-1} y$ differ in the first ternary digit. So one is in T_0 and the other in T_1. In particular,

$$\left| S^{N-1} x - S^{N-1} y \right| \geq \frac{1}{3} > r.$$

We conclude that the shift map S is chaotic.

Exercises for Section 11.5

A. If T is topologically transitive on X, show that X either is infinite or consists of a single orbit.

B. Consider the tent map of Exercise 11.3.E, which has a dense set of periodic points.

(a) What is the slope of $T^n(x)$? Use this to establish sensitive dependence on initial conditions.
(b) Show that T^n maps each interval $[k2^{-n}, (k+1)2^{-n}]$ onto $[0,1]$. Use this to establish topological transitivity.
(c) Hence conclude that the tent map is chaotic.

C. Consider the **big tent map** $Sx = \begin{cases} 3x & \text{for } x \leq \frac{1}{2}, \\ 3(1-x) & \text{for } x \geq \frac{1}{2}. \end{cases}$

(a) Sketch the graphs of S, S^2, and S^3.
(b) What are the dynamics for points outside of $[0,1]$?
(c) Describe the set $I_n = \{x \in [0,1] : S^n x \in [0,1]\}$.
(d) Describe the set $X = \bigcap_{n \geq 1} I_n$.
(e) Show that T^n has exactly 2^n fixed points, and they all belong to X. Hence show that the periodic points are dense in X.
(f) Show that S is chaotic on X.   HINT: Use the idea of the previous exercise.

D. Let $f_\infty : [0,1] \to [0,1]$ be the function constructed in Exercise 11.4.G.

(a) Show that the middle thirds Cantor set C (Example 4.4.8) is mapped into itself by f_∞.
HINT: Let S_n denote the nth stage in the construction of C, consisting of 2^n intervals of length 3^{-n}. Show that $f_\infty(S_n) = S_n$.
(b) Show that if x is not periodic for f_∞, then $f_\infty^k(x)$ eventually belongs to each S_n. Hence the distance from $f_\infty^k(x)$ to C tends to zero.
(c) Show that there are no periodic points in C.
(d) Show that f_∞ permutes the 2^n intervals of S_n in a single cycle, so that the orbit of a point $x \subset S_n$ intersects all 2^n of these intervals.
(e) Use (d) to show that the orbit of every point in C is dense in C. In particular, f_∞ is topologically transitive on C.
(f) Use (d) to show that f_∞ does *not* have sensitive dependence on initial conditions.

11.6 Topological Conjugacy

In this section, we will discuss how to show that two dynamical systems, possibly on different spaces, are essentially the same. By *essentially the same*, we mean that they have the same dynamical system properties. It is convenient to introduce two new notions that allow us to express the fact that two dynamical systems are the same map up to a reparametrization.

The notion of homeomorphism encodes the fact that two spaces have the same topology, meaning roughly that convergent sequences correspond but distances between points need not correspond.

11.6.1. DEFINITION. Two subsets of normed vector spaces X and Y are said to be **homeomorphic** if there is a continuous, one-to-one, and onto map $\sigma : X \to Y$ such that the inverse map σ^{-1} is also continuous. The map σ is called a **homeomorphism**.

11.6.2. EXAMPLE. Let f be a continuous map from $[0,1]$ into itself, and consider when this is a homeomorphism. To be onto, there must be points a and b such that $f(a) = 0$ and $f(b) = 1$. By the Intermediate Value Theorem (5.6.1), f maps $[a,b]$ onto $[0,1]$. If $[a,b]$ were a proper subset of $[0,1]$, then the remaining points would have to be mapped somewhere and f would fail to be one-to-one. Hence we have either $f(0) = 0$ and $f(1) = 1$ or $f(0) = 1$ and $f(1) = 0$. For convenience, let us assume that it is the former for a moment. By the same token, f must be strictly increasing. Indeed, if there were $x < y$ such that $f(y) \le f(x)$, then the Intermediate Value Theorem again yields a point z such that $0 \le z \le x$ and $f(z) = f(y)$, destroying the one-to-one property.

Conversely, if f is a continuous strictly increasing function such that $f(0) = 0$ and $f(1) = 1$, then the same argument shows that f is one-to-one and onto. So the inverse function f^{-1} is well defined. Moreover, it is evident that f^{-1} is also strictly increasing and maps $[0,1]$ onto itself. By Theorem 5.7.6, f^{-1} is also continuous. Therefore, f is a homeomorphism of $[0,1]$.

Likewise, if f is a continuous strictly decreasing function such that $f(0) = 1$ and $f(1) = 0$, then it is a homeomorphism.

This example makes it look as though the order on the real line is crucial to establishing the continuity of the inverse. However, this result is actually more basic and depends crucially on compactness. In fact, for a bijection $f : X \to Y$ between compact subsets of $\mathbb{R}^d$ (or indeed, compact subsets of any metric space), the continuity of f implies the continuity of f^{-1}—see Exercise 5.4.L.

This result is also true if X and Y are compact subsets of a normed vector space, with the same proof; all we need do is show that each of the theorems and lemmas in the proof holds for any normed vector space.

11.6.3. EXAMPLE. Let X be a generalized Cantor set in $\mathbb{R}$ and let C be the standard middle thirds Cantor set, both given as the intersection of sets I_n and S_n, respectively, which are the disjoint union of 2^n intervals with lengths tending to zero: $X = \bigcap_{n \ge 0} I_n$ and $C = \bigcap_{n \ge 0} S_n$, where each component interval of I_n contains two component intervals of I_{n+1}. We shall show that X is homeomorphic to C. Moreover, this homeomorphism may be constructed to be monotone increasing.

For brevity, let Z be the space of all sequences with entries in $\{0,2\}$.

We label the component intervals of S_n as in Figure 11.9. A component interval of S_n is denoted by a finite sequence of 0's and 2's. When it is split into two intervals of S_{n+1} by removing the middle third, the new intervals are labeled by adding a 0 to the label of the first interval and a 2 to the second. So, for example, when $T_{202} = [20/27, 7/9]$ is split, we label the new intervals as $T_{2020} = [20/27, 31/81]$ and $T_{2022} = [32/81, 7/9]$. The formula is more transparent in base 3:

$$T_{2020} = [0.2020_{\text{base }3}, 0.2021_{\text{base }3}] \quad \text{and} \quad T_{2022} = [0.2022_{\text{base }3}, 0.2100_{\text{base }3}].$$

So the label $\alpha_1 \dots \alpha_n$ specifies the first digits in the ternary expansion of the points in the interval $T_{\alpha_1 \dots \alpha_n}$.

$$T_{00} = [0, \tfrac{1}{9}] \qquad T_{000} = [0, \tfrac{1}{27}]$$

$$T_0 = [0, \tfrac{1}{3}] \qquad\qquad T_{002} = [\tfrac{2}{27}, \tfrac{1}{9}]$$

$$T_{02} = [\tfrac{2}{9}, \tfrac{1}{3}] \qquad T_{020} = [\tfrac{2}{9}, \tfrac{7}{27}]$$

$$T_{022} = [\tfrac{8}{27}, \tfrac{1}{3}]$$

$$T_{200} = [\tfrac{2}{3}, \tfrac{19}{27}]$$

$$T_{20} = [\tfrac{2}{3}, \tfrac{7}{9}] \qquad T_{202} = [\tfrac{20}{27}, \tfrac{7}{9}]$$

$$T_2 = [\tfrac{2}{3}, 1] \qquad\qquad T_{220} = [\tfrac{8}{9}, \tfrac{25}{27}]$$

$$T_{22} = [\tfrac{8}{9}, 1] \qquad T_{222} = [\tfrac{26}{27}, 1]$$

FIG. 11.9 Component intervals of each of S_0, S_1, and S_2.

Recall that each point y of C is determined by the sequence of component intervals of S_n that contains it. Indeed, a typical point of C is given in base 3 as

$$y = (0.y_1 y_2 y_3 \dots)_{\text{base } 3} = \sum_{k \geq 1} y_k 3^{-k},$$

where (y_k) is a sequence of 0's and 2's, i.e., an element of Z. Thus, we have a bijection between C and Z. Further, y belongs to the intervals $T_{y_1 y_2 \dots y_n}$ for each $n \geq 1$, and

$$\bigcap_{n \geq 1} T_{y_1 y_2 \dots y_n} = \{y\}.$$

Indeed, since the length of the intervals goes to zero, the intersection can contain at most one point. It is easy to show that the one point must be y.

We now describe X in the same manner. Let the interval components of I_n be denoted by $J_{\alpha_1 \alpha_2 \dots \alpha_n}$ for each finite sequence $\alpha_1 \alpha_2 \dots \alpha_n$ of 0's and 2's. When this interval is split into two parts by removing an open interval from the interior, the leftmost remaining interval will be denoted by $J_{\alpha_1 \alpha_2 \dots \alpha_n 0}$ and the rightmost by $J_{\alpha_1 \alpha_2 \dots \alpha_n 2}$. By hypothesis, each interval $J_{\alpha_1 \alpha_2 \dots \alpha_n}$ is nonempty, and the lengths tend to 0 as n goes to $+\infty$.

For each sequence $\mathbf{a} = (\alpha_k)_{n=1}^{\infty}$ of 0's and 2's in Z, define a point $x_{\mathbf{a}}$ in X by

$$\{x_{\mathbf{a}}\} = \bigcap_{n \geq 1} J_{\alpha_1 \alpha_2 \dots \alpha_n}.$$

Since the lengths of the intervals tend to 0, the intersection may contain at most one point. On the other hand, using compactness, Cantor's Intersection Theorem (4.4.7) guarantees that this intersection is nonempty. So it consists of a single point, say $x_{\mathbf{a}}$.

Conversely, each point x in X determines a unique sequence $\mathbf{a} = (\alpha_k)_{n=1}^{\infty}$ in Z because there is exactly one component interval of I_n containing x, which we denote by $J_{\alpha_1 \alpha_2 \dots \alpha_n}$. So there is a bijective correspondence between X and Z.

Define a function τ from X to C by composing the bijection from X to Z with that from Z to C. That is,

$$\tau(x_{\mathbf{a}}) = \sum_{k \geq 1} \frac{2\alpha_k}{3^k}.$$

As a composition of bijections, τ is a bijection.

Next we establish the continuity of τ. Let $x_{\mathbf{a}} \in X$ and $\varepsilon > 0$ be given. Choose N so large that $3^{-N} < \varepsilon$. Now $x_{\mathbf{a}}$ belongs to $J_{\alpha_1 \alpha_2 \ldots \alpha_N}$. Note that $J_{\alpha_1 \alpha_2 \ldots \alpha_N}$ and $I_N \setminus J_{\alpha_1 \alpha_2 \ldots \alpha_N}$ are disjoint closed sets. Let δ be the positive distance between them. Suppose that $x \in X$ and $|x - x_{\mathbf{a}}| < \delta$. Then x also belongs to $J_{\alpha_1 \alpha_2 \ldots \alpha_N}$. Hence $\tau(x)$ belongs to $T_{\alpha_1 \alpha_2 \ldots \alpha_N}$. This is an interval of length 3^{-N} containing $\tau(x_{\mathbf{a}})$ as well. Hence

$$|\tau(x) - \tau(x_{\mathbf{a}})| \leq 3^{-N} < \varepsilon.$$

Finally, we invoke Exercise 5.4.L to conclude that τ is a homeomorphism. Alternatively, the continuity of τ^{-1} can be proved in the same way as for τ. Note that the map τ preserves the order on the 2^n intervals in I_n for every n. It follows easily that τ is monotone increasing.

Now we study those homeomorphisms between two spaces that carry a dynamical system on one space to a different system on the other.

11.6.4. DEFINITION. Let (X, S) and (Y, T) be dynamical systems. They are said to be **topologically conjugate** if there is a homeomorphism σ from X onto Y such that $\sigma S = T \sigma$, or equivalently, $T = \sigma S \sigma^{-1}$. The map σ is called a **topological conjugacy** between S and T.

It is clear that if σ is a topological conjugacy between S and T, then

$$\sigma S^n = T^n \sigma \quad \text{for all} \quad n \geq 1.$$

Hence if $x \in X$ is a periodic point for S with period n, then $y = \sigma(x)$ will be periodic for T of the same order. Moreover, the fact that σ is a homeomorphism means that convergent sequences in X correspond exactly to convergent sequences in Y under this map. Hence a periodic point $y = \sigma(x)$ will be attracting or repelling exactly as x is. Indeed, we have $\mathscr{O}_T(\sigma(x)) = \sigma(\mathscr{O}_S(x))$ for every point $x \in X$.

Topological conjugacy is an equivalence relation. First, if σ conjugates S onto T, then σ^{-1} conjugates T back onto S. If R is conjugate to S and S is conjugate to T, then R and T are conjugate. Evidently, the identity map $\mathrm{id} : S \to S$ is a topological conjugacy from S to itself.

We will study topological conjugacy by examining a few examples.

11.6.5. PROPOSITION. *The tent map (Exercise 11.3.E) and the quadratic map $Q_4 x = 4x - 4x^2$ on $[0, 1]$ are topologically conjugate. So Q_4 is chaotic on $[0, 1]$.*

PROOF. We will pull the appropriate homeomorphism out of the air. So the rest of this proof will be easy to understand, but it won't explain how to choose the homeomorphism.

Let $\sigma(x) = \sin^2(\frac{\pi}{2}x)$. It is easily checked that σ is strictly increasing and continuous on $[0,1]$ and that $\sigma(0) = 0$ and $\sigma(1) = 1$. Hence by Example 11.6.2, it follows that σ is a homeomorphism of $[0,1]$. Now using $Q_4x = 4x(1-x)$, compute

$$Q_4\sigma(x) = 4\sin^2(\tfrac{\pi}{2}x)\cos^2(\tfrac{\pi}{2}x) = \sin^2(\pi x).$$

Likewise,

$$\sigma(Tx) = \begin{cases} \sin^2\big(\tfrac{\pi}{2}(2x)\big) = \sin^2(\pi x) & \text{if } 0 \le x \le \tfrac{1}{2}, \\ \sin^2\big(\tfrac{\pi}{2}(2-2x)\big) = \sin^2(\pi - \pi x) & \text{if } \tfrac{1}{2} \le x \le 1. \end{cases}$$

Thus $\sigma T = Q_4\sigma$; and so σ is a topological conjugacy intertwining T and Q_4.

By Exercise 11.5.B, the tent map is chaotic. Hence Q_4 is also. ■

The goal of the rest of this section is to continue our analysis of the quadratic maps Q_a for $a > 2+\sqrt{5}$. We will establish a topological equivalence with the shift on the Cantor set. Hence these quadratic maps are all topologically equivalent to each other, so that dynamically they all behave in exactly the same way.

11.6.6. THEOREM. *For $a > 2+\sqrt{5}$, the quadratic maps Q_a on the set X_a is topologically conjugate to the shift S on the Cantor set C.*

PROOF. Recall from Example 11.5.6 that X_a is a generalized Cantor set. We will construct a homeomorphism along the lines of Example 11.6.3, except that the ordering will be determined by the dynamics rather than by the usual order on the line.

Recall the notation from Example 11.5.6. The first step in constructing X_a is the set

$$I_1 = J_0 \cup J_1 = \{x \in [0,1] : Q_a x \in [0,1]\}.$$

For each point x in X_a, $Q_a^{n-1}x$ belongs to X_a, and thus to either J_0 or J_1. Define the **itinerary** of x to be the sequence $\Gamma x = \gamma_1 \gamma_2 \ldots$ of 0's and 1's defined by the condition

$$Q_a^{n-1}x \in J_{\gamma_n} \quad \text{for all} \quad n \ge 1.$$

The interval $J_{\alpha_1 \ldots \alpha_{n-1}}$ is mapped bijectively by Q_a^{n-1} onto the whole unit interval. And $X_a \cap J_{\alpha_1 \ldots \alpha_{n-1}}$ is mapped into X_a, and in particular into $I_1 = J_0 \cup J_1$. This dichotomy determines the sets $J_{\alpha_1 \ldots \alpha_{n-1}0}$ and $J_{\alpha_1 \ldots \alpha_{n-1}1}$, as one is mapped onto J_0 by Q_a^{n-1} and the other is mapped onto J_1. The order we need to keep track of is this itinerary ordering, not the usual order on $\mathbb{R}$. This discussion shows that if x and y both belong to $X_a \cap J_{\alpha_1 \ldots \alpha_n}$, then the itineraries Γx and Γy agree for the first n terms.

Define a map σ from X_a to C by

$$\sigma(x) = y_{\Gamma x} := \sum_{k \geq 1} \frac{2\gamma_k}{3^k}.$$

We will verify that σ is a homeomorphism.

For any $x \in X_a$ and $\varepsilon > 0$, choose N such that $3^{-N} < \varepsilon$. Then x belongs to one of the Nth-level intervals $J_{\alpha_1 \ldots \alpha_N}$. Let δ be the positive distance between this interval and the remaining $I_N \setminus J_{\alpha_1 \ldots \alpha_N}$. Then any $y \in X_a$ with $|x - y| < \delta$ also belongs to $J_{\alpha_1 \ldots \alpha_N}$. Hence the itinerary of y agrees with x for the first N terms. This means that $\sigma(x)$ and $\sigma(y)$ belong to the same Nth-level interval for C. Hence

$$|\sigma(x) - \sigma(y)| \leq 3^{-N} < \varepsilon.$$

So σ is continuous.

To see that σ is a bijection, consider any point $y \in C$. As usual, we write $y = 0.y_1 y_2 \ldots_{\text{base 3}}$ in ternary using a sequence of 0's and 2's. This is the image of all points $x \in X_a$ with itinerary $\Gamma = \gamma_1 \gamma_2 \ldots$ given by $\gamma_k = y_k/2$. However, Γ determines another unique sequence $\mathbf{a} = \alpha_1 \alpha_2 \ldots$ by the relation

$$Q_a^{n-1} J_{\alpha_1 \ldots \alpha_n} = J_{\gamma_n} \quad \text{for all} \quad n \geq 1.$$

So the points x with itinerary Γ are the points in

$$\bigcap_{n \geq 1} J_{\alpha_1 \ldots \alpha_n}.$$

As we have noted before, this intersection consists of exactly one point; call it $x_{\mathbf{a}}$. So σ is onto because this set is nonempty for each Γ, and it is one-to-one because the set is always a singleton.

Now we may apply Exercise 5.4.L to see that σ is a homeomorphism.

We must show that σ intertwines Q_a and S. Suppose that $x \in X_a$ has itinerary $\Gamma = \gamma_1 \gamma_2 \gamma_3 \ldots$. Then the itinerary of $Q_a x$ is evidently $\gamma_2 \gamma_3 \gamma_4 \ldots$ because

$$Q_a^{n-1} Q_a x = Q_A^n x \in J_{\gamma_{n+1}} \quad \text{for all} \quad n \geq 1.$$

This is just saying that

$$\sigma(Q_a x) = S\sigma(x).$$

Therefore, Q_a is topologically conjugate to the shift. ∎

11.6.7. COROLLARY. *The quadratic maps Q_a for $a > 2 + \sqrt{5}$ have a dense set of transitive points in X_a.*

PROOF. This follows from our discussion of the shift in Example 11.5.16. The shift has a dense set of transitive points. So any map topologically conjugate to the shift must have such a set as well. ∎

11.6.8. REMARK. We have been studying the quadratic logistic maps in detail throughout this chapter. Our early arguments depended on specific calculations for these functions. However, all of the arguments for chaos depend only on a few fairly general properties.

Suppose that f is a function on $[0,1]$ with $f(0) = f(1) = 0$ that is **unimodal**, meaning that f increases to a maximum at a point (x_0, y_0) and then decreases back down to $(1,0)$. In order for our arguments to work, all we need is that $y_0 > 1$ and $|f'(x)| \geq c > 1$ for all x in

$$I_1 = \{x \in [0,1] : f(x) \in [0,1]\}.$$

Then the argument of Example 11.5.6 would apply to show that the set

$$X = \{x \in [0,1] : f^n(x) \in [0,1] \text{ for all } n \geq 1\}$$

is a Cantor set.

The preceding proof showing that Q_a is topologically conjugate to the shift relied only on the fact that $Q_a^n J = [0,1]$ for every component interval of the nth-level set I_n for each $n \geq 1$. It is easy to see that this property also holds in the generality of the unimodal function f. Thus, in particular, f is chaotic on X. This enables us to recognize chaos in many situations.

A simple example of this that is very similar to the quadratic family is the function $f(x) = 3x - 3x^3$. However, it is more instructive to look at the quadratic maps again.

Consider the graph of $Q_{3.75}^2$ in Figure 11.10. Notice that $Q_{3.75}$ has a fixed point at $1 - 1/3.75 = \frac{11}{15}$; and $Q_{3.75}\frac{4}{11} = \frac{11}{15}$ as well. On the interval $J = [\frac{4}{15}, \frac{11}{15}]$, $Q_{3.75}^2$ decreases from $(\frac{4}{15}, \frac{11}{15})$ to a local minimum at $(\frac{1}{2}, \frac{225}{1024})$, and increases again to the point $(\frac{11}{15}, \frac{11}{15})$. Since $\frac{225}{1024} \approx 0.22 < 0.267 \approx \frac{4}{15}$, this graph "escapes" the square $J \times J$. The qualitative behaviour of this portion of the graph is just like that of Q_a for $a > 2 + \sqrt{5}$.

To verify that our proof applies, we need to see that the absolute value of the derivative is greater than 1 on

$$J_1 = \{x \in J : Q_{3.75}^2(x) \in J\}.$$

It is notationally easier to use the generic parameter a and substitute 3.75 for a later. To compute J_1, we first solve the quadratic $Q_a(x) = \frac{1}{a}$. This has solutions

$$\frac{1}{2} \pm \frac{\sqrt{a^2 - 4}}{2a}.$$

These are roughly 0.077 and 0.923 for $a = 3.75$. The points that Q_a^2 maps to the endpoints of J_1 are seen from the graph to be solutions of

$$\frac{1}{2} + \frac{\sqrt{a^2 - 4}}{2a} = Q_a x = ax - ax^2.$$

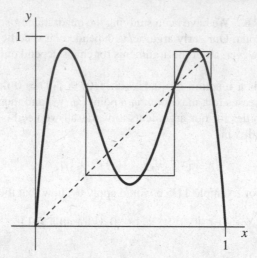

FIG. 11.10 The graph of $Q^2_{3.75}$, with $J \times J$ and $K \times K$ marked.

This quadratic has solutions

$$x_\pm := \frac{1}{2} \pm \frac{\sqrt{a^2 - 2a - 2\sqrt{a^2 - 4}}}{2a}.$$

For $a = 3.75$, we obtain $J_1 = [0.2667, 0.4377] \cup [0.5623, 0.7333]$. Now the derivative of Q^2_a is monotone increasing on J_1, changing sign at $x = 0.5$, and is symmetric about the midpoint. So the minimal slope is obtained at the two interior endpoints

$$(Q^2_a)'(x_+) = Q'_a(Q_a x_+)\, Q'_a(x_+) = a^2(1 - 2Q_a x_+)(1 - 2x_+)$$

$$= a^2 \frac{\sqrt{a^2 - 4}}{a} \frac{\sqrt{a^2 - 2a - 2\sqrt{a^2 - 4}}}{a}$$

$$= \sqrt{(a^2 - 4)(a^2 - 2a - 2\sqrt{a^2 - 4})} \approx 1.482.$$

Thus our earlier arguments apply to show that there is a Cantor set X contained in J on which $Q^2_{3.75}$ acts chaotically, and is topologically conjugate to the shift map.

Now $Q_{3.75}$ maps J_1 onto the interval $K = [.733, .923]$. The restriction of $Q^2_{3.75}$ to $K \times K$ behaves in exactly the same way, and there is another Cantor set Y on which $Q^2_{3.75}$ behaves chaotically. Moreover, $Q_{3.75}$ maps the Cantor set X into Y and vice versa. From this, it is not difficult to see that $Q_{3.75}$ acts chaotically on the Cantor set $X \cup Y$. See the exercises.

Just as Q_a is actually chaotic for $a > 4$ with a more delicate proof, the same is true for this analysis of Q^2_a. It can be shown that the preceding argument works whenever the graph in the interval $J \times J$ escapes in the middle. This occurs at about $a = 3.6786$. So once $a > 3.6786$, the quadratic map Q_a is chaotic on a Cantor set.

Exercises for Section 11.6

A. Consider the map of $[0, 2\pi)$ onto the circle $\mathbb{T}$ by wrapping around exactly once. Show that this map is continuous, one-to-one, and onto. Show that this is *not* a homeomorphism. Explain why this does not contradict Exercise 5.4.L.

B. Show that the circle $\mathbb{T}$ is not homeomorphic to the interval $[0, 1]$. HINT: If σ maps $\mathbb{T}$ onto $[0, 1]$, show that there are at least two points mapping to each $0 < y < 1$.

C. Map the circle to itself by $T\theta \equiv \theta + \frac{2\pi}{n} + \varepsilon \sin(2\pi n x)$, where $0 < \varepsilon < 1/2\pi n$.

(a) Compute $T'(\theta)$ and deduce that T is a homeomorphism.
(b) Show that 0 and $\frac{\pi}{n}$ are periodic points.
(c) If $x \notin \mathcal{O}(0)$, prove that $\text{dist}(T^k x, \mathcal{O}(\frac{\pi}{n}))$ is strictly decreasing.
(d) Show that $\mathcal{O}(0)$ is a repelling orbit and $\mathcal{O}(\frac{\pi}{n})$ is attracting, and that $\omega(x) = \mathcal{O}(\frac{\pi}{n})$ except for $x \in \mathcal{O}(0)$.

D. Show that $f(x) = 1 - 2|x|$ and $g(x) - 1 - 2x^2$ as dynamical systems on $[-1, 1]$ are topologically conjugate as follows:

(a) If φ is a homeomorphism of $[-1, 1]$ such that $\varphi(f(x)) = g(\varphi(x))$, show that φ is an odd function such that $\varphi(-1) = -1$ and $\varphi(0) = 0$.
(b) Use fixed points to show that $\varphi(1/3) = 1/2$. Deduce that $\varphi(2/3) = \sqrt{3}/2$.
(c) Guess a trig function with the properties of φ and verify that it works.

E. Let f be a homeomorphism of $[0, \infty)$ with no fixed points in $(0, \infty)$.

(a) Show that f is strictly monotone increasing, and either $f(x) < x$ or $f(x) > x$ for all $x > 0$.
(b) If $f(x) > x$, show the orbit of x converges to $+\infty$ and its orbit under f^{-1} converges to 0.
(c) Show that $(0, \infty)$ is the disjoint union of the intervals $[f^k(1), f^{k+1}(1))$ for $k \in \mathbb{Z}$.
(d) Let f and g be two homeomorphisms of $[0, \infty)$ with no fixed points in $(0, \infty)$. When are they topologically conjugate? HINT: If $f(x) > x$ and $g(x) > x$ for all $x > 0$, define φ from $[1, f(1))$ onto $[1, g(1))$. Extend this to the whole interval to obtain a conjugacy.

F. (a) Show that every quadratic function $p(x) = ax^2 + bx + c$ on $\mathbb{R}$ is topologically conjugate to some $q(x) = x^2 + d$. HINT: Use a linear map $\tau(x) = mx + e$. Compute $p(\tau(x))$ and $\tau(q(x))$ and equate coefficients to solve for m, e, and d.
(b) For which values of d is $q(x) = x^2 + d$ topologically conjugate to one of the logistic maps Q_a for $a > 0$? What are the dynamics of q when d is outside this range?

G. Suppose that $T : I \to I$ is given, and T^2 maps an infinite compact subset X into itself and is chaotic on X. Show that T is chaotic on $X \cup TX$.

11.7 Iterated Function Systems

An **iterated function system**, or **IFS**, is a multivariable discrete dynamical system. Under reasonable hypotheses, these systems have a unique compact invariant set. This invariant set exhibits certain self-similarity properties. Such sets have become known as **fractals**.

We begin with a finite set $\mathscr{T} = \{T_1, \ldots, T_r\}$ of contractions on a closed subset X of $\mathbb{R}^n$. This family of maps determines a multivariable dynamical system. The orbit of a point x will consist of the set of all points obtained by repeated application of the maps T_i in any order with arbitrary repetition. That is, for each finite *word*

$i_1 i_2 \ldots i_k$ over the alphabet $\{1, \ldots, r\}$, the point $T_{i_1} T_{i_2} \ldots T_{i_k} x$ is in the orbit $\mathscr{O}(x)$. We wish to find a compact set A with the property that

$$A = T_1 A \cup T_2 A \cup \cdots \cup T_r A.$$

Surprisingly this set turns out to be unique!

A **similitude** is a map T that is a scalar multiple of an isometry. That is, there is a constant $r > 0$ such that $\|Tx - Ty\| = r\|x - y\|$ for all $x, y \in X$. Such maps are obtained as compositions of rotations, translations, and scalings (see Exercise 11.7.F). In particular, these maps are similarities in the geometric sense that they map sets to similar sets and so preserve shape, up to a scaling factor.

In the event that each T_i is a similitude, the fact that $A = T_1 A \cup T_2 A \cup \cdots \cup T_r A$ means that each $T_i A$ is similar to A. This will be especially evident in examples in which these r sets are disjoint. The process repeats and $T_i A$ can be decomposed as $T_i A = T_i T_1 A \cup T_i T_2 A \cup \cdots \cup T_i T_r A$. After k steps, A is decomposed into r^k similar pieces. This symmetry property is called **self-similarity** and is characteristic of fractals arising from iterated function systems.

11.7.1. EXAMPLES.

(1) Let $X = \mathbb{R}^2$, and consider three affine maps $T_1 \mathbf{x} = \frac{1}{2}\mathbf{x}$, $T_2 \mathbf{x} = \frac{1}{2}\mathbf{x} + (2, 0)$, and $T_3 \mathbf{x} = \frac{1}{2}\mathbf{x} + (1, \sqrt{3})$. Notice that each T_i is a similitude with scaling factor $\frac{1}{2}$. It is easy to verify that the fixed points of these three maps are $\mathbf{v}_1 = (0, 0)$, $\mathbf{v}_2 = (4, 0)$, and $\mathbf{v}_3 = (2, 2\sqrt{3})$, respectively.

Let Δ be the solid equilateral triangle with these three vertices. A computation shows that $T_i \Delta$ for $i = 1, 2, 3$ are the three equilateral triangles with half the dimensions of the original that lie inside Δ and share the vertex $\mathbf{v}_i$ with Δ. So $\Delta_1 = T_1 \Delta \cup T_2 \Delta \cup T_3 \Delta$ equals Δ with the middle triangle removed.

Since $\Delta_1 \subset \Delta$, it follows fairly easily (see Corollary 11.7.5) that when we iterate the procedure by setting

$$\Delta_{k+1} = T_1 \Delta_k \cup T_2 \Delta_k \cup T_3 \Delta_k,$$

a decreasing sequence of compact sets is obtained. The intersection Δ_∞ of these sets is the Sierpiński triangle of Exercise 4.4.J (see Figure 4.4). It has the property that we are looking for, $\Delta_\infty = T_1 \Delta_\infty \cup T_2 \Delta_\infty \cup T_3 \Delta_\infty$.

(2) Not all fractals are solid figures. The **von Koch curve** is obtained from an IFS using the following four similitudes:

$$T_1 \begin{bmatrix} x \\ y \end{bmatrix} = \begin{bmatrix} \frac{1}{3} & 0 \\ 0 & \frac{1}{3} \end{bmatrix} \begin{bmatrix} x \\ y \end{bmatrix}, \qquad T_2 \begin{bmatrix} x \\ y \end{bmatrix} = \begin{bmatrix} \frac{1}{3} \\ 0 \end{bmatrix} + \begin{bmatrix} \frac{1}{6} & \frac{-\sqrt{3}}{6} \\ \frac{\sqrt{3}}{6} & \frac{1}{6} \end{bmatrix} \begin{bmatrix} x \\ y \end{bmatrix},$$

$$T_3 \begin{bmatrix} x \\ y \end{bmatrix} = \begin{bmatrix} \frac{1}{2} \\ \frac{\sqrt{3}}{6} \end{bmatrix} + \begin{bmatrix} \frac{1}{6} & \frac{\sqrt{3}}{6} \\ \frac{-\sqrt{3}}{6} & \frac{1}{6} \end{bmatrix} \begin{bmatrix} x \\ y \end{bmatrix}, \qquad T_4 \begin{bmatrix} x \\ y \end{bmatrix} = \begin{bmatrix} \frac{2}{3} \\ 0 \end{bmatrix} + \begin{bmatrix} \frac{1}{3} & 0 \\ 0 & \frac{1}{3} \end{bmatrix} \begin{bmatrix} x \\ y \end{bmatrix}.$$

Let $B_0 = \{(x,0) : x \in [0,1]\}$ and define $B_{k+1} = T_1 B_k \cup T_2 B_k \cup T_3 B_k \cup T_4 B_k$. Graphing these figures shows an increasingly complex curve emerging in which the previous curve is scaled by $1/3$ and used to replace the four line segments of B_1. See Figure 11.11.

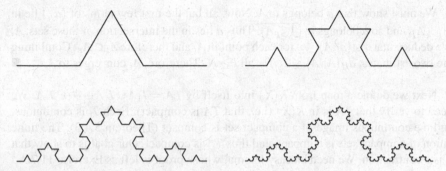

FIG. 11.11 The sets B_1 through B_4.

As we will see, this construction works in great generality, and many sets can be obtained as the invariant sets for iterated function systems. To establish these facts, we need a framework, in this case, a metric space. Let $K(X)$ denote the collection of all nonempty compact subsets of X. This is a metric space with respect to the Hausdorff metric of Example 9.1.2 5,

$$d_H(A,B) = \max \left\{ \sup_{a \in A} \text{dist}(a,B), \sup_{b \in B} \text{dist}(b,A) \right\}.$$

Our first result is the completeness of $K(X)$ in the Hausdorff metric.

11.7.2. THEOREM. *If X is a closed subset of $\mathbb{R}^n$, the metric space $K(X)$ of all compact subsets of X with the Hausdorff metric is complete.*

PROOF. Let A_n be a Cauchy sequence of compact sets in $K(X)$. Define

$$A = \bigcap_{k \geq 1} \overline{\bigcup_{i \geq k} A_i}.$$

Observe that for any $\varepsilon > 0$, there is an integer N such that $d_H(A_i, A_j) < \varepsilon$ for all $i, j \geq N$. In particular, it follows from the definition of the Hausdorff metric that $A_i \subset (A_N)_\varepsilon$ for all $i \geq N$. Consequently, $A \subset (A_N)_\varepsilon$. Now, $(A_N)_\varepsilon$ is a closed and bounded subset of $\mathbb{R}^n$, and so by the Heine–Borel Theorem is compact. It follows that A is the decreasing intersection of nonempty compact sets; by Cantor's Intersection Theorem, A is a nonempty compact set.

Having shown that $A \in K(X)$, it remains only to prove that (A_i) converges to A. We also have $A \subset (A_N)_\varepsilon \subset ((A_i)_\varepsilon)_\varepsilon = (A_i)_{2\varepsilon}$ for all $i \geq N$. Conversely, fix $a_i \in A_i$ with $i \geq N$. For each $j > i$, $A_i \subset (A_j)_\varepsilon$, and thus there is a point $a_j \in A_j$ with $\|a_i - a_j\| < \varepsilon$. Each a_j lies in the bounded set $(A_N)_\varepsilon$. By compactness, there is a convergent subsequence $\lim_{l \to \infty} a_{j_l} = a$. Clearly, $\|a_i - a\| \leq \varepsilon$.

We must show that a belongs to A. Now, all but the first few terms of (a_{j_l}) lie in $\bigcup_{j \geq k} A_j$, and so a belongs to $\overline{\bigcup_{j \geq k} A_j}$. Thus a lies in the intersection of these sets, A. We deduce that $\text{dist}(a_i, A) \leq \varepsilon$ for each point in A_i and therefore $A_i \subset A_\varepsilon$. Combining the two estimates, $d_H(A_i, A) \leq 2\varepsilon$ for all $i \geq N$. Therefore, A_i converges to A. ∎

Next we define a map from $K(X)$ into itself by $TA = T_1 A \cup T_2 A \cup \cdots \cup T_r A$. We need to verify that TA is in $K(X)$ (i.e., that TA is compact). Each T_i is continuous, and the continuous image of a compact set is compact (Theorem 5.4.3). The finite union of compact sets is compact, and thus TA is compact. Our goal is to show that it is a contraction. We need an easy lemma, whose proof is left as Exercise 11.7.A.

11.7.3. LEMMA. *Let* $A_1, \ldots, A_r$ *and* $B_1, \ldots, B_r$ *be compact subsets of* $\mathbb{R}^n$. *Then* $d_H(A_1 \cup \cdots \cup A_r, B_1 \cup \cdots \cup B_r) \leq \max \{ d_H(A_1, B_1), \ldots, d_H(A_r, B_r) \}$.

11.7.4. THEOREM. *Let* X *be a closed subset of* $\mathbb{R}^n$ *and let* $T_1, \ldots, T_r$ *be contractions of* X *into itself. Let* s_i *be the Lipschitz constants for each* T_i *and set* $s = \max\{s_1, \ldots, s_r\}$. *Then* T *is a contraction of* $K(X)$ *into itself with Lipschitz constant* s. *Hence there is a unique compact subset* A *of* X *such that*

$$A = T_1 A \cup T_2 A \cup \cdots \cup T_r A.$$

Moreover, if B *is any compact set, we have the estimates*

$$d_H(T^k B, A) \leq s^k d_H(B, A) \leq \frac{s^k}{1-s} d_H(B, TB).$$

PROOF. Let A and B be any two compact subsets of X. Observe that

$$d_H(T_i A, T_i B) \leq s_i d_H(A, B).$$

Indeed, if $a \in A$, then there is a point $b \in B$ with $\|a - b\| \leq d_H(A, B)$. Hence $\|T_i a - T_i b\| \leq s_i d_H(A, B)$. So $\sup_{a \in A} \text{dist}(T_i a, T_i B) \leq s_i d_H(A, B)$. Reversing the roles of A and B, we arrive at the desired estimate. By Lemma 11.7.3, it follows that $d_H(TA, TB) \leq s d_H(A, B)$.

Theorem 11.7.2 shows that $(K(X), d_H)$ is a complete metric space. The proof of the Contraction Principle goes through verbatim in the metric space case. Thus we may apply the Banach Contraction Principle to T. It follows that there is a unique fixed point $A = TA$. By definition of T, this is the unique compact set such that $A = T_1 A \cup T_2 A \cup \cdots \cup T_r A$. It is also an immediate consequence of the Contraction

Principle that A is obtained as the limit of iterates of T applied to any initial set B, and the estimates follow directly from the estimates in the Contraction Principle. ∎

11.7.5. COROLLARY. *Suppose that T is a contraction of $K(X)$ into itself with Lipschitz constant s. If B is a compact set such that $TB \subset B$, then the fixed point is given by $A = \bigcap\limits_{k \geq 0} T^k B$.*

PROOF. We show by induction that $T^{k+1}B \subset T^k B$ for $k \geq 0$. If $k = 0$, this is true by hypothesis. Assuming that $T^k B \subset T^{k-1}B$, we have

$$T^{k+1}B = T(T^k B) \subset T(T^{k-1}B) = T^k B.$$

By Theorem 11.7.4, the fixed point A is the limit of the sequence $T^k B$. From the *proof* of Theorem 11.7.2, this limit is given by

$$A = \bigcap_{k \geq 1} \overline{\bigcup_{i \geq k} T^i B} = \bigcap_{k \geq 1} T^k B.$$

∎

An excellent choice for the compact set B to use in computing the limit set is modeled by the Sierpiński triangle, Example 11.7.1 (1). Another good choice for an initial compact set B is a single point $\{x\}$ that happens to belong to A. Such points are easy to find. Each T_i is a contraction on X and thus has a unique fixed point x_i that may be found by iteration of T_i applied to any initial point. We give a significant strengthening of this fact, which hints at the dynamical properties of the iterated function system.

11.7.6. THEOREM. *For each word $w = i_1 i_2 \ldots i_l$ in the alphabet $\{1, \ldots, r\}$, there is a unique fixed point a_w of $T_w = T_{i_1} T_{i_2} \ldots T_{i_l}$. Each a_w belongs to the fixed set A, and the set of all of these fixed points is dense in A.*

If $a \in A$, then the orbit $\mathcal{O}(a) = \{ T_w a : w \text{ is a word in } \{1, \ldots, r\} \}$ is dense in A.

PROOF. First observe that the composition of contractions is a contraction (see Exercise 11.1.E). Therefore, each T_w is a contraction and hence has a unique fixed point a_w. Moreover, starting with any point x, the iterates $T_w^k x$ converge to a_w. Take x to be any point in A. Each T_i maps A into itself, showing that $T_w^k x$ is in A for all $k \geq 0$. Since A is closed, the limit a_w belongs to A.

Next we prove that the set of these fixed points $\{a_w : w \text{ is a word in } \{1, \ldots, r\}\}$ is dense in A. Now $A = TA = T_1 A \cup \cdots \cup T_r A$. So

$$A = T^2 A = \bigcup_{i=1}^{r} \bigcup_{j=1}^{r} T_i T_j A.$$

Repeating this N times, we obtain

$$A = T^N A = \bigcup_{\text{words } w \text{ of length } N} T_w A.$$

Fix $\varepsilon > 0$ and choose N so large that $s^N \operatorname{diam}(A) < \varepsilon$. By Exercise 11.1.E, each T_w is a contraction with Lipschitz constant no greater than s^N. Consequently, $\operatorname{diam}(T_w A) \le s^N \operatorname{diam}(A) < \varepsilon$. If $a \in A$, choose a word w of length N such that $a \in T_w A$. Since $a_w = T_w a_w$, it is clear that $a_w \in T_w A$ as well. Therefore, $\|a - a_w\| < \varepsilon$. Thus the set of these fixed points is dense in A.

If a is an arbitrary point in A, it follows that $T_w a$ belongs to A for every finite word w. Hence $\mathcal{O}(a) \subset A$. Observe that $T_i T_w a = T_{iw} a$ is another point in $\mathcal{O}(a)$. Therefore, $T \mathcal{O}(a) \subset \mathcal{O}(a)$. Let $B = \overline{\mathcal{O}(a)}$. The continuity of T implies that $TB \subset B$. By Corollary 11.7.5, $A = \bigcap_{k \ge 1} T^k B \subset B \subset A$. Therefore $\overline{\mathcal{O}(a)} = A$. ∎

This result allows us graph the fractal approximately as follows. Pick any point a. This may not lie in A. However, $b = T_1^{100} a$ will be very close to the fixed point of T_1 assuming reasonable constants. Use a computer to calculate b and then recursively plot the sets $T^k\{b\}$ for sufficiently many k. This will frequently give an excellent picture of the fractal. You can make explicit estimates to ensure good convergence.

11.7.7. EXAMPLE. Consider the maps

$$T_1 \begin{bmatrix} x \\ y \end{bmatrix} = \begin{bmatrix} .5 & -.5 \\ .5 & .5 \end{bmatrix} \begin{bmatrix} x \\ y \end{bmatrix} + \begin{bmatrix} 1 \\ 5 \end{bmatrix}, \qquad T_2 \begin{bmatrix} x \\ y \end{bmatrix} = \begin{bmatrix} .5 & -.5 \\ .5 & .5 \end{bmatrix} \begin{bmatrix} x \\ y \end{bmatrix} + \begin{bmatrix} -1 \\ 3 \end{bmatrix}.$$

A simple matrix calculation shows that $(-4, 6)$ is the fixed point of T_1 and $(-4, 2)$ is the fixed point of T_2. We use a computer to plot the sets $A_0 = \{(-4, 6), (-4, 2)\}$ and $A_{k+1} = T_1 A_k \cup T_2 A_k$ for $1 \le k \le 11$. Since A equals the closed union of all the A_k's, this yields a reasonable approximation. Look for the self-symmetry in Figure 11.12.

We finish with a simple result that shows that the sets that are fixed for iterated function schemes are extremely plentiful.

11.7.8. PROPOSITION. *Let C be a compact subset of $\mathbb{R}^n$, and let $\varepsilon > 0$. Then there is an IFS $\mathcal{T} = \{T_1, \ldots, T_r\}$ with fixed set A such that $d_H(A, C) < \varepsilon$.*

PROOF. Since C is compact, we can find a finite set of points, $C_0 = \{\mathbf{c}_1, \ldots, \mathbf{c}_r\}$ such that the union of the balls $B_\varepsilon(\mathbf{c}_i)$, call it B, contains C. Observe that B equals $(C_0)_\varepsilon$. Let R be large enough that $B_R(\mathbf{0})$ contains B. Define

$$T_i \mathbf{x} = \frac{\varepsilon}{2R}(\mathbf{x} - \mathbf{c}_i) + \mathbf{c}_i \quad \text{for} \quad 1 \le i \le r.$$

Since $T_i \mathbf{c}_i = \mathbf{c}_i$ are fixed points, C_0 is contained in the fixed set A of T. Also, $T_i B \subset T_i B_R(\mathbf{0}) \subset B_\varepsilon(\mathbf{c}_i) \subset B$ for each i. Therefore by Corollary 11.7.5, we have $A \subset B = (C_0)_\varepsilon \subset C_\varepsilon$. In addition, $C \subset (C_0)_\varepsilon \subset A_\varepsilon$. So, $d_H(A, C) < \varepsilon$ as required. ∎

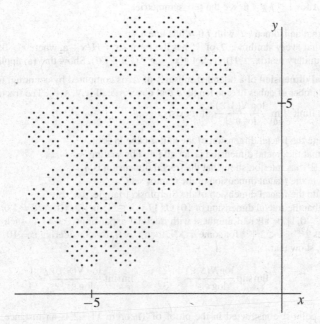

FIG. 11.12 Pointillist picture of the 'twin dragon' set.

Exercises for Section 11.7

A. (a) Let A_1, A_2, B_1, and B_2 be compact subsets of $\mathbb{R}^n$. Show that $d_H(A_1 \cup A_2, B_1 \cup B_2) \leq \max\{d_H(A_1, B_1), d_H(A_2, B_2)\}$.

(b) Use induction to prove Lemma 11.7.3.

B. (a) Let $\mathscr{T} = \{T_1, \ldots, T_r\}$ be an IFS on $\mathbb{R}^n$. Suppose that B is compact. Prove that there is a unique compact set C such that $C = B \cup TC$.   HINT: Include a constant map $T_0(X) = B$.

(b) Hence prove that there is always a compact set C containing B such that $TC \subset C$.

C. Consider the four maps on $\mathbb{R}^2$ given by $T_i x = Ax + \mathbf{b}_i$ for $i = 1, 2, 3, 4$, where $A = \frac{1}{2}\begin{bmatrix} 1 & 1 \\ -1 & 1 \end{bmatrix}$ and the vectors $\mathbf{b}_i$ are $(0,0), (1,0), (0,1)$, and $(1,1)$.

(a) Show that the fixed points of the T_i's form the vertices of a square S.

(b) Compute TS and $T^2 S$.

(c) Use a computer to generate a picture of the fixed set.

D. (a) Find an IFS on $\mathbb{R}$ that generates the Cantor set.   HINT: Identify two self maps of C with disjoint union equal to C.

(b) Find a different IFS that also generates C.

E. Consider the maps $\mathscr{T} = \{T_1, T_2, T_3, T_4\}$ given by

$$T_1 \begin{bmatrix} x \\ y \end{bmatrix} = \begin{bmatrix} 0.8 & 0 \\ 0 & 0.8 \end{bmatrix} \begin{bmatrix} x \\ y \end{bmatrix} + \begin{bmatrix} 0.1 \\ 0.04 \end{bmatrix}, \qquad T_2 \begin{bmatrix} x \\ y \end{bmatrix} = \begin{bmatrix} 0.5 & 0 \\ 0 & 0.5 \end{bmatrix} \begin{bmatrix} x \\ y \end{bmatrix} + \begin{bmatrix} 0.25 \\ 0.4 \end{bmatrix},$$

$$T_3 \begin{bmatrix} x \\ y \end{bmatrix} = \begin{bmatrix} 0.35 & -0.35 \\ 0.35 & 0.35 \end{bmatrix} \begin{bmatrix} x \\ y \end{bmatrix} + \begin{bmatrix} 0.27 \\ 0.08 \end{bmatrix}, \qquad T_4 \begin{bmatrix} x \\ y \end{bmatrix} = \begin{bmatrix} 0.35 & 0.35 \\ -0.35 & 0.35 \end{bmatrix} \begin{bmatrix} x \\ y \end{bmatrix} + \begin{bmatrix} 0.38 \\ 0.43 \end{bmatrix}.$$

(a) Use a computer to plot the maple leaf pattern A fixed by T.

(b) Plot $T_i A$ for $1 \leq i \leq 4$ to see the self-symmetries.

F. (a) Show that an isometry T with $T\mathbf{0} = \mathbf{0}$ is linear.

   (b) Show that every similitude T of $\mathbb{R}^n$ has the form $T\mathbf{x} = rU\mathbf{x} + \mathbf{a}$, where $r > 0$, $\mathbf{a} \in \mathbb{R}^n$, and U is a unitary matrix. HINT: Set $g(\mathbf{x}) = r^{-1}(T\mathbf{x} - T\mathbf{0})$. Show that (a) applies to g.

G. The **fractal dimension** of a bounded subset A of $\mathbb{R}^n$ is computed by counting, for $\varepsilon > 0$, the smallest number of cubes of side length ε that cover A; call it $N(A, \varepsilon)$. The fractal dimension is then the limit $\lim\limits_{\varepsilon \to 0^+} \dfrac{\log N(A, \varepsilon)}{\log \varepsilon^{-1}}$, if it exists.

   (a) Compute the fractal dimension of the unit n-cube in $\mathbb{R}^n$.
   (b) Show that the fractal dimension is not affected by scaling.
   (c) If $A \subset \mathbb{R}^n$ has interior, show that the fractal dimension is n.
   (d) Compute the fractal dimension of the Cantor set C.
   (e) Compute the fractal dimension of the Sierpiński triangle.
   (f) Show that the fractal dimension of $\{0\} \cup \{1/n : n \in \mathbb{N}\}$ is 1/2. HINT: $\varepsilon = 1/n - 1/(n+1)$.
   (g) Let $S \subset [0,1]$ be all real numbers with decimal expansions $0.x_1 x_2 \ldots$ such that if $i \in \mathbb{N}$ satisfies $2^{2n} \leq i < 2^{2n+1}$ for some $n \in \mathbb{N}$, then $x_i = 0$. By considering $\varepsilon = 10^{-2n}$ for n even or odd, show that

$$\limsup_{\varepsilon \to 0^+} \frac{\log N(S, \varepsilon)}{\log \varepsilon^{-1}} = \frac{2}{3}, \qquad \liminf_{\varepsilon \to 0^+} \frac{\log N(S, \varepsilon)}{\log \varepsilon^{-1}} = \frac{1}{3}.$$

H. The limit point A constructed in the proof of Theorem 11.7.2 is an instance of a general construction. For (A_k) a sequence of subsets of K, a compact subset of $\mathbb{R}^n$, define

$$\limsup A_k = \bigcap_{k \in \mathbb{N}} \overline{\bigcup_{m \geq k} A_m}.$$

   (a) Show that $a \in \limsup A_k$ if and only if there are a sequence of integers (k_i) diverging to $+\infty$ and elements $a_i \in A_{k_i}$ such that $\lim\limits_{i \to \infty} a_i = a$.
   (b) Show that if $f : K \to L$ is continuous, $L \subset \mathbb{R}^m$, then $\limsup f(A_k) = f(\limsup A_k)$.

I. The von Koch curve is the image of a one-to-one path $\gamma : [0,1] \to \mathbb{R}^2$. To show this, we use the notation of Example 11.7.1 (2) and let $B_\infty = \cap_{k \in \mathbb{N}} B_k$.

   (a) Define f from $W = \{(i_1, i_2, \ldots) : i_k \in \{1,2,3,4\}\}$ to B_∞ by $f(i_1, i_2, \ldots) = \cap_{k \in \mathbb{N}} T_{i_1 i_2 \ldots i_k} B_0$. Using base-4 expansions of real numbers, show that f induces a well-defined, continuous map $g : [0,1] \to \mathscr{B}_\infty$.
   (b) For each word w in the letters $\{1,2,3,4\}$, let C_w be the convex set containing $T_w(B_0)$. Show, for all words v that start with w, that $T_v(B_0) \subset C_w$ and hence $C_v \subset C_w$. If v and w are distinct words of the same length, show that $C_v \cap C_w$ contains at most one point. Deduce that g is one-to-one.

Chapter 12
Differential Equations

In this chapter, we apply analysis to the study of ordinary differential equations, generally called DEs or ODEs. *Ordinary* is used to indicate differential equations of a single variable, in contrast to partial differential equations (PDEs), in which several variables, and hence partial derivatives, appear. We will see some PDEs in the chapters on Fourier series, Chapters 13 and 14.

Most introductory courses on differential equations present methods for solving DEs of various special types. We will not be concerned with those techniques here except to give a few pertinent examples. Rather we are concerned with why differential equations have solutions, and why these solutions are or are not unique. This topic, crucial to a full understanding of differential equations, is often omitted from introductory courses because it requires the tools of real analysis. In particular, we will require the Banach Contraction Principle (11.1.6) which was established in the previous chapter.

12.1 Integral Equations and Contractions

We consider an example that motivates the approach of the next section. Start with an initial value problem, which consists of two parts:

$$f'(x) = \varphi(x, f(x)) \quad \text{for} \quad a \le x \le b,$$
$$f(c) = y_0.$$

The first equation is the DE and the second is an initial value condition. The function $\varphi(x, y)$ is a continuous function of two variables defined on $[a, b] \times \mathbb{R}$ and c is a given point in $[a, b]$. By solving the DE or, equivalently, solving the initial value problem, we mean finding a function $f(x)$ that is defined and differentiable on the interval $[a, b]$ and satisfies both the differential equation and the initial value condition.

K.R. Davidson and A.P. Donsig, *Real Analysis and Applications: Theory in Practice*, 293
Undergraduate Texts in Mathematics, DOI 10.1007/978-0-387-98098-0_12,
© Springer Science + Business Media, LLC 2010

This DE is of first order, since the equation involves only the first derivative. In general, the **order of a differential equation** is the highest-order derivative of the unknown function that appears in the equation.

Because of the subject's connections to physics, chemistry, and engineering, it is common in differential equations to suppress the dependence of the function f on x [i.e., to write f instead of $f(x)$]. Typically in the sciences, each variable has a physical significance, and it can be a matter of choice about which is the independent variable and which is dependent. We will sometimes do this in examples, where simplifying otherwise complicated expressions seems to be worth the extra demands this notation makes.

Our first step is to turn this problem into a fixed-point problem by integration. Indeed, from the fundamental theorem of calculus, our solution must satisfy

$$f(x) = f(c) + \int_c^x f'(t)\,dt = y_0 + \int_c^x \varphi(t, f(t))\,dt.$$

Conversely, a continuous solution of this **integral equation** is automatically differentiable by the Fundamental Theorem of Calculus, and

$$f'(x) = \frac{d}{dx}\left(y_0 + \int_c^x \varphi(t, f(t))\,dt\right) = \varphi(x, f(x))$$

and

$$f(c) = y_0 + \int_c^c \varphi(t, f(t))\,dt = y_0.$$

Thus f satisfies the DE, including the initial value condition.

This integral equation suggests studying a map from $C[a,b]$ into itself defined by

$$Tf(x) = y_0 + \int_c^x \varphi(t, f(t))\,dt.$$

The solutions to the integral equation, if any, satisfy $Tf = f$, and so correspond precisely to the fixed points of T. The Contraction Principle (11.1.6) is well suited to this kind of problem. There are also more sophisticated approaches that give weaker conclusions from weaker hypotheses. However, the Contraction Principle gives both existence and uniqueness of a solution, when it can be applied. Consider the following specific example.

12.1.1. EXAMPLE. We will solve the initial value problem

$$f'(x) = 1 + x - f(x) \quad \text{for} \quad -\tfrac{1}{2} \le x \le \tfrac{1}{2},$$
$$f(0) = 1.$$

First convert it to the integral equation for f in $C[-\tfrac{1}{2}, \tfrac{1}{2}]$:

$$f(x) = 1 + \int_0^x 1 + t - f(t)\,dt = 1 + x + \tfrac{1}{2}x^2 - \int_0^x f(t)\,dt.$$

Define a map T on $C[-\frac{1}{2}, \frac{1}{2}]$ by sending f to the function Tf given by

$$Tf(x) = 1 + x + \tfrac{1}{2}x^2 - \int_0^x f(t)\,dt.$$

The solution of the integral equation is a fixed point of T.

To use the Contraction Principle, we must show that T is a contraction. We have

$$|Tf(x) - Tg(x)| = \left| \int_0^x f(t) - g(t)\,dt \right| \leq \left| \int_0^x |f(t) - g(t)|\,dt \right|$$

$$\leq \left| \int_0^x \|f - g\|_\infty\,dt \right| = \|f - g\|_\infty \int_0^{|x|} dt \leq \frac{1}{2}\|f - g\|_\infty.$$

This estimate is independent of x in $[-\frac{1}{2}, \frac{1}{2}]$, and thus we obtain

$$\|Tf - Tg\|_\infty \leq \frac{1}{2}\|f - g\|_\infty.$$

Hence T is a contraction.

By the Contraction Principle (11.1.6), there is a unique fixed point $f_\infty = Tf_\infty$ that will solve our DE. Moreover, it shows that any sequence of functions (f_n) with $f_{n+1} = Tf_n$ will converge to f_∞ in $C[-\frac{1}{2}, \frac{1}{2}]$. For example, let us take f_0 to be the constant function 1. Then

$$f_1(x) = Tf_0(x) = 1 + x + \tfrac{1}{2}x^2 - \int_0^x 1\,dt = 1 + \tfrac{1}{2}x^2.$$

Similarly,

$$f_2(x) = Tf_1(x) = 1 + x + \tfrac{1}{2}x^2 - \int_0^x 1 + \tfrac{1}{2}t^2\,dt = 1 + \tfrac{1}{2}x^2 - \tfrac{1}{6}x^3.$$

And

$$f_3(x) = Tf_2(x) = 1 + x + \tfrac{1}{2}x^2 - \int_0^x 1 + \tfrac{1}{2}t^2 - \tfrac{1}{6}t^3\,dt = 1 + \tfrac{1}{2}x^2 - \tfrac{1}{6}x^3 + \tfrac{1}{24}x^4.$$

In general, we can establish by induction (do it yourself!) that

$$f_n(x) = 1 + \tfrac{1}{2}x^2 - \tfrac{1}{3!}x^3 + \tfrac{1}{4!}x^4 - \tfrac{1}{5!}x^5 + \cdots + \tfrac{1}{(n+1)!}(-x)^{n+1}.$$

This sequence evidently consists of the partial sums of an infinite series. The new term added at the nth stage is $\frac{1}{(n+1)!}(-x)^{n+1}$, which on $[-\frac{1}{2}, \frac{1}{2}]$ has max norm

$$\max_{|x| \leq 1/2} \left| \frac{(-x)^{n+1}}{(n+1)!} \right| = \frac{1}{2^{n+1}(n+1)!}.$$

This is summable. Therefore, this power series converges uniformly on $[-\frac{1}{2}, \frac{1}{2}]$ by the Weierstrass M-test (8.4.7). Figure 12.1 gives the graphs of f_0, f_1, f_2, and f_∞ on

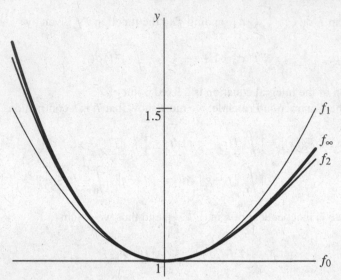

FIG. 12.1 The first three approximants and f_∞ on $[-1,1]$.

$[-1,1]$. The astute reader should notice that these are the Taylor polynomials for e^{-x} except that the term in x is missing.

We obtain

$$f_\infty(x) = x + \sum_{k=0}^{\infty} \frac{1}{k!}(-x)^k = e^{-x} + x.$$

This shows that $e^{-x} + x$ is the unique solution to our integral equation. Indeed,

$$(e^{-x} + x)' = -e^{-x} + 1 = 1 + x - (e^{-x} + x) \quad \text{and} \quad e^{-0} + 0 = 1.$$

While our reasoning shows only that the solution is valid on $[-\frac{1}{2}, \frac{1}{2}]$, $f_\infty(x) = e^{-x} + x$ is a valid solution on all of $\mathbb{R}$. One way we can deal with this is to reconsider our problem beginning at the point $x = \frac{1}{2}$, and try to extend to $[\frac{1}{2}, 1]$. We have the DE

$$f'(x) = 1 + x - f(x) \quad \text{for} \quad 0 \le x \le 1,$$
$$f(\tfrac{1}{2}) = e^{-1/2} - \tfrac{1}{2}.$$

The exact same argument on this new interval will yield the unique attractive fixed point $f_\infty(x) = e^{-x} + x$ valid on $[0,1]$. It is then easy to see that we can bootstrap our way to the unique solution on the whole line.

In the next two sections, we show how the solutions to a large family of DEs can be formulated as fixed-point problems. Then we may use the Contraction Principle to show that these DEs always have solutions.

Exercises for Section 12.1

A. Use the method of this section to solve $f'(x) = 1 + \frac{1}{2}f(x)$ for $0 \le x \le 1$ and $f(0) = 1$. You should be able to recognize the series and successfully find a closed form for the solution. Show that, in fact, this solution is valid for all real numbers.

B. For $b > 0$ and $a \in \mathbb{R}$, define T on $C[0,b]$ by $Tf(x) = a + \int_0^x f(t)xe^{-xt}\,dt$. Prove that T is a contraction. Hence show that there is a unique solution $f \in C[0,\infty)$ to the integral equation $f(x) = a + \int_0^x f(t)xe^{-xt}\,dt$.

C. Consider the DE $f'(x) = \frac{-2}{x}f(x)$ and $f(1) = 1$ for $\frac{2}{3} \le x \le \frac{3}{2}$.

  (a) Prove that the associated integral map is a contraction mapping.
  (b) Look for a solution to the DE of the form $f(x) = ax^b$.
  (c) Starting with $f_0 = 1$, find a formula for $f_n = T^n f_0$. Express this as a familiar power series of $\log x$, and hence evaluate it.   HINT: Use the closed form from (b) to guide you.

D. Consider the DE $y' = 1 + y^2$ and $y(0) = 0$.

  (a) Solve the DE directly.
  (b) Show that the associated integral map T is not a contraction on $C[-r,r]$ for any $r > 0$.
  (c) Find an $r > 0$ such that T maps the unit ball of $C[-r,r]$ into itself and is a contraction mapping on this ball.
  (d) Hence show that there is a unique solution on $[-r,r]$.

E. Consider a ball falling to the ground. The downward force of gravity is counteracted by air resistance proportional to the velocity. Find a formula for the velocity of the ball if it is at rest at time 0 [i.e., $v(0) = 0$] as follows:

  (a) Show that the velocity satisfies $v'(t) = g - cv(t)$, where g is the gravity constant and c is the constant of air resistance. What is the initial condition?
  (b) Construct the associated integral map T. Starting with $v_0(t) = 0$, iterate T and obtain a formula for the solution v_∞.
  (c) Show that this solution is valid on $[0,\infty)$ (or at least until the ball hits the ground). Compute $\lim_{t\to\infty} v_\infty(t)$. This is known as the **terminal velocity**.

12.2 Calculus of Vector-Valued Functions

In this section, we develop differentiation and integration for vector-valued functions. We will need this material in the next section, to convert an nth-order DE for a real-valued function into a first-order DE for a vector-valued function.

Consider a function $f : [a,b] \to \mathbb{R}^n$, where $f(x) = (f_1(x), f_2(x), \ldots, f_n(x))$ and each coordinate function f_i maps $[a,b]$ to $\mathbb{R}$. Although much of this is done by looking at the coordinate functions, there are some crucial differences between $\mathbb{R}^n$ and $\mathbb{R}$. Most notably, we do not have a total order in $\mathbb{R}^n$, so we cannot take the supremum or infimum of the set $f([a,b]) \subset \mathbb{R}^n$.

12.2.1. DEFINITION. We say that a vector-valued function $f : [a,b] \to \mathbb{R}^n$ is **differentiable** at a point $x_0 \in (a,b)$ if

$$\lim_{h\to 0} \frac{f(x_0+h) - f(x_0)}{h}$$

exists in $\mathbb{R}^n$. As usual, we write $f'(x_0)$ for the limit and call it the **derivative** of f at x_0. Notice that the numerator is in $\mathbb{R}^n$, while the denominator is a scalar, so $f'(x_0)$ is a vector in $\mathbb{R}^n$.

We can define **left differentiable** and **right differentiable** in the natural way, and we say that f is **differentiable** on the interval $[a, b]$ if it is differentiable at every point of (a, b) in the sense just stated and left or right differentiable at the endpoints.

The reader should verify that if $f : [a, b] \to \mathbb{R}^n$ is differentiable at $x_0 \in [a, b]$, then f is continuous at x_0.

12.2.2. PROPOSITION. *Suppose that* $f : [a, b] \to \mathbb{R}^n$ *is written as* $f(x) = (f_1(x), \ldots, f_n(x))$, *where each* f_i *maps* $[a, b]$ *into* $\mathbb{R}$. *Then* f *is differentiable at* $x_0 \in [a, b]$ *if and only if each* f_i *is differentiable at* x_0. *Moreover,*

$$f'(x_0) = (f_1'(x_0), \ldots, f_n'(x_0)).$$

PROOF. Fix $x_0 \in [a, b]$ and consider the function

$$g(h) = \frac{f(x_0 + h) - f(x_0)}{h},$$

defined for those h such that $x_0 + h \in [a, b]$. If $g(h) = (g_1(h), \ldots, g_n(h))$, then we have that $g_i(h) = (f_i(x_0 + h) - f_i(x_0))/h$ for each i. Thus f is differentiable at x_0 if and only if the limit of $g(h)$ exists as $h \to 0$, and each f_i is differentiable at x_0 if and only if the limit of $g_i(h)$ exists as $h \to 0$. But a sequence of vectors converges if and only if its coordinates converge. The result follows.   ■

The reader should verify that sums and scalar multiples of differentiable functions are differentiable. Since products of real-valued differentiable functions are differentiable, it follows easily that if f and g are vector-valued differentiable functions, then the dot product $f \cdot g$ is differentiable. Notice that $f \cdot g$ is real-valued.

On the other hand, not all results carry over from the real-valued setting. For example, consider the function $f : [0, 2\pi] \to \mathbb{R}^2$ given by $f(x) = (\cos x, \sin x)$. It is easy to see that $f(2\pi) = f(0)$, but there is no $x \in [0, 2\pi]$ such that $f'(x)$ is the zero vector. However, we do have the following result. The analaguous fact for real-valued functions is a corollary of the Mean Value Theorem.

12.2.3. THEOREM. *Suppose that* $f : [a, b] \to \mathbb{R}^n$ *is continuous on* $[a, b]$ *and differentiable on* (a, b). *Then there is* $c \in (a, b)$ *such that*

$$\|f(b) - f(a)\| \le (b - a)\|f'(c)\|.$$

PROOF. Define $\mathbf{v} = f(b) - f(a)$ and a function $g : [a, b] \to \mathbb{R}$ by $g(x) = \mathbf{v} \cdot f(x)$. Notice that if $\mathbf{v}$ is the zero vector, then we are done. So we may assume $\|\mathbf{v}\| \neq 0$.

A brief calculation shows that g is differentiable and $g'(x) = \mathbf{v} \cdot f'(x)$. Applying the Mean Value Theorem to g shows that there is $c \in (a, b)$ such that

$$g(b) - g(a) = (b-a)(\mathbf{v} \cdot f'(c)).$$

Using the definition of g and rearranging shows that $g(b) - g(a) = \|\mathbf{v}\|^2$. The Schwarz inequality (4.1.1) now gives

$$\|\mathbf{v}\|^2 = (b-a)\mathbf{v} \cdot f'(c) \leq (b-a)\|\mathbf{v}\| \|f'(c)\|.$$

Dividing by the nonzero quantity $\|\mathbf{v}\|$ gives the inequality. ∎

12.2.4. COROLLARY. *For a function $f : [a,b] \to \mathbb{R}^n$ that is C^1 on $[a,b]$, we have $\|f(b) - f(a)\| \leq (b-a)\|f'\|_\infty$.*

Next, we consider integration for vector-valued functions. The proof is straightforward, so will be omitted.

12.2.5. PROPOSITION. *Fix $f : [a,b] \to \mathbb{R}^n$ with $f(x) = (f_1(x),\ldots,f_n(x))$ for all $x \in [a,b]$. Then f is Riemann integrable if and only if each coordinate function f_i is Riemann integrable. In this case,*

$$\int_a^b f(x)\, dx = \left(\int_a^b f_1(x)\, dx, \int_a^b f_2(x)\, dx, \ldots, \int_a^b f_n(x)\, dx \right).$$

Using this proposition, we can carry over many properties of integration for real-valued functions to vector-valued functions. For example, linearity of integration and the Fundamental Theorem of Calculus carry over in this way.

12.2.6. FUNDAMENTAL THEOREM OF CALCULUS II.
Let $f : [a,b] \to \mathbb{R}^n$ be a bounded Riemann integrable function and define

$$F(x) = \int_a^x f(x)\, dx \quad for\ a \leq x \leq b.$$

Then $F : [a,b] \to \mathbb{R}^n$ is a continuous function. If f is continuous at a point x_0, then F is differentiable at x_0 and $F'(x_0) = f(x_0)$.

Finally, we have a higher-dimensional analogue of taking the absolute value inside the integral. While the proof is similar in spirit to the scalar case, the argument requires inner products and the Schwarz inequality (4.1.1).

12.2.7. LEMMA. *Let $F : [a,b] \to \mathbb{R}^n$ be continuous. Then*

$$\left\| \int_a^b F(x)\, dx \right\| \leq \int_a^b \|F(x)\|\, dx.$$

PROOF. Let $\mathbf{y} = \int_a^b F(x)\,dx$. If $\|\mathbf{y}\| = 0$, then the inequality is trivial. Otherwise, using the standard inner product on $\mathbb{R}^n$,

$$\|\mathbf{y}\|^2 = \left\langle \int_a^b F(x)\,dx, \mathbf{y} \right\rangle = \int_a^b \langle F(x), \mathbf{y} \rangle\,dx \leq \int_a^b |\langle F(x), \mathbf{y} \rangle|\,dx$$

$$\leq \int_a^b \|F(x)\|\,\|\mathbf{y}\|\,dx = \|\mathbf{y}\| \int_a^b \|F(x)\|\,dx.$$

The second line follows by the Schwarz inequality. Dividing through by $\|\mathbf{y}\|$ gives the result.   ∎

Exercises for Section 12.2

A. If $f : [a,b] \to \mathbb{R}^n$ is differentiable at $x_0 \in [a,b]$, show that f is continuous at x_0.

B. Show that if α, β are real numbers and f, g are Riemann integrable functions from $[a,b]$ to $\mathbb{R}^n$, then $\alpha f + \beta g$ is Riemann integrable and

$$\alpha \int_a^b f(x)\,dx + \beta \int_a^b g(x)\,dx = \int_a^b (\alpha f(x) + \beta g(x))\,dx.$$

C. Derive Theorem 12.2.6 from the one-variable version.

D. Prove that every continuous function $f : [a,b] \to \mathbb{R}^n$ is Riemann integrable.

E. Define a **regular curve** to be a differentiable function $f : [a,b] \to \mathbb{R}^n$ such that $f'(x)$ is never the zero vector.

(a) Given two regular curves $f : [a,b] \to \mathbb{R}^n$ and $g : [c,d] \to \mathbb{R}^n$, we say that g is a **reparametrization** of f if there is a differentiable function $h : [c,d] \to [a,b]$ such that $h'(t) \neq 0$ for all t and $g = f \circ h$. Show that $g'(t) = f'(h(t))h'(t)$.

(b) Define the **length of a regular curve** f to be $L(f) = \int_a^b \|f'(x)\|\,dx$. Show that the length is not affected by reparametrization.   HINT: Consider reparametrizations where $h'(t)$ is always positive or always negative.

(c) Given a regular curve f, show there is a reparametrization g with $\|g'(t)\| = 1$ for all t. Such a curve has **unit speed**. HINT: Show that $x \mapsto \int_a^x \|f(t)\|\,dt$ has an inverse function.

12.3 Differential Equations and Fixed Points

The goal of this section is to start with a DE of order n, and convert it to the problem of finding a fixed point of an associated integral operator.

The first step is to take a fairly general form of a higher-order differential equation and turn it into a first-order DE at the expense of making the function vector-valued. We define an **initial value problem**, for functions on $[a,b]$ and a point $c \in [a,b]$, as

$$f^{(n)}(x) = \varphi(x, f(x), f'(x), \ldots, f^{(n-1)}(x)), \qquad (12.3.1)$$

$$f(c) = \gamma_0, \; f'(c) = \gamma_1, \; \ldots, \; f^{(n-1)}(c) = \gamma_{n-1}, \qquad (12.3.2)$$

where φ is a real-valued continuous function on $[a,b] \times \mathbb{R}^n$. This is not quite the most general situation, but it includes most important examples. The first equation is referred to as a **differential equation of nth order**, and the second set is called the **initial conditions**.

A standard trick reduces this to a first-order differential equation with values in $\mathbb{R}^n$. As we shall see, this has computational advantages. Replace the function f by the vector-valued function $F : [a,b] \to \mathbb{R}^n$ given by

$$F(x) = (f(x), f'(x), \ldots, f^{(n-1)}(x)).$$

Then the differential equation becomes

$$F'(x) = \Big(f'(x), \ldots, f^{(n-1)}(x), \varphi\big(x, f(x), \ldots, f^{(n-1)}(x)\big) \Big),$$

and the initial data become

$$F(c) = (\gamma_0, \gamma_1, \ldots, \gamma_{n-1}).$$

We further simplify the notation by introducing a function Φ from $[a,b] \times \mathbb{R}^n$ to $\mathbb{R}^n$ by

$$\Phi(x, y_0, \ldots, y_{n-1}) = (y_1, y_2, \ldots, y_{n-1}, \varphi(x, y_0, \ldots, y_{n-1})),$$

and the vector

$$\Gamma = (\gamma_0, \gamma_1, \ldots, \gamma_{n-1}).$$

Note that Φ is continuous, since φ is continuous. Then (12.3.1) becomes the first-order initial value problem with vector values

$$F'(x) = \Phi(x, F(x)), \tag{12.3.3}$$
$$F(c) = \Gamma.$$

It is easy to see that a solution of (12.3.1) gives a solution of (12.3.3). To go the other way, suppose (12.3.3) has a solution

$$F(x) = (f_0(x), f_1(x), \ldots, f_{n-1}(x)).$$

Then (12.3.3) means

$$\begin{aligned}
F'(x) &= (f_0'(x), f_1'(x), \ldots, f_{n-2}'(x), f_{n-1}'(x)) \\
&= \Phi(x, f_0(x), f_1(x), \ldots, f_{n-1}(x)) \\
&= \Big(f_1(x), \ldots, f_{n-1}(x), \varphi\big(x, f(x), \ldots, f^{(n-1)}(x)\big) \Big).
\end{aligned}$$

By identifying each coordinate, we obtain

$$f_0'(x) = f_1(x), \ f_1'(x) = f_2(x), \ \ldots, \ f_{n-2}'(x) = f_{n-1}(x),$$
$$f_{n-1}'(x) = \varphi(x, f_0(x), f_1(x), \ldots, f_{n-1}(x)),$$

Thus $f_1 = f_0'$, $f_2 = f_1' = f_0^{(2)}$, $\ldots$, $f_{n-1} = f_{n-2}' = f_0^{(n-1)}$, and

$$f_0^{(n)}(x) = f_{n-1}'(x) = \varphi\big(x, f_0(x), f_0'(x), \ldots, f_0^{(n-1)}(x)\big).$$

The initial data become $\Gamma = F(c) = (f_0(c), \ldots, f_{n-1}(c))$, so

$$f_0(c) = \gamma_0, \quad f_0'(c) = \gamma_1, \quad \ldots, \quad f_0^{(n-1)}(c) = \gamma_{n-1}.$$

Thus (12.3.1) is satisfied by $f_0(x)$. Consequently, (12.3.3) is an equivalent formulation of the original problem.

12.3.4. EXAMPLE. We will express the unknown function as y, instead of $f(x)$. Consider the differential equation

$$(1 + (y')^2)y^{(3)} = y'' - xy'y + \sin x \quad \text{for} \quad -1 \le x \le 1,$$
$$y(0) = 1,$$
$$y'(0) = 0,$$
$$y''(0) = 2.$$

It is first necessary to reformulate this DE to express the highest-order derivative, $y^{(3)}$, as a function of lower-order terms. This yields

$$y^{(3)} = \frac{y'' - xy'y + \sin x}{1 + (y')^2}.$$

Then the vector function Φ defined from $[-1, 1] \times \mathbb{R}^3$ into $\mathbb{R}^3$ is given by

$$\Phi(x, y_0, y_1, y_2) = \left(y_1, y_2, \frac{y_2 - xy_1y_0 + \sin x}{1 + y_1^2}\right).$$

The initial vector is $\Gamma = (1, 0, 2)$. The DE is now reformulated as a first-order vector-valued DE looking for a function $F(x) = (f_0(x), f_1(x), f_2(x))$ defined on $[-1, 1]$ with values in $\mathbb{R}^3$ such that

$$F'(x) = \left(f_1(x), f_2(x), \frac{f_2(x) - xf_1(x)f_0(x) + \sin x}{1 + f_1(x)^2}\right) \quad \text{for} \ -1 \le x \le 1,$$
$$F(0) = (1, 0, 2).$$

Now we can integrate this example as before. Define a mapping T from the space $C([-1, 1], \mathbb{R}^3)$ of functions on $[-1, 1]$ with vector values in $\mathbb{R}^3$ into itself by sending the vector-valued function $F(x) = (f_0(x), f_1(x), f_2(x))$ to

$$TF(x) = \Gamma + \int_0^x \Phi(t, F(t)) \, dt$$
$$= \left(1 + \int_0^x f_1(t) \, dt, \ \int_0^x f_2(t) \, dt, \ 2 + \int_0^x \frac{f_2(t) - tf_1(t)f_0(t) + \sin t}{1 + f_1(t)^2} \, dt\right).$$

This converts the differential equation into the integral equation $TF = F$. An argument similar to the scalar case shows that this fixed-point problem is equivalent to the differential equation.

Returning to the general case, we need to find a suitable framework (i.e., a complete normed vector space) in which to solve the problem in general. Since the coordinate functions $f_0, \ldots, f_{n-1}$ of F are all continuous functions on $[a,b]$, we can think of F as an element of $C([a,b], \mathbb{R}^n)$, the vector space of continuous functions from $[a,b]$ into $\mathbb{R}^n$ with the sup norm. This space is complete, by Theorem 8.2.2. Recall that the norm is given by

$$\|F\|_\infty = \max_{a \le x \le b} \|F(x)\| = \max_{a \le x \le b} \left(\sum_{i=0}^{n-1} |f_i(x)|^2 \right)^{1/2}.$$

A sequence $F^k = (f_0^k, \ldots, f_{n-1}^k)$ converges to $F^* = (f_0^*, \ldots, f_{n-1}^*)$ in the max norm if and only if each of the coordinate functions f_i^k converges uniformly to f_i^* for $0 \le i \le n-1$. To see this, notice that for each coordinate i,

$$\|f_i^k - f_i^*\| = \max_{a \le x \le h} |f_i^k(x) - f_i^*(x)|$$

$$\le \left(\sum_{j=0}^{n-1} |f_j^k(x) - f_j^*(x)|^2 \right)^{1/2} = \|F^k - F^*\|_\infty.$$

Therefore, the convergence of F^k to F^* implies that f_i^k converges to f_i^* for each $0 \le i \le n-1$. Conversely,

$$\|F^k - F^*\|_\infty = \max_{a \le x \le b} \left(\sum_{j=0}^{n-1} |f_j^k(x) - f_j^*(x)|^2 \right)^{1/2}$$

$$\le \left(\sum_{j=0}^{n-1} \max_{a \le x \le b} |f_j^k(x) - f_j^*(x)|^2 \right)^{1/2} = \left(\sum_{j=0}^{n-1} \|f_j^k - f_j^*\|_\infty^2 \right)^{1/2}.$$

So if $\|f_i^k - f_i^*\|_\infty$ tends to zero for each i, then $\|F^k - F^*\|_\infty$ also converges to zero.

We have found a setting for our problem (12.3.3) in which the Contraction Principle (11.1.6) is valid, and so we can formulate the problem in terms of a fixed point. To do this, integrate (12.3.3) from a to x to obtain

$$F(x) = F(c) + \int_c^x F'(t)\, dt = \Gamma + \int_c^x \Phi(t, F)\, dt.$$

This suggests defining a map on $C([a,b], \mathbb{R}^n)$ by

$$TF(x) = \Gamma + \int_c^x \Phi(t, F(t))\, dt.$$

A solution of (12.3.3) is clearly a fixed point of T. Conversely, by the Fundamental Theorem of Calculus, a fixed point of T is a solution of (12.3.3). So the problem (12.3.1) can be solved by finding the fixed point(s) of the mapping T from $C([a,b],\mathbb{R}^n)$ into itself.

Exercises for Section 12.3

For each of the following three differential equations, convert the DE into a first-order vector-valued DE and then into a fixed-point problem.

A. $y^{(3)} + y'' - x(y')^2 = e^x$, $\quad y(0) = 1, y'(0) = -1$, and $y''(0) = 0$.

B. $f''(x) = \left(f(x)^2 + x^2\right)^{1/2} - f'(x)^2$, $\quad f(1) = 0$, and $f'(1) = 1$.

C. $\dfrac{d}{dx}\left(x\dfrac{dv}{dx}\right) + 2x^2v = 0$, $\quad v(1) = 1$, and $v'(1) = 0$.

D. (a) Solve the DE $y' = xy$ and $y(0) = 1$. Deduce that the solution is valid on the whole line. HINT: Integrate $y'/y = x$.
  (b) Define $Tf(x) = 1 + \displaystyle\int_0^x tf(t)\,dt$. Start with $f_0(x) = 1$ and compute $f_n = Tf_{n-1}$ for $n \geq 1$. Prove that this converges to the same solution. Where is this convergence uniform?

E. Let φ be a continuous positive function on $\mathbb{R}$. Consider the DE $f'(x) = \varphi(f(x))$ and $f(c) = \gamma$. Let $F(y) = c + \displaystyle\int_\gamma^y \frac{dt}{\varphi(t)}$. Show that $f(x) = F^{-1}(x)$ is the unique solution.
  HINT: Integrate $f'(t)/\varphi(f(t))$ from c to x.

F. Consider the DE: $xyy' = y^2 - 1$ and $y(1) = 1/\sqrt{2}$.

  (a) Solve the DE.   HINT: Integrate $\frac{yy'}{y^2-1} = \frac{1}{x}$.
  (b) The solution exists only for a finite interval containing 1. Explain why the solution cannot extend further.

12.4 Solutions of Differential Equations

In this section, we use a modification of the Contraction Principle to demonstrate the existence and uniqueness of solutions to a large class of differential equations. The basic idea of contraction mappings is that starting with any function, iteration of the map T leads inevitably to the solution. As we have seen for Newton's method in Section 11.2, it may well be the case that there is a contraction, provided that we start near enough to the solution. Even when T is not a contraction, it may still have an attractive fixed point. In this case, if we start at a reasonable initial approximation, the structure of the integral mapping will force convergence, which is *eventually* contractive.

To analyze the map T, we need a computational result.

12.4.1. DEFINITION. A function $\Phi(x,y)$ is **Lipschitz in the y variable** if there is a constant L such that for all (x,y) and (x,z) in the domain of Φ,

$$\|\Phi(x,y) - \Phi(x,z)\| \leq L\|y - z\|.$$

Note that both the y variable and the range may be vectors rather than elements of $\mathbb{R}$. While this estimate does not concern variation in the x variable, it does require that the constant L be independent of x.

12.4.2. LEMMA. *Let Φ be a continuous function from $[a,b] \times \mathbb{R}^n$ into $\mathbb{R}^n$ that is Lipschitz in y with Lipschitz constant L. Let*

$$TF(x) = \Gamma + \int_a^x \Phi(t, F(t)) \, dt.$$

If $F, G \in C([a,b], \mathbb{R}^n)$ satisfy $\|F(x) - G(x)\| \le \dfrac{M(x-a)^k}{k!}$, then

$$\|TF(x) - TG(x)\| \le \frac{LM(x-a)^{k+1}}{(k+1)!}.$$

In particular, T is uniformly continuous.

PROOF. Compute

$$
\begin{aligned}
\|TF(x) - TG(x)\| &= \left\| y_0 + \int_a^x \Phi(t, F(t)) \, dt - y_0 - \int_a^x \Phi(t, G(t)) \, dt \right\| \\
&= \left\| \int_a^x \Phi(t, F(t)) - \Phi(t, G(t)) \, dt \right\| \\
&\le \int_a^x \|\Phi(t, F(t)) - \Phi(t, G(t))\| \, dt \\
&\le \int_a^x L \|F(t) - G(t)\| \, dt \\
&\le \frac{LM}{k!} \int_a^x (t-a)^k \, dt = \frac{LM}{(k+1)!} (x-a)^{k+1}.
\end{aligned}
$$

In particular, $\|F - G\|_\infty = \|F - G\|_\infty \frac{(x-a)^0}{0!}$. It follows that

$$\|TF - TG\|_\infty \le \|F - G\|_\infty L \|x - a\|_\infty = \|F - G\|_\infty L(b - a).$$

So T has Lipschitz constant $L(b-a)$ and therefore is uniformly continuous. ∎

12.4.3. GLOBAL PICARD THEOREM.

Suppose that Φ is a continuous function from $[a,b] \times \mathbb{R}^n$ into $\mathbb{R}^n$ that is Lipschitz in y. Then the differential equation

$$F'(x) = \Phi(x, F(x)), \quad F(a) = \Gamma$$

has a unique solution.

PROOF. Let T map $C([a,b], \mathbb{R}^n)$ into itself by

$$TF(x) = \Gamma + \int_a^x \Phi(t, F(t)) \, dt.$$

As discussed at the beginning of this section, a function F in $C([a,b], \mathbb{R}^n)$ is a fixed point for T if and only if it is a solution of (12.3.3). Define a sequence of functions by

$$F_0(x) = \Gamma \quad \text{and} \quad F_{k+1} = TF_k \quad \text{for} \quad k \geq 0.$$

Let L be the Lipschitz constant of Φ, and set $M = \max\limits_{a \leq x \leq b} \{\|\Phi(x, \Gamma)\|\}$. We have the inequality

$$\|F_1(x) - F_0(x)\| = \left\| \int_a^x \Phi(t, \Gamma) \, dt \right\| \leq \frac{M(x-a)}{1!}.$$

Therefore, by Lemma 12.4.2,

$$\|F_2(x) - F_1(x)\| \leq \frac{ML(x-a)^2}{2!},$$

$$\|F_3(x) - F_2(x)\| \leq \frac{ML^2(x-a)^3}{3!}.$$

and, by induction, we get

$$\|F_{k+1}(x) - F_k(x)\| \leq \frac{ML^k(x-a)^{k+1}}{k+1!}.$$

As in the proof of the Banach Contraction Principle (11.1.6),

$$\|F_{n+m}(x) - F_n(x)\| \leq \sum_{k=n}^{m+n-1} \|F_{k+1}(x) - F_k(x)\|$$

$$\leq \sum_{k=n}^{m+n-1} \frac{ML^k(x-a)^{k+1}}{k+1!} \leq \frac{M}{L} \sum_{k=n+1}^{\infty} \frac{(L(b-a))^k}{k!}.$$

But the series $\sum\limits_{k=0}^{\infty} (L(b-a))^k / k!$ converges to $e^{L(b-a)}$. So given any $\varepsilon > 0$, there is an integer N such that the tail of the series satisfies

$$\frac{M}{L} \sum_{k=N}^{\infty} \frac{(L(b-a))^k}{k!} < \varepsilon.$$

Thus if $n, n+m \geq N$, it follows that $\|F_{n+m} - F_n\|_\infty < \varepsilon$. This means that the sequence (F_k) is Cauchy in $C([a,b], \mathbb{R}^n)$, and hence converges to a limit function, call it F^*.

Since T is continuous,

$$TF^* = \lim_{k \to \infty} TF_k = \lim_{k \to \infty} F_{k+1} = F^* \,.$$

If G is another solution, then G satisfies $TG = G$. An argument similar to the previous paragraph shows that

$$\|F^* - G\|_\infty = \|T^k F^* - T^k G\|_\infty \le \|F^* - G\|_\infty \frac{(L(b-a))^k}{k!} \,.$$

The left-hand side is constant and nonnegative, while the right-hand side converges to 0 as k tends to infinity. Therefore, $\|F^* - G\|_\infty = 0$. In other words, $G = F^*$ and the solution is unique. ∎

12.4.4. EXAMPLE. Consider the initial value problem

$$y'' + y + \sqrt{y^2 + (y')^2} = 0, \qquad y(0) = \gamma_0, \quad \text{and} \quad y'(0) = \gamma_1.$$

We set this up by letting $Y = (y_0, y_1)$ and

$$\Phi(x, Y) = \Phi(x, y_0, y_1) = \left(y_1, -y_0 - \sqrt{y_0^2 + y_1^2}\right) = \left(y_1, -y_0 - \|Y\|\right).$$

Then the DE becomes

$$Y'(x) = \Phi(x, Y) \qquad \text{and} \qquad Y(0) = \Gamma := (\gamma_0, \gamma_1).$$

Let us verify that Φ is Lipschitz. Let $Z = (z_0, z_1)$. Recall that the triangle inequality implies that $\big| \|Z\| - \|Y\| \big| \le \|Z - Y\|$:

$$\|\Phi(x, Y) - \Phi(x, Z)\| = \big\| \left(y_1 - z_1, z_0 - y_0 + \|Z\| - \|Y\|\right) \big\|$$
$$\le \big\| (y_1 - z_1, z_0 - y_0) \big\| + \big| \|Z\| - \|Y\| \big| \le 2 \|Z - Y\|.$$

Therefore, Picard's Theorem applies, and this equation has a unique solution that is valid on the whole real line.

It may be surprising that this DE actually can be solved explicitly. We may rearrange this equation to look like

$$\frac{y + y''}{\sqrt{y^2 + (y')^2}} = -1.$$

There is no obvious way to integrate the left-hand side. The key observation is that

$$\left(y^2 + (y')^2\right)' = 2yy' + 2y'y'' = 2y'(y + y'').$$

We may multiply both sides by y', known as an **integrating factor**, which makes both sides easily integrable in closed form:

$$\frac{2yy' + 2y'y''}{2\sqrt{y^2 + (y')^2}} = -y'.$$

The left-hand side is the derivative of $\sqrt{y^2 + (y')^2}$. Since y is a function of x and we are about to introduce constants independent of x, it is helpful to switch notation and use $y(x)$ and $y'(x)$ in place of y and y'.

Integrate both sides with respect to x from 0 to x to obtain

$$\sqrt{y(x)^2 + y'(x)^2} - \|\Gamma\| = \gamma_0 - y(x).$$

Take $\|\Gamma\|$ to the other side, square, and simplify to obtain

$$y'(x)^2 = c^2 - 2cy(x), \quad \text{where } c = \|\Gamma\| + \gamma_0.$$

Hence

$$\frac{y'(x)}{\sqrt{c^2 - 2cy(x)}} = \pm 1.$$

Integrating again from 0 to x yields

$$\frac{-1}{c}\sqrt{c^2 - 2cy(x)} + \frac{1}{c}\sqrt{c^2 - 2c\gamma_0} = \pm x.$$

It looks like we may get multiple solutions (which we know isn't the case), but let us persevere. Notice that

$$\sqrt{c^2 - 2c\gamma_0} = \sqrt{\|\Gamma\|^2 - \gamma_0^2} = \pm\gamma_1.$$

Use this identity and simplify to obtain

$$c^2 - 2cy(x) = (\pm cx + \gamma_1)^2 = c^2 x^2 \pm 2c\gamma_1 x + c^2 - 2c\gamma_0.$$

Solving for y produces

$$y(x) = \frac{-c}{2}x^2 \pm \gamma_1 x + \gamma_0.$$

The condition that $y'(0) = \gamma_1$ shows that the sign is $+$ and the unique solution is

$$y(x) = \left(\frac{-\sqrt{\gamma_0^2 + \gamma_1^2} - \gamma_0}{2}\right)x^2 + \gamma_1 x + \gamma_0.$$

Notice that this solution depends continuously on the initial data (γ_0, γ_1). This is a general phenomenon that we explore in Section 12.7.

Exercises for Section 12.4

A.   Consider the DE $y' = 1 + xy$ and $y(0) = 0$ on $[-1, 1]$.

   (a) Show that the associated integral operator is a contraction mapping.

   (b) Find a convergent power series expansion for the unique solution.

   (c) Use the Global Picard Theorem to show that there is a unique solution on $[-b,b]$ for any $b < \infty$. Hence deduce that there is a unique solution on $\mathbb{R}$.

B.   Consider the DE $y'' = y' + xy + 3x^2$, $y(0) = 2$, and $y'(0) = 1$ for $x \in [0,2]$.

   (a) Find the function Φ and vector Γ to put this DE in standard form.

   (b) Calculate the constants involved in the proof of the Global Picard Theorem, and hence find an integer N such that $\|F^* - F_N\|_\infty < 10^{-3}$.

C.   Consider the DE: $xyy' = (2 - x)(y + 2)$ and $y(1) = -1$.

   (a) Separate variables and deduce that the solution y satisfies $\frac{e^{y/2}}{y+2} = xe^{x/2}$.

   (b) Prove that both x and y are bounded.

   HINT: In part (a), minimize the left-hand side and maximize the right-hand side.

D.   Consider the DE $f'(x) = xf(x) + 1$ and $f(0) = 0$.

   (a) Use the Global Picard Theorem to show that there is a unique solution on $[-b,b]$ for any $b < \infty$. Hence deduce that there is a unique solution on $\mathbb{R}$.

   (b) Find an explicit power series that solves the DE.   HINT: Look for a solution of the form $f(x) = \sum_{n=0}^\infty a_n x^n$. Plug this into the DE and find a recurrence relation for the a_n.

   (c) Show that this series converges uniformly on the whole real line. Validate the term-by-term differentiation to verify that this power series is the unique solution.

E.   (a) Suppose φ is C^∞ function on $[a,b] \times \mathbb{R}$, and $Tf(x) = c + \int_a^x \varphi(t, f(t))\, dt$. Show by induction that if $f_0 \in C[a,b]$, then $T^n f_0$ has n continuous derivatives.

   (b) Hence conclude that a fixed point $f = Tf$ must be C^∞.

F.   (a) In the previous question, suppose that φ is C^n. Prove that a solution to $f = Tf$ has $n+1$ continuous derivatives.

   (b) Let $\varphi(t) = t$ for $t \le 1$ and $\varphi(t) = 2 - t$ for $t \ge 1$. Solve the DE $y' = \varphi(y)$ and $y(0) = 1.5$ on $\mathbb{R}$. Verify that the solution is C^1 but is not twice differentiable.

G.   Let $p > 0$. Suppose that $u(x)$ is a solution of $u(x) = \int_0^x \sin(u(t))u(t)^p\, dt$. Prove that $u = 0$.

   HINT: For $a > 0$ and $M = \sup\{|u(x)| : |x| \le a\}$, show $|u(x)| \le M^{pn}|x|^n/n!$ on $[-a,a]$ for $n \ge 0$.

H.   Suppose that Φ and Ψ are Lipschitz functions defined on $[a,b] \times \mathbb{R}$. Let f and g be solutions of $f' = \Phi(x, f(x))$ and $g' = \Psi(x, g(x))$, respectively. Also suppose that $f(a) \le g(a)$ and $\Phi(x,y) \le \Psi(x,y)$ for all $(x,y) \in [a,b] \times \mathbb{R}$. Show that $f(x) \le g(x)$ for all $x \in [a,b]$.

   HINT: If $f(x) = g(x)$, what about $f'(x)$ and $g'(x)$?

I.   **Predator–Prey Equation.** A system that models the populations of a predator $p(t)$ and quarry $q(t)$ at time t is given by the DE $\begin{bmatrix} p' \\ q' \end{bmatrix} = \begin{bmatrix} (bq - a)p \\ (c - dp)q \end{bmatrix}$ where $a,b,c,d > 0$. The quarry grows proportionally to q decreased by a factor proportional to pq that measures interactions with the predator. Likewise p will decline proportionally to p without prey, but is increased proportional to the interactions with the quarry.

   (a) Eliminate t by computing dp/dq. Separate variables and integrate to relate p and q.

   (b) Prove that both populations remain bounded and do not die off.   HINT: Exercise 12.4.C.

12.5 Local Solutions

The stipulation that Φ has to be Lipschitz over all of $\mathbb{R}^n$ is quite restrictive. However, many functions satisfy a Lipschitz condition in y on a set of the form $[a,b] \times \overline{B_R(\Gamma)}$

for $R < \infty$. For example, if Φ has continuous partial derivatives of first order, this follows from the Mean Value Theorem and the Extreme Value Theorem (see the exercises). In this case, the proof would go through as long as $F_n(x)$ always stays in this ball. While this isn't usually possible for all x, it is possible to verify this condition on a small interval $[a, a+h]$. In this way, we get a **local solution**. It is then often possible to piece these local solutions together to extend the solution to all of $[a, b]$. We shall see in a few examples that such an extension is not always possible.

12.5.1. LOCAL PICARD THEOREM.

Suppose that Φ is a continuous function from $[a,b] \times \overline{B_R(\Gamma)}$ into $\mathbb{R}^n$ satisfying a Lipschitz condition in y. Then the differential equation

$$F'(x) = \Phi(x, F(x)), \quad F(a) = \Gamma$$

has a unique solution on the interval $[a, a+h]$, where $h = \min\{b - a, R/\|\Phi\|_\infty\}$.

PROOF. The proof is the same as for the Global Picard Theorem except that we must ensure that $F_n(x)$ remains in $\overline{B_R(\Gamma)}$ so that the iterations remain defined. See Figure 12.2.

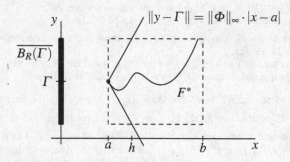

FIG. 12.2 Schematic setup for Local Picard Theorem.

This follows by induction from an easy estimate for $x \in [a, a+h]$:

$$\|F_{n+1}(x) - \Gamma\| \leq \int_a^x \|\Phi(t, F_n(t))\| \, dt \leq \|\Phi\|_\infty |x - a| \leq \|\Phi\|_\infty h \leq R.$$

Thus F_n converges uniformly on $[a, a+h]$ to a solution F^* of the differential equation. The uniqueness argument remains the same. ∎

12.5.2. EXAMPLE. Consider the differential equation

$$y' = y^2, \qquad y(0) = 1, \qquad 0 \leq x \leq 2.$$

In this case, the function is $\Phi(x, y) = y^2$. This is not Lipschitz globally because

$$\frac{\Phi(a, n + \frac{1}{n}) - \Phi(a, n)}{(n + \frac{1}{n}) - n} > 2n \quad \text{for all} \quad n \geq 1.$$

However, Φ is continuously differentiable and therefore Lipschitz on any compact set (Exercise 12.5.A), for example, on $[0, 2] \times \overline{B_{100}(1)}$. By the Local Picard Theorem, this has a unique solution beginning at the origin. The maximum of $|\Phi|$ over $[0, 2] \times B_R(1)$ is $(R + 1)^2$, which leads to $h = R/(R + 1)^2$. The optimal choice is $R = 1$ and $h = \frac{1}{4}$.

This DE may be solved by **separation of variables**. Consider $\frac{y'}{y^2} = 1$. Integrating from $t = 0$ to $t = x$, we obtain

$$x = \int_0^x 1 \, dt = \int_0^x \frac{y'(t) \, dt}{y(t)^2} = -\frac{1}{y(t)} \Big|_0^x = 1 - \frac{1}{y(x)}.$$

Therefore,

$$y(x) = \frac{1}{1 - x}.$$

Evidently, this is a solution on the interval $[0, 1)$, which has a singularity at $x = 1$. So the solution does not extend in any meaningful way to the rest of the interval. The range $[0, 1)$ is better than our estimate of $[0, \frac{1}{4}]$ but is definitely not a solution on all of $[0, 2]$.

Now let us consider how to improve on this situation. Start over at the point $\frac{1}{4}$ and try to extend the solution some more. More generally, suppose that we have used our technique to establish a unique solution on $[0, a]$, where $a < 1$. Consider the DE

$$y' = y^2, \qquad y(a) = \frac{1}{1 - a}, \qquad 0 \leq x \leq 2.$$

Again we take a ball $B_R(\frac{1}{1-a})$ and maximize $\Phi(y) = y^2$ over this ball:

$$\|\Phi\|_{B_R(1/1-a)} = \left(R + \frac{1}{1 - a}\right)^2.$$

Thus Theorem 12.5.1 applies and extends the solution to the interval $[a, a + h]$, where $h = R/\|\Phi\|$. A simple calculus argument maximizes h by taking $R = (1 - a)^{-1}$, which yields $h = (1 - a)/4$.

The thrust of this argument is that repeated use of the Local Picard Theorem extends the solution to increasingly larger intervals. Our first step produced a solution on $[0, a_1]$ with $a_1 = 1/4$. A second application extends this to $[0, a_2]$, where

$$a_2 = a_1 + \frac{1 - a_1}{4} = \frac{3a_1 + 1}{4} = \frac{7}{16}.$$

Generally, the solution on $[0, a_n]$ is extended to $[0, a_{n+1}]$, where

$$a_{n+1} = a_n + \frac{1-a_n}{4} = \frac{3a_n+1}{4}.$$

This is a monotone increasing sequence of real numbers all of which are readily seen to be less than 1. Applying the Monotone Convergence Theorem (2.6.1) to the sequence of real numbers (a_n), there is a limit $L = \lim_{n\to\infty} a_n$. Therefore,

$$L = \lim_{n\to\infty} a_{n+1} = \lim_{n\to\infty} \frac{3a_n+1}{4} = \frac{3L+1}{4}.$$

Solving yields $L = 1$.

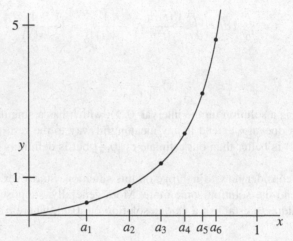

FIG. 12.3 The solution to $y' = y^2$, $y(0) = 1$ with a_n marked.

The upshot is that repeated use of the Local Picard Theorem did extend the solution until it *blew up* by going off to infinity at $x = 1$. See Figure 12.3. It is not possible to use our method further, since we are essentially following along the solution curve, which just carried us off the map.

12.5.3. EXAMPLE. Consider the differential equation

$$y' = y^{2/3}, \qquad y(0) = 0, \qquad 0 \le x \le 2.$$

The function $\Phi(x,y) = y^{2/3}$ is not Lipschitz. Indeed,

$$\lim_{h\to 0} \frac{|\Phi(0,h) - \Phi(0,0)|}{|h|} = \lim_{h\to 0} \frac{1}{|h|^{1/3}} = +\infty.$$

So the Picard Theorems do not apply.

Nevertheless, we may attempt to solve this equation as before. Separating variables and integrating the equation $y^{-2/3}y' = 1$ from 0 to x yields

$$3y^{1/3} = x + c.$$

The initial data imply that $c = 0$, and thus $y = x^3/27$. This solution is valid on the whole real line.

However, there is another nice solution that stands out, namely $y = 0$. So the solution is not unique. In fact, there are many more solutions as well. For any $a > 0$, let

$$f_a(x) = \begin{cases} 0 & \text{if } x \le a, \\ (x-a)^3/27 & \text{if } x \ge a. \end{cases}$$

Then $y = f_a(x)$ is a C^2 solution for every positive a.

We see that there can be existence of solutions without uniqueness. A result known as Peano's Theorem (see Section 12.8) establishes that the differential equation $F'(x) = \Phi(x, F(x))$ has a solution locally whenever Φ is continuous.

Now we will try to systematize what occurred in Example 12.5.2. The idea is to make repeated use of the Local Picard Theorem to extend the solution until either we reach the whole interval or the solution blows up.

12.5.4. DEFINITION. A function $\Phi(x, y)$ on $[a, b] \times \mathbb{R}^n$ is **locally Lipschitz in the y variable** if it is Lipschitz in y on each compact subset $[a, b] \times \overline{B_R}$ for all positive real numbers R.

Actually a truly local definition of locally Lipschitz in y would say that for each $(x, y) \in [a, b] \times \mathbb{R}^n$, there is a positive number $\varepsilon > 0$ such that Φ is Lipschitz in y on $[x - \varepsilon, x + \varepsilon] \times B_\varepsilon(y)$. However, a compactness argument shows that this is equivalent to the definition just given. See Exercise 12.5.J.

12.5.5. CONTINUATION THEOREM.
Suppose that $\Phi : [a, b] \times \mathbb{R}^n \to \mathbb{R}^n$ is a locally Lipschitz function. Consider the differential equation

$$F'(x) = \Phi(x, F(x)), \quad F(a) = \Gamma.$$

Then either

(1) *the DE has a unique solution $F(x)$ on $[a, b]$; or*
(2) *there is a $c \in (a, b)$ such that the DE has a unique solution $F(x)$ on $[a, c)$ and*
$$\lim_{x \to c^-} \|F(x)\| = +\infty.$$

PROOF. The Local Picard Theorem (12.5.1) establishes the fact that there is a unique solution on a nontrivial interval $[a, a + h]$. Define c to be the supremum of all values d for which the DE has a unique solution on $[a, d]$. A priori, these might be different solutions for different values of d. However, if F_1 and F_2 are the unique solutions on $[a, d_1]$ and $[a, d_2]$, respectively, for $a < d_1 < d_2$, then the restriction of F_2 to $[a, d_1]$ will also be a solution. Hence by the uniqueness, $F_2(x) = F_1(x)$, where both are defined. Therefore, we may conclude that there is a unique solution $F^*(x)$ of the DE defined on $[a,c)$ by taking the union of the solutions for $d < c$.

If F^* blows up at c, then we satisfy part (2) of the theorem; and if the solution extends to include b, we have part (1). However, it could be the case that the solution F^* remains bounded on some sequence approaching the point c, contradicting (2), yet the solution does not actually extend to or beyond the point c. The proof is complete if we can show that this is not possible. So we suppose that $c < b$, but there is a sequence x_n increasing to c and a constant K such that $\|F(x_n)\| \leq K$ for $n \geq 1$.

Consider the compact region $D = [a,b] \times \overline{B_{K+1}(0)}$. Since Φ is locally Lipschitz in y, it is Lipschitz on the region D. Set

$$M = \sup_{(x,y) \in D} \|\Phi(x,y)\| \quad \text{and} \quad \delta = \min\left\{\frac{b-c}{2}, \frac{K+1}{2M}\right\}.$$

Choose N large enough that $c - x_N < \delta$. We may apply the Local Picard Theorem (12.5.1) to the DE

$$F'(x) = \Phi(x, F(x)) \qquad F(x_N) = F^*(x_N) \quad \text{for} \quad x \in [x_N, b].$$

The result is a unique solution $F(x)$ on the interval $[x_N, x_N + h]$, where

$$h = \min\{b - x_N, (K+1)/M\} \geq 2\delta.$$

In particular, $d := x_N + h > c$. The uniqueness guarantees that $F(x)$ agrees with $F^*(x)$ on the interval $[x_N, c)$. Hence extending the definition of F^* by setting $F^*(x) = F(x)$ on $[x_N, d]$ produces a solution on $[a,d]$, contradicting the definition of c.

An easy modification of this argument is possible when $c = b$, showing that the solution always extends to the *closed* interval $[a,b]$ if the solution does not blow up. Take $\delta = (K+1)/2M$. We leave the details to the reader. ∎

The solution obtained in the Continuation Theorem is called the **maximal continuation** of the solution to the DE.

12.5.6. EXAMPLE. Not all solutions of differential equations blow up in the manner of the previous theorem. Consider the DE

$$x^4 y'' + 2x^3 y' + y = 0 \qquad \text{and} \qquad y(\tfrac{2}{\pi}) = 1, \quad y'(\tfrac{2}{\pi}) = 0 \quad \text{for} \quad x \in \mathbb{R}.$$

This looks like a reasonably nice linear homogeneous equation (see the next section). However, the coefficient of y'' is not 1, but rather a function of x that vanishes at 0. To put this in our standard form, we divide by x^4 to obtain

$$y'' + \frac{2}{x}y' + \frac{1}{x^4}y = 0.$$

The function Φ is just $\Phi(x, y_0, y_1) = \left(y_1, -\frac{2}{x}y_1 - \frac{1}{x^4}y_0\right)$. This satisfies

$$\|\Phi(x,y_0,y_1) - \Phi(x,z_0,z_1)\| = \left\|\left(y_1 - z_1, \frac{2}{x}(z_1 - y_1) + \frac{1}{x^4}(z_0 - y_0)\right)\right\|$$

$$\leq |y_1 - z_1|\left(1 + \frac{2}{|x|}\right) + \frac{1}{x^4}|z_0 - y_0|$$

$$\leq \left(1 + \frac{2}{|x|} + \frac{1}{x^4}\right)\|y - z\|.$$

So Φ satisfies a global Lipschitz condition in y, provided that x remains bounded away from 0. Thus the Global Picard Theorem (12.4.3) applies on the interval $[\varepsilon, R]$ for any $0 < \varepsilon < \frac{1}{\pi} < R < \infty$. It follows that there is a unique solution on $(0, \infty)$.

We will not demonstrate how to solve this equation. However, it is easy to check (do it!) that the solution is $f(x) = \sin(1/x)$. This solution has a very nasty discontinuity at $x = 0$, but it does not blow up—it remains bounded by 1. This does not contradict the Continuation Theorem (12.5.5). The reason is that the function Φ is not locally Lipschitz on $\mathbb{R} \times \mathbb{R}^2$. It is not even defined for $x = 0$ and has a bad discontinuity there. This shows why we cannot expect to have a global solution to a DE where the coefficient of the highest-order term $y^{(n)}$ vanishes.

Exercises for Section 12.5

A. Suppose that $\Phi(x,y)$ and $\frac{\partial}{\partial y}\Phi(x,y)$ are continuous functions on the region $[a,b] \times [c,d]$. Use the Mean Value Theorem to show that Φ is Lipschitz in y.   HINT: Let $L = \left\|\frac{\partial}{\partial y}\Phi(x,y)\right\|_\infty$.

B. Suppose that $\Phi(x,y_0,y_1)$ is C^1 on $[a,b] \times \overline{B_R(0)}$. Show that Φ is Lipschitz in y.
HINT: Estimate $\Phi(x,y_0,y_1) - \Phi(x,z_0,z_1)$ by subtracting and adding $\Phi(x,z_0,y_1)$.

C. Reformulate Theorem 12.5.1 so that it is valid when the initial conditions apply to a point c in the interior of $[a,b]$.

D. Provide the details for the proof of Theorem 12.5.5 for the case $c = b$.

E. For each of the following DEs, write down the function Φ and decide whether it satisfies (i) a global Lipschitz condition in y, (ii) a local Lipschitz condition in y, or (iii) a Lipschitz condition on a smaller region that allows a local solution.

(a) $y'' = yy'$ and $y(0) = y'(0) = 1$ for $0 \leq x \leq 10$
(b) $y' = \sqrt{1+y^2}$ and $y(0) = 0$ for $0 \leq x \leq 1$
(c) $f'(x) + f(x)^2 = 4xf(x) - 4x^2 + 2$ and $f(0) = 2$
(d) $(x^2 - x - 2)f'(x) + 3xf(x) - 3x = 0$ and $f(0) = 2$ for $-10 \leq x \leq 10$

F. Solve Exercise C (c) explicitly. Find the maximal continuation of the solution.
HINT: Find the DE satisfied by $g(x) = f(x) - 2x$ and solve it.

G. Solve Exercise C (d) explicitly. Find the maximal continuation of the solution.
HINT: Find the DE satisfied by $g(x) = f(x) - 1$ and solve it.

H. Consider $y' = \sin\left(\dfrac{x^3 + x^2 - 1}{\sqrt{101 - y^2}}\right)$ and $y(2) = 3$. Prove that there is a solution on $[-5,9]$.
HINT: Show that $|y| \leq 10$ first. Then obtain a Lipschitz condition.

I. Consider the DE $y' = 3xy^{1/3}$ for $x \in \mathbb{R}$ and $y(0) = c \geq 0$. Let $A_\varepsilon = \{y : |y| \geq \varepsilon\}$.

(a) Show that $3xy^{1/3}$ is Lipschitz in y on $[a,b] \times A_\varepsilon$ but not on $[a,b] \times [-1,1]$.

(b) Solve the DE when $c > 0$.

(c) Find at least two solutions when $c = 0$.

J. Show that the two definitions of locally Lipschitz given in Definition 12.5.4 and the subsequent paragraph are equivalent. HINT: Cover $[a,b] \times \overline{B_R(0)}$ by open sets on which Φ is Lipschitz in y. Use the Borel–Lebesgue Theorem (9.2.3).

K. Consider the DE $f(x)f'(x) = 1$ and $f(0) = a$ for $x \in \mathbb{R}$.

(a) Solve this equation explicitly.

(b) Show that there is a unique solution on an interval about 0 if $a \neq 0$ but that it extends to only a proper subset of $\mathbb{R}$, even though the solution does not blow up. Why does this not contradict the Continuation Theorem?

(c) Show that there are two solutions when $a = 0$ valid on $[0, \infty)$. Why does this not contradict the Local Picard Theorem?

L. Show that $y'' = \left(1 + (y')^2\right)^{3/2}$, $y(0) = y'(0) = 1$ has $f(x) = 1 - \sqrt{1 - x^2}$ as its unique solution on $[-1, 1]$. This solution cannot be continued beyond $x = 1$, yet $|f(x)| \leq 1$. Why does this not contradict the Continuation Theorem? HINT: How does $F(x) = \left(f(x), f'(x)\right)$ behave?

M. Consider the DE: $y' = x^2 + y^2$ and $y(0) = 0$.

(a) Show that this DE satisfies a local Lipschitz condition but not a global one.

(b) Integrate the inequality $y' \geq 1 + y^2$ for $x \geq 1$ to prove that the solution must go off to infinity in a finite time. (See Exercise 12.4.H.)

12.6 Linear Differential Equations

In this section, we explore a very important class of differential equations in greater depth. This class occurs frequently in applications and is also especially amenable to analysis. Consider the differential equation

$$f^{(n)}(x) = p(x) + q_0(x)f(x) + q_1(x)f'(x) + \cdots + q_{n-1}(x)f^{(n-1)}(x), \qquad (12.6.1)$$

$$f(c) = \gamma_0, \quad f'(c) = \gamma_1, \quad \ldots, \quad f^{(n-1)}(c) = \gamma_{n-1},$$

where $p(x)$ and $q_k(x)$ are continuous functions on $[a,b]$ and c is a point in $[a,b]$. This is called a **linear differential equation** because the function

$$\varphi(x,y) = p(x) + q_0(x)y_0 + q_1(x)y_1 + \cdots + q_{n-1}(x)y_{n-1}$$

defined on $[a,b] \times \mathbb{R}^n$ is linear in the second variable $y = (y_0, y_1, \ldots, y_{n-1})$.

Using the reduction in Section 12.3, we obtain the reformulated first-order differential equation

$$F'(x) = \Phi(x, F(x)) \qquad \text{and} \qquad F(c) = \Gamma = (\gamma_0, \gamma_1, \ldots, \gamma_{n-1}),$$

where

$$\Phi(x, y_0, \ldots, y_{n-1}) = \left(y_1, \ldots, y_{n-1}, \varphi(x,y)\right).$$

We will verify that Φ is Lipschitz in y. Set $M = \max_{0 \leq k \leq n-1} \|q_k\|_\infty$. Then for any $x \in [a,b]$ and $y = (y_0, \ldots, y_{n-1})$ and $z = (z_0, \ldots, z_{n-1})$ in $\mathbb{R}^n$,

$$\|\Phi(x,y) - \Phi(x,z)\| = \left\|\left(y_1 - z_1, \ldots, y_{n-1} - z_{n-1}, \sum_{k=0}^{n-1} q_k(x)(y_k - z_k)\right)\right\|$$

$$= \left(\sum_{i=1}^{n-1} |y_i - z_i|^2 + \left|\sum_{k=0}^{n-1} q_k(x)(y_k - z_k)\right|^2\right)^{1/2}$$

$$\leq \left(\|y - z\|^2 + (nM\|y - z\|)^2\right)^{1/2} \leq (1 + nM)\|y - z\|.$$

So Φ satisfies the Lipschitz condition with $L = 1 + nM$.

Therefore, by the Global Picard Theorem (12.4.3), this equation has a unique solution for each choice of initial values Γ.

The most important consequence of linearity is the relationship between solutions of the same DE with different initial values. Suppose that f and g are solutions of (12.6.1) with initial data Γ and $\Delta = (\delta_0, \ldots, \delta_{n-1})$, respectively. Then the function $h(x) = g(x) - f(x)$ satisfies

$$h^{(n)}(x) = p(x) + \sum_{k=0}^{n-1} q_k(x) g^{(k)}(x) - p(x) + \sum_{k=0}^{n-1} q_k(x) f^{(k)}(x) \qquad (12.6.2)$$

$$= q_0(x)h(x) + q_1(x)h'(x) + \cdots + q_{n-1}(x)h^{(n-1)}(x)$$

and satisfies the initial conditions

$$h^{(k)}(c) = \gamma_k - \delta_k \qquad \text{for} \qquad 0 \leq k \leq n-1.$$

This is a linear equation with the term $p(x)$ missing. This DE satisfied by h is called a **homogeneous linear DE**, while the equation (12.6.1) with a nonzero **forcing term** $p(x)$ is called an **inhomogeneous linear DE**.

Generally the homogeneous DE is much easier to solve. Let Γ_i denote the initial conditions $\gamma_i = 1$, $\gamma_j = 0$ for $j \neq i$, $0 \leq j \leq n-1$. In other words, the vectors Γ_i correspond to the standard basis vectors of $\mathbb{R}^n$. For each $0 \leq i \leq n-1$, let $h_i(x)$ be the unique solution of the homogeneous DE (12.6.2). Then let $\Gamma = (\gamma_0, \gamma_1, \ldots, \gamma_{n-1})$ be an arbitrary vector. Consider

$$h(x) = \sum_{i=0}^{n-1} \gamma_i h_i(x).$$

It is easy to calculate that $h^{(k)}(c) = \gamma_k$ for $0 \leq k \leq n-1$ and

$$h^{(n)}(x) = \sum_{k=0}^{n-1} q_k(x) h^{(k)}(x) = \sum_{j=0}^{n-1} \gamma_j \sum_{k=0}^{n-1} q_k(x) h_j^{(k)}(x) = 0.$$

In other words, the solutions of (12.6.2) are linear combinations of the special solutions $h_i(x)$, $0 \leq i \leq n-1$. So these n functions form a basis for the solution space of the homogeneous DE, which is an n-dimensional subspace of $C[a,b]$.

Now one searches for a single solution f_p of the inhomogeneous DE (12.6.1) without regard to the initial conditions. This solution is called a **particular solution**, to distinguish it from the general solution. We have shown that the general solution of this DE has the form $f_p + h$ for some solution h of the homogeneous DE (12.6.2).

Suppose that f_p is the solution of the inhomogeneous DE (12.6.1) with initial data Γ. Let g be the solution of (12.6.1) for initial data Δ. Then $h(x) = g(x) - f_p(x)$ is the solution of the homogeneous DE with initial data $\Delta - \Gamma$. Thus

$$g(x) = f_p(x) + h(x) = f_p(x) + \sum_{i=0}^{n-1} (\delta_i - \gamma_i) h_i(x).$$

Summing up, we have the following useful result.

12.6.3. THEOREM. *The homogeneous equation* (12.6.2) *has n linearly independent solutions* $h_0, \ldots, h_{n-1}$; *and every solution is a linear combination of them.*

If f_p *is a particular solution of the inhomogeneous DE, then every solution (for arbitrary initial conditions) is the sum of* f_p *and a solution of the homogeneous DE.*

Some techniques for solving linear DEs will be explored in the exercises. We now consider a special case in which all the functions q_k are constant.

12.6.4. EXAMPLE. Consider the second-order linear DE with constant coefficients

$$y''(x) - 5y'(x) + 6y(x) = \sin x \quad \text{for} \quad x \in \mathbb{R},$$
$$y(0) = 1, \qquad y'(0) = 0.$$

The first task is to solve the homogeneous equation $y'' - 5y' + 6y = 0$. It is useful to consider the linear map D that sends each function to its derivative: $Df = f'$. Our equation may be written as $(D^2 - 5D + 6I)y = 0$, where I is the identity map $If = f$. This quadratic may be factored as

$$D^2 - 5D + 6I = (D - 2I)(D - 3I),$$

where 2 and 3 are the roots of the quadratic equation $x^2 - 5x + 6 = 0$.

Note that the equation $(D - 2I)y = 0$ is just $y' = 2y$. We can recognize by inspection that $f(x) = e^{2x}$ is a solution. Then we may compute

$$(D^2 - 5D + 6I)e^{2x} = (4 - 10 + 6)e^{2x} = 0.$$

Similarly, e^{3x} is a solution of $(D - 3I)y = 0$ and

$$(D^2 - 5D + 6I)e^{3x} = (9 - 15 + 6)e^{3x} = 0.$$

So $h_1(x) = e^{2x}$ and $h_2(x) = e^{3x}$ are both solutions of this equation.

Let $\Gamma_1 = (h_1(0), h_1'(0)) = (1, 2)$ and $\Gamma_2 = (h_2(0), h_2'(0)) = (1, 3)$ be the initial conditions. These two vectors are evidently independent. Thus every possible vector of initial conditions is a linear combination of Γ_1 and Γ_2. From this, we see that every solution of the homogeneous DE is of the form $h(x) = ae^{2x} + be^{3x}$.

Now let us return to the inhomogeneous problem. A technique called the **method of undetermined coefficients** works well here. This is just a fancy name for good guesswork. It works for forcing functions that are (sums of) exponentials, polynomials, sines, and cosines. We look for a solution of the same type. Here we hypothesize a solution of the form

$$f(x) = c\sin x + d\cos x,$$

where c and d are constants. Plug f into our differential equation:

$$f'' - 5f' + 6f = (-c\sin x - d\cos x) - 5(c\cos x - d\sin x) + 6(c\sin x + d\cos x)$$
$$= (5c + 5d)\sin x + (5d - 5c)\cos x.$$

So we may solve the system of linear equations

$$5c + 5d = 1,$$
$$5c - 5d = 0,$$

to obtain $c = d = 0.1$. This is a particular solution.

Now the general solution to the inhomogeneous equation is of the form

$$f(x) = 0.1\sin x + 0.1\cos x + ae^{2x} + be^{3x}.$$

We compute the initial conditions

$$1 = f(0) = 0.1 + a + b,$$
$$0 = f'(0) = 0.1 + 2a + 3b.$$

Solving this linear system yields $a = 2.8$ and $b = -1.9$. Thus the solution is

$$f(x) = 0.1\sin x + 0.1\cos x + 2.8e^{2x} - 1.9e^{3x}.$$

Exercises for Section 12.6

A. Solve $y'' + 3y' - 10y = 8e^{3x}$, $y(0) = 3$, and $y'(0) = 0$.

B. Consider $y'' + by' + cy = 0$. Factor the quadratic $x^2 + bx + c = (x - r)(x - s)$.

    (a) Solve the DE when r and s are distinct real roots.

    (b) When $r = a + ib$ and $s = a - ib$ are distinct complex roots, show that $e^{ax}\sin bx$ and $e^{ax}\cos bx$ are solutions.

    (c) When r is a double real root, show that e^{rx} and xe^{rx} are solutions.

C. Observe that x is a solution of $y'' - x^{-2}y' + x^{-3}y = 0$. Look for a second solution of the form $f(x) = xg(x)$.   HINT: Find a first-order DE for g'.

D. Let A be an $n \times n$ matrix, and let $y = (y_1, \ldots, y_n)$. Consider $y' = Ay$ and $y(a) = \Gamma$.

(a) Set up the integral equation for this DE.

(b) Starting with $f_0 = \Gamma$, show that the iterates obtained are $f_k(x) = \sum_{i=0}^{k} \frac{1}{i!}(xA)^i \Gamma$.

(c) Deduce that this series converges for any matrix A. The limit is $e^{xA}\Gamma$.

E. Solve the DE of the previous exercise explicitly for $A = \begin{bmatrix} 3 & -\frac{1}{2} \\ -1 & \frac{3}{2} \end{bmatrix}$ and $\Gamma = (1, 2)$.
HINT: Find a basis that diagonalizes A.

F. Consider the DE $y'' + xy' + y = 0$.

(a) Look for a power series solution $y = \sum_{n \geq 0} a_n x^n$. That is, plug this series into the DE and solve for a_n in terms of a_0 and a_1.

(b) Find the radius of convergence of these solutions.

(c) Identify one of the resulting series in closed form $h_1(x)$. Look for a second solution of the form $f(x) = h_1(x)g(x)$. Obtain a DE for g and solve it.

(d) Use the power series expansion to find a particular *polynomial* solution of $y'' + xy' + y = x^3$.

G. **Bessel's DE.** Consider the DE $x^2 y'' + xy' + (x^2 - N^2)y = 0$, where $N \in \mathbb{N}$.

(a) Find a power series solution $y = \sum_{n \geq 0} a_n x^n$. HINT: Show $a_n = 0$ for $n < N$ and $n - N$ odd.

(b) Find the radius of convergence of this power series.

(c) The Bessel function (of the first kind) of order N, denoted by $J_N(x)$, is the multiple of this solution with $a_N = 1/(2^N N!)$. Find a concise expression for the power series expansion J_N in terms of $x/2$ and factorials.

H. **Variation of Parameters.** Let h_1 and h_2 be a basis for the solutions of the homogeneous linear DE $y'' + q_1(x)y' + q_0(x)y = 0$. Define the **Wronskian** determinant to be the function $W(x) = h_1(x)h_2'(x) - h_1'(x)h_2(x)$.

(a) Show that $W'(x) + q_1(x)W(x) = 0$ and $W(c) \neq 0$.

(b) Solve for $W(x)$. Hence show that $W(x)$ is never 0.   HINT: Integrate $W'/W = -q_1$.

(c) Let $p(x)$ be a forcing function. Show that

$$f(x) = -h_1(x) \int_c^x \frac{h_2(t)p(t)}{W(t)}\,dt + h_2(x) \int_c^x \frac{h_1(t)p(t)}{W(t)}\,dt$$

is a particular solution of $y'' + q_1(x)y' + q_0(x)y = p(x)$.

I. Use variation of parameters to solve $y'' - 5y' + 6y = 4xe^x$, $y(0) = 0$, and $y'(0) = -6$.

12.7 Perturbation and Stability of DEs

Another point of interest that can be readily achieved by our methods is the continuous dependence of the solution F on the initial data $F(a) = \Gamma$. It is important in applications that nearby initial values should lead to nearby solutions, and nearby equations have solutions that are also close. This is a variation on the notion of sensitive dependence on initial conditions, which we studied in Section 11.5.

The main theorem of this section is a bit complicated because of the explicit estimates. You should interpret this theorem more qualitatively: If two DEs are close and at least one is Lipschitz, then their solutions are close.

12.7.1. PERTURBATION THEOREM.

Let $\Phi(x,y)$ be a continuous function on a region $D = [a,b] \times \overline{B_R(\Gamma)}$ satisfying a Lipschitz condition in y with constant L. Suppose that Ψ is another continuous function on D such that $\|\Psi - \Phi\|_\infty \leq \varepsilon$. (Note that Ψ is not assumed to be Lipschitz.) Let F and G be the solutions of the differential equations

$$F'(x) = \Phi(x, F(x)), \quad F(a) = \Gamma \qquad and \qquad G(x)' = \Psi(x, G(x)), \quad G(a) = \Delta$$

respectively, such that $(x, F(x))$ and $(x, G(x))$ belong to D for $a \leq x \leq b$. Also, suppose that $\|\Delta - \Gamma\| \leq \delta$. Then, for all $x \in (a,b)$,

$$\|G(x) - F(x)\| \leq \delta e^{L|x-a|} + \frac{\varepsilon}{L}(e^{L|x-a|} - 1).$$

Thus

$$\|G - F\|_\infty \leq \delta e^{L|b-a|} + \frac{\varepsilon}{L}(e^{L|b-a|} - 1).$$

PROOF. Define

$$\tau(x) = \|G(x) - F(x)\| = \left(\sum_{i=0}^{n-1} (g_i(x) - f_i(x))^2 \right)^{1/2}.$$

In particular, $\tau(a) = \|\Delta - \Gamma\| < \delta$. Then by the Cauchy–Schwarz inequality,

$$2\tau(x)\tau'(x) = (\tau(x)^2)' = \sum_{i=0}^{n-1} 2(g_i(x) - f_i(x))(g_i'(x) - f_i'(x))$$

$$\leq 2\left(\sum_{i=0}^{n-1} (g_i(x) - f_i(x))^2 \right)^{1/2} \left(\sum_{i=0}^{n-1} (g_i'(x) - f_i'(x))^2 \right)^{1/2}$$

$$= 2\tau(x)\|G'(x) - F'(x)\|.$$

Now compute

$$\|G'(x) - F'(x)\| = \|\Psi(x, G(x)) - \Phi(x, F(x))\|$$
$$\leq \|\Psi(x, G(x)) - \Phi(x, G(x))\| + \|\Phi(x, G(x)) - \Phi(x, F(x))\|$$
$$\leq \varepsilon + L\|G(x) - F(x)\| = \varepsilon + L\tau(x).$$

Combining these two estimates, we obtain a differential inequality

$$\tau'(x) \leq \|G'(x) - F'(x)\| \leq \varepsilon + L\tau(x).$$

Hence

$$x - a = \int_a^x dt \geq \int_a^x \frac{\tau'(t)}{\varepsilon + L\tau(t)} dt = \frac{1}{L} \log(L\tau + \varepsilon) \Big|_a^x = \frac{1}{L} \log\left(\frac{L\tau(x) + \varepsilon}{L\tau(a) + \varepsilon} \right).$$

Solving for $\tau(x)$ yields

$$L\tau(x) + \varepsilon \leq e^{L(x-a)}(L\tau(a) + \varepsilon) \leq e^{L(x-a)}(L\delta + \varepsilon),$$

whence

$$\|G(x) - F(x)\| = \tau(x) \leq \delta e^{L|x-a|} + \frac{\varepsilon}{L}(e^{L|x-a|} - 1). \qquad \blacksquare$$

An immediate and important consequence of this result is continuous dependence of the solution of a DE (with Lipschitz condition) as a function of the parameter Γ. For simplicity only, we assume a global Lipschitz condition.

12.7.2. COROLLARY. *Suppose that Φ satisfies a global Lipschitz condition in y on $[a,b] \times \mathbb{R}^n$. Then the solution F_Γ of*

$$F'(x) = \Phi(x, F(x)), \qquad F(a) = \Gamma$$

is a continuous function of Γ.

PROOF. Let L be the Lipschitz constant. Since the Lipschitz condition is global, there is no need to check whether the values of $F(x)$ remain in the domain. Also, there is no need to keep δ small.

In this application of Theorem 12.7.1, we take $\varepsilon = 0$, since the function Φ is used for both functions. Hence we obtain

$$\|F_\Gamma - F_\Delta\|_\infty \leq \|\Gamma - \Delta\| e^{L|b-a|}.$$

In particular, it follows that F_Δ converges uniformly to F_Γ as Δ converges to Γ. This is referred to as **continuous dependence on parameters**. $\qquad \blacksquare$

12.7.3. EXAMPLE. Consider the linear DE (12.6.1) of Section 12.6. We showed there that linear DEs satisfy a global Lipschitz condition in y. The solution is a function f_Γ of the initial conditions. However, the estimates are expressed in terms of the vector-valued function $F_\Gamma = \left(f_\Gamma, f_\Gamma', \ldots, f_\Gamma^{(n-1)}\right)$. By Corollary 12.7.2, the solution is a continuous function of the initial data:

$$\|F_\Gamma - F_\Delta\|_\infty \leq \|\Gamma - \Delta\| e^{L|b-a|}.$$

Hence we obtain that

$$\|f_\Gamma^{(k)} - f_\Delta^{(k)}\|_\infty \leq \|\Gamma - \Delta\| e^{L|b-a|} \quad \text{for all} \quad 0 \leq k \leq n-1.$$

So the first $n - 1$ derivatives also depend continuously on the initial data. Consequently, $f_\Gamma^{(n)} = \varphi(x, F_\Gamma)$ is also a continuous function of Γ. Therefore, f_Γ is a continuous function of Γ in the $C^n[a,b]$ norm.

In this case, this is evident from the form of the solution. Recall that we let f_0 be the particular solution with $\Gamma = 0$ and found a basis of solutions h_i for the homogeneous equation (12.6.2) for initial data Γ_i. The general solution is given by

$$f_\gamma(x) = f_0(x) + \sum_{i=0}^{n-1} \gamma_i h_i(x).$$

From this, the continuous dependence of f_γ and its derivatives on Γ is evident.

Theorem 12.7.1 can be interpreted as a stability result. If the differential equation and initial data are measured empirically, then this theorem assures us that the approximate solution based on the measurements remains reasonably accurate. It is rare that differential equations that arise in practice can be explicitly solved in closed form. However, general behaviour can be deduced if the DE is close to a nice one.

12.7.4. EXAMPLE. Suppose that $g(x)$ is a solution of

$$y'' + y = e(x,y,y'), \qquad y(0) = 0, \quad \text{and} \quad y'(0) = 1 \quad \text{for} \quad x \in [-2\pi, 2\pi],$$

where $e(x,y,y')$ is a small function bounded by ε. Then g should be close to the solution of

$$y'' + y = 0, \qquad y(0) = 0, \quad \text{and} \quad y'(0) = 1 \quad \text{for} \quad x \in [-2\pi, 2\pi],$$

which is known to be $f(x) = \sin x$. This unperturbed DE corresponds to the function $\Phi(x,y_0,y_1) = (y_1, -y_0)$, which has Lipschitz constant 1. We apply Theorem 12.7.1, with $\delta = 0$, to obtain

$$\left\| (g(x) - \sin x,\ g'(x) - \cos x) \right\| \le \varepsilon(e^{|x|} - 1).$$

It follows that $g(x)$ is bounded between $\sin(x) - \varepsilon(e^{|x|} - 1)$ and $\sin(x) + \varepsilon(e^{|x|} - 1)$; see Figure 12.4.

By using the bound on g', we can describe g more precisely. Let us assume that $\varepsilon < 0.01$. Since $\cos x > 0.02$ on $[-1.55, 1.55]$, it follows that $g'(x) > 0$ on this range, and hence g is strictly increasing. So 0 is the only zero of g in this range. Similarly, g is strictly decreasing on $[1.59, 4.69]$. We see that

$$g(2.9) > \sin(2.9) - 0.01(e^{2.9} - 1) > 0.06.$$

Similarly, $g(3.5) < 0$. It follows that g has a single zero in the interval $(2.9, 3.5)$. So on $[0, 3.5]$, g oscillates much like the sine function.

Exercises for Section 12.7

A. It is known that $f(x)$ is an exact solution of $y'' + x^2 y' + 2^x y = \sin x$ on $[0,1]$. However, the initial data must be measured experimentally. How accurate must the measurements of $f(0)$ and $f'(0)$ be in order to be able to predict $f(1)$ and $f'(1)$ to within an accuracy of 0.00005?

B. Let $f(x)$ be the solution of the DE $y' = e^{xy}$ and $y(0) = 1$ for $x \in [-1,1]$. Suppose that $f_n(x)$ is the solution of $y' = \sum_{k=0}^{n} \frac{(xy)^k}{k!}$ and $y(0) = 1$ for $x \in [-1,1]$.

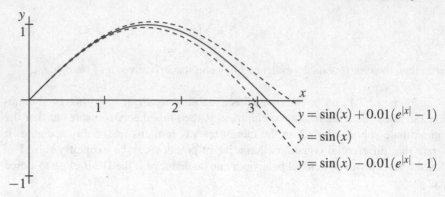

FIG. 12.4 The bounds for $g(x)$.

(a) Show that $f_n(x)$ converges to $f(x)$ uniformly on $[-1,1]$.
(b) Find an N such that $\|f - f_N\|_\infty < 0.0001$.

C. Let $A_\varepsilon = \begin{bmatrix} 1 & \varepsilon \\ \varepsilon & 1 \end{bmatrix}$. Consider the DE: $F_\varepsilon'(x) = A_\varepsilon F(x)$ and $F(0) = (2,3)$ on $[0,1]$.

(a) Solve explicitly for F_ε.   HINT: Exercise 12.6.E
(b) Compute $\|F_\varepsilon - F_0\|_\infty$.
(c) Compare (b) with the bound provided by Theorem 12.7.1.

D. For each $n \geq 1$, define a piecewise linear function $g_n(x)$ on $[0,1]$ by setting $g_n(0) = 1$ and
then defining $g_n(x) = g_n(\frac{k}{n})\left(1 + 3(x - \frac{k}{n})\right)$ for $\frac{k}{n} < x \leq \frac{k+1}{n}$ and $0 \leq k \leq n-1$.

(a) Sketch $g_{10}(x)$ and e^{3x} on the same graph.
(b) Use the fact that g_n is an approximate solution to the DE $y' = 3y$ and $y(0) = 1$ on $[0,1]$ to
estimate $\|g_n - e^{3x}\|_\infty$.
(c) How does this estimate compare with the exact value?
(d) Show that $g_n(x)$ converges uniformly to e^{3x}.

E. Let F satisfy $F'(x) = \Phi(x, F(x))$ and $F(a) = \Gamma$, where Φ is continuous on $[a,b] \times \mathbb{R}^n$ and Lip-
schitz in y with constant L. Suppose G is differentiable and satisfies $\|G'(x) - \Phi(x, G(x))\| \leq \varepsilon$
and $\|G(a) - \Gamma\| \leq \delta$. Show that $\|G(x) - F(x)\| \leq \delta e^{L|x-a|} + \frac{\varepsilon}{L}(e^{L|x-a|} - 1)$.
HINT: Apply the *proof* of Theorem 12.7.1.

12.8 Existence Without Uniqueness

So far, all of our theorems establishing the existence of solutions required a Lipschitz condition. While this is frequently the case in applications, there do exist
common situations for which there is no Lipschitz condition near a critical point
of some sort. As we saw in Example 12.5.3, this might result in the existence of
multiple solutions. It turns out that by merely assuming continuity of the function
Φ (which is surely not too much to ask), the existence of a solution is guaranteed
in a small interval. We could again use continuation methods to obtain solutions on

larger intervals. However, we will not do that here. The additional tool we need in order to proceed is a compactness theorem for functions, the Arzelà–Ascoli Theorem (8.6.9).

This is the key tool that leads to the main result of this section.

12.8.1. PEANO'S THEOREM.

Suppose that $\Gamma \in \mathbb{R}^n$ and Φ is a continuous function from $D = [a,b] \times \overline{B_R(\Gamma)}$ into $\mathbb{R}^n$. Then the differential equation

$$F'(x) = \Phi(x, F(x)), \qquad f(a) = \Gamma \quad for \quad a \leq x \leq b$$

has a solution on $[a, a+h]$, where $h = \min\{b-a, R/M\}$ and $M = \|\Phi\|_\infty$ is the max norm of Φ over the set D.

PROOF. As in Picard's proof, we convert the problem to finding a fixed point for the integral mapping

$$TF(x) = \Gamma + \int_a^x \Phi(t, F(t))\, dt.$$

For each $n \geq 1$, we define a function $F_n(x)$ on $[a, a+h]$ as follows:

$$F_n(x) = \begin{cases} \Gamma & \text{for} \quad a \leq x \leq a+\frac{1}{n}, \\ \Gamma + \int_a^{x-1/n} \Phi(t, F_n(t))\, dt & \text{for} \quad a+\frac{1}{n} \leq x \leq a+h. \end{cases}$$

Notice that the integral defines $F_n(x)$ in terms of the values of $F_n(x)$ in the interval $[u, x - \frac{1}{n}]$. Since F_n is defined to be the constant Γ on $[a, a+\frac{1}{n}]$, the definition of F_n as an integral makes sense on the interval $[a+\frac{1}{n}, a+\frac{2}{n}]$. Once this is accomplished, it then follows that the integral definition makes sense on the interval $[a+\frac{2}{n}, a+\frac{3}{n}]$. Proceeding in this way, we see that the definition makes sense on all of $[a, a+h]$, provided that we verify that $F_n(x)$ remains in $\overline{B_R(\Gamma)}$. This is an easy estimate:

$$\|F_n(x) - \Gamma\| \leq \int_a^{x-1/n} \|\Phi(t, F_n(t))\|\, dt \leq M|x-a| \leq Mh \leq R.$$

It is also easy to show that F_n is an approximate solution to the fixed-point problem. For $a \leq x \leq a+\frac{1}{n}$,

$$\|TF_n(x) - F_n(x)\| = \left\| \Gamma + \int_a^x \Phi(t, F_n(t))\, dt - \Gamma \right\|$$

$$\leq \int_a^x \|\Phi(t, F_n(t))\|\, dt \leq M(x-a) \leq \frac{M}{n}.$$

For $a+\frac{1}{n} \leq x \leq a+h$,

$$\|TF_n(x) - F_n(x)\| = \left\| \int_{x-1/n}^x \Phi(t, F_n(t))\, dt \right\| \leq \int_{x-1/n}^x \|\Phi(t, F_n(t))\|\, dt \leq \frac{M}{n}.$$

So $\|TF_n - F_n\|_\infty \le M/n$.

We will show that the family $\{F_n : n \ge 1\}$ is equicontinuous. Indeed, given $\varepsilon > 0$, let $\delta = \varepsilon/M$. If $a \le x_1 < x_2 \le a+h$ and $|x_2 - x_1| < \delta$, then

$$\|F_n(x_2) - F_n(x_1)\| \le \int_{x_1-1/n}^{x_2-1/n} \|\Phi(t, F_n(t))\| \, dt$$

$$\le M\|(x_1 - 1/n) - (x_2 - 1/n)\| < M\delta = \varepsilon.$$

Therefore we may apply the Arzelà–Ascoli Theorem. The family of functions $\{F_n : n \ge 1\}$ is bounded by $\|\Gamma\| + R$ and is equicontinuous. So its closure is compact. Thus we can extract an increasing sequence n_k such that the F_{n_k} converge uniformly on $[a, a+h]$ to a function $F^*(x)$. We will show that F^* is a fixed point of T, and hence the desired solution. Compute

$$\|F^*(x) - TF^*(x)\| \le \|F^*(x) - F_{n_k}(x)\| + \|F_{n_k}(x) - TF_{n_k}(x)\| + \|TF_{n_k}(x) - TF^*(x)\|$$

$$\le \|F^* - F_{n_k}\|_\infty + \frac{M}{n_k} + \int_a^{a+h} \|\Phi(t, F^*(t)) - \Phi(t, F_{n_k}(t))\| \, dt.$$

Now Φ is uniformly continuous on the compact set D by Theorem 5.5.9. Since F_{n_k} converges uniformly to F^* on $[a, a+h]$, it follows that $\Phi(x, F_{n_k}(x))$ converges uniformly to $\Phi(x, F^*(x))$. Hence by Theorem 8.3.1,

$$\lim_{k \to \infty} \int_a^{a+h} \|\Phi(t, F^*(t)) - \Phi(t, F_{n_k}(t))\| \, dt = 0.$$

Putting all of our estimates together and letting k tend to ∞, we obtain $TF^* = F^*$, completing the argument.   ∎

Exercises for Section 12.8

A. Show that the DE $y^{(4)} = 120y^{1/5}$ and $y(0) = y'(0) = y^{(2)}(0) = y^{(3)}(0) = 0$ has infinitely many solutions on the whole real line.   HINT: Compare with Example 12.5.3.

B. Let $\gamma \in \mathbb{R}$ and let Φ be a continuous real-valued function on $[a, b] \times [\gamma - R, \gamma + R]$. Consider the DE $y'(x) = \Phi(x, y)$ and $y(a) = \gamma$, and suppose that Peano's Theorem guarantees a solution on $[a, a+h]$. If f and g are both solutions on $[a, a+h]$, show that their maximum $f \vee g(x) = \max\{f(x), g(x)\}$ and minimum $f \wedge g(x) = \min\{f(x), g(x)\}$ are also solutions.   HINT: Verify the DE in $U = \{x : f(x) > g(x)\}$, $V = \{x : f(x) < g(x)\}$, and $X = \{x : f(x) = g(x)\}$ separately.

C. Let Φ be a continuous function on $[a, b] \times \mathbb{R}$ such that $\Phi(x, y)$ is a decreasing function of y for each fixed x.

(a) Suppose that f and g are solutions of $y'(x) = \Phi(x, y(x))$. Show that $|f(x) - g(x)|$ is a decreasing function of x.   HINT: If $f(x) > g(x)$ on an interval I and $x_1 < x_2 \in I$, express $(f(x_2) - g(x_2)) - (f(x_1) - g(x_1))$ as an integral.

(b) Show that this DE has a unique solution for the initial condition $y(a) = \gamma$.

D. Show that the set of all solutions on $[a, a+h]$ to the DE of Peano's Theorem is closed, bounded, and equicontinuous.

E. Consider the setup of Exercise 12.8.B. Prove that the set of all solutions on $[a, a+h]$ has a largest and smallest solution.   HINT: Use Exercise 12.8.D to obtain a countable dense subset $\{f_n\}$ of the set of solutions. Let $g_k = \max\{f_1, \ldots, f_k\}$ for $k \geq 1$. Show that g_k converges to the maximal solution $f_{\max}$.

F. Again the setup as in Exercise 12.8.B. Let $x_0 \in [a, a+h]$. Show that $\{f(x_0) : f$ solves the DE$\}$ is a closed interval.   HINT: If $c \in [f_{\min}(x_0), f_{\max}(x_0)]$, show that $f'(x) = \Phi(x, f(x))$ and $f(x_0) = c$ has a solution f on $[a, c]$. Consider $g(x) = (f_{\max} \wedge f) \vee f_{\min}$.

Chapter 13
Fourier Series and Physics

Fourier series were first developed to solve partial differential equations that arise in physical problems, such as heat flow and vibration. We will look at the physics problem of heat flow to see how Fourier series arise and why they are useful. Then we will proceed with the solution, which leads to a lot of very interesting mathematics. We will also see that the problem of a vibrating string leads to a different PDE that requires similar techniques to solve.

While these problems sound very applied, the infinite series that arise as solutions forced mathematicians to delve deeply into the foundations of analysis. When d'Alembert proposed his solution for the motion of a vibrating string in 1754, there were no clear, precise definitions of limit, function, or even of the real numbers—all things taken for granted in most calculus courses today. D'Alembert's solution has a closed form, and thus did not really challenge deep principles. However, the solution to the heat problem that Fourier proposed in 1807 required notions of convergence that mathematicians of that time did not have. Fourier won a major prize in 1812 for this work, but the judges, Laplace, Lagrange, and Legendre, criticized Fourier for lack of rigour. Work in the nineteenth century by many now famous mathematicians eventually resolved these questions by developing the modern definitions of limit, continuity, and uniform convergence. These tools were developed not because of some fetish for finding complicated things, but because they were essential to understanding Fourier series.

13.1 The Steady-State Heat Equation

The purpose of this section is to derive from physical principles the partial differential equation satisfied by heat flow on a surface.

Consider the problem of determining the temperature on a thin metal disk given that the temperature on the boundary circle is fixed. We assume that there is no heat loss in the third dimension. Perhaps this disk is placed between two insulating pads.

K.R. Davidson and A.P. Donsig, *Real Analysis and Applications: Theory in Practice*,
Undergraduate Texts in Mathematics, DOI 10.1007/978-0-387-98098-0_13,
© Springer Science + Business Media, LLC 2010

We also assume that the system is at equilibrium. As a consequence, the temperature at each point remains constant over time. This is the **steady-state heat problem**.

It is convenient to work in polar coordinates in order to exploit the symmetry. The disk will be given as the set

$$\overline{\mathbb{D}} = \{(x,y) : x^2 + y^2 \leq 1\} = \{(r,\theta) : 0 \leq r \leq 1, \ -\pi \leq \theta \leq \pi\},$$

where $(0,\theta)$ represents the origin for all values of θ; and we identify $(r,-\pi) = (r,\pi)$ for all $r \geq 0$. More generally, we allow (r,θ) for any real value of θ and make the identification $(r,\theta + 2\pi) = (r,\theta)$.

Let us denote the temperature distribution over the disk by a function $u(r,\theta)$, and let the given function on the boundary circle be $f(\theta)$.

As usual in physical problems such as this, we need to know a mathematical form of the appropriate physical law in order to determine a differential equation that governs the behaviour of the system. In this case, the law is that the heat flow across a boundary is proportional to the temperature difference between the two sides of the curve. Of course, our temperature distribution function will be continuous. So we must deal with the infinitesimal version of temperature change, which is the derivative of the temperature in the direction perpendicular to the boundary, known as the **normal derivative**.

With this assumption, we can write down a formula expressing the fact that given any region R in the disk with piecewise smooth boundary $\mathscr{C}$, the total amount of heat crossing the boundary $\mathscr{C}$ is 0. This yields the heat conservation equation

$$0 = \int_{\mathscr{C}} \frac{\partial u}{\partial n} \, ds.$$

Here $\partial u / \partial n$ denotes the normal derivative in the outward direction perpendicular to the tangent, and ds indicates integration over arc length along the curve.

Those students comfortable with multivariable calculus will recognize a version of the Divergence Theorem:

$$\int_{\mathscr{C}} \frac{\partial u}{\partial n} \, ds = \int_{R} \Delta u \, dA,$$

where $\Delta u = u_{xx} + u_{yy}$ is the **Laplacian** and dA represents integration with respect to area. Whenever a continuous function integrates to 0 over *every* nice region (say squares or disks), then the function must be 0 everywhere, which leads to the equation $\Delta u = 0$. This is the desired differential equation, except that it is necessary to express the Laplacian in polar coordinates. A (routine but nontrivial) exercise using the multivariate chain rule shows that

$$\Delta u = u_{rr} + \frac{1}{r} u_r + \frac{1}{r^2} u_{\theta\theta}.$$

For the convenience of students unfamiliar with ideas in the previous paragraph, we show how to derive this equation directly. This has the advantage that we can work in polar coordinates and avoid the need for a messy change of variables. The idea is to take R to be the region graphed in Figure 13.1, namely

$$R = \{(r, \theta) : r_0 \le r \le r_1, \, \theta_0 \le \theta \le \theta_1\}.$$

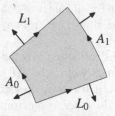

FIG. 13.1 The region R with boundary and outward normal vectors.

The boundary $\mathscr{C}$ of R consists of two radial line segments

$$L_0 = \{(r, \theta_0) : r_0 \le r \le r_1\} \quad \text{and} \quad L_1 = \{(r, \theta_1) : r_0 \le r \le r_1\}$$

and two arcs

$$A_0 = \{(r_0, \theta) : \theta_0 \le \theta \le \theta_1\} \quad \text{and} \quad A_1 = \{(r_1, \theta) : \theta_0 \le \theta \le \theta_1\}.$$

Taking orientation into account, $\mathscr{C} = L_0 + A_1 - L_1 - A_0$. Along L_1, the outward normal at (r, θ_1) is in the θ direction, and arc length along the circle with angle h is rh; thus

$$\frac{\partial u}{\partial n}(r, \theta_1) = \lim_{h \to 0} \frac{u(r, \theta_1 + h) - u(r, \theta_1)}{rh} = \frac{1}{r} u_\theta(r, \theta_1).$$

Along the arc A_1, the outward normal is the radial direction, and thus the normal derivative is $u_r(r_1, \theta)$. The arc length ds along the radii L_0 and L_1 is just dr, while arc length along an arc of radius r is $r \, d\theta$. Thus by adding over the four pieces of the boundary, we obtain the conservation law

$$0 = \int_{r_0}^{r_1} \frac{1}{r} \left(u_\theta(r, \theta_1) - u_\theta(r, \theta_0) \right) dr + \int_{\theta_0}^{\theta_1} u_r(r_1, \theta) r_1 \, d\theta - \int_{\theta_0}^{\theta_1} u_r(r_0, \theta) r_0 \, d\theta$$

$$= \int_{r_0}^{r_1} \frac{1}{r} \left(u_\theta(r, \theta_1) - u_\theta(r, \theta_0) \right) dr + \int_{\theta_0}^{\theta_1} \left(r_1 u_r(r_1, \theta) - r_0 u_r(r_0, \theta) \right) d\theta.$$

Divide by $\theta_1 - \theta_0$ and take the limit as θ_1 decreases to θ_0. The first term is integrated with respect to r, which is independent of θ, and thus the limit is evaluated using the Leibniz rule (8.3.4). For the second term, the limit follows from the Fundamental Theorem of Calculus:

$$0 = \int_{r_0}^{r_1} \frac{1}{r} \lim_{\theta_1 \to \theta_0} \frac{u_\theta(r,\theta_1) - u_\theta(r,\theta_0)}{\theta_1 - \theta_0} dr + \lim_{\theta_1 \to \theta_0} \frac{1}{\theta_1 - \theta_0} \int_{\theta_0}^{\theta_1} (r_1 u_r(r_1,\theta) - r_0 u_r(r_0,\theta)) d\theta$$

$$= \int_{r_0}^{r_1} \frac{1}{r} u_{\theta\theta}(r,\theta_0) dr + r_1 u_r(r_1,\theta_0) - r_0 u_r(r_0,\theta_0).$$

Now divide by $r_1 - r_0$ and take the limit as r_1 decreases to r_0. We obtain

$$0 = \lim_{r_1 \to r_0} \frac{1}{r_1 - r_0} \int_{r_0}^{r_1} \frac{1}{r} u_{\theta\theta}(r,\theta_0) dr + \frac{r_1 u_r(r_1,\theta_0) - r_0 u_r(r_0,\theta_0)}{r_1 - r_0}$$

$$= \frac{1}{r_0} u_{\theta\theta}(r_0,\theta_0) + \frac{\partial}{\partial r}(r u_r)(r_0,\theta_0)$$

$$= \frac{1}{r_0} u_{\theta\theta}(r_0,\theta_0) + u_r(r_0,\theta_0) + r_0 u_{rr}(r_0,\theta_0) = r_0 \Delta u(r_0,\theta_0).$$

Thus our differential equation and boundary condition become

$$\Delta u := u_{rr} + \frac{1}{r} u_r + \frac{1}{r^2} u_{\theta\theta} = 0 \quad \text{for} \quad 0 \le r < 1, \ -\pi \le \theta \le \pi,$$

$$u(1,\theta) = f(\theta) \quad \text{for} \quad -\pi \le \theta \le \pi.$$

Exercises for Section 13.1

A. Do the change of variables calculation converting $u_{xx} + u_{yy}$ to polar coordinates.

B. Let $u(r,\theta) = \log r$. Compute Δu and $u(1,\theta)$. Explain why u is not a solution of the heat equation for the boundary function $f(\theta) = 0$.

C. Suppose that u solves the steady-state heat equation for the annulus $A = \{(r,\theta) : r_0 \le r \le r_1\}$.

    (a) If u depends only on r, and not on θ, what ODE does u satisfy? Solve the heat equation for the boundary conditions $u(r_0,\theta) = a_0$ and $u(r_1,\theta) = a_1$.

    (b) Show that if u depends only on θ, then it is constant.

D. Show that $u(r,\theta) = (3 - 4r^2 + r^4) + (8r^2 - 8r^4)\sin^2\theta + 8r^4 \sin^4\theta$ satisfies $\Delta u(r,\theta) = 0$ and $u(1,\theta) = 8\sin^4\theta$.

E. Let $S = \{(x,y) : 0 \le x \le 1, y \in \mathbb{R}\}$ and consider the steady-state heat problem on this strip.

    (a) Show that $e^{n\pi y} \sin n\pi x$ is a solution for the problem of zero boundary values.

    (b) Do you believe that this is a reasonable solution to the physical problem? Discuss.

F. Suppose that an infinite rod has a temperature distribution $u(x,t)$ at the point $x \in \mathbb{R}$ at time $t > 0$. The heat equation is $u_t = u_{xx}$.

    (a) Prove that $u(x,t) = \frac{1}{\sqrt{4\pi t}} e^{-x^2/4t}$ is a solution.

    (b) Evaluate the total heat at time t: $\int_{-\infty}^{\infty} u(x,t) dx$. HINT: Let $I = \int_{-\infty}^{\infty} e^{-x^2/2} dx$. Express I^2 as a double integral over the plane, and convert to polar coordinates. Or use Example 8.3.5.

    (c) Evaluate $\lim_{t \to 0} u(x,t)$. Can you give a physical explanation of what this limit represents?

13.2 Formal Solution

The steady-state heat equation is a difficult problem to solve. We approach it first by making the completely unjustified assumption that there will be solutions of a special form. By combining these solutions, we will obtain a quite general solution, ignoring all convergence issues. After that, we will work backward and show rigorously that these solutions in fact make good sense and are completely general. We justify our first steps as *experiments* that lead us to a likely candidate for the solution but do not in themselves constitute a proper derivation of the solution. In subsequent sections, we will use our analysis techniques to justify why it works.

Our method, called **separation of variables**, is to look for solutions of the form $u(r, \theta) = R(r)\Theta(\theta)$, where R is a function only of r and Θ is a function only of θ. This enables us to split the partial differential equation into two ordinary differential equations of a single variable each. Indeed, the DE $\Delta u = 0$ becomes

$$R''(r)\Theta(\theta) + \frac{1}{r}R'(r)\Theta(\theta) + \frac{1}{r^2}R(r)\Theta''(\theta) = 0.$$

Manipulate this by taking all dependence on r to one side of the equation and all dependence on θ to the other to obtain

$$\frac{r^2 R''(r) + rR'(r)}{R(r)} = \frac{-\Theta''(\theta)}{\Theta(\theta)}.$$

The left-hand side does not depend on θ and the right-hand side does not depend on r. Since they are equal, they are both independent of all variables and so are equal to a constant c:

$$\frac{r^2 R''(r) + rR'(r)}{R(r)} = c = \frac{-\Theta''(\theta)}{\Theta(\theta)}.$$

These equations can now be rewritten as

$$\Theta''(\theta) + c\Theta(\theta) = 0$$

and

$$r^2 R''(r) + rR'(r) - cR(r) = 0.$$

The first equation is a well-known linear DE with constant coefficients. We know from Section 12.6 that this DE has a two-parameter space of solutions corresponding to the possible initial values. We can solve the equation $(D^2 + cI)y = 0$ making use of the quadratic equation $x^2 + c = 0$, which has roots $\pm\sqrt{-c}$ if $c < 0$, a double root at 0 for $c = 0$, and two imaginary roots $\pm\sqrt{c}i$ when $c > 0$. Hence by Exercise 12.6.B, we obtain the solutions

$$\Theta(\theta) = A\cos(\sqrt{c}\,\theta) + B\sin(\sqrt{c}\,\theta) \qquad \text{for} \quad c > 0,$$
$$\Theta(\theta) = A + B\theta \qquad\qquad\qquad\qquad \text{for} \quad c = 0,$$
$$\Theta(\theta) = Ae^{\sqrt{-c}\,\theta} + Be^{-\sqrt{-c}\,\theta} \qquad\quad \text{for} \quad c < 0.$$

However, not all of these solutions fit our problem. Our solutions must be 2π-periodic because $(r, -\pi)$ and (r, π) represent the same point; and more generally (r, θ) and $(r, \theta + 2\pi)$ represent the same point. Hence

$$\Theta(-\pi) = \Theta(\pi) \quad \text{and} \quad \Theta'(-\pi) = \Theta'(\pi).$$

This eliminates the case $c < 0$ and limits the $c = 0$ case to the constant functions. For $c > 0$, this forces $\sqrt{c}$ to be an integer. Hence we obtain

$$\Theta(\theta) = \begin{cases} A\cos n\theta + B\sin n\theta & \text{for} \quad c = n^2 \geq 1, \\ A & \text{for} \quad c = 0. \end{cases}$$

Now for each $c = n^2$, $n \geq 0$, we must solve the equation

$$r^2 R''(r) + r R'(r) - n^2 R(r) = 0.$$

This is not as easy to solve, but a trick, the substitution $r = e^t$, leads to the answer. Differentiation yields

$$\frac{dR}{dt} = \frac{dR}{dr}\frac{dr}{dt} = R'r$$

and

$$\frac{d^2R}{dt^2} = \frac{d}{dr}(R'r)\frac{dr}{dt} = (R''r + R')r = r^2 R'' + rR'.$$

Hence our DE becomes $\dfrac{d^2R}{dt^2} = n^2 R$. This is a linear DE with constant coefficients, which has the solutions

$$R = ae^{nt} + be^{-nt} = ar^n + br^{-n} \quad \text{for} \quad n \geq 1$$
$$R = a + bt = a + b\log r \quad \text{for} \quad n = 0.$$

Again physical considerations demand that R be continuous at $r = 0$. This eliminates the solutions r^{-n} and $\log r$. That leaves the solutions $R(r) = ar^n$ for each $n \geq 0$.

Combining these two solutions for each $c = n^2$ provides the solutions

$$u(r, \theta) = A_n r^n \cos n\theta + B_n r^n \sin n\theta \quad \text{for} \quad n \geq 0.$$

The case $n = 0$ is special and yields $u(r, \theta) = A_0$. Since the sum of solutions for a homogeneous DE such as ours will also be a solution, we obtain a formal solution (ignoring convergence issues)

$$u(r, \theta) = A_0 + \sum_{n=1}^{\infty} A_n r^n \cos n\theta + B_n r^n \sin n\theta.$$

Continuing to ignore the question of convergence, we let $r = 1$ and use our boundary condition to obtain

$$f(\theta) = A_0 + \sum_{n=1}^{\infty} A_n \cos n\theta + B_n \sin n\theta.$$

This is a **Fourier series**, which we considered in Section 7.6.

Exercises for Section 13.2

A. (a) Verify that $\Delta u = 0 = \Delta v$ implies that $\Delta (au + bv) = 0$ for all scalars $a, b \in \mathbb{R}$.
   (b) Solve the DE $y' = -y^2$. Show that the sum of any two solutions can never be a solution.
   (c) Explain the difference in these two situations.

B. Adapt the method of this section (i.e., separation of variables) to find the possible solutions of $\Delta u(r, \theta) = 0$ on the region $U = \{(r, \theta) : r > 1\}$ that are continuous on $\overline{U}$ and are **continuous at infinity** in the sense that $\lim_{r \to \infty} u(r, \theta) = L$ exists, independent of θ.

C. Find the Fourier expansion for the function $f(\theta) = 8 \sin^4 \theta$ of Exercise 13.1.D.

D. Let $HS = \{(x, y) : 0 \le x \le 1, y \ge 0\}$, and consider the steady-state heat problem on HS with boundary conditions $u(0, y) = u(1, y) = 0$ and $u(x, 0) = x - x^2$.

   (a) Use separation of variables to obtain a family of basic solutions.
   (b) Show that the conditions on the two infinite bounding lines restrict the possible solutions. If in addition, you stipulate that the solution must be bounded, express the resulting solution as a formal series.
   (c) What does the boundary condition on $[0, 1]$ imply for this formal series?

E. Consider a circular drum membrane of radius 1. At time t, the point (r, θ) on the surface has a vertical deviation of $u(r, \theta, t)$. The wave equation for the motion is $u_{tt} = c^2 \Delta u$, where c is a constant. In this exercise, we will consider only solutions that have radial symmetry (no dependence on θ).

   (a) What boundary condition should apply to $u(1, \theta, t)$?
   (b) Look for solutions to the PDE of the form $u(r, \theta, t) = R(r)T(t)$. Use separation of variables to obtain ODEs for R and T. An unknown constant must be introduced.
   (c) What conditions on the ODE for T are needed to guarantee that T remains bounded (a reasonable physical hypothesis)?
   (d) The DE for R is called Bessel's DE. What degeneracy of the DE requires us to add another condition that R remain bounded at $r = 0$?

F. In Exercise 13.2.D, find the Fourier coefficients necessary to satisfy the boundary condition on the unit interval.

13.3 Convergence in the Open Disk

It is now time to return to the original problem and systematically analyze our proposed solutions. The behaviour of the function u on the open disk

$$\mathbb{D} = \{(r, \theta) : 0 \le r < 1, \, -\pi \le \theta \le \pi\}$$

is much easier than the analysis of the boundary behaviour. We deal with that first, and we will find that it leads to a method for understanding the Fourier series of f.

Often results about the behaviour of u on $\mathbb{D}$ will follow from a stronger result on each smaller closed disk

$$\overline{\mathbb{D}}_R = \{(r,\theta) : 0 \le r \le R, -\pi \le \theta \le \pi\} \quad \text{for} \quad R < 1.$$

13.3.1. PROPOSITION. *Let f be an absolutely integrable function on $[-\pi, \pi]$ with Fourier series $f \sim A_0 + \sum_{n=1}^{\infty} A_n \cos n\theta + B_n \sin n\theta$. Then the series*

$$A_0 + \sum_{n=1}^{\infty} A_n r^n \cos n\theta + B_n r^n \sin n\theta$$

converges uniformly on $\overline{\mathbb{D}}_R$ for any $R < 1$ and thus converges everywhere on the open disk $\mathbb{D}$ to a continuous function $u(r, \theta)$.

PROOF. This follows from the Weierstrass M-test (8.4.7). First,

$$\left\| A_n r^n \cos n\theta + B_n r^n \sin n\theta \right\|_{\overline{\mathbb{D}}_R} = \max_{(r,\theta) \in \overline{\mathbb{D}}_R} \left| A_n r^n \cos n\theta + B_n r^n \sin n\theta \right|$$

$$\le (|A_n| + |B_n|)R^n \le 4\|f\|_1 R^n,$$

where the last inequality follows from Proposition 7.6.4. Since

$$\sum_{n=0}^{\infty} 4\|f\|_1 R^n = \frac{4\|f\|_1}{1-R} < \infty,$$

the M-test guarantees that the series converges uniformly on $\overline{\mathbb{D}}_R$. The uniform limit of continuous functions is continuous by Theorem 8.2.1. Thus, $u(r, \theta)$ is continuous on $\overline{\mathbb{D}}_R$. This is true for each $R < 1$, so u is continuous on the whole *open disk* $\mathbb{D}$. ■

Now we extend this argument to apply to the various partial derivatives of u. This procedure justifies **term-by-term differentiation** under appropriate conditions on the convergence.

13.3.2. LEMMA. *Suppose that $u_n(x,y)$ are C^1 functions on an open set R for $n \ge 0$ such that $\sum_{n=0}^{\infty} u_n(x,y)$ converges uniformly to $u(x,y)$ and $\sum_{n=0}^{\infty} \frac{\partial}{\partial x} u_n(x,y)$ converges uniformly to $v(x,y)$. Then $\frac{\partial}{\partial x} u(x,y) = v(x,y)$.*

PROOF. It is enough to verify the theorem on a small square about an arbitrary point (x_0, y_0) in R. By limiting ourselves to a square, the whole line segment from (x_0, y) to (x, y) will lie in R. We define functions on this small square by

$$w_n(x,y) = \sum_{k=0}^{n} u_k(x,y) = \sum_{k=0}^{n} u_k(x_0, y) + \int_{x_0}^{x} \frac{\partial}{\partial x} \sum_{k=0}^{n} u_k(t, y)\, dt$$

and

$$w(x,y) = u(x_0,y) + \int_{x_0}^{x} v(t,y)\, dt.$$

Since the integrands $\frac{\partial}{\partial x} \sum_{k=0}^{n} u_n(t,y)$ converge uniformly to $v(t,y)$, Corollary 8.3.2 shows that $w_n(x,y)$ converges uniformly to $w(x,y)$. But this limit is $u(x,y)$. Therefore, by the Fundamental Theorem of Calculus,

$$\frac{\partial}{\partial x} u(x,y) = \frac{\partial}{\partial x} w(x,y) = \frac{\partial}{\partial x}\left(u(x_0,y) + \int_{x_0}^{x} v(t,y)\, dt \right)(x,y) = v(x,y). \quad\blacksquare$$

We may apply this to our function $u(r,\theta)$.

13.3.3. THEOREM. *Let f be an absolutely integrable function with associated function $u(r,\theta) = A_0 + \sum_{n=1}^{\infty} A_n r^n \cos n\theta + B_n r^n \sin n\theta$. Then u satisfies the heat equation $\Delta u(r,\theta) = 0$ in the open disk $\mathbb{D}$.*

PROOF. Let $u_n(r,\theta) = A_n r^n \cos n\theta + B_n r^n \sin n\theta$. Then

$$\frac{\partial}{\partial r} u_n(r,\theta) = n A_n r^{n-1} \cos n\theta + n B_n r^{n-1} \sin n\theta,$$

$$\frac{\partial^2}{\partial r^2} u_n(r,\theta) = n(n-1) A_n r^{n-2} \cos n\theta + n(n-1) B_n r^{n-2} \sin n\theta,$$

$$\frac{\partial}{\partial \theta} u_n(r,\theta) = -n A_n r^n \sin n\theta + n B_n r^n \cos n\theta,$$

$$\frac{\partial^2}{\partial \theta^2} u_n(r,\theta) = -n^2 A_n r^n \cos n\theta - n^2 B_n r^n \sin n\theta.$$

We will apply the M-test to the series $\sum_{n=0}^{\infty} u_n(r,\theta)$ and to each series of partial derivatives. Indeed, on the disk $\overline{\mathbb{D}_R}$ for $0 \le R < 1$, each of the preceding terms is uniformly bounded by $4\|f\|_1 n^2 R^{n-2}$. Apply the Ratio Test (Exercise 3.2.I) to the series $\sum_{n=0}^{\infty} 4\|f\|_1 n^2 R^{n-2}$:

$$\lim_{n \to \infty} \frac{4\|f\|_1 (n+1)^2 R^{n-1}}{4\|f\|_1 n^2 R^{n-2}} = R < 1.$$

Thus this series converges; and therefore it follows from the Weierstrass M-test that each series of partial derivatives converges uniformly on $\overline{\mathbb{D}_R}$.

Therefore, by Lemma 13.3.2, the partial derivative of the sum equals the sum of the partial derivatives. This means that

$$\frac{\partial}{\partial r} u(r,\theta) = \sum_{n=0}^{\infty} \frac{\partial}{\partial r} u_n(r,\theta), \qquad \frac{\partial^2}{\partial r^2} u(r,\theta) = \sum_{n=0}^{\infty} \frac{\partial^2}{\partial r^2} u_n(r,\theta),$$

$$\frac{\partial}{\partial \theta} u(r,\theta) = \sum_{n=0}^{\infty} \frac{\partial}{\partial \theta} u_n(r,\theta), \qquad \frac{\partial^2}{\partial \theta^2} u(r,\theta) = \sum_{n=0}^{\infty} \frac{\partial^2}{\partial \theta^2} u_n(r,\theta).$$

This convergence is uniform on every disk $\overline{\mathbb{D}_R}$ for $R < 1$.

Using $\Delta u = u_{rr} + \frac{1}{r}u_r + \frac{1}{r^2}u_{\theta\theta}$, we deduce from Lemma 13.3.2 that

$$\Delta u(r,\theta) = \sum_{n=0}^{\infty} \Delta u_n(r,\theta)$$

and that this convergence is uniform on any disk $\overline{\mathbb{D}}_R$. However, we constructed the functions $r^n \cos n\theta$ and $r^n \sin n\theta$ as solutions of $\Delta u = 0$. Thus the right-hand side of this equation is zero. Hence $\Delta u(r,\theta) = 0$ everywhere in the open disk $\mathbb{D}$. ∎

13.3.4. DEFINITION. A function u such that $\Delta u = 0$ is called a **harmonic function**. The function $u(r,\theta)$ determined by the Fourier series of a 2π-periodic f is called the **harmonic extension** of f.

Exercises for Section 13.3

A.   Find the Fourier series of $f(\theta) = \theta^2$ for $-\pi \le \theta \le \pi$.

    (a) Show that the series for $u(r,\theta)$ converges uniformly on the *closed* disk $\overline{\mathbb{D}}$.
    (b) Show that the series for $u_{\theta\theta}$ does not converge *uniformly* on $\mathbb{D}$.

B.   Let A_n and B_n be the Fourier coefficients of a continuous function $f(\theta)$.

    (a) Show that $v_t(r,\theta) = u(rt,\theta) - A_0 + \sum_{n=1}^{\infty} A_n(rt)^n \cos n\theta + B_n(rt)^n \sin n\theta$ converges uniformly on the closed disk $\overline{\mathbb{D}}$ for $0 < t < 1$.
    (b) Show that v_t converges uniformly to u on $\mathbb{D}$ as $t \to 1^-$ if and only if u extends to a continuous function on $\overline{\mathbb{D}}$.

C.   (a) Show that the Fourier series of a 2π-periodic C^1 function f satisies $|A_n| + |B_n| \le C/n$ for some constant C.   HINT: Integrate by parts and compare with Proposition 7.6.4.
    (b) Show by induction that if f is C^k, then $|A_n| + |B_n| \le Cn^{-k}$ for some constant C.

D.   Show that $u(r,\theta)$ has partial derivatives $\frac{\partial^{j+k}}{\partial r^j \partial \theta^k} u$ of all orders. Hence show that u is C^∞.
    HINT: First verify this for the functions $u_n(r,\theta)$. Then show that the sequence of partial derivatives converges uniformly on each $\mathbb{D}_R$.

E.   Suppose that D_R is the disk of radius $R > 0$ about (x_0, y_0), and $f(\theta)$ is a continuous 2π-periodic function. Let $u(x_0 + r\cos\theta, y_0 + r\sin\theta)$ be the solution to the steady-state heat problem on D_R such that $u(x_0 + R\cos\theta, y_0 + R\sin\theta) = f(\theta)$. The formal solution is a series $u(x_0 + r\cos\theta, y_0 + r\sin\theta) = A_0 + \sum_{n\ge 1} A_n r^n \cos n\theta + B_n r^n \sin n\theta$.

    (a) What bounds do you obtain for A_n and B_n from f?
    (b) State the analogue of Proposition 13.3.1 for D_R.

13.4 The Poisson Formula

We seek a formula for $u(r,\theta)$ in terms of the boundary function $f(\theta)$. The basic idea is to substitute the formula for each Fourier coefficient and interchange the order of the summation and integration. This again requires uniform convergence.

Compute $u(r,\theta)$ for a *bounded* integrable function f:

$$u(r,\theta) = A_0 + \sum_{n=1}^{\infty} A_n r^n \cos n\theta + B_n r^n \sin n\theta$$

$$= \frac{1}{2\pi} \int_{-\pi}^{\pi} f(t)\,dt + \frac{1}{\pi} \sum_{n=1}^{\infty} \int_{-\pi}^{\pi} f(t)\cos nt\,dt\, r^n \cos n\theta + \int_{-\pi}^{\pi} f(t)\sin nt\,dt\, r^n \sin n\theta$$

$$= \frac{1}{2\pi} \int_{-\pi}^{\pi} f(t)\,dt + \sum_{n=1}^{\infty} \frac{1}{\pi} \int_{-\pi}^{\pi} f(t) r^n (\cos nt \cos n\theta + \sin nt \sin n\theta)\,dt$$

$$= \frac{1}{2\pi} \int_{-\pi}^{\pi} f(t)\,dt + \sum_{n=1}^{\infty} \frac{1}{\pi} \int_{-\pi}^{\pi} f(t) r^n \cos n(\theta - t)\,dt.$$

Yet again, we apply the M-test. Since

$$\|f(t) r^n \cos n(\theta - t)\| \le \|f\|_\infty R^n \quad \text{for} \quad 0 \le r \le R,$$

and $\sum_{n=0}^{\infty} \|f\|_\infty R^n = \|f\|_\infty / (1 - R) < \infty$, the series converges uniformly on $\overline{\mathbb{D}}_R$. Therefore, by Theorem 8.3.1, we can interchange the order of the summation and the integral

$$u(r,\theta) = \int_{-\pi}^{\pi} f(t) \frac{1}{2\pi} \left(1 + 2 \sum_{n=1}^{\infty} r^n \cos n(\theta - t)\right) dt$$

$$= \int_{-\pi}^{\pi} f(t) P(r, \theta - t)\,dt = \int_{-\pi}^{\pi} f(\theta - u) P(r, u)\,du.$$

We have introduced the function

$$P(r,\theta) = \frac{1}{2\pi} \left(1 + 2 \sum_{n=1}^{\infty} r^n \cos n\theta\right) \quad \text{for} \quad 0 \le r < 1, \theta \in \mathbb{R}.$$

Notice that the last step is a change of variables in which we make use of the fact that both f and P are 2π-periodic in θ.

The function $P(r, \theta)$ is known as the **Poisson kernel**. The purpose of this section is to develop its basic properties.

We will use the fact that exponentiation and trigonometric functions are related using complex variables. (See Appendix 13.9.)

13.4.1. THE POISSON FORMULA.

Let f be a bounded integrable function on $[-\pi, \pi]$. For $0 \le r < 1$ and $-\pi \le \theta \le \pi$, the harmonic extension of f is given by

$$u(r,\theta) = \int_{-\pi}^{\pi} f(\theta - t) P(r, t)\,dt,$$

where $P(r,t) = \dfrac{1}{2\pi} \dfrac{1 - r^2}{1 - 2r\cos t + r^2}.$

PROOF. From the discussion preceding the theorem, it remains only to evaluate the series for $P(r,t)$. Changing the cosines to complex exponentials allows us to sum a geometric series:

$$2\pi P(r,t) = 1 + 2\sum_{n=1}^{\infty} r^n \cos nt = 1 + \sum_{n=1}^{\infty} r^n(e^{int} + e^{-int})$$

$$= 1 + \sum_{n=1}^{\infty}(re^{it})^n + \sum_{n=1}^{\infty}(re^{-it})^n = 1 + \frac{re^{it}}{1 - re^{it}} + \frac{re^{-it}}{1 - re^{-it}}$$

$$= \frac{(1 - re^{it} - re^{-it} + r^2) + (re^{it} - r^2) + (re^{-it} - r^2)}{(1 - re^{it})(1 - re^{-it})}$$

$$= \frac{1 - r^2}{1 - r(e^{it} + e^{-it}) + r^2} = \frac{1 - r^2}{1 - 2r\cos t + r^2} \qquad \blacksquare$$

The Poisson kernel has a number of very nice properties. The most important are that $P(r,t)$ is positive and that it integrates to 1. Hence $u(r,\theta)$ is a *weighted average* of the values $f(\theta - t)$. The function $P(r,t)$ peaks dramatically at $t = 0$ when r is close to 1, and thus eventually $u(r,\theta)$ depends mostly on the values of $f(u)$ for u close to θ. See Figure 13.2. The graph of $P(15/16,t)$ has been truncated to fit the y-axis range from 0 to 3. In fact, $P(15/16,0)$ is about 4.93.

13.4.2. PROPERTIES OF THE POISSON KERNEL.
For $0 \le r < 1$ and $-\pi \le t \le \pi$,

(1) $P(r,t) > 0$.
(2) $P(r,-t) = P(r,t)$.
(3) $\displaystyle\int_{-\pi}^{\pi} P(r,t)\,dt = 1$.
(4) $P(r,t)$ *is decreasing in t on* $[0,\pi]$ *for fixed r.*
(5) *For any $\delta > 0$,* $\displaystyle\lim_{r \to 1^-} \max_{\delta \le |t| \le \pi} P(r,t) = 0$.

PROOF. Statements (1) and (2) are routine. For (3), take $f = 1$. The Fourier series of f is $A_0 = 1$ and $A_n = B_n = 0$ for $n \ge 1$. Hence $u(r,\theta) = 1$. Plugging this into the Poisson formula, we obtain

$$1 = u(r,\theta) = \int_{-\pi}^{\pi} P(r,t)\,dt.$$

For fixed r, the numerator $1 - r^2$ of $P(r,t)$ is constant, while the denominator is the monotone increasing function $1 - 2r\cos t + r^2$ on $[0,\pi]$, whence $P(r,t)$ is monotone decreasing in t. This verifies (4). Finally, from (2) and (4), we see that

$$\lim_{r \to 1^-} \max_{\delta \le |t| \le \pi} P(r,t) = \lim_{r \to 1^-} P(r,\delta) = \lim_{r \to 1^-} \frac{1 - r^2}{1 - 2r\cos\delta + r^2} = \frac{0}{2(1 - \cos\delta)} = 0. \qquad \blacksquare$$

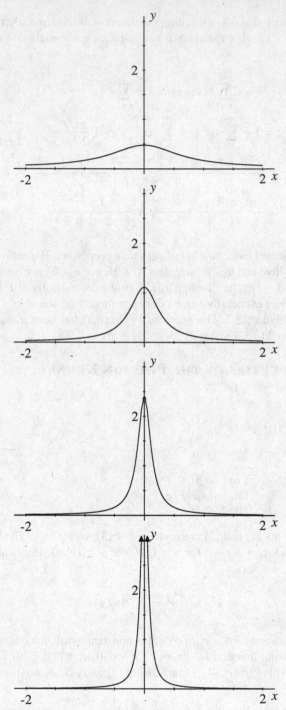

FIG. 13.2 Graphs of $P(r,t)$ for $r = 1/2,\ 3/4,\ 7/8,$ and $15/16$.

In the next chapter, we will develop two other integral kernels, the Dirichlet kernel and the Fejér kernel. The reader may want to compare their properties and compare Figures 13.2, 14.1, and 14.3.

Exercises for Section 13.4

A. Compute the Fourier series of $f(\theta) = P(s,\theta)$ for $0 \le s < 1$.

B. Find an explicit value of $r < 1$ for which $\displaystyle\int_{-0.01}^{0.01} P(r,t)\,dt > 0.999$.

C. Prove that $\dfrac{1-r}{1+r} \le 2\pi P(r,\theta) \le \dfrac{1+r}{1-r}$.

D. Let f be a *positive* continuous 2π-periodic function with harmonic extension $u(r,\theta)$.

    (a) Prove that $u(r,\theta) \ge 0$.

    (b) Prove **Harnack's inequality**: $\dfrac{1-r}{1+r}u(0,0) \le u(r,\theta) \le \dfrac{1+r}{1-r}u(0,0)$.

E. Suppose that f is a 2π-periodic integrable function such that $L \le f(\theta) \le M$ for $-\pi \le \theta \le \pi$. Let $u(r,\theta)$ be the harmonic extension of f. Show that $L \le u(r,\theta) \le M$.

F. Show that if $f(\theta)$ is absolutely integrable on $[-\pi,\pi]$ and $g_n(\theta)$ are continuous functions that converge uniformly to $g(\theta)$ on $[-\pi,\pi]$, then $\displaystyle\lim_{n\to\infty} \int_{-\pi}^{\pi} f(\theta)g_n(\theta)\,d\theta = \int_{-\pi}^{\pi} f(\theta)g(\theta)\,d\theta$.
HINT: Look at the proof of Theorem 8.3.1.

G. Use the previous exercise to show that the Poisson formula is valid for absolutely integrable functions on $[-\pi,\pi]$.

H. Prove that $\displaystyle\int_{-\pi}^{\pi} P(r,\theta-t)P(s,t)\,dt = P(rs,\theta)$.   HINT: Use the series expansion.

I. Let $f(\theta)$ be a continuous 2π-periodic function. Let $u(r,\theta)$ be the harmonic extension of f. Let $0 < s < 1$, and define $g(\theta) = u(s,\theta)$. Prove that the harmonic extension of g is $u(rs,\theta)$.
HINT: Use the Poisson formula twice to obtain the harmonic extension of g as a double integral, and interchange the order of integration.

13.5 Poisson's Theorem

Using the properties of the Poisson kernel, it is now possible to show that $u(r,\theta)$ approaches $f(\theta)$ uniformly as r tends to 1. This means that our proposed solution to the heat problem is continuous on the closed disk and has the desired boundary values. This puts us very close to solving the steady-state heat problem. It also provides a stronger reason for calling $u(r,\theta)$ an extension of f.

13.5.1. POISSON'S THEOREM.
Let f be a continuous 2π-periodic function on $[-\pi,\pi]$ with harmonic extension $u(r,\theta)$. Then $f_r(\theta) := u(r,\theta)$ converges uniformly to f on $[-\pi,\pi]$ as $r \to 1^-$. So $u(r,\theta)$ may be defined on the boundary of $\mathbb{D}$ by $u(1,\theta) = f(\theta)$ to obtain a continuous function on $\overline{\mathbb{D}}$ that is harmonic on $\mathbb{D}$ and agrees with f on the boundary.

PROOF. Let $\varepsilon > 0$ be given. We will find an $r_0 < 1$ such that

$$|f(\theta) - f_r(\theta)| \leq \varepsilon \quad \text{for all} \quad r_0 \leq r < 1, \ -\pi \leq \theta \leq \pi,$$

which will establish uniform convergence.

Let $M = \|f\|_\infty$ and set $\varepsilon_0 = (4\pi M + 1)^{-1}\varepsilon$. By Theorem 5.5.9, f is uniformly continuous. Therefore, there is a $\delta > 0$ such that

$$|f(\theta) - f(t)| \leq \varepsilon_0 \quad \text{for all} \quad |\theta - t| \leq \delta.$$

By property (5), there is an $R < 1$ such that

$$\max_{\delta \leq |t| \leq \pi} P(r,t) \leq \varepsilon_0 \quad \text{for all} \quad R \leq r < 1.$$

Now, using the Poisson formula, compute

$$f(\theta) - f_r(\theta) = f(\theta) \int_{-\pi}^{\pi} P(r,t)\,dt - \int_{-\pi}^{\pi} f(\theta - t)P(r,t)\,dt$$

$$= \int_{-\pi}^{\pi} \big(f(\theta) - f(\theta - t)\big) P(r,t)\,dt.$$

For $R \leq r < 1$, split this integral into two pieces to estimate $|f(\theta) - f_r(\theta)|$

$$\leq \int_{-\delta}^{\delta} |f(\theta) - f(\theta - t)| P(r,t)\,dt + \int_{-\pi}^{-\delta} + \int_{\delta}^{\pi} |f(\theta) - f(\theta - t)| P(r,t)\,dt$$

$$\leq \int_{-\delta}^{\delta} \varepsilon_0 P(r,t)\,dt + \int_{-\pi}^{-\delta} + \int_{\delta}^{\pi} 2M\varepsilon_0\,dt$$

$$\leq \varepsilon_0 + 2\pi(2M\varepsilon_0) = (4\pi M + 1)\varepsilon_0 = \varepsilon.$$

This establishes that $\|f - f_r\|_\infty < \varepsilon$ for $R \leq r < 1$. Hence $f_r(\theta)$ converges uniformly to $f(\theta)$ on $[-\pi, \pi]$.

Extend the definition of $u(r, \theta)$ to the boundary of the closed disk by setting $u(1, \theta) = f(\theta)$. The previous paragraphs show that this function is continuous on the closed disk $\overline{\mathbb{D}}$. By Theorem 13.3.3, u is harmonic on the interior of the disk. ∎

This result yields several important consequences quite easily. The first is the existence of a solution to the heat problem. The question of uniqueness of the solution will be handled in the next section.

13.5.2. COROLLARY. *Let f be a continuous 2π-periodic function. Then the steady-state heat equation $\Delta u = 0$ and $u(1, \theta) = f(\theta)$ has a continuous solution.*

PROOF. By Theorem 13.3.3, the function $u(r, \theta)$ is differentiable on the open disk $\mathbb{D}$ and satisfies $\Delta u = 0$. By Poisson's Theorem, u extends to be continuous on the closed unit disk $\overline{\mathbb{D}}$ and attains the boundary values $u(1, \theta) = f(\theta)$. ∎

This next corollary shows that the Fourier series of a continuous function determines the function uniquely. Continuity is essential for this result. For example, the function that is zero except for a discontinuous point $f(0) = 1$ will have the zero Fourier series but is not the zero function.

13.5.3. COROLLARY. *If two continuous 2π-periodic functions on $\mathbb{R}$ have equal Fourier series, then they are equal functions.*

PROOF. Suppose that f and g have the same Fourier series. Since the harmonic extension $u(r, \theta)$ is defined only in terms of the Fourier coefficients, this is the same function for both f and g. Now, by Poisson's Theorem,

$$f(\theta) = \lim_{r \to 1^-} u(r, \theta) = g(\theta) \quad \text{for} \quad -\pi \le \theta \le \pi.$$

Thus $f = g$. ∎

The final application concerns the case in which the Fourier series itself converges uniformly. This is not always the case, and the delicate question of the convergence of the Fourier series will be dealt with later.

13.5.4. COROLLARY. *Suppose that the Fourier series of a continuous function f converges uniformly. Then the series converges uniformly to f.*

PROOF. Let $f \sim A_0 + \sum\limits_{n=1}^{\infty} A_n \cos n\theta + B_n \sin n\theta$, and let the uniform limit of the series be g. By Theorem 8.2.1, g is continuous; and it is 2π-periodic because it is the limit of 2π-periodic functions. Compute the Fourier series of g using Theorem 8.3.1:

$$\frac{1}{\pi} \int_{-\pi}^{\pi} g(t) \sin nt \, dt = \lim_{N \to \infty} \frac{1}{\pi} \int_{-\pi}^{\pi} \left(A_0 + \sum_{k=1}^{N} A_k \cos kt + B_k \sin kt \right) \sin nt \, dt$$

$$= \lim_{N \to \infty} B_n = B_n.$$

The other coefficients are computed in the same way. It follows that f and g have the same Fourier series. Thus by the previous corollary, they are equal. That is, the Fourier series converges uniformly to f. ∎

13.5.5. EXAMPLE. Recall Example 7.6.3. It was shown that

$$|\theta| \sim \frac{\pi}{2} - \frac{4}{\pi} \sum_{k=0}^{\infty} \frac{\cos((2k+1)\theta)}{(2k+1)^2}.$$

Since $\frac{\pi}{2} + \frac{4}{\pi} \sum_{k=0}^{\infty} \frac{1}{(2k+1)^2}$ converges (by the integral test or by comparison with $\frac{1}{n^2}$), it follows from the Weierstrass M-test (8.4.7) that this series converges uniformly. By Corollary 13.5.4, it follows that the Fourier series converges uniformly to $|\theta|$.

Let us evaluate this at $\theta = 0$:

$$0 = \frac{\pi}{2} - \frac{4}{\pi} \sum_{k=0}^{\infty} \frac{1}{(2k+1)^2}.$$

Therefore,

$$\sum_{k=0}^{\infty} \frac{1}{(2k+1)^2} = \frac{\pi^2}{8}.$$

A little manipulation yields that

$$\sum_{k=1}^{\infty} \frac{1}{k^2} = \sum_{k=0}^{\infty} \frac{1}{(2k+1)^2} + \sum_{k=1}^{\infty} \frac{1}{(2k)^2} = \frac{\pi^2}{8} + \frac{1}{4} \sum_{k=1}^{\infty} \frac{1}{k^2}.$$

Solving, we obtain a famous identity of Euler (by different methods): $\sum_{k=1}^{\infty} \frac{1}{k^2} = \frac{\pi^2}{6}$.

We conclude this section with an important application of Poisson's Theorem to approximation by trigonometric polynomials. Another proof of this fact will be given in Theorem 14.4.5.

13.5.6. COROLLARY. *Every continuous 2π-periodic function is the uniform limit of a sequence of trigonometric polynomials.*

PROOF. Let f be a continuous 2π-periodic function. Given $n \geq 1$, Poisson's Theorem (13.5.1) shows that there is an $r < 1$ such that $\|f - f_r\|_\infty < \frac{1}{2n}$. Let $M = \|f\|_1$ and choose K so large that $\sum_{k=K+1}^{\infty} 4Mr^k < \frac{1}{2n}$. Then the trigonometric polynomial $t_n(\theta) = A_0 + \sum_{k=1}^{K} A_k r^k \cos k\theta + B_k r^k \sin k\theta$ satisfies

$$\|f - t_n\|_\infty \leq \|f - f_r\|_\infty + \|f_r - t_n\|_\infty$$

$$\leq \frac{1}{2n} + \sum_{k=K+1}^{\infty} \|A_k r^k \cos k\theta + B_k r^k \sin k\theta\|$$

$$\leq \frac{1}{2n} + \sum_{k=N+1}^{\infty} 4Mr^k < \frac{1}{n}.$$

Therefore, $(t_n)_{n=1}^{\infty}$ converges uniformly to f.   ∎

Exercises for Section 13.5

A. (a) Compute the Fourier series of $f(\theta) = \theta^2$ for $-\pi \leq \theta \leq \pi$.

(b) Hence evaluate $\sum_{n \geq 1} \frac{(-1)^n}{n^2}$.

B. (a) Compute the Fourier series of the function $f(\theta) = |\sin \theta|$.

(b) Hence evaluate $\sum_{n \geq 1} \dfrac{(-1)^n}{4n^2 - 1}$.

C. Suppose $Q(r, \theta)$ is a function on $\mathbb{D}$ satifying properties (1), (3), and (5) of Proposition 13.4.2. If f is a continuous function on the unit circle, define $v(r, \theta) = \displaystyle\int_{-\pi}^{\pi} f(\theta - t)Q(r,t)\,dt$. Prove that $f_r(\theta) = v(r, \theta)$ converges to f uniformly on $[-\pi, \pi]$.

D. Use rectangular coordinates for the disk $\mathbb{D} = \{(x, y) : x^2 + y^2 < 1\}$, and define a function $u(x, y) = \arctan\left(\frac{y}{1+x}\right)$.

 (a) Show that $\Delta u = u_{xx} + u_{yy} = 0$.
 (b) Show that u is continuous on $\overline{\mathbb{D}}$ except at $(-1, 0)$.
 (c) Show that u is constant on straight line segments through $(-1, 0)$. Hence evaluate $f(\theta) := \lim_{r \to 1^-} u(r \cos \theta, r \sin \theta)$.
 (d) Find the Fourier series for f, and hence find an expression for u in polar coordinates.

E. Suppose $f \sim A_0 + \sum_{n \geq 1} A_n \cos n\theta + B_n \sin n\theta$ is a 2π-periodic continuous function and $\sum_{n \geq 1} |A_n| + |B_n| < \infty$. Prove that the Fourier series converges uniformly to f.

F. A 2π-periodic continuous function $f \sim A_0 + \sum_{n \geq 1} A_n \cos n\theta + B_n \sin n\theta$ satisfies $\sum_{n \geq 1} n|A_n| + n|B_n| < \infty$. Show that $\sum_{n \geq 1} -nA_n \sin n\theta + nB_n \cos n\theta$ converges uniformly to f'.   Hint: Lemma 13.3.2.

G. The Fourier coefficients of a 2π-periodic continuous function f satisfy $|A_n| + |B_n| \leq Cn^{-k}$ for an integer $k \geq 2$. Show that f is in the class C^{k-2}.   Hint: Term-by-term differentiation and the M-test.

H. Give necessary and sufficient conditions for $A_0 + \sum_{n=1}^{\infty} A_n \cos n\theta + B_n \sin n\theta$ to be the Fourier series of a C^∞ function.   Hint: Combine Exercises 13.3.C and 13.5.G.

13.6 The Maximum Principle

The remaining point to be dealt with in the heat problem is the question of uniqueness of solutions. Physically, it is intuitively clear that a fixed temperature distribution on the boundary circle will result in a uniquely determined distribution over the whole disk. We will show this to be the case by establishing a maximal principle showing that a harmonic function must attain its maximum on the boundary circle.

13.6.1. MAXIMUM PRINCIPLE.
Suppose that u is continuous on the closed disk $\overline{\mathbb{D}}$ and $\Delta u = 0$ on the open disk $\mathbb{D}$. Then

$$\max_{(r,\theta) \in \mathbb{D}} u(r, \theta) = \max_{-\pi \leq \theta \leq \pi} u(1, \theta).$$

PROOF. First suppose that $v(r, \theta)$ satisfies $\Delta v \geq \varepsilon > 0$. If v attained its maximum at an interior point (r_0, θ_0), then the first-order partial derivatives would be zero,

$$v_r(r_0, \theta_0) = 0 = v_\theta(r_0, \theta_0),$$

and the second-order derivatives would be negative,

$$v_{rr}(r_0, \theta_0) \leq 0 \quad \text{and} \quad v_{\theta\theta}(r_0, \theta_0) \leq 0.$$

Therefore,

$$\varepsilon \leq \Delta v(r_0, \theta_0) = v_{rr}(r_0, \theta_0) + \frac{1}{r_0} v_r(r_0, \theta_0) + \frac{1}{r_0^2} v_{\theta\theta}(r_0, \theta_0) \leq 0.$$

This contradiction shows that v attains its maximum only on the boundary.

Now consider u. Let $v_n(r, \theta) = u(r, \theta) + \frac{1}{n} r^2$. A simple computation shows that

$$\Delta v_n = \Delta u + \frac{1}{n} \Delta r^2 = \frac{1}{n}\left((r^2)'' + \frac{1}{r}(r^2)' + \frac{1}{r^2}(r^2)_{\theta\theta}\right) = \frac{4}{n} > 0.$$

So v_n attains it maximum only on the boundary circle. Since v_n converges uniformly to u, we obtain

$$\begin{aligned}
\max_{(r,\theta) \in \overline{\mathbb{D}}} u(r, \theta) &= \lim_{n \to \infty} \max_{(r,\theta) \in \overline{\mathbb{D}}} v_n(r, \theta) \\
&= \lim_{n \to \infty} \max_{-\pi \leq \theta \leq \pi} v_n(1, \theta) = \max_{-\pi \leq \theta \leq \pi} u(1, \theta).
\end{aligned}$$

$\blacksquare$

13.6.2. COROLLARY. *Suppose that u is continuous on the closed disk $\overline{\mathbb{D}}$ and $\Delta u = 0$ on the open disk $\mathbb{D}$ and $u(1, \theta) = 0$ for $-\pi \leq \theta \leq \pi$. Then $u = 0$.*

PROOF. By the Maximum Principle, $u(r, \theta) \leq 0$ on $\overline{\mathbb{D}}$. However, $-u$ is also a continuous harmonic function, and thus the Maximum Principle implies that $u(r, \theta) \geq 0$ on $\overline{\mathbb{D}}$. Hence $u = 0$. $\blacksquare$

All the ingredients are now in place for a complete solution to the heat problem on the disk.

13.6.3. THEOREM. *Let $f(\theta)$ be a continuous 2π-periodic function. There exists a unique solution to the steady-state heat problem*

$$\Delta u = 0, \qquad u(1, \theta) = f(\theta),$$

given by the Poisson integral of f.

PROOF. By Corollary 13.5.2, the Poisson integral of f provides a solution $u(r, \theta)$ to the heat problem. It remains to discuss uniqueness. Suppose that $v(r, \theta)$ is another solution. Then consider $w(r, \theta) = u(r, \theta) - v(r, \theta)$. It follows that

$$\Delta w = \Delta u - \Delta v = 0 \quad \text{and} \quad w(1, \theta) = u(1, \theta) - v(1, \theta) = 0.$$

Thus by Corollary 13.6.2, $w = 0$, and so $v = u$ is the only solution. $\blacksquare$

Exercises for Section 13.6

A. Suppose $u(x,y)$ is a solution of the heat problem on $\overline{\mathbb{D}}$ (in rectangular coordinates). Let $D_R(x_0,y_0)$ be a small disk contained inside $\mathbb{D}$. Establish the **mean value property**:

$$u(x_0,y_0) = \frac{1}{2\pi} \int_0^{2\pi} u(x_0 + R\cos\theta, y_0 + R\sin\theta)\, d\theta.$$

HINT: The restriction of u to $\overline{D_R(x_0,y_0)}$ is the solution to the heat problem on this disk. Use the Poisson formula for the value at the centre of the disk.

B. (a) Suppose that $u(x,y)$ is a continuous function on $\overline{\mathbb{D}}$ that satisfies the mean value property of the previous exercise. Prove that u attains its maximum on the boundary.

    (b) Prove that if u attains its maximum value at an interior point, then it must be constant.

C. Prove that a continuous function on $\mathbb{D}$ that satisfies the mean value property is harmonic.
HINT: Fix a point (x_0,y_0) in $\mathbb{D}$ and let $D_R(x_0,y_0)$ be a small disk contained inside $\mathbb{D}$. Let $v(x,y)$ be the solution of the steady-state heat problem on $D_R(x_0,y_0)$ that agrees with u on the boundary circle. Show that $u = v$, and hence deduce that $\Delta u(x_0,y_0) = 0$.

D. Let $u(x,y)$ be a positive harmonic function on an open subset Ω of the plane. Suppose that $\overline{D_R(x_0,y_0)}$ is contained in Ω. Use Exercise 13.4.D to prove that

$$\frac{R-r}{R+r}u(x_0,y_0) \le u(x_0 + r\cos\theta, y_0 + r\sin\theta) \le \frac{R+r}{R-r}u(x_0,y_0) \quad \text{for} \quad 0 \le r < R.$$

E. Let Ω be a bounded open subset of the plane with smooth boundary Γ. A function v on Ω is harmonic if $\Delta v = 0$ on Ω.

    (a) Show that if v is continuous on $\overline{\Omega}$, then it must attain its maximum value on Γ.

    (b) Hence show that if f is a continuous function on Γ, there is at most one continuous function on $\overline{\Omega}$ which is harmonic on Ω and has boundary values equal to f.

F. Let $u(r,\theta) = \dfrac{r(1-r^2)\sin\theta}{(1-2r\cos\theta + r^2)^2}$ on $\mathbb{D}$.

    (a) Prove that u is harmonic on $\mathbb{D}$.

    (b) Show that $\lim\limits_{r\to 1^-} u(r,\theta) = 0$ for all values of θ.

    (c) Why does this not contradict the Maximum Principle?

    (d) Is u bounded?

13.7 The Vibrating String (Formal Solution)

The mathematics of a vibrating string was one of the first problems studied using Fourier series. It arose in a discussion on the oscillations of a violin string by d'Alembert in 1747, twenty-two years before Fourier was born.

Most readers will be familiar with swinging a skipping rope. The simplest mode is a single lobe oscillating between the two fixed ends. However, it is possible to set up a wave with two lobes or even three. These vibrations with more lobes are called **harmonics**. They exist in the vibration of any stringed instrument and tend to be characteristic of the instrument, giving it a distinctive sound. For example, violins have significant order-five harmonics.

Our problem is to describe the motion of a vibrating string. We imagine a uniform string stretched between two fixed endpoints under tension. We further assume that the oscillations are small compared with the length of the string. This is a reasonable assumption for a stiff string like one found on a violin or piano. This leads to the simplifying assumption that each point on the string moves only in a vertical direction. We ignore all forces other than the effect of string tension, such as the weight of the string and air resistance.

Orient the string along the x-axis of the plane, and choose units such that the endpoints are $(0,0)$ and $(\pi,0)$. The vertical displacement of the string will be given by a function $y(x,t)$, the function giving the horizontal position at the point $x \in [0,\pi]$ at the time t. For convenience, we assume that time begins at time $t = 0$. Let τ denote the tension force, and let ρ be the density of the string.

Fix x and t and consider the forces acting on the string segment between the nearby points $A = (x,y) = (x,y(x,t))$ and $B = (x+\Delta x, y+\Delta y)$, where $y + \Delta y = y(x+\Delta x, t)$. The tension τ on the string results in forces acting on both ends of the segment in the direction of the tangent, as shown in Figure 13.3.

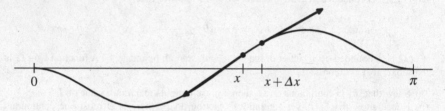

FIG. 13.3 Forces acting on a segment of the string.

The force at A is

$$-\tau \frac{\left(1, \frac{\partial y}{\partial x}(x,t)\right)}{\sqrt{1 + \frac{\partial y}{\partial x}(x,t)^2}} \approx \left(-\tau, -\tau\frac{\partial y}{\partial x}(x,t)\right).$$

The approximation is reasonable since y and $\frac{\partial y}{\partial x}$ are assumed to be small relative to 1. Likewise, the tensile force at B is approximately

$$\left(\tau, \tau\frac{\partial y}{\partial x}(x+\Delta x, t)\right).$$

The horizontal forces cancel, while the combined vertical force is

$$\Delta V(x,t) = \tau\frac{\partial y}{\partial x}(x+\Delta x,t) - \tau\frac{\partial y}{\partial x}(x,t) \approx \tau\Delta x\frac{\partial^2 y}{\partial x^2}(x,t).$$

By Newton's law, we have $F = ma$, where we have a segment of mass $\rho\Delta x$ and acceleration equal to the second derivative of $y(x,t)$ with respect to t. Substitute this in, divide by $\rho\Delta x$, and take the limit as Δx tends to 0 to obtain the linear partial differential equation

$$\frac{\partial^2 y}{\partial t^2}(x,t) = \frac{\tau}{\rho}\frac{\partial^2 y}{\partial x^2}(x,t).$$

Set $\omega^2 = \tau/\rho$. This is known as the one-dimensional **wave equation**:

$$\frac{\partial^2 y}{\partial t^2}(x,t) = \omega^2\frac{\partial^2 y}{\partial x^2}(x,t). \tag{13.7.1}$$

Since the endpoints are fixed, there are boundary conditions

$$y(0,t) = y(\pi,t) = 0 \quad \text{for all} \quad t \ge 0. \tag{13.7.2}$$

Finally, there are initial conditions: Imagine that the string is initially stretched to some (continuous) shape $f(x)$ and is moving with initial velocity $g(x)$. This gives the conditions

$$y(x,0) = f(x), \quad \frac{\partial y}{\partial t}(x,0) = g(x) \quad \text{for all} \quad x \in [0,\pi]. \tag{13.7.3}$$

These boundary conditions, together with the wave equation governing subsequent motion of the string, determine a unique solution. We shall see that it can be solved in a manner similar to our analysis of the steady-state heat problem.

As before, we begin by using separation of variables to look for solutions of the special form $y(x,t) = X(x)T(t)$. There is no way to know in advance that there are solutions of this type, but in fact there are many such solutions, which can then be combined to exhaust all possibilities. Substituting $y(x,t) = X(x)T(t)$ into the wave equation gives

$$X(x)T''(t) = \omega^2 X''(x)T(t).$$

Isolating the variables x and t, we obtain

$$\frac{T''(t)}{T(t)} = \omega^2\frac{X''(x)}{X(x)}.$$

The left-hand side of this equation is independent of x and the right-hand side is independent of t. Thus both sides are independent of both variables and therefore are equal to some constant c.

This results in two ordinary differential equations:

$$X''(x) - \frac{c}{\omega^2}X(x) = 0 \quad \text{and} \quad T''(t) - cT(t) = 0.$$

The boundary condition (13.7.2) simplifies to yield $X(0) = X(\pi) = 0$. At this stage, we must ignore the initial shape conditions (13.7.3).

The equation for X is essentially the same as the equation for Θ in Section 13.2. Depending on the sign of c, the solutions are sinusoidal, linear, or exponential:

$$X(x) = \begin{cases} A\cos\left(\frac{\sqrt{|c|}}{\omega}x\right) + B\sin\left(\frac{\sqrt{|c|}}{\omega}x\right) & \text{for } c < 0, \\ A + Bx & \text{for } c = 0, \\ Ae^{\sqrt{c}x/\omega} + Be^{-\sqrt{c}x/\omega} & \text{for } c > 0. \end{cases}$$

However, the boundary conditions eliminate both the linear and the exponential solutions. Thus the constant $-c/\omega^2$ is strictly positive, say γ^2. The solutions are

$$X(x) = a\cos\gamma x + b\sin\gamma x.$$

Since $X(0) = 0$, this forces $a = 0$. And $X(\pi) = 0$ yields $b\sin\gamma\pi = 0$. Thus a nonzero solution is possible only if γ is an integer n. Therefore, the possible solutions are

$$X_n(x) = b\sin nx \quad \text{for} \quad n \geq 1.$$

Now return to the equation for T. Since $c = -\gamma^2\omega^2 = -n^2\omega^2$, the DE for T becomes

$$T''(t) + (n\omega)^2 T(t) = 0.$$

Again this has solutions

$$T(t) = A\sin n\omega t + B\cos n\omega t.$$

Putting these together, we obtain solutions for $y(x,t)$ of the form

$$y_n(x,t) = A_n\sin nx\,\cos n\omega t + B_n\sin nx\,\sin n\omega t \quad \text{for} \quad n \geq 1.$$

The functions $y_n(x,t)$ correspond to the modes of vibration of the string. For $n = 1$, we have a string shape of a single sinusoidal loop oscillating up and down. This is the **fundamental vibration** mode of the string with the lowest frequency ω. However, for $n = 2$, we obtain a function that has two "arches" that swing back and forth. The frequency is twice the fundamental frequency. For general n, we have higher-frequency oscillations with frequency $n\omega$ that oscillate n times between the two fixed endpoints at n times the rate. As we mentioned before, these higher frequencies are called harmonics.

The differential equation (13.7.1) is linear, so linear combinations of solutions are solutions. Thinking of solutions as waves, this combination of solutions is called **superposition**; see Figure 13.4. Ignoring convergence questions, we have a large family of possible solutions, all of the form

$$y(x,t) = \sum_{n=1}^{\infty} A_n\sin nx\,\cos n\omega t + B_n\sin nx\,\sin n\omega t.$$

Now consider the initial condition (13.7.3). Substituting $t = 0$ into this, we arrive at the boundary condition

$$f(x) = y(x,0) = \sum_{n=1}^{\infty} A_n \sin nx.$$

Likewise, allowing term-by-term differentiation with respect to t, we obtain

$$\frac{\partial y}{\partial t}(x,t) = \sum_{n=1}^{\infty} -n\omega A_n \sin nx \, \sin n\omega t + n\omega B_n \sin nx \, \cos n\omega t.$$

Substituting in the boundary condition at $t = 0$ yields $g(x) = \sum_{n=1}^{\infty} n\omega B_n \sin nx$.

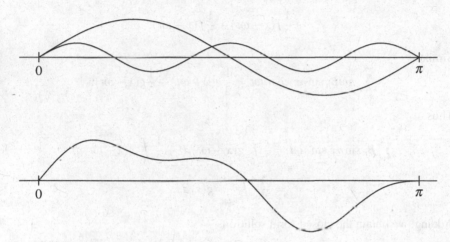

FIG. 13.4 Superposition of two sine waves.

These are Fourier series. In fact, they are sine series because f and g are defined only on $[0, \pi]$. We have seen that summing them requires a certain amount of delicacy. Notice that since the string is fixed at 0 and π, the boundary functions satisfy

$$f(0) = f(\pi) = g(0) = g(\pi) = 0.$$

Let us extend f and g to $[-\pi, \pi]$ as *odd* functions by setting $f(-x) = -f(x)$ and $g(-x) = -g(x)$. Then extend them to be 2π-periodic functions on the whole real line. The values of the coefficients A_n and B_n for $n \geq 1$ are read off from the Fourier (sine) coefficients of f and g.

$$A_n = \frac{1}{\pi} \int_{-\pi}^{\pi} f(t) \sin nt \, dt = \frac{2}{\pi} \int_0^{\pi} f(t) \sin nt \, dt$$

and

$$B_n = \frac{1}{n\omega\pi} \int_{-\pi}^{\pi} g(t) \sin nt \, dt = \frac{2}{n\omega\pi} \int_0^{\pi} g(t) \sin nt \, dt.$$

This form of the solution was proposed by Euler in 1748, and the same idea was advanced by D. Bernoulli in 1753 and Lagrange in 1759. However, we want to find a closed-form solution discovered by d'Alembert in 1747.

Returning to our proposed series solution, use the trig identity

$$2\sin nx \, \cos n\omega t = \sin n(x+\omega t) + \sin n(x-\omega t)$$

and substitute into

$$\sum_{n=1}^{\infty} A_n \sin nx \, \cos n\omega t = \frac{1}{2} \sum_{n=1}^{\infty} A_n \sin n(x+\omega t) + \frac{1}{2} \sum_{n=1}^{\infty} A_n \sin n(x-\omega t)$$

$$= \frac{1}{2} f(x+\omega t) + \frac{1}{2} f(x-\omega t).$$

Similarly,

$$\sum_{n=1}^{\infty} n\omega B_n \sin nx \, \cos n\omega t = \frac{1}{2} g(x+\omega t) + \frac{1}{2} g(x-\omega t).$$

Thus

$$\sum_{n=1}^{\infty} B_n \sin nx \, \sin n\omega t = \frac{1}{2} \int_0^t g(x+\omega t) \, dt + \frac{1}{2} \int_0^t g(x-\omega t) \, dt$$

$$= \frac{1}{2\omega} \int_{x-\omega t}^{x+\omega t} g(s) \, ds.$$

Adding, we obtain the closed-form solution

$$y(x,t) = \frac{1}{2} f(x+\omega t) + \frac{1}{2} f(x-\omega t) + \frac{1}{2\omega} \int_{x-\omega t}^{x+\omega t} g(s) \, ds.$$

We have obtained d'Alembert's form of the solution. It is worth noting that this form no longer involves a series. It doesn't matter what method was used to get here.

The role of ω becomes apparent in this formulation of the solution. Notice that $y(x,t)$ depends on the values of f and g only in the range $[x-\omega t, x+\omega t]$. We should think of ω as the speed of propagation of the signal. In particular, if $g = 0$ and f is a small "bump" supported on $[\frac{\pi}{2}, \frac{\pi}{2}+\varepsilon]$, then for $\frac{\varepsilon}{\omega} \leq t \leq 1 - \frac{\varepsilon}{\omega}$, the function $y(x,t)$ consists of two identical bumps moving out toward the ends of the string at speed ω. When they reach the end, they bounce back. When the bumps overlap, they are superimposed (added). But the shape of the bumps is not lost; the message of the bump is transmitted accurately forever. In particular, the wave equation for light has the constant c, the speed of light, in place of ω.

13.8 The Vibrating String (Rigorous Solution)

We may check directly that d'Alembert's formula yields the solution to the vibrating string problem without returning to the Fourier series expansion. Our assumption that y satisfies a second-order PDE forces f to be C^2 and g to be C^1. We will consider more general initial data afterward.

13.8.1. THEOREM. *Consider the vibrating string equation*

$$\frac{\partial^2 y}{\partial t^2}(x,t) = \omega^2 \frac{\partial^2 y}{\partial x^2}(x,t)$$

with initial conditions

$$y(x,0) = f(x) \qquad and \qquad \frac{\partial y}{\partial t}(x,0) = g(x)$$

such that f is C^2 and g is C^1 and $f(0) = g(0) = f(\pi) = g(\pi) = 0$. This has a unique solution for all $t > 0$ given by

$$y(x,t) = \frac{1}{2}f(x+\omega t) + \frac{1}{2}f(x-\omega t) + \frac{1}{2\omega}\int_{x-\omega t}^{x+\omega t} g(s)\,ds,$$

where f and g have been extended (uniquely) to the whole real line as 2π-periodic odd functions.

PROOF. First, we verify that this is indeed a valid solution. Compute

$$\frac{\partial y}{\partial x}(x,t) = \frac{1}{2}f'(x+\omega t) + \frac{1}{2}f'(x-\omega t) + \frac{1}{2\omega}g(x+\omega t) - \frac{1}{2\omega}g(x-\omega t),$$

$$\frac{\partial^2 y}{\partial x^2}(x,t) = \frac{1}{2}f''(x+\omega t) + \frac{1}{2}f''(x-\omega t) + \frac{1}{2\omega}g'(x+\omega t) - \frac{1}{2\omega}g'(x-\omega t),$$

$$\frac{\partial y}{\partial t}(x,t) = \frac{\omega}{2}f'(x+\omega t) - \frac{\omega}{2}f'(x-\omega t) + \frac{1}{2}g(x+\omega t) + \frac{1}{2}g(x-\omega t),$$

$$\frac{\partial^2 y}{\partial t^2}(x,t) = \frac{\omega^2}{2}f''(x+\omega t) + \frac{\omega^2}{2}f''(x-\omega t) + \frac{\omega}{2}g'(x+\omega t) - \frac{\omega}{2}g'(x-\omega t).$$

Thus $\frac{\partial^2 y}{\partial t^2}(x,t) = \omega^2 \frac{\partial^2 y}{\partial x^2}(x,t)$ and $y(x,0) = f(x)$ and $\frac{\partial y}{\partial t}(x,0) = g(x)$.

The question of uniqueness of solutions remains. As with the heat equation, physical properties yield a clue. The argument in this case is based on conservation of energy. Without actually justifying the physical interpretation, we introduce a quantity called total energy obtained by adding the potential and kinetic energies of the string:

$$E(t) = E(y,t) = \int_0^\pi \left(\frac{\partial y}{\partial x}(x,t)\right)^2 + \frac{1}{\omega^2}\left(\frac{\partial y}{\partial t}(x,t)\right)^2 dx.$$

This is defined for any solution y of our vibration problem. For convenience, we switch to the notation y_x, y_t, and so on for partial derivatives. Use the Leibniz rule (8.3.4) to compute

$$E'(t) = \frac{\partial}{\partial t} \int_0^\pi \left(y_x(x,t)\right)^2 + \frac{1}{\omega^2} \left(y_t(x,t)\right)^2 dx = \int_0^\pi 2y_x y_{xt}(x,t) + \frac{1}{\omega^2} 2y_t y_{tt}(x,t)\, dx.$$

Substitute $y_{tt} = \omega^2 y_{xx}$:

$$= \int_0^\pi 2y_x y_{xt}(x,t) + 2y_t y_{xx}(x,t)\, dx = \int_0^\pi \frac{\partial}{\partial x}\left(y_x y_t(x,t)\right) dx = y_x y_t(x,t)\Big|_{x=0}^{x=\pi} = 0.$$

The last equality follows since the string is fixed at both endpoints, forcing the relation $y_t(0,t) = y_t(\pi,t) = 0$. This shows that the energy is preserved (constant).

Now suppose that $y_2(x,t)$ is another solution to the wave problem. Then the difference $z(x,t) = y(x,t) - y_2(x,t)$ is also a solution of the wave equation and initial boundary conditions:

$$z_{tt} = \omega^2 z_{xx}, \qquad z(x,0) = z_t(x,0) = 0.$$

It follows that $z_x(x,0) = 0$ too, and hence the energy of the system is

$$E = \int_0^\pi z_x(s,0)^2 + z_t(s,0)^2\, ds = 0.$$

By conservation of energy, we deduce that

$$\int_0^\pi z_x(s,t)^2 + z_t(s,t)^2\, ds = 0 \quad \text{for all} \quad t \geq 0.$$

In other words, $z_x(s,t) = z_t(s,t) = 0$ for all s and t. Therefore, $z = 0$, establishing uniqueness. ∎

We know from real-world experience that a string may be bent into a non-C^2 shape. What happens then? In this case, we may approximate f uniformly by a sequence of C^2 functions f_n. Likewise, g may be approximated uniformly by a sequence g_n of C^1 functions. The wave equation with initial data $y(x,0) = f_n(x)$ and $y_x(x,0) = g_n(x)$ yields the solution

$$y_n(x,t) = \tfrac{1}{2} f_n(x+\omega t) + \tfrac{1}{2} f_n(x-\omega t) + \frac{1}{2\omega} \int_{x-\omega t}^{x+\omega t} g_n(s)\, ds.$$

This sequence of solutions converges uniformly to

$$y(x,t) = \tfrac{1}{2} f(x+\omega t) + \tfrac{1}{2} f(x-\omega t) + \frac{1}{2\omega} \int_{x-\omega t}^{x+\omega t} g(s)\, ds.$$

Thus, even though this function y may not be differentiable (so that it does not make sense to plug y into our PDE), we obtain a reasonable solution to the wave problem.

One very interesting aspect of this solution is the fact that $y(x,t)$ is not any smoother for large t than it is at time $t = 0$. Indeed, for every integer $n \geq 0$,

$$y(x, \tfrac{2n\pi}{\omega}) = \tfrac{1}{2}f(x+2n\pi) + \tfrac{1}{2}f(x-2n\pi) + \frac{1}{2\omega} \int_{x-2n\pi}^{x+2n\pi} g(s)\,ds = f(x).$$

This uses the fact that g has been extended to an odd 2π-periodic function and thus g integrates to 0 over any interval of length (a multiple of) 2π. Similarly,

$$y_t\left(x, \tfrac{2n\pi}{\omega}\right) = \frac{\omega}{2}f'(x+2n\pi) - \frac{\omega}{2}f'(x-2n\pi) + \frac{1}{2}g(x+2n\pi) + \frac{1}{2}g(x-2n\pi) = g(x).$$

Thus if the initial data f fails to be smooth at x_0, this property persists along the lines $x(t) = x_0 \pm \omega t$. This is known as **propagation of singularities**. And because our solutions are odd 2π-periodic functions, these singularities recur within our range. Following the solution only within the interval $[0, \pi]$, these singularities appear to reflect off the boundary and reenter the interval. This property is distinctly different from the solution of the heat equation, which becomes C^∞ for $t > 0$ (see Exercise 13.3.D) because the initial heat distribution gets averaged out over time.

The lack of averaging or damping in the wave equation is very important in real life. It makes it possible for us to see, and to transmit radio and television signals over long distances without significant distortion.

The Fourier series approach still has more to tell us. The Fourier coefficients in the expansion of the solution $y(x,t)$ decompose the wave into a sum of harmonics of order n for $n \geq 1$. In fact, the term

$$y_n(x,t) = A_n \sin nx \, \cos n\omega t + B_n \sin nx \, \sin n\omega t$$

may be rewritten as

$$y_n(x,t) = C_n \sin nx \, \sin(n\omega t + \tau_n),$$

where $C_n = \sqrt{A_n^2 + B_n^2}$ and the phase shift τ_n is chosen so that $\sin \tau_n = A_n/C_n$ and $\cos \tau_n = B_n/C_n$. Thus as t increases, y_n modulates through multiples of $\sin nx$ from C_n down to $-C_n$ and back.

The combination of different harmonics gives a wave its shape. In electrical engineering, one often attempts to break down a wave into its component parts or build a new wave by putting harmonics together. This amounts to finding a Fourier series whose sum is a specified function. That is the problem we will investigate further in the next chapter.

Exercises for Section 13.8

A. Using the series for $y(x,t)$ and the orthogonality relations, show that

$$E = \int_0^\pi \frac{1}{2}f'(x)^2 + \frac{1}{2\omega^2}g(x)^2 \, dx = \frac{\pi}{2} \sum_{n=1}^\infty n^2(|A_n|^2 + |B_n|^2).$$

B. Consider a guitar string that is plucked in the centre to a height h starting at rest. Assume that the initial position is piecewise linear with a sharp cusp in the centre.

  (a) What is the odd 2π-periodic extension of the initial position function f? Sketch it.
  (b) Plot the graph of the solution $y(x,t)$ for $t = 0, \frac{\pi}{3\omega}, \frac{\pi}{2\omega}, \frac{\pi}{\omega}, \frac{2\pi}{\omega}$.
  (c) How does the cusp move? Find a formula.
  (d) Compute the energy for this string.

C. Verify the formula $A_n \sin nx \cos n\omega t + B_n \sin nx \sin n\omega t = C_n \sin nx \sin(n\omega t + \tau_n)$, where $C_n = \sqrt{A_n^2 + B_n^2}$ and τ_n is chosen such that $\sin \tau_n = A_n/C_n$ and $\cos \tau_n = B_n/C_n$.

D. Let F and G be C^2 functions on the line.

  (a) Show that $y(x,t) = F(x + \omega t) + G(x - \omega t)$ for $x \in \mathbb{R}$ and $t \geq 0$ is a solution of the wave equation the whole line.
  (b) What are the initial position f and velocity g in terms of F and G? Express F and G in terms of f and g.
  (c) Show that the value of $y(x,t)$ depends only on the values of f and g in $[x - \omega t, x + \omega t]$.
  (d) Explain the physical significance of (c) in terms of the speed of propagation of the wave.

E. Consider the wave equation on $\mathbb{R}$. Let $u = x + \omega t$ and $v = x - \omega t$.

  (a) Show that $\frac{\partial}{\partial u} = \frac{1}{2}\frac{\partial}{\partial x} + \frac{1}{2\omega}\frac{\partial}{\partial t}$ and $\frac{\partial}{\partial v} = \frac{1}{2}\frac{\partial}{\partial x} - \frac{1}{2\omega}\frac{\partial}{\partial t}$. Hence deduce that after a change of variables, the wave equation becomes $\frac{\partial}{\partial u}\frac{\partial}{\partial v}y = 0$. •
  (b) Hence show that every solution has the form $y = F(u) + G(v)$.
  (c) Combine this with the previous exercise to show that the wave equation has a unique solution on the line.

F. Let $w(x)$ be a strictly positive function on $[0,1]$. Consider the PDE

$$\frac{\partial^2 y}{\partial t^2}(x,t) = w(x)^2 \frac{\partial^2 y}{\partial x^2}(x,t) + H(x,t)$$

defined for $0 \leq x \leq 1$ and $t \geq 0$ with boundary conditions $y(x,0) = f(x)$, $y_t(x,0) = g(x)$, and $y(0,t) = y(1,t) = 0$. Suppose that $y(x,t)$ and $z(x,t)$ are two solutions, and let $u = y - z$. Consider the quantity

$$E(t) = \int_0^1 u_x^2(x,t) + \frac{u_t^2(x,t)}{w(x)^2}\, dx.$$

Show that E is the zero function, and hence deduce that the solution of the PDE is unique.

13.9 Appendix: The Complex Exponential

Our goal in this section is to extend the definition of the exponential function to all complex numbers. We quickly review the basic ideas of complex numbers $\mathbb{C}$.

A complex number may be written uniquely as $a + ib$ for $a, b \in \mathbb{R}$. Addition is just given by the rule $(a + ib) + (c + id) = (a + c) + i(b + d)$. Multiplication uses distributivity and the rule $i^2 = -1$. So

$$(a + ib)(c + id) = (ac - bd) + i(ad + bc).$$

The **conjugate** of a complex number $z = a + ib$ is the number $\bar{z} := a - ib$. The absolute value or **modulus** is given by $|z| = (z\bar{z})^{1/2} = (a^2 + b^2)^{1/2}$. The set of all complex numbers is closed under addition and subtraction, multiplication, and division by nonzero elements. In particular, if $z = a + ib \neq 0$, then

$$\frac{1}{z} = \frac{\bar{z}}{z\bar{z}} = \frac{a - ib}{a^2 + b^2} = \frac{a}{a^2 + b^2} + i\frac{-b}{a^2 + b^2}.$$

This makes $\mathbb{C}$ into a field properly containing the reals. Define the **real part** and **imaginary part** of a complex number by $\text{Re}(a + ib) = a$ and $\text{Im}(a + ib) = b$.

Distance is defined between points w and z by $|w - z|$. Observe that under addition, $\mathbb{C}$ is a vector space over $\mathbb{R}$ with basis 1 and i. The natural map of $\mathbb{C}$ onto $\mathbb{R}^2$ sending $x + iy$ to (x, y) preserves the distance. So we can talk about convergence by using the topology of $\mathbb{R}^2$.

The exponential function can be defined as a power series by $e^x = \sum_{n \geq 0} \frac{1}{n!} x^n$. This converges absolutely for all $x \in \mathbb{R}$ and is uniform on $\overline{B_r(0)}$ for any $r > 0$. We use this same formula to define e^z.

13.9.1. THEOREM. *The power series* $e^z = \sum_{n \geq 0} \frac{1}{n!} z^n$ *converges absolutely for all* $z \in \mathbb{C}$ *and this convergence is uniform on* $\{z : |z| \leq r\}$ *for all* $r > 0$. *Moreover,*

(1) $e^w e^z = e^{w+z}$ *for all* $w, z \in \mathbb{C}$.
(2) $e^{x+iy} = e^x \cos y + i e^x \sin y$ *for all* $x, y \in \mathbb{R}$.
(3) $|e^z| = e^{\text{Re}\, z}$. *In particular,* $|e^{iy}| = 1$ *for* $y \in \mathbb{R}$.

PROOF. Observe that for $z \in \mathbb{C}$ on the set $\{z : |z| \leq r\}$, $\|z^n\|_\infty = r^n$. So

$$\sum_{n \geq 0} \frac{1}{n!} \|z^n\|_\infty = \sum_{n \geq 0} \frac{1}{n!} r^n = e^r.$$

Therefore this series converges absolutely and uniformly on $\{z : |z| \leq r\}$ by the Weierstrass M-test (8.4.7). So e^z is well defined for all $z \in \mathbb{C}$.

Hence the double series for $e^w e^z$ converges absolutely. So the terms may be rearranged in any order, and the same sum results by Theorem 3.3.5. So we may calculate for $w, z \in \mathbb{C}$ by collecting terms $w^k z^l$ for $k + l = n$ and using the binomial theorem:

$$e^w e^z = \sum_{k=0}^{\infty} \sum_{l=0}^{\infty} \frac{1}{k! l!} w^k z^l = \sum_{n=0}^{\infty} \frac{1}{n!} \sum_{k=0}^{n} \frac{n!}{k!(n-k)!} w^k z^{n-k} = \sum_{n=0}^{\infty} \frac{1}{n!} (w+z)^n = e^{w+z}.$$

We may calculate e^{iy} directly noting that $i^{2n} = (-1)^n$ and $i^{2n+1} = (-1)^n i$:

$$e^{iy} = \sum_{n=0}^{\infty} \frac{1}{n!} (iy)^n = \sum_{n=0}^{\infty} \frac{(-1)^n}{(2n)!} y^{2n} + i \sum_{n=0}^{\infty} \frac{(-1)^n}{(2n+1)!} y^{2n+1} = \cos y + i \sin y,$$

where the last equality comes from recognizing the power series for $\cos y$ and $\sin y$. Therefore we obtain $e^{x+iy} = e^x \cos y + i e^x \sin y$ for all $x, y \in \mathbb{R}$. Hence

$$|e^{x+iy}| = \left(e^{2x} (\cos^2 y + \sin^2 y) \right)^{1/2} = e^x = e^{\text{Re}(x+iy)}. \qquad \blacksquare$$

If $z = a + ib$ is any complex number, let $r = |z|$. Then $z/r = a/r + ib/r$ has modulus 1 and hence lies on the unit circle. There is an angle θ, unique up to a multiple of 2π, such that $z/r = \cos\theta + i\sin\theta = e^{i\theta}$. So $z = re^{i\theta}$. This is called the **polar form**, since z is represented as (r, θ) in the polar coordinates of the plane. Note that $\bar{z} = re^{-i\theta}$.

Manipulating $e^{in\theta} = \cos n\theta + i\sin n\theta$ and $e^{-in\theta} = \cos n\theta - i\sin n\theta$ yields

$$\cos n\theta = \operatorname{Re}(e^{in\theta}) = \frac{e^{in\theta} + e^{-in\theta}}{2} \quad \text{and} \quad \sin n\theta = \operatorname{Im}(e^{in\theta}) = \frac{e^{in\theta} - e^{-in\theta}}{2i}.$$

Therefore the pairs $\{\cos n\theta, \sin n\theta\}$ and $\{e^{in\theta}, e^{-in\theta}\}$ both span the same two dimensional vector space (using complex coefficients). Moreover, in the (complex) inner product on $C[-\pi, \pi]$ given by

$$\langle f, g \rangle = \frac{1}{2\pi} \int_{-\pi}^{\pi} f(e^{i\theta}) \overline{g(e^{i\theta})} \, d\theta,$$

we have

$$\langle e^{im\theta}, e^{in\theta} \rangle = \frac{1}{2\pi} \int_{-\pi}^{\pi} e^{im\theta} \overline{e^{in\theta}} \, d\theta \frac{1}{2\pi} \int_{-\pi}^{\pi} e^{i(m-n)\theta} \, d\theta = \delta_{m,n}.$$

So each $e^{in\theta}$ is a unit vector, and they form an orthonormal set.

It follows that a Fourier series of a function f on $[-\pi, \pi]$ may be rewritten using complex exponentials to obtain the **complex Fourier series**

$$f(\theta) \sim \sum_{n=-\infty}^{\infty} \langle f, e^{in\theta} \rangle e^{in\theta} = \sum_{n=-\infty}^{\infty} c_n e^{in\theta}, \tag{13.9.2}$$

where the **complex Fourier coefficients** are given by

$$c_n = \langle f, e^{in\theta} \rangle = \frac{1}{2\pi} \int_{-\pi}^{\pi} f(\theta) e^{-in\theta} \, d\theta. \tag{13.9.3}$$

We have already seen that we could sum the Poisson kernel using complex exponentials and the formula for summing geometric series. This idea has many applications in Fourier series. Here is another example.

13.9.4. LEMMA. $1 + 2 \sum_{k=1}^{n} \cos kt = \begin{cases} 2n+1 & \text{if } t = 2\pi m \text{ for } m \in \mathbb{Z}, \\ \dfrac{\sin(n+1/2)t}{\sin t/2} & \text{if } t \neq 2\pi m. \end{cases}$

PROOF. When $t = 2\pi m$, $\cos kt = \cos 2mk\pi = 1$ for all $k \in \mathbb{Z}$, so the sum is $2n+1$. Otherwise $e^{it} \neq 1$, and we sum of a geometric series:

$$1+2\sum_{k=1}^{n-1}\cos kt = 1+\sum_{k=1}^{n}e^{ikt}+e^{-ikt} = \sum_{k=-n}^{n}e^{ikt} = \frac{e^{i(n+1)t}-e^{-int}}{e^{it}-1}$$

$$=\frac{e^{i(n+1/2)t}-e^{-i(n+1/2)t}}{2i}\cdot\frac{2i}{e^{it/2}-e^{-it/2}} = \frac{\sin(n+1/2)t}{\sin t/2}.\qquad\blacksquare$$

The function $D_n(t) = \frac{1}{2\pi}\left(1+2\sum_{k=1}^{n}\cos kt\right)$ is called the Dirichlet kernel. It plays a central role in summing Fourier series. The next example is known as the Fejér kernel, $K_n(t) = \frac{1}{n}\sum_{k=0}^{n-1}D_k(t) = \frac{1}{2\pi}\left(1+\sum_{k=1}^{n}\left(1-\frac{k}{n}\right)\cos kt\right)$.

13.9.5. LEMMA. $\quad 1+\displaystyle\sum_{k=1}^{n-1}\left(1-\frac{k}{n}\right)\cos kt = \begin{cases} n & \text{if } t=2m\pi, \\ \dfrac{\sin^2(nt/2)}{n\sin^2(t/2)} & \text{if } t\neq 2m\pi. \end{cases}$

PROOF. This is clear when $t=2m\pi$. As observed above, and using Lemma 13.9.4

$$1+\sum_{k=1}^{n}\left(1-\frac{k}{n}\right)\cos kt = \frac{1}{n}\sum_{j=0}^{n-1}\left(1+\sum_{k=1}^{j}\cos kt\right) = \frac{1}{n}\sum_{j=0}^{n-1}\frac{\sin(j+1/2)t}{\sin t/2}$$

$$=\frac{1}{2in\sin(t/2)}\sum_{j=0}^{n-1}e^{i(j+1/2)t}-e^{-i(j+1/2)t}$$

$$=\frac{1}{2in\sin(t/2)}\left(\frac{e^{i(n+1/2)t}-e^{it/2}}{e^{it}-1}-\frac{e^{-i(n+1/2)t}-e^{-it/2}}{e^{-it}-1}\right)$$

$$=\frac{1}{n\sin(t/2)}\frac{(e^{int}-1)+(e^{-int}-1)}{(2i)^2}\cdot\frac{2i}{e^{it/2}-e^{-it/2}}$$

$$=\frac{1}{n\sin^2(t/2)}\left(\frac{e^{int/2}-1}{2i}\right)^2 = \frac{\sin^2(nt/2)}{n\sin^2(t/2)}.\qquad\blacksquare$$

Exercises for Section 13.9

A. Use trig identities to show that $(\cos x+i\sin x)(\cos y+i\sin y) = \cos(x+y)+i\sin(x+y)$.

B. (a) Graph the image of a line parallel to the y-axis under the exponential map.
(b) Graph the image of a line parallel to the x-axis under the exponential map.
(c) Show that the strip $\{z=x+iy \mid 0\leq y<2\pi\}$ is mapped by the exponential function one-to-one and onto the whole complex plane except for the point 0.

C. Express $f(\theta)=A_n\cos n\theta+B_n\sin n\theta$ as $c_n e^{in\theta}+c_{-n}e^{-in\theta}$. Hence show that the Fourier series of a continuous function $f(\theta)$ is converted to the complex form (13.9.2).

D. Sum the Fourier series $\displaystyle\sum_{n=0}^{\infty}2^{-n}\cos n\theta$ and $\displaystyle\sum_{n=1}^{\infty}2^{-n}\sin n\theta$.   HINT: Compute $\sum_{n=0}^{\infty}2^{-n}e^{in\theta}$.

E. Define $\cos z=(e^{iz}+e^{-iz})/2$ and $\sin z=(e^{iz}-e^{-iz})/2i$ for all $z\in\mathbb{C}$.

(a) Prove that $\sin(w+z)=\sin w\cos z+\cos w\sin z$.
(b) Find all solutions of $\sin z=2$.

F. (a) Evaluate $S_n(\theta)=\sum_{k=1}^{n}\sin k\theta$.
(b) Show that $|S_n(\theta)|\leq\pi\varepsilon^{-1}$ on $[\varepsilon,2\pi-\varepsilon]$ for all $n\geq 1$.

Chapter 14
Fourier Series and Approximation

A natural problem is to take a wave output and decompose it into its harmonic parts. Engineers are able to do this with an oscilloscope. A real difficultly occurs when we try to put the parts back together. Mathematically, this amounts to summing up the series obtained from decomposing the original wave. In this chapter, we examine this delicate question: Under what conditions does a Fourier series converge?

We start by looking at the behaviour of the Fourier coefficients. It turns out that they go to zero; and the smoother the function, the faster they go. Then we turn to the more subtle questions of pointwise and uniform convergence. The idea of kernel functions, analogous to the Poisson kernel from the previous chapter, provides an elegant method for understanding these notions of convergence. Then we turn to the L^2 norm, where there is a very clean answer. Nice applications of this include the isoperimetric inequality and sums of various interesting series. Finally, we consider applications to polynomial approximation.

14.1 The Riemann–Lebesgue Lemma

An important step in establishing convergence is the apparently modest goal of showing that at least the Fourier coefficients converge to 0. This is clearly a necessary condition for any kind of convergence. This was by no means clear in the 1850s when Riemann did his fundamental work. In fact, he introduced the modern notion of integral in order to address the question of convergence of Fourier series.

Although we need the result only for piecewise continuous functions, it is true much more generally, for absolutely integrable functions in fact.

14.1.1. THE RIEMANN–LEBESGUE LEMMA.
If f is piecewise continuous on $[a,b]$ and $\tau \in \mathbb{R}$, then

$$\lim_{n \to \infty} \int_a^b f(x) \sin(nx + \tau)\, dx = 0.$$

K.R. Davidson and A.P. Donsig, *Real Analysis and Applications: Theory in Practice*, Undergraduate Texts in Mathematics, DOI 10.1007/978-0-387-98098-0_14, © Springer Science + Business Media, LLC 2010

PROOF. We may assume that $b - a \leq \pi$. For otherwise, we just chop the interval $[a,b]$ into pieces of length at most π and prove the lemma on each piece separately. Translation by a multiple of π does not affect anything except possibly the sign of the integral. So we may assume that $[a,b]$ is contained in $[-\pi,\pi]$. Now extend the definition of f to all of $[-\pi,\pi]$ by setting $f(x) = 0$ outside of $[a,b]$.

By Proposition 7.6.1, $\{1, \sqrt{2}\cos n\theta, \sqrt{2}\sin n\theta : n \geq 1\}$ is an orthonormal set in the inner product space of piecewise continuous functions with the L^2 inner product. Applying Bessel's inequality (7.7.1), we have

$$A_0^2 + \frac{1}{2}\sum_{n=1}^{\infty} A_n^2 + B_n^2 \leq \|f\|_2^2 < \infty.$$

In particular, the series on the left-hand side must converge and its terms must tend to zero. That is,

$$\lim_{n \to \infty} A_n^2 + B_n^2 = 0.$$

So we compute

$$\left| \int_{-\pi}^{\pi} f(x)\sin(nx + \tau)\,dx \right| = \left| \int_{-\pi}^{\pi} f(x)\cos nx \sin\tau + \sin nx \cos\tau\,dx \right|$$

$$= \left| A_n \sin\tau + B_n \cos\tau \right| \leq \sqrt{A_n^2 + B_n^2} \longrightarrow 0.$$

The last estimate (included only for elegance) used the Schwarz inequality. ∎

14.1.2. COROLLARY. *If f is a piecewise-continuous 2π-periodic function with Fourier series*

$$f \sim A_0 + \sum_{n=1}^{\infty} A_n \cos n\theta + B_n \sin n\theta,$$

then $\lim\limits_{n \to \infty} A_n = \lim\limits_{n \to \infty} B_n = 0$.

PROOF. Take the definitions of A_n and B_n and apply the Riemann–Lebesgue Lemma for the interval $[-\pi,\pi]$ and displacements $\tau = \pi/2$ and $\tau = 0$, respectively. ∎

The more derivatives a function has, the faster the Fourier coefficients go to zero. First, we need to connect the two Fourier series for the function and its derivative.

14.1.3. DEFINITION. A function f is **piecewise** C^k on $[a,b]$ if f is k-times differentiable except at finitely many points, and $f^{(k)}$ is piecewise continuous.

For example, the Heaviside function H from Example 5.2.2 is piecewise C^1, as is the function $f(x) = x - \lfloor x \rfloor$, where $\lfloor x \rfloor$ indicates the largest integer $n \leq x$.

We need to show that integration by parts is valid if one of the functions is continuous but only piecewise C^1.

14.1.4. LEMMA. *If f is continuous and piecewise C^1 on $[a,b]$ and $g \in C^1[a,b]$, then*

$$\int_a^b f'(t)g(t)\,dt = f(t)g(t)\Big|_a^b - \int_a^b f(t)g'(t)\,dt.$$

PROOF. Let $a = a_0 < a_1 < \cdots < a_n = b$ be chosen such that f is C^1 on $[a_{i-1}, a_i]$ for $1 \le i \le n$. Then integration by parts is valid on each interval, so

$$\int_a^b f'(t)g(t)\,dt = \sum_{i=1}^n \int_{a_{i-1}}^{a_i} f'(t)g(t)\,dt$$

$$= \sum_{i=1}^n f(t)g(t)\Big|_{a_{i-1}}^{a_i} - \int_{a_{i-1}}^{a_i} f(t)g'(t)\,dt$$

$$= f(t)g(t)\Big|_a^b - \int_a^b f(t)g'(t)\,dt$$

because $\sum_{i=1}^n f(t)g(t)\big|_{a_{i-1}}^{a_i}$ is a telescoping sum, since f is continuous. ∎

14.1.5. LEMMA. *If $f : \mathbb{R} \to \mathbb{R}$ is a 2π-periodic continuous function that is piecewise C^1, and $f \sim A_0 + \sum_{n=1}^\infty A_n \cos n\theta + B_n \sin n\theta$, then f' has Fourier series*

$$f' \sim \sum_{n=1}^\infty nB_n \cos n\theta - nA_n \sin n\theta.$$

PROOF. Integrating by parts yields

$$\frac{1}{\pi}\int_{-\pi}^{\pi} f'(t)\cos nt\,dt = \frac{1}{\pi}f(t)\cos nt\Big|_{-\pi}^{\pi} + \frac{1}{\pi}\int_{-\pi}^{\pi} f(t)n\sin nt\,dt = nB_n.$$

Similarly,

$$\frac{1}{\pi}\int_{-\pi}^{\pi} f'(t)\sin nt\,dt = \frac{1}{\pi}f(t)\sin nt\Big|_{-\pi}^{\pi} - \frac{1}{\pi}\int_{-\pi}^{\pi} f(t)n\cos nt\,dt = -nA_n.$$

And

$$\frac{1}{2\pi}\int_{-\pi}^{\pi} f'(t)\,dt = \frac{1}{2\pi}f(t)\Big|_{-\pi}^{\pi} = \frac{1}{2\pi}\big(f(\pi) - f(-\pi)\big) = 0. \qquad \blacksquare$$

14.1.6. THEOREM. *If $f : \mathbb{R} \to \mathbb{R}$ is a 2π-periodic C^{k-1} function that is piecewise C^k, and $f \sim A_0 + \sum_{n=1}^\infty A_n \cos n\theta + B_n \sin n\theta$, then there is a constant M such that for all $n \ge k$,*

$$|A_n| \le \frac{M}{n^k} \quad and \quad |B_n| \le \frac{M}{n^k}.$$

PROOF. We use induction on k. For $k = 1$, let $M = 2\|f'\|_1$. By Lemma 14.1.5, the Fourier coefficients of f' are nB_n and $-nA_n$. Applying Proposition 7.6.4 to f' shows that $|nA_n|$ and $|nB_n|$ are bounded by M. Thus $|A_n|$ and $|B_n|$ are bounded by M/n.

For general k, applying the result for $k - 1$ to f' and using Lemma 14.1.5 gives that $|nA_n|$ and $|nB_n|$ are bounded by M/n^{k-1}. Dividing by n gives the result. ∎

14.1.7. EXAMPLE. Consider the function $f(\theta) = \theta^3 - \pi^2\theta$ for $-\pi \le \theta \le \pi$. Notice that $f(-\pi) = f(\pi) = 0$, whence f is a continuous 2π-periodic function. Moreover, $f'(\theta) = 3\theta^2 - \pi^2$ and we have $f'(-\pi) = f'(\pi) = 2\pi^2$. So f is C^1. Finally, $f''(\theta) = 6\theta$. Since $f''(-\pi) \ne f''(\pi)$, the function f is piecewise C^2 but not C^2.

By Theorem 14.1.6, the Fourier coefficients are bounded by M/n^2 for $n \ge 2$. Since $\sum_{n\ge1} n^{-2}$ converges, we know that the Fourier series converges uniformly by the Weierstrass M-test.

Let us compute the Fourier coefficients of f. Since f is odd, we need only compute the sine terms. We integrate by parts three times:

$$
\begin{aligned}
B_n &= \frac{1}{\pi}\int_{-\pi}^{\pi} (\theta^3 - \pi^2\theta)\sin n\theta\, d\theta \\
&= \frac{(\theta^3 - \pi^2\theta)}{\pi}\frac{-\cos n\theta}{n}\bigg|_{-\pi}^{\pi} + \int_{-\pi}^{\pi}\frac{3\theta^2 - \pi^2}{n\pi}\cos n\theta\, d\theta \\
&= 0 + \frac{3\theta^2 - \pi^2}{n^2\pi}\sin n\theta\bigg|_{-\pi}^{\pi} - \int_{-\pi}^{\pi}\frac{6\theta}{n^2\pi}\sin n\theta\, d\theta \\
&= 0 + \frac{6\theta}{n^3\pi}\cos n\theta\bigg|_{-\pi}^{\pi} - \int_{-\pi}^{\pi}\frac{6}{n^3\pi}\cos n\theta\, d\theta \\
&= \frac{6\pi}{n^3\pi}(-1)^n - \frac{-6\pi}{n^3\pi}(-1)^n - 0 = (-1)^n\frac{12}{n^3}.
\end{aligned}
$$

So we could have applied the M-test directly to the Fourier series. We do not yet know that the limit of the Fourier series is f.

Exercises for Section 14.1

A. Let f be a monotone function on $[-\pi, \pi]$ with $f \sim A_0 + \sum_{n=1}^{\infty} A_n\cos n\theta + B_n\sin n\theta$. Prove that the Fourier coefficients satisfy $\max\{|A_n|, |B_n|\} \le 2M/n\pi$, where $M = |f(\pi) - f(-\pi)|$.
HINT: Express B_n as the sum of integrals over intervals on which $\sin n\theta$ has constant sign, and combine into a single integral.

B. Consider the DE $y'' + 4y = g$, where g is an odd C^2 function with Fourier series $\sum_{n\ge1} B_n\sin nx$.

(a) If $B_2 = 0$, find the Fourier series of the solution.
(b) Verify that this series and its second derivative converge uniformly and provide a solution.
(c) Show that $y = -\frac{1}{4}x\cos 2x$ is the solution for $g(x) = \sin 2x$.

C.   Show that if f is a Lipschitz function on $[-\pi, \pi]$ with Lipschitz constant L, then the Fourier coefficients satisfy $|A_n| \le \frac{2L}{n}$ and $|B_n| \le \frac{2L}{n}$ for $n \ge 1$.   HINT: Split the integral into n pieces and replace each integral by $\int_{c_k - \pi/n}^{c_k + \pi/n} \left(f(x) - f(c_k) \right) \sin nx \, dx$. Then estimate each piece.

D.   Suppose that $f \sim A_0 + \sum\limits_{n=1}^{\infty} A_n \cos n\theta + B_n \sin n\theta$ satisfies $f \in \operatorname{Lip} \alpha$ (see Exercise 5.5.K).

  (a) Prove that $(A_n^2 + B_n^2)^{1/2} \le \omega(f; \frac{\pi}{n})$, where $\omega(f; \delta)$ is the modulus of continuity (see Definition 10.4.2)   HINT: Show that $A_n = \frac{1}{2\pi} \int_{-\pi}^{\pi} \left(f(t) - f(t + \frac{\pi}{n}) \right) \cos(nt) \, dt$.
  (b) If $f \in \operatorname{Lip} \alpha$, prove that there is a constant C such that $(A_n^2 + B_n^2)^{1/2} \le Cn^{-\alpha}$. Hence $|A_n|$ and $|B_n|$ are bounded by $Cn^{-\alpha}$.
  (c) If f is C^p and $f^{(p)} \in \operatorname{Lip} \alpha$, show that $(A_n^2 + B_n^2)^{1/2} \le Cn^{-p-\alpha}$.

E.   Let g be an odd function such that $g(\theta) \ge 0$ on $[0, \pi]$.

  (a) Show by induction that $|\sin n\theta| < n \sin \theta$ on $(0, \pi)$.
  (b) Prove that the Fourier coefficients satisfy $|B_n| < nB_1$.
  (c) Show by a series of examples that the previous inequality cannot be improved in general.

F.   (a) Show that $\liminf\limits_{n \to \infty} \int_a^b |\cos(nx + \alpha_n)| \, dx \ge \frac{b-a}{2}$ for any $a < b$ and any real $\alpha_n, n \ge 1$.
  (b) Show that if $r_n > 0$ satisfy $\sum_{n=1}^{\infty} r_n |\cos(nx + \alpha_n)| \le C$ for $a \le x \le b$, then $\sum_{n=1}^{\infty} r_n < \infty$.

G.   Prove the Riemann–Lebesgue Lemma for absolutely integrable functions.
     HINT: Approximate the function by a step function in the L^1 norm.

14.2 Pointwise Convergence of Fourier Series

Given a piecewise-continuous 2π-periodic function $f : \mathbb{R} \to \mathbb{R}$ with Fourier series $f \sim A_0 + \sum_{k=1}^{\infty} A_k \cos k\theta + B_k \sin k\theta$, we denote the nth partial sum by

$$S_n f(\theta) = A_0 + \sum_{k=1}^{N} A_k \cos k\theta + B_k \sin k\theta.$$

It follows from the Projection Theorem (7.5.11) that $S_n f$ is the best approximation to f in the subspace spanned by 1 and $\{\cos kx, \sin kx : 1 \le k \le n\}$, with respect to the L^2 norm. But this is not same thing as saying that for each number x, $S_n f(x)$ coverges to $f(x)$. In fact, there are continuous functions f such that there is a number x where $S_n f(x)$ goes to infinity. Such an example was first found by du Bois Reymond in 1876. We outline the construction of such a function in Exercise 14.7.E later. Further examples have been found by Fejér and by Lebesgue. Much more recently, in 1966, Carleson solved a long-standing problem conjectured 50 years earlier by Lusin. He showed that the Fourier series of a continuous function (and indeed any L^2 function) converges for all θ except for a 'small set', precisely, except for a set of measure zero. These examples and results are beyond the scope of this text.

    If we require the function to be piecewise Lipschitz, then we can establish pointwise convergence of the Fourier series. First, a better method is needed for computing the partial sums. Again there is an integral formula using a kernel. While it is not nearly as nicely behaved as the Poisson kernel of Section 13.4, this kernel still provides a better estimate than looking at terms individually.

14.2.1. DEFINITION. The sequence of functions $D_n : \mathbb{R} \to \mathbb{R}$ given by

$$D_n(t) = \frac{\sin(n+\frac{1}{2})t}{2\pi \sin t/2} \quad \text{if } t \neq 2\pi m \quad \text{and} \quad D_n(2\pi m) = \frac{2n+1}{2\pi} \quad \text{for } m \in \mathbb{Z}$$

for $n = 1, 2, \ldots$ is called the **Dirichlet kernel**.

To see that D_n is continuous at 0 (and hence also at $2\pi m$), compute

$$\lim_{t \to 0} D_n(t) = \lim_{t \to 0} \frac{\sin(n+\frac{1}{2})t}{t} \frac{t}{2\pi \sin t/2} = \frac{n+\frac{1}{2}}{2\pi/2} = \frac{2n+1}{2\pi}.$$

Another motivation for the definition of $D_n(2\pi m)$ is the following trig identity, whose proof we leave as an exercise. It can also be established using complex exponentials (see Lemma 13.9.4).

14.2.2. LEMMA. $D_n(t) = \frac{1}{2\pi}\left(1 + 2\sum_{k=1}^{n} \cos kt\right).$

The following theorem connects D_n to Fourier series.

14.2.3. THEOREM. *Let f be a piecewise-continuous 2π-periodic function. For each $x \in \mathbb{R}$,*

$$S_n f(x) = \int_{-\pi}^{\pi} f(x+t)D_n(t)\,dt.$$

PROOF. Substituting the formulae for A_k and B_k into the definition of $S_n f$ yields

$$S_n f(x) = A_0 + \sum_{k=1}^{n} A_k \cos kx + B_k \sin kx$$

$$= \frac{1}{2\pi}\int_{-\pi}^{\pi} f(t)\,dt + \sum_{k=1}^{n} \frac{1}{\pi}\int_{-\pi}^{\pi} f(t)\cos kt\,dt \cos kx + \sum_{k=1}^{n} \frac{1}{\pi}\int_{-\pi}^{\pi} f(t)\sin kt\,dt \sin kx$$

$$= \frac{1}{2\pi}\int_{-\pi}^{\pi} f(t)\left(1 + 2\sum_{k=1}^{n} \cos kt \cos kx + \sin kt \sin kx\right)dt$$

$$= \frac{1}{2\pi}\int_{-\pi}^{\pi} f(t)\left(1 + 2\sum_{k=1}^{n} \cos k(t-x)\right)dt$$

using the identity $\cos(A - B) = \cos A \cos B + \sin A \sin B$. Then, substituting $u = t - x$ and using the 2π-periodicity of f, we have

$$= \int_{-\pi}^{\pi} f(x+u)\left(\frac{1}{2\pi} + \frac{1}{\pi}\sum_{k=1}^{n} \cos ku\right)du.$$

Invoking Lemma 14.2.2 completes the proof. ∎

The Dirichlet kernel is inferior to the Poisson kernel because it is not positive. Indeed, properties (2) and (3) of the next result together show that significant cancellation must occur in integrating D_n for large n. Compare Figure 14.1 with Figure 13.2.

14.2.4. PROPERTIES OF THE DIRICHLET KERNEL.
The Dirichlet kernel has the following properties:

(1) *For each n, D_n is a continuous, 2π-periodic, even function.*

(2) $\displaystyle\int_{-\pi}^{\pi} D_n(t)\,dt = 1.$

(3) *For each n, $0.28 + 0.4\log n \le \displaystyle\int_{-\pi}^{\pi} |D_n| \le 2 + \log n$*

PROOF. By periodicity, it suffices to show that D_n is continuous at 0, which we did just after the definition. Lemma 14.2.2 shows that D_n is even and 2π-periodic.

For (2), taking $f = 1$ in Theorem 14.2.3, we have,

$$\int_{-\pi}^{\pi} D_n(t)\,dt = S_n f(0) = 1.$$

The reader can check that $\int_{-\pi}^{\pi} |D_1(t)|\,dt = \int_{-\pi}^{\pi} (|1 + 2\cos t|/(2\pi)\,dt < 1.5$. To provide the upper bound on $\int |D_n|$ for $n \ge 2$, we first observe that

$$\int_{-\pi}^{\pi} |D_n(t)|\,dt = 2\int_0^{\pi} |D_n(t)|\,dt = \int_0^{1/n} 2|D_n(t)|\,dt + \int_{1/n}^{\pi} \left|\frac{\sin(n+\frac{1}{2})t}{\pi\sin t/2}\right|\,dt.$$

The first integral is estimated for $n \ge 2$ as

$$\int_0^{1/n} 2|D_n(t)|\,dt = \frac{1}{\pi}\int_0^{1/n}\left|1 + 2\sum_{k=1}^{n}\cos kt\right|\,dt \le \frac{1}{\pi}\frac{1}{n}(1+2n) \le \frac{2.5}{\pi} < 0.8.$$

Next, use use the inequalities $\pi\sin(t/2) \ge t$ for $t \ge 0$ and $|\sin(n+\frac{1}{2})t| \le 1$ to obtain

$$\int_{1/n}^{\pi}\left|\frac{\sin(n+1/2)t}{\pi\sin t/2}\right|\,dt \le \int_{1/n}^{\pi}\frac{1}{t}\,dt = \log\pi - \log\frac{1}{n} < 1.2 + \log n.$$

Combining these two integrals, we have

$$\int_{-\pi}^{\pi} |D_n(t)|\,dt \le 2 + \log n.$$

For the lower bound, first note that

$$\int_{k\pi/(2n+1)}^{(k+1)\pi/(2n+1)} |\sin(n+1/2)t|\,dt = \int_0^{\pi/(2n+1)} \sin(n+1/2)t\,dt$$

$$= \frac{2}{2n+1}\int_0^{\pi/2} \sin t\,dt = \frac{2}{2n+1}.$$

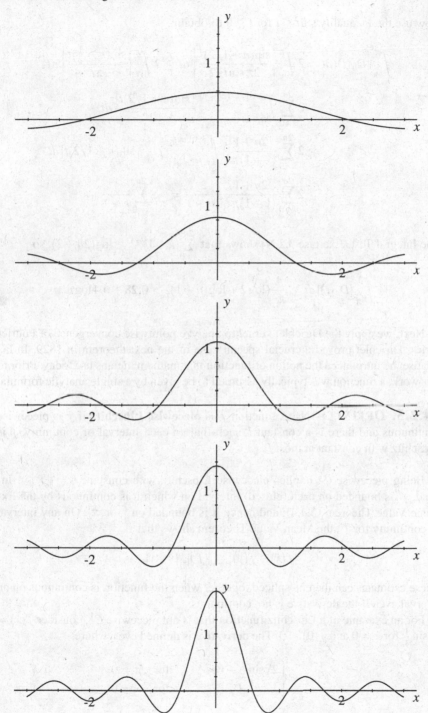

FIG. 14.1 The graphs of D_1 through D_5.

Now use the inequality $\sin t \leq t$ for $t \geq 0$ to obtain

$$
\int_{-\pi}^{\pi} |D_n(t)| \, dt = 2 \int_0^{\pi} \left| \frac{\sin(n+1/2)t}{2\pi \sin t/2} \right| dt \geq 2 \int_0^{\pi} \left| \frac{\sin(n+\frac{1}{2})t}{\pi t} \right| dt
$$

$$
= 2 \sum_{k=0}^{2n} \int_{k\pi/(2n+1)}^{(k+1)\pi/(2n+1)} \frac{|\sin(n+1/2)t|}{\pi t} \, dt
$$

$$
\geq 2 \sum_{k=0}^{2n} \frac{2n+1}{(k+1)\pi^2} \int_{k\pi/(2n+1)}^{(k+1)\pi/(2n+1)} |\sin(n+1/2)t| \, dt
$$

$$
\geq 2 \sum_{k=0}^{2n} \frac{2n+1}{(k+1)\pi^2} \frac{2}{2n+1} = \frac{4}{\pi^2} \sum_{k=0}^{2n} \frac{1}{k+1}.
$$

The Integral Test (Exercise 3.2.N) shows that $\sum_{k=0}^{2n} (k+1)^{-1} \geq \log(2n+2)$, so

$$
\int_{-\pi}^{\pi} |D_n(t)| \, dt \geq \frac{4}{\pi^2} (\log 2 + \log(n+1)) > 0.28 + 0.4 \log n. \qquad \blacksquare
$$

Next, we apply the Dirichlet kernel to analyze pointwise convergence of Fourier series. Dirichlet proved a crucial special case of the next theorem in 1829. In his treatise, he introduced the notion of function that mathematicians use today. Prior to this work, a function was typically assumed to be given by a single analytic formula.

14.2.5. DEFINITION. A function f is **piecewise Lipschitz** if f is piecewise continuous and there is a constant L such that on each interval of continuity, f is Lipschitz with constant at most L.

Being piecewise C^1 implies piecewise Lipschitz with constant $L = \|f'\|_{\infty}$. Indeed, f' is bounded on each (closed) interval on which it is continuous by the Extreme Value Theorem (5.4.4); and hence it is bounded on $[-\pi, \pi]$. On any interval of continuity for f', the Mean Value Theorem shows that

$$
|f(x) - f(y)| \leq \|f'\|_{\infty} |x - y|.
$$

These estimates can then be spliced together when the function is continuous on an interval even if the derivative is not continuous.

For an example of a Lipschitz function that is not piecewise C^1, consider $f(x) = x^2 \sin \frac{1}{x}$ for $x \neq 0$ and $f(0) = 0$. The derivative is defined everywhere:

$$
f'(x) = \begin{cases} 2x \sin \frac{1}{x} - \cos \frac{1}{x} & \text{for} \quad x \neq 0, \\ 0 & \text{for} \quad x = 0. \end{cases}
$$

This is bounded by 3 on $\mathbb{R}$, so the Mean Value Theorem argument is valid. However, f' has a nasty discontinuity at the origin. So f' is not piecewise continuous.

To prove the following result, we will use the integral formula for D_n. The argument has some similarity to the proof of Poisson's Theorem (13.5.1), but property (3) of the previous proposition forces us to be circumspect. At a certain point, we will need to combine the Lipschitz condition with the Riemann–Lebesgue Lemma to obtain the desired estimate..

We write $f(\theta^+)$ and $f(\theta^-)$ for the one-sided limits of f at θ.

14.2.6. THE DIRICHLET–JORDAN THEOREM.

If $f : \mathbb{R} \to \mathbb{R}$ is piecewise Lipschitz and 2π-periodic, then

$$\lim_{n\to\infty} S_n f(\theta) = \frac{f(\theta^+) + f(\theta^-)}{2}.$$

In particular, if f is continuous at θ, then $\lim_{n\to\infty} S_n f(\theta) = f(\theta)$.

PROOF. By Theorem 14.2.4 (2),

$$f(\theta^+) = f(\theta^+) \int_{-\pi}^{\pi} D_n(t)\, dt = 2 \int_0^{\pi} f(\theta^+) D_n(t)\, dt,$$

and a similar equality holds for $f(\theta^-)$. Using Theorem 14.2.3,

$$S_n f(\theta) = \int_{-\pi}^{\pi} f(\theta - t) D_n(t)\, dt = \int_0^{\pi} \big(f(\theta + t) + f(\theta - t)\big) D_n(t)\, dt.$$

Using this formula for $S_n f(\theta)$ with the previous two equalities gives

$$S_n f(\theta) - \frac{f(\theta^+) + f(\theta^-)}{2}$$

$$= \int_0^{\pi} \big(f(\theta + t) + f(\theta - t)\big) D_n(t)\, dt - \int_0^{\pi} \big(f(\theta^+) + f(\theta^-)\big) D_n(t)\, dt$$

$$= \int_0^{\pi} \big(f(\theta + t) - f(\theta^+)\big) D_n(t)\, dt + \int_0^{\pi} \big(f(\theta - t) - f(\theta^-)\big) D_n(t)\, dt.$$

We now consider these two integrals separately. First we prove that

$$\lim_{n\to\infty} \int_0^{\pi} \big(f(\theta + t) - f(\theta^+)\big) D_n(t)\, dt = 0.$$

An entirely similar argument will show that the second integral also goes to zero as n goes to infinity. Combining these two results proves the theorem.

Let L be the Lipschitz constant for f, and let $\varepsilon > 0$ be given. Choose a positive $\delta < \varepsilon/L$ that is so small that f is continuous on $[\theta, \theta + \delta]$. Hence

$$|f(\theta + t) - f(\theta^+)| < Lt \quad \text{for all} \quad \theta + t \in [\theta, \theta + \delta].$$

Since $|\sin(t/2)| \geq |t|/\pi$ for all t in $[-\pi, \pi]$, it follows that

$$|D_n(t)| = \left| \frac{\sin(n+\frac{1}{2})t}{2\pi \sin t/2} \right| \le \frac{1}{2|t|} \quad \text{for all} \quad t \in [-\pi, \pi].$$

Therefore,

$$\left| \int_0^\delta \left(f(\theta+t) - f(\theta^+) \right) D_n(t) \, dt \right| \le \int_0^\delta Lt \frac{1}{2t} \, dt \le \frac{L\delta}{2} < \frac{\varepsilon}{2}.$$

For the integral from δ to π, we have

$$\int_\delta^\pi \left(f(\theta+t) - f(\theta^+) \right) D_n(t) \, dt = \int_\delta^\pi \frac{f(\theta+t) - f(\theta^+)}{2\pi \sin t/2} \sin(n+\tfrac{1}{2})t \, dt$$

$$= \int_\delta^\pi g(t) \sin(n+\tfrac{1}{2})t \, dt,$$

where

$$g(t) = \frac{f(\theta+t) - f(\theta^+)}{2\pi \sin t/2}.$$

Since g is piecewise continuous on $[\delta, \pi]$, we can apply the Riemann–Lebesgue Lemma (14.1.1) to this last integral. Thus, for all n sufficiently large,

$$\left| \int_\delta^\pi \left(f(\theta+t) - f(\theta^+) \right) D_n(t) \, dt \right| = \left| \int_\delta^\pi g(t) \sin(n+\tfrac{1}{2})t \, dt \right| < \frac{\varepsilon}{2}.$$

Combining these two estimates, we have

$$\left| \int_0^\pi \left(f(\theta+t) - f(\theta^+) \right) D_n(t) \, dt \right| < \varepsilon$$

for all n sufficiently large, and thus the limit is zero. ∎

14.2.7. EXAMPLE. Let h be the following variant on the Heaviside step function:

$$h(x) = \begin{cases} -1 & \text{if} \quad -\pi < x < 0, \\ 1 & \text{if} \quad 0 \le x \le \pi. \end{cases}$$

Evidently, this function is piecewise C^1. Since h is odd, its Fourier series has only sine terms, which we compute as follows:

$$B_n = \frac{1}{\pi} \int_{-\pi}^\pi h(x) \sin nx \, dx = \frac{2}{\pi} \int_0^\pi \sin nx \, dx = \frac{-2}{n\pi} \cos nx \Big|_0^\pi = \begin{cases} 0 & \text{if } n \text{ is even,} \\ \frac{4}{n\pi} & \text{if } n \text{ is odd.} \end{cases}$$

Thus $h \sim \sum_{k=0}^\infty \frac{4}{(2k+1)\pi} \sin(2k+1)x$. See Figure 14.4 for a graph of $S_{29}h$.

The Dirichlet–Jordan Theorem tells us that this series converges to 1 on $(0, \pi)$ and to -1 for $(-\pi, 0)$. At the points of discontinuity, it converges to the average, 0. This latter fact is clear, since $\sin k\pi = 0$ for all integers k. Let us plug in a few points. For example, take $x = \pi/2$. Since $\sin(2k+1)\pi/2 = (-1)^k$,

$$1 = h\left(\frac{\pi}{2}\right) = \frac{4}{\pi}\sum_{k=0}^{\infty}\frac{(-1)^k}{2k+1}.$$

Therefore,

$$1 - \frac{1}{3} + \frac{1}{5} - \frac{1}{7} + \cdots = \frac{\pi}{4}.$$

Similarly, plugging in $x = 1$, we obtain

$$\sin 1 + \frac{1}{3}\sin 3 + \frac{1}{5}\sin 5 + \cdots = \frac{\pi}{4}.$$

Both of these series converge exceedingly slowly, so they have no real computational value. We will consider this function again in Example 14.4.6.

14.2.8. EXAMPLE. Consider the function f of Example 14.1.7. The Fourier series of f converges pointwise to f by the Dirichlet–Jordan theorem. That is,

$$\theta^3 - \pi^2\theta = \sum_{n=1}^{\infty}\frac{(-1)^n 12}{n^3}\sin n\theta \quad \text{for all} \quad \theta \in [-\pi, \pi].$$

For example, let $\theta = \pi/2$. Since $\sin(2k+1)\pi/2 = (-1)^k$ and $\sin(2k)\pi/2 = 0$,

$$\left(\frac{\pi}{2}\right)^3 - \pi^2\frac{\pi}{2} = \sum_{k=0}^{\infty}\frac{(-1)^{2k+1}12}{(2k+1)^3}(-1)^k.$$

Solving, we find that

$$\sum_{k=0}^{\infty}\frac{(-1)^k}{(2k+1)^3} = \frac{\pi^3}{32}.$$

Now consider the derivative $f'(\theta) = 3\theta^2 - \pi^2$. The Fourier series of f may be differentiated term by term, since the differentiated series converges uniformly by the M-test (8.4.7), since $\sum_{n=1}^{\infty}n|B_n| = \sum_{n=1}^{\infty}\frac{12}{n^2} < \infty$. Thus, for all $\theta \in [-\pi, \pi]$,

$$3\theta^2 - \pi^2 = \sum_{n=1}^{\infty}\frac{(-1)^n 12}{n^2}\cos n\theta.$$

Let us substitute $\theta = \frac{\pi}{2}$ here as well. Here $\cos\frac{(2k+1)\pi}{2} = 0$ and $\cos\frac{2k\pi}{2} = (-1)^k$. So

$$-\frac{\pi^2}{4} = \sum_{k=1}^{\infty}\frac{12}{4k^2}(-1)^k = -3\sum_{k=1}^{\infty}\frac{(-1)^{k-1}}{k^2}.$$

Therefore,

$$\frac{\pi^2}{12} = \sum_{k=1}^{\infty} \frac{(-1)^{k-1}}{k^2} = \sum_{k=1}^{\infty} \frac{1}{k^2} - 2\sum_{k=1}^{\infty} \frac{1}{(2k)^2} = \frac{1}{2}\sum_{k=1}^{\infty} \frac{1}{k^2}.$$

So we again obtain Euler's sum $\sum_{k=1}^{\infty} \frac{1}{k^2} = \pi^2/6$.

In fact the series for f and f' converge uniformly, but this needs further work. A sufficient condition on a function f to conclude that $S_n f$ converges uniformly to f is given in Theorem 14.7.3.

Exercises for Section 14.2

A. Prove Lemma 14.2.2.   HINT: If $t \neq 0$, multiply by $\sin t/2$ and use a trig identity.

B. Compute the Fourier series for $f(x) = x$ for $-\pi \leq x \leq \pi$, and sum of the series.

C. (a) Find the Fourier series for $f(\theta) = \sinh(\theta)$ for $|\theta| \leq \pi$.
     (b) Find a constant c such that the Fourier series for $f(\theta) - c\theta$ converges uniformly on $[-\pi, \pi]$.

D. Show that $\displaystyle\sum_{n=2}^{\infty} \frac{(-1)^n 2n^3}{n^4 - 1} \sin nx$ is the Fourier series of a piecewise C^1 function.

   HINT: Use Exercise C to subtract a multiple of x, leaving the Fourier series of a C^1 function.

E. Sum the series $\displaystyle\sum_{n=1}^{\infty} \frac{\sin n\theta}{n}$.   HINT: Exercise B and Exercise 7.6.F.

F. (a) Show that the function $h(x) = \cos(x/2)$ for $-\pi \leq x \leq \pi$ is continuous and piecewise C^1.
     (b) Find the Fourier series for h. HINT: $A_n = (-1)^{n-1} 4/(\pi(4n^2 - 1))$.
     (c) Sum the series at the point $x = 0$ in two ways, and show that they yield the same result.
        HINT: $2/(4n^2 - 1) = 1/(2n - 1) - 1/(2n + 1)$.

G. Prove **Dini's Test**: If f is a piecewise-continuous 2π-periodic function such that

$$\int_0^{\pi} \frac{1}{t} \left| \frac{f(\theta_0 + t) + f(\theta_0 - t)}{2} - s \right| dt < \infty,$$

   then $\displaystyle\lim_{n \to \infty} S_n f(\theta_0) = s$.   HINT: Look for a spot in the proof of the Dirichlet–Jordan Theorem where this integral condition may be used instead of the Lipshitz condition.

14.3 Gibbs's Phenomenon

In this section, we show that pointwise convergence of Fourier series, which we established in the previous section, is not good enough for many applications. In particular, a sequence of functions can converge pointwise without "looking like" their limit.

We have seen in Section 8.1 that pointwise convergence of functions allows surprisingly bad behaviour. Such a phenomenon arises for the functions $S_n f$ near any jump discontinuity of f. This was first discovered by an English mathematician,

Wilbraham, in 1848. Around the turn of the century, it was rediscovered by Michel-son and then explained by Gibbs, a (now) famous American physicist, in a letter to the journal *Nature*. For a discussion of this history, see [39]. Put simply, whenever f has a jump discontinuity, the graphs of $S_n f$ overshoot f near the discontinuity and increasing n does *not* reduce the error; it only pushes the overshoot nearer to the discontinuity. See Figure 14.2, for example.

As an example, we demonstrate the phenomenon for the 2π-periodic function given by

$$f(x) = \begin{cases} x & \text{if } x \in (-\pi, \pi), \\ 0 & \text{if } x = \pm\pi. \end{cases}$$

The Dirichlet–Jordan Theorem (14.2.6) shows that $\lim_{n \to \infty} S_n f(x) = f(x)$, for all $x \in \mathbb{R}$. Nonetheless, $S_n f(x)$ always overshoots $f(x)$ at some point near the discontinuity by about 9% of the gap (which is 2π in this case).

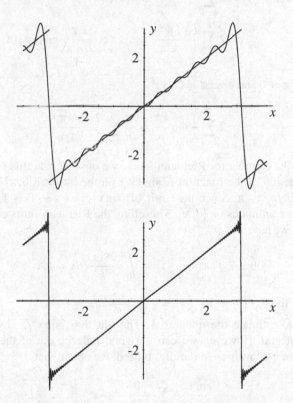

FIG. 14.2 The graphs of $S_{10}f$ and $S_{100}f$, each plotted with f.

14.3.1. THEOREM. *Let* $A = \dfrac{2}{\pi} \displaystyle\int_0^\pi \dfrac{\sin(x)}{x}\,dx \approx 1.178979744$. *For the function f just defined, we have*

$$\lim_{n\to\infty} S_n f\big(\pi(1-\tfrac{1}{n})\big) = A\pi \quad \text{and} \quad \lim_{n\to\infty} S_n f\big(-\pi(1-\tfrac{1}{n})\big) = -A\pi.$$

PROOF. Note that f is an odd function and hence has a sine series. An integration by parts argument (see Exercise 14.2.B) shows that

$$f(x) \sim 2 \sum_{k=1}^{\infty} \frac{(-1)^{k+1}}{k} \sin kx.$$

Thus,

$$S_n f\big(\pi(1-\tfrac{1}{n})\big) = 2 \sum_{k=1}^{n} \frac{(-1)^{k+1}}{k} \sin\!\left(k\pi - \frac{k\pi}{n}\right)$$

$$= 2 \sum_{k=1}^{n} \frac{(-1)^{k+1}}{k} \left(\sin k\pi \cos \frac{k\pi}{n} - \cos k\pi \sin \frac{k\pi}{n}\right),$$

and since $\sin k\pi = 0$ and $\cos k\pi = (-1)^k$,

$$= 2 \sum_{k=1}^{n} \frac{1}{k} \sin \frac{k\pi}{n} = \frac{\pi}{n} \sum_{k=1}^{n} \frac{2\sin(k\pi/n)}{k\pi/n}.$$

Remembering the formula for Riemann sums, we observe that this is the Riemann sum for the integral of the function $(2\sin x)/x$ on the interval $[0, \pi]$ using the partition $0, \pi/n, 2\pi/n, \ldots, \pi$. Since the limit of $(\sin x)/x$ as $x \to 0$ is 1, this function is bounded and continuous on $[0, \pi]$. Therefore, the Riemann sums converge to the integral. Thus, we have

$$\lim_{n\to\infty} S_n f\big(\pi(1-\tfrac{1}{n})\big) = \int_0^\pi \frac{2\sin x}{x}\,dx = \pi A$$

and, similarly, $\lim_{n\to\infty} S_n f\big(-\pi(1-\tfrac{1}{n})\big) = -\int_0^\pi \frac{2\sin x}{x}\,dx = -\pi A$.

It remains to estimate the integral A. The function $(\sin x)/x$ does not have a closed-form integral. However, we can get good mileage out of the Taylor series for $\sin x$ because it converges so rapidly. Indeed, we obtain that

$$\frac{\sin x}{x} = \sum_{k=0}^{\infty} \frac{(-1)^k}{(2k+1)!} x^{2k}$$

for all real x. Since this converges uniformly on $[0, \pi]$, we may integrate term by term by Theorem 8.3.1. Therefore,

$$A = \frac{2}{\pi} \int_0^\pi \frac{\sin(x)}{x}\, dx = \frac{2}{\pi} \int_0^\pi \sum_{k=0}^\infty \frac{(-1)^k}{(2k+1)!} x^{2k}\, dx$$

$$= \sum_{k=0}^\infty \frac{2}{\pi} \int_0^\pi \frac{(-1)^k}{(2k+1)!} x^{2k}\, dx = \sum_{k=0}^\infty \frac{2}{\pi} \frac{(-1)^k}{(2k+1)!} \frac{\pi^{2k+1}}{(2k+1)}$$

$$= 2 \sum_{k=0}^\infty \frac{(-1)^k \pi^{2k}}{(2k+1)!(2k+1)} = 2 - \frac{\pi^2}{9} + \frac{\pi^4}{300} - \frac{\pi^6}{17640} + \cdots .$$

This is an alternating series in which the terms decrease monotonically to 0, so

$$1.17357 \approx 2 - \frac{\pi^2}{9} + \frac{\pi^4}{300} - \frac{\pi^6}{17640} < A$$

$$< 2 - \frac{\pi^2}{9} + \frac{\pi^4}{300} - \frac{\pi^6}{17640} + \frac{\pi^8}{1632960} \approx 1.17938.$$

This is enough for our purposes. In fact, A is approximately 1.178979744. ∎

Gibbs's phenomenon is not special to Fourier series. Similar behaviour can be constructed using piecewise linear functions, as considered in Section 10.8, in place of trigonometric functions. See [40] for an example involving a rescaling of the function h of Example 14.2.7 and an equally spaced partition. The crucial point is that best approximations in the L^2-norm are, near jump discontinuities, going to behave badly in the uniform norm.

Exercises for Section 14.3

A. (a) Suppose h is a C^2 function on $[-\pi, \pi]$ with $h(\pi) = h(-\pi)$ but possibly $h'(\pi) \neq h'(-\pi)$. Show that $S_n h$ converges uniformly to h.

(b) Suppose that g is a C^2 function on $[-\pi, \pi]$ but $g(\pi) \neq g(-\pi)$. Subtract a multiple of the function f used in this section from g to obtain a function h as in part (a). Hence show that g also exhibits Gibbs's phenomenon.

B. (a) Following the proof of Gibbs's phenomenon, show that $\lim_{n\to\infty} S_n f\left(\pi - \frac{a}{n}\right) = \int_0^a \frac{\sin x}{x}\, dx$.

(b) Let $t_n = \int_{(n-1)\pi}^{n\pi} \frac{\sin x}{x}\, dx$. Show that t_n alternates in sign, $|t_{n+1}| < |t_n|$, and $\lim_{n\to\infty} t_n = 0$.

(c) Hence show that $\sup_{a>0} \left| \int_0^a \frac{\sin x}{x}\, dx \right| = \int_0^\pi \frac{\sin x}{x}\, dx$.

(d) Establish the existence of the improper Riemann integral $\int_0^\infty \frac{\sin x}{x}\, dx = \lim_{a\to\infty} \int_0^a \frac{\sin x}{x}\, dx$.

C. (a) Use Lemma 14.2.2 to show that $\int_0^{\pi-a} 2\pi D_n(x)\, dx = \pi - a + S_n f(a)$, where $0 < a < \pi$ and $f(x) = x$ for $-\pi \leq x \leq \pi$.

(b) Hence show that $|S_n f(a) - a| = \left| \int_0^{(n+\frac{1}{2})(\pi-a)} \frac{2\sin x}{(2n+1)\sin \frac{x}{2n+1}}\, dx - \pi \right|$.

(c) Use the Riemann–Lebesgue Lemma to show that

$$\int_0^{(n+\frac{1}{2})\frac{\pi}{2}} \frac{2\sin x}{(2n+1)\sin \frac{x}{2n+1}} - \frac{2\sin x}{x}\, dx = \int_0^{\frac{\pi}{2}} g(x)\sin(n+\tfrac{1}{2})x\, dx$$

(for a certain continuous function g) tends to 0 as n goes to $+\infty$.

(d) Use the Dirichlet–Jordan Theorem to deduce that $\int_0^\infty \dfrac{\sin x}{x}\, dx = \dfrac{\pi}{2}$.

14.4 Cesàro Summation of Fourier Series

It is natural to try to recombine the harmonics of a function f simply by adding the first several terms. However, proving that such approximations converge point-wise to f as the number of terms goes to infinity requires a Lipschitz condition (Dirichlet-Jordan Theorem (14.2.6)). Even worse is Gibbs's phenomenon, although we cannot hope for continuous approximants to converge uniformly to a discontinuous function. In this section we consider a new sequence of approximations, built from the Fourier coefficients of f, that converges uniformly to the function f for all continuous functions f. The results of this section were found about 1900 by Fejér, a Hungarian mathematician, at the age of 19.

In order to obtain this better behaviour, we replace the sequence of functions $S_n f$ with their averages, known as **Cesàro means**:

$$\sigma_n f(x) = \frac{1}{n+1} \sum_{k=0}^{n} S_k f(x).$$

This is defined whenever the Fourier coefficients of f are defined, which includes all absolutely integrable functions. Our primary interest will be for continuous functions. This new sequence of functions has an associated kernel that is much better behaved than the Dirichlet kernel. It shares many of the good properties of the Poisson kernel. In fact, it is better than the Poisson kernel, since computing $\sigma_n f$ does not require an infinite sum. Indeed,

$$\sigma_n f(x) = \frac{1}{n+1} \sum_{k=0}^{n} \left(A_0 + \sum_{j=1}^{k} A_j \cos jx + B_j \sin jx \right)$$

$$= A_0 + \sum_{j=1}^{n} \left(1 - \frac{j}{n+1} \right) \left(A_j \cos jx + B_j \sin jx \right).$$

One deficiency of $\sigma_n f$ compared to $S_n f$ is that for f a trig polynomial of degree at most n, $S_n f$ equals f but $\sigma_n f$ may not. This is more than compensated by superior convergence of the sequence.

Our first result is to turn this summation into an integral formula . The following result can be deduced using trig identities or complex exponentials (Lemma 13.9.5).

14.4.1. LEMMA. *The **Fejér kernel** is the sequence $K_n(t)$ for $n \geq 1$ given by*

$$\frac{1}{2\pi \sin^2 t/2} \sum_{k=0}^{n} \sin(k + \tfrac{1}{2})t = \begin{cases} \dfrac{n+1}{2\pi} & \text{if } t = 2m\pi, \ m \in \mathbb{Z}, \\[2mm] \dfrac{1}{2\pi(n+1)} \left(\dfrac{\sin \frac{n+1}{2} t}{\sin t/2} \right)^2 & \text{if } t \neq 2m\pi. \end{cases}$$

Each K_n is continuous at zero. This follows from $\lim_{t \to 0} (\sin at)/t = a$. We leave it as an exercise.

You should compare Figure 14.3 with the graphs of the Poisson and Dirichlet kernels, Figures 13.2 and 14.1. The key difference between the kernels K_n and D_n is that the K_n are positive. Moreover, for $t \notin 2\pi\mathbb{Z}$, $K_n(t) \to 0$ as n goes to infinity, unlike $D_n(t)$. However, $K_n(0) \to \infty$ as n goes to infinity, exactly like D_n. It is helpful to think of the K_n as functions that become more and more like spikes (i.e., large at zero and small elsewhere), as n goes to infinity. All of these properties are shared with the Poisson kernel, which leads us to define a general positive kernel below.

14.4.2. THEOREM. *If $f : \mathbb{R} \to \mathbb{R}$ is piecewise continuous and 2π-periodic, then*

$$\sigma_n f(x) = \int_{-\pi}^{\pi} f(x+t) K_n(t)\, dt.$$

PROOF. Using Theorem 14.2.3, we have

$$\sigma_n f(x) = \frac{1}{n+1} \sum_{k=0}^{n} \int_{-\pi}^{\pi} f(t+x) \frac{\sin(k+\frac{1}{2})t}{2\pi \sin t/2}\, dt$$

$$= \int_{-\pi}^{\pi} f(t+x) \frac{1}{2\pi(n+1)\sin t/2} \sum_{k=0}^{n} \sin\left(k+\frac{1}{2}\right) t\, dt,$$

and applying Lemma 14.4.1 completes the proof. ∎

14.4.3. PROPERTIES OF THE FEJÉR KERNEL.

(1) *For each n, K_n is a positive, continuous, 2π-periodic, even function.*

(2) $\displaystyle \int_{-\pi}^{\pi} K_n(t)\, dt = 1.$

(3) *For $\delta \in (0, \pi)$, K_n converges uniformly to zero on $[-\pi, -\delta] \cup [\delta, \pi]$.*

(4) *For $\delta \in (0, \pi)$, $\displaystyle \lim_{n \to \infty} \left(\int_{-\pi}^{-\delta} K_n + \int_{\delta}^{\pi} K_n \right) = 0.$*

PROOF. It is evident from the formula that K_n is positive, even, 2π-periodic, and continuous except possibly at multiples of 2π. Because of the periodicity, it suffices to check continuity at 0, which we have asserted is true.

For (2), taking $f = 1$ in Theorem 14.4.2, we have

$$\int_{-\pi}^{\pi} K_n(t)\, dt = \sigma_n f(0) = 1.$$

For (3), we let $\varepsilon > 0$. Observe that $|\sin t/2| \geq \sin \delta/2$ for t such that $\delta \leq |t| \leq \pi$. Thus,

$$|K_n(t)| \leq \frac{1}{2(n+1)} \frac{1}{|\sin \delta/2|} \quad \text{for all} \quad t \in [-\pi, -\delta] \cup [\delta, \pi].$$

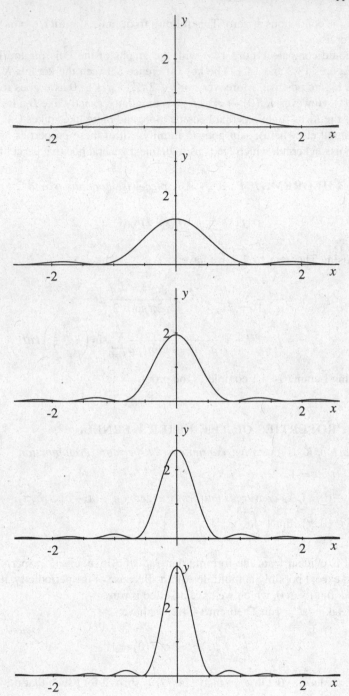

FIG. 14.3 The graphs of K_1 through K_5.

Since δ is fixed, if we choose any $N \geq \varepsilon/(2\sin\delta/2)$, then for all $n \geq N$,

$$|K_n(t)| \leq \varepsilon \quad \text{for all} \quad t \in [-\pi, -\delta] \cup [\delta, \pi].$$

That is, K_n converges uniformly to zero on $[-\pi, -\delta] \cup [\delta, \pi]$.

Finally, (4) is an immediate consequence of (3). ∎

Since we have two examples, it seems appropriate to introduce the following general definition and prove the main theorem of this section for positive kernels.

14.4.4. DEFINITION. We call a family of 2π-periodic continuous functions k_n, $n = 1,2,3\ldots$, a **positive kernel** if

(1) For all $t \in \mathbb{R}$, $k_n(t) \geq 0$.

(2) $\displaystyle\int_{-\pi}^{\pi} k_n(t)\, dt = 1$.

(3) For $\delta \in (0,\pi)$, k_n converges uniformly to zero on $[-\pi,-\delta] \cup [\delta,\pi]$.

We will write $\Sigma_n f(x)$ for $\displaystyle\int_{-\pi}^{\pi} f(x+t)k_n(t)\, dt$.

Examples include K_n, a subsequence of the Poisson kernel, such as $P(1/n,t)$, $n \geq 1$, and the de la Vallée Poussin kernel, which is given in the exercises.

We can now prove the main result.

14.4.5. FEJÉR'S THEOREM.
If f is continuous and 2π-periodic and k_n is a positive kernel, then $\Sigma_n f$ converges uniformly to f.

PROOF. Conceptually, this proof is much like the proof of Poisson's Theorem. We write $\Sigma_n f(x)$ as an integral from $-\pi$ to π and then split the integral into two parts: the interval $[-\delta,\delta]$, and second, the rest, namely $[-\pi,-\delta] \cup [\delta,\pi]$. We control the first integral using the uniform continuity of f; we control the second using the uniform convergence of the k_n to zero.

Let $M = \|f\|_\infty = \max\{|f(x)| : x \in [-\pi,\pi]\}$ and let $\varepsilon > 0$. Since f is continuous on the compact set $[-\pi,\pi]$, it is uniformly continuous by Theorem 5.5.9. Hence there is some $\delta > 0$ such that

$$|f(x) - f(y)| < \frac{\varepsilon}{2} \quad \text{whenever} \quad |x-y| < \delta.$$

With this δ fixed, we apply Definition 14.4.4 (3), to conclude that there is an integer N such that

$$k_n(x) < \frac{\varepsilon}{8\pi M} \quad \text{for all} \quad x \in [-\pi,-\delta] \cup [\delta,\pi] \quad \text{and} \quad n \geq N.$$

Using the definition of $\Sigma_n f$ and Definition 14.4.4 (2), we have

$$|\sigma_n f(x) - f(x)| = \left| \int_{-\pi}^{\pi} f(x+t) k_n(t)\, dt - f(x) \int_{-\pi}^{\pi} k_n(t)\, dt \right|$$

$$= \left| \int_{-\pi}^{\pi} \left(f(x+t) - f(x) \right) k_n(t)\, dt \right|$$

$$\leq \int_{-\pi}^{\pi} |f(x+t) - f(x)| k_n(t)\, dt.$$

Now, we split this integral into two parts, as promised above. Let $I_1 = [-\delta, \delta]$ and $I_2 = [-\pi, -\delta] \cup [\delta, \pi]$. If $t \in I_1$, then $|f(x+t) - f(x)| < \varepsilon/2$, and so

$$\int_{I_1} |f(x+t) - f(x)| k_n(t)\, dt \leq \int_{I_1} \frac{\varepsilon}{2} k_n(t)\, dt \leq \frac{\varepsilon}{2} \int_{-\pi}^{\pi} k_n(t)\, dt = \frac{\varepsilon}{2}.$$

If $t \in I_2$, then $|k_n(t)| \leq \varepsilon/(8\pi M)$, and so

$$\int_{I_2} |f(x+t) - f(x)| k_n(t)\, dt \leq \int_{I_2} 2M k_n(t)\, dt \leq 2M \int_{I_2} \frac{\varepsilon}{8\pi M}\, dt < \frac{\varepsilon}{4\pi} \int_{-\pi}^{\pi} dt = \frac{\varepsilon}{2}.$$

Adding these two results, we have

$$|\Sigma_n f(x) - f(x)| \leq \int_{-\pi}^{\pi} |f(x+t) - f(x)| k_n(t)\, dt \leq \frac{\varepsilon}{2} + \frac{\varepsilon}{2} = \varepsilon$$

for all $x \in [-\pi, \pi]$ and all $n \geq N$. Thus, by the definition of uniform convergence, $\sigma_n f$ converges uniformly to f. ∎

This proof has a strong resemblance to our proof of the Weierstrass Approximation Theorem. The common underlying technique here is controlling the integral of a product by splitting the integral into two parts, where one factor of the product is well behaved on each part. For more examples, review the proofs of the inequalities in Theorem 14.2.4 (3).

In fact, it is possible to prove the Weierstrass Approximation Theorem using Fejér's Theorem. One proof is outlined in the exercises. Another will be given in Section 14.9.

14.4.6. EXAMPLE. Consider the function h introduced in Example 14.2.7,

$$h(x) = \begin{cases} 1 & \text{for} \quad 0 \leq x \leq \pi, \\ -1 & \text{for} \quad -\pi < x < 0. \end{cases}$$

The sequence $S_n h$ will exhibit Gibbs's phenomenon (see Exercise 14.3.A) as shown in Figure 14.4. We will compute the Cesàro means for this function using the Fejér kernel. Since h is an odd function, the approximants $S_n h$ and $\sigma_n h$ are also all odd. Also, $h(\pi - x) = h(x)$, and so $S_n h$ and $\sigma_n h$ also have this symmetry. Consider a point x in $[0, \pi/2]$:

$$\sigma_n h(x) = \int_{-\pi}^{\pi} h(x+t) K_n(t)\, dt$$
$$= -\int_{-\pi}^{-x} K_n(t)\, dt + \int_{-x}^{\pi-x} K_n(t)\, dt - \int_{\pi-x}^{\pi} K_n(t)\, dt$$
$$= \int_{-x}^{x} K_n(t)\, dt - \int_{\pi-x}^{\pi+x} K_n(t)\, dt.$$

Here we have exploited the fact that K_n is even to cancel the integral from x to $\pi - x$ with the integral from $x - \pi$ to $-x$.

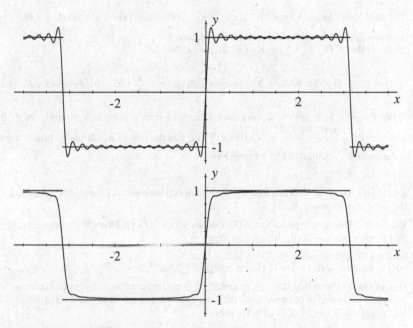

FIG. 14.4 The graphs of $S_{29}h$ and $\sigma_{29}h$, each plotted with h.

By Proposition 14.4.3, the first integral converges to 1 for x in $(0, \pi/2]$, while the second term tends to 0. On the other hand, $\sigma_n h(0) = 0$ for any n. Rewrite $\sigma_n h(x)$ as $2\int_{0}^{x} K_n(t) - K_n(\pi-t)\, dt$. Since K_n is positive on $[0, \pi]$, it follows that $\sigma_n h(x)$ is monotone increasing on $[0, \pi/2]$, then decreases symmetrically back down to 0 at π. Also, $0 < \sigma_n h(x) < 1$ here because $\int_{-x}^{x} K_n(t)\, dt < 1$. Likewise, by symmetry, $\sigma_n h(x)$ converges to -1 on $(-\pi, 0)$ and $\sigma_n h(-\pi) = 0$ for all n.

From the monotonicity, we can deduce that this convergence is uniform on intervals $[\varepsilon, \pi - \varepsilon] \cup [\varepsilon - \pi, -\varepsilon]$ for $\varepsilon > 0$. Since the function h has jump discontinuities at the points 0 or $\pm\pi$, continuous functions cannot converge uniformly to it near these points.

There are some important consequences of Fejér's Theorem (14.4.5) that have already been deduced from Poisson's Theorem (13.5.1) in the previous chapter. If you are reading this chapter without the previous one, then look at Exercises G, H, and I below.

Exercises for Section 14.4

A. Show that K_n is continuous at zero.

B. Prove Lemma 14.4.1.   HINT: Multiply through by $\sin^2 t/2$ and use a trig identity.

C. Show that if f is an absolutely integrable function with $\lim_{n\to\infty} S_n f(\theta) = a$, then $\lim_{n\to\infty} \sigma_n f(\theta) = a$.

D. The de la Vallée Poussin kernel is $V_n(x) = 2K_{2n-1}(x) - K_{n-1}(x)$ for $x \in \mathbb{R}$ and $n \in \mathbb{N}$.

   (a) Show that the functions V_n form a positive kernel.
   (b) Show that if $f \sim A_0 + \sum_k A_k \cos kx + B_k \sin kx$, then

$$\int_{-\pi}^{\pi} f(x+t)V_n(t)\,dt = A_0 + \sum_{k=1}^{n} A_k \cos kx + B_k \sin kx + \sum_{k=n+1}^{2n} \left(2 - \tfrac{k}{n}\right)(A_k \cos kx + B_k \sin kx).$$

E. Show that if f is a piecewise continuous function with a jump discontinuity at θ, then $\lim_{n\to\infty} \sigma_n f(\theta) = \dfrac{f(\theta^+) + f(\theta^-)}{2}$.   HINT: Write f as the sum of a continuous function g and a piecewise C^1 function h. Use Example 14.4.6.

F. Show that $\|\sigma_n f\|_\infty \le \|f\|_\infty$.

G. Use Fejér's Theorem to prove that if f and g are two continuous 2π-periodic functions with the same Fourier series, then $f = g$.

H. Use Fejér's Theorem to prove that if the Fourier series $S_n f$ of a 2π-periodic continuous function f converges uniformly, then it converges to f.

I. (a) Find the Fourier series of $f(t) = |t|^3$ on $[-\pi, \pi]$.
   (b) Evaluate the series at $t = 0$. Hence compute $\sum_{n=1}^{\infty} n^{-4}$.

J. (Localization) Prove that if f is a bounded 2π-periodic function that is continuous on $[a,b]$ including two-sided at the endpoints, then $\sigma_n(f)$ converges uniformly to f on $[a,b]$.
   HINT: Follow the proof of Fejér's Theorem.

K. Let $\sum_{j\ge 0} a_j$ be an infinite series. Define $s_n = \sum_{j=0}^{n} a_j$ and $\sigma_n = \frac{1}{n}\sum_{j=0}^{n-1} s_j$.

   (a) If $\lim_{n\to\infty} s_n = L$, show that $\lim_{n\to\infty} \sigma_n = L$.
   (b) Show by example that the converse of (a) is false.
   (c) **Hardy's Tauberian Theorem**: Show that if $\lim_{n\to\infty} na_n = 0$ and $\lim_{n\to\infty} \sigma_n = L$, then $\lim_{n\to\infty} s_n = L$.
   HINT: Verify that $s_N - \sigma_{N+1} = \frac{1}{N+1}\sum_{j=1}^{N} ja_j$.

L. Suppose f is a 2π-periodic function and $|A_n| + |B_n| \le C/n$ for $n \ge 1$ and some constant C.

   (a) Find a bound for $S_n f(\theta) - \sigma_n f(\theta) = \sum_{k=1}^{n} \frac{k}{n+1} A_k \cos k\theta + \frac{k}{n+1} B_k \sin k\theta$. Hence show that $\|S_n f\|_\infty \le \|f\|_\infty + C$.
   (b) Apply this to obtain a uniform bound for $S_n f$ for the function f used in our example of Gibbs's phenomenon.

M. Prove Weierstrass's Approximation Theorem for a continuous function f on $[0, \pi]$ as follows:

   (a) Set $g(\theta) = f(|\theta|)$ for $\theta \in [-\pi, \pi]$. This is an even, continuous, 2π-periodic function. Use Fejér's Theorem to approximate g within $\varepsilon/2$ by a trig polynomial.
   (b) Use the fact that the Taylor polynomials for $\cos n\theta$ converge uniformly on $[0, \pi]$ to approximate the trigonometric polynomial by actual polynomials within $\varepsilon/2$.

14.5 Least Squares Approximations

Approximation in the L^2 norm is important because it is readily computable. Also, the partial sums of the Fourier series are well behaved in this norm, unlike pointwise and uniform convergence. We will show how the Hilbert space theory from Chapter 7 applies.

Proposition 7.6.1 shows that $\{1, \sqrt{2}\cos n\theta, \sqrt{2}\sin n\theta : n \geq 1\}$ forms an orthonormal set in $C[-\pi, \pi]$. The partial sum $S_n f = A_0 + \sum_{k=1}^{n} A_k \cos k\theta + B_k \sin k\theta$ of the Fourier series of f is the orthogonal projection of f onto the subspace spanned by 1 and $\{\cos kx, \sin kx : 1 \leq k \leq n\}$. The Projection Theorem (7.5.11) implies that $S_n f$ is the best approximation to f in this subspace with respect to the L^2 norm. That is, if $t(\theta)$ is a trigonometric polynomial of degree at most n, then by inequality (7.5.12),

$$\|f - t\|_2^2 = \|f - S_n f\|_2^2 + \|S_n f - t\|_2^2 \geq \|f - S_n f\|_2^2. \tag{14.5.1}$$

Also, Lemma 7.5.7 shows that

$$\|S_n f\|_2^2 = A_0^2 + \frac{1}{2}\sum_{k=1}^{n} A_k^2 + B_k^2 \leq \|f\|_2^2.$$

The main result of this section shows that the sequence of approximants $(S_n f)$ converges to f in the L^2 norm, which requires a bit more work. The key step in the proof is to show that the trigonometric polynomials are dense in $PC[-\pi, \pi]$ in the L^2 norm. The proof given here depends on Fejér's Theorem (14.4.5); alternatively, one could use the Stone–Weierstrass Theorem, as in Corollary 10.10.7.

14.5.2. LEMMA. *Every piecewise continuous 2π-periodic function f is the limit in the $L^2(-\pi, \pi)$ norm of a sequence of trigonometric polynomials.*

PROOF. Let $x_0 = -\pi < x_1 < \cdots < x_N = \pi$ be a partition of f into continuous segments. Fix $n \geq 1$. Let $M = \|f\|_\infty$, and let

$$\delta = \min\left\{\frac{1}{32N(Mn)^2}, \frac{x_{i+1} - x_i}{2} : 0 \leq i < N\right\}.$$

Define a continuous function g_n on $[-\pi, \pi]$ as follows. Let $g_n(x) = f(x)$ for $x \in [x_i + \delta, x_{i+1} - \delta]$, $0 \leq i < N$. Also, set $g_n(-\pi) = g(\pi) = 0$. Finally, make g_n linear and continuous on each segment $J_0 = [-\pi, -\pi + \delta]$, $J_i = [x_i - \delta, x_i + \delta]$, $1 \leq i < N$, and $J_N = [\pi - \delta, \pi]$. See Figure 14.5 for an example.

Observe that $\|g_n\|_\infty \leq \|f\|_\infty = M$ and therefore $|f(x) - g_n(x)| \leq 2M$. Moreover, the two functions agree except on the intervals J_i for $0 \leq i \leq N$. The total length of these intervals is $2N\delta$. Therefore, we can estimate

$$\|f - g_n\|_2^2 \leq \sum_{i=0}^{N} \int_{J_i} (2M)^2 \, dx \leq 8NM^2\delta \leq \frac{1}{4n^2}.$$

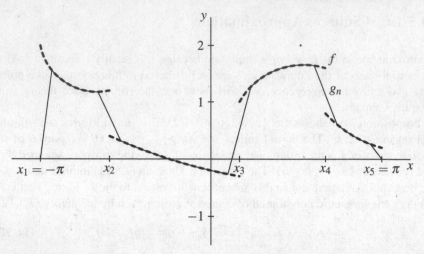

FIG. 14.5 Piecewise continuous f with continuous approximation g_n.

By Fejér's Theorem (14.4.5) (or, if you prefer, Corollary 13.5.6), the 2π-periodic continuous function g_n is a uniform limit of trig polynomials. So there is a trig polynomial t_n such that $\|g_n - t_n\|_\infty < \frac{1}{2n}$. Then $\|g_n - t_n\|_2 \le \|g_n - t_n\|_\infty < \frac{1}{2n}$ as well. Thus

$$\|f - t_n\|_2 \le \|f - g_n\|_2 + \|g_n - t_n\|_2 < \frac{1}{2n} + \frac{1}{2n} = \frac{1}{n}.$$

Therefore f is an L^2 limit of trig polynomials.   ∎

The main import of the following theorem is that the partial sums $S_N f$ converge to f in the L^2 norm. Since $S_N f$ is a trigonometric polynomial, it is continuous (and in fact C^∞). In particular, our result shows that discontinuous functions can be L^2 limits of continuous functions. Since the *uniform* limit of continuous functions remains continuous, L^2 convergence is a weaker notion.

14.5.3. LEAST SQUARES THEOREM.

If $f : \mathbb{R} \to \mathbb{R}$ is piecewise continuous and 2π-periodic, then $\lim\limits_{N \to \infty} \|f - S_N f\|_2 = 0$.
Moreover, if $f \sim A_0 + \sum_k A_k \cos kx + B_k \sin kx$, then

$$\frac{1}{2\pi} \int_{-\pi}^{\pi} |f(\theta)|^2 \, d\theta = \|f\|_2^2 = A_0^2 + \frac{1}{2} \sum_{n=1}^{\infty} A_n^2 + B_n^2. \qquad (14.5.4)$$

Note that this last equality does not follow from our statement of Parseval's Theorem (7.7.5) because $PC[-\pi, \pi]$ is not a Hilbert space.

PROOF. By Lemma 14.5.2, f is the limit of trigonometric polynomials in the L^2 norm. Thus given $\varepsilon > 0$, choose a trig polynomial t with $\|f - t\|_2 < \varepsilon$. So it follows from (14.5.1) that for n at least the degree of t,

$$\|f - S_n f\|_2 \le \|f - t\|_2 < \varepsilon.$$

The triangle inequality implies that $\big|\|f\|_2 - \|S_n f\|_2\big| \le \|f - S_n f\|_2$, and so

$$\|f\|_2^2 = \lim_{n \to \infty} \|S_n f\|_2^2 = \lim_{n \to \infty} A_0^2 + \frac{1}{2} \sum_{k=1}^{n} A_k^2 + B_k^2 = A_0^2 + \frac{1}{2} \sum_{n=1}^{\infty} A_n^2 + B_n^2. \qquad \blacksquare$$

14.5.5. EXAMPLE. In Example 7.6.3, we showed that

$$|\theta| \sim \frac{\pi}{2} - \frac{4}{\pi} \sum_{k=0}^{\infty} \frac{\cos(2k+1)\theta}{(2k+1)^2}.$$

Compute the L^2 norm using our formula:

$$\frac{1}{2\pi} \int_{-\pi}^{\pi} |\theta|^2 \, d\theta = \left(\frac{\pi}{2}\right)^2 - \frac{1}{2}\left(\frac{4}{\pi}\right)^2 \sum_{k=0}^{\infty} \left(\frac{1}{(2k+1)^2}\right)^2$$

$$= \frac{\pi^2}{4} + \frac{8}{\pi^2} \sum_{k=0}^{\infty} \frac{1}{(2k+1)^4}.$$

The integral is easily found to be $\pi^2/3$, from which we deduce that

$$\sum_{k=0}^{\infty} \frac{1}{(2k+1)^4} - \frac{\pi^2}{8}\left(\frac{\pi^2}{3} - \frac{\pi^2}{4}\right) = \frac{\pi^4}{96}.$$

Hence

$$\sum_{k=1}^{\infty} \frac{1}{k^4} = \sum_{k=0}^{\infty} \frac{1}{(2k+1)^4} + \sum_{k=1}^{\infty} \frac{1}{(2k)^4} = \frac{\pi^4}{96} + \frac{1}{16} \sum_{k=1}^{\infty} \frac{1}{k^4}.$$

Therefore, $\displaystyle \sum_{k=1}^{\infty} \frac{1}{k^4} = \frac{\pi^4}{90}.$

An immediate consequence of this theorem is that the sines and cosines span all of $PC[-\pi, \pi]$ and hence all of $C[-\pi, \pi]$. Since they are orthogonal by Proposition 7.6.1, it follows that they form an orthonormal basis.

Although we have worked with piecewise continuous functions here, there is a natural Hilbert space lurking in the background, $L^2(-\pi, \pi)$, which is the completion of $C[-\pi, \pi]$ in the L^2 norm. This Hilbert space has an elegant connection to the Hilbert space $\ell^2(\mathbb{Z})$ from Section 7.7.

First, we need a working definition of $L^2(-\pi, \pi)$. Using the Lebesgue integral, one can describe $L^2(-\pi, \pi)$ as (essentially) functions on $[-\pi, \pi]$ such that $|f|^2$ has a finite Lebesgue integral. Constructing the Lebesgue integral is an enormous effort, well beyond the scope of this book, so instead we describe $L^2(-\pi, \pi)$ as a completion of $PC[-\pi, \pi]$. To start with, take Cauchy sequences of functions (f_n) in $PC[-\pi, \pi]$. We put an equivalence relation on this collection by writing $(f_n) \approx (g_n)$

if $\|f_n - g_n\|_2 \to 0$ as $n \to \infty$. (By Lemma 14.5.2, each sequence is equivalent to a Cauchy sequence of trig polynomials.) Then $L^2(-\pi, \pi)$ is the collection of equivalence classes of Cauchy sequences. Notice that a function $f \in PC[-\pi, \pi]$ can be associated with the equivalence class of the constant sequence $(f, f, \ldots)$ in $L^2(-\pi, \pi)$. We will not need the exact sense in which these equivalence classes are bona fide functions.

The crucial property is that the inner product is well defined on $L^2(-\pi, \pi)$ as a limit. If (f_n) and (g_n) are two Cauchy sequences, then we set

$$\langle (f_n), (g_n) \rangle := \lim_{n \to \infty} \langle f_n, g_n \rangle.$$

The fact that this makes sense is left as an exercise. In particular, we can compute the Fourier coefficients and the Fourier series of an element of $L^2(-\pi, \pi)$.

14.5.6. COROLLARY. *The functions $\{1, \sqrt{2}\cos n\theta, \sqrt{2}\sin n\theta : n \geq 1\}$ form an orthonormal basis for $L^2(-\pi, \pi)$. The map sending a sequence $\mathbf{a} = (a_n)$ in $\ell^2(\mathbb{Z})$ to $F\mathbf{a} := f(\theta) = a_0 + \sum_{n=1}^{\infty} \sqrt{2}a_n \cos n\theta + \sqrt{2}a_{-n} \sin n\theta$ is a unitary map. That is, F maps $\ell^2(\mathbb{Z})$ one-to-one and onto $L^2(-\pi, \pi)$, and $\|F\mathbf{a}\|_2 = \|\mathbf{a}\|_2$ for all $\mathbf{a} \in \ell^2(\mathbb{Z})$.*

PROOF. We first define F just on the space ℓ_0 of all sequences $\mathbf{a}$ with only finitely many nonzero terms. Then $F\mathbf{a}$ is a trigonometric polynomial, and F maps ℓ_0 onto the set of all trig polynomials. Theorem 14.5.3 shows that $\|F\mathbf{a}\|_2 = \|\mathbf{a}\|_2$ for each $\mathbf{a} \in \ell^2(\mathbb{Z})$. Thus F is one-to-one because $\|F\mathbf{a} - F\mathbf{b}\| = \|\mathbf{a} - \mathbf{b}\| \neq 0$ when $\mathbf{a} \neq \mathbf{b}$.

Theorem 7.7.4 shows that $\ell^2(\mathbb{Z})$ is complete and thus is a Hilbert space. Every vector $\mathbf{a}$ is a limit of the sequence $P_n\mathbf{a} = \sum_{k=-n}^{n} a_k\mathbf{e}_k$ of vectors in ℓ_0. In particular, this sequence is Cauchy. Therefore, for each $\varepsilon > 0$, there is an integer N such that $\|P_n\mathbf{a} - P_m\mathbf{a}\|_2 < \varepsilon$ for all $n, m \geq N$. Consequently, the sequence of functions $FP_n\mathbf{a} = a_0 + \sum_{k=1}^{n} \sqrt{2}a_k \cos k\theta + \sqrt{2}a_{-k} \sin k\theta$ is also Cauchy, because

$$\|FP_n\mathbf{a} - FP_m\mathbf{a}\|_2 = \|P_n\mathbf{a} - P_m\mathbf{a}\|_2 < \varepsilon \quad \text{for all} \quad n, m \geq N.$$

So this sequence converges in the L^2 norm to an element f in $L^2(-\pi, \pi)$ (because our definition of L^2 is the set of all such limits).

This function f has a Fourier series, and, for example,

$$A_k = 2\langle f, \cos k\theta \rangle = \lim_{n \to \infty} 2\langle FP_n\mathbf{a}, \tfrac{1}{\sqrt{2}}F\mathbf{e}_k \rangle = \sqrt{2} \lim_{n \to \infty} \langle P_n\mathbf{a}, \mathbf{e}_k \rangle = \sqrt{2}a_k.$$

Hence $f \sim a_0 + \sum_{k=1}^{\infty} \sqrt{2}a_k \cos k\theta + \sqrt{2}a_{-k} \sin k\theta$. Moreover,

$$\|f\|_2^2 = \lim_{n \to \infty} \|FP_n\mathbf{a}\|_2 = \lim_{n \to \infty} \|P_n\mathbf{a}\|_2 = \sum_{-\infty}^{\infty} |a_n|^2 = \|\mathbf{a}\|_2^2.$$

This establishes a map F that maps $\ell^2(\mathbb{Z})$ into $L^2(-\pi, \pi)$ and preserves the norm. If $(F\mathbf{a}_n)$ is Cauchy in $L^2(-\pi, \pi)$, then since $\|\mathbf{a}_m - \mathbf{a}_n\| = \|F\mathbf{a}_m - F\mathbf{a}_n\|$, it follows that $(\mathbf{a}_n)$ is Cauchy in $\ell^2(\mathbb{Z})$. If $\mathbf{a}$ is its limit, then $F\mathbf{a} = \lim_{n\to\infty} F\mathbf{a}_n$ belongs to the range. So the image space is complete. The range has been defined as the completion of the trigonometric polynomials in the L^2 norm, and thus is a subspace of $L^2(-\pi, \pi)$. Equation (14.5.4) of Theorem 14.5.3 shows that the range of this map contains every continuous function. Since the range is complete, it also contains every L^2 limit of continuous functions. So the range is exactly all of $L^2(-\pi, \pi)$. This completes the proof. ∎

Exercises for Section 14.5

A. Compute the Fourier series of $f(\theta) = \theta^3 - \pi^2\theta$ for $-\pi \le \theta \le \pi$. Hence evaluate the sums $\sum_{n=0}^{\infty} \dfrac{(-1)^n}{(2n+1)^3}$ and $\sum_{n=1}^{\infty} \dfrac{1}{n^6}$.

B. Evaluate $\sum_{n=1}^{\infty} \dfrac{1}{n^8}$.

C. Show that $\{e^{in\theta} : n \in \mathbb{Z}\}$ forms an orthonormal basis for $C[-\pi, \pi]$.

D. Show that if (f_n) and (g_n) are sequences in $PC[-\pi, \pi]$ that are Cauchy in the $L^2(-\pi, \pi)$ norm, show that $\langle (f_n), (g_n) \rangle := \lim_{n\to\infty} \langle f_n, g_n \rangle$ exists.   HINT: Cauchy–Schwarz inequality.

E. Use Exercise 7.6.K to find an orthonormal basis for $C[0, \pi]$ in the given inner product.

F. (a) Compute the Fourier series of $f(\theta) = e^{a\theta}$ for $-\pi \le \theta \le \pi$ and $a > 0$.
   (b) Evaluate $\|f\|_2$ in two ways, and use this to show that

$$\frac{1}{a^2} + 2\sum_{n=1}^{\infty} \frac{1}{a^2 + n^2} = \frac{\pi}{a}\left(\frac{e^{a\pi} + e^{-a\pi}}{e^{a\pi} - e^{-a\pi}}\right) = \frac{\pi}{a}\coth(a\pi).$$

G. Recall the Chebyshev polynomials $T_n(x) = \cos(n \arccos x)$. Make a change of variables in Exercise D to show that the set $\{T_0, \sqrt{2}T_n : n \ge 1\}$ is an orthonormal basis for $C[-1, 1]$ for the inner product $\langle f, g \rangle_T = \dfrac{1}{\pi}\displaystyle\int_{-1}^{1} f(x)g(x)\,\dfrac{dx}{\sqrt{1-x^2}}$.

H. Show that the map F of Corollary 14.5.6 preserves the inner product.

14.6 The Isoperimetric Problem

In this section, we provide an interesting and nontrivial application of least squares approximation. The isoperimetric problem asks, *What is the largest area that can be surrounded by a continuous closed curve of a given length?* The answer is the circle, but a method for demonstrating this rigorously is not at all obvious.

Indeed, the Greeks were aware of the isoperimetric inequality. However, little was done in the way of a rigorous proof until the work of Steiner in 1838. Steiner gave at least five different arguments, but each one had a flaw. He could not establish the *existence* of a curve with the greatest area among all continuous curves of fixed perimeter. This difficulty was not resolved for another 50 years. In 1901, Hurwitz published the first strictly analytic proof. It is this proof that is essentially given here.

For convenience, we shall fix the length of the curve $\mathscr{C}$ to be 2π. This is the circumference of the circle of radius 1 and area π. We shall show that the circle is the optimal choice subject to the mild hypothesis that $\mathscr{C}$ is piecewise C^1. The argument to remove this differentiability requirement is left to the exercises.

Points on the curve $\mathscr{C}$ may be parametrized by the arc length s as $(x(s), y(s))$ for $0 \leq s \leq 2\pi$. This is a closed curve, and thus $x(2\pi) = x(0)$ and $y(2\pi) = y(0)$. Since the differential of arc length is $ds = \left(x'(s)^2 + y'(s)^2\right)^{1/2} ds$, we have the condition

$$x'(s)^2 + y'(s)^2 = 1.$$

The area $A(\mathscr{C})$ is given by Green's Theorem (see Exercise 14.6.B) as

$$A(\mathscr{C}) = \int_0^{2\pi} x(s) y'(s)\, ds = 2\pi \langle x, y' \rangle.$$

Since x and y are continuous and piecewise C^1, they have Fourier series

$$x(s) \sim A_0 + \sum_{n=1}^{\infty} A_n \cos ns + B_n \sin ns \quad \text{and} \quad y(s) \sim C_0 + \sum_{n=1}^{\infty} C_n \cos ns + D_n \sin ns;$$

and by Lemma 14.1.5, we have

$$x'(s) \sim \sum_{n=1}^{\infty} -nA_n \sin ns + nB_n \cos ns \quad \text{and} \quad y'(s) \sim \sum_{n=1}^{\infty} -nC_n \sin ns + nD_n \cos ns.$$

Let us integrate the condition $x'(s)^2 + y'(s)^2 = 1$ to get

$$1 = \frac{1}{2\pi} \int_0^{2\pi} x'(s)^2 + y'(s)^2\, ds = \|x'\|_2^2 + \|y'\|_2^2 = \frac{1}{2} \sum_{n=1}^{\infty} n^2 (A_n^2 + B_n^2 + C_n^2 + D_n^2).$$

The area formula yields

$$A(\mathscr{C}) = 2\pi \langle x, y' \rangle = \pi \sum_{n=1}^{\infty} n(A_n D_n - B_n C_n).$$

Therefore,

$$\pi - A(\mathscr{C}) = \frac{\pi}{2} \sum_{n=1}^{\infty} n^2 (A_n^2 + B_n^2 + C_n^2 + D_n^2) - \pi \sum_{n=1}^{\infty} n(A_n D_n - B_n C_n)$$

$$= \frac{\pi}{2} \sum_{n=1}^{\infty} (n^2 - n)(A_n^2 + B_n^2 + C_n^2 + D_n^2) +$$

$$+ \frac{\pi}{2} \sum_{n=1}^{\infty} n(A_n^2 - 2A_n D_n + D_n^2 + B_n^2 + 2B_n C_n + C_n^2)$$

$$= \frac{\pi}{2} \sum_{n=1}^{\infty} (n^2 - n)(A_n^2 + B_n^2 + C_n^2 + D_n^2) + \frac{\pi}{2} \sum_{n=1}^{\infty} n(A_n - D_n)^2 + n(B_n + C_n)^2.$$

The right-hand side of this expression is clearly a sum of squares and thus is positive. The minimum value 0 is attained only if

$$D_1 = A_1, \quad C_1 = -B_1, \quad \text{and} \quad A_n = B_n = C_n = D_n = 0 \quad \text{for} \quad n \geq 2.$$

Moreover, the arc length condition yields

$$1 = \frac{1}{2}(A_1^2 + B_1^2 + C_1^2 + D_1^2) = A_1^2 + B_1^2.$$

Therefore there is a real number θ such that $A_1 = \cos\theta$ and $B_1 = \sin\theta$. Thus the optimal solutions are

$$x(s) = A_0 + \cos\theta \cos s + \sin\theta \sin s = A_0 + \cos(s - \theta),$$
$$y(s) = C_0 - \sin\theta \cos s + \cos\theta \sin s = C_0 + \sin(s - \theta).$$

Clearly, this is the parametrization of the unit circle centred at (A_0, C_0).

Finally, we should relate this proof to the historical issues discussed at the beginning of this section. Hurwitz's proof, as we just saw, results in an inequality for *all* piecewise smooth curves in which the circle evidently attains the minimum. It does not assume the existence of an extremal curve, avoiding this problematic assumption of earlier proofs.

Exercises for Section 14.6

A. (a) Show that if f is an *odd* 2π-periodic C^1 function, then $\|f\|_2 \leq \|f'\|_2$.

(b) If $f \in C^1[a, b]$ with $f(a) = f(b) = 0$, show that $\int_a^b |f(x)|^2 \, dx \leq \left(\frac{b-a}{\pi}\right)^2 \int_a^b |f'(x)|^2 \, dx$.
    HINT: Build an odd function g on $[-\pi, \pi]$ by identifying $[0, \pi]$ with $[a, b]$.

B. (a) Let $x(t)$ and $y(t)$ be C^1 functions on $[0, 1]$ such that $x'(t) \geq 0$. Prove that the area under the curve $C = \{(x(t), y(t)) : 0 \leq t \leq 1\}$ is $\int_0^1 y(t) x'(t) \, dt$.

(b) Now suppose that C is a closed curve [i.e., $(x(0), y(0)) = (x(1), y(1))$] that doesn't intersect itself and that $x'(t)$ changes sign only a finite number of times. Prove that the area enclosed by C is $\left| \int_0^1 y(t) x'(t) \, dt \right|$.

C. Let f be a C^2 function that is 2π-periodic. Prove that $\|f'\|_2^2 \leq \|f\|_2 \|f''\|_2$.
    HINT: Use the Fourier series and Cauchy–Schwarz inequality.

D. (a) What is the maximum area which can be surrounded by a curve $\mathscr{C}$ of length 1 mile that begins and ends on a straight fence a mile long.   HINT: Reflect the curve in the fence.

(b) Suppose that two rays make an angle $\alpha \in (0, \pi)$. A curve of length 1 connects one ray to the other. Find the maximum area enclosed.   HINT: Apply a transformation in polar coordinates sending (r, θ) to $(r, \pi\theta/\alpha)$. What is the effect on area and arc length?

E. Use approximation by piecewise continuous functions to extend the solution of the isoperimetric problem to arbitrary continuous curves.

14.7 Best Approximation by Trigonometric Polynomials

We return to the theme of Chapter 10, uniform approximation. Here we are interested in approximating 2π-periodic functions by (finite) linear combinations of trigonometric functions. In this section, we obtain some reasonable estimates. Later we will establish the Jackson and Bernstein Theorems, which yield optimal estimates relating the rate of approximation with smoothness. It will turn out that there are close connections with approximation by polynomials which we also explore.

Recall that $\mathbb{TP}_n$ denotes the subspace of $C[-\pi, \pi]$ consisting of all trigonometric polynomials of degree at most n.

14.7.1. DEFINITION. The **error of approximation** to a 2π-periodic function $f : \mathbb{R} \to \mathbb{R}$ by trigonometric polynomials of degree n is

$$\widetilde{E}_n(f) = \inf\{\|f - q\|_\infty : q \in \mathbb{TP}_n\}.$$

For example, for any 2π-periodic function $f \in C[-\pi, \pi]$, both $S_n f$ and $\sigma_n f$ are in $\mathbb{TP}_n$. The subspace $\mathbb{TP}_n$ has dimension $2n + 1$ because it is spanned by the linearly independent functions $\{1, \cos kx, \sin kx : 1 \le k \le n\}$. Theorem 7.3.5 shows that there is a best approximation (in the uniform norm!) in $\mathbb{TP}_n$ to any function f. That is, given f in $C[-\pi, \pi]$, there is a trig polynomial p in $\mathbb{TP}_n$ such that

$$\|f - p\|_\infty = \inf\{\|f - q\|_\infty : q \in \mathbb{TP}_n\} = \widetilde{E}_n(f).$$

In a certain sense, the functions $S_n f$ and $\sigma_n f$ are natural approximants to f in $\mathbb{TP}_n$. The Projection Theorem (7.5.11) shows that $S_n f$ is the best L^2-norm approximant to f in $\mathbb{TP}_n$. However, it has some undesirable wildness when it comes to the uniform norm. The following theorem gives bounds on how close $S_n f$ is to the best approximation in $\mathbb{TP}_n$. It says that $S_n f$ can be a relatively bad approximation for large n. Nevertheless, the degree of approximation by $S_n f$ is sufficiently good to yield reasonable approximations if the Fourier series decays at a sufficient rate. The reason that S_n works in the next theorem is the fact that $S_n p = p$ for all p in $\mathbb{TP}_n$.

14.7.2. THEOREM. *If $f : \mathbb{R} \to \mathbb{R}$ is a continuous 2π-periodic function, then*

$$\|f - S_n f\|_\infty \le (3 + \log n)\widetilde{E}_n(f).$$

PROOF. Applying Theorems 14.2.3 and 14.2.4 (3), we have

$$\|S_n f\|_\infty \le \int_{-\pi}^{\pi} |f(x+t)D_n(t)|\,dt \le \|f\|_\infty \int_{-\pi}^{\pi} |D_n(t)|\,dt \le (2 + \log n)\|f\|_\infty.$$

Observe that if $p \in \mathbb{TP}_n$, then $S_n p = p$. In particular, if we let $p \in \mathbb{TP}_n$ be the best approximation to f, then

$$\|f - S_n f\|_\infty \le \|(f - p) - S_n(f - p)\|_\infty \le \|f - p\|_\infty + \|S_n(f - p)\|_\infty$$
$$\le \|f - p\|_\infty + (2 + \log n)\|f - p\|_\infty \le (3 + \log n)\|f - p\|_\infty.$$

Since $\|f - p\|_\infty = \widetilde{E}_n(f)$ for our choice of p, we obtain the desired estimate. ∎

Fejér's Theorem (14.4.5) suggests that $\sigma_n f$ is a reasonably good approximant in the uniform norm. On the other hand, consider $f(x) = \sin nx$. Then $S_n f = f$ is the best approximation with $\widetilde{E}_n(f) = 0$, while $\sigma_n f(x) = \frac{1}{n+1}\sin nx$ is a rather poor estimate. In spite of this, good general results can be obtained from the obvious estimate

$$\widetilde{E}_n(f) \le \|f - \sigma_n f\|_\infty.$$

We can now apply this estimate and Theorem 14.7.2 to show that the Dirichlet–Jordan Theorem (14.2.6) actually yields uniform convergence when the piecewise Lipschitz function is continuous.

14.7.3. THEOREM. *If f is a 2π-periodic Lipschitz function with Lipschitz constant L, then for $n \ge 2$,*

$$\|f - \sigma_n f\|_\infty \le \frac{(1 + 2\log n)L}{2n} \quad and \quad \|f - S_n f\|_\infty \le \frac{2\pi(1 + \log n)^2 L}{n}.$$

Hence $S_n f$ converges to f uniformly.

PROOF. We need a decent estimate for the Fejér kernel. For our purposes, the following is enough:

$$K_n(t) = \frac{1}{2\pi(n+1)}\left(\frac{\sin\frac{n+1}{2}t}{\sin\frac{1}{2}t}\right)^2 \le \min\left\{\frac{n+1}{2\pi}, \frac{\pi}{2(n+1)t^2}\right\}.$$

Indeed, $K_n(t) \le K_n(0) = \frac{n+1}{2\pi}$ yields the first upper bound. And the inequality $|\sin t/2| \ge (2/\pi)|t/2| = |t|/\pi$ on $[-\pi, \pi]$ is enough to show that

$$\frac{1}{2\pi(n+1)}\left(\frac{\sin\frac{n+1}{2}t}{\sin\frac{1}{2}t}\right)^2 \le \frac{1}{2\pi(n+1)}\left(\frac{1}{|t|/\pi}\right)^2 = \frac{\pi}{2(n+1)t^2}.$$

The first bound is better for small $|t|$, and the second becomes an improvement at the point $\delta = \pi/(n+1)$.

This proof follows the proof of Fejér's Theorem using the additional information contained in the Lipschitz condition to sharpen the error estimate. Looking back at that proof, we obtain an estimate by splitting $[-\pi, \pi]$ into two pieces. We use the Lipschitz estimate $|f(x + t) - f(x)| \le L|t|$, and use $\delta = \pi/(n+1)$:

$$|\sigma_n f(x) - f(x)| \leq \int_{-\pi}^{\pi} |f(x+t) - f(x)| K_n(t)\, dt$$

$$\leq \int_{-\delta}^{\delta} L|t| \frac{n+1}{2\pi}\, dt + \int_{-\pi}^{-\delta} + \int_{\delta}^{\pi} L|t| \frac{\pi}{2(n+1)t^2}\, dt$$

$$= \frac{(n+1)L}{\pi} \int_0^{\pi/n+1} t\, dt + \frac{\pi L}{(n+1)} \int_{\pi/n+1}^{\pi} t^{-1}\, dt$$

$$= \frac{(n+1)L}{\pi} \frac{\pi^2}{2(n+1)^2} + \frac{\pi L}{(n+1)} \log(n+1)$$

$$= \frac{L\pi}{2(n+1)} \big(1 + 2\log(n+1)\big).$$

Finally, a little calculus shows that $f(x) = (1 + 2\log x)/x$ is decreasing for $x > \sqrt{e}$. So for $n \geq 2$ we obtain

$$\|\sigma_n f - f\|_\infty \leq \frac{L\pi}{2n}(1 + 2\log n).$$

Now apply Theorem 14.7.2 to obtain

$$\|f - S_n f\|_\infty \leq (3 + \log n)\, \widetilde{E}_n(f) \leq (3 + \log n)\, \|f - \sigma_n f\|_\infty$$

$$\leq (3 + \log n) \frac{L\pi}{2n}(1 + 2\log n) \leq \frac{L\pi}{2n} 4(1 + \log n)^2.$$

Now $\displaystyle \lim_{n \to \infty} \frac{2\pi L(1 + \log n)^2}{n} = 0$. Therefore, $S_n f$ converges uniformly to f. ∎

Exercises for Section 14.7

A. Recall the de la Valée Poussin kernel $P_n f$ from Exercise 14.4.D.

(a) Show that $\|P_n(f)\|_\infty \leq 3\|f\|_\infty$.   HINT: Write $P_n f$ in terms of $\sigma_k f$ for various k.
(b) Show that $\|f - P_n(f)\|_\infty \leq 4\widetilde{E}_n(f)$.   HINT: Show $P_n p = p$ for $p \in \mathbb{TP}_n$ and imitiate the proof of Theorem 14.7.2.

B. Use the previous exercise to obtain a *lower bound* $\widetilde{E}_n(|\sin\theta|) > C/n$.
HINT: See Exercise 13.5.B. Show: $\big|P_{2n}(|\sin\theta|)(0)\big| \geq \frac{4}{\pi}\sum_{k=n}^{\infty} \frac{1}{4k^2-1} > \frac{1}{\pi n}$.

C. Suppose that a 2π-periodic function f is of class Lip α for $0 < \alpha < 1$ (see Exercise 5.5.K).

(a) Show that there is a constant C such that $\|f - \sigma_n f\|_\infty \leq Cn^{-\alpha}$.
(b) Hence show that $S_n f$ converges uniformly to f.
HINT: Follow the proof of Theorem 14.7.3 using the new estimate.

D. Let f and g be 2π-periodic functions with Fourier series $f \sim A_0 + \sum_{n \geq 1} A_n \cos n\theta + B_n \sin n\theta$ and $g \sim C_0 + \sum_{n \geq 1} C_n \cos n\theta + D_n \sin n\theta$. Prove that if f is absolutely integrable and g is Lipschitz, then $\frac{1}{2\pi}\int_{-\pi}^{\pi} f(\theta)g(\theta)\, d\theta = A_0 C_0 + \frac{1}{2}\sum_{n \geq 1} A_n C_n + B_n D_n$.   HINT: Use Theorem 14.7.3.

E. (a) Find the Fourier series of the function $g(x) = -x - \pi$ on $[-\pi, 0)$ and $= -x + \pi$ on $[0, \pi]$.
(b) For $n \geq 1$, define $g_n(x) = \sin(2nx) S_n g(x)$. Find the Fourier series for g_n.
HINT: $2\sin A \sin B = \cos(A - B) - \cos(A + B)$.

(c) Show that $\|g_n\|_\infty \le \pi + 2$.   HINT: Use Exercise 14.4.L.

(d) Show that $S_{2n}g_n(0) > \log n$.   HINT: $\sum_{k=1}^n \frac{1}{n}$ is an upper Riemann sum for $\int_1^{n+1} \frac{1}{x}\,dx$.

(e) Hence prove that $\|g_n - S_{2n}g_n\|_\infty \ge \frac{\log n}{\pi + 2}\widetilde{E}_{2n}(g_n)$. So $\log n$ is needed in Theorem 14.7.2.

(f) Show that $h(x) = \sum\limits_{n=1}^\infty \frac{1}{n^2}g_{2\cdot 3^{n^3}}(x)$ is a continuous function such that $S_{3^{n^3}}h(0)$ diverges.
    NOTE: This is difficult.

14.8 Connections with Polynomial Approximation

To connect approximation by trigonometric polynomials with approximation by polynomials, we exploit an important link between trig polynomials and the Chebyshev polynomials of Section 10.7.

The idea is to relate each function f in $C[-1,1]$ to an even function Φf in $C[-\pi, \pi]$ in such a way that the set $\mathbb{P}_n$ of polynomials of degree n is carried into $\mathbb{TP}_n$. This map is defined by

$$\Phi f(\theta) = f(\cos\theta) \quad \text{for} \quad -\pi \le \theta \le \pi.$$

Notice immediately that since $\cos\theta$ takes values in $[-1,1]$, the right-hand side is always defined. Also, Φf is an even function because

$$\Phi f(-\theta) = f(\cos(-\theta)) = f(\cos\theta) = \Phi f(\theta).$$

Crucially, the map Φ is linear. For f, g in $C[-1,1]$ and α, β in $\mathbb{R}$,

$$\begin{aligned}\Phi(\alpha f + \beta g)(\theta) &= (\alpha f + \beta g)(\cos\theta) = \alpha f(\cos\theta) + \beta g(\cos\theta) \\ &= \alpha\Phi f(\theta) + \beta\Phi g(\theta) = (\alpha\Phi f + \beta\Phi g)(\theta).\end{aligned}$$

So $\Phi(\alpha f + \beta g) = \alpha\Phi f + \beta\Phi g$, which is linearity.

Recall Definition 10.7.1 defining the Chebyshev polynomials on the interval $[-1,1]$ by $T_n(x) = \cos(n\arccos x)$. Since T_n is a polynomial of degree n, every polynomial can be expressed as a linear combination of the T_n's. In particular, $\mathbb{P}_n$ is spanned by $\{T_0, \dots, T_n\}$. Also recall that there is an inner product on $C[-1,1]$ given by

$$\langle f, g\rangle_T = \frac{1}{\pi}\int_{-1}^1 f(x)g(x)\frac{dx}{\sqrt{1-x^2}} \quad \text{for} \quad f, g \in C[-1,1].$$

The Chebyshev polynomials form an orthonormal set by Lemma 10.7.5.
    Notice that

$$\Phi T_n(\theta) = \cos(n\arccos(\cos\theta)) = \cos(n\theta).$$

It follows that Φ maps $\mathbb{P}_n$ onto the span of $\{1, \cos\theta, \dots, \cos n\theta\}$, which consists of the even trig polynomials in $\mathbb{TP}_n$.

This establishes the first parts of the following theorem. Let $E[-\pi, \pi]$ denote the closed subspace of $C[-\pi, \pi]$ consisting of all even continuous functions on $[-\pi, \pi]$.

14.8.1. THEOREM. *The map* $\Phi : C[-1,1] \to E[-\pi,\pi]$ *satisfies the following:*

(1) Φ *is linear, one-to-one, and onto.*
(2) $\Phi T_n(\theta) = \cos(n\theta)$ *for all* $n \geq 0$.
(3) $\Phi(\mathbb{P}_n) = E[-\pi,\pi] \cap \mathbb{TP}_n$.
(4) $\|\Phi f - \Phi g\|_\infty = \|f - g\|_\infty$ *for all* $f,g \in C[-1,1]$.
(5) $E_n(f) = \widetilde{E}_n(\Phi f)$ *for all* $f \in C[-1,1]$.
(6) $\langle f,g \rangle_T = \langle \Phi f, \Phi g \rangle$ *for all* $f,g \in C[-1,1]$.

PROOF. For (1), linearity has already been established. To see that Φ is one-to-one, suppose that $\Phi f = \Phi g$ for functions f,g in $C[-1,1]$. Then $f(\cos\theta) = g(\cos\theta)$ for all θ in $[-\pi,\pi]$. Since $\cos$ maps $[-\pi,\pi]$ onto $[-1,1]$, it follows that $f(x) = g(x)$ for all x in $[-1,1]$. Thus $f = g$ as required.

To show that Φ is surjective, we construct the inverse map from $E[-\pi,\pi]$ to $C[-\pi,\pi]$. For each even function g in $E[-\pi,\pi]$, define a function Ψg in $C[-1,1]$ by

$$\Psi g(x) = g(\arccos x).$$

(Notice that arccos takes all values in $[0,\pi]$. So Ψg depends only on $g(\theta)$ for θ in $[0,\pi]$. This is fine because g is even.) Compute

$$\Phi \Psi g(\theta) = \Psi g(\cos\theta) = g(\arccos(\cos\theta)) = g(|\theta|) = g(\theta).$$

So Φ maps Ψg back onto g, showing that Φ maps onto $E[-\pi,\pi]$.

We proved (2) and (3) in the discussion before the theorem. Consider (4). Note that

$$\|\Phi f\|_\infty = \sup_{\theta \in [-\pi,\pi]} |f(\cos\theta)| = \sup_{x \in [-1,1]} |f(x)| = \|f\|_\infty.$$

Hence by linearity,

$$\|\Phi f - \Phi g\|_\infty = \|\Phi(f-g)\|_\infty = \|f-g\|_\infty.$$

Applying this, we obtain

$$\begin{aligned}
E_n(f) &= \inf\{\|f - p\|_\infty : p \in \mathbb{P}_n\} \\
&= \inf\{\|\Phi f - \Phi p\|_\infty : p \in \mathbb{P}_n\} \\
&= \inf\{\|\Phi f - q\|_\infty : q \in \mathbb{TP}_n \cap E[-\pi,\pi]\}.
\end{aligned}$$

However, since Φf is even, the trig polynomial in $\mathbb{TP}_n$ closest to Φf is also even. To see this, suppose that $r \in \mathbb{TP}_n$ satisfies $\|\Phi f - r\| = \widetilde{E}_n(\Phi f)$, and let

$$q(\theta) = \frac{r(\theta) + r(-\theta)}{2}.$$

Note that if $r(\theta) = a_0 + \sum_{k=1}^{n} a_n \cos k\theta + b_n \sin\theta$, then $q(\theta) = a_0 + \sum_{k=1}^{n} a_n \cos k\theta$. Therefore q belongs to $\mathbb{TP}_n \cap E[-\pi,\pi]$ and

$$|\Phi f(\theta) - q(\theta)| = \left| \frac{\Phi f(\theta) + \Phi f(-\theta)}{2} - \frac{r(\theta) + r(-\theta)}{2} \right|$$

$$\leq \frac{1}{2}|\Phi f(\theta) - r(\theta)| + \frac{1}{2}|\Phi f(-\theta) - r(-\theta)|$$

$$\leq \frac{1}{2}\tilde{E}_n(\Phi f) + \frac{1}{2}\tilde{E}_n(\Phi f) = \tilde{E}_n(\Phi f).$$

Hence $\|\Phi f - q\|_\infty = \tilde{E}_n(\Phi f)$. Putting this information into the preceding inequality, we obtain that $E_n(f) = \tilde{E}_n(\Phi f)$.

Finally, to prove (6), we make the substitution $x = \cos\theta$ in the integral:

$$\langle f, g \rangle_T = \frac{1}{\pi} \int_{-1}^{1} f(x)g(x) \frac{dx}{\sqrt{1-x^2}}$$

$$= \frac{1}{\pi} \int_{\pi}^{0} f(\cos\theta)g(\cos\theta) \frac{-\sin\theta\, d\theta}{\sin\theta}$$

$$= \frac{1}{2\pi} \int_{-\pi}^{\pi} f(\cos\theta)g(\cos\theta)\, d\theta = \langle \Phi f, \Phi g \rangle. \qquad \blacksquare$$

In summary, this theorem shows that $C[-1,1]$ with the inner product $\langle \cdot, \cdot \rangle_T$ and $E[-\pi,\pi]$ with the inner product $\langle \cdot, \cdot \rangle$ are, as inner product spaces, the same.

Approximation questions for Fourier series are well studied, and this allows a transference to polynomial approximation. The reason we obtain estimates more readily in the Fourier series case is that the periodicity allows us to obtain nice integral formulas for our approximations.

Part (6) of this theorem shows that the Chebyshev series for $f \in C[-1,1]$ correspond to the Fourier series of Φf as $f \sim \sum_{k=0}^{\infty} a_k T_k$, where

$$a_0 = \langle f, 1 \rangle_T \quad \text{and} \quad a_n = 2\langle f, T_n \rangle_T \quad \text{for} \quad n \geq 1.$$

Let us define two series corresponding to the Dirichlet and Cesàro series

$$C_n f = \sum_{k=0}^{n} a_k T_k \quad \text{and} \quad \Sigma_n f = \sum_{k=0}^{n} \left(1 - \frac{k}{n+1}\right) a_k T_k.$$

Now it is just a matter of reinterpreting the Fourier series results for polynomials using Theorem 14.8.1.

14.8.2. THEOREM. *Let f be a continuous function on $[-1,1]$ with Chebyshev series $\sum_{k=0}^{\infty} a_k T_k$. Then $(\Sigma_n f)_{n=1}^{\infty}$ converges uniformly to f on $[-1,1]$. If f is Lipschitz, then $(C_n f)_{n=1}^{\infty}$ also converges uniformly to f. In any event,*

$$\|f - C_n f\|_\infty \leq (3 + \log n)E_n(f).$$

PROOF. The map Φ converts the problem of approximating f by polynomials of degree n to the problem of approximating Φf by trig polynomials of degree n. Part (6) of Theorem 14.8.1 shows that $\Phi C_n f = S_n \Phi f$ and $\Phi \Sigma_n f = \sigma_n \Phi f$. So the fact that $\Sigma_n f$ converge uniformly to f on $[-1, 1]$ is a restatement of Fejér's Theorem (14.4.5).

If f has a Lipschitz constant L, then

$$|\Phi f(\alpha) - \Phi f(\beta)| = |f(\cos\alpha) - f(\cos\beta)|$$
$$\leq L|\cos\alpha - \cos\beta| \leq L|\alpha - \beta|.$$

The last step follows from the Mean Value Theorem (6.2.2), since the derivative of $\cos\theta$ is $-\sin\theta$, which is bounded by 1. Thus Φf is Lipschitz with the same constant. (Warning: this step is not reversible.) By Theorem 14.7.3, the sequence $S_n \Phi f$ converges uniformly to Φf, whence $C_n f$ converges uniformly to f by Theorem 14.8.1 (4).

Theorem 14.7.2 and Theorem 14.8.1 (4) provide the estimate

$$\|f - C_n f\|_\infty = \|\Phi f - S_n f\|_\infty \leq (3 + \log n)\widetilde{E}_n(\Phi f) = (3 + \log n)E_n(f). \quad \blacksquare$$

14.8.3. EXAMPLE. Let us try to approximate $f(x) = |x|$ on $[-1, 1]$. We convert this to the function

$$g(\theta) = \Phi f(\theta) = |\cos\theta| \quad \text{for} \quad -\pi \leq \theta \leq \pi.$$

This is an even function and thus has a cosine series. Also, $g(\pi - \theta) = g(\theta)$. The functions $\cos 2n\theta$ have this symmetry, but

$$\cos\big((2n+1)(\pi - \theta)\big) = -\cos(2n+1)\theta.$$

So $A_{2n+1} = 0$ for $n \geq 0$. Compute

$$A_0 = \frac{1}{2\pi}\int_{-\pi}^{\pi} |\cos\theta|\, d\theta = \frac{2}{\pi}$$

and for $n \geq 1$,

$$A_{2n} = \frac{1}{\pi}\int_{-\pi}^{\pi} |\cos\theta|\cos 2n\theta\, d\theta = \frac{4}{\pi}\int_0^{\pi/2} \cos\theta\cos 2n\theta\, d\theta$$

$$= \frac{2}{\pi}\int_0^{\pi/2} \cos(2n-1)\theta + \cos(2n+1)\theta\, d\theta$$

$$= \frac{2}{\pi}\frac{\sin(2n-1)\theta}{2n-1} + \frac{2}{\pi}\frac{\sin(2n+1)\theta}{2n+1}\Big|_0^{\pi/2}$$

$$= \frac{2}{\pi}\left(\frac{(-1)^{n-1}}{2n-1} + \frac{(-1)^n}{2n+1}\right) = \frac{(-1)^{n-1}4}{\pi(4n^2-1)}.$$

These coefficients are absolutely summable, since they behave like the series $1/n^2$. Thus $S_n g$ converges uniformly to g by the Weierstrass M-test (see Exercise 13.5.E). Hence by Theorem 14.8.1, the sequence of polynomials

$$C_{2n}f(x) = \frac{2}{\pi} - \frac{4}{\pi} \sum_{k=1}^{n} \frac{(-1)^k}{4k^2 - 1} T_{2k}(x)$$

converges to $|x|$ uniformly on $[-1, 1]$.

We can make a crude estimate of the error by summing the remaining terms:

$$\big\| \, |x| - C_{2n}f(x) \big\| \leq \frac{4}{\pi} \sum_{k=n+1}^{\infty} \frac{1}{4k^2 - 1} \|T_n\|_\infty$$

$$= \frac{2}{\pi} \sum_{k=n+1}^{\infty} \frac{1}{2k - 1} - \frac{1}{2k + 1} = \frac{2}{\pi(2n + 1)} < \frac{1}{\pi n}.$$

This may not look very good, since $(\pi n)^{-1}$ tends to 0 so slowly. However, in 1913, Bernstein showed that $E_n(|x|) > (10n)^{-1}$ for all n. So the Chebyshev series actually gives an approximation of the correct order of magnitude.

Exercises for Section 14.8

A.   Let $f \in C[-1, 1]$ and $g(\theta) = f(\cos\theta)$. Show that $\omega(g; \delta) = \omega(g|_{[0,\pi]}; \delta) \leq \omega(f; \delta)$.

B.   Find a sequence of polynomials converging uniformly to $f(x) = \begin{cases} -x^2 & \text{for } -1 \leq x \leq 0, \\ x^2 & \text{for } \ 0 \leq x \leq 1. \end{cases}$

C.   Find $f(x) \in C[-1, 1]$ that is not Lipschitz, but Φf is Lipschitz in $C[-\pi, \pi]$.

D.   Show that there is a constant C such that $E_n(|x|) > C/n$.   HINT: See Exercise 14.7.B. Show that $\widetilde{F}_n(|\cos\theta|) > (4\pi n)^{-1}$ by a change of variables. Now use Theorem 14.8.1.

14.9 Jackson's Theorem and Bernstein's Theorem

The goal of this section is to obtain Jackson's Theorem, which provides a good estimate of the error of approximation in terms of the smoothness of the function as measured by the modulus of continuity. First, we will establish a dramatic converse (for trig polynomials) due to Bernstein in 1912 that the growth of the error function provides a good measure of the smoothness of the function. Several times in this section, we will use the complex exponential function to simplify calculations.

Recall that for $\alpha \in (0, 1]$, we defined the class $\operatorname{Lip} \alpha$ as the functions f in $C[a, b]$ for which there is a constant C with

$$|f(x) - f(y)| \leq C|x - y|^\alpha \quad \text{for all} \quad x, y \in [a, b].$$

In particular, $\operatorname{Lip} 1$ is the class of Lipschitz functions. Observe that $f \in \operatorname{Lip} \alpha$ if and only if $\omega(f; \delta) \leq C\delta^\alpha$, where $\omega(f; \delta)$ is the modulus of continuity.

We will use this class to illustrate just how tight these two theorems are. The following corollary will be deduced from our two main theorems.

14.9.1. COROLLARY. *Let f be a 2π-periodic function and let $0 < \alpha < 1$. Then f is in $\mathrm{Lip}\,\alpha$ if and only if $\widetilde{E}_n(f) \leq Cn^{-\alpha}$ for $n \geq 1$ and some constant C.*

Bernstein's Theorem will be proved first because it is more straightforward. We begin with an easy lemma. This natural proof uses complex numbers.

14.9.2. LEMMA. *Suppose that $f, g \in \mathbb{TP}_n$ and $f(\theta) = g(\theta)$ for $2n + 1$ distinct points in $(-\pi, \pi]$. Then $f = g$.*

PROOF. Let $\theta_1, \ldots, \theta_{2n+1}$ be the common points. Using complex exponentials, we may express $(f - g)(\theta)$ as

$$(f - g)(\theta) = \sum_{k=-n}^{n} a_k e^{ik\theta} = e^{-in\theta} \sum_{j=0}^{2n} a_{j-n} e^{ij\theta}.$$

Now let $p(z) = \sum_{j=0}^{2n} a_{j-n} z^j$. Observe that $(f - g)(\theta) = e^{-in\theta} p(e^{i\theta})$. Thus the polynomial p of degree $2n$ has the roots $z_i = e^{i\theta_k}$ for $1 \leq k \leq 2n + 1$. These points are distinct because $|\theta_j - \theta_k| < 2\pi$ if $j \neq k$. Therefore, $p = 0$ and so $f = g$. ∎

The key to Bernstein's Theorem is an elegant inequality. The trig polynomial $p(\theta) = \sin n\theta$ shows that the inequality is sharp (meaning that the constant cannot be improved).

14.9.3. BERNSTEIN'S INEQUALITY.
Let p be a trigonometric polynomial of degree n. Then $\|p'\|_\infty \leq n\|p\|_\infty$.

PROOF. Suppose to the contrary that $p \in \mathbb{TP}_n$ but $\|p'\|_\infty > n\|p\|_\infty$. Choose a scalar λ such that $\|\lambda p\|_\infty < 1$ yet $n < \|\lambda p'\|_\infty$, and then rename λp as p. Choose θ_0 such that $\|p'\|_\infty = p'(\theta_0)$. Choose the angle $\gamma \in [-\frac{\pi}{n}, \frac{\pi}{n}]$ such that $\sin n(\theta_0 - \gamma) = p(\theta_0)$ and the derivative $n\cos n(\theta_0 - \gamma) > 0$.

Define a trigonometric polynomial in $\mathbb{TP}_n$ by

$$r(\theta) = \sin n(\theta - \gamma) - p(\theta).$$

Set $\alpha_k = \gamma + \frac{\pi}{n}(k + \frac{1}{2})$ for $-n \leq k \leq n$. Observe that $r(\alpha_k) = (-1)^k - p(\alpha_k)$. Since $|p(\alpha_k)| < 1$, the sign of $r(\alpha_k)$ is $(-1)^k$. By the Intermediate Value Theorem, there are $2n$ points β_k with $\alpha_k < \beta_k < \alpha_{k+1}$ such that $r(\beta_k) = 0$ for $-n \leq k < n$. The interval (α_s, α_{s+1}) containing θ_0 is special. By choice of γ, $\sin n(\theta - \gamma)$ is increasing from -1 to 1 on this interval. So $r(\alpha_s) < 0 < r(\alpha_{s+1})$. In addition, $r(\theta_0) = 0$ and

$$r'(\theta_0) = n\cos n(\theta_0 - \gamma) - p'(\theta_0) < 0.$$

Therefore there are small positive numbers ε_1 and ε_2 such that $r(\theta_0 - \varepsilon_1) > 0$ and $r(\theta_0 + \varepsilon_2) > 0$. Look at Figure 14.6. Therefore, we may apply the Intermediate Value Theorem three times in this interval. Consequently, we can find two additional zeros,

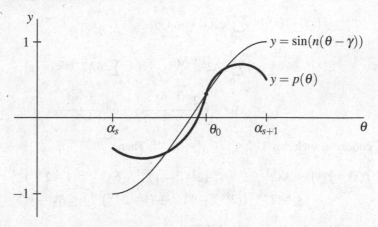

FIG. 14.6 The graphs of $\sin n(\theta - \gamma)$ and $p(\theta)$ on $[\alpha_s, \alpha_{s+1}]$.

so that r has at least $2n + 2$ zeros in (α_{-n}, α_n), which is an interval of length 2π. Therefore, by Lemma 14.9.2, we reach the absurd conclusion that r is identically 0. We conclude that our assumption was incorrect, and in fact $\|p'\|_\infty \leq n\|p\|_\infty$. ∎

We are now in a position to state and prove the desired result. The situation for $\alpha = 1$ is more complicated, and we refer the reader to [15]. The meaning of an error of the form $An^{-\alpha}$ for $\alpha > 1$ will be developed in the exercises.

14.9.4. BERNSTEIN'S THEOREM.
Let f be a 2π-periodic function such that $\widetilde{E}_n(f) \leq An^{-\alpha}$ for $n \geq 1$, where A is a constant and $0 < \alpha < 1$. Then f is in Lip α.

PROOF. Choose $p_n \in \mathbb{TP}_n$ such that $\|f - p_n\| \leq An^{-\alpha}$ for $n \geq 1$. Define $q_0 = p_1$ and $q_n = p_{2^n} - p_{2^{n-1}}$ for $n \geq 1$. Note that

$$\sum_{n \geq 0} q_n(x) = \lim_{n \to \infty} p_{2^n}(x) = f(x)$$

uniformly on $\mathbb{R}$. Compute for $n \geq 1$,

$$\|q_n\| \leq \|p_{2^n} - f\| + \|f - p_{2^{n-1}}\| \leq A2^{-n\alpha} + A2^{-(n-1)\alpha} \leq 3A2^{-n\alpha}.$$

By the Mean Value Theorem and Bernstein's inequality, we can estimate

$$|q_n(x) - q_n(y)| \leq \|q_n'\| |x - y| \leq 2^n \|q_n\| |x - y| \leq 3A2^{n(1-\alpha)} |x - y|.$$

On the other hand, a simple estimate is just

$$|q_n(x) - q_n(y)| \leq |q_n(x)| + |q_n(y)| \leq 2\|q_n\| \leq 6A2^{-n\alpha}.$$

Splitting the sum into two parts, we obtain

$$|f(x) - f(y)| \leq \sum_{n \geq 0} |q_n(x) - q_n(y)|$$

$$\leq \sum_{n=0}^{m-1} 3A2^{n(1-\alpha)}|x-y| + \sum_{n \geq m} 6A2^{-n\alpha}$$

$$\leq 3A|x-y|\frac{2^{m(1-\alpha)}-1}{2^{1-\alpha}-1} + 6A\frac{2^{-m\alpha}}{1-2^{-\alpha}}.$$

Finally, choose m such that $2^{-m} \leq |x-y| < 2^{1-m}$. Then

$$|f(x) - f(y)| \leq 3A2^{1-m}2^{m(1-\alpha)}(2^{1-\alpha}-1)^{-1} + 6A2^{-m\alpha}(1-2^{-\alpha})^{-1}$$

$$\leq 6A2^{-m\alpha}\big((2^{1-\alpha}-1)^{-1} + (1-2^{-\alpha})^{-1}\big) \leq B|x-y|^{\alpha},$$

where $B = 6A\big((2^{1-\alpha}-1)^{-1} + (1-2^{-\alpha})^{-1}\big)$.   ∎

On the other hand, Jackson's Theorem shows that smooth functions have better approximations.

14.9.5. JACKSON'S THEOREM.
Let f belong to $C[-1,1]$. Then

$$E_n(f) \leq 6\omega(f; \tfrac{1}{n}).$$

Similarly, if g is a continuous 2π-periodic function, $\widetilde{E}_n(g) \leq 6\omega(g; \tfrac{1}{n})$.

Notice that we obtain several interesting consequences immediately. This theorem shows that Proposition 10.4.4 was the correct order of magnitude, and that this result is best possible except possibly for improving the constants.

ANOTHER PROOF OF THE WEIERSTRASS APPROXIMATION THEOREM.
Not only does Jackson's Theorem prove that every continuous function is the limit of polynomials, it tells you how fast this happens. It suffices to prove the theorem for the interval $[-1,1]$ (see Exercise 10.2.A). In Section 10.4, we used the *uniform* continuity of f to show that $\lim_{n \to \infty} \omega(f; \tfrac{1}{n}) = 0$. Hence

$$0 \leq \lim_{n \to \infty} E_n(f) \leq \lim_{n \to \infty} 6\omega(f; \tfrac{1}{n}) = 0.$$

This completes the proof.   ∎

Let us apply Jackson's Theorem to the classes of functions that we have been discussing.

14.9.6. COROLLARY.
Let $\mathscr{S}$ be the class of functions in $C[0,1]$ with Lipschitz constant 1. Then $E_n(\mathscr{S}) \leq 3/n$. More generally, if $\mathscr{S}[a,b]$ is the class of functions in $C[a,b]$ with Lipschitz constant 1, then $E_n(\mathscr{S}[a,b]) \leq 3(b-a)/n$.

PROOF. First consider the interval $[-1,1]$. For any f in $\mathscr{S}[-1,1]$, we have the inequality $\omega(f;\frac{1}{n}) \leq 1/n$. Thus by Jackson's Theorem, $E_n(f) \leq 6/n$.

Now the map

$$\gamma(x) = \frac{a+b+(b-a)x}{2} \quad \text{for} \quad -1 \leq x \leq 1$$

maps $[-1,1]$ onto $[a,b]$. So for any f in $C[a,b]$, the function $\Gamma f = f(\gamma(x))$ belongs to $C[-1,1]$. Moreover, if $f \in \mathscr{S}[a,b]$, then Γf has Lipschitz constant $(b-a)/2$ (see the exercises). By the first paragraph, choose a polynomial p of degree n such that

$$\|\Gamma f - p\|_\infty \leq 6\frac{b-a}{2} = 3(b-a).$$

Then

$$q(x) = \Gamma^{-1}(q) = q(\gamma^{-1}(x)) = p\left(\frac{2x-a-b}{b-a}\right)$$

is the polynomial of degree n such that $p = \Gamma q$. Thus

$$\|f-q\|_{[a,b]} = \sup_{x \in [a,b]} |f(x) - q(x)| = \sup_{t \subset [-1,1]} |f(\gamma(t)) - q(\gamma(t))|$$

$$= \|\Gamma f - p\|_\infty \leq 3(b-a). \quad \blacksquare$$

We complete the proof of Corollary 14.9.1 with the following:

14.9.7. COROLLARY. *Suppose that f is a 2π-periodic function of class* Lip α *for any $0 < \alpha < 1$. Then $\widetilde{E}_n(f) \leq Cn^{-\alpha}$ for $n \geq 1$ and some constant C.*

PROOF. It is immediate from $|f(x) - f(y)| \leq C|x-y|^\alpha$ that $\omega(f;\frac{1}{n}) \leq Cn^{-\alpha}$. Thus Jackson's Theorem yields $\widetilde{E}_n(f) \leq 6Cn^{-\alpha}$. $\quad \blacksquare$

The proof of Jackson's Theorem is difficult, and it requires a better method of approximation that is suited to the specific function. The key idea is convolution: to integrate the function f against an appropriate sequence of polynomials to obtain the desired approximations. These ideas work somewhat better for periodic functions, so we will use the results of the last section and consider approximation by trigonometric polynomials instead. In effect, we are building a 'special purpose' kernel with the specific properties we need.

Let $\psi(\theta) = 1 + c_1 \cos\theta + \cdots + c_n \cos n\theta$. It will be important to choose the constants c_i so that ψ is positive. We will then try to make a good choice for these constants. For each 2π-periodic function f, define a function

$$\Psi f(\theta) = \frac{1}{2\pi} \int_{-\pi}^{\pi} f(\theta - t)\psi(t)\,dt. \tag{14.9.8}$$

We capture the main properties in the following lemma.

14.9.9. LEMMA. *Suppose that ψ is a trig polynomial of degree n that is positive on $[-\pi, \pi]$. Define Ψ as in equation (14.9.8). Then*

(1) $\Psi 1 = 1$.
(2) Ψ *is linear:* $\Psi(\alpha f + \beta g) = \alpha \Psi f + \beta \Psi g$ *for all* $f, g \in C[-1, 1]$ *and* $\alpha, \beta \in \mathbb{R}$.
(3) Ψ *is monotone:* $f \geq g$ *implies that* $\Psi f \geq \Psi g$.
(4) $\Psi f \in \mathbb{TP}_n$ *for all* $f \in C[-1, 1]$.

PROOF. Part (1) uses the fact that $\cos n\theta$ has mean 0 on $[-\pi, \pi]$:

$$\Psi 1(\theta) = \frac{1}{2\pi} \int_{-\pi}^{\pi} \psi(t) \, dt = \frac{1}{2\pi} \int_{-\pi}^{\pi} 1 \, dt + \sum_{k=1}^{n} c_k \int_{-\pi}^{\pi} \cos nt \, dt = 1.$$

For (2), linearity follows easily from the linearity of the integral.

For (3), if $h \geq 0$, then since $\psi(t) \geq 0$, it follows that $\Psi h(\theta) \geq 0$ as well. Thus if $f \geq g$, then

$$\Psi f - \Psi g = \Psi(f - g) \geq 0.$$

For (4), we make a change of variables by substituting $u = \theta - t$ and using the 2π-periodicity to obtain

$$\begin{aligned}
\Psi f(\theta) &= \frac{1}{2\pi} \int_{-\pi}^{\pi} f(\theta - t) \psi(t) \, dt \\
&= \frac{1}{2\pi} \int_{\theta - \pi}^{\theta + \pi} f(u) \psi(\theta - u) \, du = \frac{1}{2\pi} \int_{-\pi}^{\pi} f(u) \psi(\theta - u) \, du \\
&= \frac{1}{2\pi} \int_{-\pi}^{\pi} f(u) \, du + \sum_{k=1}^{n} c_k \int_{-\pi}^{\pi} f(u) \cos k(\theta - u) \, du \\
&= \frac{1}{2\pi} \int_{-\pi}^{\pi} f(u) \, du + \sum_{k=1}^{n} c_k \int_{-\pi}^{\pi} f(u) \big(\cos k\theta \cos ku + \sin k\theta \sin ku \big) \, du \\
&= A_0 + \sum_{k=1}^{n} 2 c_k A_k \cos k\theta + 2 c_k B_k \sin k\theta,
\end{aligned}$$

where A_k and B_k are the Fourier coefficients of f. This shows that Ψf is a trig polynomial of degree at most n. ∎

The positivity of the kernel means that the error estimates can be obtained from the proof of Fejér's Theorem (14.4.5). However, a clever choice of the weights c_i will result in a better estimate.

14.9.10. LEMMA. *If f is a continuous function on $[a, b]$, then for any $t > 0$,*

$$\omega(f; t) \leq \left(1 + \frac{t}{\delta} \right) \omega(f; \delta).$$

PROOF. Suppose that $(n - 1)\delta < t \leq n\delta$. Then if $|x - y| \leq t$, we may find points $y = x_0 < x_1 < \cdots < x_n = x$ such that $|x_k - x_{k-1}| \leq \delta$ for $1 \leq k \leq n$. Hence

$$|f(x) - f(y)| \leq \sum_{k=1}^{n} |f(x_k) - f(x_{k-1})| \leq n\omega(f; \delta) \leq \left(1 + \frac{t}{\delta}\right) \omega(f; \delta).$$

Taking the supremum over all pairs x, y with $|x - y| \leq t$ yields

$$\omega(f; t) \leq \left(1 + \frac{t}{\delta}\right) \omega(f; \delta).$$

■

14.9.11. LEMMA. *Suppose that $f \in C[-\pi, \pi]$ is a 2π-periodic function. Let $\psi = \sum_{k=0}^{n} c_k \cos k\theta$ be a positive trig polynomial of degree n.. Then*

$$\|\Psi f - f\|_\infty \leq \omega(f; \tfrac{1}{n}) \left(1 + \frac{\pi n}{2} \sqrt{2 - c_1}\right).$$

PROOF. Apply the previous lemma with $\delta = 1/n$ to obtain

$$|f(\theta - t) - f(\theta)| \leq \omega(f; |t|) \leq (1 + n|t|) \omega(f; \tfrac{1}{n}).$$

Now using the integral formula for Ψf,

$$\begin{aligned}
|\Psi f(\theta) - f(\theta)| &= \left| \frac{1}{2\pi} \int_{-\pi}^{\pi} f(\theta - t) \psi(t) \, dt - f(\theta) \frac{1}{2\pi} \int_{-\pi}^{\pi} \psi(t) \, dt \right| \\
&< \frac{1}{2\pi} \int_{-\pi}^{\pi} |f(\theta - t) - f(\theta)| \psi(t) \, dt \\
&\leq \frac{1}{2\pi} \int_{-\pi}^{\pi} (1 + n|t|) \, \omega(f; \tfrac{1}{n}) \psi(t) \, dt \\
&= \omega(f; \tfrac{1}{n}) \left(1 + \frac{n}{2\pi} \int_{-\pi}^{\pi} |t| \psi(t) \, dt\right).
\end{aligned}$$

To estimate this last term, use the Cauchy–Schwarz inequality for integrals (7.4.5):

$$\begin{aligned}
\frac{1}{2\pi} \int_{-\pi}^{\pi} |t| \psi(t) \, dt &= \frac{1}{2\pi} \int_{-\pi}^{\pi} \left(|t| \psi(t)^{1/2}\right) \psi(t)^{1/2} \, dt \\
&\leq \left(\frac{1}{2\pi} \int_{-\pi}^{\pi} t^2 \psi(t) \, dt\right)^{1/2} \left(\frac{1}{2\pi} \int_{-\pi}^{\pi} \psi(t) \, dt\right)^{1/2}.
\end{aligned}$$

The second integral is just 1.

Recall the easy estimate $\sin \theta \geq 2\theta/\pi$ for $0 \leq \theta \leq \pi/2$. This yields

$$1 - \cos t = 2 \sin^2 \frac{t}{2} \geq 2 \frac{4}{\pi^2} \left(\frac{t}{2}\right)^2 = \frac{2}{\pi^2} t^2 \quad \text{for} \quad -\pi \leq t \leq \pi.$$

Substitute this back into our integral:

$$\frac{1}{2\pi} \int_{-\pi}^{\pi} t^2 \psi(t)\, dt \leq \frac{1}{2\pi} \int_{-\pi}^{\pi} \frac{\pi^2}{2} (1 - \cos t) \psi(t)\, dt$$

$$= \frac{\pi^2}{2} \left(\frac{1}{2\pi} \int_{-\pi}^{\pi} \psi(t)\, dt - \frac{1}{2\pi} \int_{-\pi}^{\pi} \psi(t) \cos t\, dt \right) = \frac{\pi^2}{2} \left(1 - \frac{c_1}{2} \right).$$

Therefore, we obtain

$$\left| \Psi f(\theta) - f(\theta) \right| \leq \left(1 + \frac{n\pi}{2} \sqrt{2 - c_1} \right) \omega(f; \tfrac{1}{n}).$$ ∎

PROOF OF JACKSON'S THEOREM. This lemma suggests that we try to minimize $2 - c_1$ over all positive kernel functions of degree n. The most straightforward method is to write down a kernel that gets excellent estimates. Consider the complex trig polynomial

$$p(\theta) = \sum_{k=0}^{n} a_k e^{ik\theta} = \sum_{k=0}^{n} \sin \frac{(k+1)\pi}{n+2} e^{ik\theta}.$$

Our kernel will be $\psi(\theta) = c|p(\theta)|^2$. The constant c is chosen to make the constant coefficient equal to 1. Positivity of ψ is automatic, since it is the square of the modulus of p. It remains merely to do a calculation to determine the coefficients.

First, we compute

$$|p(\theta)|^2 = \sum_{j=0}^{n} a_j e^{ij\theta} \sum_{k=0}^{n} a_k e^{-ik\theta} = \sum_{j=0}^{n} \sum_{k=0}^{n} a_j a_k e^{i(j-k)\theta}$$

$$= \sum_{j=0}^{n} a_j^2 + \sum_{s=1}^{n} \sum_{k=0}^{n-s} a_k a_{k+s} \left(e^{is\theta} + e^{-is\theta} \right) = b_0 + \sum_{s=1}^{n} 2b_s \cos s\theta,$$

where $b_s = \sum_{k=0}^{n-s} a_k a_{k+s}$ for $0 \leq s \leq n$. We choose $c = 1/b_0 = \left(\sum_{j=0}^{n} a_j^2 \right)^{-1}$. It is clear that ψ is a positive cosine polynomial of degree n with constant coefficient equal to 1. So ψ satisfies the hypotheses of Lemmas 14.9.9 and 14.9.11.

So now we compute the coefficient b_1. We use some clever manipulations and the identity $\sin A + \sin B = 2 \sin \frac{A+B}{2} \cos \frac{A-B}{2}$. Notice that the periodicity of the coefficients is used in the second line.

$$2b_1 = 2 \sum_{k=0}^{n-1} a_k a_{k+1} = \sum_{k=0}^{n-1} 2 \sin \frac{(k+1)\pi}{n+2} \sin \frac{(k+2)\pi}{n+2} = \sum_{k=1}^{n} 2 \sin \frac{k\pi}{n+2} \sin \frac{(k+1)\pi}{n+2}$$

$$= \sum_{k=0}^{n-1} 2 \sin \frac{k\pi}{n+2} \sin \frac{(k+1)\pi}{n+2} = \sum_{k=0}^{n-1} \sin \frac{(k+1)\pi}{n+2} \left(\sin \frac{(k+2)\pi}{n+2} + \sin \frac{k\pi}{n+2} \right)$$

$$= 2 \sum_{k=0}^{n-1} \sin^2 \frac{(k+1)\pi}{n+2} \cos \frac{\pi}{n+2} = 2b_0 \cos \frac{\pi}{n+2}.$$

Hence

$$c_1 = \frac{2b_1}{b_0} = 2 \cos \frac{\pi}{n+2}.$$

So

$$1 + \frac{n\pi}{2}\sqrt{2-c_1} = 1 + \frac{n\pi}{2}\sqrt{2 - 2\cos\frac{\pi}{n+2}} = 1 + \frac{n\pi}{2}2\sin\frac{\pi}{2(n+2)}$$

$$\leq 1 + n\pi\frac{\pi}{2(n+2)} < 1 + \frac{\pi^2}{2} < 6.$$

The proof is now completed by appealing to Lemma 14.9.11. ∎

Exercises for Section 14.9

A. Suppose that f in $C[a,b]$ has Lipschitz constant L. Let $\gamma(x) = Ax + B$. Show that $f(\gamma(x))$ has Lipschitz constant AL.

B. Suppose f is a 2π-periodic function such that $\widetilde{E}_n(f) \leq Cn^{-p-\alpha}$, where $p \in \mathbb{N}$ and $0 < \alpha < 1$. Prove that f is C^p and that $f^{(p)}$ is in the class $\mathrm{Lip}\,\alpha$ as follows.

   (a) Write f as a sum of the polynomials q_n as in the proof of Bernstein's Theorem. Apply Bernstein's inequality p times to the q_n's. Show that the resulting series of derivatives still converges, and deduce that f is C^p.
   (b) Use this series to show that $\widetilde{E}_n(f^{(p)}) \leq C'n^{-\alpha}$. Then finish the argument.

C. (a) Suppose that f is C^1 on $[-1,1]$ and $\|f'\|_\infty = M$. Show that $E_n(f) \leq 6M/n$.
   (b) Choose a polynomial p of degree $n-1$ such that $\|f' - p\|_\infty = E_{n-1}(f')$. Let $q(x) = \int_0^x p(t)\,dt$. Show that $E_n(f) = E_n(f-q) \leq \frac{6}{n}E_{n-1}(f')$.

D. (a) Use induction on the previous exercise to show that if f has k continuous derivatives on $[-1,1]$, then $E_n(f) \leq \frac{6^k}{n(n-1)\cdots(n+1-k)}E_{n-k}(f^{(k)})$ for $n > k$.
   (b) Hence show that there is a constant C_k such that $E_n(f) \leq \frac{C_k}{n^k}\,\omega(f^{(k)}; \frac{1}{n-k})$ for $n > k$. Find this constant explicitly for $k = 2$.

E. Do a change of variables in the previous exercise to show that if f has k continuous derivatives on $[a,b]$, then $E_n(f) \leq \frac{C_k(b-a)^k}{n^k}\,\omega\left(f^{(k)}; \frac{b-a}{2(n-k)}\right)$ for $n > k$.

F. Show that if f is a 2π-periodic function and $\widetilde{E}_n(f) \leq C/n$, then $\omega(f;\delta) \leq B\delta|\log\delta|$.
   HINT: Study the proof of Bernstein's Theorem using $\alpha = 1$.

G. Prove the **Dini–Lipschitz Theorem**: If f is a continuous function on $[-1,1]$ such that $\lim_{n\to\infty}\omega(f; \frac{1}{n})\log n = 0$, then the Chebyshev series $C_n f$ converges uniformly to f.
   HINT: Combine Jackson's Theorem with Theorem 14.8.2.

H. Let $0 < \alpha < 1$.

   (a) Show that $f(x) = |x|^\alpha$ belongs to $\mathrm{Lip}\,\alpha$.
   (b) Modify the proof of Proposition 10.4.4 to obtain a lower bound for $E_n(\mathrm{Lip}\,\alpha)$.
   HINT: Piece together translates of $|x|^\alpha$.

Chapter 15
Wavelets

15.1 Introduction

In this chapter we develop an important variation on Fourier series, replacing the sine and cosine functions with new families of functions, called wavelets. The strategy is to construct wavelets so that they have some of the good properties of trig functions but avoid the failings of Fourier series that we have seen in previous chapters. With such functions, we can develop new versions of Fourier series methods that will work well for problems where traditional Fourier series work poorly.

What are the good properties of trig functions? First and foremost, we have an orthogonal basis in L^2, namely the set of functions $\sin(nx)$ and $\cos(nx)$ as n runs over $\mathbb{N}_0$. This leads to the idea of breaking up a wave into its harmonic constituents, as the sine and cosine functions appear in the solution of the wave equation. We want to retain some version of this orthogonality.

Fix a positive integer n and consider the span of $\{\sin(nx), \cos(nx)\}$, call it A_n. If $f(x)$ is in the subspace A_n then so is the translated function $f(x-a)$ for any $a \in \mathbb{R}$; and for a positive integer k, the dilated function $f(kt)$ is in A_{kn} (see Exercise 15.1.A). That is, translation leaves each subspace A_n invariant and dilation by k carries A_n to A_{kn} for each n. Moreover, these orthogonal subspaces together span all of $L^2[-\pi, \pi]$. There is a similar decomposition for wavelets, called a multiresolution, and it is central to the study of wavelets.

What are the problems with Fourier series that we would like to fix? Fourier coefficients, and hence the Fourier series approximation, depend on all values of the function. For example, if you change a function f a small amount on the interval $[0, 0.01]$, it is possible that every Fourier coefficient changes. This will then have an effect on the partial sums $S_n f(\theta)$ for all values of θ. Although these changes may be small, there are many subtleties in analyzing Fourier series approximations, as we have seen.

Further, for a badly behaved function, such as a nondifferentiable or discontinuous one, the coefficients decrease slowly. Exercises 13.5.G and 13.3.C show that the Fourier coefficients of a function go rapidly to 0 only when the functions has several

K.R. Davidson and A.P. Donsig, *Real Analysis and Applications: Theory in Practice*,
Undergraduate Texts in Mathematics, DOI 10.1007/978-0-387-98098-0_15,
© Springer Science + Business Media, LLC 2010

continuous derivatives. Thus, we may need many terms to get a close approximation, even at a point relatively far away from the discontinuity, as in Example 14.2.7.

The partial sums $S_n f(\theta)$ do not always converge to $f(\theta)$ when f is merely continuous. Thanks to Gibbs's phenomenon, $S_n f(\theta)$ will always exhibit bad behaviour near discontinuities, no matter how large n is. While we can get better approximations by using $\sigma_n f(\theta)$ instead of $S_n f(\theta)$, this will not resolve such problems as slowly decreasing Fourier coefficients.

This suggests looking for a series expansion with better local properties, meaning that coefficients reflect the local behaviour of the function and a small change on one interval affects only a few of the series coefficients and leaves unchanged the partial sums elsewhere in the domain. It may seem unlikely that there are useful wavelet bases with this local approximation property that still have nice behaviour under translation and dilation. However, they do exist, and they were developed in the 1980s. The discovery has provoked a vast literature of both theoretical and practical importance. No one family of wavelets is ideal for all problems, but we can develop different wavelets to solve specific problems. Developing such wavelets is an important practical problem.

In this chapter, we will illustrate some of the general features of wavelets. The basic example is the Haar wavelet, a rather simple case that is not the best for applications but illuminates the general theory. We construct one of the most used wavelets, the Daubechies wavelet, although we don't prove that it is continuous. This requires tools we don't have, most notably, the Fourier transform. We establish the existence of another continuous wavelet, the Franklin wavelet, but this requires considerable work. Our focus is the use of real analysis in the foundational theory. We leave the development of efficient computational strategies to more specialized treatments, such as those in the bibliography.

Most of the literature deals with bases for functions on the whole real line rather than for periodic functions, so we will work in this context. This means that we will be looking for special orthonormal bases for $L^2(\mathbb{R})$, the Hilbert space of all square integrable functions on $\mathbb{R}$ with the norm

$$\|f\|_2^2 = \int_{-\infty}^{+\infty} |f(x)|^2 \, dx.$$

We let $L^2(\mathbb{R})$ denote the completion of $C_c(\mathbb{R})$, the continuous functions of compact support on $\mathbb{R}$, in the L^2 norm.

15.1.1. DEFINITION. A **wavelet** is a function $\psi \in L^2(\mathbb{R})$ such that the set

$$\{\psi_{kj}(x) = 2^{k/2}\psi(2^k x - j) : j, k \in \mathbb{Z}\}$$

forms an orthonormal basis for $L^2(\mathbb{R})$. Sometimes ψ is called the **mother wavelet**.

This is more precisely called a dyadic wavelet to stress that dilations are taken to be powers of 2. This is a common choice but is not the most general one. Notice that the wavelet basis has two parameters, whereas the Fourier basis for $L^2(\mathbb{T})$ has only one, given by dilation alone. From the complex point of view, sines and cosines are written in terms of the exponential function $\psi(\theta) = e^{i\theta}$, and the functions $\psi(k\theta) = e^{ik\theta}$ for $k \in \mathbb{Z}$ form an orthonormal basis for $L^2(-\pi, \pi)$. A singly generated family of this form cannot have the local behaviour we are seeking.

Exercises for Section 15.1

A. (a) Given a function f with Fourier series $f(\theta) \sim A_0 + \sum_{n=1}^{\infty} A_n \cos n\theta + B_n \sin n\theta$, consider the function $g(\theta) = f(k\theta)$. If $g(\theta) \sim A_0 + \sum_{i=1}^{\infty} C_i \cos i\theta + D_i \sin i\theta$, find the formula for C_i and D_i in terms of the A_n, the B_n, and k.

   (b) Similarly, if $h(\theta) = f(\theta - x) \sim A_0 + \sum_{i=1}^{\infty} E_i \cos i\theta + F_i \sin i\theta$, find the formula for E_i and F_i in terms of the A_n, the B_n, and x.

B. Show that if ψ is a function in $L^2(\mathbb{R})$ such that $\{\psi_{0j} : j \in \mathbb{Z}\}$ is an orthonormal set, then $\{\psi_{kj} : j \in \mathbb{Z}\}$ is an orthonormal set for each $k \in \mathbb{Z}$. .

C. A map U from a Hilbert space $\mathscr{H}$ to itself is **unitary** if $\|Ux\| = \|x\|$ for all vectors $x \in \mathscr{H}$ and $U\mathscr{H} = \mathscr{H}$. Define linear maps on $L^2(\mathbb{R})$ by $Tf(x) = f(x-1)$ and $Df(x) = \sqrt{2}f(2x)$. Show that these maps are unitary.

D. Let ψ be a wavelet, and let T and D be the unitary maps defined in Exercise C. What is the relationship between the subspaces spanned by $\{T^n D\psi : n \in \mathbb{Z}\}$ and $\{DT^n \psi : n \in \mathbb{Z}\}$?

E. Let ψ be a function in $L^2(\mathbb{R})$ such that $\{\psi_{0j} : j \in \mathbb{Z}\}$ is an orthonormal set. Let χ be the characteristic function of the set $\{x \in \mathbb{R} : x - [x] < \frac{1}{2}\}$, where $[x]$ is the greatest integer $n \leq x$. Define $\varphi(x) = \chi(x)\psi(-x - \frac{1}{2}) - (1 - \chi(x))\psi(\frac{1}{2} - x)$.

   (a) Show that φ is orthogonal to ψ_{0j} for all $j \in \mathbb{Z}$.
   (b) Hence deduce that there is no function ψ in $L^2(\mathbb{R})$ such that the set of integer translates $\{\psi_{0j} : j \in \mathbb{Z}\}$ is an orthonormal basis for $L^2(\mathbb{R})$.
   (c) Show that $\{\psi_{0j}, \varphi_{0j} : j \in \mathbb{Z}\}$ is an orthonormal set.

F. For $t \in \mathbb{R}$, define $T_t f(x) = f(x - t)$ for $f \in L^2(\mathbb{R})$. Show that if $\lim_{n \to \infty} t_n = t$, then $\lim_{n \to \infty} T_{t_n} f = T_t f$ for every $f \in L^2(\mathbb{R})$. HINT: If f is continuous with compact support, use the fact that it is uniformly continuous. Next, approximate an arbitrary f.

15.2 The Haar Wavelet

To get started, we describe the **Haar system** for $L^2(0, 1)$. This will then lead to a wavelet basis for $L^2(\mathbb{R})$. For $a < b$, let $\chi_{[a,b)}$ denote the characteristic function of $[a, b)$. Set $\varphi = \chi_{[0,1)}$ and $\psi = \chi_{[0,0.5)} - \chi_{[0.5,1)}$. Then define

$$\psi_{kj}(x) = 2^{k/2}\psi(2^k x - j) \quad \text{for all} \quad k, j \in \mathbb{Z}.$$

We use only those functions that are supported on $[0, 1)$, namely $0 \leq j < 2^k$ for each $k \geq 0$. The others will be used later. The Haar system is the family

$$\{\varphi, \psi_{kj} : k \geq 0 \text{ and } 0 \leq j < 2^k\}.$$

See Figure 15.1 for examples of elements of the Haar system.

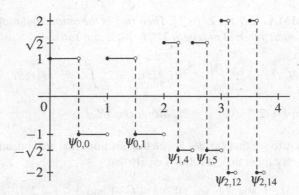

FIG. 15.1 Some elements of the Haar system.

15.2.1. LEMMA. *The Haar system is orthonormal.*

PROOF. It is straightforward to check that each of these functions has norm 1. Now ψ_{kj} and $\psi_{kj'}$ for $j \neq j'$ have disjoint supports and thus are orthogonal. More generally, if $k < k'$, then φ and ψ_{kj} are constant on the support of $\psi_{k'j'}$. Since $\int_0^1 \psi_{kj}(x)\,dx = 0$ for all j and k, it now follows that these functions are pairwise orthogonal. ∎

We may consider the inner product expansion with respect to this orthonormal set. It is natural to sum all terms of the same order at the same time to obtain a series approximant. Therefore, we define

$$H_n f(x) = \langle f, \varphi \rangle \varphi(x) + \sum_{k=0}^{n-1} \sum_{j=0}^{2^k-1} \langle f, \psi_{kj} \rangle \psi_{kj}(x).$$

The **Haar coefficients** are the inner products $\langle f, \psi_{kj} \rangle$ used in this expansion.

While we have some work yet to see that this orthogonal system spans the whole space, we can see that it has some nice properties. The local character is seen by the fact that these functions have smaller and smaller supports. If f and g agree except on the interval $[3/8, 1/2)$, then the Haar coefficients are the same for about '7/8 of the terms' in the sense that $\langle f, \psi_{kj} \rangle = \langle g, \psi_{kj} \rangle$ if $k \geq 3$ and $j/2^k \notin [3/8, 1/2)$.

This is the kind of local property we are seeking. The functions ψ_{kj} also have the translation and dilation properties that we want. However, we will have to eliminate φ somehow. We shall see that φ is not needed for a basis of $L^2(\mathbb{R})$ when we add in dilations of ψ by negative powers of 2. On the other hand, φ reappears in a central role in the next section as the scaling function.

We need a more explicit description of $H_n f$. By a **dyadic interval** of length 2^{-n}, we mean one of the form $[j2^{-n}, (j+1)2^{-n})$ for some integer j.

15.2.2. LEMMA. *Let $f \in L^2(0,1)$. Then $H_n f$ is the unique function that is constant on each dyadic interval of length 2^{-n} in $[0,1]$ and satisfies*

$$H_n f(x) = 2^n \int_{j2^{-n}}^{(j+1)2^{-n}} H_n f(t)\,dt = 2^n \int_{j2^{-n}}^{(j+1)2^{-n}} f(t)\,dt$$

for $x \in [j2^{-n}, (j+1)2^{-n})$, $0 \le j < 2^n$. Moreover, $\|H_n f\|_2 \le \|f\|_2$.

PROOF. It is easy to see that $\{\varphi, \psi_{00}\}$ span the functions that are constant on $[0,1/2)$ and on $[1/2,1)$. By induction, it follows easily that

$$M_n := \mathrm{span}\{\varphi, \psi_{kj} : 0 \le k \le n-1 \text{ and } 0 \le j < 2^k\}$$

is the subspace of all functions that are constant on each of the dyadic intervals $[j2^{-n}, (j+1)2^{-n})$ for $0 \le j < 2^n$. Notice that M_n is also spanned by the characteristic functions $\chi_{n,j} = \chi_{[j2^{-n},(j+1)2^{-n})}$ for $0 \le j < 2^n$.

Now $H_n f$ is contained in this span, and therefore is constant on these dyadic intervals. Thus $H_n f$ is the unique function of this form that satisfies $\langle H_n f, \varphi \rangle = \langle f, \varphi \rangle$ and $\langle H_n f, \psi_{kj} \rangle = \langle f, \psi_{kj} \rangle$ for $0 \le k \le n-1$ and $0 \le j < 2^k$. But this basis for M_n may be replaced by the basis of characteristic functions. Since $\|\chi_{n,j}\|_2^2 = 2^{-n}$, $H_n f$ is the unique function in M_n such that

$$H_n f(x) = 2^n \langle H_n f, \chi_{n,j} \rangle = 2^n \langle f, \chi_{n,j} \rangle$$

for all $x \in [j2^{-n}, (j+1)2^{-n})$ and $0 \le j < 2^n$, which is what we wanted.

The map H_n is the orthogonal projection of $L^2(0,1)$ onto M_n. The inequality $\|H_n f\|_2 \le \|f\|_2$ follows from the Projection Theorem (7.5.11). An elementary direct argument is outlined in Exercise 15.2.C. ∎

We can now prove that the Haar system is actually a basis. Moreover, we show that it does an excellent job of uniform approximation for continuous functions as well, even though the basis functions are not themselves continuous. In this respect, we obtain superior convergence to the convergence of Fourier series.

15.2.3. THEOREM. *Let $f \in L^2(0,1)$. Then $H_n f$ converges to f in the L^2 norm. Consequently, the Haar system is an orthonormal basis for $L^2(0,1)$. Moreover, if f is continuous on $[0,1]$, then $H_n f$ converges uniformly to f.*

PROOF. We prove the last statement first. By Theorem 5.5.9, f is uniformly continuous on $[0,1]$. Recall from Definition 10.4.2 that the modulus of continuity is $\omega(f;\delta) = \sup\{|f(x) - f(y)| : |x-y| \le \delta\}$. The remarks there also show that the uniform continuity of f implies that $\lim_{n\to\infty} \omega(f; 2^{-n}) = 0$.

For $x \in [j2^{-n}, (j+1)2^{-n})$, compute

$$|H_n f(x) - f(x)| = \left| 2^n \int_{j2^{-n}}^{(j+1)2^{-n}} f(t)\,dt - 2^n \int_{j2^{-n}}^{(j+1)2^{-n}} f(x)\,dt \right|$$

$$\leq 2^n \int_{j2^{-n}}^{(j+1)2^{-n}} |f(t) - f(x)|\,dt$$

$$\leq 2^n \int_{j2^{-n}}^{(j+1)2^{-n}} \omega(f;2^{-n})\,dt = \omega(f;2^{-n}).$$

Hence $\|H_n f - f\|_\infty \leq \omega(f;2^{-n})$ tends to 0. Therefore, $H_n f$ converges to f uniformly on $[0,1]$.

Now

$$\|H_n f - f\|_2 \leq \left(\int_0^1 \|H_n f - f\|_\infty dt \right)^{1/2} = \|H_n f - f\|_\infty.$$

So we obtain convergence in the $L^2(0,1)$ norm as well.

Next suppose that f is an arbitrary L^2 function, and let $\varepsilon > 0$ be given. Since f is the L^2 limit of a sequence of continuous functions, we may find a continuous function g with $\|f - g\|_2 < \varepsilon$. Now choose n so large that $\|H_n g - g\|_2 < \varepsilon$. Then

$$\|H_n f - f\|_2 \leq \|H_n f - H_n g\| + \|H_n g - g\|_2 + \|g - f\|_2$$
$$\leq \|H_n(f-g)\|_2 + \varepsilon + \varepsilon \leq \|f - g\|_2 + 2\varepsilon < 3\varepsilon.$$

So $H_n f$ converges to f in L^2.

Since the orthogonal expansion of f in the Haar system sums to f in the L^2 norm, we deduce that this orthonormal set spans all of $L^2(0,1)$ and thus is a basis. ∎

15.2.4. DEFINITION. The **Haar wavelet** is the function $\psi = \chi_{[0,0.5)} - \chi_{[0.5,1)}$. The **Haar wavelet basis** is the family $\{\psi_{kj} : k, j \in \mathbb{Z}\}$.

Lemma 15.2.1 can be easily modified to show that the Haar wavelet basis is orthonormal. It remains to verify that it spans $L^2(\mathbb{R})$.

15.2.5. THEOREM. *The Haar wavelet basis spans all of $L^2(\mathbb{R})$.*

PROOF. It is enough to show that any continuous function of bounded support is spanned by the Haar wavelet basis. Each such function is the finite sum of (piecewise) continuous functions supported on an interval $[m, m+1)$. But our basis is invariant under integer translations. So it is enough to show that a function on $[0,1)$ is spanned by the Haar wavelet basis. But Theorem 15.2.3 shows that the functions ψ_{kj} supported on $[0,1)$ together with φ span $L^2(0,1)$. Consequently, it is enough to approximate φ alone.

Consider the functions $\psi_{-k,0} = 2^{-k/2}\chi_{[0,2^{k-1})} - \chi_{[2^{k-1},2^k)}$ for $k \geq 1$. An easy computation shows that

$$h_N := \sum_{k=1}^{N} 2^{-k/2} \psi_{-k,0} = (1 - 2^{-N}) \chi_{[0,1)} - 2^{-N} \chi_{[1,2^N)}.$$

Thus $\|\varphi - h_N\|_2 = \|2^{-N} \chi_{[0,2^N)}\|_2 = 2^{-N/2}$. Hence φ is in the span of the wavelet basis. Therefore, the Haar wavelet basis spans all of $L^2(\mathbb{R})$. ∎

Exercises for Section 15.2

A. Let $f(x) = \sum_{k=0}^{2^n-1} s_{n,j} \chi_{[j2^{-n},(j+1)2^{-n})}$ and let $\mathbf{s}_n = (s_{n,0}, \ldots, s_{n,2^n-1})$.

(a) Define $\mathbf{a}_k = (a_{k,0}, \ldots, a_{k,2^k-1})$ and $\mathbf{s}_k = (s_{k,0}, \ldots, s_{k,2^k-1})$ by $a_{k,j} = (s_{k+1,2j} - s_{k+1,2j+1})/\sqrt{2}$
and $s_{k,j} = (s_{k+1,2j} + s_{k+1,2j+1})/\sqrt{2}$ for $0 \le k < n$ and $0 \le j < 2^k$.
Show that $f = s_{0,0}\varphi + \sum_{k=1}^{n-1} \sum_{j=0}^{2^k-1} a_{k,j} \psi_{k,j}$.
(b) Explain how to reverse this process and obtain $\mathbf{s}_n$ from the wavelet expansion of f.

B. Let f be a continuous function with compact support $[0,1]$. Fix $n \ge 1$, and define $s_j = f(j/2^n)$ for $0 \le j < 2^n$. Show that $\left\| H_n f - \sum_{k=0}^{2^n-1} s_j \chi_{[j2^{-n},(j+1)2^{-n})} \right\|_\infty \le \omega(f; 2^{-n})$.

C. Prove that $\|H_n f\|_2 \le \|f\|_2$.   HINT: Show that $\left| \int_a^{a+2^{-k}} f(x)\, dx \right|^2 \le 2^{-k} \int_a^{a+2^{-k}} |f(x)|^2\, dx$.

D. Show that $\{\psi_{kj} : k > -N, -2^{k+N} \le j < 2^{k+N}\}$ together with $\varphi_{-N,0} = 2^{-N/2} \varphi(2^{-N}x)$ and $\varphi_{-N,-1} = 2^{-N/2} \varphi(2^{-N}x + 1)$ form an orthonormal basis for $L^2(-2^N, 2^N)$.

E. Suppose that f is a continuous function with compact support contained in $[-2^N, 2^N]$ for some $N \in \mathbb{N}$. Define $P_n f(x) = \sum_{k=-n}^{n} \sum_{j=-\infty}^{\infty} \langle f, \psi_{kj} \rangle \psi_{kj}(x)$.

(a) Show that $P_n f$ is the sum of only finitely many nonzero terms.
(b) If $\int_0^{2^N} f(x)\, dx = 0 = \int_{-2^N}^0 f(x)\, dx$, then the only nonzero terms are for $k \ge -N$. Verify this.
Show that $P_n f$ converges to f uniformly.   HINT: Modify Theorem 15.2.3.
(c) Show that $P_n f$ converges uniformly to f without the integral conditions.   HINT: Prove uniform convergence for $\varphi_{-N,0}$ and $\varphi_{-N,-1}$.

F. Show that for $f \in L^2(\mathbb{R})$ that $\sum_{k=-\infty}^{\infty} \sum_{j=0}^{\infty} \langle f, \psi_{kj} \rangle \psi_{kj}(x) = \chi_{[0,\infty)} f$.

15.3 Multiresolution Analysis

Motivated by the Haar wavelet, we develop a general framework that applies to a wide range of wavelet systems. We will use this framework to construct other wavelet systems.

As before, $\varphi = \chi_{[0,1)}$ and we define the translations and dilations

$$\varphi_{kj}(x) = 2^{k/2} \varphi(2^k x - j) \quad \text{for all} \quad k, j \in \mathbb{Z}.$$

This is not an orthonormal system. But for each k, the family $\{\varphi_{kj} : j \in \mathbb{Z}\}$ consists of multiples of the characteristic functions of the dyadic intervals of length 2^{-k}. In particular, these families are orthonormal.

Define $V_k = \text{span}\{\varphi_{kj} : j \in \mathbb{Z}\}$. This is the space of L^2 functions that are constant on each dyadic interval of length 2^{-k}. Consequently $V_k \subset V_{k+1}$ for $k \in \mathbb{Z}$. That is, the V_k form a nested sequence of subspaces:

$$\cdots \subset V_{-2} \subset V_{-1} \subset V_0 \subset V_1 \subset V_2 \subset V_3 \subset \cdots.$$

It is also immediate that $f(x)$ belongs to V_k if and only if $f(2x)$ belongs to V_{k+1}. We state and prove the important (but basically easy) properties of this decomposition.

15.3.1. LEMMA. *Let* $\varphi = \chi_{[0,1)}$ *and define* V_k *as above. Then we have*

(1) **orthogonality:** $\{\varphi(x - j) : j \in \mathbb{Z}\}$ *is an orthonormal basis for* V_0.
(2) **nesting:** $V_k \subset V_{k+1}$ *for all* $k \in \mathbb{Z}$.
(3) **scaling:** $f(x) \in V_k$ *if and only if* $f(2x) \in V_{k+1}$.
(4) **density:** $\overline{\bigcup_{k \in \mathbb{Z}} V_k} = L^2(\mathbb{R})$.
(5) **separation:** $\bigcap_{k \in \mathbb{Z}} V_k = \{0\}$.

PROOF. We have already established (1), (2), and (3).

As in the proof of Theorem 15.2.3, we know that every continuous function with compact support $[-N, N]$ is the uniform limit of functions that also have support $[-N, N]$ and are constant on dyadic intervals of length 2^{-k} (i.e., functions in V_k). These functions therefore converge in L^2 as well. Consequently, the closed union of the V_k's contains all continuous functions of compact support, and thus all of $L^2(\mathbb{R})$.

Notice that ψ_{kj} is orthogonal to $\varphi_{k'j'}$ provided that $k \geq k'$ because $\psi_{k'j'}$ will be constant on the support of ψ_{kj}, and ψ_{kj} integrates to 0. So any function f belonging to the intersection $\bigcap_{k \in \mathbb{Z}} V_k$ must be orthogonal to every ψ_{kj}. By Theorem 15.2.5, it follows that f is orthogonal to every function in $L^2(\mathbb{R})$, including itself. Therefore, $\|f\|_2 = \langle f, f \rangle = 0$, whence $f = 0$. ∎

This leads us to formalize these properties in greater generality.

15.3.2. DEFINITION. A **multiresolution** of $L^2(\mathbb{R})$ with **scaling function** φ is the sequence of subspaces

$$V_j = \text{span}\{\varphi_{kj}(x) = 2^{k/2}\varphi(2^k x - j) : j \in \mathbb{Z}\}$$

provided that the sequence satisfies the five properties—orthogonality, nesting, scaling, density, and separation—described in the preceding lemma.

The function φ is sometimes called a **father wavelet**.

Notice that by a change of variables $t = 2^k x$, we obtain

$$\langle \varphi_{ki}, \varphi_{kj} \rangle = \int_{-\infty}^{\infty} 2^k \varphi(2^k x - i)\varphi(2^k x - j)\,dx = \int_{-\infty}^{\infty} \varphi(t - i)\varphi(t - j)\,dt = \delta_{ij}.$$

So $\{\varphi_{kj} : j \in \mathbb{Z}\}$ forms an orthonormal basis of V_k for each $k \in \mathbb{Z}$.

Once we have a nested sequence V_k with these properties, we can decompose $L^2(\mathbb{R})$ into a direct sum of subspaces. Set $W_k = \{f \in V_{k+1} : f \perp V_k\}$. This is the **orthogonal complement** of V_k in V_{k+1}. We write $V_{k+1} = V_k \oplus W_k$, where the $\oplus$ indicates that this is a **direct sum**, that is, a sum of orthogonal subspaces. So each vector $f \in V_{k+1}$ can be written uniquely as $f = g + h$ with $g \in V_k$ and $h \in W_k$. Since $\langle f, g \rangle = 0$, we have the Pythagorean identity

$$\|f\|_2^2 = \langle g+h, g+h \rangle = \langle g,g \rangle + \langle g,h \rangle + \langle h,g \rangle + \langle h,h \rangle = \|g\|_2^2 + \|h\|_2^2.$$

Since V_k has an orthonormal basis $\{\varphi_{kj} : j \in \mathbb{Z}\}$, Corollary 7.7.6 of Parseval's Theorem provides an orthogonal projection P_k of $L^2(\mathbb{R})$ onto V_k given by

$$P_k f = \sum_{j=-\infty}^{\infty} \langle f, \varphi_{kj} \rangle \varphi_{kj}$$

and we have the important identity

$$\|f\|_2^2 = \|P_k f\|_2^2 + \|f - P_k f\|_2^2.$$

15.3.3. LEMMA. $Q_k = P_{k+1} - P_k$ is the orthogonal projection onto W_k.

PROOF. To verify this, we will show that Q_k is an idempotent with range W_k and kernel $W_k^\perp$. Note that $P_k P_{k+1} = P_{k+1} P_k = P_k$ because V_k is contained in V_{k+1}. Hence

$$Q_k^2 = P_{k+1}^2 - P_k P_{k+1} - P_{k+1} P_k + P_k^2 = P_{k+1} - P_k = Q_k.$$

So Q_k is a projection.

We claim that $W_k^\perp = V_{k+1}^\perp + V_k$. Indeed, $P_{k+1}^\perp f$ is orthogonal to V_{k+1} and so is also orthogonal to W_k. So f is orthogonal to W_k if and only if $P_{k+1} f$ is orthogonal to W_k, which is the same as saying that $P_{k+1} f \in V_k$. This latter statement is equivalent to

$$P_{k+1} f = P_k P_{k+1} f = P_k f \qquad \text{or} \qquad (P_{k+1} - P_k) f = 0.$$

So $W_k^\perp = \ker Q_k$.

If $f \in W_k$, then $P_{k+1} f = f$, since $f \in V_{k+1}$. Also, $P_k f = 0$, since $f \perp V_k$. Hence $Q_k f = f$. Conversely, if $f = Q_k g$, then

$$P_{k+1} f = P_{k+1}^2 g - P_{k+1} P_k g = (P_{k+1} - P_k) g = f.$$

So f belongs to V_{k+1}. And

$$P_k f = P_k P_{k+1} g - P_k^2 g = (P_k - P_k) g = 0.$$

Thus f is orthogonal to V_k, and so f is in W_k. Therefore, the range of Q_k is exactly W_k. Consequently, Q_k is the orthogonal projection onto W_k. ∎

We may repeat the decomposition $V_{k+1} = V_k \oplus W_k$ finitely often to obtain

$$V_n = V_0 \oplus W_0 \oplus \cdots \oplus W_{n-1} \quad \text{and} \quad V_0 = V_{-n} \oplus W_{-n} \oplus \cdots \oplus W_{-1}.$$

Repetition of this procedure suggests that there is a decomposition of $L^2(\mathbb{R})$ as an infinite direct sum

$$\bigoplus_{k \in \mathbb{Z}} W_k = \cdots \oplus W_{-2} \oplus W_{-1} \oplus W_0 \oplus W_1 \oplus W_2 \oplus \cdots .$$

What we mean by this is that every function f in $L^2(\mathbb{R})$ should decompose uniquely as an infinite sum

$$f = \sum_{k=-\infty}^{\infty} f_k, \quad \text{where} \quad f_k \in W_k \quad \text{and} \quad \|f\|_2^2 = \sum_{k \in \mathbb{Z}} \|f_k\|_2^2.$$

We shall prove that this is indeed the case.

15.3.4. LEMMA. *Suppose that $V_k \subset V_{k+1}$ for $k \in \mathbb{Z}$ is the nested sequence of subspaces from a multiresolution of $L^2(\mathbb{R})$. Then*

$$\lim_{k \to \infty} \|f - P_k f\|_2 = 0, \quad \text{and} \quad \lim_{k \to -\infty} \|P_k f\|_2 = 0.$$

PROOF. The limit $\lim_{k \to \infty} \|f - P_k f\|_2$ is a consequence of density. For any $\varepsilon > 0$, there are an integer n and a function $g \in V_n$ such that $\|f - g\|_2 < \varepsilon$. Then for $k \geq n$, Parseval's Theorem (7.7.5) shows that

$$\|f - P_k f\|_2 = \|(f - g) - P_k(f - g)\|_2 \leq \|f - g\|_2 < \varepsilon.$$

The second limit $\lim_{k \to -\infty} \|P_k f\|_2 = 0$ is a consequence of separation. We will show that it actually follows from the first part. Let $V_k^\perp$ denote the orthogonal complement of V_k, and note that $I - P_k$ is the orthogonal projection onto it. Notice that these subspaces are also nested in the reverse order $V_{k+1}^\perp \subset V_k^\perp$. We claim that $N = \overline{\bigcup_{k \in \mathbb{Z}} V_k^\perp}$ is all of $L^2(\mathbb{R})$. Indeed, if N were a proper subspace of $L^2(\mathbb{R})$, then there would be a nonzero function $g \perp N$. Thus, in particular, $g \perp V_k^\perp$, so that g belongs to $V_k^{\perp\perp} = V_k$ for every $k \in \mathbb{Z}$. Consequently, g belongs to $\bigcap_{k \in \mathbb{Z}} V_k = \{0\}$. So $N = L^2(\mathbb{R})$. Since $\|P_k f\|_2 = \|f - (I - P_k)f\|_2$, the desired limit follows from the first part. ∎

We now are ready to derive the infinite decomposition.

15.3.5. THEOREM. *Suppose that $V_k \subset V_{k+1}$ for $k \in \mathbb{Z}$ is the nested sequence of subspaces from a multiresolution of $L^2(\mathbb{R})$. Then $L^2(\mathbb{R})$ decomposes as the infinite direct sum $\bigoplus_{k \in \mathbb{Z}} W_k$.*

PROOF. The finite decompositions are valid. So, in particular,

$$V_n = V_{-n} \oplus W_{-n} \oplus \cdots \oplus W_{n-1}.$$

Thus if f belongs to V_n and is orthogonal to V_{-n}, then f decomposes uniquely as $f = \sum_{k=-n}^{n-1} f_k$ for $f_k \in W_k$, namely $f_k = Q_k f$. Moreover, Parseval's Theorem shows that $\|f\|_2^2 = \sum_{k=-n}^{n-1} \|f_k\|_2^2$.

If f is an arbitrary function in $L^2(\mathbb{R})$ and $\varepsilon > 0$, then the lemma provides a positive integer n such that $\|(I - P_n)f\|_2^2 + \|P_{-n}f\|_2^2 < \varepsilon^2$. Thus, $g_n := P_n f - P_{-n} f$ belongs to V_n and is orthogonal to V_{-n}. Consequently, we may write $g_n = \sum_{k=-n}^{n-1} f_k$ for $f_k \in W_k$, where $f_k = Q_k g_n = Q_k f$. By Parseval's Theorem,

$$\|f - g_n\|_2^2 = \|(I - P_n)f + P_{-n}f\|_2^2 = \|(I - P_n)f\|_2^2 + \|P_{-n}f\|_2^2 < \varepsilon^2.$$

Since ε is arbitrary, it follows that g_n converges to f. That is,

$$f = \lim_{n \to \infty} \sum_{k=-n}^{n-1} f_k = \sum_{k=-\infty}^{\infty} f_k$$

and

$$\|f\|_2^2 = \lim_{n \to \infty} \|g_n\|_2^2 = \lim_{n \to \infty} \sum_{k=-n}^{n-1} \|f_k\|_2^2 = \sum_{k=-\infty}^{\infty} \|f_k\|_2^2.$$

To establish uniqueness, suppose that $f = \sum_k f_k = \sum_k h_k$ are two decompositions with f_k and h_k in W_k. Then $0 = \sum_k f_k - h_k$. The norm formula from the previous paragraph shows that $0 = \sum_k \|f_k - h_k\|_2^2$. Therefore, $h_k = f_k$ for all $k \in \mathbb{Z}$. ∎

Exercises for Section 15.3

A.   Let V_0 be the span of integer translates of the Haar scaling function φ. Suppose $f \in V_0$ has bounded support and $\{f(x - j) : j \in \mathbb{Z}\}$ is orthonormal. Prove that $f(x) = \pm \varphi(x - n)$ for some integer n.   HINT: Compute $\langle f(x), f(x - j) \rangle$ when the supports overlap on a single interval.

B.   Suppose that $\varphi \in L^2(\mathbb{R})$ is such that the subspaces V_k satisfy orthogonality, nesting, and scaling. Let $M = \overline{\bigcup_{k \in \mathbb{Z}} V_k}$. Show that if $f \in M$, then $f(x - t) \in M$ for every $t \in \mathbb{R}$.
HINT: First prove this for $t = j2^{-k}$. Then apply Exercise 15.1.F.

C.   Suppose φ is continuous with compact support $[a, a + M]$, and $\{\varphi_{0j} : j \in \mathbb{Z}\}$ are orthonormal.

(a) Suppose that $f \in V_k$, and express $f(x) = \sum_j c_j 2^{k/2} \varphi(2^k x - j)$. Use the Cauchy–Schwarz inequality to show that $|f(x)| \le 2^{k/2} M \|\varphi\|_\infty \|f\|_2$.

(b) Show that $\bigcap_{k \in \mathbb{Z}} V_k = \{0\}$.   HINT: Let $f \in \bigcap_{k \in \mathbb{Z}} V_k$. Use part (a) to estimate $\int_{-N}^{N} |f(x)|^2 \, dx$.

15.4 Recovering the Wavelet

Let us look at the decomposition obtained in the previous section in the case of the Haar system. Notice that $\varphi = \chi_{[0,1)}$ satisfies the identity

$$\varphi = \chi_{[0,0.5)} + \chi_{[0.5,1)} = \tfrac{1}{\sqrt{2}} \varphi_{10} + \tfrac{1}{\sqrt{2}} \varphi_{11}.$$

On the other hand, we can write φ_{10} and φ_{11} in terms of φ and ψ. Recalling that $\psi = \chi_{[0,0.5)} - \chi_{[0.5,1)}$, we have $\varphi_{10} = \tfrac{1}{\sqrt{2}} \varphi + \tfrac{1}{\sqrt{2}} \psi$ and $\varphi_{11} = \tfrac{1}{\sqrt{2}} \varphi - \tfrac{1}{\sqrt{2}} \psi$. In general,

$$\varphi_{kj} = \tfrac{1}{\sqrt{2}}\varphi_{k+1,2j} + \tfrac{1}{\sqrt{2}}\varphi_{k+1,2j+1}$$

and

$$\varphi_{k+1,2j} = \tfrac{1}{\sqrt{2}}\varphi_{kj} + \tfrac{1}{\sqrt{2}}\psi_{kj} \qquad \text{and} \qquad \varphi_{k+1,2j+1} = \tfrac{1}{\sqrt{2}}\varphi_{kj} - \tfrac{1}{\sqrt{2}}\psi_{kj}.$$

The subspace V_k consists of those $L^2(\mathbb{R})$ functions that are constant on the dyadic intervals of length 2^{-k}. Now ψ_{kj} belongs to V_{k+1}, it is supported on one interval of length 2^{-k}, and integrates to 0. Thus $\langle \psi_{kj}, \varphi_{kj'} \rangle = 0$ for all $j, j' \in \mathbb{Z}$. In particular, ψ_{kj} lies in W_k. So $W_k' = \mathrm{span}\{\psi_{kj} : j \in \mathbb{Z}\}$ is a subspace of W_k.

On the other hand, the identities show that every basis vector $\varphi_{k+1,j}$ belongs to $V_k + W_k'$, and thus $V_{k+1} = V_k \oplus W_k' = V_k \oplus W_k$. This forces the identity $W_k' = W_k$. So we have shown that for the Haar system, we have $W_k = \mathrm{span}\{\psi_{kj} : j \in \mathbb{Z}\}$.

There is a systematic way to construct a wavelet from a multiresolution. That is the goal of this section. Let $\{V_k\}$ be a multiresolution with scaling function φ. The construction begins with the fact that $\varphi \in V_0 \subset V_1$. Since φ_{1j} form an orthonormal basis for V_1, we may expand φ as

$$\varphi(x) = \sum_{j=-\infty}^{\infty} a_j \varphi(2x - j) = \sum_{j=-\infty}^{\infty} \frac{a_j}{\sqrt{2}} \varphi_{1j}(x), \qquad (15.4.1)$$

where $a_j = 2\langle \varphi(x), \varphi(2x - j) \rangle$. By Parseval's Theorem, $\|\varphi\|_2^2 = \tfrac{1}{2} \sum_{j=-\infty}^{\infty} |a_j|^2$. Thus (a_j) is a sequence in ℓ^2. Equation (15.4.1) is known as the **scaling relation** for φ.

15.4.2. THEOREM. *Let φ be the scaling function generating a multiresolution $\{V_k\}$ of $L^2(\mathbb{R})$ with scaling relation $\varphi(x) = \sum_{j=-\infty}^{\infty} a_j \varphi(2x - j)$. Define*

$$\psi(x) = \sum_{j=-\infty}^{\infty} (-1)^j a_{1-j}\, \varphi(2x - j).$$

Then ψ is a wavelet that generates the wavelet basis $\{\psi_{kj} : k, j \in \mathbb{Z}\}$ such that $W_k = \mathrm{span}\{\psi_{kj} : j \in \mathbb{Z}\}$ for each $k \in \mathbb{Z}$.

PROOF. Since this proof basically consists of several long computations, we provide a brief overview of the plan. The orthonormality of $\{\varphi(x - j) : j \in \mathbb{Z}\}$ will yield conditions on the coefficients a_j. Then we show that $\{\psi(x - k) : k \in \mathbb{Z}\}$ is an orthonormal set that is orthogonal to the $\varphi(x - j)$'s. Finally, we show that V_1 is spanned by V_0 and the $\psi(x - k)$'s.

In this proof, all summations are from $-\infty$ to $+\infty$, but for notational simplicity, only the index will be indicated. We define δ_{0n} to be 1 if $n = 0$ and 0 otherwise. To begin, we have

$$\delta_{0n} = \langle \varphi(x), \varphi(x - n) \rangle = \left\langle \sum_i a_i \varphi(2x - i), \sum_j a_j \varphi(2x - 2n - j) \right\rangle$$

$$= \sum_i \sum_j a_i a_j \langle \varphi(2x-i), \varphi(2x-2n-j) \rangle = \frac{1}{2} \sum_j a_{j+2n} a_j.$$

The orthonormality of $\{\psi(x-j) : j \in \mathbb{Z}\}$ follows because the coefficients of ψ are obtained from φ by reversing, shifting by one place, and alternating sign. A bit of thought will show that each of these steps preserves the property of orthogonality of translations. Here we provide the direct computation:

$$\langle \psi(x), \psi(x-n) \rangle = \left\langle \sum_i (-1)^i a_{1-i} \varphi(2x-i), \sum_j (-1)^j a_{1-j} \varphi(2x-2n-j) \right\rangle$$

$$= \sum_i \sum_j (-1)^{i+j} a_{1-i} a_{1-j} \langle \varphi(2x-i), \varphi(2x-2n-j) \rangle$$

$$= \frac{1}{2} \sum_j (-1)^{2j+2n} a_{1-j-2n} a_{1-j} = \frac{1}{2} \sum_i a_i a_{i+2n} = \delta_{0n}.$$

So $\{\psi(x-j) : j \in \mathbb{Z}\}$ is orthonormal.

The fact that the ψ's and φ's are orthogonal is more subtle. Calculate

$$\langle \psi(x-m), \varphi(x-n) \rangle = \left\langle \sum_i (-1)^i a_{1-i} \varphi(2x-2m-i), \sum_j a_j \varphi(2x-2n-j) \right\rangle$$

$$= \sum_i \sum_j (-1)^i a_{1-i} a_j \langle \varphi(2x-2m-i), \varphi(2x-2n-j) \rangle,$$

but the inner product is 0 unless $2m+i = 2n+j$,

$$= \frac{1}{2} \sum_j (-1)^{j+2n-2m} a_{1-j-2n+2m} a_j = \frac{1}{2} \sum_j (-1)^j a_{p-j} a_j,$$

where $p = 2m+1-2n$ is a fixed *odd* integer. Thus by substituting $i = p-j$, we may rearrange this sum:

$$\frac{1}{2} \sum_j (-1)^j a_{p-j} a_j = \frac{1}{2} \sum_i (-1)^{p-i} a_i a_{p-i} = -\left(\frac{1}{2} \sum_i (-1)^i a_i a_{p-i} \right).$$

Thus the sum must be 0. Hence the family $\{\psi(x-k) : k \in \mathbb{Z}\}$ is orthogonal to the family $\{\varphi(x-j) : j \in \mathbb{Z}\}$. Notice that the shift by 1 of the coefficients in the definition of ψ was to make p odd in this calculation.

Now we wish to express $\varphi_{1p}(x) = \sqrt{2}\varphi(2x-p)$ as a linear combination of these two families. To see what the coefficients should be, we compute

$$\langle \varphi_{1p}(x), \varphi(x-n) \rangle = \left\langle \sqrt{2}\varphi(2x-p), \sum_j a_j \varphi(2x-2n-j) \right\rangle = \frac{1}{\sqrt{2}} a_{p-2n},$$

since the inner product is 0 except when $2n+j = p$; and similarly

$$\langle \varphi_{1p}(x), \psi(x-n) \rangle = \left\langle \sqrt{2}\varphi(2x-p), \sum_j (-1)^j a_{1-j}\varphi(2x-2n-j) \right\rangle$$

$$= \frac{(-1)^p}{\sqrt{2}} a_{1-p+2n}.$$

Now it is a matter of adding up the series to recover $\varphi_{1p}(x)$. Compute

$$\sum_n a_{p-2n}\varphi(x-n) = \sum_n \sum_i a_{p-2n}a_i\varphi(2x-2n-i)$$

$$= \sum_k \left(\sum_n a_{p-2n}a_{k-2n} \right) \varphi(2x-k)$$

$$= \sum_k \left(\sum_n a_{p+2n}a_{k+2n} \right) \varphi(2x-k)$$

and

$$\sum_n (-1)^p a_{1-p+2n}\psi(x-n) = \sum_n \sum_i (-1)^p a_{1-p+2n}(-1)^i a_{1-i}\varphi(2x-2n-i)$$

$$= \sum_k \left((-1)^{p+k} \sum_n a_{1+2n-p}a_{1+2n-k} \right) \varphi(2x-k).$$

When $p+k$ is odd,

$$(-1)^{p+k} \sum_n a_{1+2n-p}a_{1+2n-k} = -\sum_m a_{2m+k}a_{2m+p},$$

while if $p+k$ is even,

$$(-1)^{p+k} \sum_n a_{1+2n-p}a_{1+2n-k} = \sum_m a_{1+2m+k}a_{1+2m+p}.$$

When these sums over translates of $\varphi(2x)$ and $\psi(2x)$ are added together, the coefficients of $\varphi(2x-k)$ are canceled when $p+k$ is odd, while for $p+k$ even the two sums conveniently merge to yield the sums from the orthogonality relation for the $\varphi(x-k)$. Hence the sum obtained is

$$\sum_n \frac{1}{\sqrt{2}} a_{p-2n}\varphi(x-n) + \sum_n \frac{(-1)^p}{\sqrt{2}} a_{1-p+2n}\psi(x-n)$$

$$= \sum_{k \equiv p \bmod 2} \left(\frac{1}{2}\sum_n a_{p+n}a_{k+n} \right) \sqrt{2}\varphi(2x-k) = \sqrt{2}\varphi(2x-p) = \varphi_{1p}(x).$$

Set $W = \mathrm{span}\{\psi(x-j) : j \in \mathbb{Z}\}$. Let us recap what we have established. We have shown that $\{\psi(x-j) : j \in \mathbb{Z}\}$ is an orthonormal basis for W, that W is orthogonal to V_0, and that $\varphi(2x-i)$ belongs to $V_0 + W$ for all $i \in \mathbb{Z}$. Since each $\psi(x-j)$ is expressed in terms of the $\varphi(2x-i)$, it is clear that W is a subspace of V_1. On the other hand, since each $\varphi(2x-i)$ belongs to $V_0 + W$, it follows that $V_1 = V_0 \oplus W$. Hence we deduce that W is the orthogonal complement of V_0 in V_1; that is, $W = W_0$.

It now follows from dilation that

$$\text{span}\{\psi_{kj} : j \in \mathbb{Z}\} = \{2^{k/2} f(2^k x) : f \in W_0\} = W_k.$$

Hence $\{\psi_{kj} : j \in \mathbb{Z}\}$ is an orthonormal basis for W_k for each $k \in \mathbb{Z}$. Since $L^2(\mathbb{R}) = \bigoplus_{k=-\infty}^{+\infty} W_k$, it follows that together the collection $\{\psi_{kj} : k, j \in \mathbb{Z}\}$ is an orthonormal basis for $L^2(\mathbb{R})$. Therefore, ψ is a wavelet. ∎

Exercises for Section 15.4

A. If φ is a scaling function with compact support, show that the scaling relation is a finite sum.

B. Let $\{e_k : k \in \mathbb{Z}\}$ be an orthonormal set in a Hilbert space $\mathscr{H}$. Show that the vectors $x = \sum_n a_n e_n$ and $y = \sum_n (-1)^n a_{p-n} e_n$ are orthogonal if p is odd.

C. Given a scaling relation $\varphi(x) = \sum_j a_j \varphi(2x - j)$, define the **filter** to be the complex function $m_\varphi(\theta) = \sum_j a_j e^{ij\theta}$. Prove that $|m_\varphi(\theta)|^2 + |m_\varphi(\theta + \pi)|^2 = 1$.
HINT: Compute the Fourier series of this sum, and compare with the proof of Theorem 15.4.2.

D. Suppose that φ is a scaling function that is bounded, has compact support, and satisfies $\int_{-\infty}^{\infty} \varphi(x)\,dx \neq 0$. Let $\varphi(x) = \sum_j a_j \varphi(2x - j)$ be the scaling relation.

(a) Show that $\sum_j a_j = 2$.   HINT: Integrate over $\mathbb{R}$.
(b) Show that $\sum_j (-1)^j a_j = 0$.   HINT: Use the previous exercise for $\theta = 0$.

15.5 Daubechies Wavelets

The multiresolution analysis developed in the last two sections can be used to design a continuous wavelet. We start by explaining the properties we want. The only example we have so far of a wavelet system and multiresolution analysis is the Haar wavelet system. The Haar wavelet ψ satisfies

$$\int \psi(x)\,dx = 0$$

and the multiresolution analysis uses subspaces of functions that are constant on dyadic intervals of length 2^k, $k \in \mathbb{Z}$. As a result, Haar wavelets do a good job of approximating functions that are locally constant.

It is possible to do a better job of approximating continuous functions if we use a wavelet that also satisfies

$$\int x\psi(x)\,dx = 0.$$

If you computed moments of inertia in calculus, you won't be surprised to learn that this is called the **first moment** of ψ.

Our goal in this section is to construct a continuous wavelet with this property. To be honest, our construction is not quite complete. At one crucial point, we will *assume* the uniform convergence of a sequence of functions to a continuous function. The full construction of this wavelet requires considerable work, although in the next section we provide a proof that the sequence converges in L^2. Later in this chapter, we give a full proof of the existence of another continuous wavelet, known as the Franklin wavelet.

This is part of a general family of wavelets constructed by Ingrid Daubechies in 1988. Hence these wavelets are called **Daubechies wavelets**.

15.5.1. THEOREM. *There is a continuous function φ of compact support in $L^2(\mathbb{R})$ that generates a multiresolution of $L^2(\mathbb{R})$ such that the associated wavelet ψ is continuous, has compact support, and satisfies $\int \psi(x)\,dx = \int x\psi(x)\,dx = 0$.*

PROOF. As in the last section, all of our summations are from $-\infty$ to $+\infty$; so only the index is given. We will look for a function φ with norm 1 and integral 1, that is,

$$\|\varphi\|_2^2 = \int_{-\infty}^{+\infty} |\varphi(x)|^2\,dx = 1 \qquad \text{and} \qquad \int_{-\infty}^{+\infty} \varphi(x)\,dx = 1.$$

Beyond these normalizing assumptions, we use the crucial idea of the previous section by assuming that φ satisfies a scaling relation $\varphi(x) = \sum_j a_j\,\varphi(2x-j)$. Since we wish φ to have compact support, this must be a finite sum.

Compute what follows from our assumptions:

$$1 = \langle \varphi, \varphi \rangle = \left\langle \sum_j a_j\,\varphi(2x-j), \sum_k a_k\varphi(2x-k) \right\rangle = \frac{1}{2}\sum_j |a_j|^2,$$

where we use the fact that $\{\varphi(2x-j) : j \in \mathbb{Z}\}$ is an orthogonal set of vectors in $L^2(\mathbb{R})$ with norm $1/\sqrt{2}$. Similarly,

$$1 = \int_{-\infty}^{+\infty} \varphi(x)\,dx = \int_{-\infty}^{+\infty} \sum_k a_k\varphi(2x-k)\,dx = \sum_k a_k \int_{-\infty}^{+\infty} \varphi(2x-k)\,dx = \frac{1}{2}\sum_k a_k.$$

From Theorem 15.4.2, if we can find a suitable sequence (a_k), then there is a wavelet ψ, also of norm 1, which is given by

$$\psi(x) = \sum_j (-1)^j a_{1-j}\,\varphi(2x-j). \tag{15.5.2}$$

Consider the consequences of the integral conditions on ψ, namely $\int \psi(x)\,dx = 0$ and $\int x\psi(x)\,dx = 0$, to obtain two more relations that the sequence (a_n) must satisfy:

$$0 = \int \psi(x)\,dx = \sum_j (-1)^j a_{1-j} \int \varphi(2x-j)\,dx = \frac{1}{2}\sum_j (-1)^j a_{1-j}.$$

Replace $1-j$ with j and use $(-1)^{1-j} = -(-1)^j$ to obtain $\sum_j (-1)^j a_j = 0$.

Similarly,

$$0 = \int x\psi(x)\,dx = \sum_j (-1)^j a_{1-j} \int x\varphi(2x-j)\,dx = \sum_j (-1)^j a_{1-j} \frac{1}{4} \int (t+j)\varphi(t)\,dt$$

$$= \Big(\frac{1}{4}\sum_j (-1)^j a_{1-j}\Big) \int t\varphi(t)\,dt + \Big(\frac{1}{4}\sum_j (-1)^j j a_{1-j}\Big) \int \varphi(t)\,dt$$

$$= \frac{1}{4}\sum_j (-1)^j j a_{1-j} = \frac{1}{4}\sum_k (-1)^{1-k}(1-k)a_k$$

$$= \frac{1}{4}\sum_k (-1)^k k a_k - \frac{1}{4}\sum_k (-1)^k a_k = \frac{1}{4}\sum_k (-1)^k k a_k.$$

Summarizing, we have the following equations:

$$\sum_j |a_j|^2 = \sum_j a_j = 2 \quad \text{and} \quad \sum_j (-1)^j a_j = \sum_j (-1)^j j a_j = 0.$$

As you can verify directly, one solution to these equations is given by

$$a_0 = \frac{1+\sqrt{3}}{4}, \quad a_1 = \frac{3+\sqrt{3}}{4}, \quad a_2 = \frac{3-\sqrt{3}}{4}, \quad a_3 = \frac{1-\sqrt{3}}{4}$$

with $a_j = 0$ for all other $j \in \mathbb{Z}$.

Substituting these values back into the scaling relation, we want the scaling function to satisfy

$$\varphi(x) = \frac{1+\sqrt{3}}{4}\varphi(2x) + \frac{3+\sqrt{3}}{4}\varphi(2x-1) + \frac{3-\sqrt{3}}{4}\varphi(2x-2) + \frac{1-\sqrt{3}}{4}\varphi(2x-3)$$

$$= a_0\varphi(2x) + a_1\varphi(2x-1) + a_2\varphi(2x-2) + a_3\varphi(2x-3).$$

It is not immediately clear why there should be a continuous function satisfying this equation.

We can construct such a function as the limit of a sequence of functions (φ_n), defined by $\varphi_0 = \chi_{[0,1)}$ and for $n \geq 0$,

$$\varphi_{n+1}(x) = a_0\varphi_n(2x) + a_1\varphi_n(2x-1) + a_2\varphi_n(2x-2) + a_3\varphi_n(2x-3).$$

From the first few φ_n, graphed in Figure 15.2, it is plausible that the sequence (φ_n) converges to a continuous function. However, proving this requires careful arguments using the Fourier transform, and so is beyond the scope of this book. We content ourselves with stating the following theorem.

15.5.3. THEOREM. *The sequence of functions (φ_n) converges uniformly to a continuous function φ.*

We will prove convergence in $L^2(\mathbb{R})$ in the next section. The other properties of the Daubechies wavelet can now be deduced. Note that except for the continuity of φ and ψ, all of the other properties follow from convergence in $L^2(\mathbb{R})$.

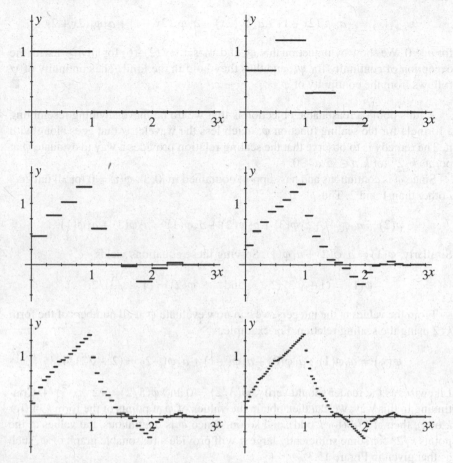

FIG. 15.2 The graphs of φ_0 through φ_5.

15.5.4. COROLLARY. *Daubechies' scaling function φ and wavelet ψ satisfy:*

(1) $\varphi(x) = a_0\varphi(2x) + a_1\varphi(2x-1) + a_2\varphi(2x-2) + a_3\varphi(2x-3)$.

(2) φ is supported on $[0,3]$.

(3) $\|\varphi\|_2 = 1$ and $\int \varphi(x)\,dx = 1$.

(4) $\psi(x) = -a_0\varphi(2x-1) + a_1\varphi(2x) - a_2\varphi(2x+1) + a_3\varphi(2x+2)$
    is continuous with support in $[-1,2]$.

(5) $\int \psi(x)\,dx = \int x\psi(x)\,dx = 0$.

(6) $\{\varphi(x-j), \psi(x-j) : j \in \mathbb{Z}\}$ *is orthonormal.*

PROOF. The proof will be left as an exercise using the following outline.

From the definition of φ_{n+1} in terms of φ_n and the convergence to φ, it follows immediately that φ satisfies the scaling relation. The Haar function $\varphi_0 = \chi_{[0,1)}$ satisfies (2) and (3) and (6a): $\{\varphi_0(x-j) : j \in \mathbb{Z}\}$ is orthonormal. We also introduce the functions

$$\psi_{n+1}(x) = -a_0\varphi_n(2x-1) + a_1\varphi_n(2x) - a_2\varphi_n(2x+1) + a_3\varphi_n(2x+2)$$

for $n \geq 0$. We show by induction that φ_n and ψ_n satisfy (2)–(6) for all $n \geq 1$ with the exception of continuity for ψ_n, and thus they hold in the limit. The continuity of ψ follows from the continuity of φ. ∎

At this point, a reasonable objection is that we do not have anything resembling a formula for the scaling function φ, much less the wavelet ψ that goes along with it. The remedy is to observe that the scaling relation provides a way to evaluate φ at points $k/2^n$ for $k, n \in \mathbb{Z}, n \geq 0$.

Since φ is continuous and has support contained in $[0,3]$, $\varphi(i) = 0$ for all integers i other than 1 and 2. Thus,

$$\varphi(2) = a_0\varphi(4) + a_1\varphi(3) + a_2\varphi(2) + a_3\varphi(1) = a_2\varphi(2) + a_3\varphi(1).$$

Similarly, $\varphi(1) = a_0\varphi(2) + a_1\varphi(1)$. Solving these equations yields

$$\varphi(1) = (1+\sqrt{3})/2 \qquad \text{and} \qquad \varphi(2) = (1-\sqrt{3})/2.$$

From the values at the integers, we can now evaluate φ at all numbers of the form $k/2$ using the scaling relation. For example,

$$\varphi\left(\tfrac{1}{2}\right) = a_0\varphi(1) + a_1\varphi(0) + a_2\varphi(-1) + a_3\varphi(-2) = (2+\sqrt{3})/4.$$

Likewise, as the reader should verify, $\varphi(3/2) = 0$ and $\varphi(5/2) = (2-\sqrt{3})/4$. Continuing in this way, we can then obtain the values of φ at points of the form $k/4$ (for k odd), then at $k/8$ (for k odd), and so on. Since φ is continuous, the values at the points $k/2^n$ for some sufficiently large n will provide a reasonable graph of φ, such as that given in Figure 15.3.

Similarly, using the relation (15.5.2), we can find the values of ψ at points $k/2^n$ and graph ψ; see Figure 15.4.

Exercises for Section 15.5

A. Prove Corollary 15.5.4 following the outline given there.

B. Evaluate the Daubechies wavelet ψ at the points $k/4$ for $k \in \mathbb{Z}$.

C. Which Daubechies wavelet coefficients are nonzero for the function given by $f(x) = x$ on $[0,2]$ and 0 elsewhere?

D. (a) Let $\mathbb{Q}[\sqrt{3}] = \{x + \sqrt{3}y : x, y \in \mathbb{Q}\}$. If $a + \sqrt{3}b = x + \sqrt{3}y$ in $\mathbb{Q}[\sqrt{3}]$, show $a = x$ and $b = y$.
  (b) Find a formula for multiplication and division in $\mathbb{Q}[\sqrt{3}]$.
  (c) Explain the significance of this fact for the efficiency of the algorithm evaluating φ and ψ given at the end of this section.

E. Let φ be the Daubechies scaling function. Let $\mathbb{D}_n = \{a/2^n : a \in \mathbb{Z}\}$ and $\mathbb{D} = \cup_{n \geq 1}\mathbb{D}_n$. Let $\mathbb{D}_n[\sqrt{3}] = \{a + \sqrt{3}b : a, b \in \mathbb{D}_n\}$ and likewise define $\mathbb{D}[\sqrt{3}]$.

  (a) Show that for $d \in \mathbb{D}$, $\varphi(d) \in \mathbb{D}[\sqrt{3}]$.
  (b) If $\overline{a + \sqrt{3}b}$ is defined as $a - \sqrt{3}b$, show that for $d \in \mathbb{D}$, $\varphi(3 - d) = \overline{\varphi(d)}$.
  HINT: Prove it for $\mathbb{D}_n$ using induction on n.

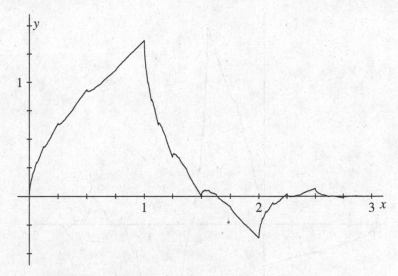

FIG. 15.3 The Daubechies scaling function.

(c) Show that for $d \in \mathbb{D}$, $1 = \sum_{k \in \mathbb{Z}} \varphi(d - k)$.   HINT: See the previous hint.

(d) Show that $\sum_{k=0}^{3} k\varphi(k) = \sum_{j=0}^{2} 2ja_{2j} = \frac{3-\sqrt{3}}{2}$.

(e) Show that for $d \in \mathbb{D}$, $d = \sum_{k \in \mathbb{Z}} \left(\frac{3-\sqrt{3}}{2} + k \right) \varphi(d - k)$.   HINT: Using part (c), reduce to

   proving $d - \frac{3-\sqrt{3}}{2} = \sum_{k \in \mathbb{Z}} k\varphi(d - k)$ Use induction, part (d), and considerable calculation.

(f) Use continuity to deduce that $1 = \sum_{k \in \mathbb{Z}} \varphi(x - k)$ and $x = \sum_{k \in \mathbb{Z}} \left(\frac{3-\sqrt{3}}{2} + x \right) \varphi(x - k)$.

F. Suppose that φ is any scaling function of compact support arising from a multiresolution analysis and that its scaling relation is $\varphi(x) = \sum_{j} a_{j} \varphi(2x - j)$.

(a) Show that $\sum_{n} \varphi(2^{-k}n) = 2^{k} \sum_{n} \varphi(n)$ for all $k \geq 1$.

(b) Show that $\int \varphi(x)\,dx = \sum_{n} \varphi(n)$.   HINT: Use Riemann sums and part (a).

(c) Show that $\sum_{n} \varphi(x - n)$ is constant.

G. Consider the sequence of functions $\varphi_{k0}(x) = 2^{k/2} \varphi(2^{k}x)$ for $k \geq 0$, where φ is the Daubechies scaling function. Compare this family to the Fejér kernel (Lemma 14.4.1). Precisely, which properties of Fejér kernel (Theorem 14.4.3) carry over directly to the functions φ_{k0}? If a property does not carry over directly, is there an analogous property that holds?

H. The construction used in this section can be extended to wavelets of higher order (i.e., with more moments vanishing).

(a) Use equation (15.5.2) and $\int x^{2} \psi(x)\,dx = \int x^{3} \psi(x)\,dx = 0$ to derive two additional conditions on the sequence (a_{k}).

(b) Show that the values $a_{0} = 0.470467$, $a_{1} = 1.141117$, $a_{2} = 0.650365$, $a_{3} = -0.190934$, $a_{4} = -0.120832$, and $a_{5} = 0.049817$ give an approximate solution to these conditions.

(c) Using the appropriate scaling relation, plot these higher-order wavelets and scaling functions. Despite the graph, this scaling function is actually differentiable.

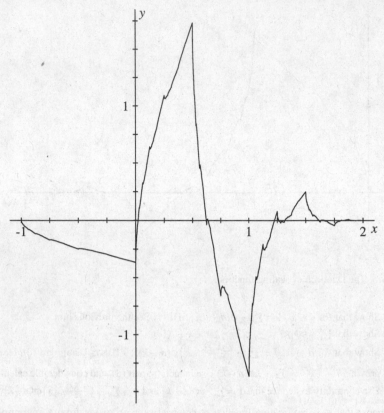

FIG. 15.4 The Daubechies wavelet.

15.6 Existence of the Daubechies Wavelet

The purpose of this section is to establish the following:

15.6.1. THEOREM. *The sequence of functions (φ_n) converges in the $L^2(\mathbb{R})$ norm to a function φ.*

We need two computational lemmas that enable us to estimate $\|\varphi_{n+1} - \varphi_n\|_2$. First, we compute $\langle \varphi_{n+1}(x), \varphi_n(x-k) \rangle$.

15.6.2. LEMMA. *Define $c_n(k) = \langle \varphi_{n+1}, \varphi_n(x-k) \rangle$ for $k \in \mathbb{Z}$ and $n \geq 0$. Then $c_n(k) = 0$ if $|k| > 2$, and the sequence of 5-tuples*

$$\mathbf{c}_n = \big(c_n(-2), c_n(-1), c_n(0), c_n(1), c_n(2)\big) \quad for \quad n \geq 0$$

satisfies $\mathbf{c}_0 = \Big(0, 0, \frac{2+\sqrt{3}}{4}, \frac{2-\sqrt{3}}{4}, 0\Big)$ and $\mathbf{c}_{n+1} = T\mathbf{c}_n$, where

$$T = \frac{1}{16} \begin{bmatrix} 0 & -1 & 0 & 0 & 0 \\ 16 & 9 & 0 & -1 & 0 \\ 0 & 9 & 16 & 9 & 0 \\ 0 & -1 & 0 & 9 & 16 \\ 0 & 0 & 0 & -1 & 0 \end{bmatrix}.$$

PROOF. Since $\varphi_0 = \chi_{[0,1)}$ and

$$\varphi_1 = \sum_{i=0}^{3} a_i \varphi_0(2x - i) = a_0 \chi_{[0,0.5)} + a_1 \chi_{[0.5,1)} + a_2 \chi_{[1,1.5)} + a_3 \chi_{[1.5,2)},$$

we easily compute $c_0(0) = \dfrac{a_0 + a_1}{2} = \dfrac{2 + \sqrt{3}}{4}$ and $c_0(1) = \dfrac{a_2 + a_3}{2} = \dfrac{2 - \sqrt{3}}{4}$ and $c_0(k) = 0$ in all other cases. Proceed by induction on n:

$$c_{n+1}(k) = \langle \varphi_{n+1}, \varphi_n(x-k) \rangle = \Big\langle \sum_{i=0}^{3} a_i \varphi_n(2x-i), \sum_{j=0}^{3} a_j \varphi_{n-1}(2x - 2k - j) \Big\rangle$$

$$= \sum_{i=0}^{3} \sum_{j=0}^{3} a_i a_j \langle \varphi_n(2x-i), \varphi_{n-1}(2x - 2k - j) \rangle.$$

Notice that making the substitution $y = 2x - i$ in the inner product results in a factor of $1/2$ from the change of variable of integration:

$$= \frac{1}{2} \sum_{i=0}^{3} \sum_{j=0}^{3} a_i a_j \langle \varphi_n(y), \varphi_{n-1}(y + i - 2k - j) \rangle.$$

Now set $l = j - i$ to obtain

$$= \frac{1}{2} \sum_{i=0}^{3} \sum_{l=-3}^{3} a_i a_{i+l} \langle \varphi_n(y), \varphi_{n-1}(y - 2k - l) \rangle$$

$$= \sum_{l=-3}^{3} \Big(\frac{1}{2} \sum_{i=0}^{3} a_i a_{i+l} \Big) c_{n-1}(2k + l).$$

Observe immediately that if $|k| \geq 3$ and $|l| \leq 3$, then $|2k + l| \geq 3$. Therefore, $c_{n+1}(k)$ for $|k| \geq 3$ depend linearly on $c_n(k)$ for $|k| \geq 3$. However, $c_0(k) = 0$ for $|k| \geq 3$, so $c_n(k) = 0$ for all $n \geq 0$ and $|k| \geq 3$. So we need only be concerned with the 5-tuple $c_n = (c_n(-2), c_n(-1), c_n(0), c_n(1), c_n(2))$.

A routine calculation yields

$$\frac{1}{2}\sum_{i=0}^{3} a_i a_{i+l} = \begin{cases} 1 & \text{when} \quad l = 0, \\ 9/16 & \text{when} \quad l = \pm 1, \\ -1/16 & \text{when} \quad l = \pm 3, \\ 0 & \text{otherwise.} \end{cases}$$

Plugging this into our identity and writing the five relations as a matrix, we obtain $\mathbf{c}_{n+1} = T\mathbf{c}_n$. ∎

Next we compute the Jordan form of T. Note immediately that $T\mathbf{e}_0 = \mathbf{e}_0$ is an obvious eigenvector.

15.6.3. LEMMA. *The matrix T factors as $T = VJV^{-1}$, where*

$$V = \begin{bmatrix} 0 & -1 & 2 & -1 & 4 \\ 0 & 8 & -4 & 4 & 0 \\ 1 & 0 & 0 & -6 & -8 \\ 0 & -8 & 4 & 4 & 0 \\ 0 & 1 & -2 & -1 & 4 \end{bmatrix} \quad and \quad J = \begin{bmatrix} 1 & 0 & 0 & 0 & 0 \\ 0 & 1/2 & 0 & 0 & 0 \\ 0 & 0 & 1/8 & 0 & 0 \\ 0 & 0 & 0 & 1/4 & 1 \\ 0 & 0 & 0 & 0 & 1/4 \end{bmatrix}.$$

PROOF. We leave it to the reader to show that $VJ = TV$ and to check that V is invertible (see the formula for V^{-1} given in the proof of Theorem 15.6.4). ∎

15.6.4. THEOREM. $\sum\limits_{n \geq 0} \|\varphi_{n+1} - \varphi_n\|_2 < \infty$ *and thus* $\lim\limits_{n \to \infty} \varphi_n$ *exists in* $L^2(\mathbb{R})$.

PROOF. The first step is to compute $c_n(0)$. Notice that $\mathbf{c}_n = T^n\mathbf{c}_0 = VJ^nV^{-1}\mathbf{c}_0$. We find that

$$J^n = \begin{bmatrix} 1 & 0 & 0 & 0 & 0 \\ 0 & 2^{-n} & 0 & 0 & 0 \\ 0 & 0 & 8^{-n} & 0 & 0 \\ 0 & 0 & 0 & 4^{-n} & 4^{1-n}n \\ 0 & 0 & 0 & 0 & 4^{-n} \end{bmatrix} \quad and \quad V^{-1} = \begin{bmatrix} 1 & 1 & 1 & 1 & 1 \\ \frac{1}{6} & \frac{1}{12} & 0 & -\frac{1}{12} & -\frac{1}{6} \\ \frac{1}{3} & \frac{1}{24} & 0 & -\frac{1}{24} & -\frac{1}{3} \\ 0 & \frac{1}{8} & 0 & \frac{1}{8} & 0 \\ \frac{1}{8} & \frac{1}{32} & 0 & \frac{1}{32} & \frac{1}{8} \end{bmatrix}.$$

Thus a straightforward multiplication yields

$$\langle \varphi_{n+1}, \varphi_n \rangle = c_n(0) = 1 - 4^{-n-2}(2 - \sqrt{3})(3n + 4).$$

Set $\varepsilon_n = 4^{-n-2}(2 - \sqrt{3})(3n + 4)$. Notice that

$$\|\varphi_{n+1} - \varphi_n\|^2 = \langle \varphi_{n+1} - \varphi_n, \varphi_{n+1} - \varphi_n \rangle$$
$$= \|\varphi_{n+1}\|_2^2 - 2\langle \varphi_{n+1}, \varphi_n \rangle + \|\varphi_n\|_2^2 = 2 - 2(1 - \varepsilon_n) = 2\varepsilon_n.$$

Consequently,

$$\sum_{n \geq 0} \|\varphi_{n+1} - \varphi_n\|_2 = \sum_{n \geq 0} \frac{\sqrt{2(2-\sqrt{3})(3n+4)}}{2^{n+2}} < \sum_{n \geq 0} \frac{n+2}{2^{n+2}} = \frac{3}{2}.$$

By Exercise 4.2.B, this implies that (φ_n) is a Cauchy sequence. Thus the limit function φ is defined in $L^2(\mathbb{R})$. ■

Exercises for Section 15.6

A. Verify that $VJ = TV$ and that the formula for V^{-1} is correct.

B. Do the calculation to compute $c_n(0)$ as indicated in Theorem 15.6.4.

C. Show that φ_n is supported on $[0,3]$ for $n \geq 0$.

D. Verify that $\sum_{n \geq 0} \frac{n+2}{2^{n+2}}$ converges. Do not compute its exact value.

E. Observe that the columns of T all sum to 1. Interpret this as saying that a related matrix has a certain eigenvector.

F. Find $\lim_{n \to \infty} T^n$.   HINT: Use the Jordan form.

15.7 Approximations Using Wavelets

In this section, we approximate functions using the Daubechies wavelet system. Our goal is to show how properties of the wavelet basis result in better (or worse) approximations. Given that the point of wavelets is to use different kinds of wavelets for different problems, it is worthwhile to see how to use properties of the wavelet and scaling function. We have already devoted Chapter 14 to approximation by Fourier series and Chapter 10 to approximation by polynomials, so we can compare approximation by wavelets to these alternatives.

Recall from the discussion of Haar wavelets that we may construct approximants to functions by first computing the projection $P_k f$ of f onto V_k. Let φ and $\{\psi_{kj}\}$ denote the Daubechies wavelet. Let us define the projections D_n onto $V_n = \text{span}\{\varphi_{n,j} : j \in \mathbb{Z}\}$ by

$$D_n f(x) = \sum_{j \in \mathbb{Z}} \langle f, \varphi_{n,j} \rangle \varphi_{n,j}.$$

For $n \geq 1$, we also realize this as

$$D_n f(x) = D_0 f(x) + \sum_{k=0}^{n-1} \sum_{j \in \mathbb{Z}} \langle f, \psi_{kj} \rangle \psi_{kj}(x).$$

If f has compact support, then only finitely many of these coefficients are nonzero at each level. For example, if the support is $[0,1]$, then there are at most three nonzero terms in the computation of D_0 and at most $2^{n-1} + 2$ additional terms to compute $D_n f$ knowing $D_{n-1} f$.

For any wavelet arising from a multiresolution analysis, the approximants $P_k f$ converge to f in $L^2(\mathbb{R})$. So, in particular, this is true for the Daubechies wavelets. In this case, we can also establish uniform convergence when f is uniformly continuous. We need a variation on the first part of the proof of Theorem 15.2.3.

15.7.1. LEMMA. *Consider the Daubechies wavelets. Fix $k \in \mathbb{N}$, $j \in \mathbb{Z}$, and $x \in [j/2^k, (j+3)/2^k]$. For any continuous function f on $\mathbb{R}$,*

$$\left| f(x) - 2^{k/2} \langle f, \varphi_{kj} \rangle \right| \le \sqrt{3}\, \omega(f, 3 \cdot 2^{-k}).$$

PROOF. Using the substitution $z = 2^k t - j$ and $\int \varphi(t)\, dt = 1$, we have

$$\left| f(x) - 2^{k/2} \langle f, \varphi_{k0} \rangle \right| = \left| f(x) - 2^k \int f(t) \varphi(2^k t - j)\, dx \right| = \left| f(x) - \int_0^3 f\!\left(\tfrac{z+j}{2^k}\right) \varphi(z)\, dz \right|$$

$$= \left| \int_0^3 \left[f(x) - f\!\left(\tfrac{z+j}{2^k}\right) \right] \varphi(z)\, dz \right| \le \omega(f; 3 \cdot 2^{-k}) \int_0^3 |\varphi(z)|\, dz,$$

since for $z \in [0,3]$, we have $|x - (z+j)/2^k| \le 3/2^k$. Finally, the Cauchy–Schwarz inequality shows that

$$\int_0^3 |\varphi(z)|\, dz \le \left(\int_0^3 |\varphi(z)|^2\, dz \right)^{1/2} \left(\int_0^3 1\, dz \right)^{1/2} = \sqrt{3}. \qquad \blacksquare$$

15.7.2. THEOREM. *If $f \in L^2(\mathbb{R})$ is uniformly continuous on $\mathbb{R}$, then the approximants $D_k f$ by Daubechies wavelets converge uniformly to f.*

PROOF. From Exercise 15.5.F, we have $\sum_{j \in \mathbb{Z}} \varphi(2^k x - j) = 1$. Multiplying this by $f(x)$, we have

$$|f(x) - D_k f(x)| = \left| \sum_{j \in \mathbb{Z}} f(x) \varphi(2^k x - j) - \sum_{j \in \mathbb{Z}} \langle f, \varphi_{kj} \rangle \varphi_{kj}(x) \right|$$

$$\le \sum_{j \in \mathbb{Z}} \left| f(x) - 2^{k/2} \langle f, \varphi_{kj} \rangle \right| \left| \varphi(2^k x - j) \right|$$

$$\le \sqrt{3}\, \omega(f, 3 \cdot 2^{-k}) \sum_{j \in \mathbb{Z}} |\varphi(2^k x - j)| \le 3\sqrt{3} \|\varphi\|_\infty \omega(f, 3 \cdot 2^{-k}),$$

because for any x, there are at most three j such that $\varphi(2^k x - j) \ne 0$. By the uniform continuity of f, $\lim_{k \to \infty} \omega(f, 3 \cdot 2^{-k}) = 0$. Thus $\lim_{k \to \infty} \|f - D_k f(x)\|_\infty = 0$. $\blacksquare$

15.7.3. REMARK. This proof does something even better because of the local nature of the Daubechies wavelets. If f is not continuous everywhere, but is continuous on a neighbourhood $[a - \delta, a + \delta]$, the same argument shows that the series converges uniformly on $[a - \varepsilon, a + \varepsilon]$ for $\varepsilon < \delta$. We will use this in Example 15.7.4.

Theorem 15.2.3 likewise shows that when f is continuous, the Haar wavelet approximants $H_n f$ converge to f uniformly. In fact, $\|H_n f - f\|_\infty \leq \omega(f; 2^{-n})$. So if f has Lipschitz constant L, the error is at most $2^{-n}L$. Since the number of coefficients in $H_n f$ doubles when n increases by 1, this is not surprising. We now have a comparable rate of convergence for Daubechies wavelets.

In approximating a function, a reasonable measure of the size of the approximant is the number of coefficients used. For $f \in L^2(0,1)$, we have

$$H_n f(x) = \langle f, \varphi \rangle \varphi(x) + \sum_{k=0}^{n-1} \sum_{j=0}^{2^k-1} \langle f, \psi_{kj} \rangle \psi_{kj}(x)$$

and so $H_n f$ uses 2^n coefficients. If there is a bound on the number of coefficients that we can use, due to storage limitations, for example, one choice is to use the largest value of n such that $H_n f$ does not have too many coefficients. It is frequently better to use a larger value of n and then replace small coefficients with zero. This has the advantage that if f has large irregularities at small resolution, these will appear in the approximation.

15.7.4. EXAMPLE. Consider the function given by

$$f(x) = \begin{cases} x & \text{if } x \in (-\pi, \pi), \\ 0 & \text{if } |x| \geq \pi. \end{cases}$$

In Section 14.3, we analyzed how the partial sums $S_n f$ of the Fourier series approximate f near the discontinuity at π. The Fourier series for f on $(-\pi, \pi)$ is

$$f(x) \sim 2 \sum_{k=1}^{\infty} \frac{(-1)^{k+1}}{k} \sin kx,$$

which converges very slowly. Even at a point well away from the discontinuity, convergence is slow. For example, if $x = \pi/2$, then we get

$$f(\pi/2) = 2 \sum_{k=0}^{\infty} \frac{(-1)^k}{2k+1}.$$

To get $2 \sum_{k=0}^{n} (-1)^k/(2k+1)$ within 10^{-6} of the exact sum of the series, we need $n \geq 500,000$. (To be fair, we can do better with Fejér kernels, but the behaviour is not optimal even then.)

The Haar wavelet approximations $H_n f$ are the step functions taking the average value of f over each interval of length 2^{-n}. It is not difficult to see that except for two intervals about the discontinuities $\pm \pi$, the convergence is uniform. The maximum error is 2^{-n-1}. Since the number of coefficients needed is roughly $2^{n+1}\pi$, we see that this convergence is not much more efficient than the Fourier series *globally*. However, to compute $f(\pi/2)$, note that only n terms in the expansion $H_n f$ are nonzero at $\pi/2$. Thus to compute this value within 10^{-6}, we only need 19 terms, since $2^{-20} < 10^{-6}$.

Even better, the vanishing moments of the Daubechies wavelets ensure that so long as the support of ψ_{kj} does not contain $\pm\pi$, the coefficient $\langle f, \psi_{kj} \rangle$ will be zero. Since ψ has support in $[0,3]$, each ψ_{kj} has support in an interval of length $3/2^k$, namely $[j/2^k, (j+3)/2^k]$. So for each k, there are only six nonzero coefficients for ψ_{kj} to contribute to the whole series.

Returning to the point $\pi/2$, notice that for the support of ψ_{kj} to contain both $\pi/2$ and π, we must have $3 \cdot 2^{-k} > \pi/2$ or $k \le 0$. Thus $D_0 f(\pi/2) = D_n f(\pi/2)$ for all $n \in \mathbb{N}$. Indeed by Remark 15.7.3, this series converges uniformly to f on any interval around $\pi/2$ that is bounded away from $\pm\pi$. In particular, we have $f(\pi/2) = \lim_{n\to\infty} D_n f(\pi/2) = D_0 f(\pi/2)$. So only three nonzero coefficients are involved in recovering this value exactly.

We can use the vanishing of the moments to obtain a better bound when the function f is C^2. Thus as for Fourier series, the Daubechies wavelet coefficients will die off quickly if the function is smooth. Moreover, because of the local nature of these wavelets, if f is smooth on some small interval, the same analysis shows that the wavelet series converges rapidly on that interval.

15.7.5. THEOREM. *If f is twice differentiable on $[(j-2)/2^k, (j+2)/2^k]$, where $j, k \in \mathbb{Z}$, and f'' bounded by B on this interval, then*

$$|\langle f, \psi_{kj} \rangle| \le \frac{4B}{2^{5k/2}}.$$

PROOF. Substituting $t = 2^k x - j$, we obtain

$$\langle f, \psi_{kj} \rangle = \int_{-\infty}^{+\infty} f(x) 2^{k/2} \psi(2^k x - j)\, dx = 2^{-k/2} \int_{-\infty}^{+\infty} f\left(\frac{t+j}{2^k}\right) \psi(t)\, dt.$$

This integral may be limited to $[-1, 2]$, the support of ψ.

On $[(j-2)/2^k, (j+2)/2^k]$, we have a Taylor series expansion for f centred at the point $b = j/2^k$, namely

$$f(x) = f(b) + f'(b)(x-b) + \frac{f''(c)}{2}(x-b)^2$$

for some point c between x and b. The vanishing moment conditions on ψ imply that $\int (mx + d)\psi(x)\, dx = 0$. Since $x - b = (t+j)/2^k - j/2^k = t/2^k$, we end up with

$$\langle f, \psi_{kj} \rangle = 2^{-k/2} \frac{1}{2} f''(c) \int_{-1}^{2} \left(\frac{t}{2^k}\right)^2 \psi(t)\, dt.$$

Since $t^2 \le 4$, we obtain

$$|\langle f, \psi_{kj} \rangle| \le \frac{2B}{2^{5k/2}} \int_{-1}^{2} |\psi(t)| \, dt.$$

Finally, the Cauchy–Schwarz inequality shows that

$$\int_{-1}^{2} |\psi(t)| \, dt \le \left(\int_{-1}^{2} |\psi(t)|^2 \, dt \right)^{1/2} \left(\int_{-1}^{2} 1 \, dt \right)^{1/2} = \sqrt{3} < 2. \qquad \blacksquare$$

Exercises for Section 15.7

A. Show that $\left\| b_0 \varphi + \sum_{k=0}^{n-1} \sum_{j=0}^{2^k-1} a_{kj} \psi_{kj} \right\|_\infty \le |b_0| + \sum_{k=0}^{n-1} 3 \cdot 2^{k/2} \max \{ |a_{kj}| : j \in \mathbb{Z} \}$.
   HINT: For each x and k, how many $\psi_{kj}(x) \ne 0$?

B. Recall the Cantor function f on $[0,1]$ from Example 5.7.8. Find the zero coefficients $\langle f, \psi_{kj} \rangle$ for $k = 1, 2, 3, 4$. What can you conclude about the functions $D_n f$?

C. (a) For the function f in Example 15.7.4, find the least $k \in \mathbb{N}$ such that $f(3) = D_k f(3)$.
   (b) In general, find a function $K(\delta)$ such that for $k \ge K(\delta)$, $f(\pi - \delta) = D_k f(\pi - \delta)$.

D. For the wavelets of Exercise 15.5.H, state and prove a version of Theorem 15.7.5.

E. We can represent $P_n f$ in two ways: $b_0 \varphi + \sum_{k=0}^{n-1} \sum_{j \in \mathbb{Z}} a_{kj} \psi_{kj}$ or $\sum_{j \in \mathbb{Z}} c_j \varphi_{nj}$.

   (a) For the Haar wavelets, what is the significance of the coefficients b_0 and a_{kj}?
   (b) For the Haar wavelets, describe how to obtain one set of coefficients from the other.
   HINT: Exercise 15.2.A.
   (c) For the Daubechies wavelets, describe how to obtain one set of coefficients from the other.

15.8 The Franklin Wavelet

It is not easy to just write down a wavelet or a scaling function. However, it is much easier to find a multiresolution with a scaling function that does not generate an orthonormal basis but does something a bit weaker. The goal is to construct a continuous piecewise linear wavelet by starting with such a system. The technique that we describe here can be adapted and refined to obtain wavelets with greater smoothness and/or with compact support. In this section, we restrict our attention to a single example known as the **Franklin wavelet**.

In these last sections, we need to take a more sophisticated view of linear maps between Hilbert spaces.

Consider the subspaces V_k of $L^2(\mathbb{R})$ consisting of continuous functions in $L^2(\mathbb{R})$ that are linear on each interval $[(j-1)2^{-k}, j2^{-k}]$. These subspaces satisfy most of the requirements of a multiresolution. It is immediately evident that V_k is contained in V_{k+1} for all $k \in \mathbb{Z}$. Also by definition, $f(x)$ belongs to V_k exactly when it is continuous and linear on each dyadic interval of length 2^{-k}, which holds precisely when $f(2x)$ is continuous and linear on each dyadic interval of length 2^{-k-1}, which means that $f(2x)$ belongs to V_{k+1}. So the V_k satisfy scaling.

The union of the V_k's is dense in all of $L^2(\mathbb{R})$. To see this, note that any continuous function g with bounded support, say contained in $[-2^N, 2^N]$, may be uniformly approximated to any desired accuracy by a piecewise linear continuous function f that is linear on dyadic intervals of length 2^{-k}, provided that k is sufficiently large. Given $\varepsilon > 0$, choose k and $f \in V_k$ so that f is also supported on $[-2^N, 2^N]$ and $\|g - f\|_\infty < 2^{-(N+1)/2}\varepsilon$. It is easy to see that

$$\|g - f\|_2^2 = \int_{-2^N}^{2^N} |g(x) - f(x)|^2\,dx \leq \int_{-2^N}^{2^N} \left(2^{-(N+1)/2}\varepsilon\right)^2 dx = \varepsilon^2.$$

Because $C_c(\mathbb{R})$ is dense in $L^2(\mathbb{R})$, it follows that we have the density property $\overline{\bigcup_{k \in \mathbb{Z}} V_k} = L^2(\mathbb{R})$.

Finally, we will demonstrate the separation property $\bigcap_{k \in \mathbb{Z}} V_k = \{0\}$. Suppose that f belongs to this intersection. Let $f(i) = a_i$ for $i = -1, 0, 1$. Since $f \in V_{-k}$ for each $k > 0$, it is linear on $[0, 2^k]$ and on $[-2^k, 0]$. So $f(x) = a_0 + (a_1 - a_0)x$ on $[0, 2^k]$ and $f(x) = a_0 - (a_{-1} - a_0)x$ on $[-2^k, 0]$. Thus

$$\|f\|_2^2 \geq \int_0^{2^k} |f(x)|^2\,dx = \tfrac{1}{3}(a_1 - a_0)^2 x^3 + a_0(a_1 - a_0)x^2 + a_0^2 x \Big|_0^{2^n}$$

$$= \tfrac{1}{3}(a_1 - a_0)^2 2^{3n} + a_0(a_1 - a_0)2^{2n} + a_0^2 2^n$$

$$= 2^n \left(\tfrac{1}{3}\left((a_1 - a_0)2^n + \tfrac{3}{2}a_0\right)^2 + \tfrac{1}{4}a_0^2\right).$$

As n tends to infinity, the right-hand side must remain bounded by $\|f\|_2^2$. This forces $a_0 = a_1 = 0$. Likewise, integration from -2^k to 0 shows that $a_{-1} = 0$. Consequently $f = 0$ on $[-2^k, 2^k]$ for every $k \geq 0$. So $f = 0$ and the intersection is trivial.

We have constructed a multiresolution except for the important scaling function. There is a function h, known as the **hat function**, which has all but one of the properties of a scaling function. It is supported on $[-1, 1]$, has $h(0) = 1$ and $h(-1) = h(1) = 0$, and is linear in between. Figure 15.5 gives its simple graph.

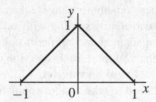

FIG. 15.5 The graph of the hat function h.

Notice that $f(x) = \sum_{j=-n}^{n} a_j h(x - j)$ is the piecewise linear function in V_0 supported on $[-n - 1, n + 1]$ that satisfies $f(j) = a_j$ for $-n \leq j \leq n$ and $f(j) = 0$ for $|j| > n$. So V_0 is spanned by translates of the hat function. Likewise, V_k is spanned by $\{2^{k/2}h(2^k x - j) : j \in \mathbb{Z}\}$. The problem with h is that these translates are not orthogonal to each other. But it does have a weaker property that serves as a substitute.

15.8.1. DEFINITION. A subset $\{\mathbf{x}_n : n \in \mathbb{Z}\}$ of a Hilbert space $\mathscr{H}$ is a **Riesz basis** if $\overline{\operatorname{span}}\{\mathbf{x}_n : n \in \mathbb{Z}\} = \mathscr{H}$ and there are constants $A > 0$ and $B < \infty$ such that

$$A\Big(\sum_n |a_n|^2\Big)^{1/2} \le \Big\|\sum_n a_n \mathbf{x}_n\Big\| \le B\Big(\sum_n |a_n|^2\Big)^{1/2}.$$

for all sequences (a_n) with only finitely many nonzero terms.

15.8.2. THEOREM. *The translates $\{h(x-j) : j \in \mathbb{Z}\}$ of the hat function form a Riesz basis for V_0.*

PROOF. Consider the inner product of two compactly supported functions in V_0, say $f(x) = \sum_{j=-n}^{n} a_j h(x-j)$ and $g(x) = \sum_{j=-n}^{n} b_j h(x-j)$. For convenience, set $a_j = b_j = 0$ for $|j| > n$:

$$\langle f, g \rangle = \int_{-\infty}^{+\infty} f(x)g(x)\,dx = \sum_{j=-n-1}^{n} \int_j^{j+1} f(x)g(x)\,dx$$

$$= \sum_{j=-n-1}^{n} \int_0^1 \big(a_j + (a_{j+1}-a_j)x\big)\big(b_j + (b_{j+1}-b_j)x\big)\,dx$$

$$= \sum_{j=-n-1}^{n} a_j b_j + \tfrac{1}{2}a_j(b_{j+1}-b_j) + \tfrac{1}{2}(a_{j+1}-a_j)b_j + \tfrac{1}{3}(a_{j+1}-a_j)(b_{j+1}-b_j)$$

$$= \frac{1}{6}\sum_{j=-n-1}^{n} 2a_j b_j + a_j b_{j+1} + a_j b_{j+1} + 2a_{j+1}b_{j+1}.$$

It is convenient to rearrange this sum further by moving the term $2a_{j+1}b_{j+1}$ to the next index to obtain

$$\langle f, g \rangle = \frac{1}{6}\sum_{j=-n-1}^{n} 4a_j b_j + a_j b_{j+1} + a_j b_{j+1}.$$

To make sense of this, we introduce two linear transformations.

Recall that $\ell^2(\mathbb{Z})$ is the Hilbert space of all square summable doubly indexed sequences $\mathbf{a} = (a_n)_{-\infty}^{+\infty}$, where we have

$$\langle \mathbf{a}, \mathbf{b} \rangle = \langle (a_n), (b_n) \rangle = \sum_{n=-\infty}^{+\infty} a_n b_n \quad \text{and} \quad \|\mathbf{a}\|_2 = \langle \mathbf{a}, \mathbf{a} \rangle^{1/2} = \Big(\sum_{n=-\infty}^{+\infty} a_n^2\Big)^{1/2}.$$

Let $\mathbf{e}_k$ for $k \in \mathbb{Z}$ denote the standard basis for $\ell^2(\mathbb{Z})$. Define the **bilateral shift** on $\ell^2(\mathbb{Z})$ by $U\mathbf{e}_k = \mathbf{e}_{k+1}$ or $(U\mathbf{a})_n = a_{n-1}$. It is easy to see that $\|U\mathbf{a}\|_2 = \|\mathbf{a}\|_2$ for all vectors $\mathbf{a} \in \ell^2(\mathbb{Z})$. Also, it is clear that U maps $\ell^2(\mathbb{Z})$ one-to-one and onto itself. Thus, U is a unitary map.

Recall from linear algebra that the adjoint (or transpose when working over the real numbers) of a linear transformation T is the linear map T^* such that

$$\langle T^*x,y\rangle = \langle x,Ty\rangle \quad \text{for all vectors } x,y \in \mathcal{H}.$$

For the unitary operator U, we have that $U^* = U^{-1}$ is the backward bilateral shift $U^*\mathbf{e}_k = \mathbf{e}_{k-1}$ or $(U^*\mathbf{a})_n = a_{n+1}$.

Second, we define a linear map H from $\ell^2(\mathbb{Z})$ onto V_0 by

$$H\mathbf{a} = \sum_{n=-\infty}^{+\infty} a_n h(x-n).$$

Looking back at our formula for the inner product in V_0, we see that

$$\langle H\mathbf{a}, H\mathbf{b}\rangle = \frac{1}{6}\sum_{j=-n-1}^{n} 4a_jb_j + a_jb_{j+1} + a_jb_{j+1} = \frac{1}{6}\langle(4I+U+U^*)\mathbf{a},\mathbf{b}\rangle. \quad (15.8.3)$$

By the Cauchy–Schwarz inequality (7.4.4), since $\|U\mathbf{a}\|_2 = \|\mathbf{a}\|_2 = \|U^*\mathbf{a}\|_2$,

$$|\langle U\mathbf{a},\mathbf{a}\rangle| = |\langle U^*\mathbf{a},\mathbf{a}\rangle| \leq \|\mathbf{a}\|_2^2.$$

Hence we obtain

$$\|f\|_2^2 = \langle H\mathbf{a}, H\mathbf{a}\rangle = \frac{1}{6}\langle(4I+U+U^*)\mathbf{a},\mathbf{a}\rangle$$
$$\leq \frac{1}{6}\left(4\|\mathbf{a}\|_2^2 + |\langle U\mathbf{a},\mathbf{a}\rangle| + |\langle U^*\mathbf{a},\mathbf{a}\rangle|\right) \leq \|\mathbf{a}\|_2^2.$$

Similarly, we obtain a lower bound

$$\|f\|_2^2 = \langle H\mathbf{a}, H\mathbf{a}\rangle = \frac{1}{6}\langle(4I+U+U^*)\mathbf{a},\mathbf{a}\rangle$$
$$\geq \frac{1}{6}\left(4\|\mathbf{a}\|_2^2 - |\langle U\mathbf{a},\mathbf{a}\rangle| - |\langle U^*\mathbf{a},\mathbf{a}\rangle|\right) \geq \frac{1}{3}\|\mathbf{a}\|_2^2.$$

So while the translates of h are not an orthonormal set, we find that they do form a Riesz basis for V_0.   ∎

Our problem now is to replace the hat function h by a scaling function φ. This function φ must have the property that the translates $\varphi(x-j)$ form an orthonormal set spanning V_0. We make use of another property relating the maps U and H. If $f(x) = H\mathbf{a} = \sum_n a_n h(x-n)$, then

$$HU\mathbf{a} = \sum_n a_{n-1}h(x-n) = \sum_n a_n h(x-n-1)$$
$$= f(x-1) = (Tf)(x) = TH\mathbf{a},$$

where $Tg(x) = g(x-1)$ is the translation operator. So $HU = TH$, namely H carries a translation by U in $\ell^2(\mathbb{Z})$ to translation by 1 in V_0. This suggests that if we can find a vector $\mathbf{c}$ such that the translates $U^n\mathbf{c}$ form an orthonormal basis with respect to the inner product

$$[\mathbf{a}, \mathbf{b}] = \frac{1}{6}\langle (4I + U + U^*)\mathbf{a}, \mathbf{b}\rangle,$$

then $\varphi = H\mathbf{c}$ will be the desired scaling function. Indeed, equation (15.8.3) becomes $\langle H\mathbf{a}, H\mathbf{b}\rangle = [\mathbf{a}, \mathbf{b}]$.

We make use of the correspondence between $\ell^2(\mathbb{Z})$ and $L^2(-\pi, \pi)$ provided by complex Fourier series. We identify the basis $\mathbf{e}_k$ with $e^{ik\theta}$, which is an orthonormal basis for $L^2(-\pi, \pi)$. This identifies the sequence $\mathbf{a}$ in $\ell^2(\mathbb{Z})$ with the function in $L^2(-\pi, \pi)$ given by $f(\theta) = \sum\limits_{n=-\infty}^{+\infty} a_n e^{in\theta}$. Now compute

$$Uf(\theta) = U\sum_n a_n e^{in\theta} = \sum_n a_n e^{i(n+1)\theta} = e^{i\theta} f(\theta).$$

Thus U is the operator that multiplies $f(\theta)$ by $e^{i\theta}$. We will write M_g to denote the operator on $L^2(-\pi, \pi)$ that multiplies by g. Such operators are called **multiplication operators**. For example, $U = M_{e^{i\theta}}$. Hence

$$\tfrac{1}{6}(4I + U + U^*) = \tfrac{1}{6}(4I + M_{e^{i\theta}} + M_{e^{i\theta}}^*) = M_{\frac{1}{6}(4 + e^{i\theta} + e^{-i\theta})} = M_{\frac{1}{3}(2 + \cos\theta)}.$$

15.8.4. THEOREM. *There is an ℓ^2 sequence (c_j) such that the function $\varphi(x) = \sum_j c_j h(x - j)$ is a scaling function for V_0.*

PROOF. Define the operator $X = M_g$, where $g(\theta) = \sqrt{3}(2 + \cos\theta)^{-1/2}$. Notice that $X = X^*$, $XU = UX$ and $X^2 = 3M_{2 + \cos\theta}^{-1} = 6(4I + U + U^*)^{-1}$. Now define $\mathbf{c} = X\mathbf{e}_0$ and $\varphi = H\mathbf{c}$. Compute

$$\begin{aligned}
\langle \varphi(x-j), \varphi(x-k)\rangle &= \langle T^j H\mathbf{c}, T^k H\mathbf{c}\rangle = \langle HU^j\mathbf{c}, HU^k\mathbf{c}\rangle \\
&= [U^j\mathbf{c}, U^k\mathbf{c}] = \frac{1}{6}\langle (4I + U + U^*)U^j\mathbf{c}, U^k\mathbf{c}\rangle \\
&= \langle X^{-2}U^j X\mathbf{e}_0, U^k X\mathbf{e}_0\rangle = \langle X^{-2}XU^j\mathbf{e}_0, XU^k\mathbf{e}_0\rangle \\
&= \langle XX^{-2}XU^j\mathbf{e}_0, U^k\mathbf{e}_0\rangle = \langle U^j\mathbf{e}_0, U^k\mathbf{e}_0\rangle = \delta_{jk}.
\end{aligned}$$

This shows that the translates of φ form an orthonormal set in the subspace V_0.

Since $\mathbf{e}_0$ is identified with the constant function 1,

$$\varphi(x) = HX1 = Hg.$$

To compute Hg, we need to find the (complex) Fourier series $g \sim \sum_n c_n e^{in\theta}$. Now g is an even function and thus $c_{-n} = c_n$; and so $g \sim c_0 + \sum\limits_{n=1}^{\infty} 2c_n \cos n\theta$. Moreover,

$$c_n = c_{-n} = \frac{\sqrt{3}}{2\pi} \int_{-\pi}^{\pi} \frac{\cos n\theta}{\sqrt{2 + \cos\theta}}\, d\theta \quad \text{for} \quad n \geq 0.$$

Hence

$$\varphi(x) = \sum_{n=-\infty}^{+\infty} c_n h(x-n). \tag{15.8.5}$$

This is the continuous piecewise linear function with nodes at the integers taking the values $\varphi(n) = c_n$.

A similar argument shows that the translates of φ span all of V_0. Indeed, note that

$$He^{ij\theta}g = HU^j g = T^j Hg = \varphi(x-j).$$

Now $h = He_0 = HXg^{-1}$. Express $g^{-1}(x) = \sqrt{(2+\cos\theta)/3}$, which is continuous, as a complex Fourier series $g^{-1} \sim \sum_j b_j e^{ij\theta}$. Then

$$h = HXg^{-1} = H\sum_j b_j e^{ij\theta} g = \sum_j b_j \varphi(x-j). \tag{15.8.6}$$

This expresses h as an ℓ^2 combination of the orthonormal basis of translates of φ, and thus h lies in their span. Evidently, this span also contains all translates of h, and so they span all of V_0. Therefore, φ is the desired scaling function. ∎

Using two formulas from the previous proof, we can write the scaling relation for φ, in terms of the sequences (b_n) and (c_n). Verify that the hat function satisfies the simple scaling relation.

$$h(x) = \tfrac{1}{2}h(2x-1) + h(2x) + \tfrac{1}{2}h(2x+1).$$

Equation (15.8.6) gives

$$h(2x) = \sum_j b_j \varphi(2x-j)$$

and similar formulas hold for $h(2x-1)$ and $h(2x+1)$. Putting these formulas into the scaling relations gives $h(x)$ as an infinite series involving $\varphi(2x-k)$ as k ranges over the integers. Substituting this series for h in equation (15.8.5), we obtain

$$\varphi(x) = \sum_{l\in\mathbb{Z}} \left(\sum_{j\in\mathbb{Z}} c_j \left[b_{2j-l} + \tfrac{1}{2}b_{2j-l+1} + \tfrac{1}{2}b_{2j-l-1} \right] \right) \varphi(2x-l).$$

This formula does not appear tractable, but the sequences (c_n) and (b_n) decay quite rapidly, so it is possible to obtain reasonable numerical results by taking sums over relatively small ranges of j, say -10 to 10.

We can then apply Theorem 15.4.2 to obtain a formula for the Franklin wavelet itself. This is plotted in Figure 15.6, along with the scaling relation. Notice that the wavelet is continuous and piecewise linear with nodes at the half-integers, as we would expect, since Theorem 15.4.2 implies that the wavelet is in V_1. Incidentally, the numerical values of the first few a_n in the scaling relation for φ are

$$a_0 = 1.15633, \quad a_1 = a_{-1} = 0.56186, \quad a_2 = a_{-2} = -0.09772,$$
$$a_3 = a_{-3} = -0.07346 \quad \text{and} \quad a_4 = a_{-4} = -0.02400.$$

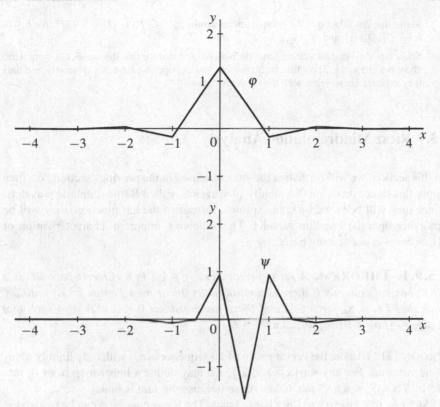

FIG. 15.6 The Franklin scaling function and wavelet.

Exercises for Section 15.8

These exercises are all directed toward the analysis of a different wavelet, the **Strömberg wavelet**, which has the same multiresolution subspaces V_k as the Franklin wavelet. See [32] for more details. Let $PL(X)$ denote the space of $L^2(\mathbb{R})$ functions that are continuous and piecewise linear with nodes on a discrete subset X of $\mathbb{R}$. As usual, h is the hat function.

A. Show that $PL(-\frac{1}{2}\mathbb{N}_0 \cup \mathbb{N})$ is spanned by $\{h(2x+k+1), h(x-k) : k \geq 0\}$.

B. Show that $PL(-\frac{1}{2}\mathbb{N}_0 \cup \{\frac{1}{2}\} \cup \mathbb{N})$ is spanned by $PL(-\frac{1}{2}\mathbb{N}_0 \cup \mathbb{N})$ and $h(2x)$. Hence show that there is a norm-1 function ψ in $PL(-\frac{1}{2}\mathbb{N}_0 \cup \{\frac{1}{2}\} \cup \mathbb{N})$ that is orthogonal to $PL(-\frac{1}{2}\mathbb{N}_0 \cup \mathbb{N})$.

C. Show that ψ is orthogonal to $2^{-k/2}\psi(2^{-k}x - j)$ for all $k < 0$ and $j \in \mathbb{Z}$ and to $\psi(x+j)$ for $j > 0$. Hence deduce that $\{\psi_{kj} : k, j \in \mathbb{Z}\}$ is orthonormal.
    HINT: Some are in $PL(-\frac{1}{2}\mathbb{N}_0 \cup \mathbb{N})$. Do a change of variables for the rest.

D. Let $V_k = PL(2^{-k}\mathbb{Z})$ and let W_k be the orthogonal complement of V_k in V_{k+1}. Show that $\mathrm{span}\{\psi_{kj} : j \in \mathbb{Z}\} = W_k$. Deduce that ψ is a wavelet.   HINT: Show that the orthogonal complement of $PL(\mathbb{Z} \cup \{k/2 : k \leq -2n-1\})$ in $PL(\mathbb{Z} \cup \{k/2 : k \leq 2n+1\})$ is $\mathrm{span}\{\psi_{0j} : |j| \leq n\}$.

E. ψ is determined by its values at the nodes, $\psi(\frac{k}{2}) = a_k$ for $k \leq 0$, $\psi(\frac{1}{2}) = b$, and $\psi(k) = c_k$ for $k \geq 1$. Show that the orthogonality relations coming from the fact that ψ is orthogonal to the basis of $PL(-\frac{1}{2}\mathbb{N}_0 \cup \mathbb{N})$ yield equations $a_{k-1} + a_k + a_{k+1} = 0$ for $k \leq -1$, and for $k \geq 2$, $c_{k-1} + c_k + c_{k+1} = 0$. Plus $a_{-2} + 6a_{-1} + 10a_0 + 6b + c_1 = 0$ and $a_0 + 6b + 13c_1 + 4c_2 = 0$.

Verify that the solution is the one-parameter family $a_k = C(2\sqrt{3}-2)(\sqrt{3}-2)^{|k|}$ for $k \leq 0$, $b = -C(\sqrt{3}+\frac{1}{2})$, and $c_k = C(\sqrt{3}-2)^{k-1}$ for $k \geq 1$.

F. Show that the Franklin scaling function is even, and deduce that the wavelet is symmetric about the line $x = \frac{1}{2}$. Show that the Strömberg wavelet does not have this symmetry, and thus they are different wavelets with the same resolution.

15.9 Riesz Multiresolution Analysis

In this section, we will formalize the structure used in the previous section. We then apply this to construct another family of wavelets, called Battle–Lemarié wavelets. Since they will be based on cubic splines, instead of the hat function, they will be smoother than the Franklin wavelet. The following important characterization of Riesz bases is our starting point.

15.9.1. THEOREM. *A set of vectors* $\{x_n : n \in \mathbb{Z}\}$ *in a Hilbert space* $\mathscr{H}$ *is a Riesz basis if and only if there is a continuous linear map* T *from* $\ell^2(\mathbb{Z})$ *onto* $\mathscr{H}$ *such that* $Te_n = x_n$ *for* $n \in \mathbb{Z}$ *and there are constants* $0 < A < B < \infty$ *such that* $A\|a\|_2 \leq \|Ta\| \leq B\|a\|_2$ *for all* $a \in \ell^2(\mathbb{Z})$.

PROOF. Let ℓ_0 denote the vector space of all sequences (a_n) with only finitely many nonzero terms. For any set $\{x_n : n \in \mathbb{Z}\}$, we may define a linear map from ℓ_0 into $\mathscr{H}$ by $Ta = \sum_n a_n x_n$, which makes sense because the sum is finite.

Suppose that $\{x_n : n \in \mathbb{Z}\}$ is a Riesz basis. The Riesz condition can be restated as

$$A\|a\|_2 \leq \|Ta\| \leq B\|a\|_2 \quad \text{for all} \quad a \in \ell_0.$$

In particular, the map T satisfies the Lipschitz condition

$$\|Ta - Tb\| = \|T(a-b)\| \leq B\|a-b\|_2$$

and thus T is uniformly continuous.

Suppose that $a \in \ell^2(\mathbb{Z})$. Then we may choose a sequence a_n in ℓ_0 that converges to a in the ℓ^2 norm. Consequently, (a_n) is a Cauchy sequence in $\ell^2(\mathbb{Z})$. Therefore, since $\|Ta_n - Ta_m\| \leq B\|a_n - a_m\|_2$, it follows that (Ta_n) is a Cauchy sequence in $\mathscr{H}$. Since $\mathscr{H}$ is complete, we may define $Ta = \lim_n Ta_n$. See the exercises for the argument explaining why this definition does not depend on the choice of the sequence. So the definition of T has been extended to all of $\ell^2(\mathbb{Z})$. Moreover, we obtain that

$$\|Ta\| = \lim_{n \to \infty} \|Ta_n\| \leq B \lim_{n \to \infty} \|a_n\|_2 = B\|a\|_2.$$

So T is (uniformly) continuous on all of $\ell^2(\mathbb{Z})$.

We similarly obtain

$$\|Ta\| = \lim_{n \to \infty} \|Ta_n\| \geq A \lim_{n \to \infty} \|a_n\|_2 = A\|a\|_2.$$

Clearly T maps ℓ_0 onto the set of all finite linear combinations of $\{\mathbf{x}_n : n \in \mathbb{Z}\}$. So the range of T is dense in $\mathscr{H}$ by hypothesis.

Let $\mathbf{y} \in \mathscr{H}$. Choose vectors $\mathbf{y}_n \in \mathrm{span}\{\mathbf{x}_n : n \in \mathbb{Z}\}$ that converge to $\mathbf{y}$. Then $(\mathbf{y}_n)$ is a Cauchy sequence. Since $\mathbf{y}_n$ belongs to the range of T, there are vectors $\mathbf{a}_n \in \ell_0$ with $T\mathbf{a}_n = \mathbf{y}_n$. Therefore,

$$\|\mathbf{a}_n - \mathbf{a}_m\|_2 \le A^{-1}\|T(\mathbf{a}_n - \mathbf{a}_m)\| = A^{-1}\|\mathbf{y}_n - \mathbf{y}_m\|.$$

Consequently, $(\mathbf{a}_n)$ is Cauchy. Since $\ell^2(\mathbb{Z})$ is complete by Theorem 7.7.4, we obtain a vector $\mathbf{a} = \lim_n \mathbf{a}_n$. The continuity of T now ensures that

$$T\mathbf{a} = \lim_{n \to \infty} T\mathbf{a}_n = \lim_{n \to \infty} \mathbf{y}_n = \mathbf{y}.$$

So T maps $\ell^2(\mathbb{Z})$ onto $\mathscr{H}$.

Conversely, if the operator T exists, then the Riesz norm condition holds (by restricting T to ℓ_0). Because T is continuous and ℓ_0 is dense in $\ell^2(\mathbb{Z})$, it follows that $\mathrm{span}\{\mathbf{x}_n : n \in \mathbb{Z}\} = T\ell_0$ is dense in $T\ell^2(\mathbb{Z}) = \mathscr{H}$; i.e., $\overline{\mathrm{span}\{\mathbf{x}_n : n \in \mathbb{Z}\}} = \mathscr{H}$. ∎

15.9.2. COROLLARY. *If $\{\mathbf{x}_n : n \in \mathbb{Z}\}$ is a Riesz basis for a Hilbert space $\mathscr{H}$, then every vector $\mathbf{y} \in \mathscr{H}$ may be expressed in a unique way as $\mathbf{y} = \sum_n a_n \mathbf{x}_n$ for some $\mathbf{a} = (a_n)$ in $\ell^2(\mathbb{Z})$.*

PROOF. The existence of $\mathbf{a} \subset \ell^2(\mathbb{Z})$ such that $T\mathbf{a} = \mathbf{y}$ follows from Theorem 15.9.1. Suppose that $T\mathbf{b} = \mathbf{y}$ as well. Then $T(\mathbf{a} - \mathbf{b}) = 0$. But $0 = \|T(\mathbf{a} - \mathbf{b})\| \ge A\|\mathbf{a} - \mathbf{b}\|_2$. Hence $\mathbf{b} = \mathbf{a}$. So T is one-to-one and $\mathbf{a}$ is uniquely determined. ∎

15.9.3. REMARK. Note that the norm estimates show that the linear map T has a continuous inverse. Indeed, Corollary 15.9.2 shows that $T^{-1}\mathbf{y} = \mathbf{a}$ is well defined. It is easy to show that the inverse of a linear map is linear. Now Theorem 15.9.1 shows that $A\|\mathbf{a}\|_2 \le \|T\mathbf{a}\|$. Substituting $\mathbf{y} = T\mathbf{a}$, we obtain $\|T^{-1}\mathbf{y}\|_2 \le A^{-1}\|\mathbf{y}\|$ for all $\mathbf{y} \in \mathscr{H}$. This shows that T^{-1} is Lipschitz and hence uniformly continuous.

In fact, a linear map is continuous if and only if it is **bounded**, meaning that $\|T\| = \sup\{\|Tx\| : \|x\| = 1\}$ is finite. (See Exercise 15.9.D.) A basic theorem of functional analysis known as **Banach's Isomorphism Theorem** states that a continuous linear map between complete normed spaces that is one-to-one and onto is invertible. Consequently, the existence of the constants A and B required in Theorem 15.9.1 are automatic if we can verify that T is a bijection.

Now we specialize these ideas to a subspace V_0 of $L^2(\mathbb{R})$ spanned by the translates of a single function h. To obtain a nice condition, we need to know some easy facts about multiplication operators. Let $g \in C[-\pi, \pi]$. Recall that the linear map M_g on $L^2(-\pi, \pi)$ is given by $M_g f(\theta) = g(\theta)f(\theta)$.

15.9.4. PROPOSITION. *Suppose that the complex Fourier series of g is given by $g \sim \sum_k t_k e^{ik\theta}$. Then the matrix $[a_{ij}]$ of M_g with respect to the orthonormal basis $\{e^{ik\theta} : k \in \mathbb{Z}\}$ for $L^2(-\pi, \pi)$ is given by $a_{jk} = t_{j-k}$.*

PROOF. This is an easy computation. Indeed,

$$a_{jk} = \langle M_g e^{ik\theta}, e^{ij\theta} \rangle = \frac{1}{2\pi} \int_{-\pi}^{\pi} g(\theta) e^{ik\theta} \overline{e^{ij\theta}} \, d\theta$$

$$= \frac{1}{2\pi} \int_{-\pi}^{\pi} g(\theta) \overline{e^{i(j-k)\theta}} \, d\theta = t_{j-k}. \qquad \blacksquare$$

We also need these norm estimates for M_g.

15.9.5. THEOREM. *If $g \in C[-\pi, \pi]$, then M_g is a continuous linear map on $L^2(-\pi, \pi)$ such that $\|M_g f\|_2 \le \|g\|_\infty \|f\|_2$. Moreover, $\|g\|_\infty$ is the smallest constant B such that $\|M_g f\|_2 \le B\|f\|_2$ for all f.*

Similarly, if $A = \inf\{|g(\theta)| : \theta \in [-\pi, \pi]\}$, then A is the largest constant such that $\|M_g f\|_2 \ge A\|f\|_2$ for all $f \in L^2(-\pi, \pi)$.

PROOF. A straightforward calculation shows that

$$\|M_g f\|_2^2 = \frac{1}{2\pi} \int_{-\pi}^{\pi} |g(\theta)|^2 |f(\theta)|^2 \, d\theta \le \frac{1}{2\pi} \int_{-\pi}^{\pi} \|g\|_\infty^2 |f(\theta)|^2 \, d\theta$$

$$= \|g\|_\infty^2 \frac{1}{2\pi} \int_{-\pi}^{\pi} |f(\theta)|^2 \, d\theta = \|g\|_\infty^2 \|f\|_2^2.$$

In particular, M_g is Lipschitz and hence continuous. Similarly,

$$\|M_g f\|_2^2 = \frac{1}{2\pi} \int_{-\pi}^{\pi} |g(\theta)|^2 |f(\theta)|^2 \, d\theta \ge \frac{1}{2\pi} \int_{-\pi}^{\pi} A^2 |f(\theta)|^2 \, d\theta = A^2 \|f\|_2^2.$$

On the other hand, suppose that $B < C < \|g\|_\infty$. Then there is a nonempty open interval (a, b) on which $|g(\theta)| > C$. Let f be a continuous function on $[-\pi, \pi]$ such that $\|f\|_2 = 1$ and the support of f is contained in $[a, b]$. Then

$$\|M_g f\|_2^2 = \frac{1}{2\pi} \int_{-\pi}^{\pi} |g(\theta)|^2 |f(\theta)|^2 \, d\theta = \frac{1}{2\pi} \int_a^b |g(\theta)|^2 |f(\theta)|^2 \, d\theta$$

$$\ge \frac{1}{2\pi} \int_a^b C^2 |f(\theta)|^2 \, d\theta = C^2 \frac{1}{2\pi} \int_{-\pi}^{\pi} |f(\theta)|^2 \, d\theta > B^2 \|f\|_2^2.$$

Therefore, $\|M_g\| > B$. So $\|g\|_\infty$ is the optimal choice.

Likewise, if $D > C > A$, there is a continuous function f with $\|f\|_2 = 1$ supported on an interval (c, d) on which $|g(\theta)| < C$. The same calculation above shows that $\|M_g f\| \le C\|f\|_2 < D\|f\|_2$. Therefore, A is the optimal lower bound. $\qquad \blacksquare$

We are now ready to obtain a practical characterization of when the translates of h form a Riesz basis. For the following results, we will make the simplifying assumption that the coefficients arising are the Fourier coefficients of a continuous function. The Riesz basis property requires that they be in $\ell^2(\mathbb{Z})$, which shows that the function is always in $L^2(-\pi, \pi)$. Theorem 15.9.5 is actually valid in this generality. However, it is often continuous in applications.

15.9.6. THEOREM. *Let $h \in L^2(\mathbb{R})$ and $V_0 = \overline{\text{span}}\{h(x - j) : j \in \mathbb{Z}\}$. Define $t_j = \langle h(x), h(x - j) \rangle$ for $j \in \mathbb{Z}$. Assume that there is a continuous function g with Fourier series $t_0 + \sum_{j=1}^{\infty} 2t_j \cos j\theta$. Then $\{h(x - j) : j \in \mathbb{Z}\}$ is a Riesz basis for V_0 if and only if there are constants $0 < A \leq B$ such that $A^2 \leq g(\theta) \leq B^2$ on $[-\pi, \pi]$.*

PROOF. Let $T\mathbf{a} = \sum_n a_n h(x - n)$ for all $\mathbf{a} \in \ell_0$. It follows from Theorem 15.9.1 that $\{h(x - j) : j \in \mathbb{Z}\}$ is a Riesz basis for V_0 if and only if there are constants $0 < A^2 \leq B^2 < \infty$ such that

$$A^2 \|\mathbf{a}\|_2^2 \leq \|T\mathbf{a}\|^2 = \left\| \sum_n a_n h(x - n) \right\|_2^2 \leq B^2 \|\mathbf{a}\|_2^2.$$

Now

$$\|T\mathbf{a}\|^2 = \langle T\mathbf{a}, T\mathbf{a} \rangle = \langle T^*T\mathbf{a}, \mathbf{a} \rangle.$$

The linear map T^*T has matrix $[t_{ij}]$ with respect to the orthonormal basis $\mathbf{e}_n$, where

$$t_{ij} = \langle T^*T\mathbf{e}_j, \mathbf{e}_i \rangle = \langle T\mathbf{e}_j, T\mathbf{e}_i \rangle = \langle h(x - j), h(x - i) \rangle = \langle h(x), h(x - i + j) \rangle = t_{i-j}.$$

So the matrix of T^*T is constant on diagonals. Note that by symmetry, $t_{-k} = t_k$.

All orthonormal bases are created equal, so we can identify $\mathbf{e}_n$ with $e^{in\theta}$ in $L^2(-\pi, \pi)$, so that ℓ_0 corresponds to all finite *complex* Fourier series. With this identification, we see from Proposition 15.9.4 that $T^*T = M_g$, where

$$g(\theta) = \sum_k t_k e^{ik\theta} = t_0 I + \sum_{k=1}^{\infty} 2t_k \cos k\theta.$$

We used $t_{-k} = t_k$ to obtain a real function g, which returns us to the real domain from this brief foray into complex vector spaces.

Next we observe that $g(\theta) \geq 0$. Indeed, we have

$$\frac{1}{2\pi} \int_{-\pi}^{\pi} g(\theta) |f(\theta)|^2 \, d\theta = \langle T^*Tf, f \rangle = \|Tf\|_2^2 \geq 0.$$

Suppose that g were not positive. Then by the continuity of g, we may choose an interval (a, b) on which $g(\theta) < -\varepsilon < 0$. Then as in the proof of Theorem 15.9.5, we deduce that for any continuous function f supported on $[a, b]$ we have

$$\frac{1}{2\pi} \int_{-\pi}^{\pi} g(\theta) |f(\theta)|^2 \, d\theta \leq -\varepsilon \|f\|_2^2 < 0,$$

which contradicts the previous inequality. So g is positive.

This allows us to define the multiplication operator $M_{\sqrt{g}}$. Since $\sqrt{g} \geq 0$, we find that $M_{\sqrt{g}}^* = M_{\sqrt{g}}$, and

$$T^*T = M_g = M_{\sqrt{g}}^2 = M_{\sqrt{g}}^* M_{\sqrt{g}}.$$

(Be warned that this does *not* show that T is equal to $M_{\sqrt{g}}$. They do not even map into the same Hilbert space.) Consequently,

$$\|Tf\|_2^2 = \langle T^*Tf, f \rangle = \langle M_g f, f \rangle = \langle M_{\sqrt{g}} f, M_{\sqrt{g}} f \rangle = \|M_{\sqrt{g}} f\|_2^2.$$

Finally, an application of Theorem 15.9.5 shows that

$$A^2 \|f\|_2^2 \leq \|M_{\sqrt{g}} f\|_2^2 \leq \|\sqrt{g}\|_\infty^2 \|f\|_2^2 = \|g\|_\infty \|f\|_2^2$$

for all $f \in L^2(-\pi, \pi)$ if and only if

$$A^2 \leq \inf_{\theta \in [-\pi, \pi]} |\sqrt{g}(\theta)|^2 = \inf_{\theta \in [-\pi, \pi]} |g(\theta)|.$$

Thus $A^2 > 0$ is possible only if g is bounded away from 0. ∎

We are now ready to modify a Riesz basis of translations of h to obtain an orthonormal basis of translates. This is the key to finding wavelets by the machinery we have already developed.

15.9.7. THEOREM. *Let $h \in L^2(\mathbb{R})$ be a function such that the set of translates $\{h(x - j) : j \in \mathbb{Z}\}$ is a Riesz basis for its span V_0. Assume that $t_0 + \sum_{j=1}^\infty 2t_j \cos j\theta$ is the Fourier series of a continuous function, where $t_k = \langle h(x), h(x - k) \rangle$. Then there is a function $\varphi \in L^2(\mathbb{R})$ such that $\{\varphi(x - j) : j \in \mathbb{Z}\}$ is an orthonormal basis for V_0.*

PROOF. By Theorem 15.9.1, there is a continuous, invertible linear map T from $\ell^2(\mathbb{Z})$ onto V_0 given by $Ta = \sum_n a_n h(x - n)$. By Theorem 15.9.6, the (continuous) function g with Fourier series $t_0 + \sum_{j=1}^\infty 2t_j \cos j\theta$ satisfies $T^*T = M_g$ and there are constants $0 < A^2 \leq B^2 < \infty$ such that

$$A^2 \leq g(\theta) \leq B^2 \quad \text{for all} \quad -\pi \leq \theta \leq \pi.$$

In particular, $g^{-1/2}$ is bounded above by $1/A$.

Now compute the Fourier series of $g^{-1/2}$, say

$$g^{-1/2} \sim \sum_n c_n e^{-in\theta} = c_0 I + \sum_{k=1}^\infty 2c_k \cos k\theta,$$

where

$$c_k = c_{-k} = \frac{1}{2\pi} \int_{-\pi}^\pi \frac{\cos n\theta}{\sqrt{g(\theta)}} \, d\theta.$$

We claim that the orthogonal generator is obtained by the formula

$$\varphi(x) = Tg^{-1/2} = \sum_{n=-\infty}^{\infty} c_n h(x-n).$$

Indeed,

$$\begin{aligned}
\langle \varphi(x-j), \varphi(x-k) \rangle &= \langle Te^{-ij\theta}g^{-1/2}(\theta), Te^{-ik\theta}g^{-1/2}(\theta) \rangle \\
&= \langle T^*Te^{-ij\theta}g^{-1/2}(\theta), e^{-ik\theta}g^{-1/2}(\theta) \rangle \\
&= \langle M_g M_g^{-1/2}e^{-ij\theta}, M_g^{-1/2}e^{-ik\theta} \rangle \\
&= \langle M_g^{-1/2}M_g M_g^{-1/2}e^{-ij\theta}, e^{-ik\theta} \rangle \\
&= \langle e^{-ij\theta}, e^{-ik\theta} \rangle = \delta_{ij}.
\end{aligned}$$

So $\{\varphi(x-k) : k \in \mathbb{Z}\}$ is orthonormal.

It is clear that each $\varphi(x-k)$ belongs to V_0, since they are in the span of the $h(x-j)$'s. Conversely, observe that $\varphi(x-k) = TM_g^{-1/2}e^{ik\theta}$. Thus

$$\text{span}\{\varphi(x-k) : k \subset \mathbb{Z}\} = TM_g^{-1/2}L^2(\mathbb{T}) = TL^2(\mathbb{T}) = V_0$$

because M_g is invertible and so $M_g^{-1/2}L^2(\mathbb{T}) = L^2(\mathbb{T})$, and T maps $L^2(\mathbb{T})$ onto V_0 by Corollary 15.9.2. So $\{\varphi(x-k) : k \in \mathbb{Z}\}$ is an orthonormal basis for V_0. ∎

The second notion that arose in our construction of a continuous wavelet was a weaker notion of a multiresolution using Riesz bases.

15.9.8. DEFINITION. A **Riesz multiresolution** of $L^2(\mathbb{R})$ with scaling function h is the sequence of subspaces $V_j = \text{span}\{h(2^k x - j) : j \in \mathbb{Z}\}$ provided that the sequence satisfies the following properties:

(1) Riesz basis: $\{h(x-j) : j \in \mathbb{Z}\}$ is a Riesz basis for V_0.
(2) nesting: $V_k \subset V_{k+1}$ for all $k \in \mathbb{Z}$.
(3) scaling: $f(x) \in V_k$ if and only if $f(2x) \in V_{k+1}$.
(4) density: $\overline{\bigcup_{k \in \mathbb{Z}} V_k} = L^2(\mathbb{R})$.
(5) separation: $\bigcap_{k \in \mathbb{Z}} V_k = \{0\}$.

The main result is now a matter of collecting together the previous theorems.

15.9.9. THEOREM. *Suppose that h is the scaling function for a Riesz multiresolution $V_j = \text{span}\{h(2^k x - j) : j \in \mathbb{Z}\}$. Assume that there is a continuous function with Fourier series $\|h\|_2^2 + \sum_{j=1}^{\infty} 2\langle h(x), h(n-k) \rangle \cos j\theta$. Then there exists a scaling function φ generating the same nested sequence of subspaces. Consequently, there is a wavelet basis for $L^2(\mathbb{R})$ compatible with this decomposition.*

PROOF. Theorem 15.9.7 provides an orthogonal scaling function for this resolution. Then Theorem 15.4.2 provides an algorithm for constructing the corresponding wavelet. ∎

15.9.10. EXAMPLE. We finish this section by describing some smoother examples of wavelets, known as **Battle–Lemarié wavelets** or **B-spline wavelets**. Let $N_0 = \chi_{[0,1)}$. For each $n \geq 0$, define

$$N_n(x) = N_{n-1} * N_0(x) = \int_{x-1}^{x} N_{n-1}(t)\,dt.$$

For example, $N_1(x) = h(x-1)$ is a translate of the hat function,

$$N_2(x) = \begin{cases} \frac{1}{2}x^2 & \text{for } 0 \leq x \leq 1, \\ \frac{3}{4} - (x - \frac{3}{2})^2 & \text{for } 1 \leq x \leq 2, \\ \frac{1}{2}(3-x)^2 & \text{for } 2 \leq x \leq 3, \end{cases}$$

and

$$N_3(x) = \begin{cases} \frac{1}{6}x^3 & \text{for } 0 \leq x \leq 1, \\ \frac{1}{6}x^3 - \frac{2}{3}(x-1)^3 & \text{for } 1 \leq x \leq 2, \\ \frac{1}{6}(4-x)^3 - \frac{2}{3}(3-x)^3 & \text{for } 2 \leq x \leq 3, \\ \frac{1}{6}(4-x)^3 & \text{for } 3 \leq x \leq 4. \end{cases}$$

Figure 15.7 gives the graphs of N_2 and N_3. Notice that $N_3(x)$ is a cubic spline. That is, N_3 is C^2, has compact support, and on each interval $[k, k+1]$ it is represented by a cubic polynomial. See the exercises for hints on establishing similar properties for general n.

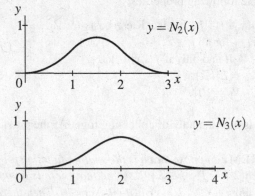

FIG. 15.7 The graphs of N_2 and N_3.

Let $S_k^{(n)}$ denote the subspace of $L^2(\mathbb{R})$ consisting of **splines of order** n with nodes at the points $2^{-k}\mathbb{Z}$. These are the L^2 functions that have $n-1$ continuous derivatives such that the restriction to each dyadic interval $[2^{-k}j, 2^{-k}(j+1)]$ is a polynomial of degree n. Clearly for each fixed n, the sequence

$$\cdots \subset S_{-2}^{(n)} \subset S_{-1}^{(n)} \subset S_0^{(n)} \subset S_1^{(n)} \subset S_2^{(n)} \subset \cdots$$

is nested and satisfies the scaling property.

In fact, for each n, this forms a Riesz multiresolution of $L^2(\mathbb{R})$ with scaling function $N_n(x)$. We will establish this for $n = 3$.

Theorem 10.9.1 showed that every continuous function f on the closed interval $[-2^N, 2^N]$ is the uniform limit of a sequence of cubic splines h_k with nodes at $2^{-k}\mathbb{Z}$. These cubic splines may be chosen to have support in $[-2^N, 2^N]$ as well. Thus

$$\lim_{k\to\infty} \|f - h_k\|_2^2 = \lim_{k\to\infty} \int_{-2^n}^{2^n} |f(x) - h_k(x)|^2\, dx \leq 2^{N+1} \lim_{k\to\infty} \|f - h_k\|_\infty^2 = 0.$$

So f is the limit in L^2 of a sequence of cubic splines. It follows that $\bigcup_k S_k^{(3)}$ is dense in $L^2(\mathbb{R})$.

The separation property is established in much the same way as the piecewise linear case. Suppose that f is a function in $\bigcap_k S_k^{(3)}$. For $k \leq 0$, functions in $S_k^{(3)}$ are cubic polynomials on $[0, 2^{|k|}]$ and on $[-2^{|k|}, 0]$. Hence the restrictions of f to $[0, \infty)$ and to $(-\infty, 0]$ agree with cubic polynomials. A nonzero cubic polynomial p has

$$\int_0^\infty |p(x)|^2\, dx = +\infty = \int_{-\infty}^0 |p(x)|^2\, dx.$$

So the only $L^2(\mathbb{R})$ function that is cubic on both half-lines is the zero function. Thus $f = 0$ is the only point in the intersection.

Finally, we will show that translates of $N_3(x)$ form a Riesz basis for $L^2(\mathbb{R})$. By Theorem 15.9.6, we must compute $t_j = \langle N_3(x), N_3(x-j)\rangle$. By symmetry, $t_{-k} = t_k$ and the fact that N_3 is supported on $[0,4]$ means that $t_k = 0$ for $|k| \geq 4$. Therefore, it suffices to compute t_0, t_1, t_2, and t_3. We spare the reader the tedious calculation and use *Maple* to obtain

$$t_0 = \tfrac{151}{315}, \quad t_1 = t_{-1} = \tfrac{397}{1680}, \quad t_2 = t_{-2} = \tfrac{1}{42}, \quad t_3 = t_{-3} = \tfrac{1}{5040}.$$

Thus we are led to consider the function

$$g(\theta) = \tfrac{1208}{2520} + \tfrac{1191}{2520}\cos\theta + \tfrac{60}{2520}\cos 2\theta + \tfrac{1}{2520}\cos 3\theta.$$

An easy calculation shows that this takes its minimum when $\cos\theta = -1$ and the minimum value is $\tfrac{76}{2520} > 0$. Since this function is positive, Theorem 15.9.9 shows that there is a wavelet basis consisting of cubic splines.

Exercises for Section 15.9

A. Let $\varphi = \chi_{[0,2)}$. Show that $\{\varphi(x-j) : j \in \mathbb{Z}\}$ is not a Riesz basis for its span.

B. Show that if $\{h(x-j) : j \in \mathbb{Z}\}$ is a Riesz basis for V_0, then $\{2^{k/2}h(2^kx-j) : j \in \mathbb{Z}\}$ forms a Riesz basis for V_k for each $k \in \mathbb{Z}$.

C. Show that if T is an invertible linear map, then T^{-1} is linear.

D. Let T be a linear map from one Hilbert space $\mathscr{H}$ to itself. Prove that T is continuous if and only if it is bounded.   HINT: If not bounded, find x_n with $\|x_n\| \to 0$ while $\|Tx_n\| \to \infty$.

E. Recall that ℓ_0 is the nonclosed subspace of $\ell^2(\mathbb{Z})$ of elements with only finitely many nonzero entries. Show that if T is a linear map from ℓ_0 into a Hilbert space $\mathscr{H}$ with $\|T\mathbf{a}\| \le B\|\mathbf{a}\|_2$, then T extends uniquely to a continuous function on $\ell^2(\mathbb{Z})$ into $\mathscr{H}$.
HINT: Fix $\mathbf{a} \in \ell^2(\mathbb{Z})$ and $\varepsilon > 0$. Show that if $\mathbf{b},\mathbf{c} \in \ell_0$ and both lie in the $\varepsilon/(2B)$ ball about $\mathbf{a}$, then $\|T\mathbf{b} - T\mathbf{c}\| < \varepsilon$. Hence deduce that if $(\mathbf{b}_i)$ and $(\mathbf{c}_j)$ are two sequences in ℓ_0 converging to $\mathbf{a}$, then $\lim_i T\mathbf{b}_i = \lim_j T\mathbf{c}_j$. Consequently, show that setting $T\mathbf{a}$ to be this limit determines a continuous function on $\ell^2(\mathbb{Z})$ extending T.

F. Suppose that $\{\mathbf{x}_n : n \in \mathbb{Z}\}$ is a Riesz basis for $\mathscr{H}$.
   (a) Show that there is a unique vector $\mathbf{y}_n$ orthogonal to the subspace $M_n = \operatorname{span}\{\mathbf{x}_j : j \ne n\}$ such that $\langle \mathbf{x}_n, \mathbf{y}_n \rangle = 1$.
   (b) Show that if $\mathbf{z} \in \mathscr{H}$, then $\mathbf{z} = \sum_n \langle \mathbf{z}, \mathbf{y}_n \rangle \mathbf{x}_n$.
   (c) Show that $\{\mathbf{y}_n : n \in \mathbb{Z}\}$ is a Riesz basis for $\mathscr{H}$.
   HINT: If $(a_n) \in \ell^2$, there is another sequence (b_n) such that $\sum_n a_n \mathbf{y}_n = \sum_n b_n \mathbf{x}_n$. Apply the Cauchy–Schwarz inequality to both $\langle \sum_j a_j \mathbf{x}_j, \sum_k a_k \mathbf{y}_k \rangle$ and $\langle \sum_j b_j \mathbf{x}_j, \sum_k a_k \mathbf{y}_k \rangle$.

G. Show that $\{N_2(x-k) : k \in \mathbb{Z}\}$ is a Riesz basis for its span.

H. Prove by induction on $n \ge 1$ that
   (a) N_n is $C^{(n-1)}$.
   (b) $N_n|_{[j,j+1]}$ is a polynomial of degree n for each $j \in \mathbb{Z}$.
   (c) $\{x \in \mathbb{R} : N_n(x) > 0\} = (0, n+1)$.
   (d) $\sum_k N_n(x-k) = 1$ for all $x \in \mathbb{R}$.

Chapter 16
Convexity and Optimization

Optimization is a central theme of applied mathematics that involves minimizing or maximizing various quantities. This is an important application of the derivative tests in calculus. In addition to the first and second derivative tests of one-variable calculus, there is the powerful technique of Lagrange multipliers in several variables. This chapter is concerned with analogues of these tests that are applicable to functions that are not differentiable. Of course, some different hypothesis must replace differentiability, and this is the notion of convexity. It turns out that many applications in economics, business, and related areas involve convex functions. As in other chapters of this book, we concentrate on the theoretical underpinnings of the subject. The important aspect of constructing algorithms to carry out our program is not addressed. However, the reader will be well placed to read that material. Results from both linear algebra and calculus appear regularly.

The study of convex sets and convex functions is a comparatively recent development. Although convexity appears implicitly much earlier (going back to work of Archimedes, in fact), the first papers on convex sets appeared at the end of the nineteenth century. The main theorems of this chapter, characterizations of solutions of optimization problems, first appeared around the middle of the twentieth century. Starting in the 1970s, there has been considerable work on extending these methods to nonconvex functions.

16.1 Convex Sets

Although convex subsets can be defined for any normed vector space, we concentrate on $\mathbb{R}^n$ with the Euclidean norm. For this space, we have an inner product and the Heine–Borel Theorem (4.4.6) characterizing compact sets in $\mathbb{R}^n$ as useful tools.

16.1.1. DEFINITION. A subset A of $\mathbb{R}^n$ is called a **convex set** if

$$\lambda a + (1 - \lambda)b \in A \quad \text{for all} \quad a, b \in A \text{ and } \lambda \in [0, 1].$$

K.R. Davidson and A.P. Donsig, *Real Analysis and Applications: Theory in Practice*,
Undergraduate Texts in Mathematics, DOI 10.1007/978-0-387-98098-0_16,
© Springer Science + Business Media, LLC 2010

Let $[a,b] = \{\lambda a + (1-\lambda)b : \lambda \in [0,1]\}$ denote the line segment joining points a and b in $\mathbb{R}^n$. Define (a,b), $[a,b)$, and $(a,b]$ in the analogous way. You should note that $\lambda \in (0,1]$ corresponds to $[a,b)$. Notice that A is convex if and only if $[a,b] \subset A$ whenever $a,b \in A$. See Figure 16.1

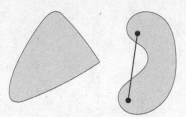

FIG. 16.1 A convex and a nonconvex set.

16.1.2. DEFINITION. A subset A of $\mathbb{R}^n$ is an **affine set** if

$$\lambda a + (1-\lambda)b \in A \quad \text{for all} \quad a,b \in A \text{ and } \lambda \in \mathbb{R}.$$

A subset C of $\mathbb{R}^n$ is a **cone** if it is a convex set that contains all of its positive scalar multiples, that is,

$$\lambda a \in C \quad \text{for all} \quad a \in C \text{ and } \lambda > 0.$$

Clearly, affine sets and cones are convex but not conversely.

There is some special terminology for this chapter: We reserve **linear function** for functions satisfying $f(\lambda a + \mu b) = \lambda f(a) + \mu f(b)$, for all a,b in the domain and all scalars λ and μ. We use **affine function** for a function g given by $g(x) = f(x) + c$, where f is a linear function and c is a constant. The reader should show that a function $f : \mathbb{R}^n \to \mathbb{R}^m$ is linear or affine if and only if the graph of f is either a subspace or an affine set, respectively (Exericse 16.1.I).

Convex sets are ubiquitous and we give a few examples, mostly without proof. Proving the following assertions is a useful warm-up exercise.

16.1.3. EXAMPLES.

(1) A subspace of $\mathbb{R}^n$ is both affine and a cone.

(2) Any ball $\overline{B_r(a)} = \{x \in \mathbb{R}^n : \|x - a\| \le r\}$ in $\mathbb{R}^n$ is convex. Indeed, if $x,y \in \overline{B_r(a)}$ and $\lambda \in [0,1]$, then

$$\begin{aligned}
\|\lambda x + (1-\lambda)y - a\| &= \|\lambda(x-a) + (1-\lambda)(y-a)\| \\
&\le \lambda\|x-a\| + (1-\lambda)\|y-a\| \\
&\le \lambda r + (1-\lambda)r = r.
\end{aligned}$$

So $\lambda x + (1-\lambda)y \in \overline{B_r(a)}$. Clearly, the ball is neither affine nor a cone.

(3) The half-space $\{(x,y) : ax + by \geq 0\}$ is a closed convex cone in $\mathbb{R}^2$.

(4) In $\mathbb{R}^n$, the positive orthant $\mathbb{R}^n_+ = \{(x_1, \ldots, x_n) : x_i > 0\}$ is a cone.

(5) If $A \subset \mathbb{R}$, then A is convex if and only if A is an interval, possibly unbounded.

(6) If $A \subset \mathbb{R}^m$ is convex and $T : \mathbb{R}^m \to \mathbb{R}^n$ is linear, then $T(A)$ is convex.

(7) If $A \subset \mathbb{R}^n$ is convex, then any translate of A (i.e., a set of the form $A + x, x \in \mathbb{R}^n$) is convex.

We now collect a number of basic properties of convex and affine sets. The proof of the first lemma is left as an exercise.

16.1.4. LEMMA. *If $\{A_i : i \in I\}$ is a collection of convex subsets of $\mathbb{R}^n$, then $\bigcap_{i \in I} A_i$ is convex. Similarly, the intersection of a collection of affine sets is affine and the intersection of a collection of cones is a cone.*

16.1.5. LEMMA. *If $A \subset \mathbb{R}^n$ is a nonempty affine set, then it is the translate of a unique subspace of $\mathbb{R}^n$.*

PROOF. Fix $a_0 \in A$, and let $L = \{a - a_0 : a \in A\}$. If $v = a - a_0 \in L$ and $t \in \mathbb{R}$, then $tv + a_0 = ta + (1-t)a_0 \in A$. Hence tv lies in $A - a_0 = L$. Suppose that $w = b - a_0$ is another element of L. Then since A is convex, $(a+b)/2$ belongs to A. Now

$$(v + w) + a_0 = a + b - a_0 = 2\left(\tfrac{a+b}{2}\right) + (1-2)a_0 \in A.$$

So $v + w$ belongs to L. This shows that L is a subspace.

To see that L is unique, suppose that $A = M + y$, where M is also a subspace of $\mathbb{R}^n$. Then $L = A - a_0 = M + (y - a_0)$. Since 0 is in L, M contains $a_0 - y$ and so $M + (y - a_0) = M$. Therefore, $L = M$. $\blacksquare$

16.1.6. DEFINITION. Suppose that S is a subset of $\mathbb{R}^n$. The **convex hull** of S, denoted by $\text{conv}(S)$, is the intersection of all convex subsets of $\mathbb{R}^n$ containing S. The **closed convex hull** of S, denoted by $\overline{\text{conv}}(S)$, is the intersection of all closed convex subsets of $\mathbb{R}^n$ containing S. The **affine hull** of S, denoted by $\text{aff}(S)$, is the intersection of all affine subsets of $\mathbb{R}^n$ containing S. Let $L(S)$ denote the unique subspace (as in Lemma 16.1.5) that is a translate of $\text{aff}(S)$. The **dimension** of S, denoted by $\dim(S)$, is the dimension of $L(S)$. Finally, if S is a subset of $\mathbb{R}^n$, we use $\text{cone}(S)$ for the intersection of all cones containing S.

By Lemma 16.1.4, $\text{conv}(S)$ and $\overline{\text{conv}}(S)$ are convex. Hence $\text{conv}(S)$ is the smallest convex set containing S. Therefore, $\text{conv}(\text{conv}(S)) = \text{conv}(S)$. The intersection of closed sets is closed, so $\overline{\text{conv}}(S)$ is the smallest closed convex set containing S. Moreover, $\overline{\text{conv}}(\overline{\text{conv}}(S)) = \overline{\text{conv}}(\text{conv}(S)) = \overline{\text{conv}}(S)$. Likewise, $\text{aff}(\text{aff}(S)) = \text{aff}(S)$ is the smallest affine set containing S and $\text{cone}(\text{cone}(S)) = \text{cone}(S)$ is the smallest cone containing S. Affine sets are closed because (finite-dimensional) subspaces are closed.

Here is a useful description of the convex hull of an arbitrary set S.

16.1.7. THEOREM. *Suppose that S is a subset of $\mathbb{R}^n$. Then a belongs to* conv(S) *if and only if there are points $s_1, \ldots, s_r$ in S and scalars $\lambda_1, \ldots, \lambda_r$ in $[0,1]$ with $\sum_{i=1}^{r} \lambda_i = 1$ such that $\sum_{i=1}^{r} \lambda_i s_i = a$.*

PROOF. We claim that

$$C = \left\{ \sum_{i=1}^{r} \lambda_i s_i : r \geq 1,\ s_i \in S,\ \lambda_i \in [0,1],\ \sum_{i=1}^{r} \lambda_i = 1 \right\}$$

is a convex set. Consider two points of C, say $a = \sum_{i=1}^{n} \mu_i s_i$ and $b = \sum_{j=1}^{m} v_j t_j$, where the s_i and t_j are in S, μ_i and v_j are in $[0,1]$, and $\sum_{i=1}^{n} \mu_i = 1 = \sum_{j=1}^{m} v_j$. Then $\lambda a + (1 - \lambda) b$ can be written as

$$\sum_{i=1}^{n} \lambda \mu_i s_i + \sum_{j=1}^{m} (1 - \lambda) v_j t_j.$$

This is a linear combination of elements of S with positive coefficients satisfying

$$\sum_{i=1}^{n} \lambda \mu_i + \sum_{j=1}^{m} (1 - \lambda) v_j = \lambda + (1 - \lambda) = 1.$$

Thus $\lambda a + (1 - \lambda) b$ also belongs to C.

Since C is convex and contains S, it follows that conv(S) is contained in C. If we show that C is contained in conv(S), then it will follow that they are equal.

Suppose that $a = \sum_{i=1}^{r} \lambda_i s_i$, where the λ_i and s_i satisfy the conditions on C. Set $\Lambda_k = \sum_{i=1}^{k} \lambda_i$ for $1 \leq k \leq r$. Let k_0 be the smallest k for which $\lambda_k > 0$. Inductively define points $a_{k_0}, \ldots, a_r$ in S by $a_{k_0} = s_{k_0}$ and

$$a_k = \frac{\Lambda_{k-1}}{\Lambda_k} a_{k-1} + \frac{\lambda_k}{\Lambda_k} s_k \quad \text{for} \quad k_0 < k \leq r.$$

Since each s_k belongs to S, it follows by induction that these convex combinations all lie in conv(S). However, we also show by induction that $a_k = \sum_{i=1}^{k} \frac{\lambda_i}{\Lambda_k} s_i$ for $k \geq k_0$. This is evident for $k = k_0$. Suppose that it is true for $k - 1$. Then

$$a_k = \frac{\Lambda_{k-1}}{\Lambda_k} \sum_{i=1}^{k-1} \frac{\lambda_i}{\Lambda_{k-1}} s_i + \frac{\lambda_k}{\Lambda_k} s_k = \sum_{i=1}^{k} \frac{\lambda_i}{\Lambda_k} s_i.$$

In particular, since $\Lambda_r = 1$, we have that $a_r = \sum_{i=1}^{r} \lambda_i s_i = a$ lies in conv(S). ∎

Since we are working in finite dimensions, this result may be sharpened so that each point in the convex hull is a combination of at most $n+1$ points.

16.1.8. CARATHÉODORY'S THEOREM.

Suppose that S is a subset of $\mathbb{R}^n$. Then each $a \in \text{conv}(S)$ may be expressed as a convex combination of $n+1$ elements of S.

PROOF. By the previous proposition, a may be written as a convex combination

$$a = \sum_{i=1}^{r} \lambda_i s_i \quad \text{where} \quad \lambda_i \geq 0, \quad \sum_{i=1}^{r} \lambda_i = 1, \text{ and } s_i \in S, \text{ for } 1 \leq i \leq r.$$

If $r \geq n+2$, we will construct another representation of a using fewer vectors. Thus we eventually reduce this to a sum of at most $n+1$ elements. We may suppose that $\lambda_i > 0$ for each i, for otherwise we reduce the list by dropping s_{i_0} if $\lambda_{i_0} = 0$.

Consider $v_i = s_i - s_r$ for $1 \leq i < r$. These are $r - 1 \geq n+1$ such vectors in an n-dimensional space, and thus they are linearly dependent. Find constants μ_i not all 0 such that

$$0 = \sum_{i=1}^{r-1} \mu_i(s_i - s_r) = \sum_{i=1}^{r} \mu_i s_i,$$

where $\mu_r = -\sum_{i=1}^{r-1} \mu_i$. Let $J = \{i : \mu_i < 0\}$, which is nonempty because $\sum_{i=1}^{r} \mu_i = 0$. Set $\delta = \min\{\lambda_i/|\mu_i| : i \in J\}$, and pick i_0 such that $\lambda_{i_0} = -\delta \mu_{i_0}$. Then

$$a = \sum_{i=1}^{r} \lambda_i s_i + \sum_{i=1}^{r} \delta \mu_i s_i = \sum_{i=1}^{r} (\lambda_i + \delta \mu_i) s_i.$$

By construction, the constants $v_i = \lambda_i + \delta \mu_i$ are at least zero and $\sum_{i=1}^{r} v_i = 1$. Plus, $v_{i_0} = 0$. So deleting s_{i_0} from the list represents a as a convex combination of fewer elements. ∎

16.1.9. COROLLARY. *If $A \subset \mathbb{R}^n$ is compact, then $\text{conv}(A)$ is compact.*

PROOF. Define a subset $X = A^{n+1} \times \Delta_{n+1}$ of $\mathbb{R}^{(n+1)^2}$ consisting of all points $x = (a_1, a_2, \dots, a_{n+1}, \lambda_1, \dots, \lambda_{n+1})$, where $a_i \in A$, $\lambda_i \in [0,1]$, and $\sum_{i=1}^{n+1} \lambda_i = 1$. It is easy to check that X is closed and bounded and therefore is compact. Consider the function $f(x) = \sum_{i=1}^{n+1} \lambda_i a_i$. This is a continuous function from X into $\text{conv}(A)$. By Carathéodory's Theorem, f maps X onto $\text{conv}(A)$. By Theorem 5.4.3, $f(X)$ is compact. Therefore, $\text{conv}(A)$ is compact. ∎

16.1.10. DEFINITION. A **hyperplane** is an affine set of codimension 1. Thus a hyperplane in $\mathbb{R}^n$ has dimension $n - 1$.

Hyperplanes are rather special affine sets, and they serve to split the whole space into two pieces. This is a consequence of the following result.

16.1.11. PROPOSITION. *A subset H of $\mathbb{R}^n$ is a hyperplane if and only if there are a nonzero vector $h \in \mathbb{R}^n$ and a scalar $\alpha \in \mathbb{R}$ such that*

$$H = \{x \in \mathbb{R}^n : \langle x, h \rangle = \alpha\}.$$

PROOF. If H is a hyperplane and $x_0 \in H$, then $L(H) = H - x_0$ is a subspace of dimension $n - 1$. Choose a nonzero vector h orthogonal to $L(H)$. This is used to define a linear map from $\mathbb{R}^n$ into $\mathbb{R}$ by $f(x) = \langle x, h \rangle$. Since the set of all vectors orthogonal to h forms a subspace of dimension $n - 1$ containing $L(H)$, it follows that $L(H) = \{h\}^{\perp} = \ker f$.

Set $\alpha = f(x_0)$. Then $f(x) = \alpha$ if and only if $f(x - x_0) = f(x) - \alpha = 0$, which occurs if and only if $x - x_0 \in L(H)$ or equivalently, when $x \in L(H) + x_0 = H$.

Conversely, the linear map $f(x) = \langle x, h \rangle$ from $\mathbb{R}^n$ into $\mathbb{R}$ maps onto $\mathbb{R}$, since $h \neq 0$. Thus $L(H) := \ker f = \{h\}^{\perp}$ is a subspace of $\mathbb{R}^n$ of dimension $n - 1$. Let x_0 be any vector with $f(x_0) = \alpha$. Then following the argument of the previous paragraph, the set $H = \{x \in \mathbb{R}^n : \langle x, h \rangle = \alpha\} = L(H) + x_0$ is a hyperplane. ∎

Note that the function f (or vector h) is not unique, but it is determined up to a scalar multiple because the subspace $H^{\perp}$ is one-dimensional. When working with a hyperplane, we will usually assume that a choice of this function has been made. This allows us to describe two **half-spaces** associated to H, which we denote by $H^{+} = \{x \in \mathbb{R}^n : f(x) \geq \alpha\}$ and $H^{-} = \{x \in \mathbb{R}^n : f(x) \leq \alpha\}$. These two subsets don't depend on the choice of f except that a sign change can interchange H^{+} with H^{-}.

Exercises for Section 16.1

A. If A is a convex subset of $\mathbb{R}^n$, show that $\overline{A}$ is convex.

B. (a) Prove Lemma 16.1.4: The intersection of convex sets is convex.
(b) State and prove the analogous result for cones and affine sets.

C. Suppose that A is a closed subset of $\mathbb{R}^n$ and whenever $a, b \in A$, the point $(a + b)/2$ is in A. Show that A is convex.   HINT: Show that $\lambda a + (1 - \lambda)b \in A$ if $\lambda = i/2^k$ for $1 \leq i < 2^k$.

D. If A is a convex subset of $\mathbb{R}^n$ such that $\text{aff}(A) \neq \mathbb{R}^n$, show that $\text{int}(A)$ is empty.

E. Let S be a subset of $\mathbb{R}^n$. Show that $\text{aff}(S) = \left\{ \sum_{i=1}^{r} \lambda_i s_i : s_i \in S, \ \lambda_i \in \mathbb{R}, \ \sum_{i=1}^{n} \lambda_i = 1 \right\}$ and $L(S) = \left\{ \sum_{i=1}^{r} \lambda_i s_i : s_i \in S, \ \lambda_i \in \mathbb{R}, \ \sum_{i=1}^{n} \lambda_i = 0 \right\}$.

F. Suppose that A is a convex subset of $\mathbb{R}^n$ and that T is a linear transformation from $\mathbb{R}^n$ into $\mathbb{R}^m$. Prove that TA is convex and $\text{aff}(TA) = T\,\text{aff}(A)$.

G. If $A \subset \mathbb{R}^m$ and $B \subset \mathbb{R}^n$, show that $\text{conv}(A \times B) = \text{conv}(A) \times \text{conv}(B)$ in $\mathbb{R}^{m+n}$.

H. (a) If S is a bounded subset of $\mathbb{R}^n$, prove that $\text{conv}(S)$ is bounded.
(b) Give an example of a closed subset S of $\mathbb{R}^2$ such that $\text{conv}(S)$ is not closed.

I. Recall that the graph of $f : \mathbb{R}^n \to \mathbb{R}^m$ is $G(f) = \{(x, y) \in \mathbb{R}^n \times \mathbb{R}^m : f(x) = y\}$. Show that f is a linear function if and only if $G(f)$ is a linear subspace of $\mathbb{R}^{m+n}$ and f is an affine function if and only if $G(f)$ is an affine set.

J. A function f on $\mathbb{R}$ is **convex** if $f(\lambda x + (1 - \lambda)y) \leq \lambda f(x) + (1 - \lambda)f(y)$ for all $x, y \in \mathbb{R}$ and $0 \leq \lambda \leq 1$. If f is any function on $\mathbb{R}$, define $\text{epi}(f) = \{(x, y) \in \mathbb{R}^2 : y \geq f(x)\}$ to be the **epigraph** of f. Show that f is a convex function if and only if $\text{epi}(f)$ is a convex set.

K.   Let A and B be convex subsets of $\mathbb{R}^n$. Prove that $\text{conv}(A \cup B)$ equals the union of all line segments $[a, b]$ such that $a \in A$ and $b \in B$.

L.   **The asymptotic cone.** Let $A \subset \mathbb{R}^n$ be closed and convex, and $x \in A$.

  (a) Show that $s(A - x) \subset t(A - x)$ if $0 < s < t$.
  (b) Show that $A_\infty(x) = \bigcap_{t>0} t(A - x) := \{d : x + td \in A \text{ for all } t > 0\}$ is a cone.
  (c) Show that $A_\infty(x)$ does not depend on the point x.   HINT: Consider $(1 - \frac{1}{k})y + \frac{1}{k}(x + ktd)$.
  (d) Prove that A is compact if and only if $A_\infty = \{0\}$.   HINT: If $\|a_k\| \to \infty$, find a cluster point d of $a_k/\|a_k\|$. Argue as in (c).

M.   A set $S \subset \mathbb{R}^n$ is a **star-shaped set** with respect to $v \in S$ if $[s, v] \subset S$ for all $s \in S$.

  (a) Show that S is convex if and only if it is star shaped with respect to every $v \in S$.
  (b) Find a set that is star shaped with respect to exactly one of its points.

N.   (a) Given a sequence of convex sets $B_1 \subset B_2 \subset B_3 \subset \cdots$ in $\mathbb{R}^n$, show that $\bigcup_{i \geq 1} B_i$ is convex.
  (b) For any sequence of convex sets $B_1, B_2, B_3, \ldots$ in $\mathbb{R}^n$, show that $\bigcup_{j \geq 1}(\bigcap_{i \geq j} B_i)$ is convex.

O.   A set $S \subset \mathbb{R}^n$ is called **affinely dependent** if there is an $s \in S$ such that $s \in \text{aff}(S \setminus \{s\})$. Show that S is affinely dependent if and only if there are distinct elements $s_1, \ldots, s_r$ of S and scalars $\mu_1, \ldots, \mu_r$, not all zero, such that $\sum_{i=1}^r \mu_i s_i = 0$ and $\sum_{i=1}^r \mu_i = 0$.   HINT: Solve for some s_i.

P.   A subset $S \subset \mathbb{R}^n$ is **affinely independent** if it is not affinely dependent. Show that an affinely independent set in $\mathbb{R}^n$ can have at most $n + 1$ points.   HINT: If $\{s_0, \ldots, s_n\} \subset S$, show that $\{s_i - s_0 : 1 \leq i \leq n\}$ is linearly independent.

Q.   **Radon's Theorem.** Suppose that $s_1, \ldots, s_r$ are distinct points in $\mathbb{R}^n$, with $r > n + 1$. Show that there are disjoint sets I and J with $I \cup J = \{1, \ldots, r\}$ such that the convex sets $\text{conv}\{s_i : i \in I\}$ and $\text{conv}\{s_j : j \in J\}$ have nonempty intersection.   HINT: Use the previous two exercises to obtain 0 as a nontrivial linear combination of these elements. Then consider those elements with positive coefficients in this formula.

R.   **Helly's Theorem.** Let C_k be convex subsets of $\mathbb{R}^n$ for $1 \leq k \leq m$. Suppose that any $n + 1$ of these sets have nonempty intersection. Prove that the whole collection has nonempty intersection.   HINT: Use induction on $m \geq n + 1$ sets. If true for $m - 1$, choose x_j in the intersection of $C_1, \ldots, C_{j-1}, C_{j+1}, \ldots, C_m$. Apply Radon's Theorem.

16.2 Relative Interior

In working with a convex subset A of $\mathbb{R}^n$, the natural space containing it is often $\text{aff}(A)$, not $\mathbb{R}^n$, which may be far too large. The affine hull is a better place to work for many purposes. Indeed, if A is a convex subset of $\mathbb{R}^n$ with $\dim A = k < n$, then thinking of A simply as a subset of $\text{aff}(A)$, which may be identified with $\mathbb{R}^k$, allows us to talk more meaningfully about topological notions such as interior and boundary. One sign that the usual interior is not useful is Exercise 16.1.D, which shows that A has empty interior whenever $\dim(A) < n$. This section is devoted to developing the properties of the interior relative to $\text{aff}(A)$.

16.2.1. DEFINITION.   If A is a convex subset of $\mathbb{R}^n$, then the **relative interior** of A, denoted by $\text{ri}(A)$, is the interior of A relative to $\text{aff}(A)$. That is, $a \in \text{ri}(A)$ if and only if there is an $\varepsilon > 0$ such that $B_\varepsilon(a) \cap \text{aff}(A) \subset A$.

  Define the **relative boundary** of A, denoted by $\text{rbd}(A)$, to be $\overline{A} \setminus \text{ri}(A)$.

16.2.2. EXAMPLE. If $\dim A = k < n$, then $\mathrm{ri}(A)$ is the interior of A when A is considered as a subset of $\mathrm{aff}(A)$, which is identified with $\mathbb{R}^k$. For example, consider the closed convex disk $A = \{(x,y,z) \in \mathbb{R}^3 : x^2 + y^2 \le 1, z = 0\}$. Then $\mathrm{aff}(A) = \{(x,y,z) \in \mathbb{R}^3 : z = 0\}$ is a plane. The relative interior of A is $\mathrm{ri}(A) = \{(x,y,z) \in \mathbb{R}^3 : x^2 + y^2 < 1, z = 0\}$, the interior of the disk in this plane, while the interior as a subset of $\mathbb{R}^3$ is empty. The relative boundary is the circle $\{(x,y,z) \in \mathbb{R}^3 : x^2 + y^2 = 1, z = 0\}$.

Notice that $A \subset B$ does not imply that $\mathrm{ri}(A) \subset \mathrm{ri}(B)$ because an increase in dimension can occur. For example, let $B = \{(x,y,z) : x^2 + y^2 \le 1, z \le 0\}$. Then $\mathrm{ri}(B) = \mathrm{int}(B) = \{(x,y,z) : x^2 + y^2 < 1, z < 0\}$. So $A \subset B$ but $\mathrm{ri}(A)$ and $\mathrm{ri}(B)$ are disjoint.

The following result shows that the phenomenon above occurs precisely because of the dimension shift. Despite its trivial proof, it will be quite useful.

16.2.3. LEMMA. *Suppose that A and B are convex subsets of $\mathbb{R}^n$ with $A \subset B$. If $\mathrm{aff}(A) = \mathrm{aff}(B)$, then $\mathrm{ri}(A) \subset \mathrm{ri}(B)$.*

PROOF. If $a \in \mathrm{ri}(A)$, then there is an $\varepsilon > 0$ with $B_\varepsilon(a) \cap \mathrm{aff}(A) \subset A$. Because $\mathrm{aff}(B) = \mathrm{aff}(A)$, we have $B_\varepsilon(a) \cap \mathrm{aff}(B) \subset A \subset B$, so $a \in \mathrm{ri}(B)$. $\blacksquare$

A **polytope** is the convex hull of a finite set. Now we obtain an analogue of Theorem 16.1.7 describing the relative interior of a polytope.

16.2.4. LEMMA. *Let $S = \{s_1, \ldots, s_r\}$ be a finite subset of $\mathbb{R}^n$. Then*

$$\mathrm{ri}\left(\mathrm{conv}(S)\right) = \left\{ \sum_{i=1}^r \lambda_i s_i : \lambda_i \in (0,1), \ \sum_{i=1}^r \lambda_i = 1 \right\}.$$

PROOF. By Theorem 16.1.7, $\mathrm{conv}(S) = \{\sum_i \lambda_i s_i : \lambda_i \in [0,1], \sum_i \lambda_i = 1\}$. The subspace $L(S)$ is given by Exercise 16.1.E as $L(S) = \{\sum_i \lambda_i s_i : \sum_i \lambda_i = 0\}$.

Let $e_1, \ldots, e_k$ be an orthonormal basis for $L(S)$. Express each e_j as a combination $e_j = \sum_i \lambda_{ij} s_i$, where $\sum_i \lambda_{ij} = 0$, and define $\Lambda = \max_i \left(\sum_j \lambda_{ij}^2\right)^{1/2}$. Suppose that x is a vector in $L(S)$. Then

$$x = \sum_{j=1}^k x_j e_j = \sum_{i=1}^r \left(\sum_{j=1}^k \lambda_{ij} x_j \right) s_i.$$

By the Schwarz inequality (4.1.1),

$$\left| \sum_{j=1}^k \lambda_{ij} x_j \right| \le \left(\sum_{j=1}^k \lambda_{ij}^2 \right)^{1/2} \left(\sum_{j=1}^k x_j^2 \right)^{1/2} \le \Lambda \|x\|.$$

Thus if $\|x\| < \varepsilon/\Lambda$, then x may be expressed as $x = \sum_i \mu_i s_i$, where $\sum_i \mu_i = 0$ and $|\mu_i| < \varepsilon$ for $1 \le i \le r$.

Let $a = \sum_i \lambda_i s_i$, where $\lambda_i \in (0,1)$ and $\sum_i \lambda_i = 1$. Then there is an $\varepsilon > 0$ such that $\varepsilon \le \lambda_i \le 1 - \varepsilon$ for $1 \le i \le r$. Note that

$$B_{\varepsilon/\Lambda}(a) \cap \operatorname{aff}(S) = a + \big(B_{\varepsilon/\Lambda}(0) \cap L(S)\big).$$

If $x \in L(S)$ and $\|x\| < \varepsilon/\Lambda$, we use the representation above to see that

$$a + x = \sum_{i=1}^{r} (\lambda_i + \mu_i) s_i.$$

Now $\sum_i \lambda_i + \mu_i = 1$ and $\lambda_i + \mu_i \in [\varepsilon, 1 - \varepsilon] + (-\varepsilon, \varepsilon) = (0,1)$. So $a + x$ belongs to $\operatorname{conv}(S)$. This shows that a belongs to $\operatorname{ri}(\operatorname{conv}(S))$.

Conversely, suppose that $a \in \operatorname{ri}(S)$. Write $a = \sum_i \lambda_i s_i$, where $\lambda_i \in [0,1]$ and $\sum_i \lambda_i = 1$. We wish to show that it has a possibly different representation with coefficients in $(0,1)$. Let $\varepsilon > 0$ be given such that $B_\varepsilon(a) \cap \operatorname{aff}(S) \subset \operatorname{conv}(S)$. If each $\lambda_i \in (0,1)$, there is nothing to prove. Suppose that $J = \{j : \lambda_j = 0\}$ is nonempty, and let k be chosen such that $\lambda_k > 0$. Pick a $\delta > 0$ so small that $|J|\delta < \lambda_k$ and $x = \delta \sum_{j \in J}(s_j - s_k)$ satisfies $\|x\| < \varepsilon$. Then $a \pm x$ belong to $\operatorname{conv}(S)$. Write $a - x = \sum_i \mu_i s_i$, where $\mu_i \in [0,1]$ and $\sum_i \mu_i = 1$. Also,

$$a + x = \sum_{j \notin J} \lambda_i s_i + \sum_{j \in J} \delta(s_j - s_k)$$

$$= \sum_{j \in J} \delta s_j + (\lambda_k - |J|\delta) s_k + \sum_{i \in (J \cup \{k\})^c} \lambda_i s_i =: \sum_{i=1}^{r} v_i s_i.$$

It is evident from our construction that each $v_i > 0$; and since $\sum_i v_i = 1$, all are less than 1. Now $a = \sum_i \dfrac{\mu_i + v_i}{2} s_i$ is expressed with all coefficients in $(0,1)$. ∎

16.2.5. COROLLARY. *If A is a nonempty convex subset of $\mathbb{R}^n$, then $\operatorname{ri}(A)$ is also nonempty. Moreover, $\operatorname{aff}(\operatorname{ri}(A)) = \operatorname{aff}(A)$.*

PROOF. Let $k = \dim \operatorname{conv}(A)$. Then there is a subset S of $k+1$ points such that $\operatorname{aff}(S) = \operatorname{aff}(A)$. Indeed, fix any element $a_0 \in A$. Since $L(A)$ is spanned by $\{a - a_0 : a \in A\}$ and has dimension k, we may choose k vectors $a_1, \ldots, a_k$ in A such that $a_i - a_0$ are linearly independent. Clearly, they span $L(A)$. So $S = \{a_0, \ldots, a_k\}$ will suffice.

Hence, by Lemma 16.2.3, $\operatorname{ri}(\operatorname{conv} A)$ contains $\operatorname{ri}(\operatorname{conv} S)$. Let $a = \frac{1}{k+1} \sum_{i=0}^{k} a_i$. By Lemma 16.2.4, $\operatorname{ri}(\operatorname{conv} S)$ is nonempty and, in particular, contains the points $b_i = (a + 2a_i)/3$ and $c_i = (2a + a_i)/3$ for $0 \le i \le k$. So $\operatorname{aff}(\operatorname{ri}(\operatorname{conv} S))$ contains $2b_i - c_i = a_i$ for $0 \le i \le k$, whence it equals $\operatorname{aff}(A)$ as claimed. ∎

The next theorem will be surprisingly useful. In particular, this theorem applies if b is in the relative boundary, $\operatorname{rbd}(A)$.

16.2.6. ACCESSIBILITY LEMMA.

Suppose that A is a convex subset of $\mathbb{R}^n$. If $a \in \mathrm{ri}(A)$ and $b \in \overline{A}$, then $[a,b)$ is contained in $\mathrm{ri}(A)$.

PROOF. We need to show that $\lambda a + (1-\lambda)b \in \mathrm{ri}(A)$ for $\lambda \in (0,1)$. Choose $\varepsilon > 0$ such that $B_\varepsilon(a) \cap \mathrm{aff}(A) \subset A$. Since $b \in \overline{A}$, pick $c \in A$ such that $x = b - c \in L(A)$ satisfies $\|x\| < \varepsilon\lambda/(2-2\lambda)$. Suppose that $z \in L(A)$ with $\|z\| < \varepsilon/2$. Then

$$\lambda a + (1-\lambda)b + \lambda z = \lambda a + (1-\lambda)(c+x) + \lambda z$$
$$= \lambda(a + z + (1-\lambda)x/\lambda) + (1-\lambda)c.$$

Since $\|z + (1-\lambda)x/\lambda\| \le \|z\| + (1-\lambda)\|x\|/\lambda < \varepsilon/2 + \varepsilon/2 = \varepsilon$, this is a vector in $B_\varepsilon(0) \cap L(A)$. Hence $d = a + z + (1-\lambda)x/\lambda$ belongs to A. So $\lambda a + (1-\lambda)b + \lambda z = \lambda d + (1-\lambda)c$ also lies in A. Therefore $a + (1-\lambda)b$ belongs to $\mathrm{ri}(A)$. ∎

Applying the preceding result when $b \in \mathrm{ri}(A)$ shows that $\mathrm{ri}(A)$ is convex. Since $\mathrm{int}(A)$ is either the empty set or equal to $\mathrm{ri}(A)$, we have the following:

16.2.7. COROLLARY. *If A is a convex subset of $\mathbb{R}^n$, so are $\mathrm{ri}(A)$ and $\mathrm{int}(A)$.*

16.2.8. THEOREM. *If A is a convex subset of $\mathbb{R}^n$, then the three sets $\mathrm{ri}(A)$, A, and $\overline{A}$ all have the same affine hulls, closures, and relative interiors.*

PROOF. By Corollary 16.2.5, $\mathrm{aff}(\mathrm{ri}(A)) = \mathrm{aff}(A)$. Since the affine hull is closed, $\overline{A} \subset \mathrm{aff}(A)$ and hence $\mathrm{aff}(A) = \mathrm{aff}(\overline{A})$.

Now $\mathrm{ri}(A) \subset A \subset \overline{A}$, and so $\overline{\mathrm{ri}(A)} \subset \overline{A} = \overline{\overline{A}}$, where the equality follows from Proposition 4.3.5. Suppose that $b \in \overline{A}$. By Corollary 16.2.5, $\mathrm{ri}(A)$ contains a point a. Hence by the Accessibility Lemma (16.2.6), $\lambda a + (1-\lambda)b$ belongs to $\mathrm{ri}(A)$ for $0 < \lambda \le 1$. Letting λ tend to 0 shows that $b \in \overline{\mathrm{ri}(A)}$. Thus $\overline{\mathrm{ri}(A)} = \overline{A}$.

For relative interiors, first observe that since the three sets have the same affine hull, Lemma 16.2.3 shows that $\mathrm{ri}(\mathrm{ri}(A)) \subset \mathrm{ri}(A) \subset \mathrm{ri}(\overline{A})$. Now $\mathrm{ri}(\mathrm{ri}(A)) = \mathrm{ri}(A)$. For if $a \in \mathrm{ri}(A)$ and $x \in B_\varepsilon(a) \cap \mathrm{aff}(A) \subset A$, then using $r = \varepsilon - \|x - a\| > 0$, it follows that $B_r(x) \cap \mathrm{aff}(A) \subset B_\varepsilon(a) \cap \mathrm{aff}(A) \subset A$. Hence $B_\varepsilon(a) \cap \mathrm{aff}(A)$ is contained in $\mathrm{ri}(A)$ and so $a \in \mathrm{ri}(\mathrm{ri}(A))$.

Suppose that $a \in \mathrm{ri}(\overline{A})$. Then there is an $\varepsilon > 0$ such that $B_\varepsilon(a) \cap \mathrm{aff}(A) \subset \overline{A}$. Pick any $b \in \mathrm{ri}(A)$, and set $x = (a - b)/\|a - b\|$. Then $c_\pm = a \pm \varepsilon x$ belong to $\overline{A}$. But $c_- = (1-\varepsilon)a + \varepsilon b \in \mathrm{ri}(A)$ by the Accessibility Lemma. Hence $a = (c_+ + c_-)/2$ belongs to $\mathrm{ri}(A)$, again by the Accessibility Lemma. Therefore, $\mathrm{ri}(\overline{A}) = \mathrm{ri}(A)$. ∎

The next two results will allow us to conveniently compute various combinations of convex sets such as sums and differences. The first is quite straightforward and is left as an exercise.

16.2.9. PROPOSITION. *If $A \subset \mathbb{R}^n$ and $B \subset \mathbb{R}^m$ are convex sets, then $A \times B$ is convex, $\mathrm{ri}(A \times B) = \mathrm{ri}(A) \times \mathrm{ri}(B)$, and $\mathrm{aff}(A \times B) = \mathrm{aff}(A) \times \mathrm{aff}(B)$.*

16.2.10. THEOREM. *If A is a convex subset of $\mathbb{R}^m$ and T is a linear map from $\mathbb{R}^m$ to $\mathbb{R}^n$, then $T\operatorname{ri}(A) = \operatorname{ri}(TA)$.*

PROOF. By Exercise 16.1.F, TA is convex. Using Theorem 16.2.8 for the first equality and the continuity of T (see Corollary 5.1.7) for the second containment, we have

$$TA \subset T\overline{A} = T\overline{\operatorname{ri}(A)} \subset \overline{T\operatorname{ri}(A)} \subset \overline{TA}.$$

Taking closures, we obtain $\overline{T\operatorname{ri}(A)} = \overline{TA}$. Using Theorem 16.2.8 again,

$$\operatorname{ri}(TA) = \operatorname{ri}(\overline{TA}) = \operatorname{ri}(\overline{T\operatorname{ri}(A)}) = \operatorname{ri}(T\operatorname{ri}(A)) \subset T\operatorname{ri}(A).$$

For the reverse containment, let $a \in \operatorname{ri}(A)$. We will show that Ta lies in $\operatorname{ri}(TA)$. By Corollary 16.2.5, $\operatorname{ri}(TA)$ is nonempty. So we may pick a point $b \in A$ with Tb in $\operatorname{ri}(TA)$. Since $a \in \operatorname{ri}(A)$, there is an $\varepsilon > 0$ such that $c = a + \varepsilon(a - b)$ belongs to A. However, $a = \frac{1}{1+\varepsilon}c + \frac{\varepsilon}{1+\varepsilon}b$, so $a \in [b, c)$. By linearity, Ta lies in $[Tb, Tc)$. Therefore, by the Accessibility Lemma (16.2.6), Ta belongs to $\operatorname{ri}(TA)$. ∎

16.2.11. COROLLARY. *Suppose that A and B are convex subsets of $\mathbb{R}^n$. Then $\operatorname{ri}(A + B) = \operatorname{ri}(A) + \operatorname{ri}(B)$ and $\operatorname{ri}(A - B) = \operatorname{ri}(A) - \operatorname{ri}(B)$.*

PROOF. For the two cases, define maps T from $\mathbb{R}^n \times \mathbb{R}^n$ to $\mathbb{R}^n$ by $T(a, b) = a \pm b$. Then $T(A \times B) = A \pm B$. By the two previous results,

$$\operatorname{ri}(A \pm B) = T\operatorname{ri}(A \times B) = T(\operatorname{ri}(A) \times \operatorname{ri}(B)) = \operatorname{ri}(A) \pm \operatorname{ri}(B). \qquad ∎$$

Exercises for Section 16.2

A. Write out a careful proof of Proposition 16.2.9.

B. Explain why the notion of relative closure of a convex set isn't needed.

C. Let C be a convex subset of $\mathbb{R}^n$ and let U be an open set that intersects $\overline{C}$. Show that $U \cap \operatorname{ri}(C)$ is nonempty.

D. Suppose that a convex subset A of $\mathbb{R}^n$ intersects every (affine) line in a closed interval. Show that A is closed.   HINT: Connect $b \in \overline{A}$ to a in $\operatorname{ri}(A)$.

E. Let A be a convex subset of $\mathbb{R}^n$ that is dense in $\operatorname{aff}(A)$. Prove that $A = \operatorname{aff}(A)$.

F. Let A be a convex subset of $\mathbb{R}^n$. Show that if B is a compact subset of $\operatorname{ri}(A)$, then $\operatorname{conv}(B)$ is a compact convex subset of $\operatorname{ri}(A)$.   HINT: Corollary 16.1.9

G. (a) If A and B are closed convex sets and A is compact, show that $A + B$ is a closed and convex.
(b) If B is also compact, show that $A + B$ is compact.
(c) Give an example of two closed convex sets with nonclosed sum.   HINT: Look for two closed convex subsets of $\mathbb{R}^2$ that lie strictly above the x-axis but 0 is a limit point of $A + B$.

H. Suppose that C and D are convex subsets of $\mathbb{R}^n$ and that $C \subset \overline{D}$. Show that if $\operatorname{ri}(D) \cap C$ is nonempty, then $\operatorname{ri}(C) \subset \operatorname{ri}(D)$.   HINT: If $d \in \operatorname{ri}(D) \cap C$ and $c \in \operatorname{ri}(C)$, extend $[d, c]$ beyond c.

I. Suppose that A and B are convex subsets of $\mathbb{R}^n$ with $\operatorname{ri}(A) \cap \operatorname{ri}(B) \neq \varnothing$.

(a) Show that $\overline{A \cap B} = \overline{A} \cap \overline{B}$.   HINT: Connect $b \in \overline{A} \cap \overline{B}$ to $a \in \operatorname{ri}(A) \cap \operatorname{ri}(B)$.
(b) Show that $\operatorname{ri}(A \cap B) = \operatorname{ri}(A) \cap \operatorname{ri}(B)$.

16.3 Separation Theorems

The goal of this section is to show that we can separate two disjoint convex sets from one another by a hyperplane. Consider a convex set $A \subset \mathbb{R}^2$ and a point $b \notin A$. Geometrically, one can see that there is a line that separates b from A. The appropriate generalization, replacing the line with a hyperplane, is true in $\mathbb{R}^n$. A general form of this theorem, called the Hahn–Banach theorem, holds for any normed vector space. We prove this result for $\mathbb{R}^n$, where we can use the Heine–Borel Theorem (4.4.6).

We begin by showing that a convex set in $\mathbb{R}^n$ always has a *unique* closest point to a given point outside. The *existence* of such a point comes from Exercise 5.4.J, and does not require convexity. In convexity theory, the map to this closest point is called a **projection**, which differs from the meaning of the term in linear algebra.

16.3.1. CONVEX PROJECTION THEOREM.
Let A be a nonempty closed convex subset of $\mathbb{R}^n$. For each point $x \in \mathbb{R}^n$, there is a unique point $P_A(x)$ in A that is closest to x. The point $P_A(x)$ is characterized by

$$\langle x - P_A(x), a - P_A(x) \rangle \leq 0 \quad \text{for all} \quad a \in A.$$

Moreover, $\|P_A(x) - P_A(y)\| \leq \|x - y\|$ *for all* $x, y \in \mathbb{R}^n$.

Geometrically, the equation says that the line segment $[x, P_A(x)]$ makes an obtuse angle with $[a, P_A(x)]$ for every point $a \in A$. The last estimate shows that P_A has Lipschitz constant 1 and, in particular, is continuous.

PROOF. Pick any vector a_0 in A, and let $R = \|x - a_0\|$. Then

$$0 \leq \inf\{\|x - a\| : a \in A\} \leq R.$$

The closest point to x in A must belong to $A \cap B_R(x)$. This is the intersection of two closed convex sets, and so is closed and convex. Moreover, $A \cap B_R(x)$ is bounded and thus compact by the Heine–Borel Theorem (4.4.6). Consider the continuous function on $A \cap B_R(x)$ given by $f(a) = \|x - a\|$. By the Extreme Value Theorem (5.4.4), there is a point $a_1 \in A$ at which f takes its minimum—a closest point.

To see that a_1 is unique, we need to use convexity. Suppose that there is another vector $a_2 \in A$ with $\|x - a_2\| = \|x - a_1\|$. We will show that $b = \frac{1}{2}(a_1 + a_2)$ is closer, contradicting the choice of a_1 as a closest point. See Figure 16.2. One way to accomplish this is to apply the parallelogram law (see Exercise 16.3.A). Instead, we give an elementary argument using Euclidean geometry.

Consider the triangle with vertices x, a_1, and a_2. Since $\|x - a_1\| = \|x - a_2\|$, this triangle is isosceles. Since $\|b - a_1\| = \|b - a_2\|$, the point b is the foot of the perpendicular dropped from x to $[a_1, a_2]$ and therefore $\|x - b\| < \|x - a_i\|$. Indeed, $\angle xba_i$ is a right angle, and the Pythagorean Theorem shows that

$$\|x - a_i\|^2 = \|x - b\|^2 + \frac{1}{4}\|a_1 - a_2\|^2.$$

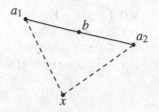

FIG. 16.2 The points x, a_1, a_2, and b.

It now makes sense to define the function P_A by setting it to be this unique closest point to x. Let $a \neq P_A(x)$ be any other point in A. Then for $0 < \lambda \leq 1$,

$$\|x - P_A(x)\|^2 < \|x - (\lambda a + (1-\lambda)P_A(x))\|^2 = \|(x - P_A(x)) - \lambda(a - P_A(x))\|^2$$
$$= \|x - P_A(x)\|^2 - 2\lambda\langle x - P_A(x), a - P_A(x)\rangle + \lambda^2\|a - P_A(x)\|^2,$$

so that $\langle x - P_A(x), a - P_A(x)\rangle \leq \frac{\lambda}{2}\|a - P_A(x)\|^2$. Let λ decrease to 0 to obtain $\langle x - P_A(x), a - P_A(x)\rangle \leq 0$.

This argument is reversible. If a_1 is any point such that $\langle x - a_1, a - a_1\rangle \leq 0$ for all $a \in A$, then

$$\|x - a\|^2 = \|x - a_1\|^2 - 2\langle x - a_1, a - a_1\rangle + \|a - a_1\|^2$$
$$\geq \|x - a_1\|^2 + \|a - a_1\|^2.$$

Hence $a_1 = P_A(x)$ is the unique closest point.

Let x and y be points in $\mathbb{R}^n$. Apply the inequality once for each of x and y:

$$\langle x - P_A(x), P_A(y) - P_A(x)\rangle \leq 0$$

and

$$\langle P_A(y) - y, P_A(y) - P_A(x)\rangle = \langle y - P_A(y), P_A(x) - P_A(y)\rangle \leq 0.$$

Adding yields $\langle (x - y) + (P_A(y) - P_A(x)), P_A(y) - P_A(x)\rangle \leq 0$. Hence

$$\|P_A(y) - P_A(x)\|^2 \leq \langle y - x, P_A(y) - P_A(x)\rangle \leq \|y - x\| \|P_A(y) - P_A(x)\|$$

by the Schwarz inequality (4.1.1). Therefore, $\|P_A(y) - P_A(x)\| \leq \|x - y\|$. ∎

16.3.2. EXAMPLE. Let A be the triangle in $\mathbb{R}^2$ with vertices $(0,0)$, $(1,0)$, and $(0,1)$ as in Figure 16.3. Then

$$P_A((x,y)) = \begin{cases} (x,y) & \text{if } (x,y) \in A, \\ (x,0) & \text{if } 0 \leq x \leq 1 \text{ and } y \leq 0, \\ (0,y) & \text{if } 0 \leq y \leq 1 \text{ and } x \leq 0, \\ (0,0) & \text{if } x \leq 0 \text{ and } y \leq 0, \\ (1,0) & \text{if } x \geq 1 \text{ and } y \leq 0, \end{cases}$$

$$P_A((x,y)) = \begin{cases} (1,0) & \text{if } 0 \leq y \leq x-1, \\ (0,1) & \text{if } y \geq 1 \text{ and } x \leq 0, \\ (0,1) & \text{if } 0 \leq x \leq y-1, \\ (s,1-s) & \text{if } x+y \geq 1 \text{ and } |y-x| \leq 1, \text{ where } s = \frac{y+1-x}{2}. \end{cases}$$

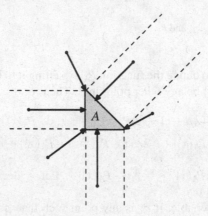

FIG. 16.3 The triangle A and action of P_A.

We can now prove the promised separation theorem, which will be fundamental to all of our later work. We provide only two consequences in this section. There are many more.

16.3.3. SEPARATION THEOREM.
Suppose that A is a closed convex set in $\mathbb{R}^n$ and $b \notin A$. Then there are a vector $h \in \mathbb{R}^n$ and scalar $\alpha \in \mathbb{R}$ such that $\langle a,h \rangle \leq \alpha$ for all $a \in A$ but $\langle b,h \rangle > \alpha$. In particular, h determines a hyperplane H such that A is contained in H^- and b is contained in the interior of H^+.

PROOF. Let $a_1 = P_A(b)$. Since $b \notin A$, $h = b - a_1$ is a nonzero vector. Define $\alpha = \langle a_1,h \rangle$, and let $H = \{x : \langle x,h \rangle = \alpha\}$. By Theorem 16.3.1, if $a \in A$, then $\langle b - a_1, a - a_1 \rangle \leq 0$. Rewriting this, we obtain $\langle a,h \rangle \leq \langle a_1,h \rangle = \alpha$. Therefore, A is contained in $H^- = \{x \in \mathbb{R}^n : \langle x,h \rangle \leq \alpha\}$. On the other hand, $\langle b - a_1, b - a_1 \rangle = \|h\|^2 > 0$, which implies that $\langle b,h \rangle > \langle a_1,h \rangle = \alpha$. ∎

16.3.4. COROLLARY.
Let A be a closed convex subset of $\mathbb{R}^n$. Then b belongs to A if and only if $\langle b,x \rangle \leq \sup\{\langle a,x \rangle : a \in A\}$ for all $x \in \mathbb{R}^n$.

PROOF. If $b \in A$, then the inequality is immediate. For the other direction, suppose that $b \notin A$. Take x to be the vector h given by the Separation Theorem,

$$\langle b,h \rangle > \alpha \geq \langle a,h \rangle \quad \text{for all} \quad a \in A.$$

Thus $\sup\{\langle a,x \rangle : a \in A\} \leq \alpha < \langle b,h \rangle$. Therefore, the inequality of the corollary is valid precisely for $b \in A$. ∎

This corollary has a beautiful geometric meaning. The proof is left as Exercise 16.3.E.

16.3.5. COROLLARY. *A closed convex set in $\mathbb{R}^n$ is the intersection of all the closed half-spaces that contain it.*

16.3.6. DEFINITION. Let A be a closed convex subset of $\mathbb{R}^n$ and $b \in \text{rbd}(A)$. A **supporting hyperplane** to A at b is a hyperplane H such that $b \in H$ and A is contained in one of the closed half-spaces determined by H.

If $\text{aff}(A)$ is a proper subspace of $\mathbb{R}^n$, then it is contained in a hyperplane that supports A. We regard this as a pathological situation and call H a **nontrivial supporting hyperplane** if it does not contain $\text{aff}(A)$.

16.3.7. SUPPORT THEOREM.

Let A be a convex subset of $\mathbb{R}^n$ and $b \in \text{rbd}(A)$. Then there is a nontrivial supporting hyperplane to A at b.

PROOF. We may suppose that A is closed by replacing A with $\overline{A}$. Let a_0 be any point in $\text{ri}(A)$. By the Accessibility Lemma (16.2.6), the interval $[a_0,b)$ is contained in $\text{ri}(A)$. By the same token, the point $b_k := b + \frac{1}{k}(b - a_0)$ is not in A for $k \geq 1$, since otherwise b would lie in $\text{ri}(A)$. Let $h_k = (b_k - P_A b_k)/\|b_k - P_A b_k\|$. The proof of the Separation Theorem (16.3.3) shows that

$$\langle a,h_k \rangle \leq \alpha_k := \langle P_A b_k, h_k \rangle < \langle b_k, h_k \rangle \quad \text{for all} \quad a \in A.$$

By the Heine–Borel Theorem (4.4.6), the unit sphere of $\mathbb{R}^n$ is compact. Thus the sequence $\{h_k : k \geq 1\}$ has a convergent subsequence $\left(h_{k_i}\right)_{i=1}^{\infty}$ with limit $h = \lim_{i \to \infty} h_{k_i}$. Then since P_A is continuous,

$$\lim_{i \to \infty} \alpha_{k_i} = \lim_{i \to \infty} \langle P_A b_{k_i}, h_{k_i} \rangle = \langle P_A b, h \rangle = \langle b,h \rangle =: \alpha.$$

If $a \in A$, then

$$\langle a,h \rangle = \lim_{i \to \infty} \langle a, h_{k_i} \rangle \leq \lim_{i \to \infty} \alpha_{k_i} = \alpha.$$

So A is contained in the half-space $H^- = \{x : \langle x,h \rangle \leq \alpha\}$.

The vectors h_k all lie in the subspace $L(A)$, and thus so does h. Consequently, $b + h$ belongs to $\text{aff}(A)$. Since $\langle b + h, h \rangle = \alpha + 1 > \alpha$, it follows that H does not contain $\text{aff}(A)$. So H is a nontrivial supporting hyperplane. ∎

Exercises for Section 16.3

A. Use the parallelogram law (Exercise 7.4.B) to give a different proof of the uniqueness portion of Theorem 16.3.1.

B. Let $A = \overline{B_r(a)}$. Find an explicit formula for $P_A(x)$.

C. If $A = B_r(a)$ and $b \in \mathrm{rbd}(A)$, show that the *unique* supporting hyperplane to A at b is $H = \{x : \langle x-a, b-a \rangle = r^2\}$.

D. Let A be a subspace. Show that $P_A(x)$ is the (linear) orthogonal projection onto A.

E. Show that a closed convex set in $\mathbb{R}^n$ is the intersection of all closed half-spaces that contain it.

F. Let $A \subset \mathbb{R}^n$ be a nonempty open convex set. Show that $a \in A$ if and only if for each hyperplane H of $\mathbb{R}^n$ containing a, the two *open* half-spaces $\mathrm{int}\, H^+$ and $\mathrm{int}\, H^-$ both intersect A.

G. (a) Suppose $A \subset \mathbb{R}^n$ is convex and $H = \{x \in \mathbb{R}^n : \langle x, h \rangle \leq \alpha\}$ is a nontrivial supporting hyperplane of A. If $a \in A$ and $\langle a, h \rangle = \alpha$, show that $a \in \mathrm{rbd}(A)$.
   (b) Show that an open convex set in $\mathbb{R}^n$ is the intersection of all the open half-spaces that contain it.

H. Suppose $S \subset \mathbb{R}^n$ and $x \in \mathrm{conv}(S) \cap \mathrm{rbd}\,\mathrm{conv}(S)$. Prove there are n points $s_1, \ldots, s_n$ in S such that $x \in \mathrm{conv}\{s_1, \ldots, s_n\}$.   HINT: Use a supporting hyperplane and Carathéodory's Theorem.

I. **Farkas Lemma.** Let $A \subset \mathbb{R}^n$ and let $C = \overline{\mathrm{cone}}(A)$ be the closed convex cone generated by A. Prove that exactly one of the following statements is valid: (1) $x \in C$ or (2) there is an $s \in \mathbb{R}^n$ such that $\langle x, s \rangle > 0 \geq \langle a, s \rangle$ for all $a \in A$.

J. Let $A = \{a_1, \ldots, a_k\}$ be a finite subset of $\mathbb{R}^n$. Prove that the following are equivalent:
   (1) $0 \in \mathrm{conv}(A)$
   (2) There is *no* $y \in \mathbb{R}^n$ such that $\langle a, y \rangle < 0$ for all $a \in A$.
   (3) $f(x) = \log\left(\sum_{i=1}^{n} e^{\langle a_i, x \rangle} \right)$ is bounded below on $\mathbb{R}^n$.
   HINT: (1) $\Leftrightarrow$ (2) separation. Not (2) $\Rightarrow$ not (3), use $f(ty)$. (2) $\Rightarrow$ (3), easy.

K. Suppose that A, B, and C are closed convex sets in $\mathbb{R}^n$ with $A + C = B + C$.
   (a) Is it true that $A = B$?
   (b) What if all three sets are compact?

L. (a) Given an arbitrary set $S \subset \mathbb{R}^n$, prove that a vector b belongs to $\overline{\mathrm{conv}}(S)$ if and only if $\langle b, x \rangle \leq \sup\{\langle s, x \rangle : s \in S\}$ for all $x \in \mathbb{R}^n$.
   (b) Show that the intersection of all closed half-spaces containing S equals $\overline{\mathrm{conv}}\, S$.

M. A hyperplane H **properly separates** A and B if $A \subset H^-$, $B \subset H^+$, and $A \cup B$ is not contained in H. Prove that convex sets A and B can be properly separated if and only if $\mathrm{ri}(A) \cap \mathrm{ri}(B) = \varnothing$. HINT: Let $C = A - B$. Show that $0 \notin \mathrm{ri}(C)$, and separate 0 from C.

16.4 Extreme Points

16.4.1. DEFINITION. Let A be a nonempty convex set. A point $a \in A$ is an **extreme point** of A if whenever $a_1, a_2 \in A$ and $a = (a_1 + a_2)/2$, then $a_1 = a_2$. The set of all extreme points of A is denoted by $\mathrm{ext}\,A$.

A **face** of A is a convex subset F of A such that whenever $a_1, a_2 \in A$ and the open interval (a_1, a_2) intersects F, then $[a_1, a_2]$ is contained in F.

Observe that a one-point face is an extreme point. Conversely, an extreme point is a one-point face. For if (a_1, a_2) contains a, then a is the average of two points of $[a_1, a_2]$ with one of them being an endpoint. Thus both those points equal a, and hence $[a_1, a_2] = \{a\}$.

Every convex set has two trivial faces: the whole set and the empty set. This could be all of them, but often there is a rich collection.

16.4.2. EXAMPLES.

(1) Consider a cube in $\mathbb{R}^3$. The whole set and $\varnothing$ are faces for trivial reasons. The extreme points are the eight vertices. Note that there are many other boundary points that are not extreme. The one-dimensional faces are the twelve edges, and the two-dimensional faces are the six sides (which are commonly called faces).

(2) On the other hand, consider the open unit ball U in $\mathbb{R}^3$. Except for the two trivial cases, there are no extreme points or faces at all. For if F is a proper convex subset, let $a \in F$ and $b \in U \setminus F$. Then since a is an interior point of U, the line segment $[b, a]$ may be extended to some point $c \in U$. Hence $(b, c) \cap F$ contains a but $[b, c]$ is not wholly contained in F.

(3) The closed unit ball B has many extreme points. A modification of the previous argument shows that no interior point is extreme. But every boundary point is extreme. This follows from the proof of Lemma 16.4.3. There are no other faces of the ball except for the two trivial cases.

16.4.3. LEMMA. *Every compact convex set A has an extreme point.*

PROOF. The norm function $f(x) = \|x\|$ is continuous on A, and thus by the Extreme Value Theorem (5.4.4) f achieves its maximum value at some point $a_0 \in A$, say $\|a_0\| = R \geq \|a\|$ for all $a \in A$. Suppose that $a_1, a_2 \in A$ and $a_0 = (a_1 + a_2)/2$. Then by the Schwarz inequality (4.1.1),

$$R^2 = \|a_0\|^2 = \left\langle \frac{a_1 + a_2}{2}, a_0 \right\rangle = \tfrac{1}{2}\langle a_1, a_0 \rangle + \tfrac{1}{2}\langle a_2, a_0 \rangle$$
$$\leq \tfrac{1}{2}\|a_1\|\,\|a_0\| + \tfrac{1}{2}\|a_2\|\,\|a_0\| \leq \tfrac{1}{2}(R^2 + R^2) = R^2.$$

Equality at the extremes forces $\langle a_i, a_0 \rangle = \|a_i\|\,\|a_0\|$ in the Schwarz inequality for $i = 1, 2$ and $\|a_1\| = \|a_2\| = R$. Hence $a_0 = a_1 = a_2$; and thus a_0 is extreme. ∎

We collect a couple of very easy lemmas that produce faces.

16.4.4. LEMMA. *If F is a face of a convex set A, and G is a face of F, then G is a face of A.*

PROOF. Suppose that $a_1, a_2 \in A$ such that (a_1, a_2) intersects G. Then a fortiori, (a_1, a_2) intersects F. Since F is a face of A, it follows that $[a_1, a_2]$ is contained in F. Therefore, $a_1, a_2 \in F$, and since G is a face of F, it follows that $[a_1, a_2]$ is contained in G. So G is a face of A. ∎

16.4.5. LEMMA. *If A is a convex set in $\mathbb{R}^n$ and H is a supporting hyperplane, then $H \cap A$ is a face of A.*

PROOF. Let $H = \{x : \langle x, h \rangle = \alpha\}$ be a hyperplane such that A is contained in $H^- = \{x : \langle x, h \rangle \le \alpha\}$. Let $F = H \cap A$. Suppose that $a_1, a_2 \in A$ such that $(a_1, a_2) \cap F$ contains a point $a = \lambda a_1 + (1 - \lambda) a_2$ for $\lambda \in (0, 1)$. Then

$$\alpha = \langle a, h \rangle = \lambda \langle a_1, h \rangle + (1 - \lambda)\langle a_2, h \rangle \le \lambda \alpha + (1 - \lambda)\alpha = \alpha.$$

Thus equality holds, so $\langle a_1, h \rangle = \langle a_2, h \rangle = \alpha$. Therefore a_1, a_2 belong to F. ∎

We have set the groundwork for a fundamental result that demonstrates the primacy of extreme points.

16.4.6. MINKOWSKI'S THEOREM.
Let C be a nonempty compact convex subset of $\mathbb{R}^n$. Then $C = \operatorname{conv}(\operatorname{ext} C)$.

PROOF. We will prove this by induction on $\dim C$. If $\dim C = 0$, then C is a single point, and it is evidently extreme. Suppose that we have established the result for compact convex sets of dimension at most $k - 1$, and that $\dim C = k$.

Let a be any point in $\operatorname{rbd}(C)$. By the Support Theorem (16.3.7), there is a nontrivial supporting hyperplane H to C at a. By Lemma 16.4.5, $F = H \cap C$ is a face of C. Also, F is compact because C is compact and H is closed, and it is nonempty since $a \in F$.

Note that $\operatorname{aff}(F)$ is contained in $\operatorname{aff}(C) \cap H$. This is properly contained in $\operatorname{aff}(C)$ because C is not contained in H. Therefore, $\dim F < \dim C = k$. By the induction hypothesis, F is the convex hull of its extreme points. However, by Lemma 16.4.4, $\operatorname{ext} F$ is contained in $\operatorname{ext} C$. So $\operatorname{conv}(\operatorname{ext} C)$ contains every boundary point of C.

Finally, let $a \in \operatorname{ri}(C)$, and fix another point $b \in C$. Let L be the line passing through a and b. In particular, L is contained in $\operatorname{aff}(C)$. Then $L \cap C$ is a closed bounded convex subset of L and thus is a closed interval that contains a in its relative interior. Let a_1, a_2 be the two endpoints. These points lie in $\operatorname{rbd}(C)$ because any ball about a_i meets L in points outside of C. By the previous paragraph, both a_1, a_2 lie in $\operatorname{conv}(\operatorname{ext} C)$. But a belongs to $\operatorname{conv}\{a_1, a_2\}$ and hence is also contained in $\operatorname{conv}(\operatorname{ext} C)$. ∎

Exercises for Section 16.4

A. Let $A \subset \mathbb{R}^n$ be convex. Show that no point in $\operatorname{ri}(A)$ is an extreme point.

B. Let $A \subset \mathbb{R}^n$ be convex. Show that $a \in A$ is an extreme point if and only if $A \setminus \{a\}$ is convex.

C. Show that if $B \subset A$ are two convex sets, then any extreme point of A that is contained in B is an extreme point of B.

D. Find a nonempty proper closed convex subset of $\mathbb{R}^2$ with no extreme points.

E.   A face of a convex set A of the form $A \cap H$, where H is a hyperplane, is called an **exposed face**. Let $A = \{(x,y) : x^2 + y^2 \leq 1\} \cup \{(x,y) : 0 \leq x \leq 1, |y| \leq 1\}$. Show that $(0,1)$ is an extreme point that is not exposed.

F.   Let $A \subset \mathbb{R}^n$ be compact and convex, and let f be an affine map of $\mathbb{R}^n$ into $\mathbb{R}$.
  (a) Show that $\{a \in A : f(a) = \sup_{x \in A} f(x)\}$ is an exposed face of A.
  (b) Let B be a subset of $\mathbb{R}^n$ such that $A = \overline{\mathrm{conv}}(B)$. Prove that $\overline{B}$ contains ext$A$.
  (c) Show that an affine function on A always takes its maximum (and minimum) value at an extreme point.

G.   Let $A = [0,1]^n = \{(x_1, \ldots, x_n) \in \mathbb{R}^n : 0 \leq x_i \leq 1, 1 \leq i \leq n\}$.
  (a) Describe ext$A$.
  (b) Explicitly show that each element of A is in the convex hull of $n+1$ extreme points.
      HINT: If $x_1 \leq x_2 \leq \cdots \leq x_n$, consider $y_j = \sum_{i=j}^{n} e_i$ for $1 \leq j \leq n+1$.

H.   A **polyhedral set** is the intersection of a finite number of closed half-spaces. Let A be a closed bounded polyhedral set determined by the intersection of closed half-spaces H_i^- for $1 \leq i \leq p$.

  (a) Show that if $a \in A$ is not in any hyperplane H_i, then a belongs to ri(A).
  (b) Show that every extreme point of A is the intersection of some collection of the hyperplanes H_i.   HINT: Use part (a) and induction on dim$A$.
  (c) Hence deduce that every closed bounded polyhedral set is a polytope.

I.   Let $A = \mathrm{conv}\{a_1, \ldots, a_r\}$ be a polytope.
  (a) Show that ext$A$ is contained in $\{a_1, \ldots, a_r\}$.
  (b) Show that every face of A is the convex hull of a subset of $\{a_1, \ldots, a_r\}$.
  (c) If F is a face of A, find a hyperplane $H \supset F$ that does not contain all of A.
      HINT: Apply the Support Theorem to a point in ri(F).
  (d) Prove that the intersection of half-spaces determined by (c) is a polyhedral set P containing A such that each face of A is contained in rbd(P).
  (e) Show that $P = A$.   HINT: If $p \in P \setminus A$ and $a \in \mathrm{ri}(A)$, consider $[a,p] \cap \mathrm{rbd}(A)$.
  (f) Hence show that every polytope is a closed bounded polyhedral set.

16.5 Convex Functions in One Dimension

Convex functions occur frequently in many applications. Generally, we are interested in minimizing these functions over a convex set determined by a number of constraints, a problem that we will discuss in later sections. The notion of convexity allows us to work with functions that need not be differentiable. The analysis of convex functions can be thought of as an extension of calculus to an important class of nondifferentiable functions.

While a few generalities are introduced here for functions on domains in $\mathbb{R}^n$, most of this section is devoted to convex functions on the line. In the next section, we extend these notions to higher dimensions.

16.5.1. DEFINITION. Suppose that A is a convex subset of $\mathbb{R}^n$. A real-valued function f defined on A is a **convex function** if

$$f(\lambda x + (1-\lambda)y) \leq \lambda f(x) + (1-\lambda)f(y) \quad \text{for all} \quad x, y \in A, \ 0 \leq \lambda \leq 1.$$

A function f is called a **concave function** if $-f$ is convex.

16.5.2. EXAMPLES.

(1) All linear functions are both convex and concave.

(2) If $\|\|\cdot\|\|$ is any norm on $\mathbb{R}^n$, the function $f(x) = \|\|x\|\|$ is convex. Indeed, the triangle inequality and homogeneity yield

$$\|\|\lambda x + (1 - \lambda)y\|\| \leq \lambda\|\|x\|\| + (1 - \lambda)\|\|y\|\| \quad \text{for} \quad x, y \in \mathbb{R}^n \text{ and } 0 \leq \lambda \leq 1,$$

which is precisely the convexity condition.

(3) $f(x) = e^x$ is convex. This is evident from an inspection of its graph. We will see that any C^2 function g with $g'' \geq 0$ is convex. Here $f''(x) = e^x > 0$.

This next result, an easy application of the Mean Value Theorem, provides many examples of convex functions on the line. It also shows that our definition is consistent with the notion introduced in calculus.

16.5.3. LEMMA. *Suppose that f is differentiable on (a, b) and f' is monotone increasing. Then f is convex. In particular, this holds if f is C^2 and $f'' \geq 0$.*

PROOF. Suppose that $a < x < y < b$ and $0 < \lambda < 1$. Let $z = \lambda x + (1 - \lambda)y$. Then there are points $c \in (a, z)$ and $d \in (z, y)$ such that

$$\frac{f(z) - f(x)}{z - x} = f'(c) \leq f'(d) = \frac{f(y) - f(z)}{y - z}.$$

Substituting $z - x = (1 - \lambda)(y - x)$ and $y - z = \lambda(y - x)$ yields

$$\frac{f(z) - f(x)}{1 - \lambda} \leq \frac{f(y) - f(z)}{\lambda}.$$

Just multiply this out to obtain the statement of convexity.

If f is C^2 and $f'' \geq 0$, then f' is an increasing function.   ■

A straightforward induction on r gives the following result from the definition of convexity. The proof is left as an exercise.

16.5.4. JENSEN'S INEQUALITY.
Suppose that $A \subset \mathbb{R}^n$ is convex and f is a convex function on A. If $a_1, \ldots, a_r$ are points in A and $\lambda_1, \ldots, \lambda_r$ are nonnegative scalars such that $\sum_{i=1}^{r} \lambda_i = 1$, then

$$f(\lambda_1 a_1 + \cdots + \lambda_r a_r) \leq \lambda_1 f(a_1) + \cdots + \lambda_r f(a_r).$$

16.5.5. EXAMPLE. In spite of the fact that Jensen's inequality is almost trivial, when it is applied we can obtain results that are not obvious. Consider the exponential function $f(x) = e^x$. Let $t_1, \ldots, t_n$ be positive real numbers and let $a_i = \log t_i$. Then

for positive values λ_i with $\sum\limits_{i=1}^{n} \lambda_i = 1$,

$$e^{\lambda_1 a_1 + \cdots + \lambda_n a_n} \leq \lambda_1 e^{a_1} + \cdots + \lambda_n e^{a_n}.$$

In other words,

$$t_1^{\lambda_1} t_2^{\lambda_2} \cdots t_n^{\lambda_n} \leq \lambda_1 t_1 + \lambda_2 t_2 + \cdots + \lambda_n t_n.$$

This is the **generalized arithmetic mean–geometric mean inequality**. If we set $\lambda_i = \frac{1}{n}$ for $1 \leq i \leq n$, we obtain

$$\sqrt[n]{t_1 t_2 \cdots t_n} \leq \frac{t_1 + t_2 + \cdots + t_n}{n}.$$

We begin by characterizing a convex function in terms of its graph, or rather its epigraph. This also serves to justify the terminology.

16.5.6. DEFINITION. Let f be a real-valued function on a convex subset A of $\mathbb{R}^n$. The **epigraph** of f is defined to be $\text{epi}(f) = \{(a,y) \in A \times \mathbb{R} : y \geq f(a)\}$.

16.5.7. LEMMA. *Let f be a real-valued function on a convex subset A of $\mathbb{R}^n$. Then f is a convex function if and only if $\text{epi}(f)$ is a convex set.*

PROOF. Suppose $p = (x,t)$ and $q = (y,u)$ belong to $\text{epi}(f)$ and $\lambda \in [0,1]$. Since A is convex, $z := \lambda x + (1-\lambda)y \subset A$. Consider $\lambda p + (1-\lambda)q = (z, \lambda t + (1-\lambda)u)$. If f is convex, then

$$f(z) \leq \lambda f(x) + (1-\lambda)f(y) \leq \lambda t + (1-\lambda)u$$

and thus $\text{epi}(f)$ is convex. Conversely, if $\text{epi}(f)$ is convex, then using $t = f(x)$ and $u = f(y)$, we see that $\text{epi}(f)$ contains the point $(z, \lambda f(x) + (1-\lambda)f(y))$, and hence $f(z) \leq \lambda f(x) + (1-\lambda)f(y)$. That is, f is convex. ∎

Now we specialize to functions on the line. We begin with a lemma about secants. In the next section, this will be applied to functions of more variables.

16.5.8. SECANT LEMMA.
Let f be a convex function on $[a,b]$, and consider three points $a \leq x < y < z \leq b$. Then

$$\frac{f(y) - f(x)}{y - x} \leq \frac{f(z) - f(x)}{z - x} \leq \frac{f(z) - f(y)}{z - y}.$$

PROOF. See Figure 16.4. Set $\lambda = \dfrac{z - y}{z - x}$, which lies in $(0,1)$. So $y = \lambda x + (1-\lambda)z$. By convexity, $f(y) \leq \lambda f(x) + (1-\lambda)f(z)$. Therefore,

$$f(y) - f(x) \leq (1-\lambda)\big(f(z) - f(x)\big).$$

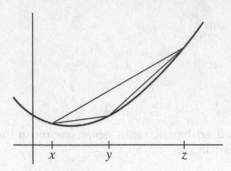

FIG. 16.4 Secants of a convex function.

Divide by $y - x = (1 - \lambda)(z - x)$ to obtain

$$\frac{f(y) - f(x)}{y - x} \leq \left(\frac{1 - \lambda}{1 - \lambda}\right) \frac{f(z) - f(x)}{z - x}.$$

The second inequality is similar. ∎

The main result about convex functions of one real variable is that convex functions are almost differentiable in a certain strong sense. The absolute value function on $\mathbb{R}$ shows that a convex function need not be differentiable, even at interior points of its domain. However, it does have left and right derivatives everywhere.

Recall that a function f has a right derivative at a if $\lim\limits_{h \to 0^+} \dfrac{f(a+h) - f(a)}{h}$ exists. It is denoted by $D_+ f(a)$. Similarly, we define the left derivative to be the limit $D_- f(a) := \lim\limits_{h \to 0^+} \dfrac{f(a) - f(a-h)}{h}$ when it exists.

16.5.9. THEOREM. *Let f be a convex function defined on (a, b). Then f has left and right derivatives at every point, and if $a < x < y < b$, then*

$$D_- f(x) \leq D_+ f(x) \leq D_- f(y) \leq D_+ f(y).$$

Therefore, f is continuous.

PROOF. Let $0 < h < k$ be small enough that $x \pm k$ belong to the interval (a, b). Apply the Secant Lemma using $x - k < x - h < x < x + h < x + k$,

$$\frac{f(x) - f(x-k)}{k} \leq \frac{f(x) - f(x-h)}{h} \leq \frac{f(x+h) - f(x)}{h} \leq \frac{f(x+k) - f(x)}{k}.$$

Thus the quotient function $d_x(t) = \dfrac{f(x+t) - f(x)}{t}$ is an increasing function of t on an interval $[-k, k]$. In particular, $\{d_x(s) : s < 0\}$ is bounded above by $d_x(t)$ for any $t > 0$. Thus by the Least Upper Bound Principle (2.3.3),

$$D_-f(x) = \lim_{s\to 0^+} \frac{f(x)-f(x-s)}{s} = \sup_{s\to 0^+} \frac{f(x)-f(x-s)}{s}$$

exists. Similarly,

$$D_+f(x) = \lim_{t\to 0^+} \frac{f(x+t)-f(x)}{t} = \inf_{t\to 0^+} \frac{f(x+t)-f(x)}{t}$$

exists. Moreover, $D_-f(x) \le D_+f(x)$, since $d_x(-s) \le d_x(t)$ for all $-s < 0 < t$. Another application of the lemma using $x < x+t < y-s < y$ shows that

$$d_x(t) = \frac{f(x+t)-f(x)}{t} \le \frac{f(y)-f(y-s)}{s} = d_y(-s)$$

if s, t are sufficiently small and positive. Thus $D_+f(x) \le D_-f(y)$.

In particular, since left and right derivatives exist, we have

$$\lim_{t\to 0^+} f(x+t) = \lim_{t\to 0^+} f(x) + t\frac{f(x+t)-f(x)}{t} = f(x) + 0D_+f(x) = f(x).$$

Similarly, $\lim_{s\to 0^+} f(x-s) = f(x)$. Therefore f is continuous. ∎

16.5.10. COROLLARY. *Let f be a convex function defined on (a,b). Then f is differentiable except on a countable set of points.*

PROOF. The right derivative $D_+f(x)$ is defined at every point of (a,b) and is a monotone increasing function. By Theorem 5.7.5, D_+f is continuous except for a countable set of jump discontinuities. At every point x where D_+f is continuous, we will show that $D_-f(x) = D_+f(x)$. From the continuity of D_+f, given any $\varepsilon > 0$, there is an $r > 0$ such that $|D_+f(x\pm r) - D_+f(x)| < \varepsilon$. Now if $0 < h < r$,

$$D_+f(x-r) \le D_-f(x-h) \le D_-f(x)$$
$$\le D_+f(x) \le D_-f(x+h) \le D_+f(x+r).$$

Thus $|D_-f(x+h) - D_+f(x)| < \varepsilon$ for all $|h| < r$. Since $\varepsilon > 0$ is arbitrary, we see that $D_-f(x) = D_+f(x)$ and $\lim_{h\to 0} D_-f(x+h) = D_+f(x)$. So D_-f is continuous at x and agrees with $D_+f(x)$. Thus, f is differentiable at each point of continuity of D_+f. ∎

16.5.11. DEFINITION. Let f be a convex function on (a,b). For each x in (a,b), the **subdifferential** of f at x is the set $\partial f(x) = [D_-f(x), D_+f(x)]$.

See Figure 16.5. The geometric interpretation is the following pretty result. When the convex function f is differentiable at c, this result says that epi(f) lies above the tangent line to f through $(c, f(c))$, while any other line through this point crosses above the graph.

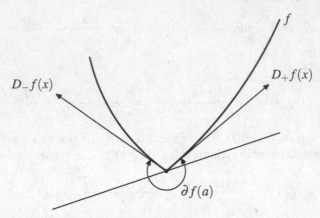

FIG. 16.5 The subdifferential and a tangent line at a nonsmooth point.

16.5.12. PROPOSITION. *Let f be a convex function on (a,b) and fix a point $c \in (a,b)$. The line $y = f(c) + m(x-c)$ is a supporting hyperplane of $\mathrm{epi}(f)$ at $(c, f(c))$ if and only if $m \in \partial f(c)$.*

PROOF. Suppose that $a < x < c < z < b$ and $m \in \partial f(c)$. From the previous proof,

$$\frac{f(c)-f(x)}{c-x} \leq D_-f(c) \leq m \leq D_+f(c) \leq \frac{f(z)-f(c)}{z-c}.$$

Hence $f(x) \geq f(c) + m(x-c)$ and $f(z) \geq f(c) + m(z-c)$. So $\mathrm{epi}(f)$ lies above the line, and thus $y = f(c) + m(x-c)$ is a support line at $(c, f(c))$.

Conversely, suppose that $m > D_+f(c)$. Since $D_+f(c) = \inf_{z>c} \frac{f(z)-f(c)}{z-c}$, there is some point $z > x$ such that $f(z) - f(c) < m(z-c)$. Thus $f(z) < f(c) + m(z-c)$ and the line intersects the interior of $\mathrm{epi}(f)$, which is a contradiction. A similar argument deals with $m < D_-f(c)$. ∎

16.5.13. EXAMPLE. There can be problems at the endpoints when f is defined on a closed interval, even if f is continuous there. For example, consider the function $f(x) = -(1-x^2)^{1/2}$ on $[-1,1]$. Then on $(-1,1)$, $f'(x) = x(1-x^2)^{-1/2}$ and $f''(x) = (1-x^2)^{-3/2}$. Since f is C^2 and $f'' > 0$ on $(-1,1)$, this function is convex. It is differentiable at every interior point. However, at the two endpoints, the graph has a vertical tangent. It is for this reason that we did not define the subdifferential at endpoints.

Exercises for Section 16.5

A. Show that if f and $-f$ are convex, then f is affine.

B. Show that the function f on $[0,1]$ given by $f(x) = 0$ for $x < 1$ and $f(1) = 1$ is a discontinuous convex function. Why does this not contradict Theorem 16.5.9?

C. Prove Jensen's inequality (16.5.4).

D. Let f be a convex function on $\mathbb{R}$ and let g be any continuous function on $[0,1]$. Show that $f\left(\int_0^1 g(x)\,dx\right) \le \int_0^1 f(g(x))\,dx$. HINT: Approximate the integrals by the Riemann sums.

E. (a) Apply the arithmetic mean–geometric mean inequality to $t_1 = \cdots = t_n = 1 + \frac{1}{n}$ and $t_{n+1} = 1$ and a second application with $t_1 = \cdots = t_{n+1} = \frac{n}{n+1}$ and $t_{n+2} = 1$.
  (b) Hence prove that $\left(1 + \frac{1}{n}\right)^n \le \left(1 + \frac{1}{n+1}\right)^{n+1} \le \left(1 + \frac{1}{n+1}\right)^{n+2} \le \left(1 + \frac{1}{n}\right)^{n+1}$.
  (c) Hence show that $\lim_{n \to \infty} \left(1 + \frac{1}{n}\right)^n$ exists.

F. Suppose that $A \subset \mathbb{R}^n$ is a convex set. Show that if $f_1, \ldots, f_k$ are convex functions on A, then the function $f(x) = \max\{f_i(x) : 1 \le i \le k\}$ is convex.

G. Suppose that $\{f_i : i \in I\}$ is a collection of convex functions such that for each $x \in \mathbb{R}$, $g(x) = \sup\{f_i(x) : i \in I\}$ is finite. Show that g is convex.

H. Suppose that f is a convex function on $\mathbb{R}$ that is bounded above. Show that f is constant.

I. (a) If f is a C^2 function on (a,b) such that $f''(x) > 0$ for all $x \in (a,b)$, prove that f is strictly convex: $f(\lambda x + (1 - \lambda)y) < \lambda f(x) + (1 - \lambda)f(y)$ for $x \ne y$ and $0 < \lambda < 1$.
  (b) Find an example of a strictly convex C^2 function on $\mathbb{R}$ such that $f''(0) = 0$.

J. Let f be a convex on (a,b). If f attains its minimum at $c \in (a,b)$, show that $0 \in \partial f(c)$.

K. **Convex Mean Value Theorem.** Consider a continuous convex function f on $[a,b]$. Show that there is $c \in (a,b)$ such that $\dfrac{f(b) - f(a)}{b - a} \in \partial f(c)$. HINT: Adapt the proof of the Mean Value Theorem. Apply Exercise J to $h(x) = f(x) - f(a) - [(f(b) - f(a))/(b - a)](x - a)$.

L. (a) If f is a convex function on $(a,b) \supset [c,d]$, prove that f is Lipschitz on $[c,d]$. HINT: Theorem 16.5.9.
  (b) Show by example that a continuous convex function on $[a,b]$ need not be Lipschitz.

M. Suppose that a function f on (a,b) and satisfies $f\left(\frac{x+y}{2}\right) \le \frac{1}{2}\left(f(x) + f(y)\right)$ for all $x, y \in (a,b)$,
  (a) Show by induction that $f\left(\frac{x_1 + \cdots + x_{2^k}}{2^k}\right) \le 2^{-k}\left(f(x_1) + \cdots + f(x_{2^k})\right)$ for $x_i \in (a,b)$.
  (b) Prove that if f is continuous, then f is convex. HINT: Take each x_i to be either x or y.

N. Prove that $f(x) = \log\left(\dfrac{\sinh ax}{\sinh bx}\right)$ is convex if $0 < b \le a$ as follows:
  (a) Show that $f''(x) > 0$ for $x \ne 0$. HINT: Show $b \sinh ax - a \sinh bx$ is increasing on $[0, \infty)$.
  (b) Find the second-order Taylor polynomial of f about 0 from your knowledge of $\sinh x$ and $\log(1 + x)$. Deduce that f is twice differentiable at 0.

O. Suppose that $A \subset \mathbb{R}^n$ is a convex set.
  (a) If f is convex on A and g is convex and increasing on $\mathbb{R}$, show that $g \circ f$ is convex on A.
  (b) Give an example of convex functions f, g on $\mathbb{R}$ such that $g \circ f$ is not convex.

16.6 Convex Functions in Higher Dimensions

In this section, we will show that convex functions are continuous on the relative interior of their domain. Then we will investigate certain sets that are defined in terms of convex functions.

This first lemma would be trivial if f were known to be continuous.

16.6.1. LEMMA. *Let f be a convex function defined on a convex set $A \subset \mathbb{R}^n$. If C is a compact convex subset of $\mathrm{ri}(A)$, then f is bounded on C.*

PROOF. Since we are only working in A, there is no loss of generality in assuming that $\mathrm{aff}(A) = \mathbb{R}^n$ and so $\mathrm{ri}(A) = \mathrm{int}(A)$.

Fix a point $a \in \mathrm{int}(A)$, and choose $\varepsilon > 0$ such that $\overline{B_\varepsilon(a)} \subset A$. We first show that there is an $r > 0$ such that f is bounded on $B_r(a)$. Let $e_1, \dots, e_n$ be an orthonormal basis for $\mathbb{R}^n$, and define $S = \{a \pm \varepsilon e_i : 1 \le i \le n\}$. Then let $M = \max\{f(s) : s \in S\}$. By Theorem 16.1.7, every b in $\mathrm{conv}(S)$ has the form $b = \sum_{i=1}^n \lambda_i(a - \varepsilon e_i) + \mu_i(a + \varepsilon e_i)$, where $\lambda_i, \mu_i \ge 0$ and $\sum_{i=1}^n \lambda_i + \mu_i = 1$. By Jensen's inequality, $f(b) \le M$.

Since $b - a \in \mathrm{conv}\{\pm \varepsilon e_i : 1 \le i \le n\}$, $2a - b = a - (b - a)$ lies in $\mathrm{conv}(S)$. Since $f(2a - b) \le M$,

$$f(a) \le \frac{f(b) + f(2a - b)}{2} \le \frac{f(b) + M}{2}.$$

Thus $f(b) \ge 2f(a) - M$. Therefore, f is bounded on S and

$$\sup_{b \in S} |f(b)| \le \max\{|M|, |2f(a) - M|\}.$$

Now $\mathrm{conv}(S)$ contains the ball $B_r(a)$, where $r = \varepsilon/\sqrt{n}$.

To complete the argument, we proceed via proof by contradiction. Suppose that there is a sequence a_k in C such that $|f(a_k)|$ tends to ∞. Since C is compact, this sequence has a convergent subsequence, say $a = \lim_{i \to \infty} a_{k_i}$. Choose $r > 0$ and M such that $|f|$ is bounded by M on $B_r(a)$. There is an integer N such that $\|a_{k_i} - a\| < r$ for all $i \ge N$. Hence $|f(a_{k_i})| \le M$ for $i \ge N$, contrary to our hypothesis. Thus f must be bounded on C.   ∎

16.6.2. THEOREM. *Let f be a convex function defined on a convex set $A \subset \mathbb{R}^n$. If C is a compact convex subset of $\mathrm{ri}(A)$, then f is Lipschitz on C. In particular, f is continuous on $\mathrm{ri}(A)$.*

PROOF. The set $X = \mathrm{aff}(A) \setminus \mathrm{ri}(A)$ is closed. Define a function on A by

$$d(a) = \mathrm{dist}(a, X) = \inf\{\|a - x\| : x \in X\}.$$

Clearly, d is a continuous function, and $d(a) = 0$ only if $a \in X$. Now C is a compact subset of $\mathrm{ri}(A)$, and thus $d(a) > 0$ for all $a \in C$. By the Extreme Value Theorem (5.4.4), the minimum value of d on C is attained. So $r = \frac{1}{2}\inf\{d(a) : a \in C\} > 0$.

Let $C_r = \{a \in \mathrm{aff}(A) : \mathrm{dist}(a, C) \le r\}$. By construction, this is a closed bounded set that is contained in $\mathrm{ri}(A)$. Thus it is compact by the Heine–Borel Theorem (4.4.6). The reader should verify that C_r is convex.

By the preceding lemma, $|f|$ is bounded on C_r by some number M. Now fix points $x, y \in C$. Let $\|y - x\| = R$ and $e = (y - x)/R$. Then the interval $[x - re, y + re]$ is contained in C_r. The function $g(t) = f(x + te)$ is a convex function on $[-r, R + r]$. Thus the Secant Lemma (16.5.8) applies:

$$\frac{g(0) - g(-r)}{r} \le \frac{g(R) - g(0)}{R} \le \frac{g(R + r) - g(R)}{r}.$$

Rewriting and using the bound M yields

$$\frac{-2M}{r} \leq \frac{f(y)-f(x)}{\|y-x\|} \leq \frac{2M}{r},$$

whence $|f(y)-f(x)| \leq L\|y-x\|$ for $L = 2M/r$.

This shows that f is Lipschitz and therefore continuous on each compact subset of ri(A). In particular, for each $a \in$ ri(A), there is a ball $\overline{B_r(a)} \cap$ aff(A) contained in ri(A), and thus f is continuous on this ball about a.   ■

A basic question in many applications of analysis is how to minimize a function, often subject to various constraints. It is frequently the case in problems from business, economics, and related fields that the functions and the constraints are convex. These functions may be differentiable, in which case the usual multivariable calculus plays a role. However, even in this case, convexity theory provides a useful perspective. When the functions are not all differentiable, this more sophisticated machinery is necessary.

16.6.3. DEFINITION. Consider a convex function f on a convex set $A \subset \mathbb{R}^n$. A point $a_0 \in A$ is a **global minimizer** for f on A if

$$f(a_0) \leq f(a) \quad \text{for all} \quad a \in A.$$

We call $a_0 \in A$ a **local minimizer** for f on A if there is an $r > 0$ such that

$$f(a_0) \leq f(a) \quad \text{for all} \quad a \in B_r(u_0) \cap A.$$

Clearly, a global minimizer is always a local minimizer. The converse is not true for arbitrary functions. For convex functions they are the same, as we prove in the next theorem. Henceforth, we drop the modifier and refer only to **minimizers**.

16.6.4. THEOREM. *Suppose that $A \subset \mathbb{R}^n$ is a convex set, and consider a convex function f on A. If $a_0 \in A$ is a local minimizer for f on A, then it is a global minimizer for f on A. The set of all global minimizers of f on A is a convex set.*

PROOF. Suppose that a_0 minimizes f on the set $B_r(a_0) \cap A$. Let $a \in A$. Since A is convex, the line $[a_0, a]$ is contained in A. Therefore, there is some $\lambda \in (0,1)$ such that $\lambda a_0 + (1-\lambda)a \in B_r(a_0) \cap A$. By the convexity of f, we have

$$f(a_0) \leq f(\lambda a_0 + (1-\lambda)a) \leq \lambda f(a_0) + (1-\lambda)f(a).$$

Solving this yields $f(a_0) \leq f(a)$. Hence a_0 is a global minimizer.

The last statement follows from Exercise 16.6.A.   ■

A **sublevel set** of a convex function f on A has the form $\{a \in A : f(a) \leq \alpha\}$ for some $\alpha \in \mathbb{R}$. For convenience, we may assume $\alpha = 0$ by the simple device of replacing f by $f - \alpha$. The constraints on the minimization problems that we will consider have this form.

16.6.5. LEMMA. *Let f be a convex function on $\mathbb{R}^n$. Suppose that there is a point $a_0 \in \mathbb{R}^n$ with $f(a_0) < 0$. Then $\mathrm{int}(\{x : f(x) \leq 0\}) = \{x : f(x) < 0\}$ and $\{x : f(x) < 0\} = \{x : f(x) \leq 0\}$.*

PROOF. By Theorem 16.6.2, f is continuous. Thus $A := \{x : f(x) \leq 0\}$ is closed and $\{x : f(x) < 0\}$ is an open subset and so is contained in $\mathrm{int}A$.

Let $a \in \mathrm{int}(A)$. Then the line segment $[a_0, a]$ extends beyond a in A to some point, say $b = (1 + \varepsilon)a - \varepsilon a_0$. Thus $a = \lambda a_0 + (1 - \lambda)b$ for $\lambda = \varepsilon/(1 + \varepsilon)$. Since f is convex,

$$f(a) \leq \lambda f(a_0) + (1 - \lambda)f(b) < 0.$$

Finally, Theorem 16.2.8 shows that $\overline{\mathrm{int}A} = A$. ∎

Exercises for Section 16.6

A. Suppose that $A \subset \mathbb{R}^n$ is convex and f is a convex function on A. Show that the sublevel set $\{a \in A : f(q) \leq \alpha\}$ is convex for each $\alpha \in \mathbb{R}$.

B. Suppose that $A \subset \mathbb{R}^n$ is convex. If $f : A \to \mathbb{R}$ is strictly convex (see Exercise 16.5.I), show that its set of minimizers is either empty or a singleton.

C. If a convex function f on a convex set $A \subset \mathbb{R}^n$ attains its maximum value at $a \in \mathrm{ri}(A)$, show that f is constant.

D. Show that $C_r = \{a \in \mathrm{aff}(A) : \mathrm{dist}(a, C) \leq r\}$ used in the proof of Theorem 16.6.2 is convex.

E. A real-valued function f defined on a cone C is called a **positively homogeneous function** if $f(\lambda x) = \lambda f(x)$ for all $\lambda > 0$ and $x \in C$. Show that a positively homogeneous function f is convex if and only if $f(x + y) \leq f(x) + f(y)$ for all $x, y \in C$.

F. A function f on $\mathbb{R}^n$ is a **sublinear function** if $f(\lambda x + \mu y) \leq \lambda f(x) + \mu f(y)$ for all $x, y \in \mathbb{R}^n$ and $\lambda, \mu \in [0, \infty)$.

(a) Prove that sublinear functions are positively homogeneous and convex.
(b) Prove that f is sublinear if and only if $\mathrm{epi}(f)$ is a cone.

G. Let f be a sublinear function on $\mathbb{R}^n$.

(a) Prove that $f(x) + f(-x) \geq 0$.
(b) Show that if $f(x) + f(-x) = 0$, then f is linear on the line $\mathbb{R}x$ spanned by x.
(c) If $f(x_j) + f(-x_j) = 0$ for $1 \leq j \leq k$, prove that f is linear on $\mathrm{span}\{x_1, \ldots, x_k\}$.
HINT: Consider $f(\pm(x_1 + x_2))$. Use induction.

H. Let f be a bounded convex function on a convex subset $A \times B$ of $\mathbb{R}^m \times \mathbb{R}^n$. Define $g(x) = \inf\{f(x, y) : y \in B\}$. Show that g is convex on A.   HINT: for $x_1, x_2 \in A$, pick y_i such that $g(x_i) = f(x_i, y_i)$. Consider the interval from (x_1, y_1) to (x_2, y_2).

I. Suppose a function $f(x, y)$ on $\mathbb{R}^2$ is a convex function of x for each fixed y and is a continuous function of y for each fixed x. Prove that f is continuous.   HINT: Fix (a, b); find $r, s > 0$ such that $f(x, y)$ is close to $f(a, b)$ on $\{(x, b) : x \in [a - r, a + r]\} \cup \{(a \pm r, y) : y \in [b - s, b + s]\}$.

J. Let $f_1, \ldots, f_r$ be convex functions on a convex set $A \subset \mathbb{R}^n$. Suppose that there is no point $a \in A$ satisfying $f_i(a) < 0$ for $1 \leq i \leq k$. Prove that there exist $\lambda_i \geq 0$ such that $\sum_i \lambda_i = 1$ and $\sum_i \lambda_i f_i \geq 0$ on A.   HINT: Separate 0 from $\{y \in \mathbb{R}^r : \text{for some } a \in A, y_i > f_i(a) \text{ for } 1 \leq i \leq r\}$.

K. Suppose that f and g are convex functions on $\mathbb{R}^n$. The **infimal convolution** of f and g is the function $h(x) := f \odot g(x) = \inf\{f(y) + g(z) : y + z = x\}$. The value $-\infty$ is allowed.

  (a) Suppose that there is an affine function $k(x) = \langle m, x \rangle + b$ for some vector $m \in \mathbb{R}^n$ and $b \in \mathbb{R}$ such that $f(x) \geq k(x)$ and $g(x) \geq k(x)$ for all $x \in \mathbb{R}^n$. Prove that $h(x) > -\infty$ for all x.
  (b) Assuming $h(x) > -\infty$ for all x, show that h is convex.

L. Let $h = f \odot g$ denote the infimal convolution of two convex functions f and g on $\mathbb{R}^n$. Prove that $\mathrm{ri}(\mathrm{epi}(h)) = \mathrm{ri}(\mathrm{epi}(f) + \mathrm{epi}(g)) = \mathrm{ri}(\mathrm{epi}(f)) + \mathrm{ri}(\mathrm{epi}(g))$.

M. Prove that if f is C^2 on a convex set $A \subset \mathbb{R}^n$, then f is convex if and only if the Hessian matrix $\nabla^2 f(x) = \left[\frac{\partial}{\partial x_i} \frac{\partial}{\partial x_j} f(x) \right]$ is positive semidefinite.   HINT: Let $x, y \in A$ and set $g(t) = f(x_t)$, where $x_t = (1-t)x + ty$. Show that $g'(t) = \langle \nabla f(x_t), y - x \rangle$ and $g''(t) = \langle (\nabla^2 f(x_t))(y-x), y-x \rangle$, where $\nabla f(a)$ is the gradient $\left(\frac{\partial}{\partial x_1} f(a), \ldots, \frac{\partial}{\partial x_n} f(a) \right)$.

N. Prove that $f(x) = -\sqrt[n]{x_1 x_2 \cdots x_n}$ is convex on $\mathbb{R}_+^n$.   HINT: Use the previous exercise.

O. Fix a convex set $A \subset \mathbb{R}^n$. Suppose that f_n are convex functions on A that converge pointwise to a function f.
  (a) Prove that f is convex.
  (b) If $S \subset \mathrm{ri}(A)$ is compact, show that f_n converges to f uniformly on S.   HINT: Show that $\sup_k f_k$ is bounded above on each ball; use the Lipschitz condition from Theorem 16.6.2.

P. A function f defined on a convex subset A of $\mathbb{R}^n$ is called a **quasiconvex function** if the sublevel set $\{a \in A : f(a) < \alpha\}$ is a convex set for each $\alpha \in \mathbb{R}$.
  (a) Show that f is quasiconvex iff $f(\lambda a + (1 - \lambda)b) \leq \max\{f(a), f(b)\}$ for all $a, b \in A$.
  (b) By part (a), a convex function is always quasiconvex. Give an example of a function f on $\mathbb{R}$ that is quasiconvex but not convex.
  (c) Let f be differentiable on $\mathbb{R}$. Show that f is quasiconvex iff $f(y) \geq f(x) + f'(x)(y - x)$ for all $x, y \in \mathbb{R}$.

16.7 Subdifferentials and Directional Derivatives

We now turn to the notions of derivatives and subdifferentials. For a convex function of one variable, we had two different notions that were useful, the left and right derivatives and the subdifferential set of supporting hyperplanes to $\mathrm{epi}(f)$. Both have natural generalizations to higher dimensions.

16.7.1. DEFINITION. Suppose that A is a convex subset of $\mathbb{R}^n$, $a \in A$, and f is a convex function on A. A **subgradient** of f at a is a vector $s \in \mathbb{R}^n$ such that

$$f(x) \geq f(a) + \langle x - a, s \rangle \quad \text{for all} \quad x \in A.$$

The set of all subgradients of f at a is the **subdifferential** and is denoted by $\partial f(a)$.

These terms are motivated by the corresponding terms for differentiable functions $f : \mathbb{R}^n \to \mathbb{R}$, so we recall them. The **gradient** of f at a, denoted by $\nabla f(a)$, is the n-tuple of partial derivatives: $\left(\frac{\partial}{\partial x_1} f(a), \ldots, \frac{\partial}{\partial x_n} f(a) \right)$. The **differential** of f at a is the hyperplane in $\mathbb{R}^{n+1}$ given by those points $(x, t) \in \mathbb{R}^n \times \mathbb{R}$ such that $t = f(a) + \langle x - a, \nabla f(a) \rangle$.

If f is a convex function on A and $a \in A$, then a vector s determines a hyperplane of $\mathbb{R}^n \times \mathbb{R}$ by

$$
\begin{aligned}
H &= \{(x,t) \in \mathbb{R}^n \times \mathbb{R} : t = f(a) + \langle x - a, s \rangle\} \\
  &= \{(x,t) \in \mathbb{R}^n \times \mathbb{R} : \langle (x,t), (-s,1) \rangle = f(a) - \langle a, s \rangle\}.
\end{aligned}
$$

The condition that s be a subgradient is precisely that the graph of f is contained in the half-space $H^+ = \{(x,t) \in \mathbb{R}^n \times \mathbb{R} : t \geq f(a) + \langle x - a, s \rangle\}$. Clearly, this is equivalent to saying that $\mathrm{epi}(f)$ be contained in H^+. Since H contains the point $(a, f(a))$, we conclude that the subgradients of f at a correspond to the supporting hyperplanes of $\mathrm{epi}(f)$ at $(a, f(a))$. It is important that the vector $(-s, 1)$ determining H have a 1 in the $(n+1)$st coordinate. This ensures that the hyperplane is nonvertical, meaning that it is not of the form $H' \times \mathbb{R}$ for some hyperplane H' of $\mathbb{R}^n$. In the case $n = 1$, this rules out vertical tangents.

An immediate and important fact is that the subdifferential characterizes minimizers for convex functions. This is the analogue of the calculus fact that minima are critical points. But with the hypothesis of convexity, it becomes a sufficient condition as well.

16.7.2. PROPOSITION. *Suppose that f is a convex function on a convex set A. Then $a \in A$ is a minimizer for f if and only if $0 \in \partial f(a)$.*

PROOF. By definition, $0 \in \partial f(a)$ if and only if $f(x) \geq f(a) + \langle x - a, 0 \rangle = f(a)$ for all $x \in A$.   ∎

16.7.3. THEOREM. *Suppose that A is a convex subset of $\mathbb{R}^n$, $a \in A$, and f is a convex function on A. Then $\partial f(a)$ is convex and closed. Moreover, if $a \in \mathrm{ri}(A)$, then $\partial f(a)$ is nonempty. If $a \in \mathrm{int}(A)$, then $\partial f(a)$ is compact.*

PROOF. Suppose that $s_1, s_2 \in \partial f(a)$ and $\lambda \in [0,1]$. Then

$$
f(x) \geq f(a) + \langle x - a, s_i \rangle \quad \text{for} \quad i = 1, 2.
$$

Hence

$$
\begin{aligned}
f(x) &\geq \lambda\big(f(a) + \langle x - a, s_1 \rangle\big) + (1 - \lambda)\big(f(a) + \langle x - a, s_2 \rangle\big) \\
     &= f(a) + \langle x - a, \lambda s_1 + (1 - \lambda)s_2 \rangle.
\end{aligned}
$$

Thus $\lambda s_1 + (1 - \lambda)s_2$ is a subdifferential of f at a, and so $\partial f(a)$ is convex.

Since the inner product is continuous, it is easy to check that $\partial f(a)$ is closed. For if (s_i) is a sequence of vectors in $\partial f(a)$ converging to $s \in \mathbb{R}^n$, then

$$
f(x) \geq \lim_{i \to \infty} f(a) + \langle x - a, s_i \rangle = f(a) + \langle x - a, s \rangle.
$$

Now suppose that $a \in \mathrm{ri}(A)$. The point $(a, f(a))$ lies on the relative boundary of $\mathrm{epi}(f)$, since if $y < f(a)$, the line segment $[(a, f(a)), (a, y)]$ meets $\mathrm{epi}(f)$ in a single point. By the Support Theorem (16.3.7), there is a nontrivial supporting hyperplane H for $\mathrm{epi}(f)$ at $(a, f(a))$, say

$$H = \{(x,t) \in \mathbb{R}^n \times \mathbb{R} : \langle (x,t), (h,r) \rangle = \alpha\},$$

where (h, r) is nonzero and $\langle a, h \rangle + r f(a) = \alpha$.

We require a nonvertical hyperplane. Suppose that $r = 0$; so $\langle a, h \rangle = \alpha$. Since H is nontrivial, $\mathrm{epi}(f)$ contains a point (b, t) such that

$$\alpha < \langle (b,t), (h,0) \rangle = \langle b, h \rangle.$$

Because a is in the relative interior, there is an $\varepsilon > 0$ such that $c = a + \varepsilon(a - b)$ belongs to A. But then

$$\alpha \leq \langle (c, f(c)), (h,0) \rangle = \langle a, h \rangle + \varepsilon \langle a - b, h \rangle = \alpha + \varepsilon(\alpha - \langle b, h \rangle).$$

Thus $\langle b, h \rangle \leq \alpha < \langle b, h \rangle$, which is absurd. Hence $r \neq 0$.

Let $s = -h/r$. Then H is nonvertical and consists of those vectors (x, t) such that $rt - \langle x, rs \rangle = \alpha = r f(a) - \langle a, rs \rangle$, or equivalently, $t = \langle x - a, s \rangle + f(a)$. Since $\mathrm{epi}(f)$ is contained in H^+, $f(x) \geq t = f(a) + \langle x - a, s \rangle$ for all $x \in A$. Therefore, $s \in \partial f(a)$, and so the subdifferential is nonempty.

Finally, suppose that $a \in \mathrm{int}(A)$ and choose $r > 0$ such that $\overline{B_r(a)} \subset \mathrm{int}(A)$. By Theorem 16.6.2, f is Lipschitz on the compact set $\overline{B_r(A)}$, say with Lipschitz constant L. If $s \in \partial f(a)$ with $s \neq 0$, then $b = a + rs/\|s\| \in \overline{B_r(A)}$. Moreover,

$$f(b) \geq f(a) + \langle b - a, s \rangle = f(a) + r\|s\|.$$

Hence

$$\|s\| \leq \frac{f(b) - f(a)}{r} \leq \frac{L\|b - a\|}{r} = L.$$

So $\partial f(a)$ is closed and bounded and thus is compact.   ∎

16.7.4. EXAMPLE.

Let g be a convex function on $\mathbb{R}$ and set $G(x,y) = g(x) - y$. Consider $\partial G(a, b)$. This consists of all vectors (s, t) such that

$$G(x,y) \geq G(a,b) + \langle (x-a, y-b), (s,t) \rangle \quad \text{for all} \quad (x,y) \in \mathbb{R}^2.$$

Rewrite this as $g(x) \geq g(a) + (x-a)s + (y-b)(t+1)$ for all $x, y \in \mathbb{R}$. If $t \neq -1$, then fixing x and letting y vary, the right-hand side will take all real values and hence will violate the inequality for certain values of y. Thus $t = -1$. The desired inequality then becomes $g(x) \geq g(a) + (x-a)s$ for all $x \in \mathbb{R}$, which is the condition that $s \in \partial g(a)$. Hence $\partial G(a, b) = \{(s, -1) : s \in \partial g(a)\}$.

Subdifferentials are closely related to directional derivatives. For simplicity, we assume that the function is defined on $\mathbb{R}^n$ instead of just a convex subset.

16.7.5. DEFINITION. Suppose that f is a convex function on $\mathbb{R}^n$. For a and $d \in \mathbb{R}^n$, we define the **directional derivative** of f at a in the direction d to be

$$f'(a;d) = \inf_{h \to 0+} \frac{f(a+hd) - f(a)}{h}.$$

It is routine to verify that this function is **positively homogeneous** in the second variable, meaning that $f'(a;td) = tf'(a;d)$ for all $t > 0$.

16.7.6. PROPOSITION. *Suppose that f is a convex function on $\mathbb{R}^n$ and fix a point a in $\mathbb{R}^n$. Then $f'(a;d) = \lim_{h \to 0+} \dfrac{f(a+hd) - f(a)}{h}$ exists for all $d \in \mathbb{R}^n$, and $f'(a; \cdot)$ is a convex function in the second variable.*

PROOF. Fix d in $\mathbb{R}^n$. Then $g(t) = f(a+td)$ is a convex function on $\mathbb{R}$. Hence by Theorem 16.5.9, $D_+g(0)$ exists. However,

$$D_+g(0) = \lim_{h \to 0+} \frac{g(h) - g(a)}{h} = \inf_{h \to 0+} \frac{g(h) - g(a)}{h}$$
$$= \lim_{h \to 0+} \frac{f(a+hd) - f(a)}{h} = \inf_{h \to 0+} \frac{f(a+hd) - f(a)}{h} = f'(a;d).$$

Fix directions $d, e \in \mathbb{R}^n$ and $\lambda \in [0,1]$. Let $c = \lambda d + (1-\lambda)e$. A short calculation using the convexity of f gives

$$\frac{f(a+hc) - f(a)}{h} = \frac{f(\lambda(a+hd) + (1-\lambda)(a+he)) - f(a)}{h}$$
$$\leq \lambda \frac{f(a+hd) - f(a)}{h} + (1-\lambda)\frac{f(a+he) - f(a)}{h}.$$

Taking the limit as h decreases to 0, we obtain

$$f'(a;c) \leq \lambda f'(a;d) + (1-\lambda)f'(a;e).$$

So $f'(a; \cdot)$ is a convex function. ∎

We are now able to obtain a characterization of the subgradient in terms of the directional derivatives, generalizing Proposition 16.5.12.

16.7.7. THEOREM. *Suppose that f is a convex function on $\mathbb{R}^n$, and $a, s \in \mathbb{R}^n$. Then $s \in \partial f(a)$ if and only if $\langle d, s \rangle \leq f'(a;d)$ for all $d \in \mathbb{R}^n$.*

PROOF. Suppose that $s \in \partial f(a)$. Then for any direction $d \in \mathbb{R}^n$ and $h > 0$, we have $f(a+hd) \geq f(a) + \langle hd, s \rangle$. Rearranging, we have

$$f'(a;d) = \lim_{h \to 0^+} \frac{f(a+hd) - f(a)}{h} \geq \langle d, s \rangle.$$

Conversely, suppose that $s \in \mathbb{R}^n$ satisfies

$$\langle d, s \rangle \leq f'(x;d) = \inf_{h \to 0^+} \frac{f(a+hd) - f(a)}{h}$$

for all $d \in \mathbb{R}^n$. Then rearranging yields $f(a) + \langle hd, s \rangle \leq f(a+hd)$. Therefore, s belongs to $\partial f(a)$. ∎

Conversely, we may express the directional derivatives in terms of the subdifferential. This requires a separation theorem. If C is a nonempty compact convex set, the **support function** of C is defined as

$$\sigma_C(x) = \sup\{\langle x, c \rangle : c \in C\}.$$

Note that σ_C is a convex function on $\mathbb{R}^n$ that is positively homogeneous.

It is a consequence of the Separation Theorem (16.3.3) that different compact convex sets have different support functions. For if D is another convex set that contains a point $d \notin C$, then there is a vector x such that

$$\sigma_C(x) = \sup\{\langle x, c \rangle : c \in C\} \leq \alpha < \langle x, d \rangle \leq \sigma_D(x).$$

The next lemma shows how to recover the convex set from a support function.

16.7.8. SUPPORT FUNCTION LEMMA.

Suppose that g is a convex, positively homogeneous function on $\mathbb{R}^n$. Let

$$C(g) = \{c \in \mathbb{R}^n : \langle x, c \rangle \leq g(x) \quad \text{for all} \quad x \in \mathbb{R}^n\}.$$

Then $C(g)$ is compact and convex, and $\sigma_{C(g)} = g$. Thus if A is a compact convex set, then $C(\sigma_A) = A$.

PROOF. Let $M = \sup\{g(x) : \|x\| = 1\}$, which is finite, since g is continuous and the unit sphere is compact. If $c \in C(g)$, then

$$\|c\| = \langle c/\|c\|, c \rangle \leq g(c/\|c\|) \leq M.$$

So $C(g)$ is bounded. Since the inner product is continuous, $\{c : \langle x, c \rangle \leq g(x)\}$ is closed for each $x \in \mathbb{R}^n$. The intersection of closed sets is closed, and hence C is closed. Thus, $C(g)$ is compact by the Heine–Borel Theorem. To see that $C(g)$ is convex, take $c, d \in C(g)$ and $x \in \mathbb{R}^n$. Then $\lambda c + (1-\lambda)d$ belongs to $C(g)$ because

$$\langle x, \lambda c + (1-\lambda)d \rangle \leq \lambda g(x) + (1-\lambda)g(x) = g(x).$$

The inequality $\sigma_{C(g)} \leq g$ is immediate from the definition.

Fix a vector a in $\mathbb{R}^n$. By Theorem 16.7.3, there is a vector s in $\partial g(a)$. Thus $g(x) \geq g(a) + \langle x - a, s \rangle$ for $x \in \mathbb{R}^n$. Rewriting this yields

$$g(x) - \langle x, s \rangle \geq g(a) - \langle a, s \rangle =: \alpha.$$

Take $x = ha$ for $h > 0$ and use the homogeneity of g to compute

$$\alpha \leq g(ha) - \langle ha, s \rangle = h(g(a) - \langle a, s \rangle) = h\alpha.$$

Hence $\alpha = 0$, $g(a) = \langle a, s \rangle$ and s belongs to $C(g)$. So $\sigma_{C(g)}(a) \geq \langle a, s \rangle = g(a)$. Therefore, $\sigma_{C(g)} = g$.

If A is compact and convex, setting $g = \sigma_A$ yields $\sigma_A = \sigma_{C(\sigma_A)}$. By the remarks preceding the proof, this implies that $C(\sigma_A) = A$.   ∎

16.7.9. COROLLARY. *Suppose that f is a convex function on $\mathbb{R}^n$. Fix $a \in \mathbb{R}^n$. Then $f'(a; \cdot)$ is the support function of $\partial f(a)$, namely*

$$f'(a; d) = \sup\{\langle d, s \rangle : s \in \partial f(a)\} \quad for \quad d \in \mathbb{R}^n.$$

PROOF. Theorem 16.7.7 shows that $\partial f(a) = C(g)$, where $g(x) = f'(a; x)$. Thus by the Support Function Lemma,

$$f'(a; d) = \sigma_{\partial f(a)}(d) = \sup\{\langle d, s \rangle : s \in \partial f(a)\}.$$ ∎

16.7.10. EXAMPLE. Consider $f(x) = \|x\|$, the Euclidean norm, on $\mathbb{R}^n$. First look at $a = 0$. A vector s is in $\partial f(0)$ if and only if $\langle x, s \rangle \leq \|x\|$ for all $x \in \mathbb{R}^n$. Thus $\partial f(0) = \overline{B_1}(0)$. Indeed, if $\|s\| \leq 1$, then the Schwarz inequality shows that

$$\langle x, s \rangle \leq \|x\| \|s\| \leq \|x\|.$$

If $\|s\| > 1$, take $x = s$ and notice that $\langle x, s \rangle = \|x\|^2 > \|x\|$.

Now we compute the directional derivatives at 0,

$$f'(0; d) = \lim_{h \to 0^+} \frac{\|hd\|}{h} = \|d\| = f(d).$$

In this case, since $f'(0; \cdot) = f$, Theorem 16.7.7 is redundant and provides no new information. Let us verify Corollary 16.7.9 directly:

$$\sup\left\{\langle d, s \rangle : s \in \overline{B_1(0)}\right\} = \|d\| = f'(0; d)$$

because $\langle d, s \rangle \leq \|d\| \|s\| = \|d\|$, while the choice $s = d/\|d\|$ attains this bound.

Now look at $a \neq 0$. We first compute the directional derivatives:

$$f'(a;d) = \lim_{h \to 0^+} \frac{\|a+hd\| - \|a\|}{h} = \lim_{h \to 0^+} \frac{\|a+hd\|^2 - \|a\|^2}{h(\|a+hd\| + \|a\|)}$$

$$= \lim_{h \to 0^+} \frac{2h\langle d,a\rangle + h^2\|d\|^2}{h(\|a+hd\| + \|a\|)} = \frac{2\langle d,a\rangle}{2\|a\|} = \left\langle d, \frac{a}{\|a\|}\right\rangle.$$

This time Theorem 16.7.7 is very useful. It says that $s \in \partial f(a)$ if and only if $\langle d,s\rangle \le \langle d, \frac{a}{\|a\|}\rangle$ for all $d \in \mathbb{R}^n$. Equivalently, $\langle d, s - \frac{a}{\|a\|}\rangle \le 0$ for all d in $\mathbb{R}^n$. But the left-hand side takes all real values unless $s = a/\|a\|$. Thus $\partial f(a) = \{a/\|a\|\}$.

Observe that $f'(a;d)$ is a linear function of d when $a \ne 0$. Thus f has continuous partial derivatives at a and $\nabla f(a) = a/\|a\|$. So $\partial f(a) = \{\nabla f(a)\}$.

Let's look more generally at the situation that occurs when a convex function f is differentiable at a. Unlike the case of arbitrary functions, the existence of partial derivatives together with convexity implies differentiability.

16.7.11. THEOREM. *Suppose that f is a convex function on an open convex set $A \subset \mathbb{R}^n$. Then the following are equivalent for $a \in A$:*

(1) $\dfrac{\partial f}{\partial x_i}(a)$ *are defined for $1 \le i \le n$.*

(2) *f is differentiable at a.*

(3) *$\partial f(a)$ is a singleton.*

In this case, $\partial f(a) = \{\nabla f(a)\}$.

PROOF. Suppose that (1) holds and consider the convex function

$$g(x) = f(a+x) - f(a) - \langle x, \nabla f(a)\rangle.$$

To show that f is differentiable, it suffices to show that $f(a) + \langle x, \nabla f(a)\rangle$ approximates $f(x)$ to first order near a, or equivalently, that $g(x)/\|x\|$ tends to 0 as $\|x\|$ tends to 0. We use $e_1, \ldots, e_n$ for the standard basis of $\mathbb{R}^n$.

The fact that the n partial derivatives exist means that for $1 \le i \le n$,

$$0 = \lim_{h \to 0} \frac{f(a+he_i) - f(a) - h\frac{\partial f}{\partial x_i}(a)}{h}$$

$$= \lim_{h \to 0} \frac{f(a+he_i) - f(a) - \langle he_i, \nabla f(a)\rangle}{h} = \lim_{h \to 0} \frac{g(he_i)}{h}.$$

Fix an $\varepsilon > 0$ and choose r so small that $|g(he_i)| < \varepsilon|h|/n$ for $|h| \le r$ and $1 \le i \le n$. Take $x = (x_1, \ldots, x_n)$ with $\|x\| \le r/n$. Then

$$g(x) = g\left(\frac{1}{n}\sum_{i=1}^n nx_ie_i\right) \le \frac{1}{n}\sum_{i=1}^n g(nx_ie_i) \le \frac{1}{n}\sum_{i=1}^n \varepsilon|x_i| \le \varepsilon\|x\|.$$

Now

$$0 = g(0) = g\left(\tfrac{x+(-x)}{2}\right) \le \tfrac{1}{2}g(x) + \tfrac{1}{2}g(-x).$$

Thus $g(x) \geq -g(-x) \geq -\varepsilon\|x\|$. Therefore, $|g(x)| \leq \varepsilon\|x\|$ on $B_r(0)$. Since $\varepsilon > 0$ is arbitrary, this proves that f is differentiable at a.

Assuming (2) that f is differentiable at a, the function $f(a) + \langle x, \nabla f(a)\rangle$ approximates $f(x)$ to first order near a with error $g(x)$ satisfying $\lim_{\|x\|\to 0} \dfrac{g(x)}{\|x\|} = 0$. Compute

$$f'(a;d) = \lim_{h\to 0^+} \frac{f(a+hd) - f(a)}{h} = \lim_{h\to 0^+} \frac{\langle hd, \nabla f(a)\rangle + g(he_i)}{h}$$

$$= \langle d, \nabla f(a)\rangle + \lim_{h\to 0} \frac{g(he_i)}{h} = \langle d, \nabla f(a)\rangle.$$

Theorem 16.7.7 now says that $s \in \partial f(a)$ if and only if $\langle d, s\rangle \leq \langle d, \nabla f(a)\rangle$ for all $d \in \mathbb{R}^n$. Equivalently, $\langle d, s - \nabla f(a)\rangle \leq 0$ for all $d \in \mathbb{R}^n$. But the left-hand side takes all real values except when $s = \nabla f(a)$. Therefore, $\partial f(a) = \{\nabla f(a)\}$.

Finally, suppose that (3) holds and $\partial f(a) = \{s\}$. Then by Corollary 16.7.9, $f'(a;d) = \langle d, s\rangle$. In particular, $f'(a; -e_i) = -f'(a; e_i)$ and thus

$$\langle e_i, s\rangle = \lim_{h\to 0^+} \frac{f(a+he_i) - f(a)}{h} = \lim_{h\to 0^-} \frac{f(a+he_i) - f(a)}{h} = \frac{\partial f}{\partial x_i}(a).$$

Hence the partial derivatives of f are defined at a, which proves (1). ∎

16.7.12. EXAMPLE. Let Q be a positive definite $n \times n$ matrix, and let $q \in \mathbb{R}^n$ and $c \in \mathbb{R}$. Minimize the quadratic function $f(x) = \langle x, Qx\rangle + \langle x, q\rangle + c$.

We compute the differential

$$f'(x;d) = \lim_{h\to 0^+} \frac{f(x+hd) - f(x)}{h}$$

$$= \lim_{h\to 0^+} \frac{\langle x+hd, Q(x+hd)\rangle + \langle x+hd, q\rangle - \langle x, Qx\rangle - \langle x, q\rangle}{h}$$

$$= \lim_{h\to 0^+} \frac{1}{h}\left(\langle hd, Qx\rangle + \langle x, Qhd\rangle + \langle hd, Qhd\rangle + \langle hd, q\rangle\right)$$

$$= \lim_{h\to 0^+} \langle d, Qx\rangle + \langle Qx, d\rangle + h\langle d, Qd\rangle + \langle d, q\rangle = \langle d, q + 2Qx\rangle.$$

Thus $\nabla f(x) = q + 2Qx$. We can solve $\nabla f(x) = 0$ to obtain the unique minimizer $x = -\frac{1}{2}Q^{-1}q$ with minimum value $f(-\frac{1}{2}Q^{-1}q) = -\frac{1}{4}\langle q, Q^{-1}q\rangle + c$.

This problem can also be solved using linear algebra. The spectral theorem for Hermitian matrices states that there is an orthonormal basis $v_1, \ldots, v_n$ of eigenvectors that diagonalizes Q. Thus there are positive eigenvalues $d_1, \ldots, d_n$ such that $Qv_i = d_i v_i$. Write q in this basis as $q = q_1 v_1 + \cdots + q_n v_n$. We also write a generic vector x as $x = x_1 v_1 + \cdots + x_n v_n$. Then

$$f(x) = c + \sum_{i=1}^n d_i x_i^2 + q_i x_i = c' + \sum_{i=1}^n d_i (x_i - a_i)^2,$$

where $a_i = -q_i/(2d_i)$ and $c' = c - \sum_{i=1}^{n} d_i a_i^2$.

Now observe by inspection that the minimum is achieved when $x_i = a_i$. To complete the circle, note that

$$x = \sum_{i=1}^{n} a_i v_i = -\frac{1}{2} \sum_{i=1}^{n} \frac{q_i}{d_i} v_i = -\frac{1}{2} Q^{-1} q$$

and the minimum value is

$$f(x) = c' = c - \sum_{i=1}^{n} d_i \left(\frac{q_i}{2d_i} \right)^2 = c - \sum_{i=1}^{n} \frac{q_i^2}{4d_i} = c - \frac{1}{4} \langle q, Q^{-1} q \rangle.$$

We finish this section with two of the calculus rules for subgradients. The proofs are very similar, so we prove the first and leave the second as an exercise.

16.7.13. THEOREM. *Suppose that* $f_1, \ldots, f_k$ *are convex functions on* $\mathbb{R}^n$ *and set* $f(x) = \max\{f_1(x), \ldots, f_k(x)\}$. *For* $a \in \mathbb{R}^n$, *set* $J(a) = \{j : f_j(a) = f(a)\}$. *Then* $\partial f(a) = \mathrm{conv}\{\partial f_j(a) : j \in J(a)\}$.

PROOF. By Theorem 16.6.2, each f_j is continuous. Thus there is an $\varepsilon > 0$ such that $f_j(x) < f(x)$ for all $\|x - a\| < \varepsilon$ and all $j \notin J(a)$. So for $x \in B_\varepsilon(a)$, $f(x) = \max\{f_j(x) : j \in J(a)\}$. Fix $d \in \mathbb{R}^n$ and note that $f(a + hd)$ depends only on $f_j(a + hd)$ for $j \in J(a)$ when $|h| < \varepsilon/d$. Thus using Corollary 16.7.9,

$$f'(a;d) = \lim_{h \to 0^+} \frac{f(a+hd) - f(a)}{h} = \lim_{h \to 0^+} \max_{j \in J(a)} \frac{f_j(a+hd) - f_j(a)}{h}$$

$$= \max_{j \in J(a)} f_j'(a;d) = \max_{j \in J(a)} \sup\{\langle d, s_j \rangle : s_j \in \partial f_j(a)\}$$

$$= \sup\{\langle d, s \rangle : s \in \cup_{j \in J(a)} \partial f_j(a)\}$$

$$= \sup\{\langle d, s \rangle : s \in \mathrm{conv}\{\partial f_j(a) : j \in J(a)\}\}.$$

This shows that $f'(a; \cdot)$ is the support function of the compact convex set $\mathrm{conv}\{\partial f_j(a) : j \in J(a)\}$. By Corollary 16.7.9, $f'(a; \cdot)$ is the support function of $\partial f(a)$. So by the Support Function Lemma (16.7.8), these two sets are equal. ∎

16.7.14. THEOREM. *Suppose that* f_1 *and* f_2 *are convex functions on* $\mathbb{R}^n$ *and* λ_1 *and* λ_2 *are positive real numbers. Then for* $a \in \mathbb{R}^n$,

$$\partial(\lambda_1 f_1 + \lambda_2 f_2)(a) = \lambda_1 \partial f_1(a) + \lambda_2 \partial f_2(a).$$

Exercises for Section 16.7

A. Show that $f'(a;d)$ is sublinear in d: $f'(a;\lambda d + \mu e) \le \lambda f'(a;d) + \mu f'(a;e)$ for all $d,e \in \mathbb{R}^n$ and $\lambda, \mu \in [0, \infty)$.

B. Suppose that f is a convex function on a convex subset $A \subset \mathbb{R}^n$. If $a,b \in A$, and $s \in \partial f(a)$, $t \in \partial f(b)$, show that $\langle s - t, a - b \rangle \ge 0$.

C. Give an example of a convex set $A \subset \mathbb{R}^n$, a convex function f on A, and a point $a \in A$ such that $\partial f(a)$ is empty.

D. Compute the subdifferential for the norm $\|x\|_\infty = \max\{|x_1|, \ldots, |x_n|\}$.

E. Let g be a convex function on $\mathbb{R}$, and define a function on $\mathbb{R}^2$ by $G(x,y) = g(x) - y$.

    (a) Compute the directional derivatives of the function G at a point (a,b).
    (b) Use Theorem 16.7.7 to evaluate $\partial G(a,b)$. Check your answer against Example 16.7.4.

F. Prove Theorem 16.7.14; compute $\partial f(a)$ when $f = \lambda_1 f_1 + \lambda_2 f_2$ as follows:

    (a) Show that $f'(a;d) = \lambda_1 f_1'(a;d) + \lambda_2 f_2'(a;d)$.
    (b) Apply Corollary 16.7.9 to both sides.
    (c) Show that $\partial(f)(a) = \lambda_1 \partial f_1(a) + \lambda_2 \partial f_2(a)$.

G. Given convex functions f on $\mathbb{R}^n$ and g on $\mathbb{R}^m$, define h on $\mathbb{R}^n \times \mathbb{R}^m$ by $h(x,y) = f(x) + g(y)$. Show that $\partial h(x,y) = \partial f(x) \times \partial g(y)$. HINT: $h = F + G$, where $F(x,y) = f(x)$ and $G(x,y) = g(y)$.

H. Suppose S is any nonempty subset of $\mathbb{R}^n$. We may still define the **support function** of S by $\sigma_S = \sup\{\langle s,x \rangle : s \in S\}$, but it may sometimes take the value $+\infty$.

    (a) If $A = \overline{\mathrm{conv}}(S)$, show that $\sigma_S = \sigma_A$. Hence show that σ_S is convex.
    (b) Show that σ_S is finite-valued everywhere if and only if S is bounded.

I. Consider a convex function f on a convex set $A \subset \mathbb{R}^n$ and distinct points $a,b \in \mathrm{ri}(A)$.

    (a) Define g on $[0,1]$ by $g(\lambda) = f(\lambda a + (1 - \lambda)b)$. If $x_\lambda = \lambda a + (1 - \lambda)b$, then show that $\partial g(\lambda) = \{\langle m, b - a \rangle : m \in \partial f(x_\lambda)\}$.
    (b) Use the Convex Mean Value Theorem (Exercise 16.5.K) and part (a) to show that there are $\lambda \in (0,1)$ and $s \in \partial f(x_\lambda)$ such that $f(b) - f(a) = \langle s, b - a \rangle$.

J. Define a **local subgradient** of a convex function f on a convex set $A \subset \mathbb{R}^n$ to be a vector s such that $f(x) \ge f(a) + \langle x - a, s \rangle$ for all $x \in A \cap B_r(a)$ for some $r > 0$. Show that if s is a local subgradient, then it is a subgradient in the usual sense.

K. (a) If f_k are convex functions on a convex subset $A \subset \mathbb{R}^n$ for $k \ge 1$ converging pointwise to f and $s_k \in \partial f_k(x_0)$ converge to s, prove that $s \in \partial f(x_0)$.
    (b) Show that in general, it is not true that every $s \in \partial f$ is obtained as such a limit by considering $f_k(x) = \sqrt{x^2 + 1/k}$ on $\mathbb{R}$.

L. Let $h = f \odot g$ be the infimal convolution (Exercise 16.6.K) of convex functions f and g on $\mathbb{R}^n$.

    (a) Suppose that $x_0 = x_1 + x_2$ and $h(x_0) = f(x_1) + g(x_2)$. Prove $\partial h(x_0) = \partial f(x_1) \cap \partial g(x_2)$. HINT: $s \in \partial h(x_0) \Leftrightarrow f(y) + g(z) \ge f(x_1) + g(x_2) + s(y + z - x_0)$. Take $y = x_1$ or $z = x_2$.
    (b) If (a) holds and g is differentiable, prove that h is differentiable at x_0.

M. **Moreau–Yosida.** Let f be a convex function on an open convex subset A of $\mathbb{R}^n$. For $k \ge 1$, define $f_k(x) = f \odot (k\| \cdot \|^2)(x) = \inf_y f(y) + k\|x - y\|^2$.

    (a) Show that $f_k \le f_{k+1} \le f$.
    (b) Show that x_0 is a minimizer for f if and only if it is a minimizer for every f_k. HINT: If $f(x_0) = f(x_1) + \varepsilon$, find $r > 0$ such that $f(x) \ge f(x_1) + \varepsilon/2$ on $B_r(x_0)$.
    (c) Prove that f_k converges to f. HINT: If L is a Lipschitz constant on some ball about x_0, estimate f_k inside and outside $B_{L/\sqrt{k}}(x_0)$ separately.
    (d) Prove that f_k is differentiable for all $k \ge 1$. HINT: Use the previous exercise.

16.8 Tangent and Normal Cones

In this section, we study two special cones associated to a convex subset of $\mathbb{R}^n$. We develop only a small portion of their theory, since our purpose is to set the stage for our minimization results, and our results are all related to that specific goal.

16.8.1. DEFINITION. Consider a convex set $A \subset \mathbb{R}^n$ and $a \in A$. Define the cone $C_A(a) = \mathbb{R}_+(A - a)$ generated by $A - a$. The **tangent cone** to A at a is the closed cone $T_A(a) = \overline{C_A(a)} = \overline{\mathbb{R}_+(A - a)}$. The **normal cone** to A at a is defined to be $N_A(a) = \{s \in \mathbb{R}^n : \langle s, x - a \rangle \leq 0 \text{ for all } x \in A\}$.

It is routine to verify that $T_A(a)$ and $N_A(a)$ are closed cones. The cone $C_A(a)$ is used only as a tool for working with $T_A(a)$. Notice that $\langle s, x - a \rangle \leq 0$ implies that $\langle s, \lambda(x - a) \rangle \leq 0$ for all $\lambda > 0$. Thus $s \in N_A(a)$ satisfies $\langle s, d \rangle \leq 0$ for all $d \in C_A(a)$. Since the inner product is continuous, the inequality also holds for $d \in T_A(a)$.

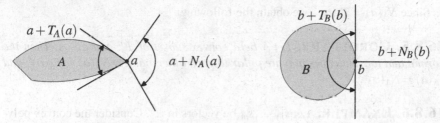

FIG. 16.6 Two examples of tangent and normal cones.

16.8.2. EXAMPLE. As a motivating example, let

$$A = \{(x,y) : x \geq 0, y > 0, x^2 + y^2 < 1\} \cup \{(0,0)\} \subset \mathbb{R}^2.$$

Then $C_A((0,0)) = \{(x,y) : x \geq 0, y > 0\} \cup \{(0,0)\}$. The tangent cone is $T_A((0,0)) = \{(x,y) : x, y \geq 0\}$. At the boundary points $(0,y)$ for $y \in (0,1)$,

$$C_A((0,y)) = T_A((0,y)) = \{(x,y) : x \geq 0\}.$$

Finally, at points $(x,y) \in \text{int} A$, $C_A((x,y)) = T_A((x,y)) = \mathbb{R}^2$.

The normal cone gets smaller as the tangent cone increases in size. Here we have $N_A((0,0)) = \{(a,b) : a, b \leq 0\}$, $N_A((0,y)) = \{(a,0) : a \leq 0\}$ for $y \in (0,1)$, and $N_A((x,y)) = \{0\}$ for $(x,y) \in \text{int} A$.

You may find it useful to draw pictures like Figure 16.6 for various points in A.

Let us formalize the procedure that produced the normal cone.

16.8.3. DEFINITION. Given a nonempty subset A of $\mathbb{R}^n$, the **polar cone** of A, denoted by A°, is $A^\circ = \{s \in \mathbb{R}^n : \langle a, s \rangle \le 0 \text{ for all } a \in A\}$.

It is easy to verify that A° is a closed cone. It is evident from the previous definition that $N_A(a) = T_A(a)^\circ$.

We need the following consequence of the Separation Theorem.

16.8.4. BIPOLAR THEOREM.
If C is a closed cone in $\mathbb{R}^n$, then $C^{\circ\circ} = C$.

PROOF. From the definition, $C^{\circ\circ} = \{d : \langle d, s \rangle \le 0 \text{ for all } s \in C^\circ\}$. This clearly includes C. Conversely, suppose that $x \notin C$. Applying the Separation Theorem (16.3.3), there are a vector s and scalar α such that $\langle c, s \rangle \le \alpha$ for all $c \in C$ and $\langle x, s \rangle > \alpha$. Since $C = \mathbb{R}_+ C$, the set of values $\{\langle c, s \rangle : c \in C\}$ is a cone in $\mathbb{R}$. Since C is bounded above by α, it follows that $C = \mathbb{R}_-$ or $\{0\}$. Hence $\langle c, s \rangle \le 0 \le \alpha < \langle x, s \rangle$ for $c \in C$. Consequently, s belongs to C°. Therefore, x is not in $C^{\circ\circ}$. So $C^{\circ\circ} = C$. ∎

Since $N_A(a) = T_A(a)^\circ$, we obtain the following:

16.8.5. COROLLARY. *Let A be a convex subset of $\mathbb{R}^n$ and $a \in A$. Then the normal and tangent cones at a are polar to each other, namely $N_A(a) = T_A(a)^\circ$ and $T_A(a) = N_A(a)^\circ$.*

16.8.6. EXAMPLE. Let $s_1, \ldots, s_m$ be vectors in $\mathbb{R}^n$. Consider the convex polyhedron given as $A = \{x \in \mathbb{R}^n : \langle x, s_j \rangle \le r_j, 1 \le j \le m\}$. What are the tangent and normal cones at $a \in A$?

Fix $a \in A$. Set $J(a) = \{j : \langle a, s_j \rangle = r_j\}$. Note that $J(a)$ is empty if $a \in \mathrm{ri}(A)$. Now

$$\begin{aligned} C_A(a) &= \{d = t(x - a) : \langle x, s_j \rangle \le r_j \text{ and some } t \ge 0\} \\ &= \{d : \langle \tfrac{d}{t} + a, s_j \rangle \le r_j \text{ and some } t \ge 0\} \\ &= \{d : \langle d, s_j \rangle \le t(r_j - \langle a, s_j \rangle) \text{ and some } t \ge 0\}. \end{aligned}$$

If $r_j - \langle a, s \rangle > 0$, this is no constraint; so $C_A(a) = \{d : \langle d, s_j \rangle \le 0, j \in J(a)\}$. This is closed, and thus $T_A(a) = C_A(a)$.

Note that $\{d : \langle d, s \rangle \le 0\}^\circ = \mathbb{R}_+ s$. Indeed, $(\mathbb{R}_+ s)^\circ = \{d : \langle d, s \rangle \le 0\}$. So the result follows from the Bipolar Theorem. Now Exercise 16.8.J tells us that

$$N_A(x) = \Big(\bigcap_{j \in J(a)} \{d : \langle d, s \rangle \le 0\} \Big)^\circ = \overline{\sum_{j \in J(a)} \mathbb{R}_+ s_j} = \mathrm{cone}\{s_j : j \in J(a)\}.$$

Indeed, $\mathrm{cone}\{s_j : j \in J(a)\}^\circ = \{d : \langle d, s_j \rangle \le 0, j \in J(a)\} = T_A(a)$. Therefore, by Corollary 16.8.5 and the Bipolar Theorem,

$$N_A(x) = T_A(a)^\circ = \mathrm{cone}\{s_j : j \in J(a)\}^{\circ\circ} = \mathrm{cone}\{s_j : j \in J(a)\}.$$

We need to compute the tangent and normal cones for a convex set A given as the sublevel set of a convex function.

16.8.7. LEMMA. *Let A be a compact convex subset of $\mathbb{R}^n$ that does not contain the origin 0. Then the cone $\mathbb{R}_+A$ is closed.*

PROOF. Suppose that $a_k \in A$ and $\lambda_k \geq 0$ and that $c = \lim\limits_{k \to \infty} \lambda_k a_k$ is a point in $\overline{\mathbb{R}_+A}$. From the compactness of A, we deduce that there is a subsequence (k_i) such that $a_0 = \lim\limits_{i \to \infty} a_{k_i}$ exists in A. Because $\|a_0\| \neq 0$,

$$\lambda_0 := \frac{\|c\|}{\|a_0\|} = \lim_{i \to \infty} \frac{\|\lambda_{k_i} a_{k_i}\|}{\|a_{k_i}\|} = \lim_{i \to \infty} \lambda_{k_i}.$$

Therefore, $c = \lim\limits_{i \to \infty} \lambda_{k_i} a_{k_i} = \lambda_0 a_0$ belongs to $\mathbb{R}_+A$. ∎

16.8.8. THEOREM. *Let g be a convex function on $\mathbb{R}^n$, and let A be the convex sublevel set $\{x : g(x) \leq 0\}$. Assume that there is a point x with $g(x) < 0$. If $a \in \mathbb{R}^n$ satisfies $g(a) = 0$, then*

$$T_A(a) = \{d \in \mathbb{R}^n : g'(a;d) \leq 0\} \quad and \quad N_A(a) = \mathbb{R}_+\partial g(a).$$

PROOF. Let $C = \{d \in \mathbb{R}^n : g'(a;d) \leq 0\}$, which is a closed cone. Suppose that d belongs to $A - a$. Then $[a, a+d]$ is contained in A and thus $g(a+hd) - g(a) \leq 0$ for $0 < h \leq 1$. So $g'(a;d) \leq 0$ and hence $d \in C$. Since C is a closed cone, it follows that C contains $\overline{\mathbb{R}_+(A - a)} = T_A(a)$.

Choose $x \in A$ with $g(x) < 0$, and set $d = x - a$. Then

$$g'(a;d) = \inf_{h>0} \frac{g(a+hd) - g(a)}{h} \leq \frac{g(a+d) - g(a)}{1} < 0.$$

Hence by Lemma 16.6.5, $\mathrm{int}\,C = \{d : g'(a;d) < 0\}$ is nonempty, and $C = \overline{\mathrm{int}\,C}$.

Let $d \in \mathrm{int}\,C$. Since $g'(a;d) < 0$, there is some $h > 0$ such that $g(a + hd) < 0$. Consequently, $a + hd$ belongs to A and $d \in \mathbb{R}_+(A - a) \subset T_A(a)$. So $\mathrm{int}\,C$ is a subset of $T_A(a)$. Thus $C = \overline{\mathrm{int}\,C}$ is contained in $T_A(a)$, and the two cones agree.

By Corollary 16.7.9,

$$g'(a;d) = \sup\{\langle d, s \rangle : s \in \partial g(a)\}.$$

Thus $d \in T_A(a)$ if and only if $\langle d, s \rangle \leq 0$ for all $s \in \partial g(a)$, which by definition is the polar cone of $\partial g(a)$. Hence by the Bipolar Theorem (16.8.4), $N_A(a)$, the polar cone of $T_A(a)$, is the closed cone generated by $\partial g(a)$. Now $0 \notin \partial g(a)$ because a is not a minimizer of g (Proposition 16.7.2). Thus, by Lemma 16.8.7, $N_A(a)$ is just $\mathbb{R}_+\partial g(a)$. ∎

Exercises for Section 16.8

A. Show that $T_A(a)$ and $N_A(a)$ are closed convex cones.

B. For a point $v \in \mathbb{R}^n$, show that $v \in N_A(a)$ if and only if $P_A(a+v) = a$.

C. If C is a closed cone, show that $T_C(0) = C$ and $N_C(0) = C^\circ$.

D. Suppose that $A \subset \mathbb{R}^n$ is convex and f is a convex function on A. Prove that $x_0 \in A$ is a minimizer for f in A if and only if $f'(x_0, d) \geq 0$ for all $d \in T_A(x_0)$.

E. Let $f(x,y) = (x - y^2)(x - 2y^2)$. Show that $(0,0)$ is not a minimizer of f on $\mathbb{R}^2$ even though $f'((0,0), d) \geq 0$ for all $d \in \mathbb{R}^2$. Why does this not contradict the previous exercise?

F. If $C_1 \subset C_2$, show that $C_2^\circ \subset C_1^\circ$.

G. If A is a subspace of $\mathbb{R}^n$, show that A° is the orthogonal complement of A.

H. Suppose that $a_1, \ldots, a_r$ are vectors in $\mathbb{R}^n$. Compute $\mathrm{conv}(\{a_1, \ldots, a_r\})^\circ$.

I. If A is a convex subset of $\mathbb{R}^n$, show that $A^\circ = \{0\}$ if and only if $0 \in \mathrm{int}(A)$.
HINT: Use the Separation Theorem and Support Theorem.

J. Suppose that C_1 and C_2 are closed cones in $\mathbb{R}^n$.
(a) Show that $(C_1 + C_2)^\circ = C_1^\circ \cap C_2^\circ$.
(b) Show that $(C_1 \cap C_2)^\circ = \overline{C_1^\circ + C_2^\circ}$.   HINT: Use the Bipolar Theorem and part (a).

K. Given a convex function f on $\mathbb{R}^n$, define g on $\mathbb{R}^{n+1}$ by $g(x, r) = f(x) - r$. Show that
(a) $T_A\big((x, f(x))\big) = \{(d, p) : f'(x; d) \leq p\}$,   HINT: Use Theorem 16.8.8.
(b) $\mathrm{int}\, T_A\big((x, f(x))\big) = \{(d, p) : f'(x; d) < p\}$, and
(c) $N_A\big((x, f(x))\big) = \mathbb{R}_+[\partial f(x) \times \{-1\}]$.
(d) For $n = 1$, explain the last equation geometrically.

L. For a convex subset $A \subset \mathbb{R}^n$, show that the following are equivalent for $x \in A$:
(1) $x \in \mathrm{ri}(A)$.
(2) $T_A(x)$ is a subspace.
(3) $N_A(x)$ is a subspace.
(4) $y \in N_A(x)$ implies that $-y \in N_A(x)$.

M. (a) Suppose $C \subset \mathbb{R}^n$ is a closed convex cone and $x \notin C$. Show that $y \in C$ is the closest vector to x if and only if $x - y \in C^\circ$ and $\langle y, x - y \rangle = 0$.   HINT: Theorem 16.3.1. Expand $\|x - y\|^2$.
(b) Hence deduce that $x = P_C(x) + P_{C^\circ} x$.

N. Give an example of two closed cones in $\mathbb{R}^3$ whose sum is not closed.
HINT: Let $C_i = \mathrm{cone}\{(x, y, 1) : (x, y) \in A_i\}$, where A_1 and A_2 come from Exercise 16.2.G (c).

O. A **polyhedral cone** in $\mathbb{R}^n$ is a set $A\mathbb{R}_+^m = \{Ax : x \in \mathbb{R}_+^m\}$ for some matrix A mapping $\mathbb{R}^m$ into $\mathbb{R}^n$. Show that $(A\mathbb{R}_+^m)^\circ = \{y \in \mathbb{R}^n : A^t y \leq 0\}$, where $z \leq 0$ means $z_i \leq 0$ for $1 \leq i \leq m$.

P. Suppose $A \subset \mathbb{R}^n$ and $B \subset \mathbb{R}^m$ are convex sets. If $(a, b) \in A \times B$, then show that $T_{A \times B}(a, b) = T_A(a) \times T_B(b)$ and $N_{A \times B}(a, b) = N_A(a) \times N_B(b)$.

Q. Suppose that A_1 and A_2 are convex sets and $a \in A_1 \cap A_2$.
(a) Show that $T_{A_1 \cap A_2}(a) \subset T_{A_1}(a) \cap T_{A_2}(a)$.
(b) Give an example where this inclusion is proper.   HINT: Find a convex set A in $\mathbb{R}^2$ such that the positive y-axis Y_+ is contained in $T_A(0)$ but $A \cap Y_+ = \{(0,0)\}$.

16.9 Constrained Minimization

The goal of this section is to characterize the minimizers of a convex function subject to constraints that limit the domain to a convex set. Generally, this set is not
explicitly described but is given as the intersection of level sets. That is, we are
interested in minimizers only in some specified convex set. The first theorem characterizes such minimizers abstractly, using the normal cone of the constraint set and
the subdifferentials of the function. If the constraint is given as the intersection of
sublevel sets of convex functions, these conditions may be described explicitly in
terms of subgradients analogous to the Lagrange multiplier conditions of multivariable calculus. Finally, we present another characterization in terms of saddle points.

We will consider only convex functions that are defined on all of $\mathbb{R}^n$, rather than a
convex subset. This is not as restrictive as it might seem. Exercise 16.9.H will guide
you through a proof that any convex function satisfying a Lipschitz condition on a
convex set A extends to a convex function on all of $\mathbb{R}^n$. There are convex functions
that cannot be extended. For example, $f(x) = -\sqrt{x-x^2}$ on $[0,1]$ is convex, but
cannot be extended to all of $\mathbb{R}$ because the derivative blows up at 0 and 1.

We begin with the problem of minimizing a convex function f defined on $\mathbb{R}^n$
over a convex subset A. A point x in A is called a **feasible point**.

16.9.1. THEOREM. *Suppose that $A \subset \mathbb{R}^n$ is convex and that f is a convex
function on $\mathbb{R}^n$. Then the following are equivalent for $a \in A$:*

(1) *a is a minimizer for $f|_A$.*
(2) *$f'(a;d) \geq 0$ for all $d \in T_A(a)$.*
(3) *$0 \in \partial f(a) + N_A(a)$.*

PROOF. First assume (3) that $0 \in \partial f(a) + N_A(a)$; so there is a vector $s \in \partial f(a)$
such that $-s \in N_A(a)$. Recall that $N_A(a) = \{v : \langle x-a, v \rangle \leq 0 \text{ for all } x \in A\}$. Hence
$\langle x-a, s \rangle \geq 0$ for $x \in A$. Now use the fact that $s \in \partial f(a)$ to obtain

$$f(x) \geq f(a) + \langle x-a, s \rangle \geq f(a).$$

Therefore, a is a minimizer for f on A. So (3) implies (1).

Assume (1) that a is a minimizer for f on A. Let $x \in A$ and set $d = x-a$. Then
$[a,x] = \{a + hd : 0 \leq h \leq 1\}$ is contained in A. So $f(a+hd) \geq f(a)$ for $h \in [0,1]$
and thus

$$f'(a;d) = \lim_{h \to 0^+} \frac{f(a+hd) - f(a)}{h} \geq 0.$$

Because $f'(a;\cdot)$ is positively homogeneous and is nonnegative on $A - a$, it follows
that $f'(a;d) \geq 0$ for d in the cone $\mathbb{R}_+(A - a)$. But $f'(a;\cdot)$ is defined on all of $\mathbb{R}^n$,
and hence is continuous by Theorem 16.6.2. Therefore, $f'(a;\cdot) \geq 0$ on the closure
$T_A(a) = \overline{\mathbb{R}_+(A - a)}$. This establishes (2).

By Theorem 16.7.3, since a is in the interior of $\mathbb{R}^n$, the subdifferential $\partial f(a)$ is
a nonempty compact convex set. Thus the sum $\partial f(a) + N_A(a)$ is closed and convex

by Exercise 16.2.G. Suppose that (3) fails: $0 \notin \partial f(a) + N_A(a)$. Then we may apply the Separation Theorem (16.3.3) to produce a vector d and scalar α such that

$$\sup\{\langle s+n,d\rangle : s \in \partial f(a), n \in N_A(a)\} \leq \alpha < \langle 0,d\rangle = 0.$$

It must be the case that $\langle n,d\rangle \leq 0$ for $n \in N_A(a)$, for if $\langle n,d\rangle > 0$, then

$$\langle s+\lambda n,d\rangle = \langle s,d\rangle + \lambda\langle n,d\rangle > 0$$

for very large λ. Therefore, d belongs to $N_A(a)^\circ = T_A(a)$ by Corollary 16.8.5. Now take $n = 0$ and apply Corollary 16.7.9 to compute

$$f'(a;d) = \sup\{\langle s,d\rangle : s \in \partial f(a)\} \leq \alpha < 0.$$

Thus (2) fails. Contrapositively, (2) implies (3).   ∎

Theorem 16.9.1 is a fundamental and very useful result. In particular, condition (3) does not depend on where a is in the set A. For example, if a is an interior point of A, then $N_A(a) = \{0\}$ and this theorem reduces to Proposition 16.7.2. Given that all we know about the constraint set is that it is convex, this theorem is the best we can do. However, when the constraints are described in other terms, such as the sublevel sets of convex functions, then we can find more detailed characterizations of the optimal solutions.

16.9.2. DEFINITION. By a **convex program**, we mean the ingredients of a minimization problem involving convex functions. Precisely, we have a convex function f on $\mathbb{R}^n$ to be minimized. The set over which f is to be minimized is not given explicitly but instead is determined by constraint conditions of the form $g_i(x) \leq 0$, where $g_1,\ldots,g_r$ are convex functions. The associated problem is

Minimize $f(x)$

subject to constraints $g_1(x) \leq 0, \ldots, g_r(x) \leq 0.$

We call $a \in \mathbb{R}^n$ a **feasible vector** for the convex program if a satisfies the constraints, that is, $g_i(a) \leq 0$ for $i = 1,\ldots,r$. A solution $a \in \mathbb{R}^n$ of this problem is called an **optimal solution** for the convex program, and $f(a)$ is the **optimal value**.

The function f is minimized over the convex set $A = \bigcap_{1 \leq i \leq r}\{x : g_i(x) \leq 0\}$. The r functional constraints may be combined into a single condition, namely $g(x) \leq 0$, where $g(x) = \max\{g_i(x) : 1 \leq i \leq r\}$. The function g is also convex. This is a useful device for technical reasons, but in practice the conditions $g_i \leq 0$ may be superior (e.g., they may be differentiable). So it is better to express optimality conditions in terms of the g_i themselves.

In order to solve this problem, we need to impose some sort of regularity condition on the constraints that allows us to use our results about sublevel sets.

16.9.3. DEFINITION. A convex program satisfies **Slater's condition** if there is point $x \in \mathbb{R}^n$ such that $g_i(x) < 0$ for $i = 1, \ldots, r$. Such a point is called a **strictly feasible point** or a **Slater point**.

16.9.4. KARUSH–KUHN–TUCKER THEOREM.

Consider a convex program that satisfies Slater's condition. Then $a \in \mathbb{R}^n$ is an optimal solution if and only if there is a vector $w = (w_1, \ldots, w_r) \in \mathbb{R}^r$ with $w_j \geq 0$ for $1 \leq j \leq r$ such that

$$0 \in \partial f(a) + w_1 \partial g_1(a) + \cdots + w_r \partial g_r(a),$$
$$g_j(a) \leq 0, \quad w_j g_j(a) = 0 \quad for \quad 1 \leq j \leq r. \tag{KKT}$$

The relations (KKT) are called the **Karush–Kuhn–Tucker conditions**. If a is an optimal solution, the set of vectors $w \in \mathbb{R}_+^r$ that satisfy (KKT) are called the (Lagrange) **multipliers**.

A slight variant on these conditions for differentiable functions was given in a 1951 paper by Kuhn and Tucker and was labeled with their names. Years later, it came to light that they also appeared in Karush's unpublished Master's thesis of 1939, and so Karush's name was added.

This definition of multipliers appears to depend on which optimal point a is used. However, the set of multipliers is in fact independent of a; see Exercise 16.9.F.

PROOF. We introduce the function $g(x) = \max\{g_i(x) : 1 \leq i \leq r\}$. Then the feasible set becomes

$$A = \{x \in \mathbb{R}^n : g_i(x) \leq 0, 1 \leq i \leq r\} = \{x \in \mathbb{R}^n : g(x) \leq 0\}.$$

Slater's condition guarantees that the set $\{x : g(x) < 0\}$ is nonempty. Hence by Lemma 16.6.5, this is the interior of A.

Assume that $a \in A$ is an optimal solution. In particular, a is feasible, so $g_j(a) \leq 0$ for all j. By Theorem 16.9.1, $0 \in \partial f(a) + N_A(a)$. When a is an interior point, we have $N_A(a) = \{0\}$ and so $0 \in \partial f(a)$. Set $w_j = 0$ for $1 \leq j \leq r$, and the conditions (KKT) are satisfied. Otherwise we may suppose that $g(a) = 0$.

The hypotheses of Theorem 16.8.8 are satisfied, and therefore $N_A(a) = \mathbb{R}_+ \partial g(a)$. When the subdifferential of g is computed using Theorem 16.7.13, it is found to be $\partial g(a) = \text{conv}\{\partial g_j(a) : j \in J(a)\}$, where $J(a) = \{j : g_j(a) = 0\}$. We claim that

$$N_A(a) = \left\{ \sum_{j \in J(a)} w_j s_j : w_j \geq 0, s_j \in \partial g_j(a) \right\}.$$

Indeed, every element of $\partial g(a)$ has this form, and multiplication by a positive scalar preserves it. Conversely, if $w = \sum_{j \in J(a)} w_j \neq 0$, then $\sum_{j \in J(a)} \frac{w_j}{w} s_j$ belongs to $\text{conv}\{\partial g_j(a) : j \in J(a)\}$, and so $\sum_{j \in J(a)} w_j s_j$ belongs to $\mathbb{R}_+ \partial g(a)$.

Thus the condition $0 \in \partial f(a) + N_A(a)$ may be restated as $s \in \partial f(a)$, $s_j \in \partial g_j(a)$, $w_j \geq 0$ for $j \in J(a)$, and $s + \sum_{j \in J(a)} w_j s_j = 0$. By definition, $g_j(a) = g(a) = 0$ for

$j \in J(a)$, whence we have $w_j g_j(a) = 0$. For all other j, we set $w_j = 0$, and (KKT) is satisfied.

Conversely, suppose that (KKT) holds. Since $g_j(a) \le 0$, a is a feasible point. The conditions $w_j g_j(a) = 0$ mean that $w_j = 0$ for $j \notin J(a)$. If a is strictly feasible, $J(a)$ is the empty set. In this event, (KKT) reduces to $0 \in \partial f(a) = \partial f(a) + N_A(a)$. On the other hand, when $g(a) = 0$, we saw that in this instance the (KKT) condition is equivalent to $0 \in \partial f(a) + N_A(a)$. In both cases, Theorem 16.9.1 implies that a is an optimal solution. ∎

Notice that if f and each g_i are differentiable, then the first part of (KKT) becomes

$$0 = \nabla f(a) + w_1 \nabla g_1(a) + \cdots + w_r \nabla g_r(a),$$

which is more commonly written as a system of linear equations

$$0 = \frac{\partial f}{\partial x_i}(a) + w_1 \frac{\partial g_1}{\partial x_i}(a) + \cdots + w_r \frac{\partial g_r}{\partial x_i}(a) \quad \text{for} \quad 1 \le i \le n.$$

This is a Lagrange multiplier problem. We adopt the same terminology here.

These conditions can be used to solve concrete optimization problems in much the same way as in multivariable calculus. Their greatest value for applications is in understanding minimization problems, which can lead to the development of efficient numerical algorithms.

16.9.5. DEFINITION. Given a convex program, define the **Lagrangian** of this system to be the function L on $\mathbb{R}^n \times \mathbb{R}^r$ given by

$$L(x, y) = f(x) + y_1 g_1(x) + \cdots + y_r g_r(x).$$

Next, we recall the definition of a saddle point from multivariable calculus. There are several equivalent conditions for saddle points given in the exercises.

16.9.6. DEFINITION. Suppose that X and Y are sets and L is a real-valued function on $X \times Y$. A point $(x_0, y_0) \in X \times Y$ is a **saddle point** for L if

$$L(x_0, y) \le L(x_0, y_0) \le L(x, y_0) \quad \text{for all} \quad x \in X, y \in Y.$$

We shall be interested in saddle points of the Lagrangian over the set $\mathbb{R}^n \times \mathbb{R}^r_+$. We restrict the y variables to the positive orthant $\mathbb{R}^r_+$ because the (KKT) conditions require nonnegative multipliers.

16.9.7. THEOREM. *Consider a convex program that admits an optimal solution and satisfies Slater's condition. Then a is an optimal solution and w a multiplier for the program if and only if (a, w) is a saddle point for its Lagrangian function $L(x, y) = f(x) + y_1 g_1(x) + \cdots + y_r g_r(x)$ on $\mathbb{R}^n \times \mathbb{R}^r_+$. The value $L(a, w)$ at any saddle point equals the optimal value of the program.*

PROOF. First suppose that $L(a,y) \leq L(a,w)$ for all $y \in \mathbb{R}^n_+$. Observe that

$$\sum_{j=1}^{r}(y_j - w_j)g_j(a) = L(a,y) - L(a,w) \leq 0.$$

Since each y_j may be taken to be arbitrarily large, this forces $g_j(a) \leq 0$ for each j. So a is a feasible point. Also, taking $y_j = 0$ and $y_i = w_i$ for $i \neq j$ yields $-w_j g_j(a) \leq 0$. Since this quantity is positive, we deduce that $w_j g_j(a) = 0$. So

$$L(a,w) = f(a) + \sum w_j g_j(a) = f(a).$$

Now turn to the condition $L(a,w) \leq L(x,w)$ for all $x \in \mathbb{R}^n$. This means that $h(x) = L(x,w) = f(x) + \sum_j w_j g_j(x)$ has a global minimum at a. Proposition 16.7.2 shows that $0 \in \partial h(a)$. We may compute $\partial h(a)$ using Theorem 16.7.14. Therefore $0 \in \partial f(a) + \sum_{j=1}^{r} w_j \partial g_j(a)$. This establishes the (KKT) conditions, and thus a is a minimizer for the convex program and w is a multiplier.

Conversely, suppose that a and w satisfy (KKT). Then $g_j(a) \leq 0$ because a is feasible, and $w_j g_j(a) = 0$ for $1 \leq j \leq r$. Therefore, $w_j = 0$ except for $j \in J(a)$. Thus $L(a,y) - L(a,w) = \sum_{j \notin J(a)} y_j g_j(a) \leq 0$. The other part of (KKT) states that the function $h(x)$ has $0 \in \partial h(a)$. Thus by Proposition 16.7.2, a is a minimizer for h. That is, $L(a,w) \leq L(x,w)$ for all $x \in \mathbb{R}^n$. So L has a saddle point at (a,w). ∎

If we have a multiplier w for the convex program, then to solve the convex program it is enough to solve the unconstrained minimization problem

$$\inf\{L(x,w) : x \in \mathbb{R}^n\}.$$

This shows one important property of multipliers: They turn constrained optimization problems into unconstrained ones. In order to use multipliers in this way, we need a method for finding multipliers without first solving the convex program. This problem is addressed in the next section.

16.9.8. EXAMPLE. Consider the following example. Let g be a convex function on $\mathbb{R}$ and fix two points $p = (x_p, y_p)$ and $q = (x_q, y_q)$ in $\mathbb{R}^2$. Minimize the sum of the distances to p and q over $A = \text{epi}(g) = \{(x,y) : G(x,y) \leq 0\}$, where $G(x,y) = g(x) - y$, as indicated in Figure 16.7.

Then the function of $v = (x,y) \in \mathbb{R}^2$ to be minimized is

$$f(v) = \|v-p\| + \|v-q\| = \sqrt{(x-x_p)^2 + (y-y_p)^2} + \sqrt{(x-x_q)^2 + (y-y_q)^2}.$$

Using Example 16.7.10, we may compute that

$$\partial f(v) = \begin{cases} \overline{B_1(0)} + \frac{p-q}{\|p-q\|} & \text{if} \quad v = p, \\ \overline{B_1(0)} - \frac{p-q}{\|p-q\|} & \text{if} \quad v = q, \\ \frac{v-p}{\|v-p\|} + \frac{v-q}{\|v-q\|} & \text{if} \quad x \neq p,q. \end{cases}$$

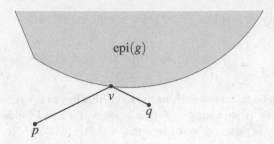

FIG. 16.7 Minimizing the sum of two distances.

Note that $0 \in \partial f(v)$ if $v = p$ or q or the two vectors $v - p$ and $v - q$ point in opposite directions, namely $v \in [p,q]$. This is obvious geometrically.

To make the problem more interesting, let us assume that A does not intersect the line segment $[p,q]$. The (KKT) conditions at the point $v = (x,y)$ become

$$g(x) \leq y, \qquad w \geq 0, \qquad w(g(x) - y) = 0,$$
$$0 \in \partial f(v) + w \partial G(v).$$

The first line reduces to saying that $v = (x, g(x))$ lies in the boundary of A and $w \geq 0$. Alternatively, we could observe that $N_A(v) = \{0\}$ when $v \in \text{int} A$, and thus $0 \notin \partial f(v) + N_A(v)$, whence the minimum occurs on the boundary.

At a point x, we know that $\partial g(x) = [D_-g(x), D_+g(x)]$. Thus by Example 16.7.4 for $v = (x, g(x))$,

$$\partial G(v) = \big\{ (s, -1) : s \in [D_-g(x), D_+g(x)] \big\}.$$

So by Theorem 16.8.8, $N_A(v) = \big\{ (st, -t) : s \in [D_-g(x), D_+g(x)], \ t \geq 0 \big\}$. Thus the second statement of (KKT) says that the sum of the two unit vectors in the directions $p - v$ and $q - v$ is an element of $N_A(v)$. Now geometrically this means that $p - v$ and $q - v$ make the same angle on opposite sides of some normal vector.

In particular, if g is differentiable at v, then $N_A(v) = \mathbb{R}_+(g'(x), -1)$ is the outward normal to the tangent line at v. So the geometric condition is just that the angles to the tangent (from opposite sides) made by $[p,v]$ and $[q,v]$ are equal. In physics, this is the following well-known law: The angle of incidence equals the angle of reflection. For light reflecting off a surface, this is explained by Fermat's principle that a beam of light will follow the fastest path (which is, in this case, the shortest path) between two points.

However, this criterion works just as well when g is not differentiable. For example, take $g(x) = |x|$ and points $p = (-1, -1)$ and $q = (2, 0)$. Then

$$\partial g(x) = \begin{cases} -1 & \text{if } x < 0, \\ [-1, 1] & \text{if } x = 0, \\ +1 & \text{if } x > 0. \end{cases}$$

We will verify the (KKT) condition at the point $0 = (0,0)$. First observe that $\partial G(0) = [-1,1] \times \{-1\}$. So

$$N_A(0) = \mathbb{R}_+ \left(\partial g(0) \times \{-1\} \right) = \{(s,t) : t \leq 0\}$$

consists of the lower half-plane. Now

$$\partial f(0) = (1,1)/\sqrt{2} + (-2,0)/2 = \left(\tfrac{1}{\sqrt{2}} - 1, \tfrac{1}{\sqrt{2}} \right)$$

lies in the upper half-plane. In particular,

$$\left(\tfrac{1}{\sqrt{2}} - 1, \tfrac{1}{\sqrt{2}} \right) + \tfrac{1}{\sqrt{2}} (1 - \sqrt{2}, -1) = (0,0).$$

So $w = 1/\sqrt{2}$ and $v = (0,0)$ satisfy (KKT), and thus $(0,0)$ is the minimizer.

Exercises for Section 16.9

A. Minimize $x^2 + y^2 - 4x - 6y$ subject to $x \geq 0$, $y \geq 0$, and $x^2 + y^2 \leq 4$.

B. Minimize $ax + by$ subject to $x \geq 1$ and $\sqrt{xy} \geq K$. Here a, b, and K are positive constants.

C. Minimize $\dfrac{1}{x} + \dfrac{4}{y} + \dfrac{9}{z}$ subject to $x + y + z = 1$ and $x, y, z > 0$.   HINT: Lagrange multipliers.

D. Suppose (x_1, y_1) and (x_2, y_2) are saddle points of a real-valued function p on $X \times Y$.
  (a) Show that (x_1, y_2) and (x_2, y_1) are also saddle points.
  (b) Show that p takes the same value at all four points.
  (c) Prove that the set of saddle points of p has the form $A \times B$ for $A \subset X$ and $B \subset Y$.

E. (a) Use the previous exercise to show that the set of multipliers for a convex program does not depend on the choice of optimal point.
  (b) Show that the set of multipliers is a closed convex subset of $\mathbb{R}_+^r$.
  (c) Show that the set of saddle points for the Lagrangian is a closed convex rectangle $A \times M$, where A is the set of optimal solutions and M is the set of multipliers.

F. Given a real-valued function p on $X \times Y$, define functions α on X and β on Y by $\alpha(x) = \sup\{p(x,y) : y \in Y\}$ and $\beta(y) = \inf\{p(x,y) : x \in X\}$. Show that for $(x_0, y_0) \in X \times Y$ the following are equivalent:
  (1) (x_0, y_0) is a saddle point for p.
  (2) $p(x_0, y) \leq p(x, y_0)$ for all $x \in X$ and all $y \in Y$.
  (3) $\alpha(x_0) = p(x_0, y_0) = \beta(y_0)$.
  (4) $\alpha(x_0) \leq \beta(y_0)$.

G. Let g be a convex function on $\mathbb{R}$ and let $p = (x_p, y_p) \in \mathbb{R}^2$. Find a criterion for the closest point to p in $A = \mathrm{epi}(g)$.
  (a) What is the function f to be minimized? Find $\partial f(v)$.
  (b) What is the constraint function G? Compute $\partial G(v)$.
  (c) Write down the (KKT) conditions.
  (d) Simplify these conditions and interpret them geometrically.

H. Suppose that A is an open convex subset of $\mathbb{R}^n$ and f is a convex function on A that is Lipschitz with constant L. Construct a convex function g on $\mathbb{R}^n$ extending f as follows:
  (a) Show if $a \in A$ and $v \in \partial f(a)$, then $\|v\| \leq L$.   HINT: Check the proof of Theorem 16.7.3.

(b) For $x \in \mathbb{R}^n$ and $a, b \in A$ and $s \in \partial f(a)$, show that $f(b) + L\|x - b\| \geq f(a) + \langle s, x - a \rangle$.

(c) Define g on $\mathbb{R}^n$ by $g(x) = \inf\{f(a) + L\|x - a\| : a \in A\}$. Show that $g(x) > -\infty$ for $x \in \mathbb{R}^n$ and that $g(a) = f(a)$ for $a \in A$.

(d) Show that g is convex.

I. Let f and $g_1, \ldots, g_m$ be C^1 convex functions on $\mathbb{R}^n$, and set $A = \{x : g_j(x) \leq 0, 1 \leq j \leq m\}$. The problem is to minimize f over A. Let $J(x) = \{j : g_j(x) = 0\}$. Prove that a feasible point x_0 is a local minimum if and only if $\lambda_0 \nabla f(x_0) + \sum_{j \in J(x_0)} \lambda_j \nabla g_j(x_0) = 0$ for constants λ_i not all 0.   HINT: Let $g(x) = \max\{f(x) - f(x_0), g_j(x) : j \in J(x_0)\}$. Compute $\partial g(x_0)$ using Theorem 16.7.13. Then use Exercise 16.3.J to deduce $g'(x_0; d) \geq 0$ for all d if and only if $0 \in \mathrm{cone}\{\nabla f(x_0), \nabla g_j(x_0) : j \in J(x_0)\}$.

J. **Duffin's duality gap.** Let $b \geq 0$, and consider the convex program

Minimize $f(x, y) = e^{-y}$ subject to $g(x, y) = \sqrt{x^2 + y^2} - x \leq b$ in $\mathbb{R}^2$.

(a) Find the feasible region. For which b is Slater's condition satisfied?

(b) Solve the problem. When is the minimum attained?

(c) Show that the solution is not continuous in b.

K. An alternative approach to solving minimization problems is to eliminate the constraint set $g_i(x) \leq 0$ and instead modify f by adding a term $h(g_i(x))$, where h is an increasing function with $h(y) = 0$ for $y \leq 0$. The quantity $h(g_i(x))$ is called a **penalty**, and this approach is the **penalty method**. Assume that f and each g_i are continuous functions on $\mathbb{R}^n$ but not necessarily convex. Let $h_k(y) = k(\max\{y, 0\})^2$. For each integer $k \geq 1$, we have the following minimization problem: Minimize $F_k(x) = f(x) + \sum_{i=1}^r h_k(g_i(x))$ for $x \in \mathbb{R}^n$. Suppose that this minimization problem has a solution a_k and the original has a solution a.

(a) Show that $F_k(a_k) \leq F_{k+1}(a_{k+1}) \leq f(a)$.

(b) Show that $\lim_{k \to \infty} \sum_{i=1}^r h_k(g_i(a_k)) = 0$.

(c) If a_0 is the limit of a subsequence of (a_k), show that it is a minimizer.

(d) If $f(x) \to \infty$ as $\|x\| \to \infty$, deduce that the minimization problem has a solution.

16.10 The Minimax Theorem

In addition to the Lagrangian of the previous section, saddle points play a central role in various other optimization problems. For example, they arise in game theory and mathematical economics. Our purpose in this section is to examine the mathematics that leads to the existence of a saddle point under quite general hypotheses. Examination of a typical saddle point in $\mathbb{R}^3$ shows that the cross sections in the xz-plane are convex functions, while the cross sections in the yz-plane are concave. See Figure 16.8. It is this trade-off that gives the saddle its characteristic shape. Hence we make the following definition:

16.10.1. DEFINITION. A function $p(x, y)$ defined on $X \times Y$ is called **convex–concave** if $p(\cdot, y)$ is a convex function of x for each fixed $y \in Y$ and $p(x, \cdot)$ is a concave function of y for each fixed $x \in X$.

The term **minimax** comes from comparing two interesting quantities:

$$p_* = \sup_{y \in Y} \inf_{x \in X} p(x, y) \quad \text{and} \quad p^* = \inf_{x \in X} \sup_{y \in Y} p(x, y),$$

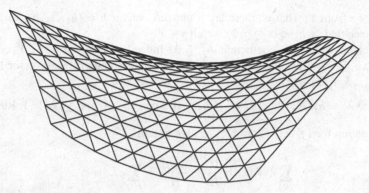

FIG. 16.8 A typical saddle point.

which are the **maximin** and **minimax**, respectively. These quantities make sense for any function p. Moreover, for $x_1 \in X$ and $y_1 \in Y$,

$$\inf_{x \in X} p(x, y_1) \leq p(x_1, y_1) \leq \sup_{y \in Y} p(x_1, y).$$

Take the supremum of the left-hand side over $y_1 \in Y$ to get $p_* \leq \sup_{y \in Y} p(x_1, y)$, since the right-hand side does not depend on y_1. Then take the infimum over all $x_1 \in X$ to obtain $p_* \leq p^*$.

Suppose that there is a saddle point $(\bar{x}, \bar{y})$, that is, $p(\bar{x}, y) \leq p(\bar{x}, \bar{y}) \leq p(x, \bar{y})$ for all $x \in X$ and $y \in Y$. Then

$$p_* \geq \inf_{x \in X} p(x, \bar{y}) \geq p(\bar{x}, \bar{y}) \geq \sup_{y \in Y} p(\bar{x}, y) \geq p^*.$$

Thus the existence of a saddle point shows that $p_* = p^*$.

We will use the following variant of Exercise 16.6.J.

16.10.2. LEMMA. *Let $f_1, \ldots, f_r$ be convex functions on a convex subset X of $\mathbb{R}^n$. For $c \in \mathbb{R}$, the following are equivalent:*

(1) *There is no point $x \in X$ satisfying $f_j(x) < c$ for $1 \leq j \leq r$.*
(2) *There exist $\lambda_j \geq 0$ such that $\sum_j \lambda_j = 1$ and $\sum_j \lambda_j f_j \geq c$ on X.*

PROOF. If (1) is false and $f_j(x_0) < c$ for $1 \leq j \leq k$, then $\sum_j \lambda_j f_j(x_0) < c$ for all choices of $\lambda_j \geq 0$ with $\sum_j \lambda_j = 1$. Hence (2) is also false.

Conversely, assume that (1) is true. Define

$$Y = \{ y \in \mathbb{R}^r : y_j > f_j(x) \text{ for } 1 \leq j \leq r \text{ and some } x \in X \}.$$

This set is open and convex (Exercise 16.10.B). By (1), $z = (c, c \ldots, c) \in \mathbb{R}^r$ is not in Y. Depending on whether z belongs to $\bar{Y}$ or not, we apply either the Support Theorem (16.3.7) or the Separation Theorem (16.3.3) to obtain a hyperplane that separates z from Y. That is, there are a nonzero vector $h = (h_1, \ldots, h_r)$ in $\mathbb{R}^r$ and $\alpha \in \mathbb{R}$ such that $\langle y, h \rangle < \alpha \le \langle z, h \rangle$ for all $y \in Y$.

We claim that each coefficient $h_j \le 0$. Indeed, for any $x \in X$, $\overline{Y}$ contains $(f_1(x), \ldots, f_r(x)) + te_j$ for any $t \ge 0$, where e_j is a standard basis vector for $\mathbb{R}^r$. Thus $\sum_{j=1}^r h_j f_j(x) + th_j \le \alpha$, which implies that $h_j \le 0$.

Define $\lambda_j = h_j / H$, where $H = \sum_{j=1}^r h_j < 0$. Then $\lambda_j \ge 0$ and $\sum_j \lambda_j = 1$. Restating the separation for $(f_1(x), \ldots, f_r(x)) \in \overline{Y}$, we obtain

$$\sum_{j=1}^r \lambda_j f_j(x) \ge \frac{\alpha}{H} \ge \frac{\langle z, h \rangle}{H} = \sum_{j=1}^r c\lambda_j = c.$$

So (2) holds.   ∎

Now we establish our saddle-point result. First we assume compactness. We will remove it later, at the price of adding a mild additional requirement.

16.10.3. MINIMAX THEOREM (COMPACT CASE).
Let X be a compact convex subset of $\mathbb{R}^n$ and let Y be a compact convex subset of $\mathbb{R}^m$. If p is a convex–concave function on $X \times Y$, then p has a nonempty compact convex set of saddle points.

PROOF. For each $y \in Y$, define a convex function on X by $p_y(x) = p(x, y)$. For each $c > p_*$, define $A_{y,c} = \{x \in X : p_y(x) \le c\}$. Then this is a nonempty compact convex subset of X.

For any finite set of points $y_1, \ldots, y_r$ in Y, we claim that $A_{y_1,c} \cap \cdots \cap A_{y_r,c}$ is nonempty. If not, then there is no point x so that $p_{y_j}(x) < c$ for $1 \le j \le r$. So by Lemma 16.10.2, there would be scalars $\lambda_i \ge 0$ with $\sum_i \lambda_i = 1$ so that $\sum_i \lambda_i p_{y_i} \ge c$ on X. Set $\bar{y} = \sum_{i=1}^r \lambda_i y_i$. Since p is concave in y,

$$c \le \sum_{i=1}^r \lambda_i p(x, y_i) \le p(x, \bar{y}) \quad \text{for all} \quad x \in X.$$

Consequently, $c \le \min_{x \in X} p(x, \bar{y}) \le p_*$, which is a contradiction.

Let $\{y_i : i \ge 1\}$ be a dense subset of Y, and set $c_n = p_* + 1/n$. Then the set $A_n = \bigcap_{i=1}^n A_{y_i,c_n}$ is nonempty closed and convex. It is clear that A_n contains A_{n+1}, and thus this is a decreasing sequence of compact sets in $\mathbb{R}^n$. By Cantor's Intersection Theorem (4.4.7), the set $A = \bigcap_{n \ge 1} A_n$ is a nonempty compact set. (It is also convex, as the reader can easily verify.)

Let $\bar{x} \in A$. Then $\bar{x} \in A_{y_i,c_n}$ for all $n \ge 1$. Thus $p(\bar{x}, y_i) \le c_n$ for all $i \ge 1$ and $n \ge 1$. So $p(\bar{x}, y_i) \le p_*$. But the set $\{y_i : i \ge 1\}$ is dense in Y and p is continuous, so $p(\bar{x}, y) \le p_*$ for all $y \in Y$. Therefore,

$$p^* = \inf_{x \in X} \max_{y \in Y} p(x, y) \le \max_{y \in Y} p(\bar{x}, y) \le p_*.$$

Since $p_* \leq p^*$ is always true, we obtain equality. Choose a point $\bar{y} \in Y$ such that $p(\bar{x}, \bar{y}) = p_*$. Then $p(\bar{x}, y) \leq p(\bar{x}, \bar{y}) \leq p(x, \bar{y})$ for all $x \in X$ and $y \in Y$. That is, $(\bar{x}, \bar{y})$ is a saddle point for p.

By Exercise 16.9.D, the set of saddle points is a rectangle $A \times B$. Moreover, the same argument required in Exercise 16.9.E shows that this rectangle is closed and convex.   ∎

Slippery things can happen at infinity if precautions are not taken. However, the requirements of the next theorem are often satisfied.

16.10.4. MINIMAX THEOREM.
Suppose that X is a closed convex subset of $\mathbb{R}^n$ and Y is a closed convex subset of $\mathbb{R}^m$. Assume that p is convex–concave on $X \times Y$, and in addition assume that

(1) *if X is unbounded, there is $y_0 \in Y$ such that $p(x, y_0) \to +\infty$ as $\|x\| \to \infty$ for $x \in X$.*

(2) *if Y is unbounded, there is $x_0 \in X$ such that $p(x_0, y) \to -\infty$ as $\|y\| \to \infty$ for $y \in Y$.*

Then p has a nonempty compact convex set of saddle points.

PROOF. We deal only with the case in which both X and Y are unbounded. The reader can find a modification that works when only one is unbounded.

By the hypotheses, $\max\limits_{y \in Y} p(x_0, y) = \alpha < \infty$ and $\min\limits_{x \in X} p(x, y_0) = \beta > -\infty$. Clearly, $\beta \leq p(x_0, y_0) \leq \alpha$. Set

$$X_0 = \{x \in X : p(x, y_0) \leq \alpha + 1\} \quad \text{and} \quad Y_0 = \{y \in Y : p(x_0, y) \geq \beta + 1\}.$$

Conditions (1) and (2) guarantee that X_0 and Y_0 are bounded, and thus they are compact and convex. Let $A \times B$ be the set of saddle points for the restriction of p to $X_0 \times Y_0$ provided by the compact case.

In particular, let $(\bar{x}, \bar{y})$ be one saddle point, and let $c = p(\bar{x}, \bar{y})$ be the critical value. Then

$$\beta \leq p(\bar{x}, y_0) \leq p(\bar{x}, \bar{y}) = c \leq p(x_0, \bar{y}) \leq \alpha.$$

Let $x \in X \setminus X_0$, so that $p(x, y_0) > \alpha + 1$. Now $p(\cdot, y_0)$ is continuous, and hence there is a point x_1 in $[x, \bar{x}]$ with $p(x_1, y_0) = \alpha + 1$. So $x_1 \in X_0$ and $x_1 \neq \bar{x}$. Thus $x_1 = \lambda x + (1 - \lambda)\bar{x}$ for some $0 < \lambda < 1$. Since $p(\cdot, \bar{y})$ is convex,

$$c \leq p(x_1, \bar{y}) \leq \lambda p(x, \bar{y}) + (1 - \lambda)p(\bar{x}, \bar{y}) = \lambda p(\bar{x}, \bar{y}) + (1 - \lambda)c.$$

Hence $p(\bar{x}, \bar{y}) = c \leq p(x, \bar{y})$. Similarly, for every $y \in Y \setminus Y_0$, we may show that $p(\bar{x}, y) \leq c = p(\bar{x}, \bar{y})$. Therefore, $(\bar{x}, \bar{y})$ is a saddle point in $X \times Y$.   ∎

Now let us see how this applies to the problem of constrained optimization. Consider the following convex programming problem: Minimize a convex function $f(x)$ over the closed convex set $X = \{x : g_j(x) \leq 0, 1 \leq j \leq r\}$. Suppose that it satisfies

Slater's condition. The Lagrangian $L(x,y) = f(x) + y_1g_1(x) + \cdots + y_rg_r(x)$ is a convex function of x for each fixed $y \in \mathbb{R}^r_+$, and is a linear function of y for each $x \in \mathbb{R}^n$. So, in particular, L is convex–concave on $\mathbb{R}^n \times \mathbb{R}^r_+$.

Now we also suppose that this problem has an optimal solution. Then we can apply Theorem 16.9.7 and the Karush–Kuhn–Tucker Theorem (16.9.4) to guarantee a saddle point (a,w) for L; and $L(a,w)$ is the solution of the convex program. By the arguments of this section, it follows that the existence of a saddle point means that the optimal value is also obtained as the maximin:

$$f(a) = \min_{x \in X} f(x) = L^* = \max_{y \in \mathbb{R}^r_+} \inf_{x \in \mathbb{R}^n} L(x,y).$$

Define $h(y) = \inf_{x \in \mathbb{R}^n} f(x) + y_1g_1(x) + \cdots + y_rg_r(x)$ for $y \in \mathbb{R}^r_+$. While its definition requires an infimum, h gives a new optimization problem, which can be easier to solve. This new problem is called the **dual program**:

$$\text{Maximize } h(y) \text{ over } y \in \mathbb{R}^r_+.$$

16.10.5. PROPOSITION. *Consider a convex program that admits an optimal solution and satisfies Slater's condition. The solutions of the dual program are exactly the multipliers of the original program, and the optimal value of the dual program is the same.*

PROOF. Suppose a is an optimal solution of the original program and w a multiplier. Then $L(a,y) \leq L(a,w) \leq L(x,w)$ for all $x \in \mathbb{R}^n$ and $y \in \mathbb{R}^r_+$, since (a,w) is a saddle point for the Lagrangian L. In particular,

$$h(w) = \inf\{L(x,w) : x \in \mathbb{R}^n\} = L(a,w).$$

Moreover, for any $y \in \mathbb{R}^r_+$ with $y \neq w$, $h(y) \leq L(a,y) \leq L(a,w) = h(w)$. So w is a solution of the dual problem, and the value is $L(a,w) = L^*$, which equals the value of the original problem.

Conversely, suppose that w' is a solution of the dual program. Let (a,w) be a saddle point. Then

$$L^* = L(a,w) \geq L(a,w') \geq h(w') = L^*.$$

Thus $h(w') = L(a,w') = L^*$. Therefore, $L(a,w') = h(w') \leq L(x,w')$ for all $x \in \mathbb{R}^n$. Also, since (a,w) is a saddle point, $L(a,y) \leq L(a,w) = L(a,w')$ for all $y \in \mathbb{R}^r_+$. Consequently, w' is a multiplier. $\blacksquare$

An important fact for computational purposes is that since these two problems have the same answer, we can obtain estimates for the solution by sampling. Suppose that we have points $x_0 \in X$ and $y_0 \in \mathbb{R}^m_+$ such that $h(y_0) - f(x_0) < \varepsilon$. Then since we know that the solution lies in $[f(x_0), h(y_0)]$, we have a good estimate for the solution even if we cannot compute it exactly.

16.10.6. EXAMPLE. Consider a **quadratic programming** problem. Let Q be a positive definite $n \times n$ matrix, and let $q \in \mathbb{R}^n$. Also let A be an $m \times n$ matrix and $a \in \mathbb{R}^m$. Minimize the quadratic function $f(x) = \langle x, Qx \rangle + \langle x, q \rangle$ over the region $Ax \leq a$.

We can assert before doing any calculation that this minimum will be attained. This follows from the global version, Example 16.7.12, where it was shown that f may be written as a sum of squares. Thus f tends to infinity as $\|x\|$ goes to infinity. Therefore, the constraint set could be replaced with a compact set. Then the Extreme Value Theorem asserts that the minimum is attained.

The constraint condition is really m linear conditions $\langle x, A^t e_j \rangle - a_j \leq 0$ for $1 \leq j \leq m$, where $a = (a_1, \ldots, a_m)$ with respect to the standard basis $e_1, \ldots, e_m$. If $\text{rank} A = m$, then $m \leq n$ and A maps $\mathbb{R}^n$ onto $\mathbb{R}^m$. Thus there are strictly feasible points and so Slater's condition is satisfied. In general this needs to be checked.

The Lagrangian is defined on $\mathbb{R}^n \times \mathbb{R}^m_+$ by

$$L(x,y) = f(x) + \sum_{j=1}^m \left(\langle x, A^t e_j \rangle - a_j \right) y_j = \langle x, Qx \rangle + \langle x, q + A^t y \rangle - \langle a, y \rangle.$$

To find a solution to the dual problem, we first must compute $h(y) = \inf_{x \in \mathbb{R}^n} L(x,y)$. This was solved in Example 16.7.12, so

$$h(y) = -\tfrac{1}{4} \langle q + A^t y, Q^{-1}(q + A^t y) \rangle - \langle a, y \rangle$$
$$= -\tfrac{1}{4} \langle y, AQ^{-1}A^t y \rangle - \langle y, a + \tfrac{1}{2} AQ^{-1} q \rangle - \tfrac{1}{4} \langle q, Q^{-1} q \rangle.$$

The dual problem is to maximize $h(y)$ over the set $\mathbb{R}^m_+$. This is now a quadratic programming problem with a simpler set of constraints, possibly at the expense of extra variables if $m > n$. The matrix $AQ^{-1}A^t$ is positive semidefinite but may not be invertible. This is not a serious problem.

Now look at a specific case:

Minimize $f(x_1, x_2) = 2x_1^2 - 2x_1 x_2 + 2x_2^2 - 6x_1$
subject to $x_1 \geq 0, x_2 \geq 0$, and $x_1 + x_2 \leq 2$.

This is a quadratic programming problem with $Q = \begin{bmatrix} 2 & -1 \\ -1 & 2 \end{bmatrix}$, $q = (-6, 0)$, $A = \begin{bmatrix} -1 & 0 \\ 0 & -1 \\ 1 & 1 \end{bmatrix}$, and $a = (0,0,2)$. Note that Slater's condition is satisfied, for example, at the point $(1/2, 1/2)$.

We can compute $Q^{-1} = \tfrac{1}{3} \begin{bmatrix} 2 & 1 \\ 1 & 2 \end{bmatrix}$ and $\tfrac{1}{4} AQ^{-1}A^t = \tfrac{1}{12} \begin{bmatrix} 2 & 1 & -3 \\ 1 & 2 & -3 \\ -3 & -3 & 6 \end{bmatrix}$, and in addition, $a + \tfrac{1}{2} AQ^{-1}q = (2, 1, -1)$ and $\tfrac{1}{4} \langle q, Q^{-1}q \rangle = 6$. Thus

$$h(y_1, y_2, y_3) = \tfrac{1}{12}(2y_1^2 + 2y_2^2 + 6y_3^2 + 2y_1 y_2 - 6y_1 y_3 - 6y_2 y_3) - 2y_1 - y_2 + y_3 - 6.$$

This problem can be solved most easily using the (KKT) conditions. It will be left to Exercise 16.10.G to show that $x = (3/2, 1/2)$ and $y = (0,0,1)$ satisfy (KKT).

Notice that $f(3/2,1/2) = -11/2 = h(0,0,1)$. Since the minimum of f and the maximum of h are equal, this is the minimax value. Hence the value of the program must be $-11/2$, the minimizer is $(3/2,1/2)$, and the multiplier is $(0,0,1)$.

Exercises for Section 16.10

A. Compute p_* and p^* for $p(x,y) = \sin(x+y)$ on $\mathbb{R}^2$.

B. Show that the set Y defined in the proof of Lemma 16.10.2 is convex and open.

C. Show that the set A constructed in the proof of the Minimax Theorem (compact case) is compact and convex.

D. Modify the proof of the Minimax Theorem to deal with the case in which X is unbounded and Y is compact.

E. Let $p(x,y) = e^{-x} - e^{-y}$.

    (a) Show that p is convex–concave.
    (b) Show that $p_* = p^*$
    (c) Show that there are no saddle points.
    (d) Why does this not contradict the Minimax Theorem?

F. Suppose that $p(x,y)$ is convex–concave on $X \times X$ for a compact subset X of $\mathbb{R}^n$ and satisfies $p(y,x) = -p(x,y)$. Prove that $p_* = p^* = 0$.

G. (a) Solve the (KKT) equations for the numerical example in Example 16.10.6.
    (b) Write down the Lagrangian and verify the saddle point.

H. Consider the following **linear programming** problem: Minimize $\langle x,q \rangle$ subject to $Ax \leq a$, where $q \in \mathbb{R}^n$, A is an $m \times n$ matrix, and $a \in \mathbb{R}^m$.

    (a) Express this problem as a convex program and compute the Lagrangian.
    (b) Find the dual program.
    (c) Show that if the original program satisfies Slater's condition and has a solution v, then the dual program has a solution w and $\langle v,q \rangle = \langle w,a \rangle$.

References

Calculus

1. G. Klambauer, *Aspects of Calculus*, Springer-Verlag, New York, 1986.
2. M. Spivak, *Calculus*, 3rd edition, Publish or Perish, Inc., Houston, 1994.
3. M. Spivak, *Calculus on Manifolds*, W. A. Benjamin, New York, 1965.

Basic Analysis

4. J. E. Marsden, *Elementary Classical Analysis*, W.H. Freeman and Co., San Francisco, 1974.
5. A. Mattuck, *Introduction to Analysis*, Prentice Hall, Upper Saddle River, N.J., 1999.
6. W. Rudin, *Principles of Mathematical Analysis*, 3rd edition, McGraw-Hill, New York, 1976.
7. W. R. Wade, *An Introduction to Analysis*, Prentice Hall, Englewood Cliffs, N.J., 1995.

Linear Algebra

8. S. H. Friedberg, A. J. Insel, and L. E. Spence, *Linear Algebra*, 2nd edition, Prentice Hall, Englewood Cliffs, N.J., 1989.
9. K. Hoffman and R. Kunze, *Linear Algebra*, 2nd ed., Prentice Hall, Englewood Cliffs, N.J., 1971.
10. G. Strang, *Linear Algebra and Its Applications*, 2nd ed., Academic Press, New York, 1980.

Advanced Analysis (Measure Theory)

11. A. M. Bruckner, J. B. Bruckner, and B. S. Thomson, *Real Analysis*, Prentice Hall, Upper Saddle River, N.J., 1997.
12. G. B. Folland, *Real Analysis*, 2nd edition, John Wiley & Sons, New York, 1999.
13. H. Royden, *Real Analysis*, 3rd edition, Macmillan Pub. Co., New York, 1988.

Approximation Theory

14. C. deBoor, *A Practical Guide to Splines*, Springer-Verlag, New York, 1978.
15. W. Cheney, *An Introduction to Approximation Theory*, McGraw-Hill, New York, 1966.
16. W. Cheney and W. Light, *A Course in Approximation Theory*, Brooks/Cole Pub. Co., Pacific Grove, Calif., 2000.
17. P. J. Davis, *Interpolation and Approximation*, Blaisdell Pub. Co., New York, 1963.
18. R. P. Feinerman and D. J. Newman, *Polynomial Approximation*, Williams & Wilkins, Baltimore, 1974.

Dynamical Systems

19. M. Barnsley, *Fractals Everywhere*, 2nd edition, Academic Press, Boston, 1993.
20. R. Devaney, *An Introduction to Chaotic Dynamical Systems*, Addison-Wesley, Redwood City, Calif., 1989.
21. K. Falconer, *Fractal Geometry: Mathematical Foundations and Applications*, John Wiley & Sons, New York, 1990.

22. R. A. Holmgren, *A First Course in Discrete Dynamical Systems*, Springer-Verlag, New York, 1996.
23. C. Robinson, *Dynamical Systems*, 2nd edition, CRC Press, Boca Raton, Fla., 1999.

Differential Equations

24. G. Birkhoff and G. C. Rota, *Ordinary Differential Equations*, 4th edition, John Wiley & Sons, New York, 1989.
25. G. F. Simmons, *Differential Equations*, 2nd edition, McGraw-Hill, New York, 1991.
26. W. Walter, *Ordinary Differential Equations*, Grad. Texts in Math., Vol. 182, Springer-Verlag, New York, 1998.

Fourier Series

27. H. Dym and H. P. McKean, *Fourier Series and Integrals*, Academic Press, New York, 1972.
28. T. Körner, *Fourier Analysis*, Cambridge University Press, Cambridge, U.K., 1988.
29. R.T. Seeley, *An Introduction to Fourier Series and Integrals*, W. A. Benjamin, New York, 1966.

Wavelets

30. G. Bachman, E. Beckenstein, and L. Narici, *Fourier and Wavelet Analysis*, Springer-Verlag, New York, 2000.
31. E. Hernández and G. Weiss, *A First Course on Wavelets*, CRC Press, Boca Raton, Fla., 1996.
32. P. Wojtaszczyk, *A Mathematical Introduction to Wavelets*, LMS Student Texts, Vol. 37, Cambridge University Press, Cambridge, U.K., 1997.

Convex Optimization

33. J. Borwein and A. Lewis, *Convex Analysis and Nonlinear Optimization*, Springer-Verlag, New York, 2000.
34. J. B. Hiriart-Urruty and C. Lemarichal, *Convex Analysis and Minimization algorithms I*, Springer-Verlag, New York, 1993.
35. A. L. Peressini, F. E. Sullivan, and J. J. Uhl, *The Mathematics of Nonlinear Programming*, Springer-Verlag, New York, 1988.
36. J. van Tiel, *Convex Analysis*, John Wiley & Sons, New York, 1984.
37. R. Webster, *Convexity*, Oxford University Press, New York, 1994.

Articles

38. J. Banks, J. Brooks, G. Cairns, G. Davis, and P. Stacey, "On Devaney's Definition of Chaos," *Amer. Math. Monthly* **99** (1992), 332–334.
39. H. Carslaw, "A Historical Note on Gibbs' Phenomenon in Fourier's Series and Integrals," *Bull. Amer. Math. Soc.* (2) **31** (1925), 420–424.
40. J. Foster and F. B. Richards, "The Gibbs Phenomenon for Piecewise-Linear Approximation," *Amer. Math. Monthly* **98** (1991), 47–49.
41. J. E. Hutchinson, "Fractals and Self-Similarity," *Indiana Univ. Math. J.* **30** (1981), 713–747.
42. T. Li, J. Yorke, "Period Three Implies Chaos," *Amer. Math. Monthly* **82** (1975), 985–992.
43. M. Vellekoop and R. Berglund, "On Intervals, Transitivity = Chaos," *Amer. Math. Monthly* **101** (1994), 353–355.

Index

A **boldface** page number indicates a definition or primary entry.